Geophysical Monograph Series

Including

IUGG Volumes
Maurice Ewing Volumes
Mineral Physics Volumes

Geophysical Monograph Series

Geophysical Monograph Volumes

144 **The Subseafloor Biosphere at Mid-Ocean Ridges** *William S. D. Wilcock, Edward F. DeLong, Deborah S. Kelley, John A. Baross, and S. Craig Cary (Eds.)*

145 **Timescales of the Paleomagnetic Field** *James E. T. Channell, Dennis V. Kent, William Lowrie, and Joseph G. Meert (Eds.)*

146 **The Extreme Proterozoic: Geology, Geochemistry, and Climate** *Gregory S. Jenkins, Mark A. S. McMenamin, Christopher P. McKay, and Linda Sohl (Eds.)*

147 **Earth's Climate: The Ocean–Atmosphere Interaction** *Chuzai Wang, Shang-Ping Xie, and James A. Carton (Eds.)*

148 **Mid-Ocean Ridges: Hydrothermal Interactions Between the Lithosphere and Oceans** *Christopher German, Jian Lin, and Lindsay Parson (Eds.)*

149 **Continent-Ocean Interactions Within East Asian Marginal Seas** *Peter Clift, Wolfgang Kuhnt, Pinxian Wang, and Dennis Hayes (Eds.)*

Maurice Ewing Volumes

1 **Island Arcs, Deep Sea Trenches, and Back-Arc Basins** *Manik Talwani and Walter C. Pitman III (Eds.)*

2 **Deep Drilling Results in the Atlantic Ocean: Ocean Crust** *Manik Talwani, Christopher G. Harrison, and Dennis E. Hayes (Eds.)*

3 **Deep Drilling Results in the Atlantic Ocean: Continental Margins and Paleoenvironment** *Manik Talwani, William Hay, and William B. F. Ryan (Eds.)*

4 **Earthquake Prediction—An International Review** *David W. Simpson and Paul G. Richards (Eds.)*

5 **Climate Processes and Climate Sensitivity** *James E. Hansen and Taro Takahashi (Eds.)*

6 **Earthquake Source Mechanics** *Shamita Das, John Boatwright, and Christopher H. Scholz (Eds.)*

IUGG Volumes

1 **Structure and Dynamics of Earth's Deep Interior** *D. E. Smylie and Raymond Hide (Eds.)*

2 **Hydrological Regimes and Their Subsurface Thermal Effects** *Alan E. Beck, Grant Garven, and Lajos Stegena (Eds.)*

3 **Origin and Evolution of Sedimentary Basins and Their Energy and Mineral Resources** *Raymond A. Price (Ed.)*

4 **Slow Deformation and Transmission of Stress in the Earth** *Steven C. Cohen and Petr Vaníček (Eds.)*

5 **Deep Structure and Past Kinematics of Accreted Terrances** *John W. Hillhouse (Ed.)*

6 **Properties and Processes of Earth's Lower Crust** *Robert F. Mereu, Stephan Mueller, and David M. Fountain (Eds.)*

7 **Understanding Climate Change** *Andre L. Berger, Robert E. Dickinson, and J. Kidson (Eds.)*

8 **Evolution of Mid Ocean Ridges** *John M. Sinton (Ed.)*

9 **Variations in Earth Rotation** *Dennis D. McCarthy and William E. Carter (Eds.)*

10 **Quo Vadimus Geophysics for the Next Generation** *George D. Garland and John R. Apel (Eds.)*

11 **Sea Level Changes: Determination and Effects** *Philip L. Woodworth, David T. Pugh, John G. DeRonde, Richard G. Warrick, and John Hannah (Eds.)*

12 **Dynamics of Earth's Deep Interior and Earth Rotation** *Jean-Louis Le Mouël, D. E. Smylie, and Thomas Herring (Eds.)*

13 **Environmental Effects on Spacecraft Positioning and Trajectories** *A. Vallance Jones (Ed.)*

14 **Evolution of the Earth and Planets** *E. Takahashi, Raymond Jeanloz, and David Rubie (Eds.)*

15 **Interactions Between Global Climate Subsystems: The Legacy of Hann** *G. A. McBean and M. Hantel (Eds.)*

16 **Relating Geophysical Structures and Processes: The Jeffreys Volume** *K. Aki and R. Dmowska (Eds.)*

17 **Gravimetry and Space Techniques Applied to Geodynamics and Ocean Dynamics** *Bob E. Schutz, Allen Anderson, Claude Froidevaux, and Michael Parke (Eds.)*

18 **Dynamics and Predictability of Geophysical Phenomena** *William I. Newman, Andrei Gabrielov, and Donald L. Turcotte (Eds.)*

Mineral Physics Volumes

1 **Point Defects in Minerals** *Robert N. Shock (Ed.)*

2 **High Pressure Research in Mineral Physics** *Murli H. Manghnani and Yasuhiko Syona (Eds.)*

3 **High Pressure Research: Application to Earth and Planetary Sciences** *Yasuhiko Syona and Murli H. Manghnani (Eds.)*

Geophysical Monograph 150
IUGG Volume 19

The State of the Planet: Frontiers and Challenges in Geophysics

Robert Stephen John Sparks
Christopher John Hawkesworth
Editors

American Geophysical Union
Washington, DC

International Union of Geodesy and Geophysics

Published under the aegis of the AGU Books Board

Jean-Louis Bougeret, Chair; Gray E. Bebout, Carl T. Friedrichs, James L. Horwitz, Lisa A. Levin, W. Berry Lyons, Kenneth R. Minschwaner, Andy Nyblade, Darrell Strobel, and William R. Young, members.

Library of Congress Cataloging-in-Publication Data

The state of the planet : frontiers and challenges in geophysics / Robert Stephen John Sparks, Christopher John Hawkesworth, editors.
    p. cm. -- (Geophysical monograph, ISSN 0065-8448 ; 150) (IUGG ; v. 19)
 Includes bibliographical references.
 ISBN 0-87590-415-7 (alk. paper)
 1. Geophysics--Research. I. Sparks, R. S. J. (Robert Stephen John), 1949- II. Hawkesworth, C. J. III. Series. IV. IUGG (Series) ; v. 19.

QC807.5.S73 2004
550'.72--dc22

2004062266

ISBN 0-87590-415-7
ISSN 0065-8448

Copyright 2004 by the American Geophysical Union
2000 Florida Avenue, N.W.
Washington, DC 20009

Figures, tables, and short excerpts may be reprinted in scientific books and journals if the source is properly cited.

Authorization to photocopy items for internal or personal use, or the internal or personal use of specific clients, is granted by the American Geophysical Union for libraries and other users registered with the Copyright Clearance Center (CCC) Transactional Reporting Service, provided that the base fee of $1.50 per copy plus $0.35 per page is paid directly to CCC, 222 Rosewood Dr., Danvers, MA 01923. 1526-8448/04/$01.50+0.35.

This consent does not extend to other kinds of copying, such as copying for creating new collective works or for resale. The reproduction of multiple copies and the use of full articles or the use of extracts, including figures and tables, for commercial purposes requires permission from the American Geophysical Union.

Printed in the United States of America.

CONTENTS

Preface
Steve Sparks and Chris Hawkesworth vii

Foreword
Masaru Kono ix

The Earth's Core: An Approach From First Principles
G. David Price, D. Alfè, L. Vočadlo, and M. J. Gillan 1

Modeling the Earth's Dynamo
Gary A. Glatzmaier, Darcy E. Ogden, and Thomas L. Clune 13

Core-Mantle Boundary Structures and Processes
Thorne Lay and Edward J. Garnero 25

Seismological Insights Into Heterogeneity Patterns in the Mantle
B. L. N. Kennett 43

Geophysical and Geochemical Models of Mantle Convection: Successes and Future Challenges
Yanick Ricard and Nicolas Coltice 59

Melting of Fertile Peridotite With Variable Amounts of H_2O
Bernard J. Wood 69

Geophysical Constraints on Slab Subduction and Arc Magmatism
Akira Hasegawa and Junichi Nakajima 81

Evolution of Arc Magmas and Their Volatiles
Richard J. Arculus 95

Volatile Controls on Magma Ascent and Eruption
Katharine V. Cashman 109

Climatic Impact of Volcanic Emissions
Alan Robock 125

High-Resolution Gravity Mapping: the Next Generation Sensors
Christopher Jekeli 135

Satellite Magnetic Field Measurements: Applications in Studying the Deep Earth
Catherine G. Constable and Steven C. Constable 147

**Global Navigation Satellite Sounding of the Atmosphere and GNSS Altimetry:
Prospects for Geosciences**
T. P. Yunck and G. A. Hajj 161

Dense GPS Array as a New Sensor of Seasonal Changes of Surface Loads
Kosuke Heki 177

Remote Sensing of Terrestrial Water Storage, Soil Moisture and Surface Waters
James S. Famiglietti 197

High-Resolution Measurement of Ocean Surface Topography by Radar Interferometry for Oceanographic and Geophysical Applications
Lee-Leung Fu and Ernesto Rodriguez 209

The Global Water Cycle
Taikan Oki, Dara Entekhabi, and Timothy Ives Harrold 225

Cryosphere During the Twentieth Century
Atsumu Ohmura 239

Climate Prediction: The Limits of Ocean Models
Peter H. Stone 259

Biosphere Dynamics: Challenges for Earth System Models
I. Colin Prentice, Corinne Le Quéré, Erik T. Buitenhuis, Joanna I. House, Christine Klass, and Wolfgang Knorr 269

The UVic Earth System Climate Model and the Thermohaline Circulation in Past, Present, and Future Climates
Andrew J. Weaver 279

Complexities in the Climate System and Uncertainties in Forecasts
Ronald G. Prinn 297

Modeling Human-Climate Interaction
Henry D. Jacoby 307

Uncertainty and Predictability in Geophysics: Chaos and Multifractal Insights
Daniel Schertzer and Shaun Lovejoy 317

Earthquake Prediction and Forecasting
David D. Jackson 335

Earthquake Prediction, Seismic Hazard, and Vulnerability
Seiya Uyeda and Kimiro Meguro 349

Volcanic Activity: Frontiers and Challenges in Forecasting, Prediction and Risk Assessment
R. S. J. Sparks and W. P. Aspinall 359

Geophysical Risk, Vulnerability, and Sustainability
Tom Beer 375

People Induced Geophysical Risks and Urban Sustainability
Ian Douglas 387

Urban Climate, Weather and Sustainability
Gerald Mills 399

PREFACE

Perhaps the most pressing scientific challenge before us concerns our understanding of the Earth and environmental change, which is occurring at a dramatic rate. Humanity can expect serious difficulties on the road ahead as a result, even to the point of threatening civilized progress itself. There is thus an urgent need to understand the state of the planet, to anticipate the effects and consequences of environmental change, and if necessary take preventive action.

Our understanding of change and its consequences can only advance, however, if we understand the Earth system well. Geophysics (*sensu lato*), one of the critical disciplines in the emerging interdisciplinary Earth System Science, forms the basis of the present book—which moves from the deepest parts of the Earth to surface environments, to a final discussion on the interface between science and society. While depicting the frontiers in each of their fields and focusing on present challenges, our contributing authors have also written for wide readership, specialists and non-specialists alike.

Remarkable advances are being made in understanding the Earth's core and mantle through theoretical studies combined with analysis of the huge amounts of data generated by observing seismic, magnetic and gravitational variations in time and space. What occurs down there is a critical part of the Earth System; fluctuations in magnetic fields can affect satellite performance and the migration of birds, and the dynamics of tectonic plates and internal mantle motions govern the large-scale cycles of the Earth and define the planet's state. Volcanism provides one of the most obvious connections between the deep interior of the Earth and the surface environment. This is particularly clear at subduction zones where tectonic processes lead to geophysical hazards such as volcanism and earthquakes, with volcanic volatiles emissions, aerosols and dusts significantly influencing the climate worldwide.

Progress in science is strongly dependent on technological innovation. This book reflects the impact of two major technologies on Earth System science: Earth Observation and computing power. Several contributions consider the extraordinary advances and prospects in Earth Observation, notably using satellites. The dense array of GPS stations in Japan, for example, monitors the motion of the colliding plates and the deformation field for the arc. Remarkably, deformation rates measured over little more than a decade agree almost exactly with plate velocities estimated over millions of years. Faults can be seen to lock and accumulate strain so that the locations of future major earthquakes can be identified. It is even possible to investigate the hypothesis that winter snow in Japan influences crustal deformation. The gravitational field of the Earth and its fluctuations can be mapped in unprecedented detail with major impacts on our ability to monitor attributes of the planet's state from deep mantle plumes to shifts in water masses across the Earth.

Implicit in many of the contributions here is the profound influence of computing power and the internet, which aids in storing, analysing and synthesizing the vast amounts of data collected by geophysical instruments and satellites. It is now possible to model the non-linear stochastic processes that characterise the geosphere, hydrosphere, cryosphere, biosphere, and atmosphere in sufficient detail to gain deep insight into how complex natural systems work. Major modelling advances are being made in almost every corner of the geophysical world. Indeed, Global Climate Models (GCMs) seek to integrate all component systems and their interactions in fairly plausible simulations of reality. With computing power ever on the rise, our ability to understand the past, present and future of climate states has expanded in response. The new Japanese Earth Simulator, for example, has already refined GCM research by allowing modellers to decrease the grid size from 300 km to 60 km. In so doing the models can now simulate large ocean gyres like the Gulf Stream.

Several papers provide cautionary tales on just how formidable the task of modelling environmental change is likely to be. GCMs are required to parameterise all the sub-grid processes, as well as account for uncertainties. Other discussions on heat transport in the oceans, ocean-atmosphere coupling, the role of the biosphere in climate principally through the carbon cycle, volcanic emissions, the global water cycle, and ice sheet dynamics parallel this caution. Because all these topics concern processes that have an intrinsic role in climate, how we understand them will measure how much confidence we might give to the parameterisations that are integrated into GCMs. Indeed, several authors emphasize that there are major challenges and unresolved problems in understanding such critical processes and, in some cases, draw attention to weaknesses in current GCM parameterisations. More philosophically, GCMs must include both uncertainties in knowledge about natural processes and uncertainties related to natural variability. At the same time, these epistemic and aleatory uncertainties contribute to models in ways that are poorly understood, and may lead to results that do not simulate nature realistically. There are also limits to the predictability of natural systems as a consequence of complexity, which provides challenges for theoreticians and modellers.

How does Earth System science interface with society? Models are now being produced that combine simplified physical models of climate with models of economic and social processes and parameters. They are run in probabilistic fashion using Monte Carlo techniques to sample from parameter uncertainties. The current MIT model is described in two papers that assess the likelihood of more extreme scenarios for the Earth's future climate and allow for comparisons of different mitigation strategies; for example, in terms of fossil fuel use. This kind of integrated model is rapidly emerging as a major thrust of modelling designed to help society form judgements and develop policies to ameliorate the effects of global change. The imperative of helping society find solutions for its global problems and the threat of climate change are also driving geophysics to collaborate ever more widely not just in the science community but also in politics, social sciences, and economics. Of course, geophysicists raised on diets of hard science may not feel comfortable here, and may shy away from problems that elude the rigor of their traditional training. Discussions on forecasting geophysical hazards, risk assessment, sustainability issues, and the science of the urban environment illustrate these issues forcibly. As central concerns of society, they provide an imperative for scientists to get involved and contribute to a sustainable and better future.

This book derives from Union Symposia and Lectures devoted to the "State of the Planet: Frontiers and Challenges," adopted as the theme for the first General Assembly of the International Union of Geodesy and Geophysics (IUGG) in the 21st century, held in Sapporo, Japan, July 2003.

We acknowledge and wish to thank our authors, symposium convenors, reviewers, and AGU Books for making this book possible.

Steve Sparks and Chris Hawkesworth
Bristol University, UK

FOREWORD

The International Union of Geodesy and Geophysics (IUGG) was established in 1919 as an international, non-governmental organization. Since then, the IUGG has organized or sponsored a wide range of activities to promote the scientific study of the Earth and its environment, and disseminate the knowledge thus gained to benefit the scientific community as well as the general public and society. The IUGG consists of seven disciplinary Associations engaged in geodesy (IAG), geomagnetism and aeronomy (IAGA), hydrology (IAHS), atmospheres (IAMAS), oceans (IAPSO), seismology (IASPEI), and volcanology (IAVCEI).

During its history, one of the most successful achievements of the IUGG was the organization of the International Geophysical Year (IGY) held between 1957 and 1958. The IGY also expanded the scope of the former two International Polar Years (1982–1983 and 1932–1933), and the entire Earth system became the focus of scientific observations. Under difficult world political conditions, the IGY succeeded in coordinating the research activities of many countries participating in the program. Examples of such international cooperation can be found in the coordinated atmospheric and oceanic research done and the scientific expeditions to the Antarctic, which many countries have continued to this date. The most notable event of the IGY period may be the launch of Sputnik, which became the first artificial satellite to orbit the Earth. This flight marked the start of the space age, which continues to provide us great wonders with the amazing discoveries reported by space missions to the Moon and to other solar system planets.

Coming into the 21st century, the Earth science pursued by the IUGG seems to gain in importance year by year. There are many reasons for this trend, including rapid urbanization and the formation of magacities, which are vulnerable to large-scale natural hazards (earthquakes, volcanic eruptions, tsunamis, etc.). At the same time, the worldwide increase in human population threatens the supply of clean water and energy resources, which are essential to support life. Most prominently, however, it has been recognized from the last quarter of the 20th century that human activity is actually triggering a very fast climate change in the form of global warming. Although the Earth system is quite large and complex, human activity has reached a level that Earth system behavior cannot be well understood unless the human factor is taken into consideration.

With these contexts as background, the IUGG sought to reform itself to better fulfill the changing needs of society and the scientific community. Opinions were sought from the National Committees and the Associations, and then discussed in several meetings of the Bureau and Executive Committee, with many action items decided on for IUGG to engage in now and in the coming years. One of the most important items here was to make the General Assembly more attractive and more effective for pursuing the goal of the IUGG. The most important problems for which Earth science is key (examples of which are listed above) are complex by nature. The climate problem, for instance, cannot be solved simply by considering the atmosphere without considering interactions with other other essential parts of the system, including the oceans, cryosphere, and the human input. Variability in solar activity also is an important factor in climate change.

The IUGG is well suited for this kind of interdisciplinary approach. Being a conglomerate of seven strong Associations with widely different disciplines, the search for cross-cutting approaches are part of the everyday business of the IUGG. As a result, in its 23rd General Assembly held in Sapporo, Japan, 2003, the IUGG took up this general theme: "The State of the Planet: Frontiers and Challenges." Four Union Lectures delivered by distinguished scientists were combined with the six Union Symposia organized with the cooperation of all the Associations. Our aim was to summarize the most up-to-date scientific achievements about the state of the Earth, and to provide this knowledge to society as basic scientific factors to be taken into consideration.

In forming these Union Symposia, Steve Sparks (IAVCEI president) acted as organizer. The series of Union Lectures and Union Symposia were a great success at the Sapporo Assembly, a success that would not have been possible without the concerted effort of the organizer. I am grateful to Steve and other Association officers for their effort in these activities. The publication of the present volume is a welcome addition to the activities in the General Assembly, and provides a reference document for the scientific community and society.

Misasa, September, 2004

Masaru Kono
President, IUGG 1999—2003

The Earth's Core: An Approach From First Principles

G. David Price, D. Alfè, L. Vočadlo

Research School of Earth Sciences, Birkbeck and University College London, UK.

M. J. Gillan

Department of Physics and Astronomy, University College London, UK.

The Earth's core is largely composed of iron (Fe), alloyed with less dense elements such as sulphur, silicon and/or oxygen. The phase relations and physical properties of both solid and liquid Fe alloys are therefore of great geophysical importance. Over the past fifty years the properties of Fe and its alloys have been extensively studied experimentally. However, achieving the extreme pressures and temperatures found in the core provide a major experimental challenge, and there are still considerable discrepancies in the results obtained by using different experimental techniques. In the past fifteen years quantum mechanical techniques have been applied to predict the properties of Fe. First principles methods used to study Fe now enable us to conclude: (i) that pure Fe adopts an hexagonal close packed structure under core conditions and melts at ~6200 K at 330 GPa, (ii) that thermodynamic equilibrium and observed seismic data are satisfied by a liquid Fe alloy outer core with a composition of ~10 mole% S (or Si) and 8 mole% O crystallising at ~ 5500 K to give an Fe alloy inner core with ~8 mole% S (or Si) and 0.2 mole % O, and (iii) that with such concentrations of S (or Si), an Fe alloy might adopt a body centred cubic structure in all or part of the inner core. In the future the roles of Ni, C, H and K in the core need to be studied, and techniques to predict the transport and rheological properties of Fe alloys need to be developed.

INTRODUCTION

Knowing about the nature of the Earth's core lies at the heart of understanding the state of our planet. Today, we believe that the core is the source of the Earth's magnetic field, and that heat-flow from the core contributes significantly to driving mantle convection, and hence ultimately contributes to plate tectonics and the resulting evolution of the planet's surface. What we know of the details of the core's structure comes largely from seismology. Although previously inferred to exist from a knowledge of the mass and moment of inertia of the Earth, the presence of the Earth's core was only firmly established from seismology by *Oldham* [1906]. Further seismological study enabled *Gutenberg* [1913] to determine the depth of the core-mantle boundary (CMB) to be ~2890 km, and in 1936 *Lehmann* discovered that there existed an inner core (with a radius now known to be ~1220 km). It was not until 35 years later that *Dziewonski and Gilbert* [1971] definitely proved that the inner core was a solid region within the surrounding liquid outer core. Seismology has subsequently shown that the inner core is anisotropic, with seismic waves travelling faster parallel to the poles than in the equatorial plane [e.g. *Creager*, 1992], and most recently, seismic measurements have been interpreted as showing that the Earth's solid inner core is even more structurally complex. The detailed interpretations of the data differ, but all workers [e.g. *Song and Helmberger*, 1998; *Ishii* and *Dziewonski*, 2002; *Beghein and Trampert*, 2003] conclude that the inner core exhibits a significant degree of layering, which may either reflect the changing history of core crystallisation and deformation, or the occurrence of an unidentified change in the core-forming phases.

The fact that the core is largely composed of Fe was firmly established as a result of *Birch's* [1952] analysis of mass-density/sound-wave velocity systematics. Today we believe that the outer core is about 6 to 10% less dense than pure liquid Fe, while the solid inner core is a few percent less dense than crystalline Fe [e.g. *Poirier*, 1994a]. From cosmochemical and other considerations, it has been suggested [e.g. *Poirier*, 1994b; *Allègre et al.*, 1995; *McDonough and Sun*, 1995] that the alloying elements in the core might include S, O, Si, H and C. It is also probable that the core contains minor amounts of other elements, such as Ni and K. The exact temperature profile of the core is still controversial [e.g. *Alfè et al.*, 2002a], but it is generally held that the inner core is crystallising from the outer core as the Earth slowly cools, and that (as a result of the work outlined below) temperatures across the core range between ~4000 at the base of the mantle to 5500K at its centre, where the pressure is ~360 GPa.

Before a full understanding of the chemically complex core can be reached, it is necessary to understand the properties and behaviour at high pressure (P) and temperatures (T) of its primary constituent, namely metallic Fe. Experimental techniques have evolved rapidly in the past years, and today using diamond anvil cells or shock experiments the study of minerals at pressures up to ~200 GPa and temperatures of a few thousand Kelvin is possible. These studies, however, are still far from routine, and results from different groups are often in conflict [see for example reviews by *Poirier*, 1994b; *Shen and Heinz*, 1998; *Stixrude and Brown*, 1998; *Boehler*, 2000]. As a result, in order to complement these existing experimental studies and to extend the range of pressure and temperature over which we can model the Earth, computational mineral physics has, in the past decade, become an established and growing discipline.

Within computational mineral physics a variety of atomistic simulation methods (developed originally in the fields of solid state physics and theoretical chemistry) are used. These techniques can be divided approximately into those that use some form of interatomic potential model to describe the energy of the interaction of atoms in a mineral as a function of atomic separation and geometry, and those that involve the approximate solution of Schrödinger's equation to calculate the energy of the mineral species by quantum mechanical techniques. For the Earth sciences, the accurate description of the behaviour of minerals as a function of temperature is particularly important, and computational mineral physics usually uses either lattice dynamics or molecular dynamics methods to achieve this important step. The relatively recent application of all of these advanced condensed matter physics methods to geophysics has only been possible by the very rapid advances in the power and speed of computer processing. Techniques, which in the past were limited to the study of structurally simple compounds, with small unit cells, can today be applied to describe the behaviour of complex, low symmetry structures (which epitomise most minerals) and liquids.

In this paper, we will focus on recent studies of Fe and its alloys, which have been aimed at predicting their geophysical properties and behaviour under core conditions. We will contrast what is known from experiment or approximate theory, with the developing insight which is coming from computational mineral physics research. Although interatomic potentials have been used to study Fe [e.g. *Matsui and Anderson*, 1997], many of its properties are very dependent upon a precise description of its metallic nature and can only be modelled accurately by quantum mechanical methods. Thus, below we briefly introduce the essential ab initio techniques used in the most recent studies of Fe alloys [see also *Stixrude et al.*, 1998; *Vočadlo et al.*, 2003; *Steinle-Neumann et al.*, 2003]. We then present a discussion of the structure of the stable phase of Fe at core pressures and temperatures, its melting behaviour at core pressures, ab initio derived estimates of the composition of the core and its predicted thermal structure. We conclude with a materials-based discussion of the interpretation of the seismic structure of the inner core, and high-light some of the issues which must be addressed in the future.

QUANTUM MECHANICAL SIMULATIONS

To investigate the microscopic properties of matter from first principles one needs to solve the Schrödinger equation of the system:

$$H\Psi = E\Psi \qquad (1)$$

where H is the Hamiltonian, and is given by $H = T_i + T_e + V_{ii} + V_{ie} + V_{ee}$, with T_i and T_e representing the kinetic energies of the nuclei and electrons respectively, V_{ii} the repulsive Coulombic energy between the nuclei, V_{ie} the attractive Coulombic energy between the nuclei and the electrons and V_{ee} the repulsive Coulombic energy between electrons, and Ψ is the many body wave function of the system that depends on the positions of all the nuclei and electrons. Since the electrons are much less massive than the nuclei, their motion adiabatically follows that of the nuclei, and it is a reasonable approximation to decouple the electronic degrees of freedom from the ionic ones. Therefore, for each ionic configuration, R, Equation 1 can be written as:

$$H(\{R\})\Psi = E(\{R\})\Psi \qquad (2)$$

where now $H(\{R\})$ depends only parametrically on the position of the ions, R, and one has to solve Equation 2 only with respect to the electronic degrees of freedom. The resulting energy, $E(\{R\})$, can then be used as the potential energy for the motion of the nuclei. Energy minimisation techniques may then be applied in order to obtain the equilibrium structure for the system under consideration.

Unfortunately, the complexity of the wave function, Ψ, for an N electron system scales as M^N, where M is the number of degrees of freedom for a single electron wave function, ψ. This type of problem cannot readily be solved for large systems due to computational limitations, and therefore the exact solution to the problem for large systems is intractable. However, there are a number of approximations that may be made to simplify the calculation, whereby good predictions of the structural and electronic properties of materials can be obtained by solving self-consistently the one-electron Schrödinger equation for the system, and then summing these individual contributions over all the electrons in the system [e.g. see *Gillan*, 1997; *Stixrude et al.*, 1998]. Fe and its alloys have been studied extensively by a number of groups [e.g., see *Vocadlo et al.*, 2003; *Steinle-Neumann et al.*, 2003], and it now appears that ab initio calculations give an accurate description of the known properties of Fe. The calculations performed by us and others [e.g. *Steinle-Neumann et al.*, 2003] are based on Density Functional Theory (DFT) generally within the Generalised Gradient Approximation (GGA). Our calculations have been performed using the VASP code [*Kresse and Furthmüller*, 1996], and most recently we have used the Projected-Augmented Wave (PAW) method [*Blöchl*, 1994] to calculate the total energy of the system. The PAW method is closely related to the ultra-soft pseudopotential method and has been shown to give results that agree accurately with all-electron methods [*Kresse and Joubert*, 1999; *Alfè et al.*, 2000, 2001].

To study Fe under core conditions, however, we need not only to explore the energetics of bonding, but we are also concerned with the effect of temperature on the system. This requires us to calculate the Gibbs free energy of the systems, which can be done either using lattice dynamic or molecular dynamic methods. Lattice dynamics is a semi-classical approach that can be used with the quasiharmonic approximation (QHA) to describe a cell in terms of independent quantised harmonic oscillators, the frequencies of which vary with cell volume, thus allowing for a description of thermal expansion [e.g. *Born and Huang*, 1954]. The motions of the individual particles are treated collectively as lattice vibrations or phonons, and the phonon frequencies, $\omega(q)$, and displacement eigenvectors, $e(q)$, are obtained by solving:

$$m\omega^2(q) e_i(q) = D(q) e_j(q) \qquad (3)$$

where m is the mass of the atom, and the dynamical matrix, $D(q)$, is defined by:

$$D(q) = \Sigma_{ij} (\partial^2 U/\partial u_i \partial u_j) \exp(iq \cdot r_{ij}) \qquad (4)$$

where U is the potential energy of the system, r_{ij} is the interatomic separation, and u_i and u_j are the displacements of atoms i and j from their equilibrium position. For a unit cell containing N atoms, there are 3N eigenvalue solutions, $\omega^2(q)$, for a given wave vector, q. There are also 3N sets of eigenvectors, $e_x(q)$, $e_y(q)$, $e_z(q)$, which describe the pattern of atomic displacements for each normal mode.

The vibrational frequencies of a lattice can be calculated ab initio, by standard methods such as the small-displacement method [e.g. *Kresse et al.*, 1995]. Having calculated the vibrational frequencies, a number of thermodynamic properties may then be calculated using standard statistical mechanical relations, which are direct functions of these vibrational frequencies. Thus, for example, the Helmholtz free energy, F, is given by:

$$F = k_B T \Sigma_i^M (x_i/2 + \ln(1 - e^{-x_i})) \qquad (5)$$

where $x_i = h\omega_i/2\pi k_B T$, h is Planck's constant and k_B is Boltzmann's constant, and the sum is over all the M=3N normal modes. Modelling the effect of pressure is essential if one is to obtain accurate predictions of phenomena such as phase transformations and anisotropic compression. This problem is now routinely being solved using codes that allow constant stress, variable geometry cells in both static and dynamic simulations. In the case of lattice dynamics, the mechanical pressure is calculated from strain derivatives, whilst the thermal kinetic pressure is calculated from phonon frequencies [e.g. *Parker and Price*, 1989]. The balance of these forces can be used to determine the variation of cell size as a function of pressure and temperature.

Molecular dynamics is routinely used for medium to high temperature simulations of minerals and in all simulations of liquids, where lattice dynamics is of course inapplicable. The method is essentially classical, and its details are presented, for example, in *Allen and Tildesley* [1987]. The interactions between the atoms within the system have traditionally been described in terms of the interatomic potential models mentioned earlier, but instead of treating the atomic motions in terms of lattice vibrations, each ion is treated individually. As the system evolves, the required dynamic properties are calculated iteratively at the specified pressure and temperature. The ions are initially assigned positions and velocities within the simulation box; their co-ordinates are usually chosen to be at the crystallographically determined sites, whilst their velocities are equilibrated such that they are compatible with the required system temperature, and such that both energy and momentum is conserved. In order to calculate subsequent positions and velocities, the forces acting on any individual ion are then calculated from the first derivative of the total energy function, and the new position and velocity of each ion may be calculated at each time-step by solving Newton's equation of motion. Both the particle positions and the volume of the system, or simulation box, can be used as dynamical variables, as is described in detail in *Parrinello and Rahman* [1980] and by *Wentzcovitch* [1991]. The kinetic energy, and therefore temperature, is obtained directly from the velocities of the individual particles. With this explicit particle motion, anharmonicity is implicitly accounted for at high temperatures.

Because of advances in computer power, it is now possible to perform ab initio molecular dynamics (AIMD), with the forces calculated fully quantum mechanically instead of relying upon the use of interatomic potentials [e.g. see *Scandolo*, 2002]. The first pioneering work in AIMD was that of *Car and Parrinello* [1985], who proposed a unified scheme to calculate ab initio forces on the ions and keep the electrons close to the Born-Oppenheimer surface while the atoms move. We have used in the work summarized below an alternative approach, in which the dynamics are performed by explicitly minimizing the electronic free energy functional at each time step. This minimization is more expensive than a single Car-Parrinello step, but the cost of the step is compensated by the possibility of making longer time steps.

PROPERTIES OF FE UNDER CORE CONDITIONS

The High P Structure of Fe

Before we can even begin to provide a materials based interpretation of the composition and structure of the core, we must understand the behaviour of its primary constituent (Fe) under core conditions. At ambient conditions, Fe is crystalline and has a body centred cubic (bcc) structure. This transforms with temperature to a face centred cubic (fcc) form, and with pressure transforms to a hexagonal close packed (hcp) phase, e-Fe. The high P/T phase diagram of pure iron itself however is still controversial (see Figure 1 and also the discussions in *Stixrude and Brown* [1998], *Anderson* [2003], and *Steinle-Neumann et al.* [2003]).

Various diamond anvil cell (DAC) based studies have been interpreted as showing that hcp Fe transforms at high temperatures to a phase that has variously been described as having a double hexagonal close packed structure (dhcp) [*Saxena et al.*, 1996] or an orthorhombicly distorted hcp structure

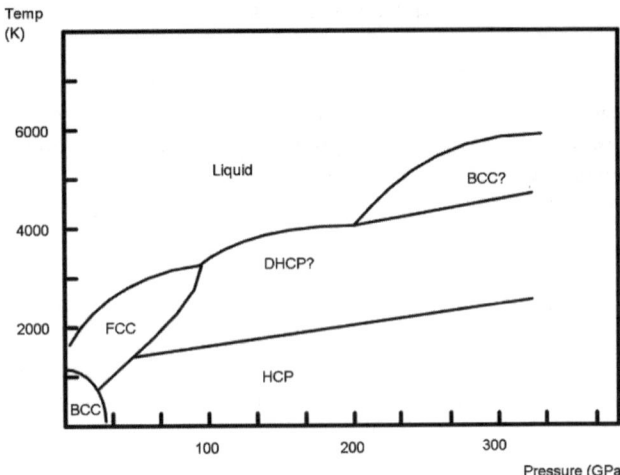

Figure 1. A hypothetical phase diagram for Fe, incorporating all the experimentally suggested high P/T phase transformations.

[*Andrault et al.*, 1997]. Furthermore, high pressure shock experiments have also been interpreted as showing a high pressure solid-solid phase transformation [*Brown and McQueen*, 1986; *Brown*, 2001], which has been suggested could be due to the development of a bcc phase [*Matsui and Anderson*, 1997]. Other experimentalists, however, have failed to detect such a post-hcp phase [e.g. *Shen et al.*, 1998; *Nguyen and Holmes*, 1999], and have suggested that the previous observations were due either to minor impurities or to metastable, strain-induced behaviour.

Further progress in interpreting the nature and evolution of the core would be severely hindered if the uncertainty concerning the crystal structure of the core's major chemical component remained unresolved. An alternative approach to this problem, however, is to use ab initio calculations, which have been shown to provide an accurate means of calculating the properties of materials at high P and T [e.g. *Gillan*, 1997]. Thus, *Vočadlo et al.* [1999] carried out a series of calculations to determine ab initio the stable phase of Fe. Here, we performed spin polarized simulations on candidate phases (including a variety of distorted bcc and hcp structures and the dhcp phase) at pressures ranging from 325 to 360 GPa. These revealed, in agreement with *Söderlind et al.* [1996], that at core pressures only bcc Fe has a residual magnetic moment and all other phases have zero magnetic moments. We found that at core pressures and zero temperature, both the bcc and the suggested orthorhombic polymorph of iron [*Andrault et al.*, 1997] are mechanically unstable. The bcc phase can be continuously transformed to the fcc phase (confirming the findings of *Stixrude and Cohen* [1995]), while the orthorhombic phase spontaneously transforms to the hcp phase, when allowed to relax to a state of isotropic stress. In contrast, hcp, dhcp and fcc Fe remain mechanically stable at core pressures, and we were therefore able to calculate their phonon frequencies and free energies. It was concluded that, on the basis of lattice dynamic calculations over the whole P-T space investigated, the hcp phase of Fe has the lowest Gibbs free energy, and is therefore the stable form of Fe under core conditions, indicating that the true phase diagram for Fe is simpler than previously suggested (Figure 1) and is better described by Figure 2.

The High P Melting of Fe

Having shown how ab initio calculations can be used to establish the sub-solidus phase relations in high P Fe, we now consider its high P/T melting behaviour. The temperature distribution in the core is poorly constrained and consequently a reliable estimate of the melting temperature of Fe at the pressure of the inner-core boundary (ICB) would be valuable. There is much controversy over the high P melting behaviour of Fe [e.g. see *Shen and Heinz*, 1998], with estimates of the T_m of Fe at ICB pressures ranging between ~4500 K to ~7500 K.

Since both our calculations and recent experiments [*Shen et al.*, 1998] suggest that Fe melts from the e-phase in the pressure range immediately above 60 GPa, we focus here on the equilibrium between hcp Fe and liquid Fe. The condition for two phases to be in thermodynamic equilibrium at a given temperature and pressure is that their Gibbs free energies, G(P,T), are equal. To determine T_m at any pressure, we calculate G for the solid and liquid phases as a function of T and determine where they are equal. In fact, we calculate the Helmholtz free energy, F(V,T), as a function of volume, V, and hence obtain the pressure through the relation $P = -(\partial F/\partial V)_T$ and G through its definition, G = F + PV.

To validate our method and to prove its accuracy, we modeled the well studied high P melting behaviour of Al [*de Wijs et al.*,

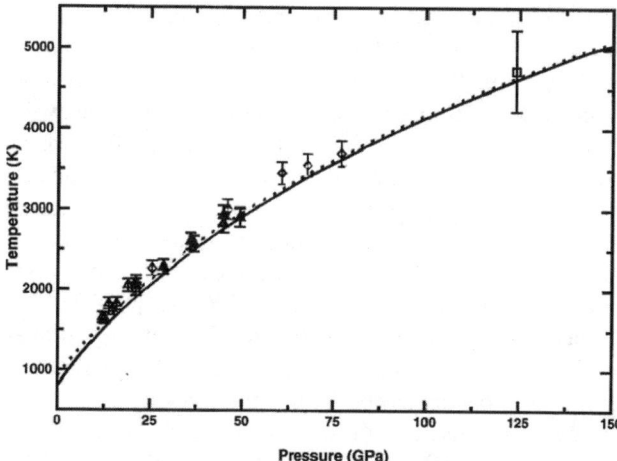

Figure 3. Our calculated high-pressure melting curve for Al [see *Vočadlo and Alfè*, 2002] is shown passing through a variety of recent high P experimental points.

1998; *Vočadlo and Alfè*, 2002]. Figure 3 shows the very good agreement obtained for this system. Our results for the T_m of Fe are shown in Figure 4, and we conclude that Fe melts at ICB pressures between 6,200 and 6,350 K [*Alfè et al.*, 1999, 2002b,d]. A full analysis of the errors and uncertainties in these calculations has been reported in *Alfè et al.* [2002b,d]. For pressures P < 200 GPa (the range covered by DAC experiments) the ab

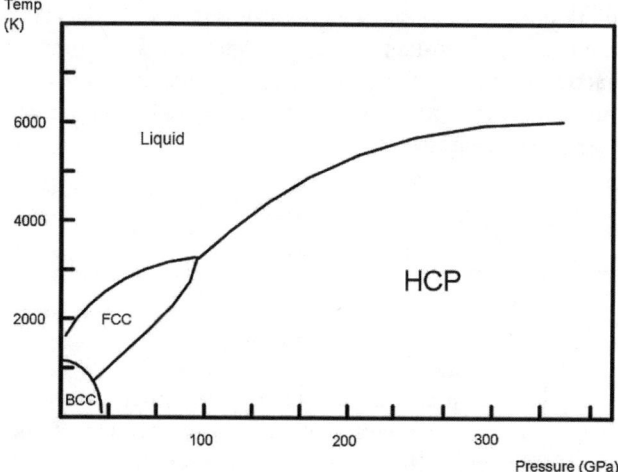

Figure 2. Schematic phase diagram for Fe based on the results of *Vočadlo et al.* [1999].

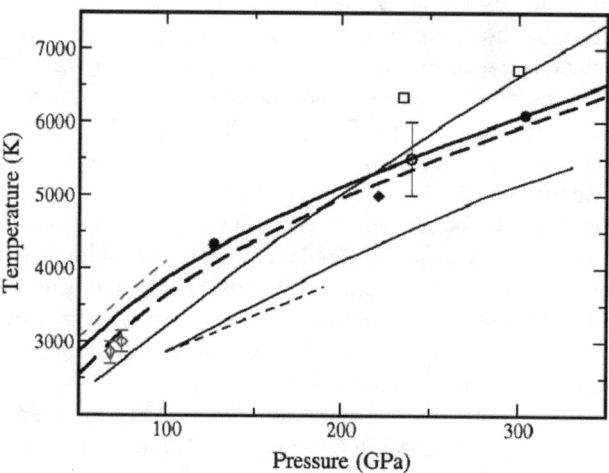

Figure 4. Our calculated high P melting curve of Fe (plotted as a solid thick black line) is shown passing through the shock wave datum (open circle) of *Brown and McQueen* [1986]. Other data shown includes: our melting curve corrected for the GGA pressure error (black dashed line), Belonoshko's melting curve (thin black line), Belonoshko's melting data corrected for errors in potential fitting (black dots), Laio et al's melting curve (gray line), Boehler's DAC curve (gray dashed line), Shen's data (gray open diamonds), Yoo's shock data (open squares), William's melting curve (pale gray dashed line).

initio curve lies ~900 K above the experimental values of *Boehler* [1993] and ~200 K above the more recent values of *Shen et al.* [1998] (who stress that their values are only a lower bound to T_m). The ab initio curve falls significantly below the shock-based estimates for the T_m of *Yoo et al.* [1993], in whose experiments temperature was deduced by measuring optical emission (however, the difficulties of obtaining temperature by this method in shock experiments are well known), but accords almost exactly with the shock data values of *Brown and McQueen* [1986] and *Nguyen and Holmes* [1999].

There are other ways of determining the melting temperature of a system by ab initio methods, including performing simulations that model co-existing liquid and crystal phases. The melting temperature of such a system can then be inferred by seeing which of the two phases grows during the course of a series of simulations at different temperatures. This intuitively attractive approach was used by *Laio et al.* [2000] and by *Belonoshko et al.* [2000] to study the melting of Fe. But because of current computer limitations, they had to modelled Fe melting using interatomic potentials fitted to ab initio surfaces. We have recently also used the co-existence method [*Alfè et al.*, 2002d], but with a model potential fitted to our own ab initio calculations. Initially, our raw model failed to give the same melting temperature as obtained from our ab initio free energy calculations, but when the results were corrected for the free energy mismatch of the model potential with respect to the ab initio energies of liquid and solid, the results for the two methods came into agreement. Thus the shortcomings of model potential co-existence calculations can be corrected by calculating the free energy differences between the model and the ab initio system for both the liquid and solid phases. This difference in free energy between liquid and solid can then be transformed into an effective temperature correction. When this is done to Belonoshko's data, there is excellent agreement with our ab initio melting curve for Fe (see Figure 4). In the future, when greater computing resources will allow full ab initio simulations, the co-existence approach may prove the more attractive strategy for modelling melting behaviour, as it is an intuitive simulation of the melting process.

Interestingly *Poirier and Shankland* [1993] obtained a value of T_m for Fe at 330 GPa of 6100 ±100 K by using the more empirical dislocation melting model, and *Anderson* [2003] finds a T_m of 5900 ± 300 K on the basis of thermodynamic analysis. Thus with our ab initio calculated value of 6200 to 6350 K having been shown to be robust, there appears to be now an emerging consensus on the high P melting of pure Fe. However fully to understand the core, we need to investigate the effect of other alloying elements on the melting behaviour of Fe. Some aspects of this problem are address below.

THE COMPOSITION AND TEMPERATURE OF THE CORE

Seismological data suggest that the outer and inner core contain some light element impurities [*Birch*, 1964]. Cosmochemical abundances of the elements, combined with models of the Earth's history, limit the possible impurities to a few candidates. Those most often discussed are S, O and Si [e.g. *Poirier*, 1994b; *Allègre et al.*, 1995; *McDonough and Sun*, 1995], and we have to date confined our studies to these three. Our strategy for constraining the impurity fractions and the temperature of the core is based on the supposition that the solid inner core is slowly crystallising from the liquid outer core, and that therefore the inner and outer core are in thermodynamic equilibrium at the ICB. This implies that the chemical potentials of Fe and of each impurity must be equal on the two sides of the ICB.

If the core consisted of pure Fe, equality of the chemical potential (Gibbs free energy in this case) would tell us only that the temperature at the ICB is equal to the melting temperature of Fe at the ICB pressure of 330 GPa. With impurities present, equality of the chemical potentials for each impurity element imposes a relation between the mole fractions in the liquid and the solid, so that with S, O and Si we have three such relations. But these three relations must be consistent with the accurate values of the mass densities in the inner and outer core deduced from seismic and free-oscillation data. Below we outline our finding [*Alfè et al.*, 2002c] that ab initio results for the densities and chemical potentials in the liquid and solid Fe-S, Fe-O and Fe-Si alloys determine with useful accuracy the probable mole fraction of O and the sum of the S and Si mole fractions in the outer and inner core, as well as enabling us to determine the temperature at the ICB.

The chemical potential, μ_x, of a solute x in a solid or liquid solution is conventionally expressed as $\mu_x = \mu_x^0 + k_B T \ln a_x$, where μ_x^0 is a constant and a_x is the activity. It is common practice to write $a_x = \gamma_x c_x$, where γ_x is the activity coefficient and c_x the concentration of x. The chemical potential can therefore be expressed as:

$$\mu_x = \mu_x^0 + k_B T \ln \gamma_x c_x \qquad (6)$$

which we rewrite as:

$$\mu_x = \mu_x^* + k_B T \ln c_x \qquad (7)$$

It is helpful to focus on the quantity μ_x^* for two reasons: first, because it is a convenient quantity to obtain by ab initio calculations [*Alfè et al.*, 2002c]; second, because at low concentrations the activity coefficient, γx, will deviate only weakly from unity by an amount proportional to

c_x, and by the properties of the logarithm the same will be true of μ_x^*.

The equality of the chemical potentials μ_x^l and μ_x^s in coexisting liquid and solid (superscripts l and s respectively) then requires that:

$$\mu_x^{*l} + k_B T \ln c_x^l = \mu_x^{*s} + k_B T \ln c_x^s \qquad (8)$$

or equivalently:

$$c_x^s / c_x^l = \exp((\mu_x^{*l} - \mu_x^{*s})/ k_B T) \qquad (9)$$

This means that the ratio of the mole fractions c_x^s and c_x^l in the solid and liquid solution is determined by the liquid and solid thermodynamic quantities μ_x^{*l} and μ_x^{*s}. Although liquid-solid equilibrium in the Fe-S and Fe-O systems has been experimentally studied up to pressures of around 60 GPa [e.g *Boehler*, 2000], there seems little prospect of obtaining experimental data for $\mu_x^{*l} - \mu_x^{*s}$ for Fe alloys at the much higher ICB pressure. However, we have shown [*Alfè et al.*, 2002c] recently that the fully ab initio calculation of μ_x^{*l} and μ_x^{*s} is technically feasible. Thus, the chemical potential, μ_x, of chemical component x can be defined as the change of Helmholtz free energy when one atom of x is introduced into the system at constant temperature, T, and volume, V. In ab initio simulations, it is awkward to introduce a new atom, but the awkwardness can be avoided by calculating $\mu_x^* - \mu_{Fe}^*$, which is the free energy change, ΔF, when an Fe atom is replaced by an x atom. For the liquid, this ΔF is computed by applying the technique of 'thermodynamic integration' to the (hypothetical) process in which an Fe atom is continuously transmuted into an x atom. We have recently performed such calculations for transmuting Fe atoms into S, O and Si [*Alfè et al.*, 2002a].

Alfè et al. [2000, 2002c] performed simulations at constant volume and temperature on systems of 64 atoms; the duration of the simulations after equilibration was typically 6 ps in order to reduce statistical errors to an acceptable level; the number of thermodynamic integration points used in transmuting Fe into x was 3, and we carefully checked the adequacy of these numbers of points. Our results reveal a major qualitative difference between O and the other two impurities. For S and Si, μ_{x^*} is almost the same in the solid and the liquid, the differences being at most 0.3 eV, i.e. markedly smaller than $k_B T \sim 0.5$ eV; but for O the difference of μ_x^* between solid and liquid is ~ 2.6 eV, which is much bigger than $k_B T$. This means that added O will partition strongly into the liquid, but added S or Si will have similar concentrations in the two phases.

Our simulations of the chemical potentials of the alloys can be combined with simulations of their densities to investigate whether the known densities of the liquid and solid core can be matched by any binary Fe/x system, with x = S, O or Si. Using our calculated partial volumes of S, Si and O in the binary liquid alloys, we find that the mole fractions required to reproduce the liquid core density are 16, 14 and 18 % respectively (Figure 5, panel (a) displays our predicted liquid density as a function of c_x compared with the seismic density). Our calculated chemical potentials in the binary liquid and solid alloys then give the mole fractions in the solid of 14, 14 and 0.2 % respectively that would be in equilibrium with these liquids (see Figure 5, panel (b)). Finally, our partial volumes in the binary solids give ICB density discontinuities of 2.7±0.5, 1.8±0.5 and 7.8±0.2 % respectively (Figure 5, panel (c)). As expected, for S and Si, the discontinuities are considerably smaller than the known value of 4.5±0.5 %;

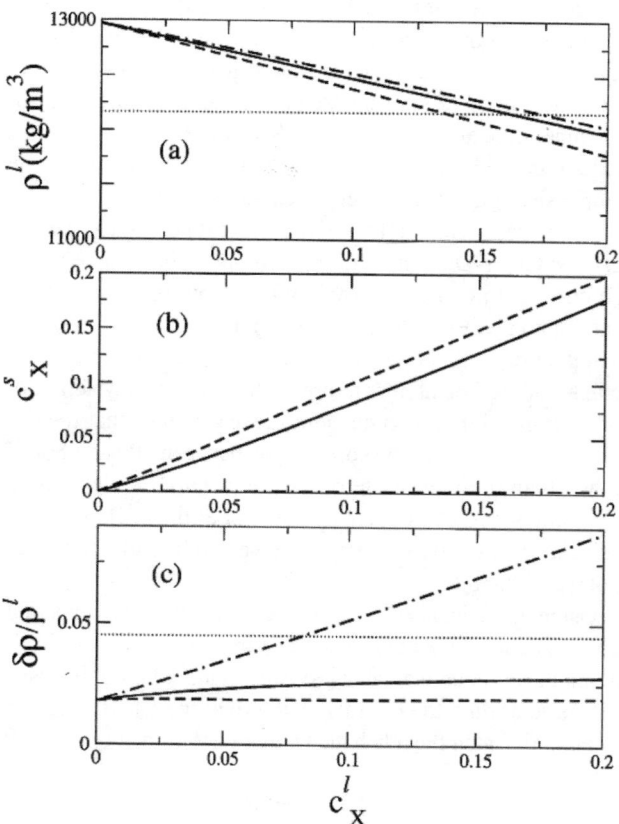

Figure 5. Liquid and solid impurity mole fractions c_x^l and c_x^s of impurities x = S, Si and O, and resulting densities of the inner and outer core predicted by ab initio simulations. Solid, dashed and chain curves represent S, Si and O respectively. (a) liquid density ρ^l (kgm-3); horizontal dotted line shows density from seismic data. (b) mole fractions in solid resulting from equality of chemical potentials in solid and liquid. (c) relative density discontinuity ($\delta\rho/\rho^l$) at the ICB; horizontal dotted line is the value from free oscillation data.

for O, the discontinuity is markedly greater than the known value. We conclude that none of the binary systems can account for the discontinuity quantitatively. However, it clearly can be accounted for by O together with either or both of S and Si. Ab initio calculations on general quaternary alloys containing Fe, S, O and Si will be feasible in the future, but currently they are computationally too demanding, so for the moment we assume that the chemical potential of each impurity species is unaffected by the presence of the others. Our estimated mole fractions needed to account for the ICB density discontinuity, were reported in *Alfè et al.* [2002a] as being 8.5±2.5 mole % S(or Si) and 0.2±0.1 % O in the inner core and 10±2.5 % S (or Si) and 8±2.5 % O in the liquid outer core. This compositional estimate was based on the value of the density discontinuity at the ICB determined by *Shearer and Masters* [1990]. Since then, *Masters and Gubbins* [2003] have reassessed the free oscillation data set, and have determined the density jump at the ICB to be 0.82±0.18 Mg m^{-3}, which is larger than the previous estimates. Using the new density data of *Masters and Gubbins* [2003] leads to a revised core composition of 7±2.5 mole % S (or Si) and 0.2±0.1 % O for the inner core and 8±2.5 % S (or Si) and 13±2.5 % O for the outer core. This change in our estimate of core composition emphasises the need for very accurate seismic data.

With the calculated impurity chemical potentials, we can use the Gibbs-Duhem relation to compute the change in the Fe chemical potential caused by the impurities in the solid and liquid phases [*Alfè et al.*, 2002a]. By requiring the chemical potential of Fe to be the same in both phases we can calculate the liquidus temperature in the iron alloy system. For our estimated liquid compositions, we calculate that the liquidus temperature of the core alloy is between 700 and 800 K lower than the melting temperature of pure Fe, depending upon the value of the density contrast at the ICB. Thus we calculate that the Earth's temperature at the ICB to be between 5400 and 5650 K.

Using this calculated value for the ICB temperature, and our ab initio values for the Grüneisen parameter, γ, for liquid Fe along the outer core adiabat [*Vočadlo et al.*, 2003], it is possible to determine the core temperature at the CMB from the relation defining the adiabatic temperature gradient:

$$\partial T/\partial r = -\gamma g T/\Phi \qquad (10)$$

where Φ is the seismic parameter and g is the acceleration due to gravity. Like *Anderson* [2003], we find γ to be ~1.5 and virtually constant in the outer core, which leads to an ab initio estimate of the core temperature at the CMB of between 3950 and 4200K. Again, this range is in excellent agreement with that inferred from thermodynamic arguments [*Anderson*, 2003].

POSSIBLE STRUCTURE OF THE INNER CORE

The conventional interpretation of the origin of the seismic anisotropy of the inner core is based on the idea of the development of partial alignment of the elastically anisotropic crystals of hcp Fe [e.g. *Song*, 1997]. The static elastic constant of hcp Fe were first calculated by *Stixrude and Cohen* [1995] and then by *Söderlind et al.* [1996]. The low temperature elastic constants of hcp Fe at 39 and 211 GPa were measured in an experiment reported by *Mao et al.* [1999]. The overall agreement between the experimental and various ab initio studies is very good, and the measured and calculated bulk and shear moduli and the seismic velocities of hcp Fe as a function of pressure are shown in Figures 6 and 7. The calculated values compare well with experimental data at higher pressures, but discrepancies at lower pressures are probably due to the neglect of magnetic effects in the simulations [see also the discussion in *Steinle-Neumann et al.*, 2003].

The effect of temperature on the elastic constants of Fe was reported by *Steinle-Neumann et al.* [2001] based on calculations using the approximate 'particle in a cell' method. With increasing temperature, they found a significant change in the c/a axial ratio of the hcp structure, which in turn caused a marked reduction in the elastic constants c_{33}, c_{44} and c_{66}. This led them to conclude that increasing temperature reverses the sense of the single crystal longitudinal anisotropy of hcp Fe, and that the anisotropy of the core should now be viewed

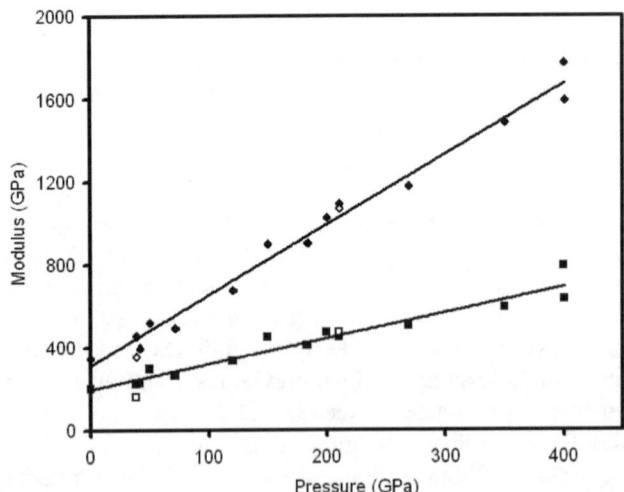

Figure 6. Plot of bulk modulus (diamonds) and shear modulus (squares) for hcp Fe as a function of pressure, with values taken from *Stixrude and Cohen* [1995], *Steinle-Neumann et al.* [1999], *Söderlind et al.* [1996], *Mao et al.* [1999] and *Vočadlo et al.* [2003]. Black diamonds and squares represent ab initio values, while white diamonds and squares represent values obtained from experimentally determined elastic constants.

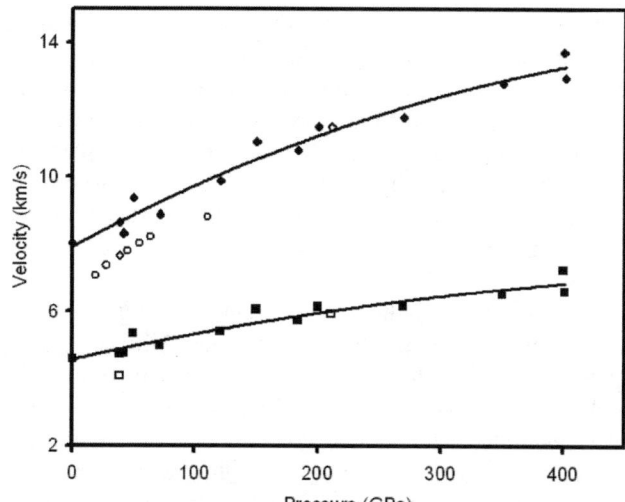

Figure 7. Plot of aggregate v_p (diamonds) and v_s (squares) wave velocity for hcp Fe as a function of pressure, with values taken from *Stixrude and Cohen* [1995], *Steinle-Neumann et al.* [1999], *Söderlind et al.* [1996], *Mao et al.* [1999] and *Vočadlo et al.* [2003]. Black diamonds and squares represent ab initio values, while white diamonds and squares represent values obtained from experimentally determined elastic constants. White circles are the experimental data of *Fiquet et al.* [2001].

as being due to hcp Fe crystals having their c-axis preferably aligned equatorially, rather than axially as originally suggested by *Stixrude and Cohen* [1995]. However, the ab initio determination of high T elastic constants is very difficult. *Karki et al.* [1999] and *Oganov et al.* [2001] have shown how to use truly ab initio methods to obtain high T elastic constants, and further work is needed to confirm the high T properties of hcp Fe. Recent calculations [*Gannarelli et al.*, 2003] have failed thus far to reproduce the strong effect of temperature on c/a seen by *Steinle-Neumann et al.* [2001], and if this result is confirmed by more precise molecular dynamic simulations, then it will have important implications for the interpretation of the seismic tomography of the inner core.

The nature of inner core anisotropy has recently been shown to be more complex than previously thought, and *Beghein and Trampert* [2003] have shown that free oscillation data cannot be simply interpreted in terms of the elastic properties of hcp Fe as reported by *Steinle-Neumann et al.* [2001]. Furthermore, the assumption that hcp Fe is the thermodynamically stable polymorph of Fe at the high temperatures found in the inner core has also been recently questioned [*Brown*, 2001]. In spite of the arguments in favour of hcp Fe, it has been proposed by a number of workers that bcc Fe might be the stable high P/T phase [*Ross et al.*, 1990; *Matsui and Anderson*, 1997]. A strong argument in favour of the stability of bcc Fe under these conditions is that a number of transition metals (e.g. Ti, Zr..) are known to transform from close-packed structures to the bcc structure at temperatures just below their melting curve. However, until recently it appeared that theoretical calculations had ruled out the bcc structure as a candidate for the stable phase of Fe in the core. As reviewed, for example, in *Steinle-Neumann et al.* [2003], this was because ab initio calculations have shown that the bcc structure becomes elastically unstable at pressures above ~150 GPa, and that the enthalpy of the perfect bcc structure is considerably higher than that of the hcp phase. However, these arguments are not conclusive because they are based on athermal or lattice dynamical ab initio calculations, which because of the dynamical instability of bcc Fe cannot be used to determine entropic effects at high pressures. In the past, computing limitations prevented a more sophisticated analysis, but recent methodological developments mean that it is now possible accurately to address these thermal effects. Thus, to resolve the controversy over the effect of temperature on the stability of bcc Fe, we recently performed ab initio molecular dynamics (AIMD) calculations on bcc Fe to simulate directly its behaviour at the high temperatures relevant to the Earth's inner core [*Vočadlo et al*, 2003].

Table 1 shows the calculated Helmholtz free energies for the bcc and hcp phases over a range of volumes and temperatures along (and below) the previously determined melting curve of

Table 1. The ab initio Helmholtz free energy per atom of the *bcc* and *hcp* phases of Fe at state points along (*and below) the calculated melting curve. In this work, the reference system is a simple inverse power potential which takes the form $U=4\varepsilon(\Gamma/r)^\alpha$ where $\varepsilon=1$ eV, $\Gamma=1.77$ Å, $\alpha=5.86$. We have previously shown that this reference potential, based on only a repulsive term, describes the ab initio system extremely well; the bonding term is almost independent of the atomic positions, depending only on the volume and temperature of the system. We stress, however, that the final result is totally independent of the choice of reference system. We performed these calculations on a 64 atom super-cell (4x4x4 primitive cells) with a 3x3x3 k-point grid. Considerable effort was spent on convergence tests in k-point sampling to reduce the error in free energies to <10 meV per atom. Cell size effects have been extensively studied in our previous work hcp-Fe [*Alfè et al.*, 2001].

V(Å³)	T(K)	F_{bcc}(eV)	F_{hcp}(eV)	ΔF(meV)
9.0	3500	-10.063	-10.109	46
8.5	3500	-9.738	-9.796	58
7.8	5000	-10.512	-10.562	50
7.2	6000	-10.633	-10.668	35
6.9	6500	-10.545	-10.582	37
6.7	6700	-10.288	-10.321	33
7.2*	3000	-7.757	-7.932	175

hcp Fe. It is clear that in all cases $F_{bcc} > F_{hcp}$, and so as previously concluded (on the basis of lattice dynamical as opposed to molecular dynamical calculations) pure Fe adopts the hcp structure at core pressures and temperatures. However, the differences in free energies are small (33–58 meV/atom along T_m). The Earth's inner core is known not to be made of pure Fe, but is expected to be alloyed with between 5 to 10 mol% of a lighter element. It is argued that, either separately or together, S and Si are two of the most probable light elements alloyed with Fe in the core [*Poirier*, 1994a; *Allègre et al.*, 1995]. Recent experiments [*Dobson et al.*, 2002] have shown that at high pressures FeSi crystallises with the CsCl-structure (i.e. has identical atomic co-ordinates to bcc-Fe), and it has been found that at low concentrations Si favours the formation of bcc Fe over the hcp polymorph [*Lin et al.*, 2002; *Dubrovinsky et al.*, 2003]. We therefore investigated the energetic effect of the substitution of S and Si in bcc and hcp Fe at representative core densities. We found that the enthalpies of the S and Si defects are respectively 1.4 and 1.2 eV per defect atom more stable in the bcc structure than in the hcp phase. Therefore, for example, a 5 mol% concentration of Si in Fe, would stabilise the bcc phase by 60 meV. Thus, we conclude that the presence of S or Si as the light impurity element in the core, at appropriate concentrations, could favour the formation of a bcc- rather than an hcp-structured Fe alloy phase at temperatures just below T_m at inner core pressure. The presence of a bcc-structured alloy may, therefore, be a candidate for explaining the observed seismic complexity of the inner core recently reported by *Beghein and Trampert* [2003], however more work on the high T elasticity of this and other candidate phases is needed before the seismic anisotropy of the inner core can be fully understood.

CONCLUSION

The past decade has seen a major advance in the application of ab initio methods in the solution of high pressure and temperature geophysical problems, thanks to the rapid developments in high performance computing. We are now in a position to calculate from first principles the free energies of solid and liquid phases, and hence to determine both the phase relations and the physical properties of planetary forming phases. So far we have been able to study the effects of S, O and Si on the behaviour of Fe, but in the future it will be essential also to consider other elements, including Ni, C, H and K. However, from what has been studied to date, it is likely that the ICB temperature lies between 5400 and 5650K and the core temperature at the CMB is between 3950 and 4200K. Using ab initio methods it has been shown the composition of the core can be constrained, but estimates are very sensitive to the seismologically inferred values of core density.

In the future, the clearer insight that ab initio modelling gives to our understanding of the properties of the core will enable the development of more quantitative descriptions of the history of the Earth, and even to the establishment of quantitative models of the geodynamo [see for example *Gubbins et al.*, 2003]. The high P and T structure and elastic properties of Fe and its alloys are still the subject of some uncertainty, but theory should soon be able to resolve these issues.

We look forward to the advent of routinely available 'terascale' computing. This will open new possibilities for geophysical modelling. It will be possible to model more complex and larger systems, to investigate for example solid state rheological problems, or physical properties such as thermal and electrical conductivity, and to model the possible chemical processes occurring at the CMB. However, it must be recognised that the DFT methods, that are currently used, are still approximate, and fail for example to describe the band structure of important phases such as FeO. In the future it will be necessary to consider using terascale facilities to implement more demanding but more accurate techniques, such as those based on quantum Monte Carlo methods. Nevertheless, significant progress has been made to date, and it is certain than in the future further insights from computational mineral physics will enhance the understanding of our planet.

Acknowledgements. DA and LV are supported by Royal Society University Research Fellowships. DA also acknowledges support from the Leverhulme Trust. MJG thanks GEC and Daresbury Laboratory for their support. Our work reported in this paper was supported by NERC grants GR3/12083 and GR9/03550. The calculations were run on the Cray T3D and Cray T3E machines at Edinburgh and the Manchester CSAR Centre, and on the Origin 2000 machine at the UCL HiPerSPACE Centre. We would like to thank Renata Wentzcovitch, Steve Sparks and Sandro Scandolo for helpful comments and corrections.

REFERENCES

Alfè, D., Program available from, http://chianti.geol.ucl.ac.uk/~dario/phon.tar.z, 1998.

Alfè, D., Gillan, M. J., and Price, G. D., The melting curve of iron at the pressures of the Earth's core from ab initio calculations. *Nature*, *401*, 462–464, 1999

Alfè, D., Gillan, M. J., and Price, G. D., Composition and temperature of the Earth's core constrained by combining ab initio calculations and seismic data. *Earth Planet. Sci. Lett.*, *195*, 91–98, 2002a.

Alfè, D., Gillan, M. J., and Price, G. D., Ab initio chemical potential of solid and liquid solutions and chemistry of the Earth's core. *J. Chemical Physics.*, *116*, 7127–7136, 2002c.

Alfè, D., Gillan, M. J., and Price, G. D., Complementary approaches to the ab initio calculation of melting properties. *J. Chemical Physics.*, *116*, 6170–6177, 2002d.

Alfè, D., Price, G. D., and Gillan, M. J., Thermodynamics of hexagonal-close-packed iron under Earth's core conditions. *Physical Review.*, B64, 045123, 2001.

Alfè, D., Price, G. D., and Gillan, M. J., Iron under Earth's core conditions: Liquid-state thermodynamics and high-pressure melting curve from ab initio calculations. *Phys. Rev.*, B65, 165118, 1–11, 2002b.

Allègre, C. J., Poirier, J. P., Humler, E., and Hofmann, A. W., The chemical composition of the Earth. *Phys. Earth Planet. Interiors.*, 134, 515–526, 1995.

Allen, M. P., and Tildesley, D. J., *Computer Simulation of Liquids.* Oxford University Press, Oxford, UK, 1987.

Anderson, O. L., *Equations of state of solids for Geophysics and ceramic science.* Oxford University Press, Oxford, UK, 1995.

Anderson, O. L., The three-dimensional phase diagram of iron. In: *Earth's Core* (eds. V. Dehant, K. C. Creager, S-i. Karato, S. Zatman). AGU, Geodynamics Series, 31, 83–104, 2003.

Andrault, D., Fiquet, G., Kunz, M., Visocekas, F., and Häusermann, D., The orthorhombic structure of iron: an in situ study at high temperature and high pressure. *Science*, 278, 831–834, 1997.

Beghein, C., and Trampert, J., Robust Normal Mode Constraints on Inner-Core Anisotropy from Model Space Search. *Science*, 299, 552–555, 2003.

Belonoshko, A. B., Ahuja, R., and Johansson, B., Quasi- ab initio molecular dynamic study of Fe melting. *Phys Rev Lett.*, 84, 3638–3641, 2000.

Birch, F., Elasticity and the constitution of the Earth's interior. *J. Geophys. Res.*, 5, 227–286, 1952.

Birch, F., Density and composition of mantle and core. *J. Geophys. Res.*, 69, 4377–4388, 1964.

Blöchl, P. E., Projector augmented-wave method. *Phys. Rev.*, B 50, 17953–17979, 1994.

Boehler, R., Temperature in the Earth's core from the melting point measurements of iron at high static pressures. *Nature*, 363, 534–536, 1993.

Boehler, R., High-pressure experiments and the phase diagram of lower mantle and core materials. *Reviews of Geophys.*, 38, 221–245, 2000.

Born, M., and Huang, K., *Dynamical Theory of Crystal Lattices.* Oxford University Press, Oxford, UK, 1954.

Brown, J. M., The equation of state of Iron to 450GPa: another high pressure solid phase? *Geophys. Res. Letts.*, 28, 4339–4342, 2001.

Brown, J. M., and McQueen, R. G., Phase transitions, Grüneisen parameter and elasticity for shocked iron between 77 GPa and 400 GPa. *J. Geophys. Res.*, 91, 7485–7494, 1986.

Car, R., and Parrinello, M., Unified Approach for Molecular Dynamics and Density Functional Theory. *Phys. Rev. Lett.*, 55, 2471–2474, 1985.

Creager, K. C., Anisotropy of the inner core from differential travel times of the phases PKP and PKIKP. *Nature*, 356, 309–314, 1992.

Dobson, D. P., Vocadlo, L., and Wood, I. G., A new high-pressure phase of FeSi. *Amer.Mineral.*, 87, 784–787, 2002.

Dubrovinsky, L., Dubrovinskaia, N., Langenhorst, F., Dobson, D., Rubie, D., Gessmann, C., Abrikosov, I.A., Johansson, B., Baykov, V .I., Vitos, L., Le Bihan, T., Crichton, W. A., Dmitriev, V., and Weber, H. P., Iron-silica interaction at extreme conditions and the electrically conducting layer at the base of Earth's mantle. *Nature*, 422, 58–61, 2003.

Dziewonski, A. M., and Gilbert, F., Solidity of the inner core of the Earth inferred from normal mode observations. *Nature*, 234, 465–466, 1971.

Fiquet, G., Badro, J., Guyot, F., Requardt, H., and Krisch, M., Sound velocities in iron to 110 gigapascals. *Science*, 291, 468–47, 2001.

Gannarelli, C. M. S., Alfè, D., and Gillan, M. J., The particle-in-cell model for ab initio thermodynamics: implications for the elastic anisotropy of the Earth's inner core. *Phys. Earth Planet. Interiors.*, 139, 243–253, 2003.

Gao, F., Johnston, R. L., and Murrell, J. N., Empirical many-body potential energy functions for Iron. *J. Phys. Chem.*, 97, 12073–12082, 1993.

Gillan, M. J., 1997. The virtual matter laboratory. *Contemp. Phys.*, 38, 115–130.

Gubbins, D., Alfè, D., Masters, G., Price, G. D., and Gillan, M. J., Can the Earth's dynamo run on heat alone? *Geophys. J. Int.*, 155, 609–622, 2003.

Gutenberg, B., Uber die Konstitution der Erdinnern, erschlossen aus Erdbebenbeobachtungen. *Phys. Zeit.*, 14, 1217, 1913.

Ishii, M., and Dziewonski, A. M., The innermost inner core of the earth: Evidence for a change in anisotropic behaviour at the radius of about 300 km. *Proceedings of the National Academy of Sciences of the United States of America*, 99, 14026–14030, 2002.

Karki, B. ., Wentzcovitch, R. M., de Gironcoli, S., Baroni, S., First-principles determination of elastic anisotropy and wave velocities of MgO at lower mantle conditions. *Science*, 286, 1705–1707, 1999.

Kresse, G., Furthmüller, J., and Hafner, J., Ab-initio force-constant approach to phonon dispersion relations of diamond and graphite. *Europhys. Lett.*, 32, 729–734, 1995.

Kresse, G., and Furthmüller, J., Efficient iterative schemes for ab-initio total-energy calculations using a plane-wave basis set. *Phys. Rev.*, B54, 11169–11186, 1996.

Kresse G., Joubert D., From ultrasoft pseudopotentials to the projector augmented-wave method. *Phys. Rev.*, B59, 1758–1775, 1999.

Laio, A., Bernard, S., Chiarotti, G.L., Scandolo, S., and Tosatti, E., Physics of iron at Earth's core conditions. *Science*, 287, 1027–1030, 2000.

Lehmann, I., P'. *Trav. Sci. Sect. Seis. U.G.G.I. (Toulouse)*, 14, 3–31, 1936.

Lin, J-F., Heinz, D. L., Campbell, A. J., Devine, J. M., and Shen, G., Iron-silicon alloy in Earth's core? *Science*, 295, 313–315, 2002.

Mao, H. K., Shu, J., Shen, G., Hemley, R. J., Li, B., and Sing, A. K., Elasticity and rheology of iron above 220GPa and the nature of the Earth's inner core. *Nature*, 399, 280–282, 1999.

Mao, H. K., Xu, J., Struzhkin, V. V., Shu, J., Hemley, R. J., Sturhahn, W., Hu, M. Y., Alp, E. E., Vocadlo, L., Alfè, D., Price, G. D., Gillan, M. J., Schwoerer-Bohning, M., Hausermann, D., Eng, P., Shen, G., Giefers, H., Lubbers, R., and Wortmann, G., Phonon density of states of iron up to 153 gigapascals. *Science*, 292, 914–916, 2001.

Masters, G., and Gubbins, D., On the resolution of density within the Earth. *Phys. Earth Planet. Interiors.*, 140, 159–167, 2003.

Matsui, M., and Anderson, O. L., The case for a body-centered cubic phase for iron at inner core conditions. *Phys. Earth Planet. Interiors.*, *103*, 55–62, 1997.

McDonough, W. F., and Sun, S.-S., The composition of the Earth. *Chem. Geol.*, *120*, 223–253, 1995.

Nguyen, J. H., and Holmes, N. C., Iron sound velocities in shock wave experiments. *Shock Compression of Condensed Matter.*,*CP505*, 81–84, 1999.

Oldham, R. D., The constitution of the Earth as revealed by earthquakes. *Quart. J. Geol. Soc.*, *62*, 456–75, 1906.

Oganov, A. R., Brodholt, J. P., Price, G. D., The elastic constants of $MgSiO_3$ perovskite at pressures and temperatures of the Earth's mantle. *Nature*, *411*, 934–937, 2001.

Parker, S. C., and Price, G. D., Computer Modelling of Phase Transitions in Minerals. *Advances in Solid State Chemistry*, *1*, 295–327, 1989.

Parrinello, M., and Rahman, A., Crystal Structure and Pair Potentials: A Molecular Dynamics Study. *Phys. Rev. Lett.*, *45,* 1196–1199, 1980.

Poirier, J. P., Light elements in the Earth's outer core: A critical review. *Phys. Earth Planet. Interiors.*, *85*, 319–337, 1994a.

Poirier, J. P., Physical-properties of the Earths core. *Cr. Acad. Sci., II. 318*, 341–350, 1994b.

Poirier, J. P., and Shankland, T. J., Dislocation Melting Of Iron and the Temperature of the Inner-Core Boundary, Revisited. *Geophys. J. Int.*, *115*, 147–151, 1993.

Ross, M., Young, D. A., and Grover, R., Theory of the iron phase diagram at Earth core conditions. *J. Geophys. Res.*, *95*, 21,713–21,716, 1990.

Saxena, S. K., Dubrovinsky, L. S., and Häggkvist, P., X-ray evidence for the new phase of ß-iron at high temperature and high pressure. *Geophys. Res. Lett.*, *23*, 2441–2444, 1996.

Shearer, P. M., and Masters, G., The density and shear velocity contrast at the inner core boundary. *Geophys. J. Int.*, *102*, 491–498, 1990.

Scandolo, S., First-principles molecular dynamics simulations at high pressure. *Proc. of the International School of Physics "E. Fermi" on "High Pressure Phenomena", Course CXLVII* (IOS, Amsterdam), 195–214, 2002.

Shen, G. Y., and Heinze, D. L., High Pressure melting of deep mantle and core materials. In: *Ultrahigh-pressure Mineralogy* (ed. R.J. Hemley). Reviews in Mineralogy, *37*, 369–398, 1998.

Shen, G. Y., Mao, H. K., Hemley, R. J., Duffy, T. S., and Rivers, M. L., Melting and crystal structure of iron at high pressures and temperatures. *Geophys. Res. Lett.*, *25*, 373–376, 1998.

Söderlind, P., Moriarty, J. A., and Wills, J. M., First-principles theory of iron up to earth-core pressures: Structural, vibrational and elastic properties. *Phys. Rev.*, *B53*, 14063–14072, 1996.

Song, X. D., Anisotropy of the Earth's Inner core. *Reviews Geophys.*, *35*, 297–313, 1997.

Song, X. D., and Helmberger, D. V., Seismic evidence for an inner core transition zone. *Science, 282*, 924–927, 1998.

Steinle-Neumann, G., Stixrude, L., and Cohen, R. E., First-principles elastic constants for the hcp transition metals Fe, Co, and Re at high pressure. *Phys. Rev.*, *B60*, 791–799, 1999.

Steinle-Neumann, G., Stixrude, L., and Cohen, R. E., Physical properties of iron in the inner core. In: *Earth's Core* (eds. V. Dehant, K. C. Creager, S-i. Karato, S. Zatman). AGU, Geodynamics Series, *31*, 137–162, 2003.

Steinle-Neumann, G., Stixrude, L., Cohen, R. E., and Gülseren, O., Elasticity of iron at the temperature of the Earth's inner core. *Nature*, *413*, 57–60, 2001.

Stixrude, L., and Brown, J.M., The Earth's Core. In: *Ultrahigh-pressure Mineralogy* (ed. R.J. Hemley). Reviews in Mineralogy, *37*, 261–283, 1998.

Stixrude, L., and Cohen, R. E., Constraints on the Crystalline Structure of the Inner Core - Mechanical Instability of bcc Iron at High-Pressure. *Geophys. Res. Lett.*, *22*, 125–128, 1995.

Stixrude, L., Cohen, R. E., and Hemley, R. J., Theory of minerals at high pressure. In: *Ultrahigh-pressure Mineralogy* (ed. R. J. Hemley). Reviews in Mineralogy, *37*, 639–671, 1998.

Stixrude, L., Wasserman, E., and Cohen, R. E., Composition and temperature of the Earth's inner core. *J. Geophys. Res.*, *102*, 24729–24739, 1997.

Vočadlo, L., and Alfè, D., The ab initio melting curve of aluminum. *Phys. Rev., B65*, 214105, 1–12, 2002.

Vočadlo, L., Alfè, D., Gillan, M. J., Wood, I. G., Brodholt, P. J., and Price, G. D., Possible thermal and chemical stabilisation of body-centred- cubic iron in the Earth's core? *Nature*, *424*, 536–53, 2003.

Vočadlo, L., Brodholt, J., Alfè, D., Price, G. D., and Gillan, M. J., The structure of iron under the conditions of the Earth's inner core. *Geophys. Res. Lett.*, *26*, 1231–1234, 1999.

Vočadlo, L., deWijs, G. A., Kresse, G., Gillan, M. J., and Price, G. D., First principles calculations on crystalline and liquid iron at Earth's core conditions. *Faraday Discussions, 106*, 205–217, 1997.

Wasserman, E., Stixrude, L., and Cohen, R. E., Thermal properties of iron at high pressures and temperatures. *Phys. Rev.*, *B53*, 8296–8309, 1996.

Wentzcovitch, R. M., Invariant Molecular-Dynamics Approach to Structural Phase-Transitions. *Phys. Rev.*, *B44*, 2358–2361, 1991.

de Wijs, G. A., Kresse, G., and Gillan, M. J., First order phase transitions by first principles free energy calculations: The melting of Al. *Phys. Rev., B57*, 8233–8234, 1998.

Yoo, C. S., Holmes, N. C., Ross, M., Webb, D. J., and Pike, C., Shock temperatures and melting of iron at Earth core conditions. *Phys. Rev. Lett., 70*, 3931–3934, 1993.

G. D. Price, Research School of Earth Sciences, Birkbeck and University College London, Gower Street, London, WC1E 6BT, UK; d.price@ucl.ac.uk

D. Alfè, Research School of Earth Sciences, Birkbeck and University College London, Gower Street, London, WC1E 6BT, UK; d.afle@ucl.ac.uk

L. Vočadlo, Research School of Earth Sciences, Birkbeck and University College London, Gower Street, London, WC1E 6BT, UK; l.vocadlo@ucl.ac.uk

M. J. Gillan, Department of Physics and Astronomy, University College London, Gower Street, London WC1E 6BT, UK; m.gillan@ucl.ac.uk

Modeling the Earth's Dynamo

Gary A. Glatzmaier and Darcy E. Ogden

Earth Sciences Department, University of California, Santa Cruz, California

Thomas L. Clune

Science Computing Branch 931, NASA Goddard Space Flight Center, Greenbelt, Maryland

For the past decade, three-dimensional time-dependent computer models have been used to predict and explain how the geomagnetic field is maintained by convection in the Earth's fluid core. Geodynamo models have simulated magnetic fields that have surface structure and time dependence similar to the Earth's field, including dipole reversals. However, no dynamo model has yet been run at the spatial resolution required to simulate a broad spectrum of turbulence, which surely exists in the Earth's fluid core. Two-dimensional simulations of magnetoconvection show how the structure and time dependence of even the large-scale features change dramatically when the solution becomes strongly turbulent. Although these two-dimensional turbulent simulations lack the important effects of three-dimensional spherical geometry, based on their results one must question how geophysically realistic the large-scale dynamo mechanism is in current three-dimensional laminar simulations. Whatever the answer, we look forward to new discoveries from the next generation of turbulent dynamo models.

1. BASIC GEODYNAMO THEORY

We spend our entire lives immersed within the Earth's magnetic field yet most people today seldom stop to ponder about what it is, why it exists, how it changes or how life would be different without it. The geomagnetic field helps to shield us from cosmic radiation and for ages has been used for navigation. However, the main reason for studying the geodynamo is to gain a better understanding and appreciation for the origin and evolution of this fascinating field we live in.

It has long been known that the geomagnetic field originates in the Earth's core. However, the temperature of the core is well above the Curie temperature for permanent magnetism; the field would decay away with a half-life of about 15,000 years if it were not continually being regenerated. The Earth's magnetic field must therefore be maintained by large electric currents within its iron-rich fluid core.

The Earth is cooling and the resulting drop in temperature with radius through the fluid core is steep enough to drive thermal convection. In addition, as the core cools, iron in the iron-alloy fluid preferentially plates onto the solid inner core releasing latent heat and leaving behind a higher concentration of light elements. Thermal and compositional buoyancy at the inner core boundary (ICB) are more effective in driving convection than the secular cooling of the entire fluid core because they originate far from the rising fluid's destination: the core–mantle boundary (CMB). Additional volumetric heating sources may exist in the fluid core due to radioactive decay, potassium-40 being the best candidate [*Chabot and Drake*, 1999; *Buffett*, 2000; *Brodholt and Nimmo*, 2002; *Gessman and Wood*, 2002; and *Murthy et al.*, 2003]; the amount is uncertain and the focus of many current studies. Its presence would have a significant effect on the age of the Earth's solid inner core and the style of convection and magnetic field generation in the early Earth.

Hot, light fluid at the ICB, however, does not rise straight up to the CMB and likewise cold heavy fluid does not sink straight down. Coriolis forces (within the frame of reference of the rotating Earth) cause the low-viscosity fluid to flow in curved trajectories. The resulting twisting and shearing that this electrically conducting fluid does to the existing magnetic field converts kinetic energy into magnetic energy. Another way of saying this is that flow of metallic fluid through magnetic field produces an electromotive force that drives large electric currents (Ohm's law). These electric currents induce new magnetic fields around them (Ampere's law) that compensate for the continual decay of the magnetic field (Faraday's law).

The fundamentals of dynamo theory, first proposed in the 1950's, are the following: that differential rotation within the fluid core shears poloidal (north–south and radial) magnetic field into toroidal (east–west) magnetic field and that three-dimensional (3D) helical fluid flow twists toroidal field into poloidal field. At the same time, the more sheared and twisted the field the faster it decays away; that is, magnetic diffusion (reconnection) continually smooths out the field. The field is self-sustaining if, on average, the generation of field is balanced by its decay. Discovering and understanding the details of how rotating convection in the Earth's fluid outer core does this to maintain the observed intensity, structure and time dependencies requires 3D computer models of the geodynamo.

2. HISTORY OF DYNAMO MODELING

Many studies of magnetic field generation have been based on kinematic models. In these models the fluid flow induces magnetic field; but the feedback on the flow via Lorentz forces, which resist the movement of conducting fluid through the magnetic field, is neglected. Some compute or prescribe a large-scale 3D flow structure [*e.g., Kumar and Roberts*, 1975]. Other models compute or prescribe only the axisymmetric (two-dimensional) part of the flow and use it with a parameterized longitudinally-averaged effect of a hypothetical 3D helical flow to compute an axisymmetric magnetic field. Much has been learned from these studies about the types of fields that can be induced by various flow structures. However, the lack of the time-dependent magnetic feedback on the flow precludes a detailed prediction or explanation of the 3D flow and field structures in the Earth's core, i.e., the geodynamo mechanism. In addition, based on rough estimates of the flow and field amplitudes in the core, the magnetic energy in the core is likely a thousand times greater than the kinetic energy of the flow that maintains it; therefore the neglect of Lorentz forces in kinematic dynamo models is not really appropriate for detailed investigations of the Earth's dynamo mechanism.

Magnetohydrodynamic (MHD) dynamo simulations, on the other hand, are dynamically self-consistent, but also more challenging. They solve a coupled set of nonlinear differential equations that describe the 3D evolution of the thermodynamic variables, the fluid velocity and the magnetic field with all the major feedbacks. These geodynamo simulations are conducted to investigate the structure and time-dependence of convection and magnetic field generation in the Earth's core. So little can be directly observed other than the surface magnetic field (today's field in detail and the paleomagnetic field in much less detail) and what can be inferred from seismic measurements and variations in the length of the day and possibly in the gravitational field. Therefore, geodynamo models are used as much to predict what has not been observed as they are used to explain what has. When an MHD computer model generates a magnetic field with structure, intensity, and time dependence qualitatively similar to the Earth's surface field, then it is plausible that the 3D flows and fields inside the model core are qualitatively similar to those in the Earth's core. Analyzing this detailed simulated data provides a physical description and explanation of the model's dynamo mechanism and, by assumption, of the geodynamo.

The first 3D global convective dynamo simulations were developed in the 1980's to study the *solar* dynamo. Gilman and Miller [1981] pioneered this style of research by constructing the first 3D MHD dynamo model. However, they simplified the problem by specifying a constant background density, i.e., they used the Boussinesq approximation of the equations of motion. Glatzmaier [1984] developed a 3D MHD dynamo model using the anelastic approximation, which accounts for the stratification of density within the sun. Zhang and Busse [1988] used a 3D model to study the onset of dynamo action within the Boussinesq approximation. However, the first 3D MHD models of the *geodynamo* that successfully produced a dominantly dipolar field at the model's surface were not published until 1995 [*Kageyama et al.*, 1995; *Jones et al.*, 1995; and *Glatzmaier and Roberts*, 1995a]. Since then several groups around the world have developed dynamo models [*e.g., Kuang and Bloxham*, 1997; *Sarson et al.*, 1997; *Kida et al.*, 1997; *Busse et al.*, 1998; *Christensen et al.*, 1998; *Sakuraba and Kono*, 1999; *Katayama et al.*, 1999; and *Hollerbach*, 2000] and several others are currently being designed. Some features of the various simulated fields are robust, like the dominance of the dipolar part of the field outside the core. Other features, like the 3D structure and time dependence of the temperature, flow and field inside the core, depend somewhat on the chosen boundary conditions, parameter space and numerical resolution. Many review articles have been written that describe and compare these models [*e.g., Hollerbach*, 1996;

Glatzmaier and Roberts, 1997; Fearn, 1998; Busse, 2000; Roberts and Glatzmaier, 2000; Dormy et al., 2000; Christensen et al., 2001; Busse, 2002; Kono and Roberts, 2002; and Glatzmaier, 2002].

Although considerable progress has been made, none of these models has been able to run in a realistic parameter regime for the Earth's core. The problem is that, because of the low fluid viscosity, a broad spectrum of turbulence likely exists in the core, from global length scales down to scales on the order of meters; whereas current simulations have about 10 to 100 km resolution at best. In addition, because of the dominance of the Coriolis and Lorentz forces, this turbulence is heterogeneous and anisotropic. Capturing a good portion of this spectrum in a 3D geodynamo simulation would require more computing resources than are currently available. Therefore, greatly enhanced diffusion coefficients have been used in geodynamo simulations, which produces smooth large-scale (laminar) flows, not turbulence.

3. MODEL DESCRIPTION

We begin by briefly describing the Glatzmaier-Roberts geodynamo model, which employs the anelastic equations of motion. These equations allow for a realistic variation of density, temperature and pressure with depth, while filtering out acoustic waves. Effectively, the sound speed is assumed infinite so the pressure distribution at each numerical time step is in equilibrium throughout the fluid core. This approximation is valid when the fluid velocity is small relative to the local sound speed and the thermodynamic variations, or perturbations, are small relative to their background reference state values. These conditions are very well satisfied for the Earth's liquid core. The reason for using this approximation is that the numerical time step can be about a million times larger since fast-moving low-energy sound waves are not simulated. All other geodynamo models have employed the Boussinesq approximation, which assumes a constant background density. The small (20%) change in density across the Earth's fluid core would seem to justify this simpler approach; however, an important source of vorticity, which helps to generate magnetic field, is consequently neglected.

The anelastic MHD equations [Glatzmaier, 1984; and Braginsky and Roberts, 1995] that define our dynamo model describe the 3D, time-dependent perturbations of the flow, field and thermodynamic variables relative to a no-flow, non-magnetic, radially-dependent thermodynamic reference state fitted to the Preliminary Reference Earth Model (PREM). The set of equations ensures mass and magnetic flux conservation. An equation of state relates perturbations in the entropy, pressure and composition to density perturbations, which determine the perturbations in the gravitational field and the buoyancy forces. Newton's second law of motion determines how the local fluid velocity changes with time due to buoyancy, pressure gradient, viscous, Coriolis and Lorentz forces and to advection of momentum by the flow. The magnetohydrodynamic equations (i.e., Maxwell's equations and Ohm's law with the extremely good assumption that the fluid velocity is small relative to the speed of light) describe how the local magnetic field changes with time due to induction by the flow and diffusion due to finite conductivity. The second law of thermodynamics dictates how advection of entropy by the flow, thermal diffusion, and Joule and viscous heating determine the local time rate of change of entropy. A final advection-diffusion equation describes the local time rate of change of composition. These equations are solved each numerical time step to obtain the evolution of the 3D fluid flow and magnetic field and the perturbations in density, pressure, specific entropy, light constituent mass fraction and gravitational potential relative to the background reference state in a frame of reference rotating with the Earth's mean angular velocity.

The thermal boundary conditions at the ICB constrain the local flux of latent heat to be proportional to the local flux of light constituents and to the local cooling rate [Glatzmaier and Roberts, 1996a]. The flux of light constituents through the CMB is set to zero. A non-zero heat flux is prescribed at the CMB; it controls the cooling rate of the core and therefore the production rate of buoyancy sources at the ICB and thus affects the intensity of the magnetic field. Normally, in this model the total heat flow out of the core is set to 7.2 TW, the generally assumed, although highly uncertain, value for the Earth. Of this, 5 TW is due to heat flow conducted down the adiabat. (Note, thermal convection occurs when the total temperature drop across the fluid outer core is greater than what it would be for an adiabatic temperature profile.)

The solid inner core and solid mantle rotate in reaction to magnetic torques, viscous torques if non-slip boundary conditions are applied, and gravitational torques if the gravitational forces between the mantle and the inner core are included [Buffett and Glatzmaier, 2000]. Viscous torques in the model represent the rate of momentum transfer between the fluid and the boundaries by small unresolved turbulence. The total angular momentum of the inner core, fluid core and mantle remains zero in the rotating frame, but the angular momentum of each of these is time dependent.

The magnetic field is generated in the model's fluid outer core and diffuses into its conducting solid inner core. Since the electrical conductivity of the Earth's mantle is very small relative to that of the core, everything above a thin layer at the base of the mantle is assumed to be a perfect insulator. The external magnetic field is therefore a potential field, albeit time dependent and determined by the dynamics inside the core.

The set of coupled nonlinear equations is solved at each numerical time step using a spectral method (spherical harmonic and Chebyshev polynomial expansions) [*Glatzmaier*, 1984; and *Glatzmaier and Roberts*, 1996a]. The nonlinear terms are computed each time step by a spectral transform method, i.e., the simulated data are transformed from wave number space to grid space where the nonlinear products are computed, which are then transformed back to wave number space where the solution is advanced a time step. Simulations that span hundreds of thousands of years, involving tens of millions of numerical time steps, have been run at very low spatial resolution (*e.g.*, 33 radial, 32 latitudinal, 64 longitudinal levels). Simulations have also been done at much higher resolution (*e.g.*, 289 radial, 384 latitudinal, 384 longitudinal levels), which are more accurate because they resolve more of the energy spectrum; but they span less simulated time, on the order of tens of thousands of years.

The code is run on massively parallel computers with the data spread over the processors according to radial grid levels and spherical harmonic wave numbers. The spectral method requires each processor to send data to and receive data from every other processor every time step. This global communication among the parallel processors quickly grows as one increases the number of processors. In addition, although fast Fourier transforms are used for the spectral transform in longitude and radius, no efficient fast transform exists in latitude, and so this operation is very time consuming at high resolution.

The review articles mentioned above describe the variations on the equations and numerical methods employed in other geodynamo models [*e.g., Glatzmaier*, 2002]. Most dynamo models use spherical harmonic expansions, which have proven to be very accurate and efficient at relatively low spatial resolution. However, to reach much greater spatial resolution in the future, new methods are being developed that do not employ spherical harmonics and therefore avoid global inter-processor communication and the latitudinal spectral transform.

4. MODEL RESULTS

Since the mid 1990's 3D computer simulations have advanced our understanding of the geodynamo. This is true even though they have been forced to use significantly enhanced diffusion coefficients that result in large-scale laminar convection. That is, the boundaries have a significant influence on the structure of the simulated flow because convective cells and plumes typically span the entire depth of the fluid outer core, unlike the turbulence that likely exists in the Earth's core. The simulations have however shown that a strong, dominantly dipolar magnetic field, not unlike the Earth's, can be maintained by convection driven by an Earth-like heat flux.

A typical snapshot of the simulated magnetic field from a geodynamo model is illustrated in Figure 1 with a set of field lines. In the fluid outer core, where the field is generated, field lines are twisted and sheared by the flow. The field that extends beyond the core is significantly weaker and dominantly dipolar at the model's surface, not unlike the geomagnetic field. For most geodynamo simulations, the non-dipolar part of the surface field, at certain locations and times, propagates westward at about 0.2° per year, as has been observed in the geomagnetic field over the past couple hundred years [*e.g., Bloxham and Jackson*, 1992].

Several dynamo models have electrically conducting inner cores that on average drift eastward relative to the mantle [*e.g., Glatzmaier and Roberts*, 1995a; *Sakuraba and Kono*, 1999; and *Christensen et al.* 2001], opposite to the propagation direction of the surface magnetic field. Inside the fluid core the simulated flow has a "thermal wind" component that, near the inner core, is predominantly eastward relative to the mantle. The magnetic field in these models that permeates both this flow and the inner core tries to drag the inner core in the direction of the flow. This magnetic torque is resisted by a gravi-

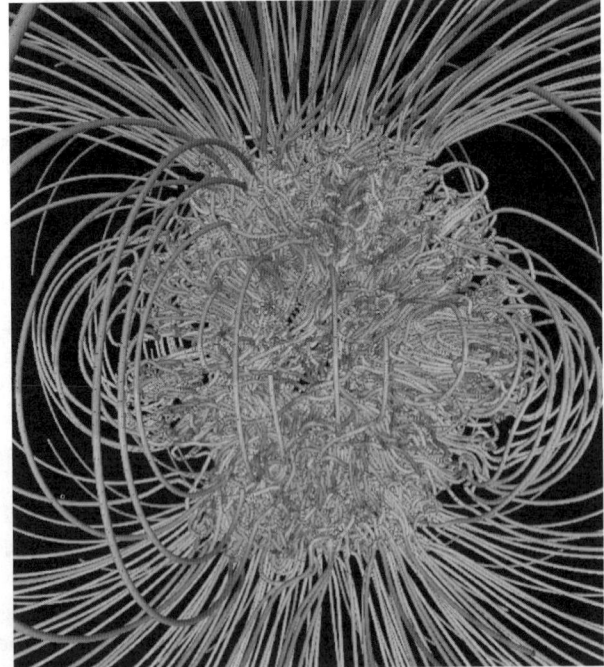

Figure 1. A snapshot of the 3D magnetic field structure generated with the Glatzmaier-Roberts geodynamo model and illustrated with a set of magnetic lines of force. The axis of rotation is vertical and centered in the image. The field is a smooth, dipole-dominated, potential field outside the core.

tational torque between the inner core and mantle [*Buffett and Glatzmaier*, 2000].

Figure 2 shows the resulting time dependent angular rotation rate of the inner core relative to the mantle during a short interval of a simulation that prescribed a vanishing viscous torque on the inner core. Most of the time the inner core is rotating slightly faster than the mantle; but the rate is highly variable. The average super-rotation of the model's inner core depends on the very poorly constrained viscosity assumed for the inner core's deformable surface layer, which by definition is near the melting temperature. The amplitude of the super-rotation rate predicted by geodynamo models depends on the model's specified parameters and assumptions; the original prediction was an average of about 2° longitude per year [*Glatzmaier and Roberts*, 1995a]. Since then the super-rotation rate of the Earth's inner core today has been inferred from several seismic analyses [*e.g., Song and Richards*, 1996; *Su et al.*, 1996; and *Xu and Song*, 2003], but is still quite uncertain [*e.g., Creager*, 1997; *Souriau*, 1998; and *Laske and Masters*, 1999]. Currently there is a spread in the inferred values, from the initial estimates of 1° to 3° eastward per year (relative to the Earth's surface) to some that are zero to within an uncertainty of 0.2° per year. The average simulated rate in Figure 2 is closer to the lower end of this range; but again this depends on how deformable the model's inner core surface is assumed to be.

On a much longer time scale, some dynamo simulations have produced spontaneous non-periodic magnetic dipole reversals [*Glatzmaier and Roberts*, 1995b; *Sarson and Jones*, 1999; *Kageyama et al.*, 1999; *Glatzmaier et al.*, 1999; and *Kutzner and Christensen*, 2002]. Periodic reversals, like the *dynamo-wave* reversals seen in early solar dynamo simulations [*Gilman*, 1983], have also occurred in recent dynamo simulations [*Kida et al.*, 1997; *Kida and Kittauchi*, 1998; *Grote et al.*, 1999, 2000; and *Simitev and Busse*, 2002]. The highly variable times between reversals seen in the paleomagnetic record are measured in hundreds of thousands of years; whereas the time to complete a reversal is typically a few thousand years, less than a magnetic dipole decay time [*e.g., Merrill and McFadden*, 1999].

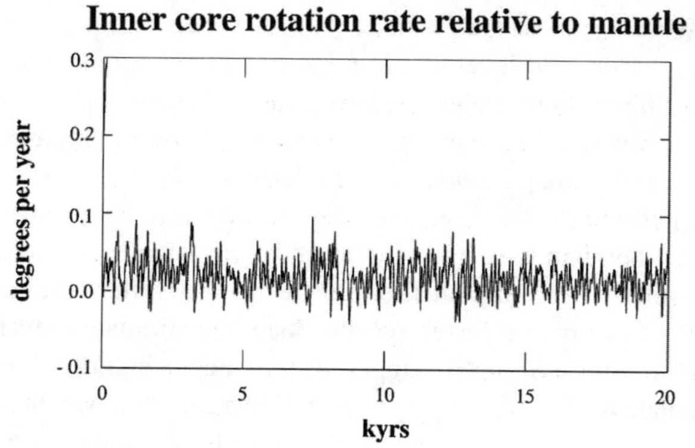

One of the simulated reversals is portrayed in Figure 3 with four snapshots spanning about 9000 years. The radial component of the field is shown at both the CMB and the surface of the model Earth. The reversal, as viewed in these surfaces, begins with reversed magnetic flux patches in both the northern and southern hemispheres. (It is interesting that a reversed flux patch is currently growing in the Earth's southern hemisphere as the dipole moment of the geomagnetic field is slowly decreasing.) In addition, the longitudinally-averaged poloidal and toroidal parts of the field inside the core are illustrated at these times. Although when viewed at the model's surface the reversal appears complete by the third snapshot, another three thousand years is required for the original field polarity to decay out of the inner core and the new polarity to diffuse in.

On an even longer time scale the frequency of reversals seen in the paleomagnetic record varies. The frequency of non-periodic reversals in geodynamo simulations has been found to depend on the magnitude of the convective driving relative to the effect of rotation [*Kutzner and Christensen*, 2002]. It also depends on the pattern of outward heat flux imposed over the CMB [*Glatzmaier, et al.* 1999]. In the Earth, the CMB heat flux distribution might be determined by the distribution of cold subducted slabs at the base of the mantle. Since this distribution is determined by mantle convection, which is a million times slower than core convection, time-independent patterns of the CMB heat flux are prescribed via thermal boundary conditions. In addition, convection in the low viscosity outer core is so efficient that the maximum temperature variation in the core fluid at the CMB is no more than 10^{-3} K, compared with temperature variations in the high viscosity mantle of the order of hundreds of degrees K. Therefore the temperature drop, and so also the heat flux, between the core and mantle is presumably greatest where the mantle is relatively cold.

Two case studies are illustrated in Figure 4. One has higher heat flux imposed in the polar regions and in the equatorial region, with minimum heat coming out at mid-latitude. The other has a CMB heat flux patterned after today's seismic tomography of the Earth's lower mantle, with high heat flux along the high seismic velocity "rim around the Pacific" where cold slabs are thought to currently reside. The latitude of the south magnetic pole and the dipole moment are plotted for 300,000 years of these two simulations, for which the average numerical time step is about 15 days. The first case is seen to be extremely stable, with the magnetic pole never wondering

Figure 2. The angular velocity of the solid inner core for a 20,000 year interval of a geodynamo simulation. Positive is eastward relative to the mantle. [From *Buffett and Glatzmaier*, 2000]

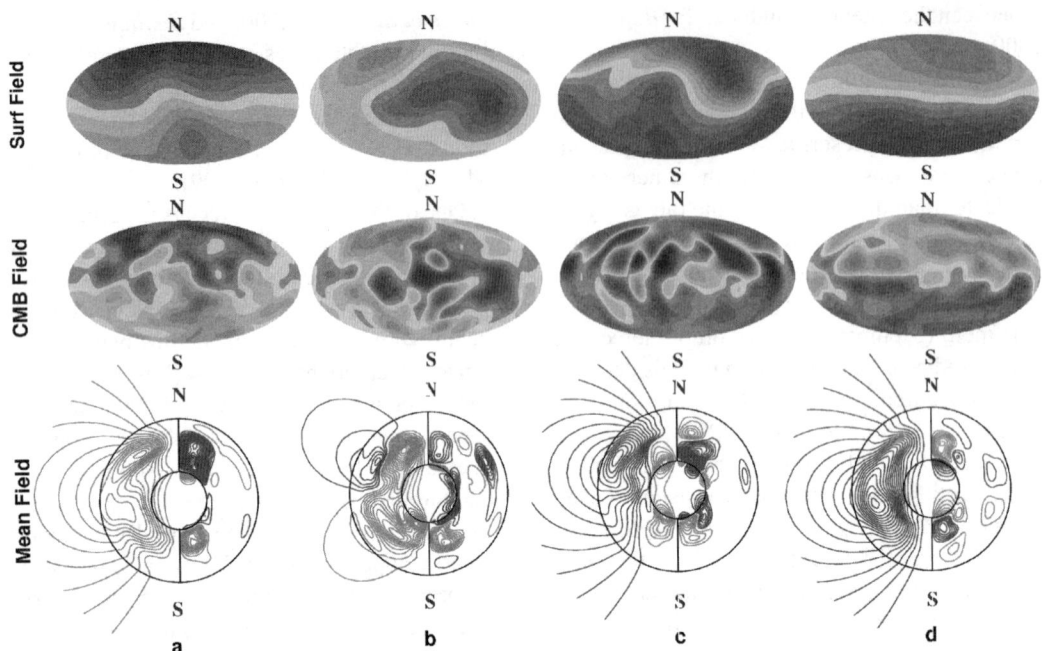

Figure 3. A sequence of snapshots of the longitudinally averaged magnetic field through the interior of the core and of the radial component of the field at the core–mantle boundary and at what would be the surface of the Earth, displayed at roughly 3000-year intervals spanning a dipole reversal from a geodynamo simulation. In the plots of the average field, the small circle represents the inner core boundary and the large circle is the core–mantle boundary. The poloidal field is shown as magnetic field lines on the left-hand sides of these plots (light shade or blue is clockwise and dark or red is counter-clockwise). The toroidal field direction and intensity are represented as contours (not magnetic field lines) on the right-hand sides (dark or red is eastward and light or blue is westward). Aitoff-Hammer projections of the entire core–mantle boundary and surface are used to display the radial component of the field (with the two different surfaces displayed as the same size). Light shades or red represent outward directed field and dark or blue represent inward field; the surface field, which is typically an order of magnitude weaker, was multiplied by 10 to enhance the color contrast. [From *Glatzmaier et al.*, 1999].

far from the geographic pole. Since the natural convective heat flux within the fluid core preferentially transports more heat to the polar and equatorial regions, the imposed pattern of heat flux at the CMB for this case is very compatible with the internal fluid dynamics. Therefore magnetic instabilities have little chance of growing to significant amplitude. The second case is more Earth-like, with several reversal attempts and two successful reversals in the 300,000 years. Also, as seen in the paleomagnetic record, the intensity of the field typically decreases by at least an order of magnitude during a reversal.

Small changes in the local flow structure continually occur in this highly nonlinear chaotic system [*e.g.*, *Olson et al.*, 1999]. These can generate local magnetic anomalies that are reversed relative to what would be the direction of the global dipolar field structure. If the thermal and compositional perturbations continue to drive the fluid flow in a way that amplifies this reversed field polarity while destroying the original polarity, the entire global field structure would eventually reverse. However, more often the local reversed polarity is not able to survive and the original polarity fully recovers because it takes a couple thousand years for the original polarity to decay out of the solid inner core [*Hollerbach and Jones*, 1993; *Glatzmaier and Roberts*, 1995b]. This is a plausible explanation for "events" [*Lund et al.*, 1998], which occur when the paleomagnetic field (as measured at the Earth's surface) reverses and then reverses back, all within about ten thousand years [*Gubbins*, 1999; *Glatzmaier et al.*, 1999].

These studies of reversals require long simulated times; and, other than setting the values of the diffusion coefficients and the pattern of heat flux out of the CMB, they are not adjusted to get preferred results. Also, the simulated reversals are not externally triggered; they occur naturally and spontaneously in this highly nonlinear system. However, these long simulated times require tens of millions of numerical time steps, which can only be afforded at relatively low spatial resolution. Alternatively, studies of the structure and time dependence of the flow and field during relatively quiet times

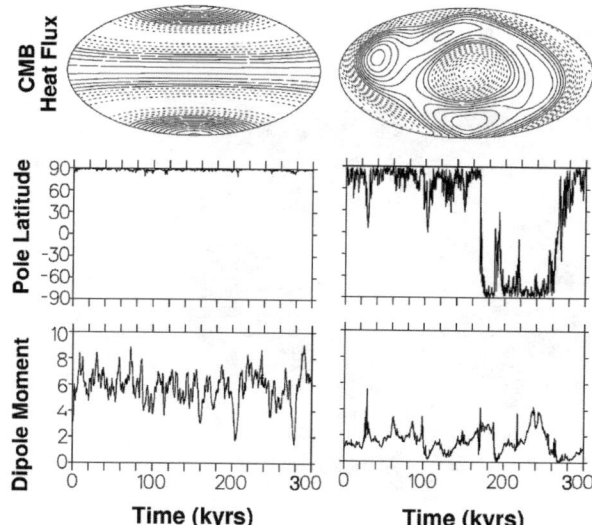

Figure 4. The latitude of the south magnetic pole and the dipole moment (in units of 10^{22} A m^2) versus time (in units of 1000 years) for two 300,000-year geodynamo simulations. The tic marks on the time axis are at intervals of 20,000 years, one dipole magnetic diffusion time. The pattern of the imposed heat flux at the core–mantle boundary (CMB) is displayed for each case: solid contours represent outward heat flux greater than the mean and broken contours represent heat flux less than the mean. [From *Glatzmaier et al.*, 1999]

between reversals requires less simulated time and therefore can be done at much greater spatial resolution.

For example, in the 1970s, there was a debate concerning the presence of radiogenic elements in the Earth's core. Due to a lack of convincing evidence to support their presence, this debate ended with a general consensus that there are no significant amounts of radioactive elements present [*e.g., Stacey*, 1992]. Recently, however, new evidence in the form of experimental studies [*e.g., Chabot and Drake*, 1999; *Gessman and Wood*, 2002; and *Murthy et al.*, 2003] and thermal evolution models [*e.g., Buffett*, 2000; and *Brodholt and Nimmo*, 2002] have shown that potassium-40 may have partitioned into the Earth's iron core during its formation; however, the predicted concentration today is still debated, ranging between 1 ppm and 300 ppm.

In an attempt to shed some light on this question, we have run a modified version of the Glatzmaier-Roberts geodynamo model to check if there are any significant effects on the simulated magnetic field due to radiogenic heating from 250 ppm of potassium-40 in the core. This corresponds to prescribing 25% of the core–mantle boundary heat flow to be coming from a volumetric heating source proportional to the radially-dependent density. Another case, the control case, was run without radiogenic heating. The other three quarters of the CMB heat flow for the internally heated case (and the total for the control case) comes from the latent heat source at the ICB, Joule and viscous heating and secular cooling. Otherwise the two cases are the same. In this model, the viscous, thermal, compositional and magnetic diffusivities are set to be constant, equal and independent of the length scale of the spherical harmonic mode. (Note, this prescription differs from previous Glatzmaier-Roberts simulations and the calculations have been performed at much greater spatial resolution.)

Snapshots of the vertical flow in the equatorial plane are shown in Figure 5. Notice that even for these more highly resolved and less diffusive simulations, the convective plumes typically span the entire depth of the outer core. That is, these simulations are still not turbulent. The results show a subtle difference in the vertical flow structure (Figure 5); the internally heated case has weaker upwellings at the ICB. This occurs because the thermal gradient at the ICB, which drives convection there, is less steep relative to the control case since the diffusive heat flux (and compositional flux) there is less. Note, that this also implies that the inner core for the internally heated case would be much older than the same size inner core of the control case.

The magnetic field structures for the two cases look surprisingly similar; a snapshot of the field for the internally heated case is displayed in Figure 1. The internally heated case does, however, maintain an average magnetic field 10–20% more intense than that of the control case, and an average kinetic energy 10–20% less (Figure 5) than that of the control case. There are also slight differences in the magnetic energy spectra between the two cases. The internally heated case, unlike the control case, usually has a greater quadrupole component than octupole at the surface, as is the case for the present day Earth. The control case however also displays this characteristic occasionally. Unfortunately, we have seen no significant difference in the surface fields of the two cases to argue for or against potassium-40 in the Earth's core. This is in agreement with a similar dynamo study [*Kutzner and Christensen*, 2002]. Higher spatial resolution and lower diffusivities may be needed to produce a noticeable difference in the dynamo.

Many other studies have been conducted via dynamo simulations to, for example, assess the effects of the size and conductivity of the solid inner core [*e.g., Sakuraba and Kono*, 1999; *Bloxham*, 2000; *Morrison and Fearn*, 2000; *Roberts and Glatzmaier*, 2001; and *Wicht*, 2002], of a stably stratified layer at the top of the core [*Glatzmaier and Roberts*, 1997], of heterogeneous thermal boundary conditions [*e.g., Sarson et al.*, 1997; *Glatzmaier et al.*, 1999; *Bloxham*, 2000; *Olson and Christensen*, 2002; and *Christensen and Olson*, 2003], of different velocity boundary conditions [*e.g., Kuang and Bloxham*, 1999; *Kuang*, 1999; and *Christensen et al.*,

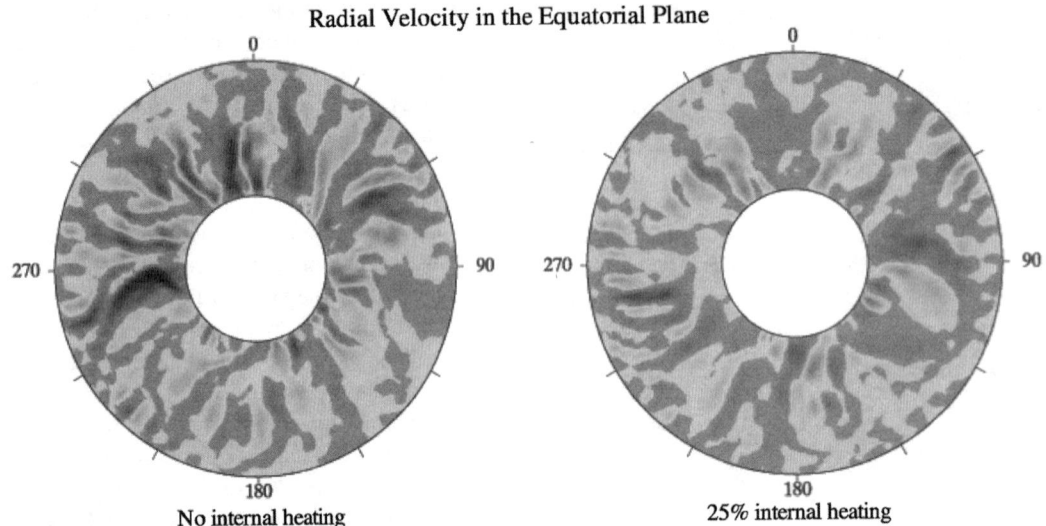

Figure 5. Snapshots of the amplitude of the radial flow in the equatorial plane for two dynamo simulations that have the same total heat flow out of the core. One has no specified internal heating; the other has one quarter of the heat flow out of the core coming from internal heating.

1999] and of computing with different parameters [*e.g.*, *Gilman*, 1983; *Olson et al.*, 1999; *Christensen et al.*, 1999, *Grote et al.*, 2000; *Simitev and Busse*, 2002]. These models differ in several respects. For example, models employ either the Boussinesq or the anelastic approximation, compositional buoyancy is usually neglected as are perturbations in the gravitational field, different boundary conditions and spatial resolutions are chosen, the inner core may be treated as an insulator instead of a conductor or may not be free to rotate. As a result, the simulated flow and field structures inside the core differ. For example, the strength of the shear flow on the "tangent cylinder" (the imaginary cylinder tangent to the inner core equator), which depends on the relative dominance of the Coriolis forces, varies among the simulations. Likewise, the vigor of the convection and the resulting magnetic field generation tends to be greater outside this tangent cylinder for some models and inside for others. But all the solutions have a westward zonal flow in the upper part of the fluid core and a dominantly dipolar magnetic field outside the core.

However, when assuming Earth values for the radius and rotation rate of the core, all models of the geodynamo have been forced (due to computational requirements) to use a greatly enhanced viscous diffusivity to account for mixing by the small-scale (unresolved) turbulence. However, the values used have been at least three or four orders of magnitude larger than estimates of what it should be for the current spatial resolutions that are employed, which is roughly equal to the Earth's actual magnetic diffusivity. One must also decide how to prescribe the thermal, compositional and magnetic diffusivities [*e.g.*, *Dormy et al.*, 2000; *Kono and Roberts*, 2002; *Simitev and Busse*, 2002; and *Glatzmaier*, 2002]. They all have units of length squared per time and serve as coefficients that relate the degree of structure in the temperature, composition and magnetic field, respectively, to the rates these quantities diffuse. One of two extremes has typically been chosen. These three diffusivities could be set equal to the Earth's actual magnetic diffusivity (2 m^2/s), making the viscous diffusivity much greater than these; this was the choice for the previous Glatzmaier-Roberts simulations. Alternatively, they could be set equal to the enhanced viscous diffusivity, making all (turbulent) diffusivities too large, but at least equal; this was the choice of most of the other models, including the internally heated version of the Glatzmaier-Roberts model presented above. Neither choice is satisfactory.

5. TWO-DIMENSIONAL MODELS

The fundamental question about geodynamo models is how well do they simulate the actual dynamo mechanism of the Earth's core. Many of us have argued, or at least suggested, that the large (global) scales of the temperature, flow and field seen in these simulations should be fairly realistic because the diffusivities may be asymptotically small enough. That is, although molecular viscosity of the fluid in the outer core may be as much as 10^{10} times smaller than what the (turbulent) viscosity is set to in current geodynamo simulations, viscous forces in most simulations (away from the boundaries) still tend to be about 10^4 times smaller than the Coriolis and Lorentz forces. No one believes geodynamo simulations need to use the actual molecular viscosity to be able to get realistic large-scale flows because the transport of momentum by the small unresolved turbulence is much more efficient than actual viscous forces. All models of the geodynamo have in fact crudely modeled this nonlinear mixing process as a simple linear diffusion process with a very enhanced viscous diffusivity. The question is how much of the turbulence spectrum needs to be numerically resolved to depict adequately the large scale flow, and therefore field, in the Earth's core, i.e., the geodynamo mechanism. Only when numerical methods and computing resources improve to the point where we can further reduce the turbulent diffusivities by several orders of magnitude and produce at least moderately turbulent 3D dynamo simulations will we be able to answer this fundamental question.

To gain some insight to what the answer might be, we examine two simulations of two-dimensional (2D) magnetoconvection. Magnetoconvection refers to an MHD simulation with an imposed background magnetic field; convection amplifies and distorts the field and Lorentz forces affect the flow, but, unlike in a convective dynamo, the field cannot decay away. Although the 2D constraint precludes realistic global flows and a dynamo mechanism, it does allow us to run at much greater spatial resolution and therefore with much smaller viscous, thermal and magnetic diffusivities. That is, we are able to run in a more realistic parameter regime, one that we would like to be able to reach someday in 3D.

We specify a box of fluid in the horizontal (x) and vertical (z) directions, with no flow, field or variations in the y direction. There is a constant gravitational acceleration in the $-z$ direction; and the frame of reference is rotating about an axis in the y direction. This box can be considered a small part of the equatorial plane, far from the axis of rotation. The top and bottom boundaries are impermeable and stress free and there is a specified drop in specific entropy across the depth of the box. There is an externally applied, uniform, background magnetic field in the z direction. The side boundaries are periodic.

We solve a system of magnetohydrodynamic equations in 2D for the thermodynamic perturbations, the fluid flow and the induced magnetic field using a numerical method that has been employed in many previous 2D Boussinesq studies [*e.g., Weiss*, 1981], but with modifications that account for the density stratification of our anelastic model [*Rogers et al.*, 2003; and *Glatzmaier*, 2004]. The numerical time step is limited by a Courant condition based on the fluid and Alfvén velocities and the grid resolution. That is, to maintain numerical stability, the time step cannot exceed the shortest time it takes fluid to flow, or magnetic waves to propagate, between any two adjacent grid points. We use 2001 (non-uniform) grid levels in the vertical direction and 4001 (uniform) grid levels in the horizontal direction.

We begin by describing a 2D simulation that is time dependent but laminar. The density stratification across the convection layer is set to that of the Earth's outer fluid core, i.e., the density at the bottom boundary is about 20% greater than that at the top. The viscous, thermal and magnetic diffusivities are constant and equal but much larger than Earth values. Buoyancy, rotational, magnetic and viscous effects on the flow are similar in magnitude to those in current 3D geodynamo simulations.

The resulting convective velocities are a thousand times greater than the diffusive velocity and ten times smaller than the rotational velocity. The induced magnetic field has an average maximum intensity about an order of magnitude greater than the applied background field intensity and the magnetic energy is comparable to the kinetic energy. Figure 6 is a snapshot of the perturbation of entropy (approximately the temperature perturbation). The flow is said to be "laminar" because thermal plumes extend from one boundary to the other, similar to large convection cells. The evolution resem-

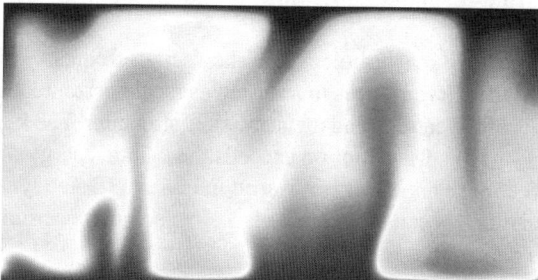

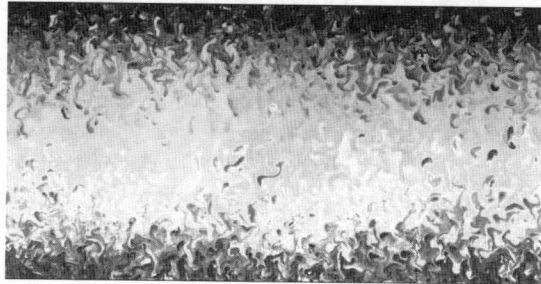

Figure 6. Snapshots of the entropy for two different 2D anelastic, rotating, magnetoconvection calculations. Gravity is downward. Hot (red) plumes rise from the bottom boundary and cold (blue) sink from the top. The turbulent case has viscous, thermal and magnetic diffusivities one thousand times smaller than those of the laminar case.

bles the flow seen in a lava lamp. This time-dependent large-scale laminar convection is typical of the style of convection simulated with current 3D geodynamo models.

Now we examine what happens when we reduce all three diffusivities, each by a factor of a thousand. Convective velocities are now almost a million times greater than the diffusive velocity and the resulting Coriolis forces are typically 10^9 times greater than the (turbulent) viscous forces, closer to the conditions in the Earth's fluid core. A snapshot of this turbulent case is also shown in Figure 6. As expected, there is much more energy in the small spatial scales compared with the laminar case. After detaching from the boundaries, turbulent plumes interact with each other, not the boundaries.

However, somewhat unexpected, deep turbulent boundary layers exist. This occurs because kinetic energy of buoyant plumes rising from the bottom boundary is converted into rotational kinetic energy. That is, as plumes rise from the bottom boundary the Coriolis forces resulting from the expansion (due to the slight density stratification) generate vortices with counter-clockwise rotation. Clockwise rotating vortices are generated in plumes sinking from the top boundary. This effect is relatively insignificant in the laminar case and in the current 3D anelastic geodynamo simulations because viscosity in these calculations is too large. It is completely absent in Boussinesq simulations, which do not account for density stratification.

A movie of this turbulent case shows high-frequency high-amplitude Alfvén waves manifested as local horizontal oscillations of the entropy and magnetic field structures. The kinetic energy of this wave motion exceeds the kinetic energy of the convection. This effect is not seen in current 3D geodynamo simulations because the large viscosity used in those simulations damps these waves.

6. NEXT GENERATION OF MODELS

Our understanding of how the geomagnetic field is generated by convection in the Earth's outer fluid core has improved during the past decade with the development of magnetohydrodynamic dynamo models. The relatively Earth-like magnetic field structure and time dependence, including dipole reversals, produced by these models is certainly encouraging. However, no dynamo model has yet simulated strongly turbulent convection, which likely exists in the Earth's core. We have assumed that the enhanced diffusivities in our laminar simulations are adequately, albeit crudely, accounting for the mixing by the unresolved turbulence. Two-dimensional magnetoconvection simulations, however, demonstrate that strongly turbulent rotating magnetoconvection is significantly different than the corresponding laminar simulations obtained with much larger diffusivities.

These findings suggest that current 3D laminar dynamo simulations may be missing critical dynamical phenomena. It is important, therefore, to strive for much greater spatial resolution in 3D models in order to significantly reduce the enhanced diffusion coefficients and actually simulate turbulence. This will require faster parallel computers and improved numerical methods, which hopefully will happen within the next decade or two. In addition, sub-grid scale models need to be added to geodynamo models to better represent the heterogeneous anisotropic transport of heat, composition, momentum and possibly also magnetic field by the part of the turbulence spectrum that remains unresolved.

How geophysically realistic is the laminar dynamo mechanism in the current geodynamo models? Whatever the answer, these findings suggest that there are still several exciting and important phenomena about the geodynamo waiting to be discovered with the next generation of models.

Acknowledgments. The computing resources were provided by the NSF (MCA99S012S) at the Pittsburgh Supercomputing Center and at UCSC via an NSF MRI grant (AST-0079757) and by the DOE at NERSC (DE-FC02-01ER41176). Support for this research was provided by grants from NSF (EAR-9902969, EAR-0221941), NASA (NAG5-11220) and from Los Alamos National Laboratory IGPP (1225A) and UCDRD (9948).

REFERENCES

Braginsky, S. I., and P. H. Roberts, Equations governing convection in Earth's core and the geodynamo, *Geophys. and Astrophys. Fluid Dyn., 79*, 1–97, 1995.

Bloxham, J., Sensitivity of the geomagnetic axial dipole to thermal core-mantle interactions, *Nature, 405*, 63–65, 2000.

Bloxham, J., and A. Jackson, Time-dependent mapping of the magnetic field at the core–mantle boundary, *J. Geophys. Res., 97*, 19537–19563, 1992.

Brodholt, J. and F. Nimmo, Core values, *Nature, 418*, 489–491, 2002.

Buffett, B. A., Estimates of heat flow in the deep mantle based on the power requirements for the geodynamo, *Geophys. Res. Lett., 29*, doi:10.1029/2001GL014649, 2002.

Buffett, B. A., and G. A. Glatzmaier, Gravitational braking of inner-core rotation in geodynamo simulations, *Geophys. Res. Lett., 27*, 3125–3128, 2000.

Busse, F. H., Homogeneous dynamos in planetary cores and in the laboratory, *Ann. Rev. Fluid Mech., 32*, 383–408, 2000.

Busse, F. H., Convective flows in rapidly rotating spheres and their dynamo action, *Phys. Fluids, 14*, 1301–1314, 2002.

Busse, F. H., E. Grote, and A. Tilgner, On convection driven dynamos in rotating spherical shells, *Studia Geoph. Geod., 42*, 1–6, 1998.

Chabot, N. L. and M. J. Drake, Potassium solubility in metal: the effects of composition at 15 kbar and 1900 C on partitioning between iron alloys and silicate melts, *Earth Planet. Sci. Lett., 172*, 323–335, 1999.

Christensen, U. R., J. Aubert, P. Cardin, E. Dormy, S. Gibbons, et al., A numerical dynamo benchmark, *Phys. Earth Planet. Inter.*, *128* 25–34, 2001.

Christensen, U. R. and P. Olson, Secular variation in numerical geodynamo models with lateral variations of boundary heat flow, *Phys. Earth Planet. Inter.*, *138*, 39–54, 2003.

Christensen, U., P. Olson, and G. A. Glatzmaier, A dynamo model interpretation of geomagnetic field structures, *Geophys. Res. Lett.*, *25*, 1565–1568, 1998.

Christensen, U., P. Olson, and G. A. Glatzmaier, Numerical modeling of the geodynamo: a systematic parameter study, *Geophys. J. Int.*, *138*, 393–409, 1999.

Creager, K., Inner core rotation rate from small scale heterogeneity and time-varying travel times, *Science*, *128*, 1284–1288, 1997.

Dormy, E., J.-P. Valet, and V. Courtillot, Numerical models of the geodynamo and observational constraints, *Geochem. Geophys. Geosyst. 1*, 2000GC000062, 2000.

Fearn, D. R., Hydromagnetic flow in planetary cores, *Rep. Prog. Phys.*, *61*, 175–235, 1998.

Gessman, C. K. and B. J. Wood, Potassium in the Earth's core?, *Earth Planet. Sci. Lett.*, *200*, 63–78, 2002.

Gilman, P. A., Dynamically consistent nonlinear dynamos driven by convection in a rotating spherical shell. II. Dynamos with cycles and strong feedbacks, *Astrophys. J. Suppl. Ser.*, *53*, 243–268, 1983.

Gilman, P. A. and J. Miller, Dynamically consistent nonlinear dynamos driven by convection in a rotating spherical shell, *Astrophys. J. Suppl. Ser.*, *46*, 211–238, 1981.

Glatzmaier, G. A., Numerical simulations of stellar convective dynamos I. The model and the method, *J. Comput. Phys.*, *55*, 461–484, 1984.

Glatzmaier, G. A., Geodynamo Simulations—How Realistic are They?, *Ann. Rev. Earth Planet. Sci.*, *30*, 237–257, 2002.

Glatzmaier, G. A., Planetary and stellar dynamos: Challenges for next generation models, in *Astrophysical and Geophysical Fluid Dynamics and Dynamos*, A. M. Soward, C. A. Jones, D. W. Hughs and N. O. Weiss eds., Taylor and Francis, 2004, in press.

Glatzmaier, G. A., R. S. Coe, L. Hongre, and P. H. Roberts, The role of the Earth's mantle in controlling the frequency of geomagnetic reversals, *Nature*, *401*, 885–890, 1999.

Glatzmaier, G. A., and P. H. Roberts, A three-dimensional convective dynamo solution with rotating and finitely conducting inner core and mantle, *Phys. Earth Planet. Inter.*, *91*, 63–75, 1995a.

Glatzmaier, G. A., and P. H. Roberts, A three-dimensional self-consistent computer simulation of a geomagnetic field reversal, *Nature*, *377*, 203–209, 1995b.

Glatzmaier, G. A., and P. H. Roberts, An anelastic evolutionary geodynamo simulation driven by compositional and thermal convection, *Physica D*, *97*, 81–94, 1996a.

Glatzmaier, G. A., and P. H. Roberts, Rotation and magnetism of Earth's inner core, *Science*, *274*, 1887–1891, 1996b.

Glatzmaier, G. A., and P. H. Roberts, Simulating the Geodynamo, *Contemp. Phys.*, *38*, 269–288, 1997.

Grote, E., F. H. Busse and A. Tilgner, Convection-driven quadrupolar dynamos in rotating spherical shells, *Phys. Rev. E*, *60*, 5025–5028, 1999.

Grote, E., F. H. Busse and A. Tilgner, Regular and chaotic spherical dynamos, *Phys. Earth Planet. Inter.*, *117*, 259–272, 2000.

Gubbins, D., The distinction between geomagnetic excursions and reversals, *Geophys. J. Int.*, *137*, F1–F3, 1999.

Hollerbach, R., On the theory of the geodynamo, *Phys. Earth Planet. Inter.*, *98*, 163–185, 1996.

Hollerbach, R., A spectral solution of the magneto-convection equations in spherical geometry, *Int. J. Numer. Meth. Fluids*, *32*, 773–797, 2000.

Hollerbach, R. and C. A. Jones, Influence of the Earth's inner core on geomagnetic fluctuations and reversals, *Nature*, *365*, 541–543, 1993.

Jones, C. A., A. Longbottom, and R. Hollerbach, A self-consistent convection driven geodynamo model, using a mean field approximation, *Phys. Earth Planet. Inter.*, *92*, 119–141, 1995.

Kageyama, A., M. Ochi, T. Sato, Flip-flop transitions of the magnetic intensity and polarity reversals in the magnetohydrodynamic dynamo, *Phys. Rev. Lett.*, *82*, 5409–5412, 1999.

Kageyama, A., T. Sato, K. Watanabe, R. Horiuchi, T. Hayashi, et al., Computer simulation of a magnetohydrodynamic dynamo II, *Phys. Plasmas*, *2*, 1421–1431, 1995.

Katayama, J. S., M. Matsushima, and Y. Honkura, Some characteristics of magnetic field behavior in a model of MHD dynamo thermally driven in a rotating spherical shell, *Phys. Earth Planet. Inter.*, *111*, 141–159, 1999.

Kida, S., K. Araki, and H. Kitauchi, Periodic reversals of magnetic field generated by thermal convection in a rotating spherical shell, *J. Phys. Soc. Japan*, *66*, 2194–2201, 1997.

Kida, S. and H. Kitauchi, Thermally driven MHD dynamo in a rotating spherical shell, *Prog. Theor. Phys.*, *130*, Suppl. 121–136, 1998.

Kono, M. and P. H. Roberts, Recent geodynamo simulations and observations of the geomagnetic field, *Rev. Geophys.*, *40*, 4-1–53, 2002.

Kuang, W., Force balances and convective state in the Earth's core, *Phys. Earth Planet. Inter.*, *116*, 65–79, 1999.

Kuang, W. and J. Bloxham, An Earth-like numerical dynamo model, *Nature*, *389*, 371–374, 1997.

Kuang, W. and J. Bloxham, Numerical modeling of magnetohydrodynamic convection in a rapidly rotating spherical shell: Weak and strong field dynamo action, *J. Comp. Phys.*, *153*, 51–81, 1999.

Kumar, S. and P. H. Roberts, A three-dimensional kinematic dynamo, *Proc. R. Soc. London, Ser. A*, *344*, 235–258, 1975.

Kutzner, C. and U. R. Christensen, From stable dipolar towards reversing numerical dynamos, *Phys. Earth Planet. Inter.*, *131*, 29–45, 2002.

Laske, G. and G. Masters, Limits on differential rotation of the inner core from an analysis of the Earth's free oscillations, *Nature*, *402*, 66–69, 1999.

Lund, S. P., G. Acton, B. Clement, M. Hastedt, M. Okada, and T. Williams, Geomagnetic field excursions occurred often during the last million years, *EOS Trans. AGU*, *79*, 178–179, 1998.

Merrill, R. T. and P. I. McFadden, Geomagnetic polarity transitions, *Rev. Geophys.*, *37*, 201–226, 1999.

Morrison, G. and D. R. Fearn, The influence of Rayleigh number, azimuthal wavenumber and inner core radius on 2D hydromagnetic dynamos, *Phys. Earth Planet. Inter.*, *117*, 237–258, 2000.

Murthy, V. R., W. van Westrenen, and Y. Fei, Experimental evidence that potassium is a substantial radioactive heat source in planetary cores, *Nature*, *423*, 163–165, 2003.

Olson, P. and U. R. Christensen, The time-averaged magnetic field in numerical dynamos with nonuniform boundary heat flow, *Geophys. J. Int.*, *151*, 809–823, 2002.

Olson, P., U. Christensen and G. A. Glatzmaier, Numerical modeling of the geodynamo: mechanisms of field generation and equilibration, *J. Geophys. Res.*, *104*, 10383–10404, 1999.

Roberts, P. H. and G. A. Glatzmaier, Geodynamo theory and simulations, *Rev. Mod. Phys.*, *72*, 1081–1123, 2000.

Roberts, P. H. and G. A. Glatzmaier, The geodynamo, past, present and future, *Geophys. Astrophys. Fluid Dyn.*, *94*, 47–84, 2001.

Rogers, T. M., G. A. Glatzmaier, and S. E. Woosley, Simulations of two-dimensional turbulent convection in a density-stratified fluid, *Phys. Rev. E*, *67*, 026315-1–6, 2003.

Sakuraba, A., and M. Kono, Effect of the inner core on the numerical solution of the magnetohydrodynamic dynamo, *Phys. Earth Planet. Inter.*, *111*, 105–121, 1999.

Sarson, G. R. and C. A. Jones, A convection driven dynamo reversal model, *Phys. Earth Planet. Inter.*, *111*, 3–20, 1999.

Sarson, G. R., C. A. Jones, and A. W. Longbottom, The influence of boundary region heterogeneities on the geodynamo, *Phys. Earth Planet. Inter.*, *101*, 13–32, 1997.

Song, X., and P. Richards, Seismological evidence for differential rotation of the Earth's inner core, *Nature*, *382*, 221–224, 1996.

Souriau, A., New seismological constraints on differential rotation of the inner from Novaya Zemlya events recorded at DRV Antarctica, *Geophys. J. Int.*, *134*, F1–F5, 1998.

Simitev, R. and F. H. Busse, Parameter dependences of convection driven spherical dynamos, in *High Performance Computing in Science and Engineering '02*, E. Krause and W. Jager, eds, pp. 15–35, Springer, 2002.

Stacey, F. D., *Physics of the Earth*, Brookfield Press, 1992.

Su, W., A. M. Dziewonski, and R. Jeanloz, Planet within a planet: Rotation of the inner core of the Earth, *Science*, *274*, 1883–1887, 1996.

Weiss, N. O., Convection in an imposed magnetic field. Part 1. The development of nonlinear convection, *J. Fluid Mech.*, *108*, 247–272, 1981.

Wicht, J., Inner-core conductivity in numerical dynamo simulations, *Phys. Earth Planet. Inter.*, *132*, 281–302, 2002.

Xu, X. and X. Song, Evidence for inner core super-rotation from time-dependent differential PKP travel times observed at Beijing Seismic Network, *Geophys. J. Int.*, *152*, 509–514, 2003.

Zhang, K. and F. H. Busse, Finite amplitude convection and magnetic field generation in a rotating spherical shell, *Geophys. Astrophys. Fluid Dyn.*, *41*, 33–53, 1988.

Thomas L. Clune, Science Computing Branch 931, NASA Goddard Space Flight Center, Greenbelt, MD 20771.

Gary A. Glatzmaier, Earth Sciences Department, University of California, Santa Cruz, CA 95064.

Darcy E. Ogden, Earth Sciences Department, University of California, Santa Cruz, CA 95064.

Core-Mantle Boundary Structures and Processes

Thorne Lay

Earth Sciences Department, University of California Santa Cruz, Santa Cruz, California

Edward J. Garnero

Department of Geological Sciences, Arizona State University, Tempe, Arizona

Seismological and geodynamical observations have established the presence of a major thermo-chemical boundary layer (TCBL) in the lowermost mantle. This boundary layer plays a critical role in regulating heat flow through the core-mantle boundary, thereby influencing the dynamo-generating core flow regime. It also plays an important role in the mantle convection system, possibly serving as a source of boundary-layer instabilities and as a reservoir for long-lived geochemical heterogeneities. Two end-member conceptual models for the TCBL have emerged, both reconcilable with current observational constraints: a global, stably-stratified, chemically distinct layer may exist in the lowermost 250 km of the mantle (the global TCBL model), or this region may be a partially mixed boundary layer involving a composite of downwelling thermo-chemical anomalies such as oceanic lithospheric slabs or eclogitic oceanic crustal components and ancient dense chemical anomalies dynamically concentrated into large agglomerations beneath upwellings (the hybrid TCBL model). For the global TCBL model, laterally varying partial melt fractions within the layer are required to account for various seismological observations, and large dynamic topography on the upper boundary of this layer is expected: there is evidence for both of these attributes of the TCBL. The hybrid TCBL model requires additional complexity such as a phase transition or structural fabric transition to account for various seismological observations: some mineralogical candidates have been proposed. The outstanding challenge, requiring multi-disciplinary advances, is to discriminate between these competing conceptual models, as they differ in implications for thermal history, chemical processing, and dynamical behavior of the TCBL.

INTRODUCTION

The past two decades have witnessed tremendous progress in seismological, geodynamical, geomagnetic, and mineral physics efforts to quantify deep Earth processes. Our understanding of structures and processes in the lowermost mantle has advanced accordingly, and there is now widespread agreement that some form of major thermo-chemical boundary layer (TCBL) exists on the mantle side of the core-mantle boundary (CMB), extending at least several hundred kilometers upward into the lower mantle. Boundary layers play critical roles in heat transport and dominant length scales of thermal convection systems, so there is a concerted effort to understand the lowermost mantle boundary layer. The end-member conceptual models now being considered for the TCBL at the base of the mantle are reviewed here along with their observational foundations, and the directions of future multi-disciplinary research required to advance our understanding of CMB structures and processes are defined.

The CMB lies about 2900 km below Earth's surface, with this interface between mantle silicate and oxide rocks and molten iron alloy core materials being the primary internal compositional contrast within the planet. With density, viscosity, convective flow, and compositional contrasts comparable to or exceeding those at the surface of the Earth, the CMB separates the two major dynamical regimes of the interior; this makes it a place where chemical heterogeneities might be expected to accumulate. As such, from the early stages of core-formation and evolution of the primary chemical stratification of the planet to the current mantle convection system driving plate tectonics and the core flow regime generating the magnetic field, the boundary layers on either side of the CMB have played key roles in the chemical and dynamical evolution of the Earth. The multi-disciplinary advances in observational constraints on structures and processes near the CMB have been eagerly greeted by all disciplines engaged in understanding deep Earth processes. An assessment of our current state of knowledge (and ignorance) of this remote region is provided here, augmenting recent reviews by *Lay et al.* [2003], *Garnero* [2000], *Lay et al.* [1998a], and many papers in *Gurnis et al.* [1998].

CONCEPTUAL MODELS OF A DEEP THERMO-CHEMICAL BOUNDARY LAYER

By 1949 seismological observations were sufficient to establish that the lowermost mantle exhibits inhomogeneity relative to the overlying lower mantle, primarily manifested as decreased gradients in seismic velocities with depth. The bottom 200 km of the mantle were designated the D" region [*Bullen*, 1949], and it has become common to associate this region with a thermal and/or chemical boundary layer above the CMB. Subsequent radially averaged Earth models, such as the Preliminary Reference Earth Model (PREM) of *Dziewonski and Anderson* [1981], have incorporated reduced velocity gradients in the deepest 150 km of the mantle to accommodate the global departure of D" velocity structure from that of the overlying mantle, where velocity gradients are generally compatible with homogeneous self-compression. The presence of a thermal boundary layer (TBL) in D" caused by heat fluxing from the core into the mantle has long been postulated, and efforts have been made to infer the properties of such a boundary layer based on the reduced velocity gradients in models like PREM [e.g., *Stacey and Loper*, 1983]. However, as established by studies over the past half century that demonstrate increased seismic wave travel time fluctuations for paths traversing D", there appears to be substantial heterogeneity in this boundary layer on a wide variety of scale-lengths; a simple TBL interpretation is not sufficient to account for all seismically inferred properties of the region, nor is there a meaningful 'average' structure for the region to guide any physically viable interpretation of the boundary layer.

In the decades following the plate tectonics revolution, the importance of boundary layers to mantle and core dynamic systems became increasingly evident. It is well-recognized that the behavior of the plate tectonics system is largely governed by the surface thermo-chemical boundary layer: strong temperature-dependent viscosity effects and chemical differentiation play key roles in development of oceanic plates, while strong compositional variations play a key role in sustaining continental masses at the surface. The notion of hot plumes ascending from the interior to feed long-lived hotspot volcanic systems raised interest in the possible role of deeper boundary layers as the source region for plume genesis. Chemical anomalies and variance associated with many hotspot magmas relative to mid-ocean ridge basalts suggest the notion of deep boundary layers containing unmixed, isolated reservoirs within the mantle [cf., *Hofmann*, 1997; *Kellogg et al.*, 1999]. While most of the ensuing work on internal boundary layers focused on the possibility of compositional and dynamical stratification of the upper and lower mantles, substantial interest was directed toward the D" region as a possible boundary layer source for thermal instabilities and chemical heterogeneities. The location of the D" boundary layer at the CMB, across which there has been a long history of chemical transport and where there is a massive density contrast, makes D" a logical site to accumulate chemically anomalous dregs from the mantle and dross from the core [e.g., *Anderson*, 1998]. Emerging notions of mantle-wide distributed chemical heterogeneities sampled by upwellings from deep boundary layers [e.g., *Davies*, 1990; *Helffrich and Wood*, 2001] sustain this interest in the possible existence of a deep mantle boundary layer, even if the mantle is not chemically stratified.

Advances in geophysical disciplines have yielded the current state of knowledge of the mantle dynamic system summarized in Figure 1a. The dynamical system near the surface is well-characterized as being comprised of a thermo-chemical boundary layer that is partially mixing, with production and recycling of oceanic lithosphere, gradual addition to the chemical heterogeneities of the buoyant and enduring continental crust, and diverse scales of upwelling beneath ridges and hotspot volcanoes. Substantial complexity of the phase equilibria of the upper mantle is recognized, particularly in the shallow mantle where abundant volatiles are likely to be present. But there is broad agreement that (a) global seismic velocity increases near depths of 410 and 660 km are likely due to phase transitions in the $(Mg,Fe)_2SiO_4$ system; (b) most com-

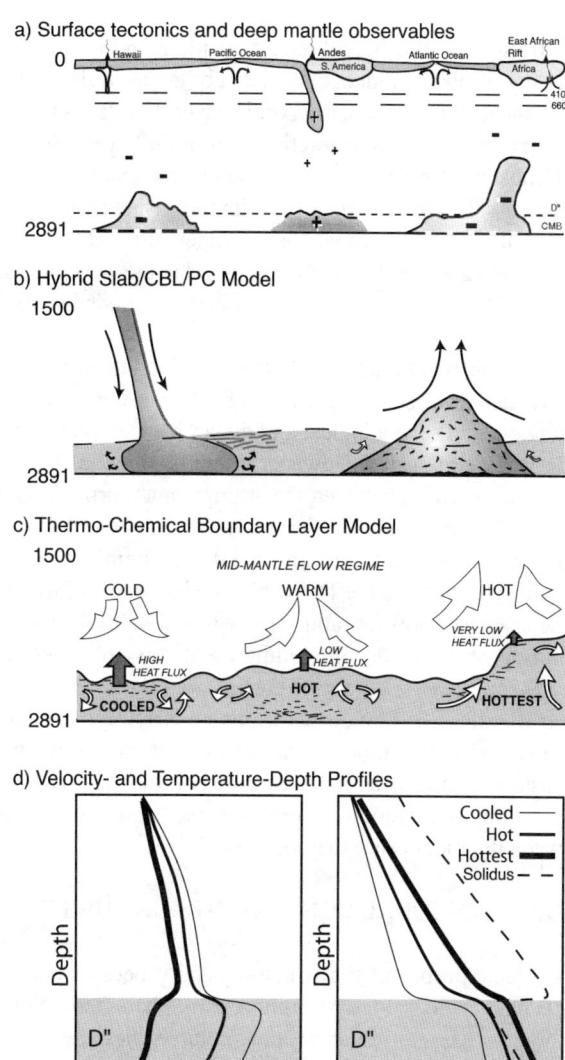

Figure 1. (a) Schematic cartoon of well-resolved attributes of the mantle convection system emphasizing the boundary layer near the surface comprised of the chemically differentiated continents, the chemically and thermally cycling oceanic slabs, and primary upper mantle phase transitions. The mid-mantle has large-scale patterns of seismic velocity heterogeneity with faster velocities (+ signs) tending to underlie regions of recent subduction of oceanic lithosphere, some of which may penetrate below the transition zone, and slow velocities (- signs) under the central Pacific and south Atlantic/Africa regions. The CMB boundary layer has large scale patterns of stronger heterogeneity which involves strong radial increases or decreases in velocity about 250 km above the CMB, with a predominant degree 2 pattern. (b) The hybrid thermo-chemical boundary layer (TCBL) concept for the deep mantle, in which subducting slabs penetrate to the CMB, providing thermal and chemical anomalies that will eventually rise back up in the mantle flow, while hot dense chemical anomalies are swept into large piles under upwellings. Either a phase change or radial gradient in structural fabric exists as well. (c) The global TCBL model, in which the lowermost mantle is a dense chemically distinct layer, possibly of primordial nature, which remains unmixed, but thermally coupled to overlying flow. Variable heat flow out of the chemical layer occurs in response to the configuration of mantle flow, leading to lateral variation in thermal structure across the boundary layer. Topography is induced on the chemical layer by the mid-mantle flow as well. (d) Schematic elastic velocity profiles across the D" region for regions of the TCBL that have relatively hot or cool thermal structures. For the global TCBL model, it is assumed that the eutectic solidus is intersected by the thermal profiles over varying depth extent depending on the regional temperature level.

Global seismic tomography has established that large-scale patterns of mid- and deep-mantle heterogeneity have some correlation with the shallow mantle structures and circulation: large regions of relatively high seismic velocity mantle (presumably lower temperature, higher density, and thus, descending) tend to underly regions of substantial oceanic lithosphere subduction over the past 200 million years or so [e.g. *Lithgow–Bertelloni and Richards*, 1998], while large regions of relatively low seismic velocity mantle (presumably higher temperature, lower density, and thus, ascending) are located beneath regions replete with surface hotspot volcanism. These large-scale seismic velocity patterns increase in strength in the deepest mantle; the D" region exhibits predominant large-scale heterogeneity at spherical harmonic degree 2. The highest velocity areas of D" tend to be accompanied by fairly strong (1–3%) increases in seismic velocity with depth about 250 km above the CMB, while the lowest velocity areas have comparable decreases in velocity near the same depth. Overall lateral variations in seismic velocity for large scales in most tomographic models are on the order of ±3% for *S* velocity and ±1% for *P* velocity within the D" region, about a factor of 2 to 3 larger than for large-scale vari-

mon upper mantle minerals undergo a dissociative phase transition to (Mg,Fe)SiO$_3$ in perovskite structure and (Mg,Fe)O between 660 and 800 km or so; and (c) (Mg,Fe)SiO$_3$ and (Mg,Fe)O are the primary components of a relatively homogeneous lower mantle (with additional Ca-perovskite and other minor components). Dynamical and seismological constraints favor the notion of flux of at least some oceanic slab material into the upper portions of the lower mantle, with seismic images in some regions suggestive of deep penetration of advectively thickened masses of cold oceanic lithosphere to depths of 2000 km or deeper [e.g., *van der Hilst and Kárason*, 1999; *Kellogg et al.*, 1999; *Fukao et al.*, 2001; *Grand*, 2002]. The mid-mantle may have chemical stratification, but as yet, there is no clear detection of global layering, so we focus on the lowermost mantle boundary layer.

ations at mid-mantle depths. The increase in velocity heterogeneity and the existence of large-scale structures in D" support the probability that this regions serves as a major boundary layer within the interior.

Efforts to reconcile the general observations of Figure 1a (and their attendant details as discussed below) with characteristics of a boundary layer at the base of the mantle have resulted in different conceptual models for the nature of D". For the purpose of focusing discussion, we define two end-member scenarios for the boundary layer, and relate them to pertinent observational details in the next section.

The first is what we will call the hybrid TCBL model, involving large-scale chemical heterogeneities embedded in a partially mixed boundary layer that has many similarities to the surface thermo-chemical boundary layer (Figure 1b). In this hybrid model, subducting slabs descend to the D" region, cool large-scale regions beneath downwellings, and help to physically sweep aside dense lowermost mantle chemical heterogeneities [e.g., *Wysession*, 1996; *Grand*, 2002; *Tan et al.*, 2002], which subsequently concentrate beneath upwellings. The upwellings include thermal boundary layer instabilities that give rise to plumes that ascend to the upper mantle. The pile of dregs can resist total entrainment if dense enough, but some chemical anomaly will be conveyed by the plume nonetheless. This model requires an additional aspect such as either a phase change or a strong vertical gradient in structural fabric imparted by shear flow to provide the abrupt radial seismic velocity increases at the top of the D" region observed in cooled areas.

An alternate scenario [*Lay et al.*, 2003] that we call the global TCBL model invokes the notion of a global chemically distinct layer in D" that remains relatively stable beneath overlying mantle upwellings and downwellings. The chemical composition is not constrained, but could involve differences in relative amounts of iron, calcium, aluminum, and/or silica relative to the overlying mantle. The composition could represent primordial differences associated with heterogeneous accretion [e.g., *Ruff and Anderson*, 1980], accumulation of ancient dense subducted products [e.g., *Anderson*, 1998], or products of chemical reactions between the core and mantle [e.g., *Knittle and Jeanloz*, 1989; *Goarant et al.*, 1992; *Dubrovinsky et al.*, 2001]. Substantial topography over a wide spectrum of wavelengths may be imposed on the layer by overlying mantle flow, including downwelling slabs in the mid-mantle and possibly plumes from the upper thermal boundary layer of D" (Figure 1c). Lateral variations in the overlying thermal system modulate heat flow out of the layer, resulting in large-scale lateral temperature gradients in the boundary layer that are thermally coupled to the mid-mantle, giving apparent continuity of seismic velocity heterogeneities. The lateral variations in temperature cause the boundary layer to either exceed or remain below the eutectic solidus at different lateral positions in the layer. The eutectic is probably reduced from that of the overlying mantle due to the distinct composition of the layer. This leads to laterally varying partial melt fraction within the layer. The melt itself must be effectively neutrally buoyant, remaining distributed across the layer but possibly with increasing melt fraction with depth. In turn, the variations in temperature and melt cause large-scale seismic velocity variations within the layer and variable velocity increases or decreases at the upper boundary of the layer (Figure 1d).

Both conceptual models explicitly involve lateral temperature variations in D", long-lasting chemical heterogeneity of the region, and seismological and dynamical complexity on a wide-variety of scales. However, there is potential for substantial differences between the strongly and partially stratified TCBL scenarios in, for example: (a) heat transport efficiency; (b) origin and nature of the chemical heterogeneities; (c) the role the TCBL plays relative to surface phenomena such as hotspots and the fate of slabs; (d) the thermal evolution of the system including the history of inner core growth; and (e) the extent of partial melting in the ancient lower mantle. The key observations underlying these competing end-member models will now be outlined, along with discussion of how each model may accommodate the observations. Future directions of research needed to discriminate between the models are then discussed.

OBSERVATIONAL CONSTRAINTS ON THE TCBL

Available probes of the structures and processes near the CMB have quite limited resolution, thus, characterization of the boundary layer must draw upon multiple lines of evidence. Given that seismological, geodynamical and geomagnetic information does not directly reveal the thermal structure of the deep mantle, even the existence of a TBL must be deduced indirectly. For example, the reduced seismic velocity gradients used to define the D" region could be a manifestation of chemical heterogeneity such as increasing iron content with depth rather than purely an effect of a superadiabatic temperature increase. When efforts are made to extrapolate experimentally constrained tie-points on temperature from the inner-core boundary and from upper mantle phase transitions in $(Mg,Fe)_2SiO_4$, the estimated superadiabatic temperature increase across the D" region is on the order of 1000–2000K [e.g., *Williams*, 1998; *Boehler*, 2000; *Anderson*, 2002]. This is comparable to the temperature contrast across the lithosphere, and favors the existence of a major TBL above the CMB. However, if superadiabatic thermal gradients are present shallower in the mantle, perhaps in the transition zone, mid-mantle or at the top of a global TCBL, this estimated temperature increase could be significantly reduced.

Another line of evidence is that the energetics of Earth's geodynamo require ongoing loss of heat from the core to the mantle, with estimates ranging from 2–10 TW annually [e.g., *Buffet* 2002, 2003; *Labrosse*, 2002]. These estimates are also somewhat dependent on assumptions about mantle stratification, the presence of which could reduce the estimates. While the uncertainties in both lines of evidence remain large, there is broad agreement, essentially now a paradigm, that a TBL with a significant overall temperature contrast exists at the base of the mantle. This raises the potential for TBL instabilities rising from the CMB, either within a stably stratified layer or as part of the larger mantle convection system. For the global TCBL model, one TBL is at the base of the chemically distinct layer (at the CMB), with a second TBL expected at its top. Such stratification gives large uncertainties in the actual temperature drop across the CMB, but does not negate the requirement of some heat fluxing through the CMB. For the hybrid TCBL model, the TBL at the base of the mantle is a more prominent feature of mantle convection because it serves as the lower boundary layer of the deep mantle convective system, but heat flow is still modified laterally by thermal and chemical heterogeneity.

Lateral Variations in the Boundary Layer

The case for lateral variations in the boundary layer, whether of thermal or chemical nature, is most compelling from the arena of global seismic tomography. There is now substantial convergence in large-scale mapping of seismic velocity heterogeneity in the deep mantle, particularly amongst shear wave models. Plate 1 presents comparisons of several recent global models for shear velocity variations (dVs) and compressional velocity variations (dVp) within the lowermost 250 km of the mantle; additional model comparisons are provided by *Garnero* [2000]. Large-scale shear velocity variations of ±3% are dominated by relatively high velocities beneath the circum-Pacific, with relatively low velocities under the central Pacific and south Atlantic/Africa [e.g., *Grand*, 2002; *Gu et al.*, 2001; *Ritsema and van Heijst*, 2000; *Mégnin and Romanowicz*, 2000; *Masters et al.*, 2000; *Kuo et al.*, 2000; *Castle et al.*, 2000]. The predominant degree 2 pattern in shear velocities is readily evident in Plate 1. The lowermost mantle portions of global compressional velocity models with ±1% velocity fluctuations tend to be less consistent, as apparent in Plate 1. Consistent features between *P* velocity models include fast regions beneath eastern Asia and Middle America, and slow regions under the South Pacific and the southern Atlantic [e.g, *Kárason and van der Hilst*, 2001; *Zhao*, 2001; *Fukao et al.*, 2001; *Boschi and Dziewonski*, 1999, 2000; *Bijwaard et al.*, 1998; *Vasco and Johnson*, 1998]. There is generally good correlation between dVp and dVs structures at very long wavelength [e.g., *Masters et al.*, 2000], although there are some well-sampled regions where the velocity variations decorrelate, such as within the Central Pacific and beneath North America (Plate 1). Several simultaneous inversions of *P* and *S* wave data have been performed with the intent of isolating bulk sound velocity variations from shear velocity variations [e.g., *Robertson and Woodhouse*, 1995; *Su and Dziewonski*; 1997, *Kennett et al.*, 1998; *Masters et al.*, 2000], but as yet there is little agreement amongst bulk sound velocity models.

The general expectation that shear velocity will be more sensitive to thermal variations than compressional velocity suggests that at least some of the large-scale regional pattern is the result of lateral temperature variations in the boundary layer of the order of several hundred degrees [e.g., *Forte and Mitrovica*, 2001; *Trampert et al.*, 2001]. However, both large- and small-scale regions are found where P and S velocity anomalies do not correlate [e.g., *Saltzer et al.*, 2001; *Wysession et al.*, 1999]. There are also regions where shear velocity variations are positively correlated with, but much stronger than, compressional velocity variations (as in the south Pacific) [*Masters et al.*, 2000; *Lay et al., 2003*]. These observations require that any thermal variations be augmented by or competing with chemical or partial melting effects. For the hybrid TCBL model, high seismic velocity regions are associated with cooled regions where slabs have descended to the CMB, while low seismic velocity regions are hot piles of chemical dregs concentrated, but largely resisting entrainment under upwellings. The global TCBL model accounts for the large-scale patterns of seismic velocity by lateral variations in temperature (hence, in partial melt volume), resulting from thermal coupling with the overlying mid-mantle convection system. In either case, one expects lateral variations in thermal gradient above the CMB, affecting both mantle and core dynamics.

An indirect line of evidence favoring a thermal contribution to the seismic velocity heterogeneity comes from the nature of Earth's magnetic field, for which a few strong flux bundles in the northern and southern hemispheres appear to sustain relative stationarity beneath high seismic velocity regions in D" [e.g., *Gubbins*, 1998]. The very low viscosity of the core ensures that the CMB is nearly isothermal; however, the probable existence of lateral temperature variations within the D" region will result in lateral variations in the thermal gradient above the CMB, introducing a variable heat flow boundary condition on the core convection regime. This variable heat flux boundary condition can drive thermal winds in the core [e.g., *Bloxham and Gubbins*, 1987; *Zhang and Gubbins*, 1993] while possibly stabilizing large magnetic flux concentrations, and perhaps even influencing preferred paths of virtual geomagnetic poles (VGP) during reversals [cf., *Gubbins*, 1998]. The complexity of quantifying mantle-core thermal

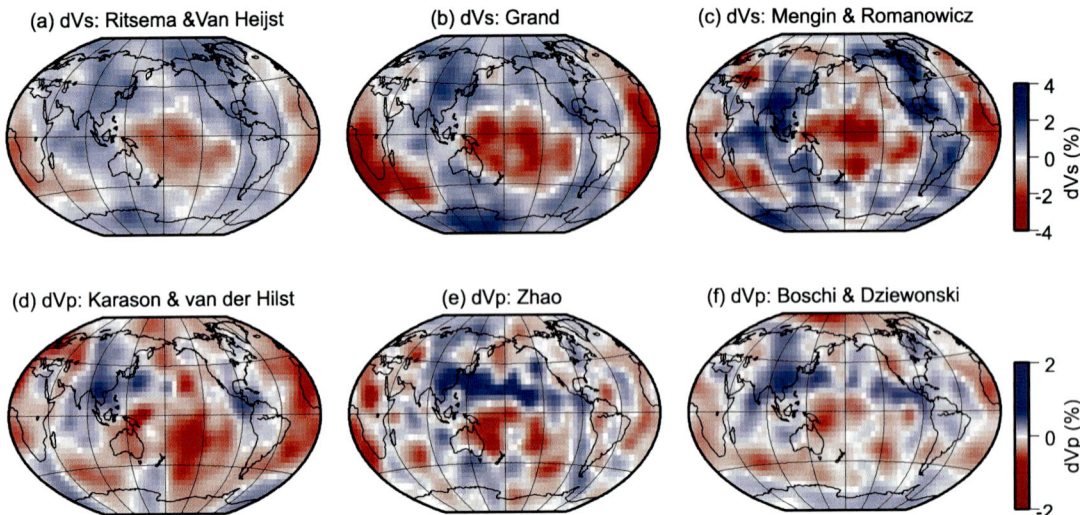

Plate 1. Maps of shear velocity (a–c) and compressional velocity (d–f) variations in the lowermost portions of several global mantle models. Shear velocity variations are all on a common scale, for the models of *Ritsema and van Heijst* [2000], *Grand* [2002], and *Mégnin and Romanowicz* [2000]. Compressional velocity variations are a different common scale for the models of *Kárason and van der Hilst* [2001], *Zhao* [2001], and *Boschi and Dziewonski* [2000].

interactions is enhanced by the possibility that chemical variations in D″ may affect both thermal and electrical conductivity in the mantle [e.g., *Buffett*, 1992], as well as by the possible effects of CMB topography (which remains poorly determined), so this continues to be an area of active research [cf., *Buffett*, 1998].

Geomagnetic observations also provide one probe of a possible boundary layer on the core side of the CMB; particularly the possibility of a stably stratified outermost core layer [*Gubbins et al.*, 1982; *Braginsky*, 1993]. Thermal buoyancy in such a thermally stratified layer would compete with effects of compositional buoyancy in the deeper core associated with expulsion of light alloying components upon solidification of core material at the inner core boundary [e.g., *Lister and Buffet*, 1998]. Such a TBL could defy seismological detection, but there have been several studies that suggest the presence of slightly anomalous compressional velocity gradients in the outermost 50–200 km of the core based on *SmKS* phases [*Lay and Young*, 1990; *Souriau and Poupinet*, 1991; *Garnero et al.*, 1993b; *Tanaka and Hamaguchi*, 1993], and this remains an open issue even in the face of increasing complexity being recognized to exist in the mantle-side boundary layer [*Garnero and Lay*, 1998]. This is an important area for study given that the existence of any inhomogeneous structure in the outer core (generally assumed to be negligible) could trade-off with models for inner core structure. On a much finer scale, core-side boundary layer structure could involve ponding of buoyant light-alloying components under topographic highs in the CMB and development of finite rigidity in a very thin (<5 km) underplating layer [e.g., *Garnero and Jeanloz*, 2000; *Buffet et al.*, 2000]. Possible observation of very localized structure less than 0.2 km thick has been presented by *Rost and Revenaugh*, [2001], sustaining interest in the possibility of a thin mushy layer of sediments accumulating on the CMB. As yet there is very little constraint on any core-side thermal/chemical boundary layer, so the remainder of this article will focus on mantle-side structure.

Local Stratification of the Boundary Layer

Complex locally layered seismic structures have also been detected on the mantle side of the CMB at both the top and bottom of the D″ region. The shallower structure, typically from 150 to 350 km above the CMB, was first detected by array analysis of *P* waves [*Wright and Lyons*, 1979; *Wright et al.*, 1985; *Weber and Davis*, 1990] and by analysis of profiles of *S* waves [*Lay and Helmberger*, 1983; *Young and Lay*, 1987b; *Young and Lay*, 1990]. *Wysession et al.* [1998] review the many subsequent studies that characterize this feature as a rapid increase in seismic velocity with depth, over a depth extent of 0–30 km, with 2–3% shear velocity increase and 0.5–3% compressional velocity increase. The depth of the velocity increase varies laterally over both large (>500 km) and short (<100 km) spatial scales [e.g., *Kendall and Shearer*, 1994; *Weber et al.*, 1996; *Lay et al.*, 1997]. This feature is often called a discontinuity, but the sharpness and lateral continuity of the increase remains important research topics. Figure 2b indicates regions where the most compelling observations (from detailed waveform analyses) of the shear velocity increase are found, relative to large scale patterns in shear velocity (Figure 2a, Plate 1). Generally, the regions with strong velocity increases are imaged by tomographic analyses as having higher than average shear velocity, as under Middle America, eastern Eurasia, and India; however, the Pacific has evidence for relatively small (0.5–1.5%) velocity increases in areas that are low velocity in the global tomographic models [*Russell et al.*, 2001].

There are intermittent or isolated regions of the lowermost mantle that are fairly well sampled by seismic waves where any shear velocity increase appears to either be very small or not present [e.g. *Weber and Davis*, 1990; *Kendall and Nangini*, 1996; *Garnero and Lay*, 2003]. When considering all regions sampled, the statistical correlation between shear and compressional velocity increases and large-scale tomographic patterns is actually quite low [*Wysession et al.*, 1998]. This requires further investigation and assessment of the reliability of isolated detections based on waveform complexity. This is particularly true for *P* waves, as array processing of large-numbers of observations is required to confidently detect, or rule-out, small velocity increases of 0.5% or so, especially if distributed over some tens of kilometers radially. This is the case even at grazing incidence where the phases are amplified by triplication effects [e.g. *Reasoner and Revenaugh*, 1999].

The relative infrequency of clear short-period reflections from the top of D" for steeply incident waves [e.g., *Persch et al.*, 2001; *Castle and van der Hilst*, 2003] tends to favor either strong lateral variations in the D" discontinuity or obscuring effects such as a gradational transition zone or small-scale topography on the feature. Procedures embedding assumptions of one-dimensional reference models and horizontal reflectors may give incorrect estimates of actual structures.

The global TCBL model interprets the observed *P* and *S* velocity increases as the upper boundary of the chemically distinct layer, which is expected to be the site of a TBL as well. The topography on the boundary is expected as a result of dynamic loading by mid-mantle flow and possibly by internal convection of the boundary layer, with depressed discontinuity depths below downwellings, and elevated discontinuities under upwellings, both with high variability. Lateral variations in observability of the discontinuity caused by the intrinsic chemical contrast of D" are explained as the result of both topography and gradient of the thermal-chemical contrast at the top of the boundary, compounded by lateral variations in degree of partial melt within the layer, which has a profound affect on seismic velocities. Cooled regions under downwellings are expected to be sub-solidus, and thus have the

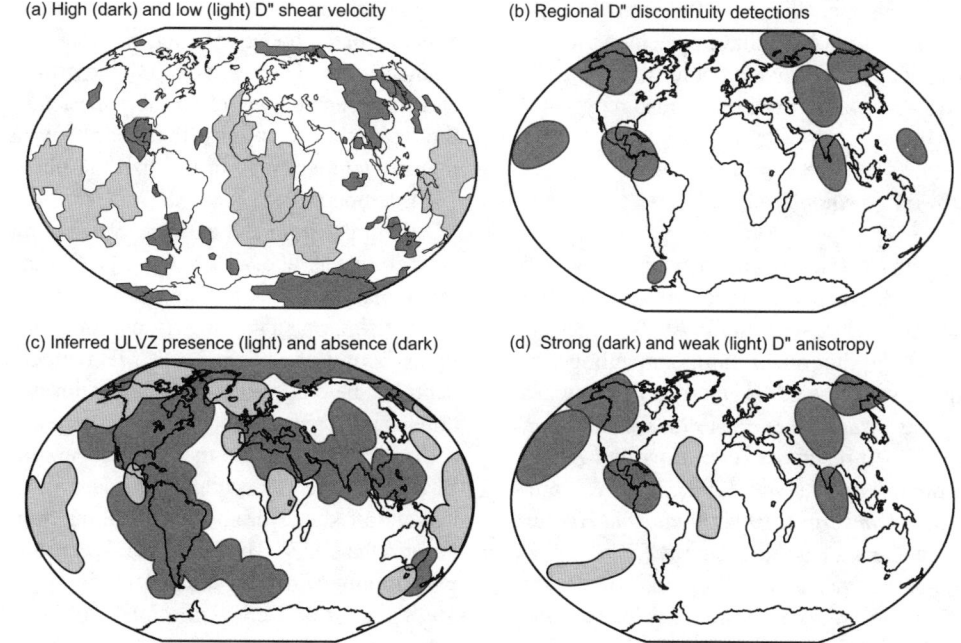

Figure 2. (a) Large-scale shear-velocity variations in the lowermost 250 km of the mantle from the global model of *Grand* [2002]. Dark shaded regions have shear velocity anomalies 1.2% or more faster than average, while light shaded regions have shear velocity anomalies 1.2% or more slower than average. (b) Shaded zones highlight regions where rapid shear velocity increases of 1.5–3% have been observed at depths near 250±50 km above the CMB. (c) Distribution of ultra-low velocity zone observations (light shade) and non-observations (dark shade). (d) Locations where relatively strong (>1%) vertical transverse isotropy have been detected in D" (dark shading) and where weak or non-existent anisotropy has been observed (light shading).

strongest positive velocity increases due to the absence of any competing partial melt component. Warmer regions under upwellings should have higher degrees of melt and hence weaker velocity increases, or possibly even decreases.

The hybrid TCBL model accounts for the velocity increases at the top of D" as the result of thermal anomalies of slab materials combined with an unspecified phase change to sharpen the velocity increase [e.g., *Sidorin et al.*, 1999]. The notion of a global phase change to account for the discontinuity has been around for a decade [*Nataf and Houard*, 1993], but a clear candidate for the phase change has not been established amongst the predominant Mg-perovskite and ferropericlase minerals expected in the lower mantle [see *Wysession et al.*, 1998]. Recent work by *Badro et al.* [2003] has demonstrated a possible change in the spin state of (Mg,Fe)O, from high-spin to low-spin for lower mantle pressures. This could favor iron enrichment in this mineral and iron depletion of silicate perovskite in the lowermost mantle, with possible attendant viscosity and thermal transport effects, but probably only a minor seismic velocity or bulk density effect. Another candidate effect for producing a radial increase in velocity in the hybrid model is a gradient in fabric into the boundary layer, with the stress increase and cooling of the mid-mantle downwelling allowing dislocation creep processes to develop lattice preferred orientation (LPO) in the (Mg,Fe)O component. *McNamara et al.* [2001] have presented models where suitable conditions are predicted below slabs, even if they do not descend all the way to the CMB. Anisotropy in D" is discussed further below.

Complex Mantle-Core Transition Zone

Evidence for very strong velocity contrasts just above the CMB dates back to work on spectra of core reflections and diffracted wave velocities in the 1960's and 1970's [see a review by *Young and Lay*, 1987a], but much more compelling evidence for a transition zone at the CMB emerged from studies of *SPdiffKS* waves [e.g., *Garnero et al.*, 1993; *Garnero and Helmberger*, 1996], *PcP* precursors [e.g. *Mori and Helmberger*, 1995; *Revenaugh and Meyer*, 1997]; and *PKP* precursors [e.g., *Wen and Helmberger*, 1998; *Vidale and Hedlin*, 1998]. These studies, and many since (see *Garnero et al.* [1998] for a review) demonstrate that in some regions, 10–40 km thick zones have 5–10% low compressional velocities and 15–30% low shear velocities right above the CMB, either in horizontally extensive layers or concentrated into blob-like domes. The magnitude of the velocity reductions and the large ratio between shear and compressional velocity variations favor an interpretation of these so-called ultra-low velocity zones (ULVZs) as regions of significant (6% to 30%) partial melt volume [*Williams and Garnero*, 1996]. The spatial extent of ULVZ detections and non-detections is indicated in Figure 2c. Non-detections are difficult to appraise, as the structure may be too thin or too gradational to give rise to clear waveform complexity, but there are regions where there is at least no evidence supporting significant ULVZ presence [e.g. *Castle and van der Hilst*, 2000; *Persch et al.*, 2001]. Figure 2 demonstrates a general correlation between major ULVZ regions and large-scale patterns of low shear velocity in D", although there are some exceptions such as under Central America. The interpretation that ULVZ material is partially molten is shared by the two end-member TCBL models, but in the hybrid model localized chemical anomalies may account for the finite extent of the ULVZ zones. For the global model it is also possible that lateral chemical heterogeneities contribute to the spatial pattern, but it may simply be that the ULVZ is seismically detectable in the hottest regions of the layer, where the steep thermal gradient in the CMB TBL exceeds the solidus over tens of kilometers rather than over only a few kilometers. In cool regions with a steep thermal gradient over a very narrow depth range, it may be very difficult to detect any ULVZ even with short-period reflected waves.

The dramatic velocity reductions invoked to explain the ULVZ observations motivate consideration of partial melting in the lowermost mantle, the presence of which has profound implications for seismic velocity heterogeneity, viscosity structure of the boundary layer, and chemical processing in the boundary layer [*Lay et al.*, 2003]. Experimental constraints on end-member perovskite and ferropericlase mineralogies of the deep mantle suggest that each may have much higher melting temperatures in D" than the likely upper bound on CMB temperatures [*Zerr and Boehler*, 1993, 1994]; however, for a multi-component eutectic system a much lower solidus temperature is likely. *Zerr et al.* [1998] and *Boehler* [2000] extrapolated a pyrolite composition solidus to CMB conditions, estimating a melting temperature of 4300 K, within the upper bound of estimates of CMB temperatures. Additional chemical heterogeneity, associated with subduction products, core-mantle reaction products, or ancient chemical stratification could lower the melting temperature within D" even further (Figure 1), accounting for laterally varying melt fraction within the layer and/or partial melting in the basal TBL to form the ULVZ. The few available constraints on liquid-solid partitioning of iron in silicates at high pressure [see *Knittle*, 1998], favor iron concentration into the melt phase, which provides a density effect that can stabilize melts in the boundary layer, or possibly lead to their accumulation in the ULVZ due to negative buoyancy [see *Lay et al.*, 2003]. At this point, partial melting of either the lowermost or all of the hottest regions of D" must be considered a possibility. Detailed investigation of seismic attenuation properties of the ULVZ could help to constrain the structure.

Low Velocity Provinces

Improvements in global distribution of seismometers have enabled increasing focus on detailed structure in the low velocity regions under the Central Pacific and Africa [e.g., *Tanaka*, 2002; *Wen et al.*, 2001; *Bréger et al.*, 2001]. Both regions appear to have strong lateral gradients, with shear velocity variations involving abrupt 3–5% reductions and low velocity zones several hundred kilometers in thickness. Extensive regions of the South Atlantic have a velocity decrease at the top of the low velocity region, with 1–3% drop near 200–250 km, at about the same average depth as the 2–3% increases in circum-Pacific regions [e.g., *Ni and Helmberger*, 2003; *Wen* 2002]. The low velocity region extends upward 500–800 km into the mid-mantle, still with sharp lateral gradients, under Africa [*Ni et al.*, 2002], suggesting large topography on the low velocity body. The Atlantic/Africa region appears to have shear/compressional velocity ratios compatible with partial melting [e.g., *Simmons and Grand*, 2002; *Tkalcic and Romanowicz*, 2001], while the central Pacific region may involve anomalous ratios requiring chemical effects as well as partial melting [e.g., *Masters et al.*, 2000; *Lay et al.*, 2003]. In the hybrid TCBL model, the large low velocity regions under the southern Pacific and southern Atlantic/Africa are considered chemical superplumes; where hot, dense chemical heterogeneities have piled up under upwellings (Figure 1b). The chemical anomaly accounts for part of the velocity decrease and the sharp edges of the structure. In the global TCBL model, these are the thickest, hottest regions of the boundary layer, with the most extensive fractional melting across the chemical layer, resulting in relatively abrupt seismic velocity decreases across the layer interface (Figure 1d). Relatively small melt fractions (0.5–2%) can produce the strong velocity reductions needed to account for these structures [*Lay et al.*, 2003]. In both cases the large topography is induced by mantle flow above hot, but dense material.

Boundary Layer Anisotropy and Scattering

Seismic anisotropy has been demonstrated to exist in the D" region for quite some time (see *Lay et al.* [1998] and *Kendall* [2000] for reviews). The general observations support widespread occurrence of anisotropy compatible with vertical transverse isotropy (VTI) (Figure 2d), primarily below circum-Pacific, relatively high shear velocity regions, with weak anisotropy beneath regions of moderate shear velocity anomalies. Intermittent azimuthal anisotropy is found beneath the low velocity central Pacific. Anisotropy appears to have increased strength and spatial coherence in the boundary layer relative to the overlying mid-mantle. Recognizing that the boundary layer is likely to have relatively strong shear flows and lateral temperature variations, a mix of possible anisotropy mechanisms have been proposed, ranging from shearing of partial-melt components in horizontal or vertical flows [e.g., *Kendall and Silver*, 1998; *Russell et al.*, 1998] to stress-induced LPO in (Mg,Fe)O in low temperature regions where dislocation glide mechanisms are active [e.g., *Yamazaki and Karato*, 2002; *McNamara et al.*, 2001, 2003]. Both end-member TCBL models can accommodate either form of anisotropy as a result of boundary layer shearing of chemical heterogeneities, partial melt blobs, or mineral alignments. Observations of coupling between the shear velocity increase at the top of D" and an onset of anisotropy [e.g., *Matzel et al.*, 1996; *Garnero and Lay*, 1997] suggest that either acquisition of preferred fabric is responsible for the discontinuity or that the chemical change giving rise to the discontinuity produces favorable conditions for the development of the fabric. Further characterization of D" anisotropy through more detailed seismic analyses is needed to establish its role in the TCBL.

A further line of evidence pertaining to structures and processes near the CMB is the presence of very small-scale seismic heterogeneity, manifested in the scattered wavefield accompanying deeply penetrating seismic phases. Scattering of short-period *P* waves by structural heterogeneity, including possible CMB topography, indicates that scale-lengths of a few to tens of kilometers have velocity fluctuations of a few to ten percent [e.g., *Bataille and Lund*, 1996; *Cormier*, 2000; *Earle and Shearer*, 1997; *Hedlin and Shearer*, 2000]. In some cases the scattering may arise from the ULVZ [e.g., *Vidale and Hedlin*, 1998; *Wen and Helmberger*, 1998 *Niu and Wen*, 2001; *Rost and Revenaugh*, 2003], and in a few cases it can be imaged by scattering migrations [e.g., *Thomas et al.*, 1999; *Rost and Thomas*, 2003]. The overall spectrum of heterogeneity of the boundary layer is not well determined yet, but it appears to be relatively red, with substantial power at long wavelengths and moderate power at short wavelengths, possibly with an anisotropic spatial distribution [e.g., *Cormier*, 1999]. The two end-member TCBL models both involve thermal and chemical heterogeneities that can account for a reddened heterogeneity spectrum, with the global model having the added degrees of freedom provided by strong velocity effects of distributed fractional melts. Both models also admit the possibility of CMB topography on a variety of scale-lengths, which could account for much of the scattering in the region due to the strong density and velocity contrasts involved. Determining CMB topography has proved very challenging and there is little agreement amongst recent models [e.g., *Garcia and Souriau*, 2000; *Sze and van der Hilst*, 2003]. Improved characterization of the spatial pattern of small-scale heterogeneities could help to constrain specific causal mechanisms within the boundary layer.

DISCUSSION AND CONCLUSIONS

Table 1 summarizes the primary observational constraints on the D" region and the various ways that these constraints have been built into the two competing models we discuss in this paper. With so few hard-constraints on viable chemistry, melting, and dynamical structures in the boundary layer, the end-member models are sufficiently flexible to accommodate most observations with reasonable degree of plausibility. Particularly challenging for the global model is to account for the reversal in sign of the velocity contrast across the top of the layer from positive in high velocity areas to negative in low velocity areas. However, the dramatic effect of small amounts of partial melt, suitably distributed in the boundary layer material, provides a possible mechanism. The absence of a universal sharp reflector is also a challenge for this model, but the demonstrated presence of strong topography, along with the viability of a gradational transition zone with superimposed effects of a TBL may reconcile this constraint. The hybrid model struggles to account for the presence of rapid velocity increases at the top of D", as such are not readily produced in thermal models of subducting slabs [e.g., *Sidorin et al.*, 1999], and this leads to the need for an additional effect, such as a phase change (of unknown type), or perhaps an

Table 1. Observational Characteristics of the CMB TCBL and How End-Member Models Account for Them

Feature	Characteristics	Hybrid TCBL Model	Global TCBL Model
High Heat Flux Values Across CMB	Thermal model based estimates of CMB heat flux of as much as 10 TW	Hot thermal boundary layer instabilities	Layering would lower heat flux estimates
Mantle/Magnetic Field Correlations	Strong localized quasi-stationary magnetic flux concentrations under regions of high seismic velocity	Slab cooling produces long-term heat flux boundary condition	Thermally modulated layer produces long-term heat flux boundary condition
Large-scale Vs and Vp Heterogeneity	±3-4% dVs, ±1% dVp, degree 2 dominance ±1% bulk sound velocity circum-Pacific high velocity C. Pacific/S. Atlantic low velocity strongest in lowermost 300-400 km	Thermal origin in large scale flow regime, slabs cause high velocity, chemical heterogeneity in slow regions, superplumes	Partial melt variations within chemical layer temperatures modulated by thermal coupling chemical layer has bulk sound velocity anomaly
D" Velocity Increases	+1.5-+3% in Vs, +0.5-+3% in Vp 0-30 km thick transition zone in high velocity regions, weaker in low velocity regions, depth varies from 130 to 300+ km above CMB	Slab related with a phase change, gradient in fabric, gradient in heterogeneity spectrum	Intrinsic to chemical contrast, varying partial melt fraction modulates, dynamic topography on chemical layer
Simple ScP Phases	Widespread observations of steeply incident CMB reflections with no high frequency precursors and simple waveforms	Pristine areas between slab piles and low velocity chemical blobs	Fuzzy upper boundary of chemical layer due to topography and/or chemical/thermal gradients, ULVZ very thin in cooled regions
ULVZ	5-10% Vp reductions, 10-30% Vs reductions, 5-40 km thickness sometimes sharp onset (< few km) best developed in low velocityregions, but present in high velocity regions	Partial melt in basal boundary layer in regions with piles of chemical heterogeneities,	Partial melt in basal boundary layer due to laterally modulated temperature profile across chemical layer
Low Velocity Blobs	3-5% slow shear velocity over 250-800 km thick zones, with sharp (-1 to –3%) decreases at top and laterally	Chemical heterogeneities, hot, dense piles under up-wellings	Hottest, highest melt fraction regions of TCBL thick under upwellings
D" Anisotropy	Large scale 1-2% in boundary layer, primarily with VTI symmetry in high velocity regions. variable strength/symmetry in low velocity regions. absent from average regions.	LPO in MgO, SPO in chemical/melt, lamellae from sheared chemical anomalies	LPO in MgO SPO in chemcial/melt lamellae from sheared chemical anomalies
D" Scattering	1-10% rms heterogeneity in boundary layer, 10-100 km scale lengths related to ULVZ, change in spectrum relative to mid-mantle	CMB topography, chemical/melt variations, thermal explanation not likely without melt, slab crustal components	CMB topography chemical/melt variation small scale convection

onset of anisotropy in the boundary layer due to the stresses and cooling of the downwelling slab, as in the models of *McNamara et al.* [2003].

The evidence for chemical heterogeneity above and beyond thermal effects, including partial melting, is rather compelling. The strong gradients laterally and radially into the low velocity zone beneath the south Atlantic/Africa are hard to account for thermally without some chemical contribution. While abundant dense partial melt could play some role, sustaining strong lateral gradients in upwelling material for hundreds of kilometers appears problematic. The anomalous bulk sound velocity measurements found for the central Pacific also appear to require a compensating bulk modulus perturbation even if partial melting accounts for the strong shear velocity reductions in the region. This could still be partially accounted for by anisotropy, so further constraints on the cause and orientation of anisotropy are needed. Small-scale fluctuations are unlikely to be due to thermal heterogeneity unless the temperatures are right at the solidus (as proposed by the global TCBL model). Of course, partial melting itself should give rise to chemical heterogeneity, such as iron fractionation into the melt, so one cannot truly separate melting from chemical heterogeneity.

Dynamical models for global and hybrid TCBL structures comprised of primordial components, core-mantle chemical reaction products, or segregated subduction products have been explored quite extensively in the past few years using two- and three-dimensional mantle convection codes. Some of the key issues are the evolution of a TCBL and the viability of chemical heterogeneities denser than normal mantle material accumulating at the base of the mantle and surviving entrainment by mid-mantle flow. Most calculations indicate that a density contrast of 3–6% relative to the overlying mantle is required to sustain a coherent global layer [e.g., *Christensen*, 1984; *Sleep*, 1988; *Kellogg and King*, 1993; *Kellogg*, 1997; *Sidorin and Gurnis*, 1998; *Montague and Kellogg*, 2000, *Montague et al.*, 1998]. The requisite density contrast for stability of the layer may actually be much lower, on the order of 0.5–1% when allowance is made for compressibility effects and strong temperature dependence of thermal expansion (the buoyancy number must be computed for the local, reduced values, which is not usually done in the literature), along with effects such as reduction of viscosity in the boundary layer caused by temperature-dependent viscosity [e.g., *Schott et al.*, 2002]. If the density increase of the chemical heterogeneities is too low to maintain a coherent layer, the material will be concentrated into patches under upwellings [e.g., *Davies and Gurnis*, 1986; *Hansen and Yuen*, 1989, *Manga and Jeanloz*, 1996; *Tackley*, 1998; *Davaille*, 1999; *Gonnermann et al.*, 2002]; a means by which an initially global TCBL situation could have evolved into a hybrid TCBL today. Various dynamical models differ in the extent to which subducting slabs reach the deepest mantle, whether crustal components can separate from the slab, and the extent of thermal and viscous disruption and induced topography of the boundary layer that takes place [e.g., *Christensen and Hofmann*, 1994; *Sidorin and Gurnis*, 1998; *Tackley*, 2000; *McNamara et al.*, 2001; *Coltice and Ricard*, 1999].

There have also been numerous geodynamic explorations of the possible role of D" as a source of thermal plumes, and their potential to bring up chemical heterogeneities from the deep boundary that can account for geochemical anomalies in ocean island basalts [*Hofmann and White*, 1982; *Albarede and van der Hilst*, 1999]. The presence of chemical heterogeneity in the boundary layer can affect the stability and distribution of thermal plumes rising from within or above the boundary layer [e.g., *Kellogg and King*, 1993; *Farnetani*, 1997; *Jellinek and Manga*, 2002] as well as the overall heat transport across the boundary layer [e.g., *Namiki and Kurita*, 2003]. The extent to which entrained boundary layer materials are mixed in ascending plume shear flows has also been examined [e.g., *Farnetani and Richards*, 1995]. Shear flow in the boundary layer likely plays a major role in the development of seismic wave anisotropy [*McNamara et al.*, 2001; *Lay et al.*, 1998], as well as possibly contributing to ULVZ formation and/or growth by shear heating [*Steinbach and Yuen*, 1999]. At present, the calculations support the viability of plumes sampling chemical heterogeneities either within or from the top of the boundary layer and bringing them to the surface, but this has not yet been demonstrated to occur.

Advancing our understanding of structures and processes near the CMB will require observational, laboratory and modeling advances across several disciplines. Figure 3 highlights some of the major boundary layer features that require improved observational constraints; including aspects of D" discontinuities, D" anisotropy, large-scale low velocity zones, ULVZs, CMB topography, and attendant dynamical issues. It is also clearly of great importance to establish whether there is any density increase in the lowermost mantle along with whether there are density anomalies associated with large low velocity provinces. Preliminary work with normal modes suggests that low velocity regions may have anomalously high density [e.g., *Ishii and Tromp*, 1999], but the resolution of normal mode approaches remains limited [e.g., *Romanowicz*, 2001; *Kuo and Romanowicz*, 2002]. Many of the topics noted in Figure 3 require improved seismological imaging and modeling, which includes better 2- and 3-dimensional wave propagation approaches as well as more advanced array methods for characterizing subtle features in the wavefields. There are tremendous limitations imposed by the geometry of sources and receivers that need to be addressed by innovative data collection strategies.

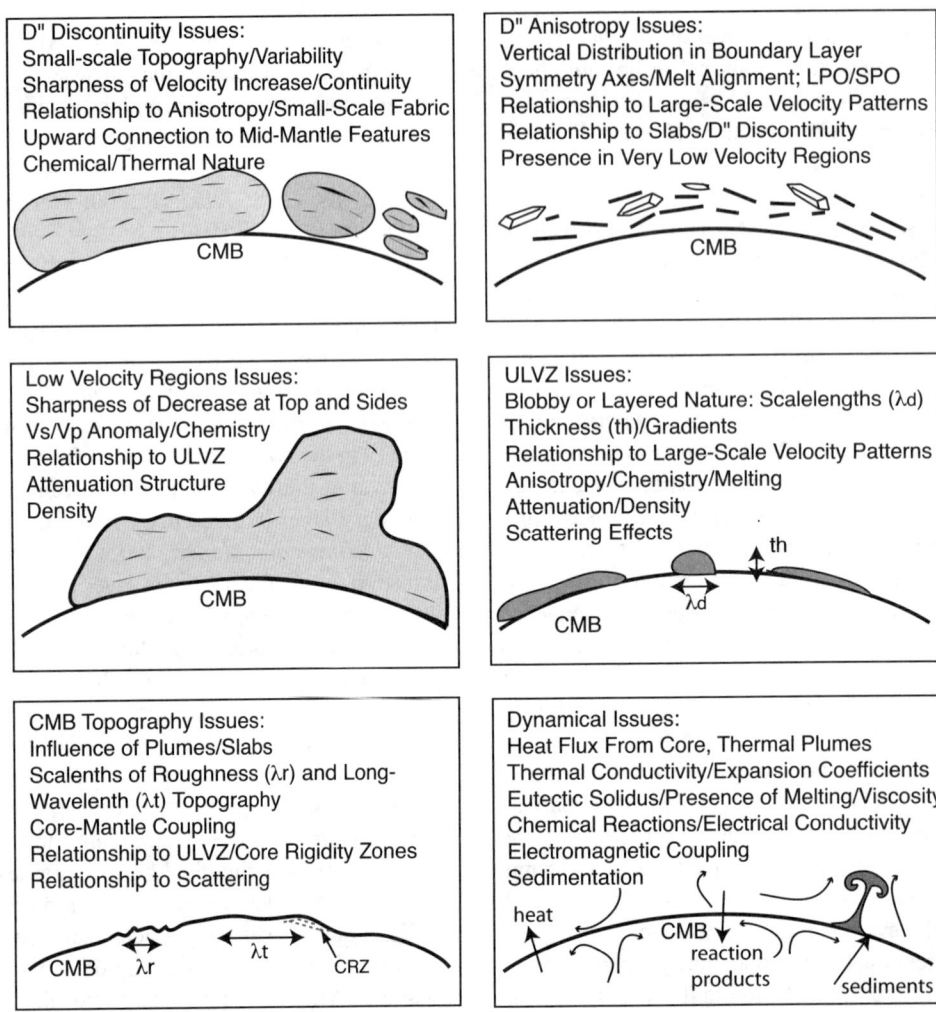

Figure 3. Summary of principal research topics associated with major features of the CMB TCBL including the D" discontinuity, D" anisotropy, large-scale low velocity regions, ULVZ, CMB topography, and associated dynamical issues. Progress on each area requires advances in seismology, mineral physics, geodynamics, and possibly geomagnetism.

While numerically challenging at present, dynamical calculations need to account for three-dimensional, compressible flow with the possibility of small-scale partial melting. The strong viscosity reduction upon melting and the multi-scale nature of partial melting present formidable challenges. Quantification of boundary layer shear flows that might account for anisotropy by mineral or melt/chemical inclusion alignment is also needed. Enhanced mineral physics experimental constraints on the stability and iron spin-state of silicate perovskite and ferropericlase for D" conditions are needed. Exploration of the eutectic melting behavior of plausible high pressure lower mantle assemblages is also needed. All of these needs are at the frontiers of current technologies, and concerted effort will be required to resolve the issue of a strongly stratified or dynamically disrupted TCBL above the CMB. Establishing the current configuration of deep mantle structure is prerequisite to extrapolating back in time to an earlier, hotter Earth system, and to understanding the thermal evolution of the core [e.g., *Buffett*, 2003], the extent of partial melting of the ancient mantle, and variations in the role played by D" through time.

Acknowledgements. We thank Geoff Davies, Brian Kennett, and Chris Hawkesworth for comments on the manuscript. This research was supported by NSF grants EAR-0125595 (TL). EAR-9996302 (EG). Contribution No. 466, Center for the Study of Imaging and Dynamics of the Earth, IGPP, UCSC.

REFERENCES

Albarede, F., and R.D. van der Hilst, New mantle convection model may reconcile conflicting evidence, *EOS, 80,* 535–539, 1999.

Anderson, D.L., The EDGES of the mantle, in *The Core-Mantle Boundary Region*, edited by M. Gurnis, M. E. Wysession, E. Knittle, and B. A. Buffett, pp. 255–271, AGU, Washington, D.C., 1998.

Anderson, O. L., The power balance at the core-mantle boundary, *Phys. Earth Planet. Inter., 131*, 1–17, 2002.

Badro, J., G. Fiquet, F. Guyot, J.-P. Rueff, V. V. Struzhkin, G. Vankó, and G. Monaco, Iron partitioning in Earth's mantle: Toward a deep lower mantle discontinuity, *Science, 300*, 789–791, 2003.

Bataille, K., and F. Lund, Strong scattering of short-period seismic waves by the core-mantle boundary and the *P*-diffracted wave, *Geophys. Res. Lett., 18*, 2413–2416, 1996.

Bijwaard, H., W. Spakman, and E. R. Engdahl, Closing the gap between regional and global travel time tomography, *J. Geophys. Res., 103*, 30055–30078, 1998.

Bloxham, J., and D. Gubbins, Thermal core-mantle interactions, *Nature, 325*, 511–513, 1987.

Boehler, R., High-pressure experiments and the phase diagram of lower mantle and core materials, *Rev. Geophys., 38*, 221–245, 2000.

Boschi, L., and A. M. Dziewonski, High- and low-resolution images of the Earth's mantle: Implications of different approaches to tomographic modeling, *J. Geophys. Res., 104*, 25,567–25,594, 1999.

Boschi, L., and A. M. Dziewonski, Whole Earth tomography from delay times of *P*, *PcP*, and *PKP* phases: Lateral heterogeneities in the outer core or radial anisotropy in the mantle?, *J. Geophys. Res., 105*, 13675–13696, 2000.

Braginsky, S. I., MAC-oscillations of the hidden ocean of the core, *J. Geomag. Geoelectr., 45*, 1517–1538, 1993.

Bréger, L., B. Romanowicz, and C. Ng, The Pacific plume as seen by *S*, *ScS*, and *SKS*, *Geophys. Res. Lett., 28*, 1859–1862, 2001.

Buffett, B. A., Constraints on magnetic energy and mantle conductivity from the forced nutations of the Earth, *J. Geophys. Res., 97*, 19,581–19,597, 1992.

Buffett, B. A., Free oscillations in the length of day: inferences on physical properties near the core-mantle boundary, in *The Core-Mantle Boundary Region*, edited by M. Gurnis, M. E. Wysession, E. Knittle, and B. A. Buffett, pp. 153–165, AGU, Washington, D.C., 1998.

Buffett, B. A., Estimates of heat flow in the deep mantle based on the power requirements for the geodynamo, *Geophys. Res. Lett., 29*, No. 12, 10.1029/2001GL014649, 2002.

Buffett, B.A., The thermal state of Earth's core, *Science, 299*, 1675–1677, 2003.

Buffett, B. A., E. J. Garnero, and R. Jeanloz, Sediments at the top of Earth's core, *Science, 290*, 1338–1342, 2000.

Bullen, K. E., Compressibility-pressure hypothesis and the Earth's interior, *Month. Not. R. Astr. Soc., Geophys. Suppl., 5*, 355–368, 1949.

Castle, J. C., and R. D. van der Hilst, The core-mantle boundary under the Gulf of Alaska: No ULVZ for shear waves, *Earth and Planet. Sci. Lett. 176*, 311–321, 2000.

Castle, J. C., and R. D. van der Hilst, Using *ScP* precursors to search for mantle structures beneath 1800 km depth, *Geophys. Res. Lett., 30*, No. 8, 1422, doi:10.1029/2002GL016023, 2003.

Castle, J.C., Creager, K.C., and J.P. Winchester, Shear wave speeds at the base of the mantle, *J. Geophys. Res., 105*, 21,543–21,557, 2000.

Christensen, U., Instability of a hot boundary layer and initiation of thermo-chemical plumes, *Annales Geophysicae, 2*, 311–319, 1984.

Christensen, U. R., and A. W. Hofmann, Segregation of subducted oceanic crust in the convecting mantle, *J. Geophys. Res., 99*, 19,867–19,884, 1994.

Coltice, N., and Y. Ricard, Geochemical observations and one layer mantle convection, *Earth Planet Sci. Lett., 174*, 125–137, 1999.

Cormier, V. F., Anisotropy of heterogeneity scale lengths in the lower mantle from *PKIKP* precursors, *Geophy. J. Int., 136*, 373–384, 1999.

Cormier, V.F., D" as a transition in the heterogeneity spectrum of the lowermost mantle, *J. Geophys. Res., 105*, 16,193–16,205, 2000.

Davaille, A., Simultaneous generation of hotspots and superswells by convection in a heterogeneous planetary mantle, *Nature, 402*, 756–760, 1999.

Davies, G., Mantle plumes, mantle stirring and hotspot chemistry, *Earth Planet. Sci. Lett., 99*, 94–109, 1990.

Davies, G., and M. Gurnis, Interaction of mantle dregs with convection—lateral heterogeneity at the core-mantle boundary, *Geophys. Res. Lett., 13*, 1517–1520, 1986.

Dubrovinsky, L., H. Annersten, N. Dubrovinskala, F. Westman, H. Harryson, O. Fabrichnaya, and S. Carlson, Chemical interactions of Fe and Al_2O_3 as a source of heterogeneity at the Earth's core-mantle boundary, *Nature, 412*, 527–529, 2001.

Dziewonski, A.M., D. L. Anderson, Preliminary reference earth model, *Phys. Earth Planet Int., 25*, 297–356, 1981.

Earle, P. S., and P. M. Shearer, Observations of *PKKP* precursors used to estimate small-scale topography on the core-mantle boundary, *Science, 277*, 667–670, 1997.

Farnetani, C. G., Excess temperature of mantle plumes: The role of chemical stratification across D", *Geophys. Res. Lett., 24*, 1583–1586, 1997.

Farnetani, C. G., and M. A. Richards, Thermal entrainment and melting in mantle plumes, *Earth Planet. Sci. Lett., 136*, 251–267, 1995.

Forte, A. M., and J. X. Mitrovica, Deep-mantle high-viscosity flow and thermochemical structure inferred from seismic and geodynamic data, *Nature, 410*, 1049–1056, 201.

Fukao, Y., S. Widiyantori, and M. Obayashi, Stagnant slabs in the upper and lower mantle transition region, *Rev. of Geophys., 39*, 291–323, 2001.

Garcia, R., and A. Souriau, Amplitude of the core-mantle boundary topography estimated by stochastic analysis of core phases, *Phys. Earth Planet. Inter., 117*, 345–359, 2000.

Garnero, E. J., Lower mantle heterogeneity, *Ann. Rev. Earth Planetary Sci., 28*, 509–37, 2000.

Garnero, E. J., and D. V. Helmberger, Seismic detection of a thin laterally varying boundary layer at the base of the mantle beneath the central-Pacific, *Geophys. Res. Lett., 23*, 977–980, 1996.

Garnero, E. J., and R. Jeanloz, Fuzzy patches on the Earth's core-mantle boundary, *Geophys. Res. Lett., 27*, 2777–2780, 2000.

Garnero, E. J., and T. Lay, Lateral variations in lowermost mantle shear wave anisotropy beneath the north Pacific and Alaska, *J. Geophys. Res., 102*, 8121–8135, 1997.

Garnero, E. J., and T. Lay, Effects of D" anisotropy on seismic velocity models of the outermost core, *Geophys. Res. Lett., 25*, 23412344, 1998.

Garnero, E. J., and T. Lay, D" shear velocity heterogeneity, anisotropy, and discontinuity structure beneath the Caribbean and Central America, *Phys. Earth Planet. Int.*, in press, 2003.

Garnero, E. J., S. P. Grand, and D. V. Helmberger, Low P wave velocity at the base of the mantle, *Geophys. Res. Lett., 20*, 1843–1846, 1993a.

Garnero, E. J., D. V. Helmberger, and S. P. Grand, Constraining outermost core velocity with SmKS waves, *Geophys. Res. Lett., 20*, 2463–2466, 1993b.

Garnero, E. J., J. S. Revenaugh, Q. Williams, T. Lay, and L. H. Kellogg, Ultralow velocity zone at the core-mantle boundary, in *The Core-Mantle Boundary Region*, edited by M. Gurnis, M. E. Wysession, E. Knittle, B. A. Buffett, pp. 319–334, AGU, Washington, DC, 1998.

Goarant, F., F. Guyot, J. Peyroneau, and J. P. Poirier, High-pressure and high-temperature reactions between silicates and liquid iron alloys in the diamond anvil cell, studied by analytical electron microscopy, *J. Geophys. Res., 97*, 4477–4487, 1992.

Gonnermann, H. M., M. Manga, and A M. Jellinek, Dynamics and longevity of an initially stratified mantle, *Geophys. Res. Lett., 29*, No. 10, 10.1029/2002GL014851, 2002.

Grand, S.P., Mantle shear-wave tomography and the fate of subducted slabs, *Phil. Trans. R. Soc. Lond., A, 360*, 2475–2491, 2002.

Gu, Y.J., A.M. Dziewonski, W. Su, and G. Ekström, Models of the mantle shear velocity and discontinuities in the pattern of lateral heterogeneities, *J. Geophys. Res.,106*, 11,169–11,199, 2001.

Gubbins, D., Interpreting the paleomagnetic field, in *The Core-Mantle Boundary Region*, edited by. M. Gurnis, M. Wysession, E. Knittle, and B. Buffet, pp. 167–182, AGU Washington D.C., 1998.

Gubbins, D., C. J. Thompson, and K. A Whaler, Stable regions in the Earth's liquid core, *Geophys. J. R. Astron. Soc., 68*, 241–251, 1982.

Gurnis, M., M. E. Wysession, E. Knittle, B. A. Buffett (editors), *The Core-Mantle Boundary Region*, AGU, Geodynamics Series, Vol. 28, 334pp, 1998.

Hansen, U., and D. A. Yuen, Dynamical influences from thermal-chemical instabilities at the core-mantle boundary, *Geophys. Res. Lett., 16*, 629–632, 1989.

Hedlin, M. A. H., and P. M. Shearer, An analysis of large-scale variations in small-scale mantle heterogeneity using Global Seismographic Network recordings of precursors to PKP, *J. Geophys. Res., 105*, 13,655–13,673, 2000.

Hellfrich, G. R., and B. J. Wood, The Earth's mantle, *Nature, 412*, 501–507, 2001.

Hofmann, A. W., Mantle geochemistry: the message from oceanic volcanism, *Nature, 385*, 219–229, 1997.

Hofmann, A. W., and W. M. White, Mantle plumes from ancient oceanic crust, *Earth Planet. Sci. Lett., 57*, 421–436, 1982.

Ishii, M., and J. Tromp, Normal-mode and free-air gravity constraints on lateral variations in velocity and density, *Science, 285*, 1231–1236, 1999.

Jellinek, A. M., and M. Manga, The influence of a chemical boundary layer on the fixity, spacing and lifetime of mantle plumes, *Nature, 418*, 760–763, 2002.

Kárason, H., and R.D. van der Hilst, Tomographic imaging of the lowermost mantle with differential times of refracted and diffracted core phases (PKP, $Pdiff$), *J. Geophys. Res., 106*, 6569–6587, 2001.

Kellogg, L. H., Growing the Earth's D" layer: Effect of density variations at the core-mantle boundary, *Geophys. Res. Lett., 24*, 2749–2752, 1997.

Kellogg, L.H., and S. D. King, Effect of mantle plumes on the growth of D" by reaction between the core and mantle, *Geophys. Res. Lett., 20*, 379–382, 1993.

Kellogg, L.H., Hager, B.H., and R.D. van der Hilst, Compositional stratification in the deep mantle, *Science, 283*, 1881–1884, 1999.

Kendall, J. M., Seismic Anisotropy in the Boundary Layers of the Mantle, in *Earth's Deep Interior: Mineral Physics and Tomography From the Atomic to the Global Scale*, edited by S. Karato, A.M. Forte, R.C. Liebermann, G. Masters, and L. Stixrude, pp. 133–159, AGU, Washington, D.C., 2000.

Kendall, J. M., and C. Nangini, Lateral variations in D" below the Caribbean, *Geophys. Res. Lett., 23*, 399–402, 1996.

Kendall, J. M., and P. M. Shearer, Lateral variations in D" thickness from long-period shear wave data, *J. Geophys. Res., 99*, 11,575–11,590, 1994.

Kendall, J. M., and P. G. Silver, Investigating causes of D" anisotropy, in *The Core-Mantle Boundary Region*, edited by M. Gurnis, M. Wysession, E. Knittle, and B. Buffet, pp. 97–118, AGU, Washington D.C., 1998.

Kennett, B. L. N., S. Widiyantoro, and R. D. van der Hilst, Joint seismic tomography for bulk sound and shear wave speed in the Earth's mantle, *J. Geophys. Res., 103*, 12,469–12,493, 1998.

Knittle, E., The solid/liquid partitioning of major and radiogenic elements at lower mantle pressures: Implications for the core-mantle boundary region, in *The Core-Mantle Boundary Region*, edited by M. Gurnis, M. Wysession, E. Knittle, and B. Buffet, pp. 119–130, AGU, Washington D.C., 1998.

Knittle, E., and R. Jeanloz, Simulating the core-mantle boundary: an experimental study of high-pressure reactions between silicates and liquid iron, *Geophys. Res. Lett., 16*, 609–612, 1989.

Kuo, B. Y., E. J. Garnero, and T. Lay, Tomographic Inversion of S-SKS times for shear wave velocity heterogeneity in D": Degree 12 and hybrid models, *J. Geophys Res., 105*, 28,139–28,157, 2000.

Kuo, C., and B. Romanowicz, On the resolution of density anomalies in the Earth's mantle using spectral fitting of normal-mode data, *Geophys. J. Int., 150*, 162–179, 2002.

Labrosse, S., Hotspots, mantle plumes and core heat loss, *Earth Planet. Inter., 199*, 147–156, 2002.

Lay, T., E. J. Garnero, and Q. Williams, Partial melting in a thermochemical boundary layer at the base of the mantle, *Phys. Earth Planet. Int.*, in press, 2003.

Lay, T., E. J. Garnero, C. J. Young, and J. B. Gaherty, Scale-lengths of heterogeneity at the base of the mantle from S-wave differential times, *J. Geophys. Res., 102*, 9887–9909, 1997.

Lay, T., and D. V. Helmberger, A lower mantle S-wave triplication and the shear velocity structure of D", *Geophys. J. R. Astron. Soc.*, 75, 799–838, 1983.

Lay, T., and C. J. Young, The stably-stratified outermost core revisited, *Geophys. Res. Lett.*, 17, 2001–2004, 1990.

Lay, T., Q. Williams, and E. J. Garnero, The core-mantle boundary layer and deep Earth dynamics, *Nature*, 392, 461–468, 1998a.

Lay, T., Q. Williams, E. J. Garnero, L. Kellogg, and M.E. Wysession, Seismic wave anisotropy in the D" region and its implications, in *The Core-Mantle Boundary Region*, edited by. M. Gurnis, M. Wysession, E. Knittle, and B. Buffet, pp. 229–318, AGU, Washington, D.C., 1998b.

Lister, J. R., and B. A. Buffett, Stratification of the outer core at the core-mantle boundary, *Phys. Earth Planet. Inter.*, 105, 5–19, 1998.

Lithgow-Bertelloni C, and M. A. Richards, The dynamics of Cenozoic and Mesozoic plate motions, *Rev. Geophys.*, 36, 27–78, 1998.

Manga, M., and R. Jeanloz, 1996. Implications of a metal-bearing chemical boundary layer in D" for mantle dynamics, Geophys. Res. Lett., 23, 3091–3094.

Masters, G., G. Laske, H. Bolton, and A.M. Dziewonski, The relative behavior of shear velocity, bulk sound speed, and compressional velocity in the mantle: implications for chemical and thermal structure, in *Earth's Deep Interior: Mineral Physics and Tomography From the Atomic to the Global Scale*, edited by S. Karato, A. M. Forte, R. C. Liebermann, G. Masters, and L. Stixrude, pp. 63–87, AGU, Washington, D.C., 2000.

Matzel, E., M. K. Sen, and S. P. Grand, Evidence for anisotropy in the deep mantle beneath Alaska, *Geophys., Lett.*, 23, 2417–2420, 1996.

McNamara, A. K., S.-I. Karato, and P. E. van Keken, Localization of dislocation creep in the lower mantle: implications for the origin of seismic anisotropy, *Earth Planet. Sci. Lett.*, 191, 85–99, 2001.

McNamara, A. K., P. E. van Keken, and S.-I. Karato, Development of finite strain in the convecting lower mantle and its implications for seismic anisotropy, *J. Geophys. Res.*, 108 (B5), 2230, doi:10.1029/2002JB001970, 2003.

Mégnin, C., and B. Romanowicz, The three-dimensional shear velocity structure of the mantle from the inversion of body, surface, and higher-mode waveforms, *Geophys. J. Int.*, 143, 709–728, 2000.

Montague, N., and L. Kellogg, Numerical models of a dense layer at the base of the mantle and implications for the dynamics of D", *J. Geophys. Res.*, 105, 11,101–11,114, 2000.

Montague, N., L. H. Kellogg, and M. Manga, High Rayleigh number thermo-chemical models of a dense boundary layer in D", *Geophys. Res. Lett.*, 25, 2345–2348, 1998.

Mori, J., and D. V. Helmberger, Localized boundary layer below the mid-Pacific velocity anomaly identified from a *PcP* precursor, *J. Geophys. Res.*, 100, 20,359–20,365, 1995.

Namiki, A., and K. Kurita, Heat transfer and interfacial temperature of two-layered convection: Implications for the D"-mantle coupling, *Geophys. Res. Lett.*, 30, 1023, doi:10.1029/2002GL015809, 2003.

Nataf, H., and S. Houard, Seismic discontinuity at the top of D": A world-wide feature?, *Geophys. Res. Lett.*, 20, 2371–2374, 1993.

Ni, S., and D. V. Helmberger, Ridge-like lower mantle structure beneath South Africa, *J. Geophys. Res.*, 108, B2, 2094, doi:10.1029/2001JB001545, 2003.

Ni, S., E. Tan, M. Gurnis, and D. Helmberger, Sharp sides to the African superplume, *Science*, 296, 1850–1852, 2002.

Niu, F., and L. Wen, Strong seismic scatterers near the core-mantle boundary west of Mexico, *Geophys. Res. Lett.*, 28, 3557–3560, 2001.

Persch, S. T., J. E. Vidale, and P. S. Earle, Absence of short-period ULVZ precursors to *PcP* and *ScP* from two regions of the CMB, *Geophys. Res. Lett.*, 28, 387–390, 2001.

Reasoner, C., and J. Revenaugh, Short-period *P* wave constraints on D" reflectivity, *J. Geophys. Res.*, 104, 955–961, 1999.

Revenaugh, J., and R. Meyer, Seismic evidence of partial melt within a possibly ubiquitous low-velocity layer at the base of the mantle, *Science*, 277, 670–673, 1997.

Ritsema, J., and H. J. van Heijst, Seismic imaging of structural heterogeneity in Earth's mantle: evidence for large-scale mantle flow, *Science Progress*, 83, 243–259, 2000.

Robertson, G. S., and J. H. Woodhouse, Ratio of relative *S* to *P* velocity heterogeneity in the lower mantle, *J. Geophys. Res.*, 101, 20,041–20,052, 1996.

Romanowicz, B., Can we resolve 3D density heterogeneity in the lower mantle?, *Geophys. Res. Lett.*, 28, 1107–1110, 2001.

Rost, S., and J. Revenaugh, Seismic detection of rigid zones at the top of the core, *Science*, 294, 1911–1914, 2001.

Rost, S., and J. Revenaugh, Small-scale ultralow-velocity zone structure imaged by *ScP*, *J. Geophys. Res.*, 108, 2056, doi:10.1029/2001JB001627, 2003.

Rost, S., and Ch. Thomas, Array seismology: methods and applications, *Rev. Geophys.*, 40, 1008, doi:10,1029/2000RG000100, 2002.

Ruff, L. J., and D. L. Anderson, Core formation, evolution and convection: A geophysical model, *Phys. Earth. Planet. Inter.*, 21, 181–201, 1980.

Russell, S. A., T. Lay, and E. J. Garnero, Seismic evidence for small-scale dynamics in the lowermost mantle at the root of the Hawaiian hotspot, *Nature*, 369, 255–257, 1998.

Russell, S. A., C. Reasoner, T. Lay, and J. Revenaugh, Coexisting shear- and compressional-wave seismic velocity discontinuities beneath the central Pacific, *Geophys. Res. Lett.*, 28, 2281–2284, 2001.

Saltzer, R.L., R.D. van der Hilst, and H. Kárason, Comparing *P* and *S* wave heterogeneity in the mantle, *Geophys. Res. Lett.*, 28, 1335–1338, 2001.

Schott, B., D. A. Yuen, and A. Braun, The influences of composition- and temperature-dependent rheology in thermal-chemical convection on entrainment of the D"-layer, *Phys. Earth Planet. Inter.*, 129, 43–65, 2002.

Sidorin, I., and M. Gurnis, Geodynamically consistent seismic velocity predictions at the base of the mantle, in *The Core-Mantle Boundary Region*, edited by M. Gurnis, M. Wysession, E. Knittle, and B. Buffett, pp. 209–230, AGU, Washington, D.C., 1998.

Sidorin, I., M. Gurnis, and D. V. Helmberger, Evidence for a ubiquitous seismic discontinuity at the base of the mantle, *Science*, 286, 1326–1331, 1999.

Simmons, N. A., and S. P. Grand, Partial melting in the deepest mantle, *Geophys. Res. Lett., 29*, No. 11, 10.1029/2001Gl013716, 2002.

Sleep, N. H., Gradual entrainment of a chemical layer at the base of the mantle by overlying convection, *Geophys. J., 95*, 437–447, 1988.

Souriau, A., and G. Poupinet, A study of the outermost liquid core using differential travel times of the *SKS, SKKS,* and *S3KS* phases, *Phys. Earth Planet. Int., 68*, 183–199, 1991.

Stacey, F. D., and D. E. Loper, The thermal boundary-layer interpretation of D″ and its role as a plume source, *Phys. Earth Planet. Int., 33*, 45–55, 1983.

Steinbach, V., and D. A. Yuen, Viscous heating: a potential mechanism for the formation of the ultralow velocity zone, *Earth Planet. Sci. Lett., 172*, 213–220, 1999.

Su, W. J., and A. M. Dziewonski, Simultaneous inversion for 3-D variations in shear and bulk velocity in the mantle, *Phys. Earth Planet. Int., 100*, 135–156, 1997.

Sze, E. K. M., and R. D. van der Hilst, Core mantle boundary topography from short period *PcP, PKP,* and *PKKP* data, *Phys. Earth Planet. Inter., 135*, 27–46, 2003.

Tackley, P. J., Three-dimensional simulations of mantle convection with a thermo-chemical basal boundary layer: D″?, in *The Core-Mantle Boundary Region*, edited by M. Gurnis, M. E. Wysession, E. Knittle, and B. A. Buffett, pp. 231–253, AGU, Washington, D. C., 1998.

Tackley, P., Mantle convection and plate tectonics: Toward an integrated physical and chemical theory, Science, 288, 2002–2007, 2000.

Tan, E., M. Gurnis, and L. Han, Slabs in the lower mantle and their modulation of plume formation, *Geochemistry, Geophysics, Geosystems, 3* No. 11, 1067, doi:10.1029/2001GC000238, 2002.

Tanaka, S., Very low shear wave velocity at the base of the mantle under the South Pacific superswell, *Earth Planet. Sci. Lett., 203*, 879–893, 2002.

Tanaka, S., and H. Hamaguchi, Velocities and chemical stratification in the outermost core, *J. Geomag. Geoelectr., 45*, 1287–1301, 1993.

Thomas, Ch., M. Weber, C. W. Wicks, and F. Scherbaum, Small scatterers in the lower mantle observed at German broadband arrays, *J. Geophys. Res., 104*, 15,073–15,088, 1999.

Tkalcic, H. and B. Romanowicz, Short scale heterogeneity in the lowermost mantle: insights from *PcP-P* and *ScS-S* data, *Earth Planet. Sci. Lett., 201*, 57–68, 2001.

Trampert, J., P. Vacher, and N. Vlaar, Sensitivities of seismic velocities to temperature, pressure and composition in the lower mantle, *Phys. Earth Planet. Inter., 124*, 255–267, 2001.

van der Hilst, R. D., and H. Kárason, Compositional heterogeneity in the bottom 1000 kilometers of Earth's mantle: Toward a hybrid convection model, *Science, 283*, 1885–1888, 1999.

Vasco, D. W., and L. R. Johnson, Whole Earth structure estimated from seismic arrival times, *J. Geophys. Res., 103*, 2633–2671, 1998.

Vidale, J.E., and M. A. H. Hedlin, Evidence for partial melt at the core-mantle boundary north of Tonga from the strong scattering of seismic waves, *Nature, 391*, 682–684, 1998.

Weber, M., and J. P. Davis, Evidence of a laterally variable lower mantle structure from *P*- and *S*-waves, *Geophys. J. Int., 102*, 231–255, 1990.

Weber, M. J. P. Davis, C. Thomas, F. Krüger, F. Scherbaum, J. Schlittenhardt, and M. Körnig, The structure of the lowermost mantle as determined from using seismic arrays, in *Seismic Modelling of the Earth's Structure*, edited by E. Boschi, G. Ekström, and A. Morelli, pp. 399–442, Istit. Naz. Di Geophys. Rome, 1996.

Wen, L., An *SH* hybrid method and shear velocity structures in the lowermost mantle beneath the central Pacific and South Atlantic Oceans, *J. Geophys. Res., 107* (B3), 10.1029/2001JB000499, 2002.

Wen, L., and D.V. Helmberger, Ultra-low velocity zones near the core-mantle boundary from broadband *PKP* precursors, *Science, 279*, 1701–1703, 1998.

Wen, L., P. Silver, D. James, and R. Kuehnel, Seismic evidence for a thermo-chemical boundary at the base of the Earth's mantle, *Earth and Planet. Sci. Lett., 189*, 141–153, 2001.

Williams, Q., The temperature contrast across D″, in *The Core-Mantle Boundary Region*, edited by. M. Gurnis, M. E. Wysession, E. Knittle, and B. A. Buffett, pp. 73–81, AGU, Washington, D. C., 1998.

Williams, Q., and E. J. Garnero, Seismic evidence for partial melt at the base of the Earth's mantle, *Science, 273*, 1528–1530, 1996.

Wright, C., and J. A. Lyons, The identification of radial velocity anomalies in the lower mantle using an interference method, *Phys. Earth Planet. Int., 18*, 27–33, 1979.

Wright, C., K J. Muirhead, and A. E. Dixon, The *P* wave velocity structure near the base of the mantle, *J. Geophys. Res., 90*, 623–634, 1985.

Wysession, M. E., Imaging cold rock at the base of the mantle: The sometimes fate of Slabs?, in *Subduction: Top to Bottom*, edited by G. E. Bebout, D. Scholl, S. Kirby, and J. P. Platt, pp. 369–384, AGU, Washington, D. C., 1996.

Wysession, M., T. Lay, J. Revenaugh, Q. Williams, E.J. Garnero, R. Jeanloz, and L. Kellogg, The D″ discontinuity and its implications, in *The Core-Mantle Boundary Region*, edited by M. Gurnis, M. E. Wysession, E. Knittle, and B. A. Buffett, pp. 273–298, AGU, Washington, D.C., 1998.

Wysession, M. E., A. Langenhorst, M. J. Fouch, K. M. Fischer, G. I. Al-Eqabi, P. J. Shore, and T. J. Clarke, Lateral variations in compressional/shear velocities at the base of the mantle, *Science, 284*, 120–125, 1999.

Yamazaki, D., and S.-I. Karato, Fabric development in (Mg,O) during large strain shear deformation: implications for seismic anisotropy in Earth's lower mantle, *Phys. Earth. Planet. Int., 131*, 251–267, 2002.

Young, C. J., and T. Lay, The core mantle boundary, *Ann Rev. Earth Planet Sci., 15*, 25–46, 1987a.

Young, C. J., and T. Lay, Evidence for a shear velocity discontinuity in the lower mantle beneath India and the Indian Ocean, *Phys. Earth Planet. Int., 49*, 37–53, 1987b.

Young, C. J., and T. Lay, Multiple phase analysis of the shear velocity structure in the D″ region beneath Alaska, *J. Geophys. Res., 95*, 17,385–17,402, 1990.

Zerr, A., and R. Boehler, Melting of (Mg,Fe)SiO$_3$-perovskite to 625 kbar: Indication of a high melting temperature in the lower mantle, *Science, 262*, 553–555, 1993.

Zerr, A., and R. Boehler, Constraints on the melting temperature of the lower mantle from high pressure experiments on MgO and magnesiowüstite, *Nature, 371*, 506–508, 1994.

Zerr, A., A. Diegeler, and R. Boehler, Solidus of the Earth's deep mantle, *Science, 281*, 243–245, 1998.

Zhang, K., and D. Gubbins, On convection in the earth's core forced by lateral temperature variations in the lower mantle, *Geophys. J. Int., 108*, 247–255, 1993.

Zhao, D., Seismic structure and origin of hotspots and mantle plumes, *Earth Planet. Sci. Lett., 192*, 251–265, 2001.

T. Lay, Earth Sciences Department, U.C. Santa Cruz, Santa Cruz, CA 95064 (e-mail:tlay@es.ucsc.edu)

E. J. Garnero, Department of Geological Sciences, Box 871404, Arizona State University, Tempe, AZ 85287-1404 (e-mail:garnero@asu.edu)

Seismological Insights Into Heterogeneity Patterns in the Mantle

B. L. N. Kennett

Research School of Earth Sciences, Australian National University, Canberra, ACT, Australia

Both geophysical and geochemical results point to pervasive 3-D heterogeneity in the Earth's mantle. Geophysical evidence presents a snapshot of current structure, whereas geochemical data contain important information on age. A major source of information on heterogeneity within the Earth comes from seismic tomography, particularly when both *P* and *S* wave data can be exploited. A powerful tool for examining the character of heterogeneity comes from the comparison of images of bulk-sound and shear wavespeed extracted in a single inversion, since this isolates the dependencies on the elastic moduli. Such studies are particularly effective when common path coverage is achieved for *P* and *S* as, e.g., when common source and receiver pairs are extracted for arrival times of the phases. The relative behavior of bulk-sound and shear wavespeed can provide a useful guide to the definition of heterogeneity regimes. For subduction zones a large part of the tomographic signal comes from *S* wavespeed variations, but in the upper mantle and transition zone there can be significant bulk-sound speed contributions for younger slabs (<85 Ma), and in stagnant slabs associated with slab roll-back. The narrow segments of fast wavespeeds in the depth range 900–1500 km in the lower mantle are dominated by *S* variations, with very little bulk-sound contribution, so *P* images are controlled by shear. Deep in the mantle there are many fast features without obvious association with subduction in the last 100 Ma, which suggests long-lived preservation of components of the geodynamic cycle. Changes in the patterns of heterogeneity occur near 1200 km and 2000 km depth in the lower mantle and indicate the complexity of processes occurring in the current Earth.

INTRODUCTION

Both geophysical and geochemical techniques contribute to our understanding of the complex nature of the Earth's mantle, but the two sources of information provide very different viewpoints. Most geophysical evidence provides an instantaneous snapshot of current structure, whilst geochemical evidence has much more information on age. Further, geophysical probes are naturally integrative since they depend on sampling by finite wavelength disturbances. Geochemical probes increasingly concentrate on detailed in-situ analysis of small samples so that they are oriented towards highly localized variations in trace elements (or their isotopes).

The mantle represents the silicate shell around the metallic core of the Earth, and much of its properties will inevitably have been strongly influenced by the complex process of accretion of the Earth and the segregation of the core (see, e.g., *O'Neill and Palme*, [1998]). The major features of mantle structure are presented schematically in Figure 1 with indications of the nature of the geophysical and geochemical evidence used to constrain the behavior. Geophysical con-

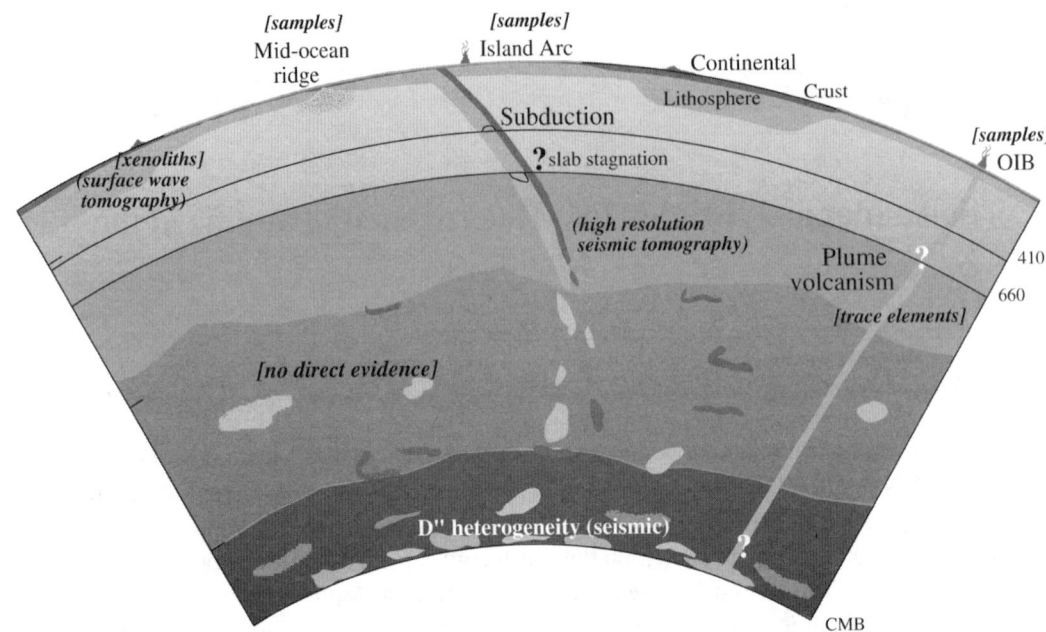

Figure 1. Schematic representation of processes and structures in the Earth's mantle, the features are drawn at approximately true scale. The sources of available information are indicated by brackets for geophysical constraints and square brackets for geochemical constraints.

straints are indicated with brackets and geochemical constraints with square brackets. Although the lower mantle has long been supposed to have rather simple properties, there is increasing evidence for complexity from seismic tomography and the schematic in Figure 1 is unlikely to represent the true variations. Some features are indicated with question marks because there is some debate about their origins. For example, some, but not all, subduction zones have a near horizontal section in the transition zone (410–660 km depth) which has been interpreted as representing stagnation of the descending slab [*Fukao et al.*, 1992]. The depth of the origin of ocean island basalts (OIB) is also somewhat contentious and the evidence suggesting deep mantle origins has been challenged by *Anderson*, [2000, 2001].

The aim of this paper is to give an overview of the current understanding of deep Earth structure and to provide some insights into the challenges to enhance knowledge of both structure and processes within the Earth.

Geophysical Probes

The dominant variation in the properties of the Earth is with depth, with superimposed 3-D variability. The main features in the radial structure of the Earth were understood by the 1930's, with the discovery of weak seismic reflections from the inner core by Inge Lehmann. The identification of the long-period free-oscillations of the Earth from the great Chilean earthquake spurred an effort to develop improved models of radial structure and this culminated in the Preliminary Reference Earth Model (PREM) of *Dziewonski and Anderson* [1981]. PREM was constructed to fit both the available observations of the frequencies of the free-oscillations of the Earth and constraints from the times of passage of seismic waves through the Earth.

Most of the geophysical probes for 3-D mantle structure depend on seismological techniques and observations. Information is available across a broad range of frequencies from the splitting of free oscillations [e.g., *Ishii and Tromp*, 1999; *Kuo and Romanowicz*, 2002], through waveform tomography using long-period observations [e.g., *Dziewonski and Woodhouse*, 1987; *Megnin and Romanowicz*, 2000; *Masters et al.*, 2000], to the use of seismic arrival times [e.g., *van der Hilst et al.*, 1997; *Kennett et al.*, 1998; *Bijwaard et al.*, 1998; *Grand*, 2002]. Some recent studies, e.g., *Ritsema et al.*, [1999]; *Masters et al.*, [2000]; *Antolik et al.*, [2003], use a wide range of different types of information to try to achieve the maximum level of sampling of the Earth's mantle. However, such studies face the complication of assigning relative weighting to different classes of information and also of combining information in different frequency bands. The properties of higher frequency seismic energy scattered back from the mantle provides a different class of information and provides direct insight into heterogeneity on small scales, typically 20 km size or less, [e.g., *Hellfrich*, 2002].

Three-dimensional models of the variation in seismic wavespeeds are normally displayed as deviations from a 1-D reference model. Frequently, the way in which these models are derived from observed data is also based on some class of perturbative analysis from a 1-D reference model (see *Kennett* [2002] for a discussion of styles of tomographic inversion). Many studies of seismic tomography have concentrated attention on a single wavetype (particularly *S*). Current *S* images derived from long-period seismic data (such as free oscillations, waveforms of multiple *S* phases), provide good definition of heterogeneity with horizontal scales larger than 1000 km across the whole mantle. The model of *Megnin and Romanowicz* [2000] based on just the use of *SH* waves (Figure 2) represents the current capabilities of waveform inversion on a global scale, using coupled normal modes to represent long-period body wave phenomena.

Higher resolution can be achieved using arrival times extracted from seismograms for both *P* and *S* waves, but at the cost of less coverage of mantle structure. The inversion of the arrival times of seismic phases derived from bulletin reports indicate rapid variations in wavespeed in parts of the mantle for both *P* and *S*. The *P* results are best known [e.g., *van der Hilst et al.*, 1997; *Bijwaard and Spakman*, 2000], but despite the difficulties of picking *S* as a later phase the resolution of features in shear wavespeed, such as the slabs in the lower mantle, is good [*Widiyantoro et al.*, 2000]. A prerequisite for high quality imaging in tomographic inversion is multidirectional sampling through any zone. The configuration of earthquake sources and, mostly, continental seismic stations restricts the sampling for body waves, unless multiply reflected body waves can be exploited. Thus, greater global coverage for *S* paths has been achieved by *Grand* [1994, 2002] by picking the arrival times of many different types of *S* phases from digital records, using comparisons with theoretical seismograms.

In those regions where different studies have achieved a comparable coverage, the major features of the tomographic images are in reasonable correspondence, even though the details vary (see, e.g., the comparisons of recent models by *Masters et al.*, [2000] and *Grand*, [2002]). The highest levels of heterogeneity are found near the Earth's surface and near the core-mantle boundary (see Figure 2). More subtle features appear in the mid-mantle, including relatively narrow zones of elevated wavespeed that are most likely associated with past subduction. In the uppermost mantle, the ancient cores of continents stand out with fast wavespeeds, while the mid-ocean ridge system and orogenic belts show slow wavespeeds. Below 400 km depth the high-wavespeed anomalies are mostly associated with subduction zones; in some regions they extend to around 1100 km depth, but in a few cases tabular fast wavespeed structures seem to extend to 2000 km or deeper. The base of the mantle shows long-wave-

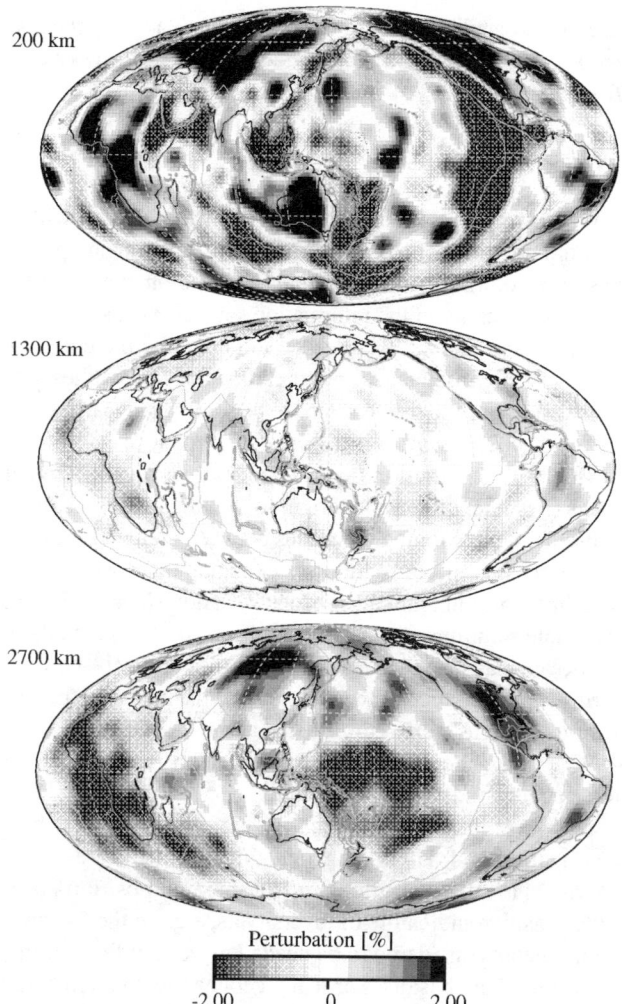

Figure 2. Maps showing the variation of shear wavespeed at depths of 200 km, 1300 km and 2700 km for the *SH* model SAW24B16 of *Megnin and Romanowicz* [2000] displayed as deviations from the PREM reference model *Dziewonski and Anderson*, [1981].

length regions of higher wavespeeds, most likely associated with past subduction, and two major regions of slow wavespeed beneath the central Pacific and southern Africa (see Figure 2) which may represent sites of upwelling of hotter material.

Despite recent advances in imaging the Earth's interior, there are still substantial gaps in our knowledge, particularly with respect to variations in density and viscosity. Information has been derived from tomographic images via scaling of shear wavespeed variations and by the use of inferred flow models. Efforts are being made to improve the situation by exploiting a wider range of geophysical observables [e.g., *Forte et al.*, 2002], but there is still little independent information. Inversions of the splitting parameters of low frequency normal modes, which depend on the density distributions, offer the possibility of improving on scaling relations [*Ishii and Tromp*, 1999; *Kuo and Romanowicz*, 2002; *Resovsky and Trompert*, 2003].

Geochemical Probes

When we turn to geochemical issues we are faced with a very different style of probes of the Earth in which a multiplicity of sources of information are brought to bear on mantle structure. The diversity of the information sources means that considerable specialisation is needed to undertake the complex analyses. In consequence interpretations of the synthesis of geochemical information can tend to be somewhat polarized [cf. *Allégre*, 2002; *Albaréde and van der Hilst*, 2002].

Much of the variation that is encountered in geochemical systems is likely to owe its origin to melting processes, with consequent partitioning of species between melt and residue. The processes of convection are expected by most workers to enhance mixing and thus reduce variation [*Davies*, 1999]. Available sampling, from limited outcrop or dredge hauls in the ocean, may tend to enhance apparent variability by nonuniform coverage. Much of the geochemical data indicate that the various systems were segregated at about 1–2 Ga [*Hofmann*, 1997] and so the geochemical information may not reflect the current condition of the mantle.

A wide range of radiogenic systems are used in the characterization and dating of geochemical systems (see, e.g., *Davies*, [1999]). Variations in isotopic compositions provide sensitive indicators of diversity as, e.g., in the 5 (or so) components considered to be needed to explain the patterns in ocean-island basalts. For many geochemical systems, the reference model is the assumed concentrations in the primitive mantle [e.g., *Hofmann*, 1997]. The deviations from the standard are then to be interpreted in terms of the action of different classes of processes which control *inter alia* the relative abundance of elements. Certainly the different classes of rocks derived from the mantle show considerable variety in elemental abundances enhanced by different degrees of melting.

The conventional view of the mantle is that a number of relatively distinct geochemical reservoirs with reasonably well-defined properties can be recognized with a reasonably well defined mean composition, even though there may be some chemical heterogeneity, as in the case of the continental crust. For example, *O'Neill and Palme*, [1998] define as reservoirs: 'the continental crust; the sub-continental lithospheric mantle; the depleted, well-stirred upper mantle, as the source for mid-ocean-ridge basalt (MORB); the enriched, heterogeneous source regions for ocean-island basalt (OIB), which may include subducted oceanic crust; and, possibly, primitive mantle that has never been differentiated.'

The subduction process with both mass and thermal transport has the potential to stir the mantle but also to leave dregs at the base (in the D″ layer). Basalts sampled from oceanic islands appear to sample from the base of the mantle, and part of this evidence comes from the properties of the various isotopes of the noble gases. The conventional interpretation based on expectations about U/Th ratios in the deep Earth requires the presence of material in the lower mantle with low gas loss, but degassing and depletion of about 50% of the mass of the mantle (see, e.g., *Allégre*, [2002]). An alternative viewpoint based on mixing of radiogenic and non-radiogenic components in the upper mantle [*Anderson*, 2000; *Meibom et al.*, 2003] does not require a deep origin for the OIB material.

Synthesis of Information on the Mantle

Contrasts in physical properties are required to generate the variations in seismic wavespeed imaged through seismic tomography. A substantial component of such wavespeed variability can be expected to have a thermal origin, for example, subduction of oceanic lithosphere transports mass and injects colder material into the mantle. When changes in seismic wavespeeds occur which are not purely due to thermal effects, then it is likely that the major element chemistry of the materials is changed. The influence of compositional variation appears to be most significant at both the top of the mantle in the continental lithosphere and in the D″zone just above the core-mantle boundary. Geochemical probes concentrate on minor element tracers, and as a result there does not have to be a simple relation between geophysical images of wavespeed heterogeneity and geochemical reservoirs.

Both geophysical and geochemical evidence requires the presence of 3-D heterogeneity permeating the mantle. But, we are still faced with the question of the significance of the patterns of heterogeneity. We anticipate that they will change with time driven by thermal processes in the mantle, but could the small-scale components be chemical remnants? There is strong spatial variation in the mantle imaged by, e.g., seismic tomography but how can we link this to time? Geochemical observations provides some help particularly when considered in connection with modeling of thermochemical convection in the mantle (see, e.g., *van Keken et al.*, [2002]; *Tackley and Xie*, [2002]; *Ballentine et al.*, [2002]). Nevertheless our constraints on mantle dynamics are rather indirect [*Davies*, 1999; *Bunge et al.*, 2002].

The advent of high performance computing has led to major advances in the modeling of the dynamics of the mantle. Convection calculations can now be run with sufficient spatial resolution to approximate the necessary mantle flow conditions see, e.g., *Bunge et al.*, [2002]. Plate-like features can be generated as the models evolve and coupling to geochemical transport is possible in, at least, 2-D models [*Tackley and Xie*, 2002]. However, a major complication is that the results of the complicated mantle convection calculations are strongly dependent on the assumed initial state. Thus it is not possible to expect to explain present-day mantle heterogeneity by running even a perfect mantle convection code forward in time [*Bunge et al.*, 2002]. A partial solution is to work with 'data assimilation' techniques in which the convection scheme is integrated over the period (~120 Ma) for which past subduction can be effectively extrapolated, and the configuration of subducted plates can be used to constrain the convection patterns. This approach helps to provide constraints on the radial distribution of mantle viscosity, but the heterogeneity in the deepest mantle is still sensitive to the initial conditions [*Bunge et al.*, 2002]. The heterogeneity patterns from seismic imaging provide the main constraints on mantle flow models, and there is some promise that back-projection in time can help to resolve the problems of defining suitable initial conditions.

Developments in Seismic Imaging

The early success of seismic tomography came from the striking images of large scale 3-D structure and, later, of the details of subduction zones. The interpretation of such images is based firmly on the variations in seismic wavespeed. The conventional color scheme for seismic images with blue-fast and red-slow reflects an expected link to thermal processes, but chemical heterogeneity could also play a significant role, particularly in the regions with strong variability at the top and bottom of the mantle.

Results for a single wavespeed are not sufficient to indicate the nature of the observed anomalies. Recent developments in seismic imaging are therefore moving towards ways of extracting multiple images in which different aspects of the physical system are isolated. This may be from P and S images (preferably from common data sources) or via the use of the bulk modulus, shear modulus and density. Such multiple images of mantle structure encourage an interpretation in terms of processes and mineral physics parameters, since the relative variation of the different parameters adds additional information to the spatial patterns.

In principle, we can derive a significant increase in understanding of heterogeneity if we can use both P and S information. The P wavespeed α depends on both the bulk modulus κ and the shear modulus μ as

$$\alpha = [(\kappa + \tfrac{4}{3}\mu)/\rho]^{1/2} \qquad (1)$$

where ρ is the density. We can thus isolate the dependence on the shear modulus μ and the bulk modulus κ, by working with the S wavespeed

$$\beta = [\mu/\rho]^{1/2}, \qquad (2)$$

and the bulk-sound speed ϕ derived from both the P wavespeed α and the S wavespeed β

$$\phi = [\alpha^2 - \tfrac{4}{3}\beta^2]^{1/2} = [\kappa/\rho]^{1/2}, \qquad (3)$$

which isolates the bulk modulus κ. This style of parameterization has been employed by *Su and Dziewonski* [1997], *Kennett et al.* [1998] and *Masters et al.* [2000]. The variation of the three wavespeeds are indicated in Figure 3 for the reference model AK135 of *Kennett et al.* [1995] designed for travel time studies.

An unfortunate complication in the combined use of P and S information comes from the uneven geographic distribution of data. Whereas S wave data are available from globe circling paths; P wave information, dominantly from travel times, is dictated by the location of seismic sources and receivers. A number of recent studies, such as that of *Masters et al.* [2000], use a wide variety of different data sets with the aim of providing sampling of the mantle in many ways, so that the final images should be more reliable. There is some component of P wave information in long period waveform data but this does not provide strong constraints on mantle structure. It therefore remains difficult to compare full global coverage from S with information derived from P waves with a much more limited geographic coverage [*Su and Dziewonski*, 1997;

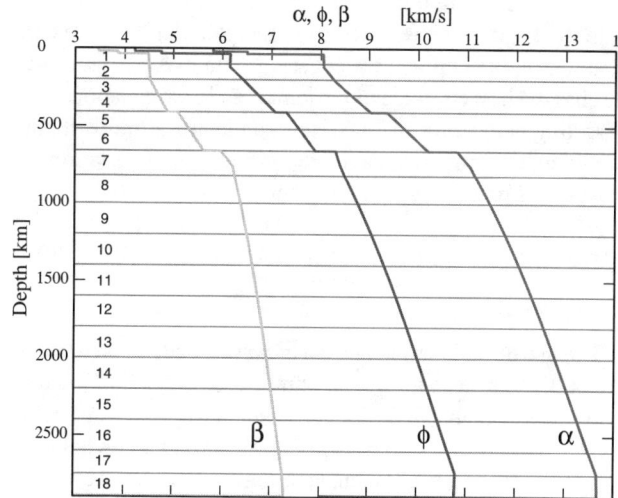

Figure 3. Variation of the P wavespeed α, S wavespeed β and bulk-sound speed ϕ with depth in the mantle for the reference model AK1315, *Kennett et al.* [1995], together with the 18 layer parameterization used in the global tomography studies illustrated in this paper.

Masters et al., 2000]. *Antolik et al.* [2003] endeavor to improve P coverage by incorporating PP-P differential times from long-period records, as well as waveforms for different frequency bands. However, their model J362D28 for the 3-D variations in P and S wavespeed, obtained from joint inversion, does not achieve the same level of resolution for the two wavetypes, because of the strong path influence from the P travel times.

An alternative approach is to restrict attention to paths for which both P and S information is available; the consequence is that sampling of the mantle is reduced, but the reliability of the images is high and direct assessment can be made of the relative behavior of either P and S wavespeed variations, or the variations in bulk-sound speed and shear wavespeed [*Kennett et al.*, 1998; *Widiyantoro et al.*, 1999; *Saltzer et al.*, 2001; *Gorbatov and Kennett*, 2003; *Kennett and Gorbatov*, 2004].

The images of seismic structure presented in this paper are derived from joint inversions of P and S arrival-time data, with light damping and a broad residual range designed to capture strong features in the uppermost and lowermost mantle. The global results are taken from the model of *Kennett and Gorbatov* [2004] with $2° \times 2°$ cells and 18 layers through the mantle, as illustrated in Figure 3. The regional results for the western Pacific are taken from the models of *Gorbatov and Kennett* [2003] which employ a 19 layer model to 1500 km depth with $1° \times 1°$ cells for the region of interest embedded in a global model with $5° \times 5°$ cells. In each case a linearized inversion is first performed for P and S separately, and then a joint inversion for ϕ and β is undertaken with 3-D ray tracing. The use of the restricted P and S data means that the data ray coverage is limited below 2500 km, and that in consequence lowermost mantle features are only weakly represented.

In the figures below, taken from the global model, we concentrate on the hemisphere covering Asia to Australia, since this has the best coverage of the joint P and S observations. The reporting pattern for the Americas is much less uniform and it is more difficult to separate the influences of variations in wavespeed properties and sampling.

WAVESPEED VARIATIONS AND HETEROGENEITY REGIMES

The use of bulk-sound and shear wavespeeds as proxies for the bulk modulus and shear modulus provides a link to laboratory experiments and *ab initio* calculations, but highlights the limitations imposed by our relatively weak knowledge of the variations in density within the Earth. When we work with the bulk-sound speed we are closer to experimental information than when P wavespeed is employed, since diamond anvil work provides direct constraints on the properties of the bulk modulus (see, e.g., *Jackson and Ridgen*, [1996]; *Liebermann*, [2000]). By contrast, our knowledge of the properties of the shear modulus at high pressures is very limited, although measurements have begun [*Sikelnikov et al.*, 1998]. For conditions at the base of the mantle, various styles of *ab initio* calculations can be attempted for mineral assemblages, but the thermal properties represent a challenge to computational techniques (see, e.g., *Stixrude*, [2000]; *Brodholt et al.*, [2002]). Comparisons between mineral physics results and seismological models need to take account of anelastic effects [*Karato and Karki*, 2001], and these will be strongly temperature dependent.

The variations of the bulk-sound and shear wavespeeds due to changes in, e.g., temperature and composition, can be expressed in terms of contributions from the elastic moduli and density

$$\frac{\delta\phi}{\phi} = \frac{1}{2}\left(\frac{\delta\kappa}{\kappa} - \frac{\delta\rho}{\rho}\right), \quad \frac{\delta\beta}{\beta} = \frac{1}{2}\left(\frac{\delta\mu}{\mu} - \frac{\delta\rho}{\rho}\right); \quad (4)$$

and will be equal for variations in density alone. The influence of temperature on the moduli κ and μ would be expected to be similar, with a decrease associated with increasing temperature that would be more pronounced for the shear modulus μ (see, e.g., *Brodholt et al.*, [2002]). The relative behavior of the two wavespeeds depends on the balance of the variations of the moduli and density. If

$$\frac{\delta\kappa}{\kappa} < \frac{\delta\rho}{\rho} < \frac{\delta\mu}{\mu}, \quad (5)$$

then $\delta\phi/\phi$ would be negative and $\delta\beta/\beta$ would be positive. Such a weak anti-correlation appears through much of the lower mantle in the model of *Masters et al.* [2000].

When the bulk-sound speed and shear wavespeed variations are large and anti-correlated we need to seek the influence of factors other than temperature, such as chemical heterogeneity or, at shallow depths, the presence of volatiles to lower shear wavespeed. Even where we are not able to assign a likely cause, the relative variations of bulk-sound speed and shear wavespeeds provide a means of characterizing different heterogeneity regimes.

The ratio of wavespeed deviations from a reference model can be used to link to mineral physics results [e.g., *Masters et al.*, 2000; *Karato and Karki*, 2001]. For example, the large slow anomalies for S at the base of the mantle (e.g., under the mid-Pacific) do not have a corresponding expression in P. The resulting heterogeneity ratios fall outside the domain for purely thermal contributions, suggesting a significant component of chemical heterogeneity.

Figure 4 displays a cross-section through the global bulk-sound and shear wavespeed model that passes through a range of different mantle features. We see the concentration of heterogeneity near the surface, but also the presence of notable

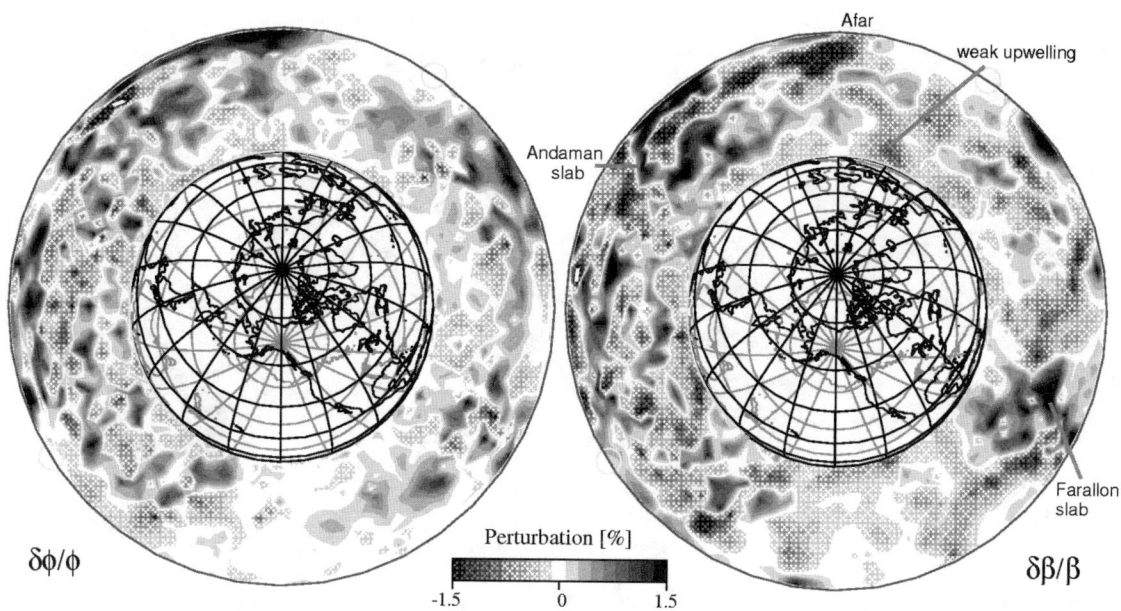

Figure 4. Cross-section through the variation of bulk-sound speed $\delta\phi/\phi$ and shear wavespeed $\delta\beta/\beta$ relative to the AK135 reference model, crossing the mantle feature beneath S. America associated with the Farallon slab, the Andaman slab and the Afar hotspot.

fast wavespeed zones penetrating into the mantle. The section cuts through the northern part of South America and passes obliquely through the high shear wavespeed zone which has been associated with past Farallon plate subduction [e.g., *Grand et al.*, 1997]. Even more distinct is the Andaman extension of the Indonesian subduction zone that is inferred to extend to about 900 km depth. There is an apparent weak connection to a further pronounced shear-wave anomaly in the mid-mantle which is likely to have been produced by subduction at the northern margin of the Tethys Ocean (cf. Figure 8). Note that neither of the subduction zone features have any significant expression in bulk-sound speed that suggests the influence of factors other than just thermal effects.

At the top of the figure, the prominent low-velocity zone for *S* reaches the surface at the Afar region. This zone appears to have a weak connection to a deep zone of lowered wavespeed beneath central Africa. Such an inclined link is consistent with a number of recent studies [e.g., *Ritsema et al*, 1999]. Possible upwelling from the core-mantle boundary can also be seen beneath the western Pacific. The blank zone in the eastern Pacific is a reflection of the limitations of the arrival-time dataset. Structure cannot be imaged unless crossing ray paths traverse the region.

The use of the common source-receiver pairs for both *P* and *S* has proved to be particularly effective for subduction related features, where we are striving to image structures which are faster than their surroundings and which contain themselves many seismic events in the upper mantle.

Regions of lowered seismic wavespeed, which are likely to be hot and thus sources of upwelling, form an important component of the geodynamic system [*Ritsema and Allen*, 2003]. However, these features are not well sampled by our restricted data set, yet some significant zones of lowered shear wavespeed can be captured (Figure 4). Recent advances in finite-frequency tomography [*Nolet et al.*, 2003] offer the potential for markedly improved imaging of low wavespeed features using travel times measured from longperiod records. An alternative approach, which examines different physical properties, is to undertake attenuation tomography [*Romanowicz*, 1995] in which upwelling material can be identified by its higher attenuation (lowered *Q*).

UPPERMOST MANTLE

For the main shield regions, e.g. Australia, there is a strong shear wavespeed anomaly (up to 6% or more) down to about 250 km accompanied by a somewhat weaker bulk-sound speed signature (Figure 5). In contrast, the major orogenic belts from southern Europe to Iran (and also western N. America) show rather slow *S* wavespeeds, accompanied by fast bulk-sound speeds. A similar behavior is evident for the Red Sea and E. African rifts. The anti-correlation of bulk-sound and shear wavespeed is pronounced, a large component can be thermal because of the very strong reduction of the shear modulus as the solidus is reached, but volatiles may also be significant.

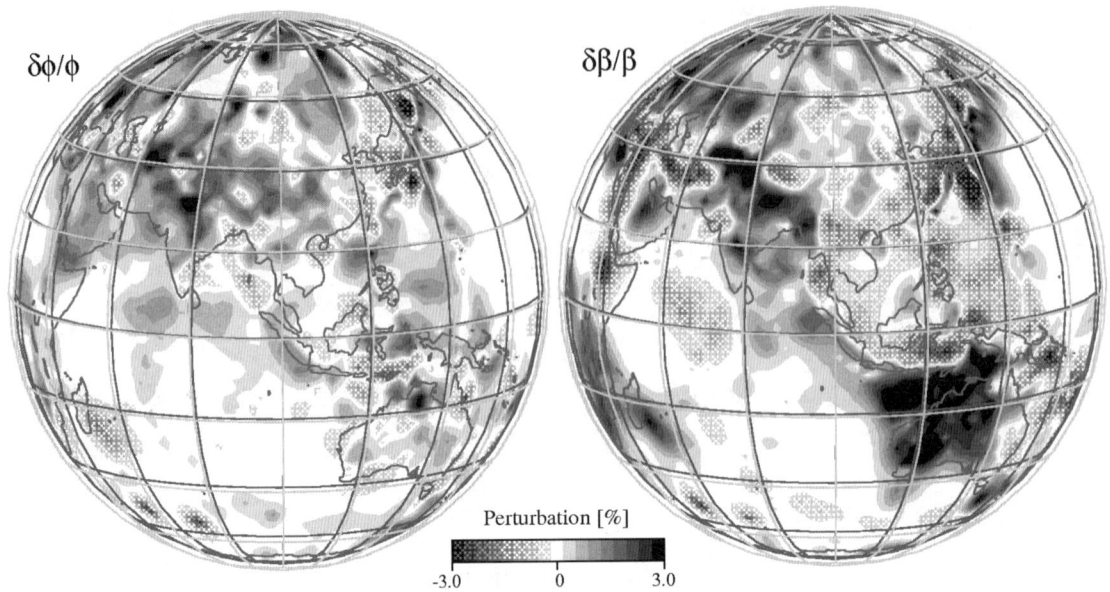

Figure 5. Variation of bulk-sound speed $\delta\phi/\phi$ and shear wavespeed $\delta\beta/\beta$ relative to the AK135 reference model for the layer from 100–200 km depth.

In eastern Asia we infer some portions of subducted slab, e.g. the Ryukyu arc, standing out very clearly by their high shear wavespeeds from the lower background; these also have some expression in bulk-sound speed. The variations in the bulk-sound speed signature along the Indonesian arc, which are clearly visible in Figure 5, correlate with the age of the slab as it enters the subduction zone. Older slab material (> 85 Ma) has a stronger shear wavespeed signature relative to bulk-sound speed than younger material [*Gorbatov and Kennett*, 2003].

A distinct lowered *S* wavespeed anomaly occurs in northwestern India, with a modest bulk-sound expression. This feature was noted by *Kennett and Widiyantoro* [1999] from *P* wave tomography, and corresponds to the region where the volcanism of the Deccan traps originated at about 63 Ma, probably due to plume breakthrough. A somewhat weaker *S* wavespeed anomaly can be recognised in southeastern Australia associated with hotspot volcanism over the Neogene.

SUBDUCTION

Seismic images of fast wavespeed in subduction zones have become quite familiar in recent years [e.g., *van der Hilst et al.*, 1991]. These images derived from inversion of the times of passage of *P* waves have helped us to get a picture of the way in which the former ocean lithosphere descends into and interacts with the mantle.

As the subduction zones begins to interact with the 660 km discontinuity, a range of behavior is found. The results of *Fukao et al.* [1992], and subsequent workers, using *P* wave tomography have demonstrated the presence of stagnant slabs remaining in the transition zone, as in the Izu-Bonin region south of Japan. Only slightly further to the south in the same subduction system in the Marianas, the slab is nearly vertical and appears to penetrate well into the lower mantle.

Some regions such as Izu-Bonin and Tonga, where there has been trench retreat [e.g., *van der Hilst and Seno*, 1993; *van der Hilst*, 1995], show indications of slab material in the transition zone. However, in many cases a significant fast wavespeed anomaly occurs in the top of the lower mantle beneath the subduction zone, suggesting ponding of former slab material [*Fukao et al.* 1992], as originally proposed by Ringwood [1991].

Insight into the nature of the subduction process can be obtained by exploiting the times of passage of both *S* and *P* waves to generate two complementary images. *Widiyantoro et al.* [1999] used *P* and *S* arrival time data with common source/station pairs to conduct separate *P* and *S* inversions for structure in the northwest Pacific along the line of section displayed in Figure 6. A cross-section through the model crossing the Ryukyu and Izu-Bonin arcs is shown in Figure 7(a). Clear slab images are found for both *P* and *S*, but the slab material lying on top of the 660 km discontinuity has a different character in the *P* and *S* images, with a stronger *P* anomaly. This suggests a difference in the physical properties of this stagnant segment.

Gorbatov and Kennett [2003] have undertaken a systematic study of western Pacific subduction zones with joint inversion for bulk-sound speed and shear wavespeed incor-

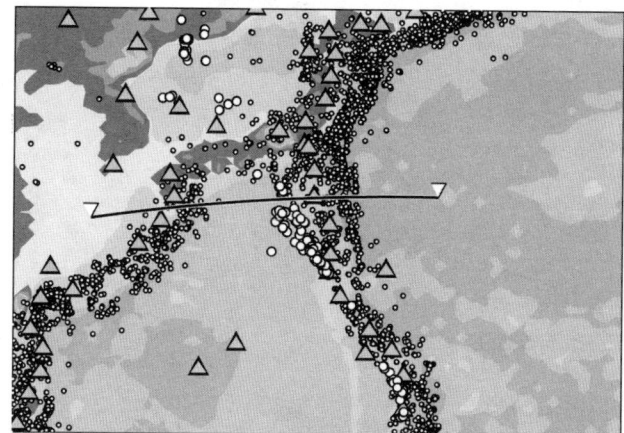

Figure 6. Line of cross-sections across the Ryukyu and IzuBonin arcs. south of Japan, displayed in fig 7. The triangles indicate volcanoes and earthquakes are shown by circles with white infill.

porating a non-linear inversion with an allowance for 3-D ray tracing. Figure 7(b) shows results for the same line of section as Figure 7(a), but now with the joint inversion. The inclusion of 3-D ray tracing provides a sharper image of the slabs. We also see a very clear contrast between the bulksound speed and shear wavespeed variations. Almost all the P wave signature down to 500 km is controlled by the variations in the shear modulus. There is little sign of the slabs in the bulksound image, except for the stagnant portion in the transition zone where it is the dominant component.

The results from global joint inversion of P and S arrival time data confirm that stagnant portions of the slab in the transition zone systematically have a stronger signature in bulksound speed than in shear wavespeed. Lower contrasts in rigidity are found in Izu-Bonin, the Philippines and Tonga where stagnant slabs are reported, but not in the Kuriles, Marianas and the Indonesian Arc where the slab penetrates into the lower mantle.

The study of *Gorbatov and Kennett* [2003] for the western Pacific subduction zones indicates a varying balance between bulk-sound speed and shear wavespeed behavior in different subduction zones. These differences in physical properties indicate there are still many puzzles to resolve as to the nature of the subduction process. There is a strong correlation of the slab properties in the uppermost mantle with the age of the subducted crust at the subduction zone. For crust older than 85 Ma, shear wavespeed anomalies dominate, whereas for younger crust the bulk-sound speed is most prominent. As noted above, the variations are clearly seen along the Indonesian arc in Figure 5; the bulk-sound anomaly changes from negative to positive at the point where younger sea floor enters the trench. Similar behavior also occurs along the South American subduction zone in northern Peru where a limited segment of older crust is being subducted. The bulk-sound anomaly switches to negative for this portion, with positive values on either side for the younger subduction. The age break comes at about the point where oceanic lithosphere is expected to achieve thermal maturity, but a physical mechanism still needs to be sought for the change in the material properties.

LOWER MANTLE

Gu et al. [2001] present evidence for a major rearrangement of the long wavelength features of the shear wavespeed distribution across the 660 km discontinuity. The changes in the short wavelength components are distinct but more subtle. We have already noted the accumulation of large positive anomalies in shear wavespeed down about 1100 km at the top of the lower mantle associated with past subduction. A number of observations have been made of seismic 'discontinuities' in the interval from 950 km to 1100 km in the regions of these high wavespeed zones (see, e.g., *Niu and Kawakatsu*, [1997]) and the locations of these S to P conversions appear to mark the base of the anomalies. The lower boundary of the slab accumulation zones might be related to a local maximum in mantle viscosity [cf. *Forte et al.*, 2002] which impedes downward progress of slab material, with either thermal assimilation in place or penetration to depth after significant delay [*Fukao et al.*, 2001].

Only in a few places are there clear indications of slab-like behavior extending below 1300 km depth (see, e.g., *Grand et al.*, [1997]), most notably in southern Asia and beneath the Americas. These two features has been linked to subduction in the past 80 Ma at the northern edge of the Tethys Ocean and of the now-extinct Farallon plate beneath the Americas [*Lithgow-Bertelloni and Richards*, 2000].

The heterogeneity regime in the mid-mantle is illustrated in Figure 8 with the variations in bulk-sound speed and shear wavespeed near 1100 km depth from the model of Kennett and Gorbatov (2004). The striking fast shear anomalies extending from Iran to Indonesia are almost absent in the bulk-sound image. This means that the 'Tethyan' anomalies imaged by P wave inversions [e.g., *van der Hilst et al.*, 1997; *Bijwaard et al.*, 1998; *Karâson and van der Hilst*, 2000] are controlled by shear variations. Away from the major features, the two wavespeeds show comparable levels of variability on intermediate scales, with a weak anti-correlation in the patterns that is compatible with minor thermal fluctuations.

The relatively narrow, slab-like structures in southern Asia and the Americas become less coherent with depth and appear in places to link with drip-like features in the lowermost mantle. Cross-sections of these structures can tend to be misleading because of the influence of oblique cuts. Indeed, it is difficult to follow the behavior in 3-D because of the various

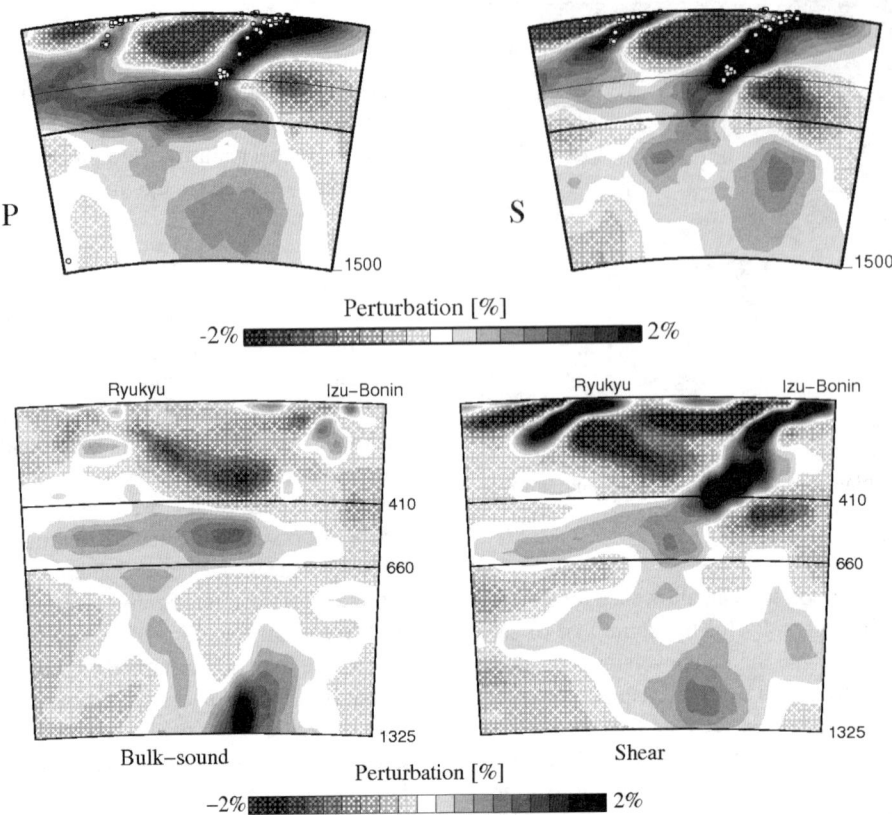

Figure 7. Cross-sections along the profile across the Ryukyu and Izu-Bonin arcs shown in fig 6: (a) variations of P and S wavespeed from inversions for paths with common source and receiver pairs for P and S phases (Widiyantoro et al, 1999.) (b) Variations of bulk-sound speed and shear wavespeed from a joint inversion with incorporation of 3-D ray tracing (Gorbatov and Kennett, 2003). In each case the variations are shown relative to the AK135 reference model. The 410 and 660 km discontinuities and the depth of the sector are indicated for each section.

factors that can influence the amplitudes of the imaged wavespeed variations. Unfortunately, limited ray path coverage reduces the apparent amplitudes, particularly for the joint inversions near the core-mantle boundary.

The character of the heterogeneity patterns in the mantle changes with depth, notably below 2000 km. In the interval around 2100 km, illustrated in Figure 9, the bulk-sound speed variations are subdued, which is consistent with the results of *Robertson and Woodhouse* [1996] and *Kennett et al.* [1998]. There is, however, significant variation in S wavespeed with striking anomalies, especially in Asia, with a very different pattern from that seen at 1100 km depth. Some aspects of the pattern appear to have links to past subduction, e.g., the feature now beneath Australia that extends down from 1800 km depth. This location corresponds to a confluence of multiple subduction episodes along the margin of the Panthalassian Ocean in the interval from 400 to 150 Ma (see, e.g., *Collins*, [2003]).

It would appear that some fragments of former plates may survive well beyond the time frame expected from simple thermal assimilation arguments. This survival may well be linked with variations in the physical and chemical properties of the subducted material, [*Kennett and Gorbatov*, 2004]; the material entering current subduction zones show complex patterns of structures arising from, e.g., seamounts and other features. *Albaréde and van der Hilst* [2002] have suggested that the presence of oceanic plateau material could enhance plate sinking in the lower mantle and hence encourage descent to the lowermost mantle.

The character of the heterogeneity regime in the mantle undergoes further change as the core-mantle boundary is approached. The amplitude of bulk-sound variation which is very low as we have seen near 2100 km increases with depth, and the pattern of variation increasingly becomes uncorrelated with shear wavespeed. The amplitude of the shear heterogeneity also increases rapidly with depth [*Masters et al.*, 2002]. Just above the core-mantle boundary, in the D″ zone, a wide range of different pieces of information indicates the presence of extensive but variable heterogeneity including variable seismic anisotropy and narrow zones with very low seismic wavespeeds (see, e.g., *Lay and Garnero*, [2004], this volume).

A striking feature of the bulk-sound speed and shear wavespeed distributions at the base of the mantle is the discordance in the patterns of variation [*Masters et al.*, 2000]. Such behavior is not compatible with a simple thermal origin and suggests the presence of widespread chemical heterogeneity. Such a class of heterogeneity would fit quite well with the geochemical information derived from ocean island basalts, which are thought to include plume components raised from the base of the mantle.

DISCUSSION

A variety of lines of evidence including the correlation properties of tomographic images, density distributions compatible with the earth's free oscillations [*Kennett*, 1998], and the variation in the ratio between bulk-sound speed and shear wavespeed variations, suggest that we need to consider three major zones in the lower mantle [cf. *van der Hilst and Karâson*, 1999]. The uppermost zone reflects a continuation of the transitions in radial mantle structure and extends to around 1100 km with an ill-defined lower boundary. The main structures in tomographic images reach only a little deeper suggesting some barrier to penetration. There seems to be little organized structure in the mid-mantle away from localized slab-like features (but this could be a function of our current probes).

The correlation properties of the bulk-sound and shear wavespeed models, (see Appendix), suggest that it is inappropriate to just segregate the zone above 660 km (which represents about 30% of the mass of the mantle). A more sensible division occurs near 1200 km depth corresponding closely to 50% of the mantle mass.

At the base of the mantle below 2100 km there is a distinct change in the heterogeneity regime with an increase in the amplitude of 3-D variation and divergence in the variations in bulk-sound and shear wavespeed. The change in the nature of the heterogeneity is most likely associated with a change in the dynamical regime in the deep mantle. *Kellogg et al.* [1999] attempted to reconcile the range of observations with a model that includes a compositionally distinct zone in the lower part of the mantle containing chemically enriched material. The intrinsically greater density (about 4%) of the lower zone is very nearly balanced by buoyancy due it to being hotter than the mantle above. A characteristic of the model is that the boundary is dynamic and irregular allowing for the formation of mega-plumes. However, no evidence for a global mid-mantle interface has been found in seismological studies [*Albaréde and van der Hilst*, 2002].

Our picture of the mantle continues to evolve as the quality of imaging improves and more integrated interpretation of the many different lines of evidence about the internal structure becomes possible. However, we definitely need improved constraints on 3-D variations in density and viscosity to provide critical input into simulations of mantle behavior.

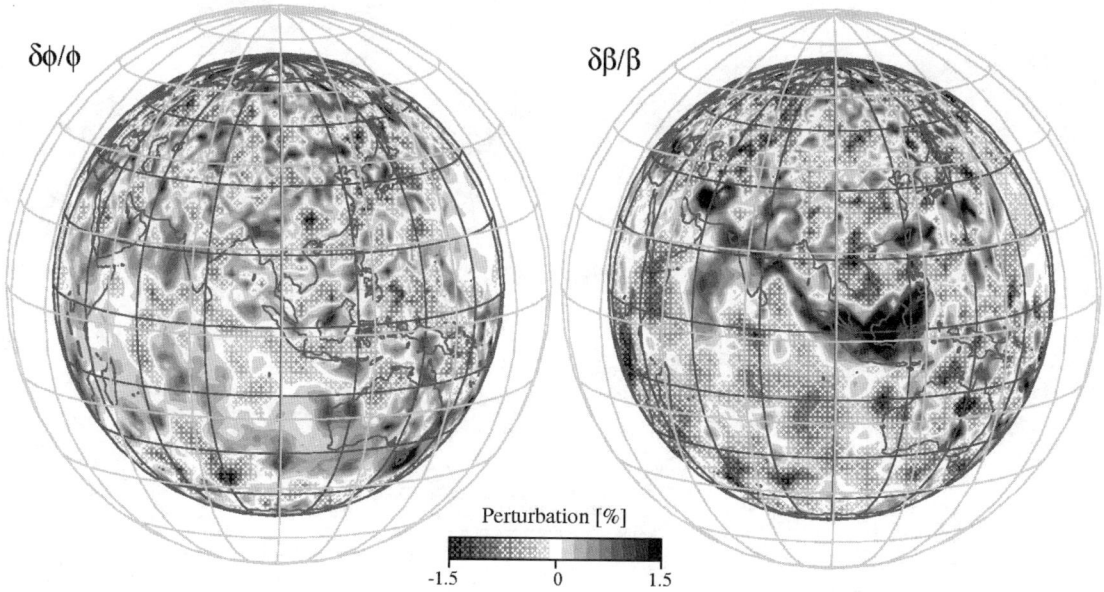

Figure 8. Variation of bulk-sound speed $\delta\phi/\phi$ and shear wavespeed $\delta\beta/\beta$ relative to the AK135 reference model for the layer from 1000–1200 km depth.

It is clear that the whole mantle is a dynamic system dominated by thermal influences. In some sense the whole mantle must convect, even though at some stages in Earth history quasi-stratification may develop. The picture of the mantle derived from high-resolution joint tomography is complex (see Figure 4), but has distinct resemblances to the configuration of geodynamic modeling in a whole mantle model with temperature dependent viscosity with long term mixing.

Even high-resolution seismic tomography can only resolve features in the mantle on scales of the order of 100 km or more, since there has to be sufficient volume to produce an observable change in the passage times of seismic waves. Smaller scale features are not seen, yet are in evidence from scattering studies [e.g., *Helffrich*, 2002]. It is likely that the entire mantle comprises a melange of features on a very broad range of scales, with the large-scale ordering reflected in global seismic tomography using long-period waves (cf. Figure 2). The resulting picture of mantle structure has some resemblances to the statistical upper mantle assemblage (SUMA) proposed on geochemical grounds by *Meibom and Anderson* [2004]. In this model, the concept of distinct geochemical reservoirs is replaced by multiple sampling of heterogeneous structure, with partial melting and blending of magmas to produce different geochemical components. However, the geophysical evidence would not confine the stochastic heterogeneity to the upper mantle.

The changes in the organization of mantle heterogeneity near 1200 km and 2100 km may well prove to have some link to the issues of the location of chemical reservoirs. Indeed it is likely that the short wavelength components of seismic structure are more relevant to geochemical signatures than the long wavelengths. The extraction of melt is a surface not a volume process, and the physical recycling of material becomes a balance between the buoyancy and thermal equilibration of different components.

APPENDIX—RADIAL CORRELATION PROPERTIES OF TOMOGRAPHIC MODELS

One of the ways in which we can investigate the geodynamic implications of the snapshots of structure provided by tomographic imaging is through the correlation properties of the model as in Figure 10. In this figure each entry in a matrix represents the weighted correlation of the wavespeed speed structure between different layers in the bulk-sound speed or shear wavespeed models, or between the two wavespeeds in the cross-wavespeed function. The amplitudes of the wavespeed variations are weighted by the sampling, to place most reliance on well resolved features of the model. The diagonal entries for the bulk-sound or shear-wave speed represent the full correlation for a layer with itself, and the bands of significant correlation extending away from the diagonal show where the structure across a group of layers has similar properties. The weighted cross-correlation between the bulk-sound speed and shear wavespeed is generally small with a tendency to anti-correlation.

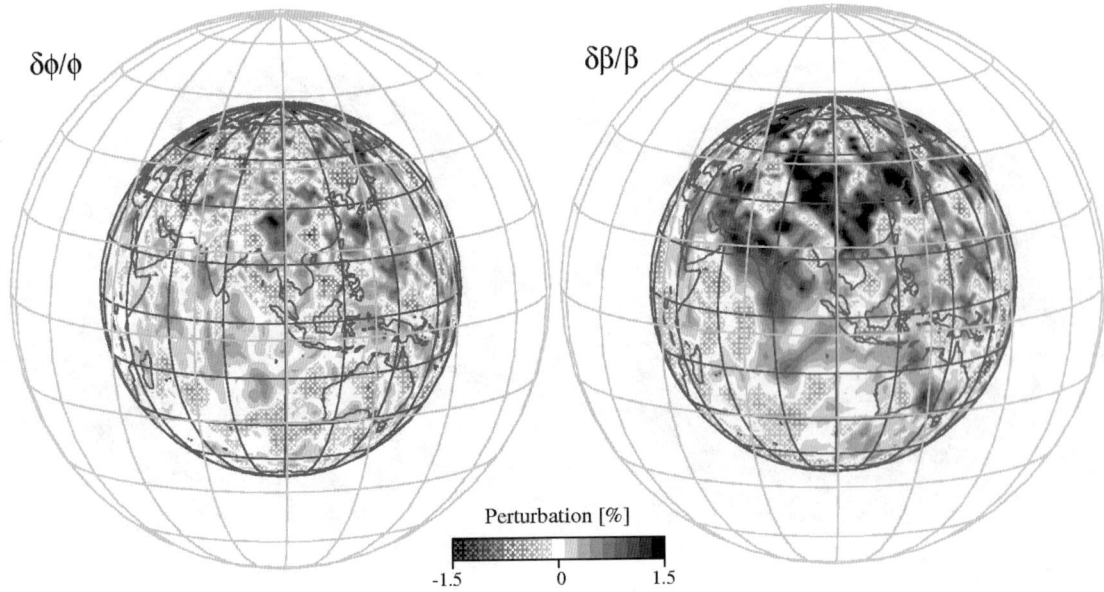

Figure 9. Variation of bulk-sound speed $\delta\phi/\phi$ and shear wavespeed $\delta\beta/\beta$ relative to the AK135 reference model for the layer from 2000–2200 km depth.

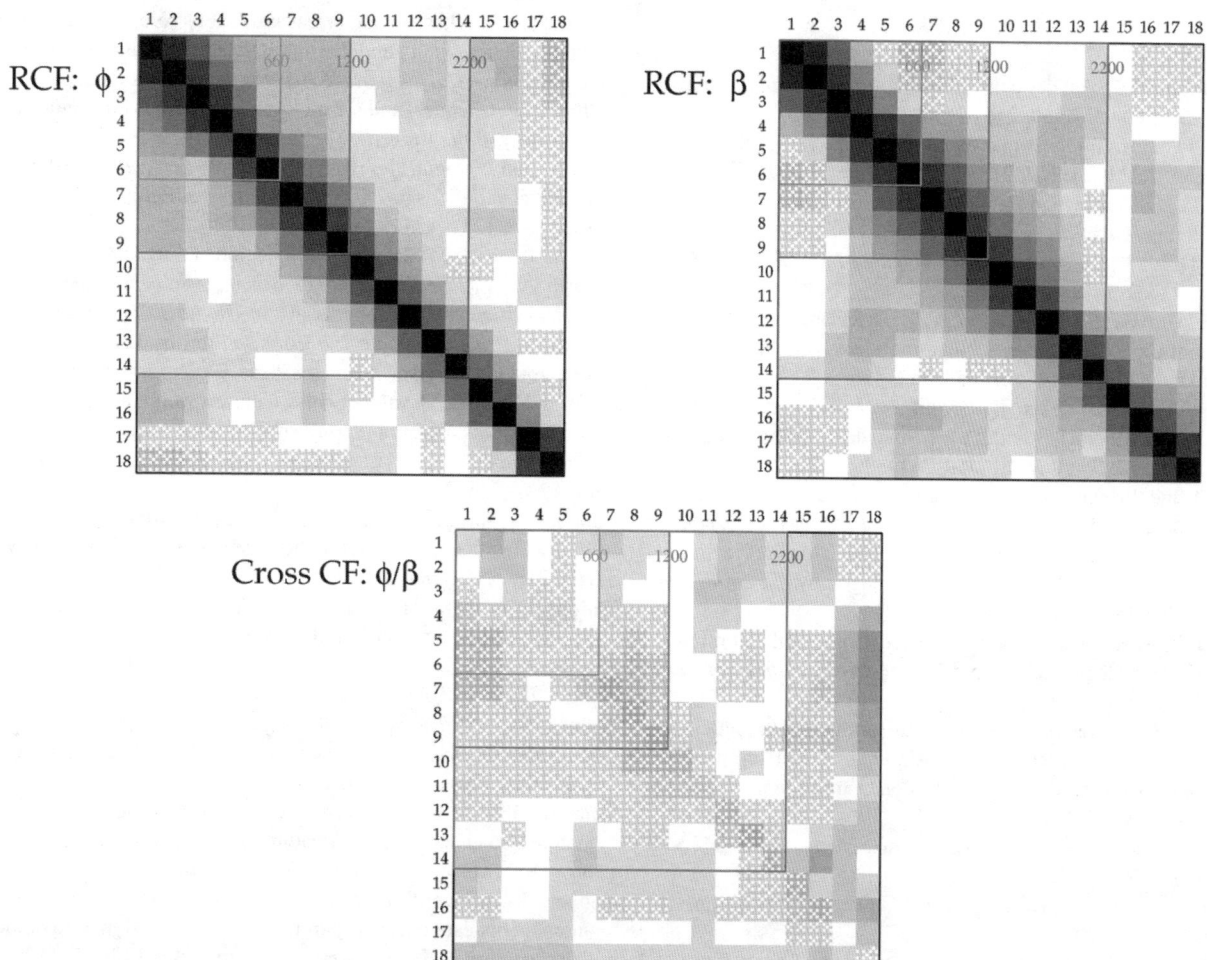

Figure 10. Radial correlation functions between the variations in wavespeed for the layers of the joint tomographic model (bulk-sound speed ϕ and shear wavespeed β) and the cross-wavespeed function (ϕ/β). Positive correlation is indicated by grey tone, with deeper tones for stronger correlation; negative correlation is indicated by patterning.

When we look at the patterns of the correlation properties we see a tendency for there to be blocks of entries with a close relation to each other. To guide the interpretation of these results the divisions corresponding to the 660 km discontinuity, 1200 km and 2200 km depth are superimposed on each correlation matrix. These guidelines correspond quite well to changes in the correlation properties of the wavespeed variations, but we might wish to shift to the boundary of layer 13 (i.e. 2000 km depth) rather than layer 14 (2200 km).

However, we see that it is difficult to separate the segment above 660 km from that down to 1200 km except through the weak anti-correlation between the bulk-sound and shear wavespeeds. From the perspective of the tomographic results we should be thinking of the region to at least 1000 km depth as representing the transition zone into the lower mantle. There is again a change in the organization of the cross-wavespeed correlation below 2200 km, but we need to recognize that ray coverage is diminishing strongly near the core-mantle boundary so that the correlations are for a rather restricted global coverage.

Nevertheless, the results suggest that the dominant character of the mantle which needs to be captured in geodynamic simulations comprises three segments of approximately 1000 km each in depth and that the major phase boundary near 660 km depth does not mark a distinct limit to the geophysical state.

Acknowledgments. I would like to acknowledge the contributions of many workers on mantle structure who have influenced this paper (even if they might feel that their views are misrepresented). In particular I would like to thank my colleagues at the Research School of Earth Sciences at the Australian National University, with a special appreciation to the late Ted Ringwood who particularly encouraged my forays into deeper mantle structure.

REFERENCES

Albaréde, F. and R. D. van der Hilst, Zoned mantle convection *Phil. Trans. R. Soc. Lond.*, A*360*, 2569–2592, 2002.

Allégre, C., The evolution of mantle mixing, *Phil. Trans. R. Soc. Lond.*, A*360*, 2411–2431, 2002.

Anderson, D. L., The thermal state of the upper mantle: no role for mantle plumes, *Geophys. Res. Lett.*, *27*, 3623–3626, 2000.

Anderson, D. L., Top-down tectonics? *Science*, *293*, 2016–2018, 2001.

Antolik, M., Yu.J. Gu, G. Ekström, and A. M. Dziewonski, J362D28: a new joint model of compressional and shear velocity in the Earth's mantle, *Geophys. J. Int.*, *153*, 443–466, 2003.

Ballentine, C.J., P.E. van Keken, D. Porcelli, and E.H. Hauri, Numerical models, geochemistry and the zero-paradox noble-gas mantle, *Phil. Trans. R. Soc. Lond.*, A*360*, 2611–2631, 2002.

Bijwaard, H. and W. Spakman, Non-linear global P wave tomography by iterated linearized inversion, *Geophys. J. Int.*, *141*, 71–82, 2000.

Bijwaard, H., W. Spakman, and E. R. Engdahl, Closing the gap between regional and global travel time tomography, *J. Geophys. Res.*, *103*, 30 055–30 078, 1998.

Brodholt, J. P., A. R. Oganov and G. D. Price, Computational mineral physics and the physical properties of Perovskite, *Phil. Trans. R. Soc. Lond.*, A*360*, 2507–2520, 2002.

Bunge, H.-P., M. A. Richards and J. R. Baumgardner, Mantlecirculation models with sequential data assimilation: inferring present-day mantle structure from plate-motion histories, *Phil. Trans. R. Soc. Lond.*, A*360*, 2545–2567, 2002.

Collins, W. J., Slab pull, mantle convection and Pangaean assembly and dispersal, *Earth Planet. Sci. Lett.*, *205*, 225–237, 2003.

Davies, G. F., *Dynamic Earth*, Cambridge University Press, Cambridge, 458 pp., 1999.

Dziewonski, A. M. and D. L. Anderson, Preliminary reference Earth model, *Phys. Earth Planet. Inter.*, *25*, 297–356, 1981.

Dziewonski, A. M. and J. H. Woodhouse, Global images of the Earth's interior, *Science*, *236*, 37–48, 1987.

Forte, A. M., J. X. Mitrovica and A. Espesset, Geodynamic and seismic constraints on the thermochemical structure and dynamics of convection in the deep mantle, *Phil. Trans. R. Soc. Lond.*, A*360*, 2521–2543, 2002.

Fukao, Y., M. Obayashi and H. Inoue, Subducting slabs stagnant in the mantle transition zone, *J. Geophys. Res.*, *97*, 4909–4922, 1992.

Fukao, Y., S. Widiyantoro and M. Obayashi, Stagnant slabs in the upper and lower mantle transition zone, *Rev. Geophys.*, *39*, 291–233, 2001.

Gorbatov, A. and B. L. N. Kennett, Joint bulk-sound and shear tomography for western Pacific subduction zones, *Earth Planet Sci Lett*, *210*, 527–543, 2003.

Grand, S. P., Mantle shear structures beneath the Americas and surrounding oceans, *J. Geophys. Res.*, *99*, 11 591–11 621, 1994

Grand, S. P., Mantle shear-wave tomography and the fate of subducted slabs, *Phil. Trans. R. Soc. Lond.*, A*360*, 2475–2491, 2002

Grand, S. P., R. D. van der Hilst and S. Widiyantoro, Global seismic tomography: a snapshot of convection in the Earth, *Geology Today*, *7*,(4) 1–7, 1997.

Gu, Yu-J., A. M. Dziewonski, W. Su and G. Ekström, Models of the mantle shear velocity and discontinuities in the pattern of lateral heterogeneities, *J. Geophys. Res.*, *106*, 11 169–11 199, 2001.

Helffrich, G., Chemical and seismological constraints on mantle heterogeneity, *Phil. Trans. R. Soc. Lond.*, A*360*, 2493–2505, 2002.

Ishii., M. and J. Tromp, Normal mode and free-air gravity constraints on lateral variations in velocity and density of Earth's mantle, *Science*, *285*, 1231–1236, 2001.

Jackson, I. and S. M. Ridgen, Composition and temperature of the Earth's mantle: Seismological models interpreted through experimental studies of Earth materials, in *The Earth's Mantle: Composition, Structure and Evolution*, edited by I. Jackson, pp. 405–460, Cambridge University Press, Cambridge, 1998.

Karâson, K. and R. D. van der Hilst, Constraints on mantle convection from seismic tomography, in *The history and dynamics of global plate motion*, American Geophysical Union Monograph, *121*, 277–288, 2000.

Karato, S.-I. and B. B. Karki, Origin of lateral variation of seismic wave velocities and density in the deep mantle, *J. Geophys. Res.*, *106*, 21 771–21 783, 2001.

Kellogg, L. H., B. H. Hager and R. D. van der Hilst, Compositional stratification in the deep mantle, *Science*, *283*, 1881–1884, 1999.

Kennett B. L. N, On the density distribution within the Earth, *Geophys. J. Int.*, *132*, 374–384, 1998.

Kennett, B. L. N., *The Seismic Wavefield II—Interpretation of Seismograms on Regional and Global Scales*, 534 pp., Cambridge University Press, Cambridge, 2002.

Kennett, B. L. N. and A. Gorbatov, Seismic heterogeneity in the mantle strong shear wave signature of slabs from joint tomography, *Phys. Earth Planet. Inter.*, *146*, 87–100.

Kennett, B. L. N. and S. Widiyantoro, A low seismic wavespeed anomaly beneath northwestern India—a seismic signature of the Deccan plume? *Earth Planet. Sci. Lett.*, *165*, 145–155, 1999.

Kennett, B. L. N., E. R. Engdahl and R. Buland, Constraints on the velocity structure in the Earth from travel times, *Geophys. J. Int.*, *122*, 108–124, 1995.

Kennett, B. L. N., S. Widiyantoro and R. D. van der Hilst, Joint seismic tomography for bulk-sound and shear wavespeed in the Earth's mantle, *J. Geophys. Res.*, *103*, 12 469–12 493, 1998.

Kuo, C. and B. Romanowicz, On the resolution of density anomalies in the Earth's mantle using spectral fitting of normal-mode data, *Geophys. J. Int.*, *150*, 162–179, 2002.

Lay, T. and E. Garnero, The core-mantle boundary region, in *State of the Planet*, edited by S. Sparks and C. Hawkesworth, American Geophysical Union, Washington D.C., 2004.

Liebermann, R. C., Elasticity of mantle minerals (experimental studies), in *Earth's Deep Interior: Mineral Physics and Tomography from the Atomic to the Global Scale*, AGU Geophysical Monograph, *117*, 181–199, 2000.

Lithgow-Bertelloni, C. and M. A. Richards, The dynamics of Cenozoic and Mesozoic plate motions, *Rev. Geophys.*, *36*, 27–78, 1998.

Masters, G., G. Laske, H. Bolton and A. Dziewonski, The relative behavior of shear velocity, bulk sound speed, and compressional velocity in the mantle: implications for chemical and thermal structure, in *Earth's Deep Interior: Mineral Physics and Tomog-

raphy from the Atomic to the Global Scale, AGU Geophysical Monograph, *117*, 63–87, 2000.

Megnin, C. and B. Romanowicz, The three-dimensional shear velocity structure of the mantle from the inversion of body, surface and higher-mode waveforms, *Geophys. J. Int.*, *143*, 709–728, 2000.

Meibom, A., D. L. Anderson, N. H. Sleep, R. Frei, C. H. Chamberlain, M.T. Hren and J. L. Wooden, Are high He/He ratios in oceanic basalts an indicator of deep mantle plume components, *Earth Planet. Sci. Lett.*, *208*, 197–204, 2003.

Meibom, A. and D. L. Anderson, The statistical upper mantle assemblage, *Earth Planet. Sci. Lett.*, *217*, 123–140, 2004.

Nolet, G., R. Montelli, G. Masters, F. A. Dahlen, and S.-H. Hung, Finite frequency tomography shows a variety of plumes, *Geophysical Research Abstracts*, *5*, 03146, 2003.

Niu, F. and H. Kawakatsu, Depth variation of the mid-mantle seismic discontinuity, *Geophys. Res. Lett.*, *24*, 429–432, 1997.

O'Neill, H. St. C. & H. Palme, Composition of the silicate Earth: implications for accretion and core formation, in *The Earth's Mantle: Structure, Composition and Evolution*, edited by I. Jackson, pp. 3–126, Cambridge University Press, Cambridge, 1998.

Resovsky, J. and J. Trampert, Reliable mantle density error bars: An application of the neighbourhood algorithm to normal mode and surface wave data, *Geophys. J. Int.*, *150*, 665–672, 2003.

Ringwood, A. E., Phase transformations and their bearing on the constitution and dynamics of the mantle, *Geochem. Cosmochem. Acta*, *55*, 2083–2110, 1991.

Ritsema, J. and R.M. Allen, The elusive mantle plume, *Earth Planet. Sci. Lett.*, *207*, 1–12, 2003.

Ritsema, J., H. J. van Heijst and J. H. Woodhouse, Complex shear velocity structure imaged beneath Africa and Iceland. *Science*, *286*, 1925–1928, 1999.

Robertson, G. S. and J. H. Woodhouse, Ratio of relative S to P velocity heterogeneities in the lower mantle, *J. Geophys. Res*, *101*, 20 041–20 052, 1996.

Romanowicz, B., A global tomographic model of shear attenuation in the upper mantle, *J. Geophys. Res.*, *100*, 12 375–12 394, 1995.

Saltzer, R. L., R. D. van der Hilst and H. Kârason, Comparing P and S wave heterogeneity in the mantle, *Geophys. Res. Lett.*, *28*, 1335–1338, 2001.

Sikelnikov, Y. D., G. Chen, D. R. Neuville, M.T. Vaughan and R C. Liebermann, Ultrasonic shear velocities of $MgSiO_3$ perovskite at 8 GPa and 800 K and lower mantle composition, *Science*, *281*, 677–679, 1998.

Stixrude, L., Elasticity of mantle phases at high pressure and temperature, in *Earth's Deep Interior: Mineral Physics and Tomography from the Atomic to the Global Scale*, AGU Geophysical Monograph, *117*, 181–199, 2000.

Su W.-J. and A. M. Dziewonski, Simultaneous inversions for 3-D variations in shear and bulk velocity in the mantle, *Phys. Earth. Planet. Inter.*, *100*, 135–156, 1997.

Tackley, P. J. and S. Xie, The thermo-chemical structure and evolution of Earth's mantle: constraints and numerical models, *Phil. Trans. R. Soc. Lond.*, A*360*, 2593–2609, 2002.

van der Hilst, R. D., Complex morphology of subducted lithosphere in the mantle beneath the Tonga trench. *Nature*, *374*, 154–157, 1995.

van der Hilst, R. D. and T. Seno, Effects of relative plate motion on the deep structure and penetration depth of slabs below the Izu-Bonin and Mariana island arcs, *Earth. Planet. Sci. Lett.*, *120*, 375–407.

van der Hilst, R. D. and K. Kârason, Compositional heterogeneity in the bottom 100 kilometers of Earth's mantle: towards a hybrid convection model, *Science*, *283*, 1885–1889, 1997.

van der Hilst, R. D., R. Engdahl, W. Spakman and G. Nolet, Tomographic imaging of subducted lithosphere below northwest Pacific island arcs, *Nature*, *353*, 733–739, 1991.

van der Hilst, R. D., S. Widiyantoro and E. R. Engdahl, Evidence for deep mantle circulation from global tomography, *Nature*, *386*, 578–584, 1997.

Van der Voo, R., W. Spakman and H. Bijwaard, Tethyan subducted slabs under India, *Earth Planet. Sci. Lett.*, *171*, 7–20, 1999.

van Keken, P. E., E. H. Hauri, and C. J. Ballentine, Mantle mixing: the generation, preservation, and destruction of chemical heterogeneity, *Ann. Rev. Earth Planet. Sci.*, *30*, 493–525, 2002.

Widiyantoro, S., B. L. N. Kennett and R. D. van der Hilst, Seismic tomography with P and S data reveals lateral variations in the rigidity of deep slabs, *Earth. Planet. Sci. Lett.*, *173*, 91–100, 1999.

Widiyantoro, S., A. Gorbatov, B. L. N. Kennett and Y. Fukao, Improving global shear wave traveltime tomography using three-dimensional ray tracing and iterative inversion, *Geophys. J. Int.*, *141*, 747–758, 2000.

Woodhouse, J. H. and A. M. Dziewonski, Mapping of the upper mantle: three-dimensional modeling of earth structure by inversion of seismic waveforms, *J. Geophys. Res.*, *89*, 5953–5986, 1984.

B. L. N. Kennett, Research School of Earth Sciences, Australian National University, Canberra ACT 0200, Australia, (e-mail: brian@rses.anu.edu.au)

Geophysical and Geochemical Models of Mantle Convection: Successes and Future Challenges

Yanick Ricard and Nicolas Coltice

Laboratoire de Sciences de la Terre, Ecole Normale Supérieure de Lyon, Lyon, France

Although more and more robust evidence for whole mantle convection comes from seismic tomography and geoid modeling, the rare gases and other isotopic or trace element signatures of ridge and hotspot basalts indicate the presence of various isolated geochemical reservoirs in the mantle. We discuss this discrepancy between fluid dynamic views of mantle convection and chemical observations. We compare the standard model of geodynamicists where the mantle behaves as a fluid mostly heated from within with the findings of seismic tomography. We suggest that a significant part of the subducted oceanic crust transforms into dense eclogitic assemblages, and partially segregates to form a layer that has grown with time above the Core Mantle Boundary (CMB) and should correspond to the D″ layer of seismological models (~280 km thick). We show how a two component marble cake mantle filling the whole mantle except D″ can account for the variability of Ocean Island Basalts (OIB) and Mid-Ocean Ridge Basalts (MORB) in rare gases. We then present the state of the art in thermochemical convection of the mantle and emphasize the numerical and conceptual progress that must be made to provide a quantitative test of the geochemical hypotheses.

1. INTRODUCTION

Geochemists, seismologists and geodynamicists try to understand the behavior of our planet by means of very different tools. The observations of geochemists provide a time integrated view. The isotopic, trace or major element concentrations and ratios that they measure are the results of 4.5 byrs of dynamics that includes major events like core segregation and the formation of the continental crust. Seismologists, on the other hand have only access to a snapshot of this evolution, namely the present-day structure of the Earth.

There are no obvious reasons to believe that the time-integrated and the instantaneous views of the Earth should be identical. Using experimentally measured parameters (like densities, viscosities) and physical laws (mass, energy and momentum conservation), geodynamicists have the difficult task of proposing a scenario that is consistent with these two viewpoints. We are far from a detailed understanding of how the mantle works, but at least we can describe where the problems are and suggest some possible solutions.

2. THE MANTLE SEEN BY GEOPHYSICISTS

The striking advances in mantle tomography in the last 20 years have made it difficult to believe that mantle flow can be stratified at any depth by any sharp discontinuity. Since the first global images 20 years ago, inversion methods have been improved by more precise location of the events, local grid refinements (Bijwaard et al., 1998), multibounce phase modeling (Grand et al., 1997), a more accurate description of wave propagation (Montelli et al., 2004), etc. Although there are still significant differences between the results, all models share sheet-like fast structures reminiscent of past subduction. These structures are very well defined under North and Central America (through most of the mantle) and below the Tethyan suture from the Mediterranean sea to the north of Australia (at least down to mid mantle).

These tomographic observations rule out a strict stratification of the mantle. This of course does not mean that slabs penetrate the lower mantle without difficulty, nor that all slabs reach the core-mantle boundary. In various places, like around the Philippine plate or Tonga, folded slabs or slabs flattening in the transition zone are observed (Fukao et al., 2001). In other places the sheet structure of the fossil slabs seems to fade away around mid-mantle depths or are replaced by finger-like downwellings. These two observations are in good agreement with geodynamic models.

The oceanic lithosphere cools over a thickness L during its thermal contraction in spreading at the sea floor. The thermal diffusion equation implies the well known relation

$$L^2 \sim 4D_{th}t, \qquad (1)$$

where D_{th} is the thermal diffusivity and t the age of the lithosphere. The same diffusion equation implies that this lithosphere reheats after a time of order $t/4$ in the deep mantle (the lithosphere is cooled only from its top but is reheated from both sides). This means that the lithosphere lasts around 30–40 myrs in the mantle before halving its temperature deficit. With a sinking velocity of 2 cm/yr in the lower mantle, this lithosphere can travel down to the mid lower mantle before being sufficiently reheated to loose its integrity. This indeed corresponds to the depth where many slabs observed by tomography in the shallower mantle seem to fade out. This simple calculation agrees with more realistic numerical simulations (Bunge et al., 1998).

The likely viscosity increase through the transition zone causes a decrease in the sinking velocity of the subducting material and in the dip angle of the descending slabs, similar to a refraction kink. The effects of phase changes from ringwoodite to perovskite plus oxides, that occur at a greater depth in the cold slabs than in the surrounding mantle, also tend to affect the slab penetration in the lower mantle. When these effects are taken into account in addition to potential trench migrations (roll-back), numerical simulations are able to reproduce in a very realistic way all the complexities of mid mantle slab trajectories, but the conclusion is that viscosity increases, phase transitions and roll back do not impede a large scale flow throughout the mantle (Christensen, 1996). A comparison between tomographic images and paleogeographic plate reconstructions shows a close agreement between the fast structures and the positions of Cenozoic and Mesozoic trenches at global (Ricard et al., 1993) and regional scales (Van der Voo et al., 1999).

Subduction removes primitive and radiogenic heat by burying cold lithosphere at great depths. There are various indications that this is the major source of buoyancy that drives the mantle (Bercovici et al., 2000). The return flow associated with these active downwellings is mostly passive and should consist of a uniform upwelling flow with an averaged velocity at least one order of magnitude lower than the slab sinking velocities (in the proportion of the surface area of the descending slabs to the surface area of the Earth). As a simple numerical example, the downgoing velocities should be of the order of 10 cm/yr in the upper mantle, and a few cm/yr in the lower mantle where viscosity increases by one or two orders of magnitude. Except for the velocities of actively rising material in plumes, the background upwelling velocities should be around a few mm/yr. This behavior is very different from what occurs with a bottom heated fluid where upwellings and downwellings have similar absolute velocities. The flow regime is such that a complete overturn in the internally heated mantle (transport from ridge to trench, subduction through the whole mantle, and back to the ridge) is controlled by the return flow, is thus very slow and the time scale is of the order of 1 byr. This simple scenario indicates that isotopic ratios involving radioactive chains like U-Pb and Rb-Sr have enough time to evolve and generate observable heterogeneities.

The role of plumes in convection models is to carry the excess heat out of a hot boundary layer. No experimental or numerical model has ever generated plumes from within of a convection cell. An obvious candidate for the source of hotspots is the core-mantle boundary where heat diffuses out from the core throught a thermal boundary. The existence of other thermal boundary layers in the mantle, at 670 km depth or more speculatively on top of an abyssal layer, has no clear observational support (Castle and van der Hilst, 2003). The plumes themselves are very difficult to observe although striking progress has been made in recent years (Nataf and VanDecar, 1993; Montelli et al., 2003). This difficulty comes from their expected very small dimensions and low excess temperature. According to geodynamicists, hotspots carry buoyancy fluxes from 7000 kg s^{-1} for Hawaii to 300 kg s^{-1} for the smaller detectable ones (Davies, 1988). Their excess temperature is only ~250 K (Sleep, 1990) and their ascent velocity should be significantly larger than a typical plate velocity to resist entrainment by the large scale mantle flow (Steinberger and O'Connell, 1998). These figures imply radii of the order of 10 km up to 100 km for the strongest plumes, indeed very small to be detected with present techniques.

This agreement between simple thermal internally heated convection and seismic observations is only first order. Seismologists have observed various structures in the deep mantle that probably have chemical origins. In D″ near the core-mantle boundary (CMB), diffractive bodies (Weber, 1996), anisotropy (Vinnik et al., 1995), and ultra-low velocity zones (Garnero and Helmberger, 1995), suggest complex small scale thermochemical processes. At a larger scale, subtle contradictions between P- and S-wave models (for example, anticorrelations of the anomalies and changes in Vp/Vs ratios) also point to chemical variations in the deep mantle (Saltzer et al., 2001). Increased complexity in the models is necessary for seismologists as well as for dynamicists. The density variations in the mantle are not only due to thermal expansion but are also related to mineralogical and chemical variations.

3. GEOCHEMICAL OBSERVATIONS

The differentiation of the mantle through melting, alteration and other processes has significantly transformed the initial distribution of chemical elements. For this reason, the geochemistry of mantle derived rocks contributes to constraining models of mantle dynamics.

The initial bulk chemical composition of the Earth in refractory elements is modeled using the chemistry of meteorites and especially carbonaceous chondrites, the supposed parent bodies of our planet (McDonough and Sun, 1995). The initial composition in volatile elements is still questioned since significant volatilization probably occurred during Earth's accretion (Tyburczy et al., 1986). During Earth's differentiation, some chemical elements are retained in the silicated mantle and crust (the lithophile elements), others form iron alloys in the core (the siderophile elements). For refractory lithophile elements and isotopes, the difference between the present-day chemistry of the mantle and the initial composition reflects the history of mantle evolution and mixing.

The simplest model of present-day distribution of lithophile elements is to suppose that the primitive mantle evolved through extraction of the crust leaving a residual mantle. As a consequence, the chemistry of the continental crust should be complementary to the present-day mantle relative to its primitive composition. A simple test can be performed with uranium (which is also valid for most other elements): the continental crust contains at most $3.9 \, 10^{16}$ kg of uranium (Rudnick and Fountain, 1995) and the whole mantle would contain $3.2 \, 10^{16}$ kg (Jochum et al., 1983) assuming that its composition is similar to that of the shallow mantle. This would make a total of $7.1 \, 10^{16}$ kg of uranium for the silicate Earth. However, the primitive mantle should contain $8.4 \, 10^{16}$ kg of uranium (McDonough and Sun, 1995). At least 15% of the uranium is therefore missing in the balance. One could store the missing uranium in the outer core, which then would have a uranium concentration similar to that of the shallow mantle of 8 ppb. This seems highly improbable as uranium is not a siderophile element. The other solution is to invoke at least one more hidden domain having a composition different from that of the shallow mantle, that unlike the shallow mantle cannot be sampled.

The composition-volume tradeoff of this reservoir can be derived from the mass balance

$$M_{pm} C_{pm} = M_{cc} C_{cc} + M_{sm} C_{sm} + M_{h} C_{h}, \qquad (2)$$

where M_i and C_i are the mass and the concentration of an element in a reservoir i. The reservoirs are the primitive mantle, pm, the continental crust, cc, the shallow mantle, sm and a hidden reservoir h. We have shaded in Figure 1 the possible range of mass and composition of this reservoir for two elements, uranium (U top) and aluminum (Al bottom). The possible concentrations of this U- and Al-rich hidden reservoir are

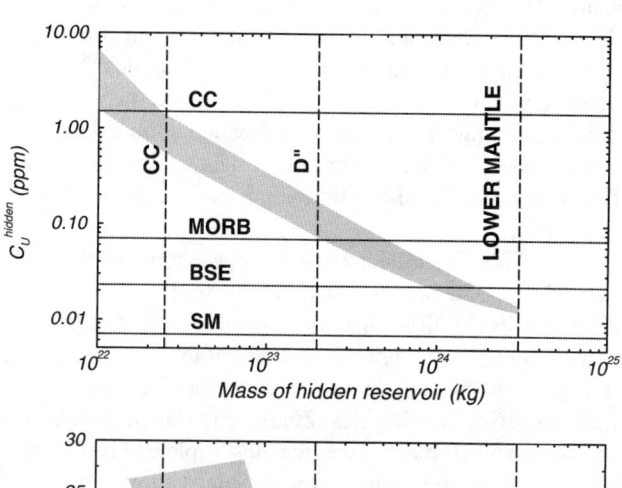

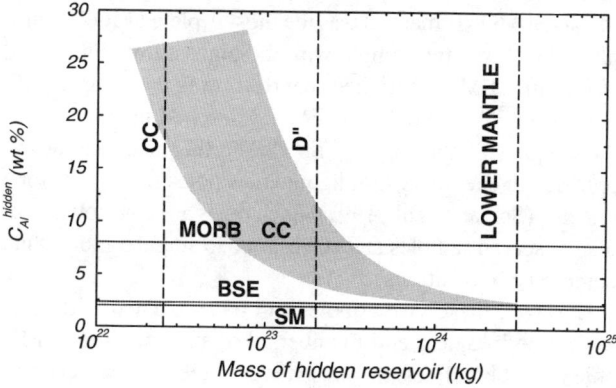

Figure 1. The possible concentrations of two lithophile incompatible elements (top uranium, bottom aluminum) in the hidden reservoir as a function of mass of this reservoir are indicated by shadowed areas. To account for the abundance of the incompatible elements in the bulk silicate Earth (BSE), the continental crust (CC) and the shallow mantle (SM) are not enough. A hidden reservoir is necessary with a lithophile incompatible concentration larger than in the shallow mantle. This reservoir could have a rather small volume (like that of D″) but be very rich in incompatible elements (with concentrations somewhat similar to that of subducted MORBs). Alternatives could be a lower mantle somewhat depleted in U and primitive in Al or a thick primitive abyssal layer (~50% of the lower mantle).

of course higher than those in the shallow mantle (SM). This graph shows that a hidden reservoir the size of the lower mantle would not be primitive, since the U and Al contents differ from those of the bulk silicate Earth, BSE. A primitive reservoir would represent only 30–60% of the mass of the lower mantle; if D″ is the hidden reservoir (~280 km thick on top of the core-mantle boundary), its U and Al contents would be roughly similar to that of MORBs. A reservoir with a volume and concentrations comparable to that of the continental crust, CC, would be acceptable for the U budget but would not contain enough Al.

Another source of information comes from isotopes. Two isotopes of the same element do not fractionate upon melting and so the erupted material has the same isotopic ratio as its source. However, melting, or other modification of rocks (e.g. dehydration, alteration and phase changes) can fractionate the parent-daughter ratio for radioactive isotopes. The elements that concentrate in the melt are called incompatible, those remaining in the residue are called compatible. As a consequence, a melt and a residue will have significantly different isotope ratios after a time comparable to the radioactive decay time.

We will not discuss the whole isotope taxonomy of mantle derived rocks, but oceanic island basalts (OIBs) and mid ocean ridge basalts (MORBs) display a specific diversity (see Hofmann, 1997 for a review). At the end of the 80's, Sr, Nd, Pb were the most studied isotope systems and a specific effort was made to define mantle poles (Zindler and Hart, 1986). DMM (depleted MORB mantle) was the most depleted MORB sample, HIMU was the sample with the highest time integrated U/Pb ratio, EM-1 (enriched mantle 1) was the sample with the lowest $^{143}Nd/Nd^{144}$ and EM-2 (enriched mantle 2) was the sample with the highest $^{87}Sr/^{86}Sr$. Further analytical developments in chemistry highlighted new poles for new isotope systems (Schiano et al., 1997). Such poles are concepts rather than observed samples and correspond to singular rather than common compositions.

However, these virtual poles can be associated with geological/petrological end-members. For example, the HIMU pole probably reflects the presence of ancient altered crust in the source (Hofmann and White, 1982). Various isotope systems can be used to identify the petrological components in the source of magmas like $\delta^{18}O$, Os, Hf and the presence of ancient oceanic crust sections has been detected in the source of MORBs and of all OIBs (Eiler et al., 1996; Schiano et al., 1997; Blichert-Toft, 1999).

These observations show that the mantle is made of a heterogeneous mixture of rocks that have different time integrated histories of differentiation. For some samples, the time needed to develop their specific signature is of the order of 2 billion years (Hofmann, 1997). In general, the homogeneity of MORBs contrasts with the heterogeneity of OIBs, but for some isotope ratios the diversity is comparable. The fact that various hotspots are associated with volcanic chains that have crossed ridges indicates that these hotspots at least, and we think all hotspots, have a deep origin (Richards et al., 1989). The heterogeneity of OIBs should reflect the heterogeneity of the deep mantle.

Volatiles provide information on degassing, especially light noble gases that are not recycled into the mantle. The $^3He/^4He$ ratio evolves through degassing of the stable isotope ^3He, and through radiogenic growth of ^4He produced by decay of U and Th. The standard interpretation of higher $^3He/^4He$ ratios in some OIBs, like Hawaii or Iceland, is that their source is rich in ^3He, hence mostly undegassed (Allégre et al., 1986). For that reason, many authors suggest that there are still some undegassed rocks within the convective mantle, and hypothesize the existence a deep primitive layer, sometimes identified with the lower mantle, which would also contain the missing incompatible elements.

Argon is a better candidate than helium to monitor the degassing of the planet since it does not escape from Earth's atmosphere, unlike helium. It is well known that about half of the argon-40 that has been produced by potassium-40 decay has not been degassed and cannot be found in the continental crust or the shallow mantle (Allègre et al., 1996). However, the argon budget is controlled by the potassium budget, a volatile element for which the bulk abundance is not well known. Assuming the frequently quoted K/U ratio of 12700 (Jochum et al., 1986), a primitive, undegassed lower mantle would close the budget. However, a whole mantle with shallow mantle composition similar to that of the source of MORBs overlying a potassium rich D″ layer made of segregated oceanic crust, is another way to close the budget if the bulk Earth K/U ratio was 20% lower (Coltice and Ricard, 2002).

In conclusion, every isotope system can potentially provide a constraint on mantle history but the signal is not easy to interpret in terms of mantle dynamics. The major observations are (a) that MORBs and OIBs often differ in regard to their isotope signature and heterogeneity, (b) subducted rocks are recycled in the mantle and dominate within some plumes, (c) noble gases suggest that rocks with various degrees of degassing coexist within a vigorously convective flow.

4. PHYSICS OF MIXING

The chemical heterogeneities of the mantle may have different origins. The initial chemical conditions of the mantle when plate tectonics, or some kind of convection started, is certainly poorly known. Various researchers have discussed the possible survival of primitive material in the form of a continuous deep layer (Kellogg et al., 1999) or of entrained lumps

(Becker et al., 1999). The heterogeneity could also be the mere consequence of plate tectonics. The oceanic ridges associated with the divergent motion of the plates entrained by subduction induce an adiabatic melting due to pressure release. The chemical elements are partitioned between the basaltic melt and the residual. This is of course a major source of mantle heterogeneities.

Whether they are primitive or recycled, heterogeneities are then entrained by convection and mixed back in the mantle. Ultimately, complete mixing is obtained when chemical diffusion has erased the heterogeneities. However, without deformation (stirring), the chemical solid-state diffusion alone is an incredibly slow process. The diffusivity of uranium for example is of order $D_{ch} \sim 10^{-19}$ m^{2}s^{-1} (Hofmann and Hart, 1978). Since the formation of the Earth (4.5 byrs), uranium would only have migrated by 20 cm. Even noble gases that are many orders of magnitude more mobile than uranium, $D_{ch} \sim 10^{-13}$ m^{2}s^{-1} ms (Trull and Kurtz, 1993), would only have migrated by 200 m. Without stirring, elements are basically frozen in the mantle.

However, the mantle is not steady and convection stirs, stretches and folds the heterogeneities until they reach small enough dimensions, where diffusion can be efficient. To provide a quantitative understanding of the interaction between deformation and diffusion, we can consider a heterogeneity localized in an infinite stripe of thickness $a(t)$ and assume that the concentration only varies in the y direction, perpendicular to the stripe. Instead of writing the diffusion equation in an Eulerian fixed frame as usual, we can write the elemental balance across the deforming stripe itself; using the Lagrangian variable $\tilde{y} = y/a(t)$, one can demonstrate that diffusion produces

$$\frac{dC}{dt} = D_{ch}\left(\frac{a_0}{a(t)}\right)^2 \frac{d^2 C}{d\tilde{y}^2}. \quad (3)$$

Stretching therefore increases the apparent diffusivity by a factor $(a_0/a(t))^2$. By analogy with the usual diffusion solution, one can guess that a stripe of initial thickness a_0 is erased by diffusion after a homogeneisation time t_h, so that

$$a_0^2 \sim 4D_{ch}\int_0^{t_h} \left(\frac{a_0}{a(u)}\right)^2 du. \quad (4)$$

Although we have not performed a rigorous demonstration of the two previous equations, the result agrees with more sophisticated approaches (e.g. Kellogg and Turcotte, 1986; Olson et al., 1984).

To perform some quantitative estimates, we can consider a heterogeneity in a strongly time dependent flow. In this case the distance between two points grows exponentially with time (Ottino, 1989) and as in a pure shear deformation, one has

$$a(t) = a_0 \exp(-At). \quad (5)$$

The parameter A has the dimension of a strain rate; it is, however, an effective strain rate computed along the Lagrangian trajectories of the flow and it is not equivalent to the instantaneous strain rate $\dot{\epsilon}$ of the flow. The quantity A is also called a finite-time Lyapunov exponent and even for very simple flows, its relationship with the average strain rate is far from being obvious (Ferrachat and Ricard, 1998). In general it also depends on the rheological properties of the heterogeneity with respect to its surroundings (du Vignaux and Fleitout, 2001). Combining (3) and (5) indicates that in time dependent flows, the apparent diffusivity depends on the Lyapunov exponent A and varies exponentially with time like D_{ch} exp$(2At)$. Equation (4) implies

$$t_h \sim \frac{1}{2A}\log\frac{A a_0^2}{D_{ch}}. \quad (6)$$

Numerical application considering the fate of an anomaly related to former oceanic crust ($a_0 = 7$ km, $A = 5\ 10^{-16}$ s^{-1}) indicates that it retains its He during 395 myrs, and its U for 835 myrs. The corresponding thickness of the crustal layer when helium and uranium diffuses out, are 14 m and 1.4 cm respectively. The very large difference between the diffusion coefficients of these two elements (6 orders of magnitude) has only a moderate impact on their retention inside the advected anomalies (a factor 2 in the retention time). The mixing (stirring+diffusion) acts in a very different way from the slow usual diffusion: for times less than the homogeneisation time t_h the elements are frozen in the flow after this time, they suddenly diffuse out exponentially.

Another important point that sometimes creates a misunderstanding between geochemists and geophysicists is the quantity of primitive material that may have remained unprocessed in the mantle. In convective flow, heterogeneities are constantly folded and stretched; as the size and the number of heterogeneities decreases, the probability of their being sampled under a ridge also decreases. The quantity of primitive material surviving convection therefore varies exponentially with time. Depending on the way the time dependence of Earth's mantle convection is parametrized, and assuming whole mantle convection, the remaining primitive material can account for 40% (even 60% in some extreme cases) to less than 3% of the mantle mass (van Keken and Ballentine, 1998; Ferrachat and Ricard, 2001; Davies, 2001). The fraction of undegassed mantle is very uncertain and depends on various assumptions regarding the history of mantle convection, the rheological properties of the mantle, the way the magma extraction is modeled and the present-day abundance of radiogenic heat sources. At least we know that a significant part of the mantle is still undegassed. The observation that 40% of the Ar produced is still in the mantle is not surprising in itself. However the primitive material should be mostly in the form of stripes, blobs, etc. of varying ages, partly erased by diffusion rather than taking the form of a homogeneous large scale reservoir. The primitive material should be intrinsically mixed with the recycled components and should also constitute a component of the shallow mantle.

5. A WHOLE MANTLE MARBLE CAKE MODEL

Subduction of oceanic crust is one of the major observations leading to plate tectonic theory. The descending slabs can be tracked within the mantle using seismology or gravity. After several tens of myrs, they are thermally equilibrated with the surrounding mantle and almost undetectable by seismology. Slabs are then folded and elongated by convective mixing and dispersed everywhere in the mantle.

The concept of a marble-cake mantle, introduced by Allègre and Turcotte (1986) explains to first order the petrological and chemical heterogeneity of the shallow mantle. In this model, the shallow mantle is made of two components: peridotite and pyroxenite. The pyroxenite represents the oceanic crust stirred by convection embedded in a peridotite matrix. This is a well accepted view of the shallow mantle, supported by field observations and geochemistry of mantle rocks and MORBs. In peridotic massifs like Beni-Boussera the pyroxenite layers have centimetric thicknesses which suggests that the 7 km thick oceanic crust has been stretched by a factor 10^5. In the marble-cake mantle, the peridotite is itself a complex mixture. It contains the remaining primitive material that, as we have seen, is likely still to exist in the convective mantle. It also includes in large proportions, ancient depleted lithospheric mantle (from fertile and depleted lherzolite to residual harzburgite). It may also include the chemical elements of ancient crustal filaments below the centimeter scale that have been wiped out by diffusion.

We propose that this model for the shallow mantle can be extended to the whole mantle and can also explain the geochemistry of OIBs and the deep mantle structure. Four observations from geochemistry need to be reproduced: a reservoir with high incompatible element concentrations, the recycling of the subducted crust within OIB sources, the higher ^{3}He/^{4}He of some OIBs, and better mixing of MORB sources compared to OIB sources. At the same time, geophysics strongly argues for whole mantle convection.

The marble-cake model can explain the existence of a deep reservoir with high incompatible element content. The subducted oceanic crust transforms into eclogitic assemblages which are likely to be a few percent denser than the surrounding mantle except in a limited depth range below 670 km depth (Hirose et al., 1999). Numerical models incorporating mineralogical transformations combined with variable viscosity produce a layer of subducted crust at the base of the mantle, fed by slabs and entrained by plumes (Christensen and Hofmann, 1994; Tackley and Xie, 2002). This layer often displays an uneven thickness with interconnected ridges commonly known as a spoke pattern, with a total volume comparable to that of D″. Hence, subduction of oceanic plates could have formed a deep reservoir, growing in time, rich in incompatible elements and poor in primitive volatiles that could account for the seismological complexities of D″. We have seen in Figure 1 that such a layer can account for the balance of U and Al. It can also account for the balance of other incompatible elements like rhenium (Hauri, 2002). This layer would explain why all plumes contain recycled slab components. The presence of piles of dense material at the base of the convective mantle could favor the relative fixity of hotspots (Davaille et al., 2002) and explain their low excess temperature (Farnetani, 1997).

The marble-cake model is also able to explain noble gas isotopes. As seen before, ^{3}He is the stable isotope. It escapes the mantle through melting and degassing. For this reason, only the peridotic component contains significant amounts of this isotope. The ^{4}He is produced from the decay of U and Th, which are concentrated in the crust. Therefore the variability of ^{3}He/^{4}He ratios could be explained from the variability of the recycled crust fraction in the mantle sources of basalts.

A two component mantle where the mass fraction of recycled crust is f, has a helium composition such that

$$\left[^3\text{He}\right]_s = (1-f)\left[^3\text{He}\right]_{per} \quad (7)$$

$$\left[^4\text{He}\right]_s = f\left[^4\text{He}\right]_c + (1-f)\left[^4\text{He}\right]_{per} \quad (8)$$

where s, per, and c stand for source, peridotite and crust. These two equations are valid for the MORB source (or shallow mantle, SM) and we assume that such a source corresponds to a mass fraction of ancient crust, f, of 8% and helium concentrations of 120 10^{-9} mol g^{-1} for ^{4}He and 1.4 10^{-12} for ^{3}He (i.e a ^{3}He/^{4}He ratio of 8 times the atmospheric ratio (Moreira et al., 1998)). We further assume that the ancient crust is 1.5 byrs old and contains 70 ppb U, similar to modern MORBs. The observed U content of the shallow mantle, 7 ppb, would come partly from ancient crust contributing 5.6 ppb (8%×70 ppb) and the rest from the peridotic component. All these estimates are subject to large uncertainties but the exact values do not affect our qualitative results.

We can now plot in Figure 2 the predicted ^{3}He/^{4}He ratio as a function of the fraction of oceanic crust in the OIB source, f. For simplicity we assume that the corresponding variabil-

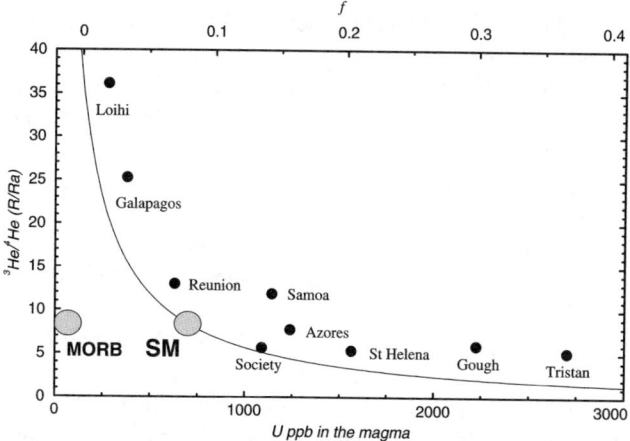

Figure 2. Various observations of ^{3}He/^{4}He as a function of U content in the magmas of various hotspots (bottom horizontal axis) (Helium data are averaged from Kurz et al. (1982), Graham et al. (1990), and Hanyu and Kaneoka (1997), U from Sims and DePaolo (1997), and Chauvel et al. (1992)). The model line shows the predicted ratio as a function of the mass fraction of ancient crust in a marble cake mantle source (top horizontal line). The presence of crust increases the uranium content but decreases the ^{3}He/^{4}He ratio. The correspondence between the U content of the source and the magma assumes a uniform enrichment in all OIBs by a factor 100. SM corresponds to a hypothetical magma obtained by enriching the shallow mantle with the same factor. With a partial melting degree of 10% the shallow mantle gives the usual MORB (^{3}He/^{4}He=8, U=70 ppb).

ity of U in the same OIBs is only due to the variability in the source and that all hotspots far from ridges, have the same degree of partial melting, 1% (U in the magma is ~100 times that in the source). This last approximation is certainly drastic but allows us to have a direct correspondence between the fraction of crust (top horizontal axis) and the U content of the magma (bottom horizontal axis). It gives a very reasonable fit to the data taken from various hotspots. The shallow mantle source corresponds by hypothesis to a mass fraction of crust of 8% (top horizontal axis); with a melt fraction of 1% it would yield a basalt with 700 ppb U (bottom horizontal axis). A melt fraction of 10%, appropriate to ridges, would produce a typical 70 ppb U MORB. Figure 2 indicates that the highest ^{3}He/^{4}He ratio, corresponding to that found for Loihi could be obtained by mixing about 2% of ancient crust with peridotite whereas the lowest ratios such as that for Tristan would imply 35% of crust in their source. This model implies that the ^{3}He concentrations in the sources of MORBs and OIBs are roughly the same as they vary with $1 - f$ which is 0.98 for Loihi, 0.92 for MORBs, 0.65 for Tristan. Only the ^{4}He concentrations differ.

This simple two end-member model does not explain all the geochemical poles, but it accounts for other geochemical isotope observations that are also related to recycling (Coltice and Ricard, 2002). Components of the EM1, EM2 and HIMU poles indicate the need to take into account both continental and oceanic sediments. However, the presence of these minor ingredients in the source of plumes are not in conflict with the model outlined here. It seems to us, at any rate, that the primary difficulty in matching geochemical and geophysical views comes from the variability of noble gas ratios.

The fact that MORBs are better mixed than OIBs may come from two mechanisms. First, the melting zones of MORBs are sampling a much larger volume than those of OIBs, second, the shallow mantle is filled with material that is more thoroughly mixed than hotspots, which come from a heterogeneous stagnant bottom boundary layer. The process of crustal segregation maintains the presence of heterogeneities near the CMB.

As a conclusion, a two component marble cake mantle can account for the existence of a deep reservoir rich in incompatible elements and for the variability of noble gas concentration in mantle magmas. Moreover, this model is consistent with seismic observations of deep slab penetration and provides an origin for D″ formed from the segregation of dense crust.

6. TOWARD THERMOCHEMICAL CONVECTION

The previous discussion in this paper shows that the only way to reconcile seismology and chemistry is explicitly to take account of petrology in geodynamic models. Clearly one of the major petrological density contrasts that exists in the mantle is that related to the difference in composition and mineralogy between the oceanic crust and the rest of the upper mantle.

A long standing model for upper mantle composition is pyrolite. Pyrolite is therefore taken to be the average composition of the marble-cake mantle made of peridotite and pyroxenite. Its transformations at high pressure gives a good fit to PREM (see Figure 3). The oceanic crust is much richer in Si than pyrolite and has a much higher Fe/Mg ratio. At depth, basalt transforms into a garnet/stishovite assemblage and at approximately 730 km depth, garnet transforms into perovskite/magnesiowüstite while the stishovite remains stable. Stishovite and Fe-rich assemblages are significantly denser than the olivine and the Fe-poor assemblages of the normal mantle. The high Al content slightly lightens the basalt. However, with the exception of depths between 650 and 730 km where the basalt remains as garnetite while the pyrolite has transformed into post spinel phases, the oceanic crust is a few percent denser (see Figure 3). The most depleted residue left after the extraction of the crust (harzburgite) is only very slightly lighter than pyrolite. From 730 to ~ 900 km depth, there is a general agreement that eclogite is about 3% denser than pyrolite (Hirose et al., 1999; Ono et al., 2001). In the

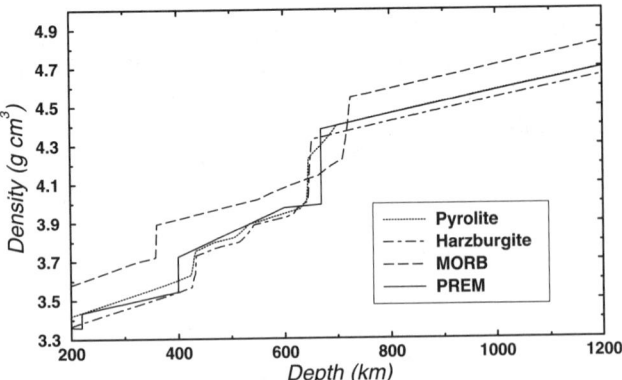

Figure 3. Densities of various petrological references as a function of depth, compared with the density in PREM. These densities have been computed from the thermodynamic properties of the various involved phases. The evolution of the phase diagrams have themselves been computed from Gibbs energy minimization (Matas, 1999). The pyrolite composition gives a rather close fit to PREM. The subducted crust always appears significantly denser than PREM except in a narrow zone on top of the lower mantle.

deepest lower mantle the density evolution and even the structures of all the phases involved are not well known (Badro et al., 2003; Gillet et al., 2000). Figure 3 was computed using a rigorous selection of the equations of state of the various mineralogical phases (Matas, 1999). Even at CMB conditions, it is inferred that eclogite would be ~ 1.5% denser. Other authors however suggest than the eclogite density may intersect the average mantle density in the lower mantle (Kesson et al., 1998; Ono et al., 2001).

The introduction of petrological components in thermal convection models is complex and the numerical simulations introduce a lot of approximations. A continuous representation of the concentrations is impossible since for affordable mesh sizes, artificial numerical diffusion would be much larger than any realistic chemical diffusion. The most common approach consists in using tracers advected by the flow and carrying chemical properties (Christensen and Hofmann, 1994; Ferrachat and Ricard, 2001; Tackley and Xie, 2002). The interpretation of the results of such models in term of petrology or geochemistry is not straightforward. The advection of tracers does not represent the real stretching of the material and diffusion cannot be taken into account. For statistical reasons a large number of tracers have to be used even in 2D (typically one million) and 3D simulations are for now out of reach. Moreover, the differences between hotspots and ridges, i.e., between cylindrical and linear upwellings does not exist in 2D simulations. Another approach, consisting in the advection of the chemical interfaces is promising in 3D (Schmalzl and Loddoch, 2003). However to avoid the treatment of too convolved interfaces, topological simplifications are needed that introduce spurious diffusion.

In addition to numerical problems, modelers also have some conceptual problems. For all thermochemical numerical codes, there is the difficulty of handling the formation of the continental crust. This rate of formation controls the chemical and thermal budget of the Earth, but all these phenomena are largely ignored in convection models. The final difficulty arises from the need for a realistic self consistent representation of surface plate motion that organizes the mantle flow and is associated with stable subduction and more erratic ridges (Bercovici et al., 2000).

Various papers have presented and discussed 2D convection models with petrological components (Christensen and Hofmann, 1994; Samuel and Farnetani, 2002; Xie and Tackley, 2003). Some of the ingredients discussed above are indeed confirmed by these models. For example the remnants of primitive material in the mantle and the delamination of ~20–30% of the eclogitized oceanic crust at the CMB (Christensen and Hofmann, 1994). This segregation occurs mostly in the hot boundary layer of the deep mantle, by Rayleigh-Taylor instability in the thickest parts of the oceanic crust. This favors the segregation of thick oceanic plateaus as also proposed from geochemical arguments (Albarède and van der Hilst, 2002). The segregation would also generate long lasting chemical heterogeneities near D″, in the sluggish lower mantle. Some basaltic components may also be trapped just below the 670 km discontinuity (Mambole and Fleitout, 2002). Without this segregation, no differences in the sizes or compositions of heterogeneities are found between the upper and lower mantle (van Keken and Ballentine, 1998). This conclusion is not affected by a moderate viscosity increase with depth in the mantle, up to 2 or 3 orders of magnitude (Ferrachat and Ricard, 2001). However, due to the exclusive use of 2D simulations, it has remained difficult to predict precisely what should be sampled in active upwellings (hotspots) and passive ridges in these models.

Ten years ago geophysicists and geochemists were working with very different tools and concepts that appear now equally naive. While geophysicists were mostly thinking in terms of purely thermal convection, geochemists were mostly intepretating their observations in terms of isolated boxes. The basis for a common language has now been set up, where the petrological components play an important role and where the topology of the reservoirs has gained much complexity. On the other hand, a simple tool to study mantle chemical evolution is still unavailable until necessary ingredients like 3-dimensionality, a correct description of petrology, and continental crust evolution are introduced in thermochemical convection models.

REFERENCES

Albarède, F., and R. D. van der Hilst, Zoned mantle convection, *Phil. Trans. Royal Soc.*, 360, 2569–2592, 2002.

Allègre, C. J., and D. L. Turcotte, Implications of a two component marble-cake mantle, *Nature*, 323, 123–127, 1986.

Badro, J., G. Fiquet, F. Guyot, J.-P. Rueff, V. V. Struzhkin, and G. Vank, Iron partitioning in Earth's mantle: toward a deep lower-mantle discontinuity, *Science*, 300, 789–791, 2003.

Becker, T. W., J. B., Kellogg, and R. J. O'Connell, Thermal constraints on the survival of primitive blobs in the lower mantle, *Earth Planet. Sci. Lett.*, 171, 351–365, 1999.

Bercovici, D., Y. Ricard, and M. A. Richards, The relation between mantle dynamics and plate tectonics: A primer, The History and Dynamics of Global Plate Motions, M. A. Richards, R. Gordon and R. Van der Hilst Ed., AGU Geophysical Monograph 21, 5–46, 2000.

Bijwaard, H., W. Spakman, and E. R. Engdahl, Closing the gap between regional and global travel time tomography, *J. Geophys. Res.*, 103, 30055–30078, 1998.

Blichert-Toft, J., F. A. Frey, and F. Albarède, Hf isotope evidence for pelagic sediments in the source of Hawaiian basalts, *Science*, 285, 879–882, 1999.

Bunge, H. P., M. A. Richards, C. Lithgow-Bertelloni, J. R. Baumgardner, S. P. Grand, and B. A. Romanowicz, Time scales and heterogeneous structure in geodynamic Earth models, *Science*, 280, 91–95, 1998.

Castle, J. C., and R. D. van der Hilst, Searching for seismic scattering of mantle interfaces between 800 and 2000 km depth, *J. Geophys. Res.*, 108, 2095–2109, 2003.

Chauvel, C., A. W. Hofmann, and P. Vidal, HIMU-EM: the French Polynesian connection, *Earth Planet. Sci. Lett.*, 110, 99–119, 1992.

Christensen, U. R., The influence of trench migration on slab penetration into the lower mantle, *Earth Planet. Sci. Lett.*, 140, 27–39, 1996.

Christensen, U. R., and A. W. Hofmann, Segregation of subducted oceanic crust in the convecting mantle, *J. Geophys. Res.*, 99, 19867–19884, 1994.

Coltice, N., and Y. Ricard, Geochemical observations and one layer mantle convection, *Earth Planet. Sci. Lett.*, 174, 125–137, 1999.

Coltice, N., and Y. Ricard, On the origin of noble gases in mantle plumes, *Phil. Trans. Royal Soc.*, 360, 2633–2648, 2002.

Davaille, A., F. Girard, and M. Le Bars, How to anchor hotspots in a convecting mantle?, *Earth Planet. Sci. Lett.*, 203, 621–634, 2002.

Davies, G. F., Ocean bathymetry and mantle convection 1. large scale flow and hotspots, *J. Geophys. Res.*, 93, 10467–10480, 1988.

Davies, G. F., Stirring geochemistry in mantle convection models with stiff plates and slabs, *Geochim. Cosmochim. Acta*, 66, 3125–3142, 2001.

du Vignaux, N. M., and L. Fleitout, Stretching and mixing of viscous blobs in Earth's mantle, *J. Geophys. Res.*, 106, 30,893–30908, 2001

Eiler, J. M., K. A. Farley, J. M. Valley, A.W. Hofmann, and E M. Stolper, Oxygen isotope constraints on the sources of Hawaiian volcanism, *Earth Planet. Sci. Lett.*, 144, 453–468, 1996.

Farnetani, C. G., Excess temperature of mantle plumes: the role of chemical stratification across D″, *J. Geophys. Res.*, 24, 1583–1586, 1997.

Ferrachat, S., and, Y. Ricard, Regular vs. chaotic mantle mixing, *Earth Planet. Sci. Lett.*, 155, 75–86, 1998.

Ferrachat, S. and Y. Ricard, Mixing properties in the Earth's mantle: Effects of the viscosity stratification and of oceanic crust segregation, *Geochem. Geophys. Geosys.*, 2, 2000GC000092, 2001.

Fukao, Y., S. Widiyantoro, and M. Obayashi, Stagnant slabs in the upper and lower mantle transition region, *Rev. Geophys.*, 39, 291–323, 2001

Garnero, E. J., and D. V. Helmberger, A very slow basal layer underlying large-scale low-velocity anomalies in the lower mantle beneath the Pacific: evidence from core phases, *Phys. Earth Planet. Inter.*, , 91, 161–176, 1995

Gillet, P., M. Chen, L. Dubrovinsky, and A. El Goresi, Natural $NaAlSi_3O_8$-hollandite in the shocked Sixiangkou meteorite, *Science*, 287, 1633–1636, 2000.

Graham, D., J. Lupton, F. Albarède, and M. Condomines, Extreme temporal heterogeneity of helium isotopes at Piton de la Fournaise, Réunion island, *Nature*, 347, 545–548, 1990.

Grand, S. P., R. D. van der Hilst, and S. Widiyantoro, Global seismic tomography: a snapshot of convection in the Earth, *GSA Today*, 7, 1–7, 1997.

Hanyu, T., and I. Kaneoka, The uniform and low $^3He/^4He$ ratios of HIMU basalts as evidence for their origin as recycled materials, *Nature*, 390, 273–276, 1997.

Hauri, E. H., Osmium isotopes and mantle convection, *Phil. Trans. R. Soc. London A*, 360, 2371–2382, 2002.

Hirose, K., Y. Fei, Y. Ma, and H.-K. Mao, The fate of subducted basaltic crust in the Earth's lower mantle, *Nature*, 397, 53–56, 1999.

Hofmann, A. W., Mantle geochemistry: the message from oceanic volcanism, *Nature*, 385, 219–229, 1997.

Hofmann, A. W., and S. R. Hart, An assessment of local and regional isotopic equilibrium in the mantle, *Earth Planet. Sci. Lett.*, 38, 44–62, 1978.

Jochum, K. P., A. W. Hofmann, E. Ito, H. M. Seufert, and W. M. White, K, U and Th in mid-ocean ridge basalt glasses and heat production, *Nature*, 306, 431–436, 1983.

Kellogg, L. H., and D. L. Turcotte, Homogeneisation of the mantle by convective mixing and diffusion, *Earth Planet. Sci. Lett.*, 81, 371–378, 1987.

Kellogg, L. H., B. H. Hager, and R. D. van der Hilst, Compositional stratification in the deep mantle, *Science*, 283, 1881–1884, 1999.

Kesson, S. E., J. D. F. Gerald, and J. M. Shelley, Mineralogy and dynamics of a pyrolite lower mantle, *Nature*, 393, 252–255, 1998.

Kurz, M. D., W. J. Jenkins, and S. R. Hart, Helium isotopic systematics of oceanic islands and mantle heterogeneity, *Nature*, 297, 43–47, 1982.

Mambole, A., and L. Fleitout, Petrological layering induced by an endothermic phase transition in the Earth's mantle, *Geophys. Res. Lett.*, , 29, 2044–2047, 2002.

Matas, J., Modélisation thermochimique des propretés de solides à hautes températures et hautes pressions. Applications géophysiques, PhD, Ecole Normale de Lyon, 1999.

Montelli, R., G. Nolet, F. A. Dahlen, G. Masters, E. R. Engdahl, and S.-H. Hung, Finite frequency tomography reveals a variety of plumes in the mantle, *Science*, 303, 338–343, 2004.

Moreira, M., J. Kunz, and C. J. Allègre, Rare gas systematics in popping rock: isotopic and elemental compositions in the upper mantle, *Science*, 279, 1178–1181, 1998.

McDonough, W. F., and S. Sun, The composition of the Earth, *Chem. Geol.* 120, 223–253, 1995.

Nataf, H.-C., and J. C. VanDecar, Seismological detection of a mantle plume?, *Nature*, 264, 115–120, 1993.

Olson, P., D. A. Yuen, and D. Balsinger, Mixing of passive heterogeneities by mantle convection, *J. Geophys. Res.*, 89, 425–436, 1984.

Ono, S., E. Ito, T., Katsura, Mineralogy of subducted basaltic crust (MORB) from 25 to 37 GPa, and chemical heterogeneity of the lower mantle, *Earth Planet. Sci. Lett.*, 190, 57–63, 2001.

Ottino, J. M., *The Kinematics of Mixing: Stretching, Chaos, and Transport*, Cambridge Univ. Press, New York, 1989.

Ricard, Y., M. A. Richards, C. Lithgow-Bertelloni, and Y. LeStunff, A geodynamic model of mantle density heterogeneity, *J. Geophys. Res.*, 98, 21895–21909, 1993.

Richards, M. A., R. A. Duncan and V. E. Courtillot, Flood basalts and hot-spot tracks: plume heads and tails, *Science*, 246, 103–107, 1989.

Saltzer, R., R. D. Van der Hilst, and H. Karason, Comparing P and S wave heterogeneity in the mantle, *Geophys. Res. Lett.*, , 28, 1335–1338, 2001.

Schmalzl, J., and A. Loddoch, Using subdivision surfaces and adaptative surface simplification algorithms for modeling chemical heterogeneities in geophysical flows, *Geochem. Geophys. Geosys.*, 4, 8303, doi:10.1029/2003GC000578, 2003.

Sims, K. W., and D. J. DePaolo, Inferences about mantle magma sources from incompatible element concentration ratios in oceanic basalts, *Geochim. Cosmochim. Acta*, 61, 765–784 , 1997.

Sleep, N. H., Hotspots and plumes: some phenomenology, *J. Geophys. Res.*, 95, 6715–6736, 1990.

Steinberger, B., and R. J. O'Connell, Advection of plumes in mantle flow; implications on hotspot motion, mantle viscosity and plume distribution, *Geophys. J. Int.*, 132, 412–434, 1998.

Tackley, P. J., and S. X. Xie, The thermochemical structure and evolution of Earth's mantle: constraints and numerical models, *Philos. Trans. Roy. Astron. Soc.*, 1800, 2593–2609, 2002.

Trull, T. W., and M. D. Kurz, Experimental measurements of He-3 and He-4 mobility in olivine and clinopyroxene at magmatic temperatures, *Geochim Cosmochim. Acta*, 57, 1313–1324, 1993.

Tyburczy, J. A., B. Frisch, and T. J. Arhens, Shock-induced volatile loss from a carbonaceous chondrite: implications for planetary accretion, *Earth Planet. Sci. Lett.*, 80, 201–207, 1986.

van der Voo R., W. Spakman, and H. Bijwaard, Tethyan subducted slabs under India, *Earth Planet. Sci. Lett.*, 171, 7–20, 1999.

van Keken, P. E., and C. J. Ballentine, Whole-mantle versus layered convection and the role of a high-viscosity lower mantle in terrestrial volatile evolution, *Earth Planet. Sci. Lett.*, 156, 19–32, 1998.

Vinnik, L., B. Romanowicz, Y. LeStunff, and L. Makeyeva, Seismic anisotropy in the D″ layer, *Geophys. Res. Lett.*, , 22, 1657–1660, 1995.

Weber, M., J. P. Davis, C. Thomas, F. Krger, F. Scherbaum, J. Schlittenhardt, and M. Krnig, The structure of the lower- most mantle as determined from using seismic arrays; In: Seismic modeling of the Earth's structure, Eds. Boschi, E., Ekstrm, G., Morelli A., Istituto Nazionale di Geophysica, Roma, 399–442, 1996.

Zindler, A., and S. Hart, Chemical geodynamics, *Annu. Rev. Earth Planet. Sci.*, 14, 493–571, 1986.

Y. Ricard, Laboratoire des Sciences de la Terre, Ecole Normale Supérieure de Lyon, 46 allée d'Italie, F-69364 Lyon, Cedex 07, France. (ricard@ens-lyon.fr)

N. Coltice, Laboratoire des Sciences de la Terre, Ecole Normale Supérieure de Lyon, 46 allée d'Italie, F-69364 Lyon, Cedex 07, France. (ricard@ens-lyon.fr)

Melting of Fertile Peridotite With Variable Amounts of H$_2$O

Bernard J. Wood

Department of Earth Sciences, University of Bristol, Bristol BS8 1 RJ, U.K.

An empirical model of the conditions of equilibration of anhydrous and hydrous melts with lherzolite has been developed from the available experimental data. Anhydrous melts in equilibrium with olivine, orthopyroxene and clinopyroxene exhibit clear compositional dependences on pressure and total alkali (Na$_2$O+K$_2$O) content. With increasing pressure and decreasing alkali content, the equilibrium SiO$_2$ content of the melt decreases while its MgO concentration increases. In the latter case the relationship for natural bulk compositions is described by:

$$MgO(weight\%) = (11.14 - 1.262 Alk) + (2.765 + 0.0945 Alk)P$$

Where Alk refers to total (Na$_2$O+K$_2$O) in the melt and P is in GPa. For a melt of given MgO content this can be rearranged to give a barometer:

$$P = \frac{MgO - 11.14 + 1.262 Alk}{2.765 + 0.0945 Alk}$$

Similarly, the temperature of equilibrium can be expressed in terms of pressure and alkali contents:

$$T(°C) = 1230 - 23.75(Alk) + (98 + 5.6[Alk])P$$

The equilibrium MgO content of the melt declines by about 0.23% for every 1 weight % H$_2$O added. This is equivalent to an apparent shift in pressure of about 0.06 GPa for every 1% H$_2$O and is accompanied by a corresponding increase in SiO$_2$ content of the melt. As H$_2$O is added, the equilibrium temperature declines approximately as:

$$\text{Liquidus depression (°C)} \cong 80(H_2Owt\%)^{0.4}$$

The above relationships were applied to a number of putative primitive magmas from arc environments, some of which had been studied experimentally to determine conditions of equilibrium with the lherzolite assemblage. The empirical relationships seem to work well, generally reproducing measured equilibrium conditions to within about 0.3 GPa and 40°C.

The compositions of primitive magmas from the Cascade arc indicate equilibrium with the mantle under conditions close to the crust–mantle boundary. High alumina olivine tholeiites, which are essentially anhydrous yield equilibrium conditions of about 0.9 GPa and 1275°C, in good agreement with experimental data. Hydrous basaltic andesites appear to have equilibrated with the mantle at about 1.4 GPa and 1170°C. The high temperatures at relatively shallow depths are consistent with melt-focussing and heat advection towards the wedge corner at the boundary of upper plate and subducting slab.

INTRODUCTION

The origin and evolution of magmas in the arc setting is a subject which, because of petrologic and tectonic diversity, has generated enormous debate since well before the advent of plate tectonics. What seems to be certain is that addition of H_2O from the downgoing slab is critical to the production of large volumes of intermediate liquids which, based on some experimental evidence (e.g *Kushiro*, 1973), may be primary melts of the mantle wedge overlying the subduction zone. It is the purpose of this paper to develop a method, based on the available experimental data, of determining whether or not specific melts could be primary and to determine the conditions of pressure, temperature and H_2O content under which they would be in equilibrium with the mantle. This will provide a means of constraining the physical conditions under specific volcanic arcs and enable them to be related to models of temperature structure and thermal history.

A common experimental approach to the identification of possible primary magmas is through the determination of the conditions of multiple saturation. This entails the performance of a series of high pressure experiments close to the liquidus until a region of pressure–temperature-H_2O space is identified where olivine, orthopyroxene and clinopyroxene are all liquidus phases (e.g *Tatsumi*, 1982; *Pichavant et. al.* 2002). Thus, if there exists a region of pressure–temperature space in which the melt is in equilibrium with the residual peridotite minerals and the latter have appropriate Mg/Fe ratios (olivine of Fo_{88-92} for example), then this melt is a plausible primary magma. In the case of supra-subduction zone volcanics the range of melt compositions, in terms of SiO_2 and MgO contents proposed as primary mantle melts is extremely large. Thus, for example, olivine basalts (~48% SiO_2, 12% MgO), basaltic andesites (~52% SiO_2, 9% MgO) and high magnesian andesites (~57% SiO_2, 7% MgO) can all obey the 'multiple saturation' criterion under certain circumstances. The big differences between subduction zone volcanics and those from mid-ocean ridges and oceanic islands lie in their H_2O contents and oxygen fugacities. Mid-ocean ridge basalts (N-MORB) are typified by H_2O contents of about 0.1% (*Sobolev* and *Chaussidon*, 1996) and oxygen fugacities about 2 log units below the Ni-NiO (NNO) buffer (e.g *Wood et. al.*, 1990; *Carmichael*, 1991). Subduction volcanics are much richer and more variable in H_2O contents, based both on measurements of glass inclusions and on inferences drawn from high pressure multiple saturation experiments. Thus, for example, glass inclusions yield H_2O values as low as 0.2% for a low-K olivine tholeiite from Medicine Lake, California and up to 6.2% for a basaltic andesite from Fuego, Guatemala (*Sisson and Layne*, 1993; *Sisson and Bronto*, 1998), while phase equilibria studies suggest 1–2% H_2O for primary magmas from NE Japan (*Tatsumi and Eggins*, 1995), 2% for St. Vincent basalts (*Pichavant et. al.* 2002), ~7% for high-Mg andesite from SW Japan (*Tatsumi*, 1982) and >6% for andesites from Mexico (*Carmichael*, 2002). The more oxidised conditions and wider range of oxygen fugacities of subduction zone volcanics relative to MORB (e.g *Carmichael*, 1991) also lead to a greater range of Mg/Fe ratios in melts which can potentially equilibrate with a lherzolite assemblage in which the mantle has olivine of composition Fo_{88-92}. Thus, the large variations in H_2O content and f_{O_2} lead to a broad spectrum of plausible primary melts.

Given these observations the important task of experiments is to determine the conditions of P, T, H_2O- content and f_{O_2} under which primary magmas are produced. While the multiple saturation experiments described above are extremely useful in this regard, they are extremely time-consuming and they are only applicable to specifically-identified melts. A more general approach that embodies all of the available data is really required. This is what I have attempted to develop here in a simple form. As might be anticipated, given the range of physical conditions and chemical composition that merit investigation, not all available data are in accord with one another. What I have done, therefore, is to attempt a best-fit approach and only leave out data when there is a clear reason for doing so. The approach involves starting with the available data on melting of lherzolite under anhydrous conditions and then adding the hydrous data in order to address explicitly the effects of H_2O. The first step is to consider the melting of peridotite under anhydrous conditions.

ANHYDROUS PERIDOTITE MELTING

In this section I will review the available data on the melting of peridotite under anhydrous conditions. Because most of the reliable data apply to residues of olivine plus orthopyroxene plus clinopyroxene (lherzolite), I will not address the conditions of equilibration of highly magnesian melts with harzburgitic (olivine, orthopyroxene) residue. The discussion therefore centres on liquids which comprise less than 20–25% melts of fertile upper mantle.

Experimentally, peridotite melting experiments have been performed in a number of different ways, each of which has advantages and disadvantages. Direct melting experiments are not useful at low melt fractions (<30%) unless the melt can be physically separated from the residual crystals. This is because, during quenching, the melt composition may be modified by back-reaction with the crystals. Microprobe analysis of the quenched glass does not, in such cases, give the equilibrium melt composition. A new technique, that of the 'diamond aggregate' (*Hirose and Kushiro*, 1993; *Baker and Stolper*, 1994) was developed in the 1990's to get around the problem of back-reaction. In this method a layer of small diamond crystals is inserted into the peridotite. The strength of the diamond means that the pore space between the crystals remains stable during pressurization and that the melts generated from the peridotite are drawn into the diamond layer. On quenching, the liquid is physically separated from the residual crystals and is therefore unable to back-react with them. The diamond host remains essentially inert during the quench. Although achieving the desired result of a glass which is the same composition as the high pressure liquid, the forced physical separation of crystals and liquid means that the latter may never be a true equilibrium melt, particularly if the starting mixture contains relatively coarse crystals (*Falloon et. al.* 1999).

The 'sandwich' technique (*Stolper*, 1980) attempts to force equilibrium with the residual peridotite mineralogy, but at the same time minimises the problem of back-reaction by increasing the ratio of melt to crystals. This method involves placing a relatively thick layer of possible equilibrium melt on top of the residual mineralogy. During the experiment the two layers react and the melt approaches equilibrium more closely. On quenching the liquid back-reacts along its interface with the crystals, but the layer is sufficiently thick that distal regions are unmodified and can be analysed by microprobe. Practically, this is a relatively simple method which gives good results. The main drawbacks arise from the case where the initial melt reacts strongly with the peridotite residuum. This can lead to one or more solid phases disappearing at the interface and/or to phase compositions zoned with respect to the interface. Then the melt is not in equilibrium with the bulk peridotite. Furthermore, this method, in which the fraction of liquid present is greater than in the mantle, does not lead directly to a determination of the dependence of melt composition on extent of melting. Robinson et. al. (1998) extended the 'sandwich' method to resolve these principal drawbacks. Firstly, they synthesized the measured melt composition produced during the sandwich experiment and re-reacted it with the peridotite mineralogy. After 2 cycles the melt did not usually react further with the peridotite and could be considered close to equilibrium. A final check on equilibrium was provided by taking the melt composition alone and testing for precipitation of olivine, orthopyroxene and clinopyroxene under the same conditions as those of the 'sandwich' experiment. Robinson et. al. (1998) also showed how, by carefully measuring the proportions of phases present before the experiment and the compositions after the experiment, the results could be converted to an apparent melt fraction for melting of the peridotite of interest. This type of (time-consuming) study seems to provide the best results currently available.

The SiO_2 contents of melts in equilibrium with olivine, orthopyroxene and clinopyroxene are, as shown by Hirschmann et. al. (1998), strongly dependent on pressure and on the total concentrations of alkali oxide ($Na_2O + K_2O$) which the melts contain. At fixed pressure, increasing alkali oxide results in increasing SiO_2 content of the melt in equilibrium with the lherzolite assemblage. The effect of alkali decreases with increasing pressure, however, such that, at 3 GPa, it is near zero. Following from Hirschmann et. al.'s observations, I took experiments from a number of studies and fitted, for each study SiO_2 (weight%) as a function of weight %(Na_2O+K_2O) in the melt at fixed pressure. I used the same approach to determine the effect of (Na_2O+K_2O) on MgO content of the melt. This approach enables comparison of different sets of data at fixed alkali content and pressure.

Figure 1 shows the results of Walter and Presnall (1994) on melts in the simplified system Na_2O-CaO-MgO-Al_2O_3-SiO_2 (NCMAS). In each case I have plotted the concentration of SiO_2 or MgO at fixed Na_2O contents in the melt of 2, 4 and 6 weight %. Construction of the 6% isopleth required, at each pressure, extrapolation slightly beyond the range of the data (5–5.5% maximum) and, for this isopleth, I excluded pressures where the maximum Na_2O content was below 4.6%.

As can be seen from Figure 1, SiO_2 contents of melts in equilibrium with the lherzolite assemblage increase with increasing Na_2O and with decreasing pressure. The dependence of SiO_2 on pressure is approximately logarithmic, as can be inferred from the curves fitted to the data. In contrast to SiO_2, the dependence of melt MgO on pressure is essentially linear (Figure 1b). MgO content (in equilibrium with olivine, orthopyroxene and clinopyroxene) increases strongly with decreasing alkali content of the melt and with increasing pressure. This simplified system, although not directly applicable to mantle melting, provides a model for anticipated dependences of SiO_2 and MgO on pressure and alkali content. Because natural peridotites have molar Mg/(Mg+Fe) of about 0.9 rather than the 1.0 of the model system, melts in equilibrium with the mantle at a given pressure are lower in MgO than in the NCMAS system. Mantle melts are also slightly lower in SiO_2 than melts in the model NCMAS system.

Figure 2 shows the SiO_2 and MgO contents of melts from different types of experiment on natural compositions ('sandwich', 'diamond aggregate' and sandwich plus multiple saturation). These experiments were aimed at determining the compositions of melts in equilibrium with the mantle assemblage at different pressures and temperatures. There is appreciable scatter at any given pressure (of the order of $\pm 1\%$ SiO_2 and MgO), but the nature of the pressure dependences are, in both cases very similar to those shown by the data of Walter and Presnall (1994) (Figure 1). The scatter exhibited by the natural system data can be ascribed to a combination of differences in bulk composition, notably Mg#, and to the equilibration problems discussed earlier. Nevertheless, the data

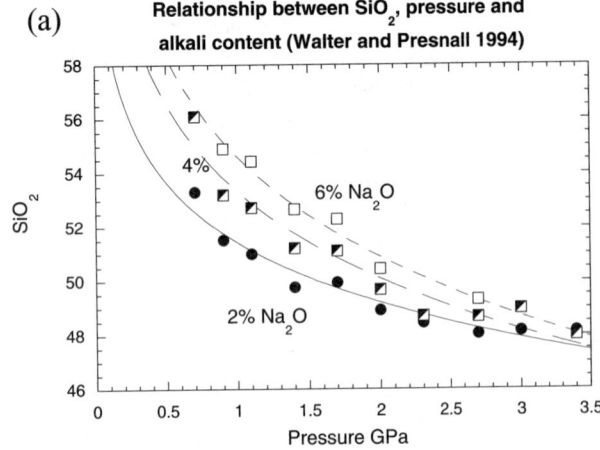

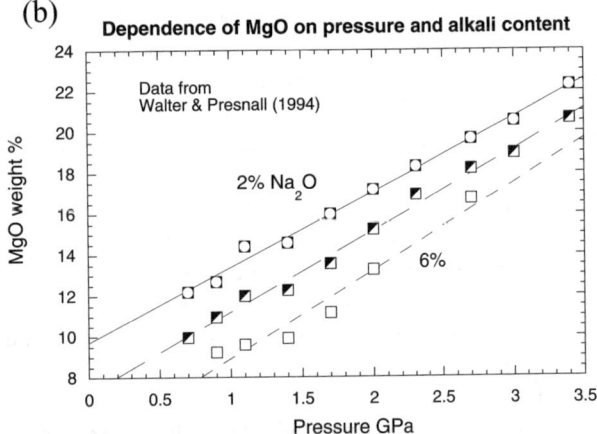

Figure 1. (a) The relationships between SiO$_2$ contents of equilibrium melts of model lherzolite in the system Na$_2$O-CaO-MgO-Al$_2$O$_3$-SiO$_2$ and pressure and Na$_2$O contents of the melts. Curves are logarithmic and shown purely for illustration. Data from Walter and Presnall (1994). (b) MgO contents of the same melts plotted against pressure at fixed Na$_2$O content.

provide the basis of a reasonable method for estimating the conditions of equilibration of melts with the mantle under anhydrous conditions. I used the experimental data on liquids coexisting with olivine, orthopyroxene and clinopyroxene to construct best-fit curves of melt composition as a function of pressure and total alkali content. Results are shown in Figures 3 and 4. Although I have removed the data to avoid confusion, there are sufficient data to construct the curves at 4% and 6% (Na$_2$O+K$_2$O) and the scatter is of similar amount to that shown in Fig 2ab.

Figures 3 and 4 show SiO$_2$ and MgO contents of melt as a function of pressure for fixed total alkali (Na$_2$O+K$_2$O) content of the melt. In the pressure range 0.1–2.4 GPa the MgO curves are given by:

$$MgO(wt\%) = (11.14 - 1.262 Alk) + (2.7653 + 0.0945 Alk)P \quad (1)$$

In this equation Alk refers to (Na$_2$O+K$_2$O) in the melt (in weight %) and P is in GPa. For a liquid of given MgO content this can be rearranged to give a barometer:

$$P = \frac{MgO - 11.14 + 1.262 Alk}{2.7653 + 0.0945 Alk} \quad (2)$$

with, assuming an uncertainty of $\pm 1\%$ in the calibration, an error of ± 0.33 GPa approximately.

At any fixed pressure, the MgO and SiO$_2$ contents of the melt depend, as shown in Figs 3 and 4, on the alkali content and hence on temperature and the degree of melting. The degree of melting may be estimated from the contents of incompatible elements such as Na in the melt (e.g *Langmuir et. al.* 1992). In this case, the data on liquid Na$_2$O content as a function of melt fraction F given by Robinson et. al. (1998) and Kushiro (1996) agree reasonably well between 0.5 and 2 GPa with the curve presented by Langmuir et. al. (1992). For

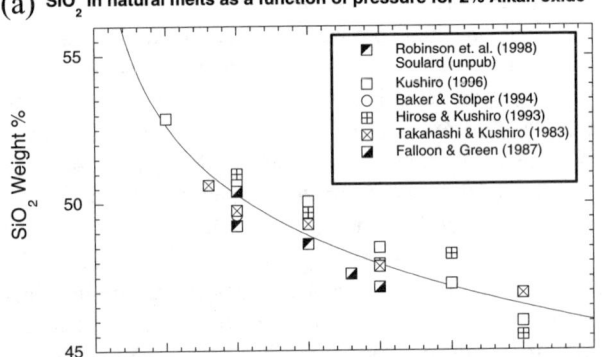

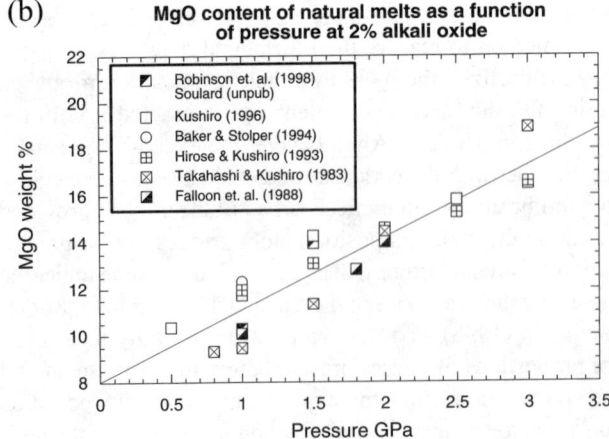

Figure 2. (a) Similar to figure 1a except for natural anhydrous melts containing Fe at, for clarity, fixed (Na$_2$O+K$_2$O) of 2%. (b) Similar to figure 1b except for natural anhydrous melts containing Fe at fixed (Na$_2$O+K$_2$O) of 2%.

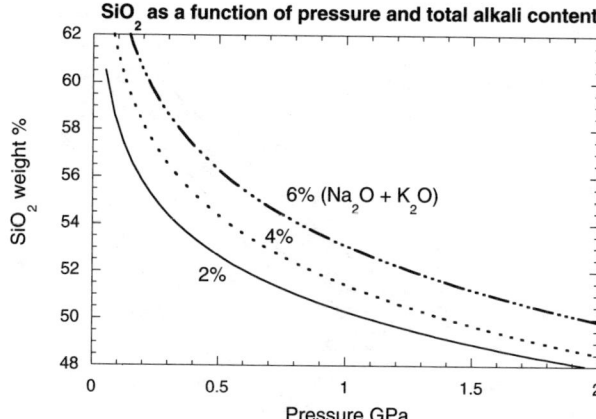

Figure 3. Best-fit curves showing SiO_2 contents of equilibrium melts of natural lherzolite as a function of pressure at fixed total alkali (Na_2O+K_2O) contents of the melts. Data sources (omitted for clarity) as in Fig 2.

F (batch melting) values between 0.02 and 0.25 they can be represented by:

$$F = 0.41 e^{\left(\frac{-0.17 Na_2O^{melt}}{Na_2O^{source}}\right)} \qquad (3)$$

where the unknowns on the right hand side refer to Na_2O contents of the melt and the initial source peridotite.

The anhydrous temperatures which are consistent with the melt compositions discussed above have been fitted as a function of alkali content and pressure, and yield the following relationship:

$$T\ (°C) = 1230 - 23.75(Alk) + (98 + 5.6[Alk])P \pm 30 \qquad (4)$$

The uncertainty of $\pm 30°$ refers to 1 standard deviation of the fit of equation (4) to the data (105 points).

The implications of equations (1)–(4) are that for any putative mantle melt we can estimate the pressure and temperature under which it would be in equilibrium with the assemblage olivine, orthopyroxene and clinopyroxene, where the olivine composition is close to Fo_{90}.

EFFECT OF H_2O ON MELTING

Since the work of Tuttle and Bowen (1958), it has been known that, even at low concentrations, H_2O dissolved in melts has profound effects on crystallisation temperatures, phase equilibria and viscosity. In both granitic and basaltic compositions addition of 1% water to the melt depresses the liquidus by 50 to 80°C and in the basaltic case, expands the stability field of olivine relative to orthopyroxene (e.g. *Kushiro*, 1973). Precise work in simple systems such as $CaMgSi_2O_6$-H_2O, show that liquidus temperature is generally a nonlinear function of water addition and these observations have led to determination of the complex activity-composition relations in hydrous silicate melts (e.g. *Burnham*, 1975). In volcanic arcs, water-rich fluids derived from dehydration of the subducted slab play a major role in reducing the solidus temperature of mantle peridotite and initiating partial melting at relatively low temperatures.

Experiments on the effect of H_2O on peridotite melting follow similar philosophies to those applied to the problem of melting under anhydrous conditions. The sandwich technique of forced saturation in the mantle phases has been used by Gaetani and Grove (1998) and Hall (1999), while direct melting of peridotite with separation of melt from residue was employed by Hirose and Kawamoto (1995). Finally, direct experimentation, using possible mantle melts with known amounts of added water, has been employed by Tatsumi (1982) and Pichavant et. al. (2002). In these cases P-T-H_2O space is explored to determine whether conditions exist under which the hypothetical melt precipitates the mantle assemblage on the liquidus.

Figures 5 and 6 show the effect of added water on the degree of melting of fertile peridotite at 1 and 1.5 GPa (*Hirose and Kawamoto*, 1995; *Hall*, 1999) respectively. Addition of 0.2% H_2O lowers the temperature at which a given melt fraction is generated from KLB-1 peridotite at 1GPa by up to 100°. The effect appears to be strongly non-linear, with 0.5% H_2O lowering temperature by 100–150°. The results of Hall (1999) at 1.5 GPa are similar. Addition of 0.5% H_2O lowers the temperature for a given melt fraction by 100–150°, while addition of 0.85% H_2O causes a temperature depression of 120–180°. Also shown in Figures 5 and 6 are calculated curves for the addition of fixed amounts of water. These were obtained from

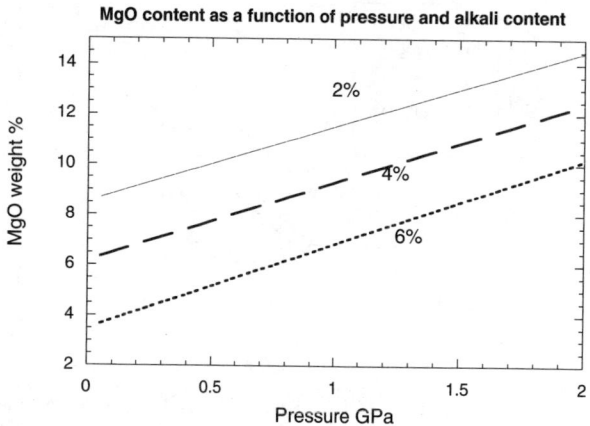

Figure 4. Linear fits to the MgO contents of the melts of Fig 3 plotted against pressure at fixed (Na_2O+K_2O) content.

a slightly modified form of the empirical expression for the depression of the olivine liquidus which was given by Falloon and Danyushevsky (2000) :

$$\text{Liquidus depression (°C)} = 80(H_2O wt\%)^{0.4} \quad (5)$$

The curves were constructed by assuming that all H_2O partitions into the melt and with the assumption that depression of the olivine liquidus is equivalent to depression of temperature of equilibration of the melt with the solid olivine-rich residue. As can be seen from Figs 5 and 6, the calculated depression of the olivine liquidus is in good agreement with the observed temperature depression at given F when up to at least 0.55% H_2O is added. Deviations from the calculated curves at 1.5 GPa and temperatures below 1150°C are due to the appearance of amphibole, which lowers the water content of the melt. At both 1 and 1.5 GPa however, the effect of water on F at fixed pressure and temperature,

$$\left(\frac{\partial F}{\partial C_{H_2O}}\right)_{P,T}$$

is strongly nonlinear, being greatest at low amounts of added water. The results agree reasonably well with estimates of the effect of water on melting made by Stolper and Newman (1994) for primitive basalts from the Mariana back-arc trough. These authors calculated that addition of 0.2% H_2O raised the degree of melting by about 0.12 which, as can be seen from Figs 5 and 6, is close to our calculated curves.

The principal uncertainties in applying equation (5) to hydrous melting lie firstly in the fact that it only strictly

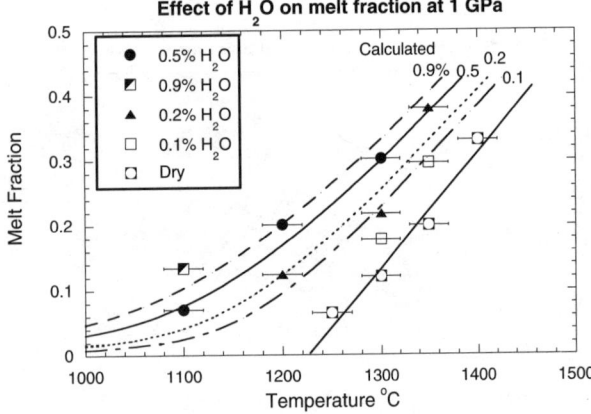

Figure 5. The effect of H_2O content of KLB-1 peridotite on melt fraction generated at 1 GPa. Data from Hirose and Kawamoto (1995). Curves calculated at 0.1, 0.2, 0.5 and 0.9% H_2O using equation 5 and the dry melting curve of Hirose and Kushiro (1993).

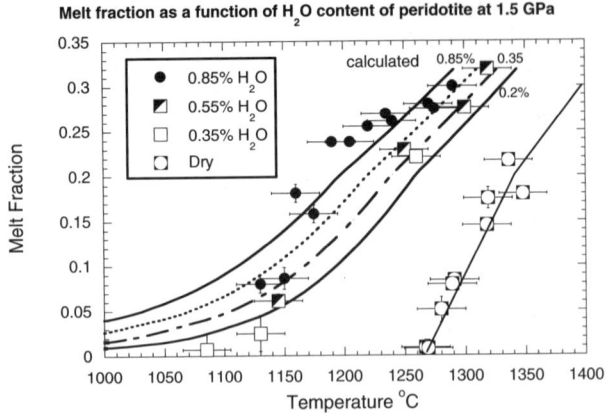

Figure 6. The effect of H_2O content of MORB-pyrolite on melt fraction generated at 1.5 GPa. Data from Hall (1999). Curves calculated at 0.2, 0.35, 0.55 and 0.85% H_2O using equation 5 and the dry melting curve of Robinson et al (1998).

applies for melts containing less than about 3% total Na_2O plus K_2O (*Falloon and Danyushevsky,* 2000) and secondly in the assumption that, apart from H_2O content, the melt composition remains unchanged relative to the dry system. It is well-known, however, that melting of peridotite under hydrous conditions leads to melts which (on an anhydrous basis) are elevated in SiO_2 and depressed in MgO relative to the corresponding dry system (*Kushiro,* 1972; *Nicholls and Ringwood,* 1972). Furthermore, this trend increases with increasing water content of the melt such that at 1GPa H_2O-saturated melts in equilibrium with peridotite may be andesitic rather than basaltic in nature (*Hirose,* 1997). Melts produced by Hirose (1997) at 1GPa contain 10–15% H_2O and on an anhydrous basis have SiO_2 and MgO contents of 60.3 and 5.8% respectively at a total alkali content of 3.4%. This contrasts strongly with dry melts which, at 1 GPa and comparable alkali content, contain about 51% SiO_2 and 10% MgO (Figures 3 and 4).

Figure 7 shows the observed isobaric decreases in MgO content for the hydrous peridotite melting data presented in Figures 5 and 6. Apparent, or expected MgO contents were calculated on an anhydrous basis for the measured alkali content and pressure using equation (1). For valid comparison with the anhydrous data, only those experiments in which clinopyroxene, olivine and orthopyroxene were present in the residuum are shown. The observed MgO contents were then recalculated on an anhydrous basis and subtracted from the calculated values. As can be seen, there is a good correlation between the difference in calculated MgO ($MgO_{calc} - MgO_{obs}$) and the H_2O content of the melt. The fitted line suggests that MgO content of the melt decreases by 0.23% for every 1 weight % water added.

As can be seen from Figure 7, addition of water decreases, at constant pressure, the MgO contents (on an anhydrous basis) of melts in equilibrium with olivine, orthopyroxene and clinopyroxene, while under dry conditions, melts in equilibrium with these phases decrease in MgO content with decreasing pressure (Figure 4). Therefore the lowering of MgO content with water addition can be converted into an apparent pressure shift given by equation (2). Figure 8 shows the difference between actual and calculated pressures ($P_{calc}-P_{obs}$) plotted as a function of the H_2O content of the silicate melt. As anticipated, the data of Hall(1999) and Hirose and Kawamoto (1995) show a good correlation between ($P_{calc}-P_{obs}$) and water content. For comparison, the data of Gaetani and Grove (1998) which were not used for calibrating the geobarometer are also shown in Figure 8. The results of Gaetani and Grove (1998) obtained under anhydrous conditions are very accurately predicted by equation (4). Some of their results under hydrous conditions are not, however, in such good agreement with the expected shift due to H_2O-addition. The reason for this may be related to the fact that Gaetani and Grove made stringent efforts to ensure that oxygen fugacity in their experiments remained low (below the FMQ buffer) under both hydrous and anhydrous conditions. In contrast, the hydrous experiments of Hall (1999) and Hirose and Kawamoto (1995) were performed in AuPd alloy capsules and had the potential to lose H_2 becoming, in the process, more oxidised. A posteriori measurements of oxygen fugacity using the olivine-orthopyroxene-spinel oxygen barometer (*Wood et. al.* 1990) demonstrates that oxygen fugacities were indeed high. Hirose and Kawamoto measured values 1–2 logf_{O_2} units above the FMQ buffer, while Hall (1999) obtained values mainly in the range 0–3 log units above FMQ. Thus it is possible that some of the apparent effect of H_2O shown in Figures 7 and 8 is actually an effect of changing oxidation state

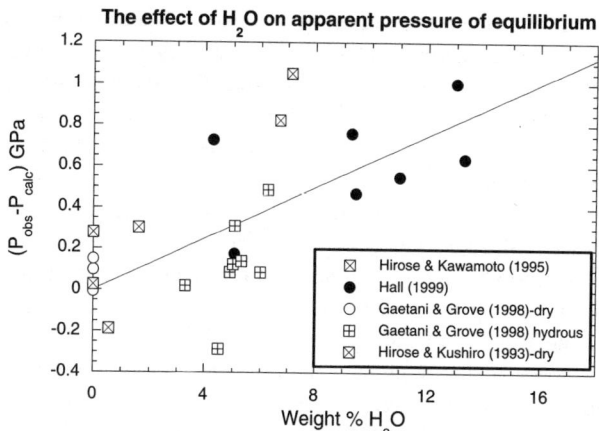

Figure 8. MgO contents of hydrous melts were recalculated to an anhydrous basis and then used to estimate equilibrium pressure using equation 2. These values are, as shown, generally lower than the actual (observed) pressure of the experiment.

since the anhydrous comparator experiments were performed in graphite capsules under relatively reducing conditions.

Despite the proviso about oxidation state, there is other evidence to suggest that H_2O shifts the apparent equilibration pressures of melts. This comes from direct experiments under hydrous conditions in which the conditions of multiple saturation were determined. The experiments of Pichavant et. al. (2002) and of Tatsumi (1982) demonstrate that when H_2O is added to melts saturated in olivine, orthopyroxene and clinopyroxene, the conditions of multiple saturation move to progressively higher pressures and lower temperatures with increasing water content. These observations are consistent with Figures 7 and 8 which indicate that the pressure shift is approximately 0.06GPa to higher pressures for every 1% H_2O added. The observation that the oxygen fugacities are substantially higher than those of oceanic basalts and in the range of the hydrous melting experiments (*Carmichael*, 1991) provides further support for the application of the pressure shift due to increasing water content which is illustrated in Figure 8.

Finally, melts containing high contents of H_2O in equilibrium with the lherzolite assemblage appear to have slightly lower CaO contents than corresponding anhydrous melts. Without being able to quantify the effect of H_2O, I find that the 3 phase assemblage produces melts in which the CaO content decreases with H_2O content. Thus, for example, the dry peridotite melting data summarised earlier have approximate CaO contents given by, at clinopyroxene saturation:

$$CaO(anhydrous) = 13.4 - 0.85(Alk) \text{ weight\%} \quad (6)$$

The hydrous melting data of Hirose and Kawamoto (1995), Hirose (1997) and Hall (1999) all show lower CaO contents

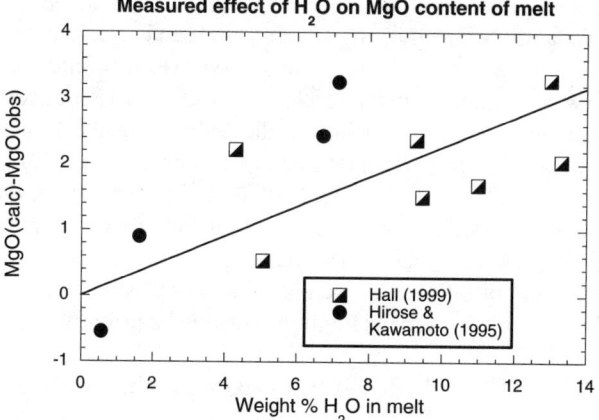

Figure 7. Dependence of MgO contents of hydrous melts on H_2O contents. Observed values were recalculated on an anhydrous basis while calculated values were obtained from equation 1.

when recalculated on an anhydrous basis. For example, the water-saturated melt produced by Hirose (1997) at 1GPa contained 8.5 % CaO as opposed to 10.5% calculated and the H_2O-undersaturated melts of Hirose and Kawamoto (1995) and Hall (1999) were up to 1.3% lower in CaO than anticipated from equation (6).

In summary then, addition of H_2O to peridotite at constant pressure shifts product melts to lower MgO and CaO contents and to higher SiO_2 concentrations than those generated during anhydrous melting. If, on the other hand, we were to fix the MgO, CaO and SiO_2 contents of the melt and to trace the pressure–temperature conditions of melting, then these would shift to lower temperature and higher pressure as water is added. These observations provide a means of constraining the conditions of equilibrium of hydrous melts in the mantle wedge.

IDENTIFICATION OF PRIMITIVE MELTS

Since low pressure evolution of basaltic and intermediate liquids takes place through the precipitation of ferromagnesian silicates plus plagioclase, the principal means of identification of primitive liquids has been through the Mg# (Mg/[Mg+Fe^{2+}]). For liquids in equilibrium with the mantle, the simplest constraint, because of the simplicity of olivine composition, is given by the Mg-Fe^{2+} partition coefficient between olivine and silicate melt:

$$Kd^{ol\text{-}liq}_{Fe\text{-}Mg} = \frac{Fe^{2+}_{ol} \cdot Mg^{2+}_{liq}}{Fe^{2+}_{liq} \cdot Mg^{2+}_{ol}} \quad (7)$$

where Fe^{2+}_{ol} refers to the atomic fraction or weight fraction of Fe in the olivine.

Ulmer (1989) determined the partition coefficient for basaltic liquids coexisting with olivine at pressures up to 3GPa. In this range he found a linear dependence of the logarithm of $Kd^{ol\text{-}liq}_{Fe\text{-}Mg}$ on pressure:

$$\log\left(Kd^{ol\text{-}liq}_{Fe\text{-}Mg}\right) = -0.5214 + 0.0323P \quad (8)$$

where P is in GPa. This means, for example that, at 1.5 GPa, melts coexisting with olivine of Fo_{90} composition should, in principle, have an Mg# of about 0.75. In practise the apparent Mg#'s of melts coexisting with mantle olivines are lower than this because of the presence of Fe^{3+} in the melt. Ulmer's experiments were performed in graphite capsules under quite reducing conditions, well below the FMQ buffer and the Fe^{3+} contents of the product melts are also likely to be very low (*Ulmer*, 1989). Under more oxidising conditions the melts contain significant amounts of Fe^{3+} an ion which does not

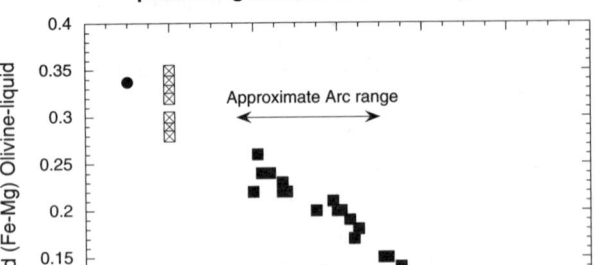

Figure 9. Measurements of apparent $Kd^{ol\text{-}liq}_{Fe\text{-}Mg} = \frac{(Fe/Mg)_{ol}}{(Fe/Mg)_{liq}}$ as a function of oxygen fugacity at 1.5 GPa. Data of Hall (1999) are from hydrous experiments and oxygen fugacity was estimated from the olivine-orthopyroxene-spinel barometer. Data of Robinson et. al. (1998) and Ulmer (1989) are for reduced conditions well below the FMQ buffer.

enter mantle olivines except in trace amounts. Thus, with increasing oxygen fugacity, the measured Mg/[Fe^{2+}+Fe^{3+}] of the melt decreases as Fe^{3+} is added to Fe^{2+} and $Kd^{ol\text{-}liq}_{Fe\text{-}Mg}$ appears to decrease.

Figure 9 shows the effect of oxygen fugacity on the apparent $Kd^{ol\text{-}liq}_{Fe\text{-}Mg}$ measured by Hall (1999). In this case the oxygen fugacity was calculated from the olivine-orthopyroxene-spinel barometer and $Kd^{ol\text{-}liq}_{Fe\text{-}Mg}$ was calculated by assuming that all Fe is Fe^{2+}. It is clear from the figure that increasing oxygen fugacity has the predicted effect on the apparent partition coefficient. If we assume that all Fe is Fe^{2+}, then Kd decreases with increasing oxygen fugacity because of the increasing concentration of Fe^{3+} in the melt. Alternatively, if we fix at the value given by Ulmer (1989) and calculate the amount of Fe^{3+} needed to give the observed (lower) value we obtain Fe^{3+} concentrations which are broadly consistent with the measured oxygen fugacities (Hall, 1999). Given the measured oxygen fugacity range for arcs (e.g Carmichael, 1991) apparent $Kd^{ol\text{-}liq}_{Fe\text{-}Mg}$ in the range 0.18–0.30 appear possible. If all Fe is reported as FeO (FeO*) then melts in equilibrium with Fo_{90} olivine could, in this oxygen fugacity range, have MgO/FeO* of 0.9–1.5 by weight. This provides us with an indication of which melts might reasonably be primary.

APPLICATION TO EQUILIBRATION OF WET MELTS WITH LHERZOLITE

A number of direct experimental studies have been made of the conditions under which specific primitive magmas would be in equilibrium with peridotitic mantle. In most such experiments (e.g *Tatsumi,* 1982; *Pichavant et. al.,* 2002) the hypothetical primary melt is held at fixed temperature, pressure and H_2O content and the identity of the liquidus phase or phases determined. By searching through pressure–temperature-H_2O space, the conditions where Mg-rich olivine, orthopyroxene and clinopyroxene are all present on the liquids can be determined to within 0.2 GPa and 20°C, in some cases.

Figure 10 shows a comparison between calculations performed using the procedures discussed above and experimentally-determined conditions of equilibrium between melts and an olivine-orthopyroxene-clinopyroxene residuum. The melts chosen for illustration differ widely in composition. For example, the basalt (STV 301) investigated by Pichavant et. al. (2002) contains 47% SiO_2 12.5% MgO and 2.7% (Na_2O+K_2O), whereas the andesite (SD-261) studied by Tatsumi (1982) contains 57% SiO_2 7.4%MgO and 5.2% total alkalis. Conditions of olivine-orthopyroxene-clinopyroxene saturation were calculated as follows: Pressure and temperature of equilibration under anhydrous conditions were estimated from equations 2 and 4 respectively. Addition of H_2O to the melt, at 3% intervals, then increases pressure by 0.06 GPa per 1% H_2O. Temperature was incremented with pressure using equation 4 and decreased using equation 5 which takes account of H_2O addition. As can be seen, calculated and observed conditions of equilibrium are in good agreement at the H_2O-contents of the melts used by the authors concerned, although my calculations appear to overestimate pressures and temperatures slightly. The final set of comparisons shown

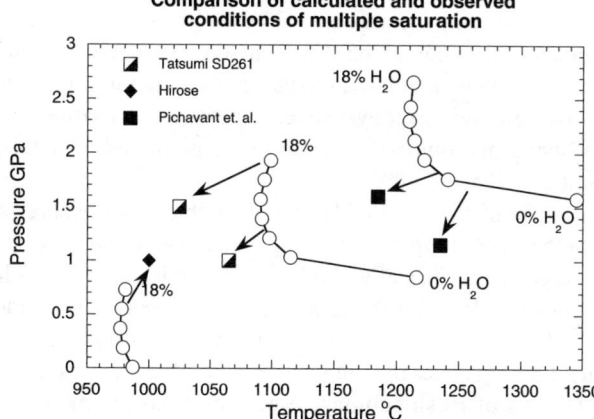

Figure 10. Conditions of equilibration of 3 melts, ranging from 47% SiO_2 12.5% MgO (highest temoerature) to 60.3% SiO_2 5.8% MgO (lowest) with lherzolite calculated for 0–18% H_2O in the melt at 3% intervals. Experimental results are connected to the calculations by arrows at the appropriate concentrations of H_2O in the melt. Amphibole is not reported to be a liquidus phase for any of the compositions studied.

in Figure 10 are for the H_2O-saturated melt generated by direct melting of KLB-1 peridotite at 1 GPa by Hirose (1997). On an anhydrous basis this melt contains 60.3% SiO_2, 5.8% MgO and 3.4% total alkalis. In this case the calculations slightly underestimate the pressure and temperature of equilibrium with peridotite, but, as with the more basic melts, agreement between observed and calculated equilibration conditions remains good.

The comparisons between calculated and observed equilibration conditions shown in Figure 10 indicate that the methodology outlined here provides reasonably accurate estimates of the conditions under which specific melts would, at fixed H_2O-content, equilibrate with the lherzolite assemblage.

As an example of the potential applications of the experimental data summarised in equations (1)–(6), I have used them to calculate the conditions under which primitive melts from the Cascades (*Bacon et. al.*,1997) would be in equilibrium with olivine, orthopyroxene and clinopyroxene.

Bacon et. al. (1997) identified three end-member primitive magma groups in the Cascades: high alumina olivine tholeiite (HAOT), arc basalt and basaltic andesite and intraplate basalt. In order to elucidate the physical conditions of melt generation in the mantle underlying the Cascade arc, I have calculated the apparent pressures and temperatures of equilibrium of primitive HAOT and basaltic andesite lavas with the olivine-orthopyroxene-clinopyroxene assemblage (Figure 11). Primitive melts were identified by Bacon et. al. (1997) on the basis of high Mg# ([Mg/ Mg+Fe^{2+}+Fe^{3+}]>0.60), high Mg content (8–10.5% MgO) and high Ni content (>140ppm). Following the measurements of Sisson and Layne (1993), the high alumina olivine tholeiites are assumed to be essentially anhydrous melts. In the absence of H_2O, calculated equilibration conditions form an olivine fractionation trend away from a point at about 0.9 GPa and 1275 °C, with the highest pressures and temperatures being given by lavas with the highest Mg#'s and Ni contents (0.67 and >200 ppm respectively). The CaO contents of the most primitive lavas (11–12 weight%) are consistent, at their total alkali content (~2.4weight%) with saturation in clinopyroxene at high pressures (equation 6) indicating an origin by partial melting of lherzolite. The results imply that parts of the sub-arc mantle is extremely hot at shallow depths (30 km) just beneath the continental crust. This conclusion is substantiated by the experiments of Baker et. al. (1994) who concluded that high alumina olivine tholeiites from Mt. Shasta would be in equilibrium with lherzolite at about 1.1 GPa and 1300°C.

Primitive basaltic andesite lavas (Mg# 0.67–0.71, Ni~160 ppm) from the Cascades also form a fractionation trend, in this case leading away from a point at about 1170°C and 1.4 GPa, with some variation due to assumed pre-eruptive H_2O content of the melt. Following measurements of melt inclusions

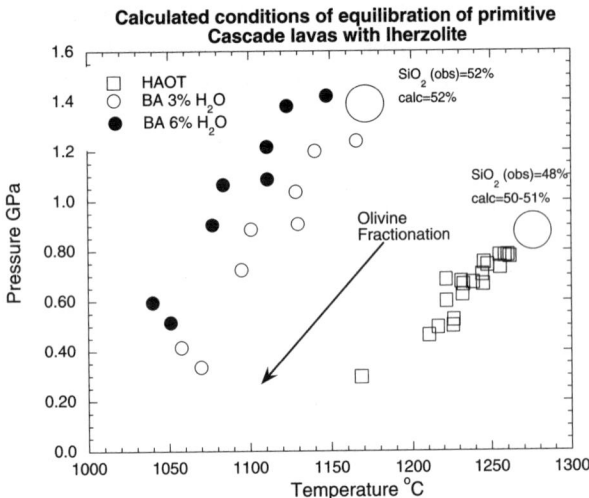

Figure 11. Calculated conditions of equlibration with lherzolite of dry high alumina olivine tholeiites (HAOT) and basaltic andesites containing 3–6% H_2O. All data, from Bacon et. al.(1997) are for Cascade volcanics.

by Sisson and Layne (1993) and phase equilibria studies of Baker et. al (1994), pre-eruptive contents of 3–6% H_2O have been assumed. Baker et. al. (1994) found equilibrium of these melts with olivine and orthopyroxene at about 1 GPa and 1200°C for similar H_2O contents to those used in the calculations. CaO contents of the lavas are slightly lower (0.2–0.5%) than those predicted by equation 6, consistent with clinopyroxene saturation at high pressures under hydrous, H_2O-undersaturated conditions.

The results summarised in Figure 11 indicate that primary basaltic andesite magma (~52% SiO_2) is in equilibrium with the mantle under the Cascades at about 45 km depth and temperatures of about 1170°C. Basaltic melts are extracted at higher temperatures and lower pressures, consistent with an inverted temperature–depth relationship, but with the zones of melt generation shifted to much shallower depths than would be expected from conductive thermal models of subduction zones (e.g *Turcotte and Schubert*, 1982). However, these thermal models take no account of thermal perturbations introduced by the advection of magma, which should be focussed towards the wedge corner at the junction of the upper plate and the subducting slab, as demonstrated by Spiegelman and McKenzie (1987). The latter authors showed that melts from all depths are drawn towards the wedge corner by a singularity in the piezometric pressure and that this effect expresses itself as a narrow region of arc magmatism. Temperatures close to the corner are raised by advection of heat and the melt extracted is a mixture of liquids initially generated at a range of depths. Despite their origins at greater depths, however, the erupted products appear, based on the calculations presented above, to have re-equilibrated with the mantle at depths just below the Moho.

CONCLUSIONS

Melting data on peridotitic bulk compositions under anhydrous conditions have been parameterised as a basis for understanding the effects of H_2O on the melting process. The anhydrous data, both in simple and in complex natural systems, show clear correlations of SiO_2 and MgO contents of the melt with pressure and with total alkali (Na_2O+K_2O) content of the melt under conditions where olivine, orthopyroxene and clinopyroxene are residual phases. When water is added under isothermal, isobaric conditions the degree of melting increases by about 0.1 for addition of 0.1% H_2O or 0.12–0.15 for addition of 0.2% H_2O. These measured effects are in good agreement with those estimated by Stolper and Newman (1994) for primitive basalts from the Mariana back-arc trough. As well as increasing the degree of melting, addition of H_2O shifts melt composition, on an anhydrous basis, to higher SiO_2 and lower MgO and CaO contents.

Using experimentally-measured degrees of melting and melt compositions from Hirose and Kawamoto (1997) (at 1 GPa) and Hall (1999) (1.5 GPa), I have made an approximate calibration of the effects of H_2O on melt composition, pressure and temperature. For a given melt the procedure involves recalculating the composition on an anhydrous basis and then calculating the apparent equilibrium pressure–temperature conditions from equations 2 and 4. Then, in the presence of H_2O the equilbrium pressure is shifted upwards by 0.06 GPa per 1% added to the melt. The depression of the equilibrium temperature under hydrous conditions is then calculated from equation5. Comparison with detailed melting studies on natural lavas shows that the conditions of multiphase (olivine, orthopyroxene, clinopyroxene) saturation can, using my approach, generally be predicted to within about 0.3 GPa and 40°C.

Because of the input of H_2O and soluble minor elements from the subducting slab, melting beneath arcs and in back-arc basins is intrinsically much more complex than beneath mid-ocean ridges. Furthermore, subaerial lavas are substantially degassed, making initial H_2O contents difficult to estimate. Recent studies of melt inclusions (e.g Sisson and Layne, 1993) and of fresh submarine glasses from the back-arc (Stolper and Newman, 1994) demonstrate, however, that pre-eruptive H_2O contents can be accurately determined in some cases. The principal questions now concern the relationships between the amount of water added, the degree of melting and the temperature–pressure conditions in the mantle wedge. Stolper and Newman (1994) studied the compositions of Mariana back-arc glasses and showed that it is possible to estimate the composition of the component added from the slab and the dependence of degree of melting on amount of added H_2O. This approach can now be extended by using concentrations of trace elements such as Ti as a measure of the degree of mantle melting F for known amounts of added H_2O (*Kelley et. al.*, 2003). Kelley et al. found that, after correction for crystal fractionation, the Mariana arc and back-arc melts form an array on a plot of H_2O versus F which intercepts at finite F for zero H_2O. This means that while most of the melting in both arc and back-arc is due to addition of H_2O, some is due to decompression melting of the mantle. Extension of this approach to other arcs will enable the relative importance of decompression and H_2O-addition to be generalised. The experimental data summarised here provide an important input to the process. It is a means to the quantification of the pressures and temperatures at which the melts are extracted from the mantle and hence to a quantification of the relationship between temperature, F and H_2O.

Application of the experimental data to primitive lavas from the Cascade arc demonstrates that they separated from the mantle at relatively shallow depths and high temperatures. When apparent pressures and temperatures of equilibration of (H_2O-free) high alumina olivine tholeiite lavas with lherzolite are calculated, the results show a fractionation trend away from a point at about 0.9 GPa and 1275°C. Hydrous basaltic andesite lavas appear to have equilibrated with the mantle at slightly higher pressures and lower temperatures, approximately 1.4 GPa and 1170°C respectively. Application of the methodology to other arcs will enable differences in subcrustal temperature structures and melting regimes to be elucidated. The remaining experimental challenges revolve around improving the empirical methodology outlined here and improving the ways in which primitive magmas are identified. The scatter shown in Figures 2–8 is partly due to small differences in peridotite bulk composition and partly to lack of equilibrium in some of the experiments plotted. These uncertainties need to be resolved in order to improve the calibration. Identification of primitive magmas is generally based on Mg# and content of compatible trace elements such as Ni. Evolved magmas are corrected back to 'primitive' compositions using experimentally measured crystal-liquid partition coefficients (*Kelley et. al.*, 2003). Generally the effects of H_2O and melt composition on the latter are poorly known and elucidation of these effects is an important and difficult challenge faced by experimentalists.

Acknowledgments. The unpublished research cited here was supported by the NERC (U.K.). I am extremely grateful for the many discussions with colleagues, particularly Steve Sparks, Jon Blundy and Chris Hawkesworth and for the review of Yoshiyuki Tatsumi.

REFERENCES

Bacon, C. R., P. E. Bruggman, R. L.Christiansen, M. A. Clynne, J. M. Donnelly-Nolan, and W.Hildreth, Primitive magmas at five Cascade volcanic fields: melts from hot, heterogeneous sub-arc mantle, *Can Mineral. 35*, 397–423, 1997.

Baker M. B., T. L.Grove and R. Price, Primitive basalts and andesites from the Mt. Shasta region, N. California: Products of varying melt fraction and H_2O content, *Contrib. Mineral. Petrol., 118*, 111–129, 1994.

Baker M. B., M. M. Hirschmann, M. S. Ghiorso, and E. Stolper, Compositions of near-solidus peridotite melts from experiments and thermodynamic calculations, *Nature, 375, 308–311*, 1995.

Baker M. B., and E. Stolper, Determining the composition of high-pressure mantle melts using diamond aggregates, *Geochim. Cosmochim. Acta , 58*, 2811–2827, 1994.

Burnham C. W., Water and magmas: A mixing model, *Geochim. Cosmochim. Acta 39,* 1077–1084, 1975.

Carmichael, I. S. E., The redox states of basic and silicic magmas- A reflection of their source regions, *Contrib. Mineral. Petrol. 106*, 129–141, 1991.

Carmichael, I. S. E. The andesite aqueduct: perspectives on the evolution of intermediate magmatism in west–central (105–99°W) Mexico, *Contrib. Mineral. Petrol 143*, 641–663, 2002.

Falloon T. J., D. H. Green, C. J. Hatton, and K. L. Harris, Anhydrous partial melting of a fertile and depleted peridotite from 2–30kb and application to basalt petrogenesis, *J. Petrol. , 29,* 1257–1282, 1988.

Falloon T. J., and L. V. Danyushevsky, Melting of refractory mantle at 1.5, 2 and 2.5 GPa under, anhydrous and H2O-undersaturated conditions: Implications for the petrogenesis of high-Ca boninites and the influence of subduction components on mantle melting, *J. Petrol. 41,* 257–283 2000.

Falloon T. J., D. H. Green, L.V . Danyushevsky, and U. H. Faul, Peridotite melting at 1.0 and 1.5 GPa: an experimental evaluation of techniques using diamond aggregates and mineral mixes for determination of near-solidus melts, *J. Petrol 40,* 1343–1375 1999.

Fujii T., and C. M. Scarfe, Composition of liquids coexisting with spinel lherzolite at 10kbar and the genesis of MORBs, *Contrib. Mineral. Petrol. , 90,* 18–28, 1985.

Gaetani G. A., and T. L. Grove, The influence of water on mantle melting, *Contrib. Mineral. Petrol., 131,* 323–345, 1998.

Green D. H., Experimental melting studies on a model upper mantle composition at high pressure under water-saturated and water-undersaturated conditions, *Earth Planet. Sci. Lett., 19,* 37–53, 1973 .

Hall, L. J. The effect of water on mantle melting, Ph.D. Thesis, University of Bristol, 159 pp, 1999.

Hirose K., Melting experiments on lherzolite KLB-1 under hydrous conditions and generation of high-magnesian andesitic melts, *Geology, 25,* 42–44, 1997.

Hirose K., and T. Kawamoto, Hydrous partial melting of lherzolite at 1GPa: The effect of H_2O on the genesis of basaltic magmas, *Earth Planet. Sci. Lett., 133,* 463–473, 1995.

Hirose K., and I. Kushiro, Partial melting of dry peridotites at high pressures: Determination of compositions using aggregates of diamond, *Earth Planet. Sci. Lett., 114,* 477–489, 1993.

Hirschmann M. M., M. B. Baker, and E. Stolper, The effect of alkalis on the silica content of mantle-derived melts, *Geochim. Cosmochim. Acta 62,* 883–902, 1998.

Hirth G., and D. L. Kohlstedt, Water in the oceanic upper mantle: Implications for rheology, melt extraction and the evolution of the lithosphere, *Earth Planet. Sci. Lett., 144,* 93–108, 1996 .

Kushiro I., Partial melting of a fertile mantle peridotite at high pressures: An experimental study using aggregates of diamond, in Basu A. S. & Hart S. (eds.), Earth Processes: Reading the Isotopic Code. *Amer. Geophys. Union, Geophys. Monogr. 95,* 109–122, 1996.

Kushiro I., Effect of water on the compositions of magmas formed at high pressures, *Journal of Petrology, 13,* 311–334, 1972.

Kushiro, I., Melting of hydrous upper mantle and possible generation of andesitic magma: an approach from synthetic systems, *Earth Planet. Sci. Lett., 22, 294–299,* 1974.

Kelley, K. A., T. Plank, S. Newman, E. Stolper, T. L. Grove, S. Parman, and E. Hauri, Mantle melting as a function of water content in arcs, *EOS, Fall Meeting Amer. Geophys. Union F1585,* 2003.

Langmuir, C. H., E. M. Klein, and T. Plank, Petrological systematics of Mid-Ocean Ridge Basalts:constraints on melt generation beneath ocean ridges, in Morgan, J. P., Blackman, D. K and Sinton, J. M. eds : Mantle flow and melt generation at Mid-Ocean Ridges, *Amer. Geophys. Union Geophys. Monogr. 71,* 183–280 1992.

Nicholls, I. A., and A. E. Ringwood, Effect of water on olivine stability in tholeiites and the production of silica saturated magmas in the island-arc environment, *Jour. Geol. 81,* 285–300, 1973.

Pichavant, M., B. O. Mysen, and R. Macdonald, Source and H_2O content of high-MgO magmas in island arc settings: An experimental study of a primitive calc-alkaline basalt from St. Vincent, Lesser Antilles arc, *Geochim. Cosmochim. Acta 66,* 2193–2209, 2002.

Robinson J. A. C., B. J. Wood, and J. D. Blundy, The beginning of melting of fertile and depleted peridotite at 1.5GPa, *Earth Planet. Sci. Lett., 155,* 97–111, 1998.

Sisson, T. W., and S. Bronto, Evidence for pressure-release melting beneath magmatic arcs from basalt at Galunggung, Indonesia, *Nature, 391,* 883–886, 1998.

Sisson, T. W and G. D. Layne, H_2O in basalt and basaltic andesite glass inclusions from 4 subduction-related volcano, *Earth Planet. Sci. Lett., 117,* 619–635, 1993.

Sobolev A. J., and M. Chaussidon, H_2O concentrations in primary melts from supra-subduction zones and mid-ocean ridges: implications for H_2O storage and recycling in the mantle, *Earth Planet. Sci. Lett., 137,* 45–55, 1996.

13–27.

Spiegelman, M., and D. McKenzie Simple 2-D models for melt extraction at mid-ocean ridges and island arcs, *Earth Planet. Sci. Lett., 83,* 137–152, 1987.

Stolper E., and S. Newman S. The role of water in the petrogenesis of Mariana trough magmas, *Earth Planet. Sci. Lett., 121,* 293–325, 1994.

Takahashi E., and I. Kushiro, Melting of a dry peridotite at high pressures and basalt magma genesis, *Amer. Mineral., 68,* 859–879, 1983.

Tatsumi Y., Origin of high-magnesian andesites in the Setouchi volcanic belt, southwest Japan, II, Melting experiments at high pressures, *Earth Planet. Sci. Lett., 60,* 305–317, 1982.

Turcotte D. L., and G. Schubert, Geodynamics: application of continuum physics to geological problems, 450 pp., John Wiley, New York 1982.

Tuttle, O. F., and N. L Bowen, Origin of granite in light of experimental studies in the system $NaAlSi_3O_8$ $KalSi_3O_8$-SiO_2-H_2O, *Geol. Soc. Amer. Memoir 74,* 1958

Ulmer P., The dependence of the Fe^{2+}-Mg cation partitioning between olivine and basaltic liquid on pressure, temperature and composition: An experimental study to 30kb, *Contrib. Mineral. Petrol., 101,* 261–273, 1989.

Walter M. J., and D. C. Presnall, Melting behaviour of simplified lherzolite in the system CaO-MgO-Al2O3-SiO2-Na2O from 7 to 35kb, *J. Petrol., 35,* 329–359, 1994.

Wood B. J., L. T. Bryndzia, and K. E. Johnson, Mantle oxidation state and its relationship to tectonic environment and fluid speciation, *Science, 248,* 337–345, 1990.

Bernard J. Wood, Department of Earth Sciences, University of Bristol, Bristol BS8 1RJ, U.K.

Geophysical Constraints on Slab Subduction and Arc Magmatism

Akira Hasegawa and Junichi Nakajima

Research Center for Prediction of Earthquakes and Volcanic Eruptions, Graduate School of Science, Tohoku University, Japan

Seismic tomography studies have revealed highly heterogeneous structures for the mantle wedge beneath several volcanic arcs. These structures are inclined seismic low-velocity and high-attenuation zones at depths shallower than ~150 km in the mantle wedge sub-parallel to the slab, which probably correspond to the upwelling-flow portion of subduction-induced convection. Seismic studies for NE Japan suggest that temperatures are higher than the wet solidus of peridotite and that melt inclusions with volume fractions of 0.1–1% exist within this upwelling flow. Aqueous fluids supplied from the underlying slab meet this hot upwelling flow at depths of 100–150 km and perhaps cause partial melting. This inclined low-velocity zone crosses the Moho at the volcanic front, suggesting that the location of the volcanic front is determined by the position of this hot upwelling flow. Observations of heat flow and seismic anisotropy also support the existence of the upwelling flow. Seismic tomography study of the mantle wedge of NE Japan has further revealed an along-arc variation of the inclined low-velocity zone: very low velocity regions periodically occur about every 80 km along the strike of the arc. Clustering of Quaternary volcanoes and topographic highs at the surface are located immediately above these very low-velocity areas in the mantle wedge, and low-frequency microearthquakes, perhaps caused by rapid movements of fluids in the lower crust, occur right above them also. These observations show the value of 3D modeling of arc magmatism.

1. INTRODUCTION

Cold and hence heavy oceanic lithosphere descends deep into the earth's mantle at subduction zones, causing the most active seismicity and volcanism on the Earth. Subduction zones are characterized geomorphologically by deep ocean trenches and island arcs or continental margins, seismically by landward dipping deep seismic zones, and magmatically by arcuate belts of volcanoes. Both subduction and arc magmatism are fundamental processes in the evolution of the Earth.

They play crucial roles in the present-day differentiation of earth's materials and are believed to be major sites of generation of the continental crust. Subduction is also significant in the water and carbon cycles.

In most subduction zones, the volcanic front is formed sub-parallel to the trench axis at the location below which the subducting slab reaches depths of 112 ± 19 km [*Tatsumi*, 1986]. The fundamental paradox of magmatism in subduction zones is high heat flow values and abundance of melt generation above the cold subducting plate. Most petrological models of subduction-related magmatism assign hydrous volatiles supplied from the descending slab to generate melting regions in the core of the hot asthenospheric wedge by lowering the solidus temperature [*e.g.*, *Tatsumi et al*, 1983; *Peacock*, 1996]. On the other hand, magmas that were derived from pressure-release (decompression) melting of hot mantle peridotite have been found in some subduction zones [*e.g., Sisson and Bronto*, 1998]. Essential details of metamorphic processes taking place in the slab and the overlying mantle wedge, and the mechanism of melt transport to the surface, are still uncertain.

In the last decade, researchers have investigated the deep structure of subduction zones and ongoing dynamics. Their observations have enhanced our knowledge of subduction zones. Thus in the present paper we discuss geophysical, particularly seismological, constraints on slab subduction and arc magmatism. We focus our discussions on the upper ~200 km depth of subduction zones, considering that the cycling of volatiles through the subduction system occurs mainly in this depth range.

2. INTERMEDIATE-DEPTH EARTHQUAKES AND SLAB STRUCTURE

Intermediate-depth earthquakes occur within subducted slabs at depths of 50–200 km. They are mostly located in the uppermost several kilometers of subducted slabs [*Abers*, 1992; *Kirby*, 1995], but they also lie in a secondary plane, 30–40 km below the slab surface. The double seismic zone (Plate 1) consists of two distinct seismic layers separated vertically by 30–40 km and has been found in many subduction zones [*e.g., Engdahl and Scholz*, 1977; *Hasegawa et al.*, 1978; *Kawakatsu*, 1986; *Abers*, 1992; *Gorvatov et al.*, 1994; *Kao and Liu*, 1995; *McGuire and Wiens*, 1995; *Comte et al.*, 1999]. Seismological studies have revealed the existence of a thin (~10 km) low-velocity layer at depths of 60–150 km in relatively cold slabs. This layer is typically 5–8% slower in seismic velocity than the surrounding mantle. In the case of young and hot slabs, this low-V layer persists only to depths of 50–60 km. The low-V layer lies at the top of the slab and is thought to correspond to the subducted oceanic crust [*e.g., Matsuzawa et al.*, 1986; *Cassidy and Ellis*, 1993; *Abers*, 2000]. The subducted crust is generally interpreted as either hydrous blueschist or as basalt and gabbro persisting metastably within the eclogite stability field [*e.g., Helffrich*, 1996]. The upper plane seismicity of the double seismic zones (or intermediate-depth seismicity of a single seismic zone) perhaps lies in this subducted oceanic crust and therefore the lower plane seismicity of the double seismic zone occurs in the middle of the subducted oceanic mantle.

The stress state of the slab is generally characterized by down-dip compression and down-dip extension along the upper and lower seismic planes, respectively, although there are some exceptions such as the Aleutians where down-dip extension is dominant in both planes [*Abers*, 1992]. Several models, such as unbending of the slab [*Engdahl and Scholz*, 1977], have been proposed as a possible mechanism of the observed stress state, but they are not sufficient to generate the double seismic zone. At depths greater than > 50 km, lithostatic pressure becomes too high for any brittle fracturing or frictional slip to be possible [*e.g., Frohlich*, 1994; *Kirby*, 1995]. Some mechanisms that reduce the strength of rocks are required for the occurrence of intermediate-depth earthquakes. Dehydration embrittlement is a possible mechanism for reducing the strength of rocks and causing intermediate-depth earthquakes. Fluid released by dehydration of hydrous minerals could induce *in situ* mechanical instability and brittle deformation at the time of the dehydration reaction [*Raleigh and Paterson*, 1965].

The upper plane earthquakes of the double seismic zones (or intermediate-depth earthquakes of the single seismic zone) occurring in the subducted oceanic crust are probably caused by dehydration embrittlement associated with the transformation of metabasalt and metagabbro to eclogite [*Kirby et al.*, 1996]. In NE Japan, there exists a thin seismic plane in the uppermost part of the subducted oceanic crust; thus forming a triple seismic zone instead of the double seismic zone [*Igarashi et al.*, 2001]. Events in this uppermost plane have P axes normal to the down-dip direction of the slab, while events below them in the upper plane are characterized by down-dip compression. Although the cause of this stress state has not yet been fully understood, densification of the uppermost portion of the crust by phase transformation may explain it. The phase transformation would gradually occur from the top since the subducted crust is warmed from the upper surface, resulting in locally denser uppermost crust. This mechanism can cause the uppermost crust to be in tension and the remaining part of the crust below it in compression. Alternatively, *Hacker et al.* [2003] suggested that the densification reactions in the upper crust at shallower levels might be driven by higher H_2O content and finer grain size than those of the lower crust; these ideas might also explain the stress state inferred from seismic observations.

The lower plane seismicity of the double seismic zones could be caused by serpentine dehydration in the subducting oceanic mantle [*Seno and Yamanaka*, 1996; *Peacock*, 2001]. *Yamasaki and Seno* [2003] determined the loci for dehydration of serpentinized oceanic mantle for six typical subduction zones using an experimentally derived phase diagram. An example for NE Japan is shown in Plate 1. They found that the lower plane seismicity is located at these dehydration loci of serpentine, which may explain why the lower plane events occur on a plane in the middle of the subducted mantle. If this is the cause of lower plane seismicity, the oceanic mantle must be hydrated prior to its subduction. Serpentinization may occur in the trench-outer rise region, where faulting causes infiltration of seawater to depths of several tens of kilometers into the oceanic plate [*Peacock*, 2001].

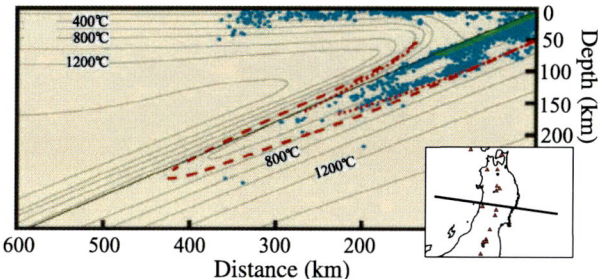

Plate 1. Vertical cross-section of earthquakes, thermal structure and dehydration loci of crust and mantle along a line in the inserted map of NE Japan [*Yamasaki and Seno*, 2003]. Dehydration loci of metamorphosed crust are shown by green lines, and those of serpentinized mantle are shown by red dashed lines or red dotted lines.

Aqueous fluids released from the subducted slab by dehydration reactions can move via at least three types of flow paths; percolation through locally, high permeability zones, flow through cracks caused by the local stress state, and post-seismic flow through fault zones [*Hacker et al.*, 2003]. Aqueous fluids expelled from the subducting slab may react with overlying mantle materials and form hydrous minerals such as serpentine, chlorite, and amphibole in the mantle wedge immediately above the slab [*e.g., Davies and Stevenson*, 1992; *Iwamori*, 1998]. Fluids hosted by hydrous minerals are then dragged down, perhaps to depths of 150–200 km, for old subduction zones such as NE Japan [*Iwamori*, 1998]. Breakdown of serpentine and chlorite at these depths could result in the formation of fluid columns through which aqueous fluids are transported upwards [*e.g., Iwamori*, 1998; *Schmit and Poli*, 1998]. This suggests that a relatively large amount of aqueous fluids can be released to the mantle wedge at depths of 150–200 km for NE Japan, although we do not yet have direct evidence from the observations. For subduction of young and hot slabs, the breakdown of serpentine and chlorite occurs at shallower depths, which would cause the formation of the hydrous columns at shallower levels.

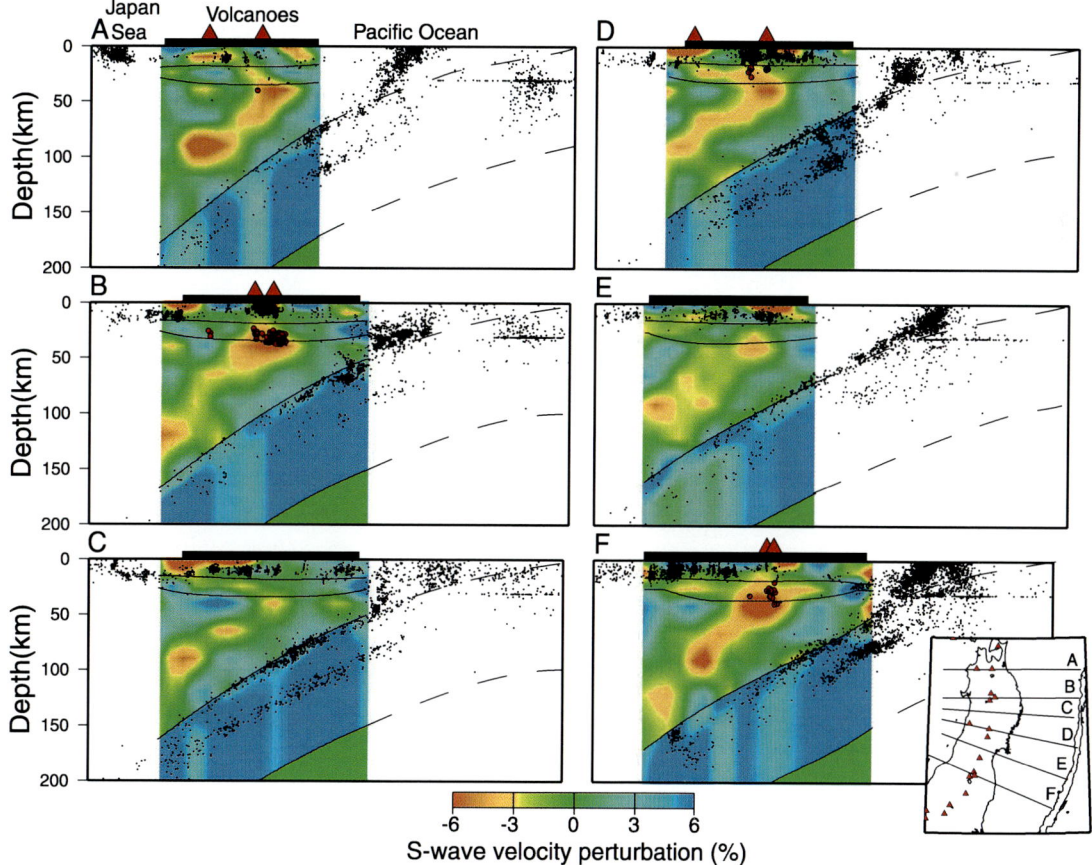

Plate 2. Vertical cross-sections of S-wave velocity perturbations along lines in the inserted map of NE Japan [*Nakajima et al.*, 2001a]. Red and blue colors represent low and high velocities, respectively, according to the scale at the bottom. Red triangles and bars on the top represent active volcanoes and the land area, respectively. Dots and red circles show earthquakes and deep, low-frequency microearthquakes, respectively.

84 SLAB SUBDUCTION AND ARC MAGMATISM

3. MANTLE WEDGE STRUCTURE

3.1 Seismic Velocity Structure

Numerous studies on three-dimensional seismic velocity structure for local and regional scales have been conducted in various subduction zones by seismic tomography, which provided important clues for understanding slab subduction and arc magmatism. One of the most detailed studies was conducted on the NE Japan subduction zone. *Nakajima et al.* [2001a] imaged inclined low-V zones for both P- and S-waves in the mantle wedge of NE Japan down to ~120 km depth, which are distributed sub-parallel to the down-dip direction of the subducting slab (Plate 2). They are observed in the cross-sections with active volcanoes and also in those without active volcanoes. These findings indicate the existence of continuous inclined low-V zones in the mantle wedge, having the form of a single inclined sheet. Plate 3 shows velocity images of vertical cross-sections beneath Kamchatka [*Gorbatov et al.*, 1999]. Inclined low-V zones sub-parallel to the subducted slab are also identified in the mantle wedge, although they are not so clear especially in Plate 3(a) because of lower resolution. Similar inclined low-velocity zones have been imaged in the mantle wedge of Tonga [*Zhao et al.*, 1997], the eastern Aleutians [*Abers*, 1994] and Alaska [*Zhao et al.*, 1995].

The subduction of the oceanic plate generates mechanically-induced secondary convection in the overlying mantle wedge [*e.g., McKenzie*, 1969]. Mantle materials near the slab are forced to flow downward. This flow then induces advection of mantle materials toward the corner of the mantle wedge. Since temperature in the mantle wedge increases with depth, this flow causes upward advection of high-temperature and low-viscosity mantle materials toward the corner of the mantle wedge. Velocity fields of upward flows in the mantle wedge are modeled to be sub-parallel to the dip of the subducting slab [*e.g., Furukawa*, 1993a; *Eberle et al.*, 2002]. Inclined low-V zones detected in the mantle wedge perhaps correspond to this upward flow of hot mantle materials.

The inclined low-V zones imaged by *Nakajima et al.* [2001a] are confined to depths shallower than ~120 km, due to source and receiver geometry used in their inversion. *Zhao and Hasegawa* [1993] and *Zhao et al.* [1994] obtained the velocity structure beneath the Japan Islands using travel time data from regional and teleseismic sources and showed no prominent inclined low-V zones in the mantle wedge of NE Japan at depths deeper than ~150 km. When large amounts of aqueous fluids are released by the dehydration reactions of serpentine and chlorite immediately above the slab at depths of 150–200 km, they may migrate upward in the mantle wedge due to buoyancy and meet the upwelling flows at around a depth of 100–150 km. Addition of aqueous fluids to hot

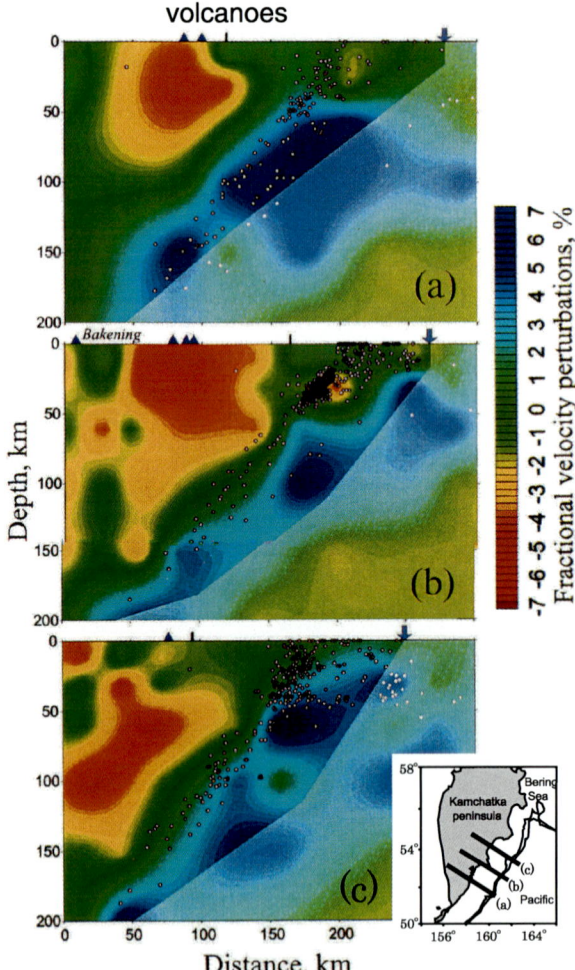

Plate 3. (a) Vertical cross-sections of P-wave velocity perturbations along lines in the inserted map of Kamchatka [*Gorbatov et al.*, 1999]. Red and blue colors represent low and high velocities, respectively, according to the scale on the right. Solid triangles show volcanoes and circles are earthquakes. Vertical black bars mark the coastline and arrows indicate the trench.

upwelling mantle flows would lower the solidus. Temperatures within the low-V zones estimated from seismic attenuation structure and laboratory experiment data are higher than the wet solidus of peridotite [*Nakajima and Hasegawa*, 2003a; *Wood*, this volume]. Observed values of dlnVs/dlnVp (velocity decrease ratio between P- and S-waves) are consistent with the existence of melt inclusions with aspect ratios of 0.01–0.1 and volume fractions of 0.1–1 % in these low-V zones [*Nakajima and Hasegawa*, 2003c]. These observations are thus consistent with the idea that the addition of aqueous fluids causes partial melting in the low-V zones, reducing the seismic velocity. Decompression effect also may contribute to the partial melting in the low-V zones [*e.g., Conder et al.*, 2002]. Such mechanisms can explain the velocity structure obtained in the mantle wedge of NE Japan, with prominent low-V zones being confined to depths shallower than ~150 km.

The confinement of low-V zones at depths shallower than ~150 km in the mantle wedge agree with the prediction of numerical simulations of mechanically-induced secondary convection in subduction zones. Plate 4(b) shows P-wave velocity anomalies predicted by a numerical simulation of the subduction–induced convection flow with temperature-dependent viscosity [*Eberle et al.*, 2002]. Inclined low-V zones sub-parallel to the slab are simulated at depths shallower than ~130 km due to the thermal effect caused by the upwelling flow, consistent with the observations.

Inferred transportation paths of aqueous fluids for NE Japan are schematically shown in Plate 5. Both the upwelling flow of hot mantle materials and the addition of aqueous fluids are postulated to cause partial melting in the inclined low-V zones. Aqueous fluids are absorbed into the melt and transported upward along the upwelling flow. This upwelling flow finally meets the Moho beneath the volcanic front. Here melt separates and distributes below the Moho along the volcanic front. Magma then migrates upward into the lower crust because of its buoyancy, where fractional crystallization and chemical reactions with the surrounding crustal rocks can occur. This portion of the crust can be imaged as seismic low-V zones [*Nakajima et al.*, 2001b]. Further upward migration of magma and repeated discharge to the surface forms the chain of volcanoes. This model envisages the location of the volcanic front as the place where the upwelling flow meets the Moho.

In the case of young and hot slabs, the breakdown of serpentine and chlorite is expected to occur at shallower depths; hydrous columns would be expected to form at shallower levels, closer to the trench. If the postulated columns do not meet the upwelling flow portion, partial melting may not occur, resulting in feeble or absent arc volcanism [*Kirby et al.*, 1996]. Slab geometry may also affect the arc volcanism, since the upwelling flow in the mantle wedge would be formed sub-parallel to the subducting slab. There exist volcano gaps in the Andes; no active volcanoes between 8°S and 13°S, and between 28°S and 33°S, where the Nazca plate subducts with ~30° dip down to ~100 km depth and then becomes nearly horizontal with a dip of only ~5° [*e.g., Hasegawa and Sacks*, 1981; *Cahill and Isacks*, 1992]. The horizontal slab probably inhibits the upwelling mantle flow induced by subduction, resulting in no active volcanoes above it.

We infer that the upward flows, mechanically induced by the slab subduction and imaged seismically as low-V zones, are common in all subduction zones. Absence of the inclined low-V zones in the mantle wedge for some subduction zones with arc volcanism is, therefore, considered to be due to insufficient

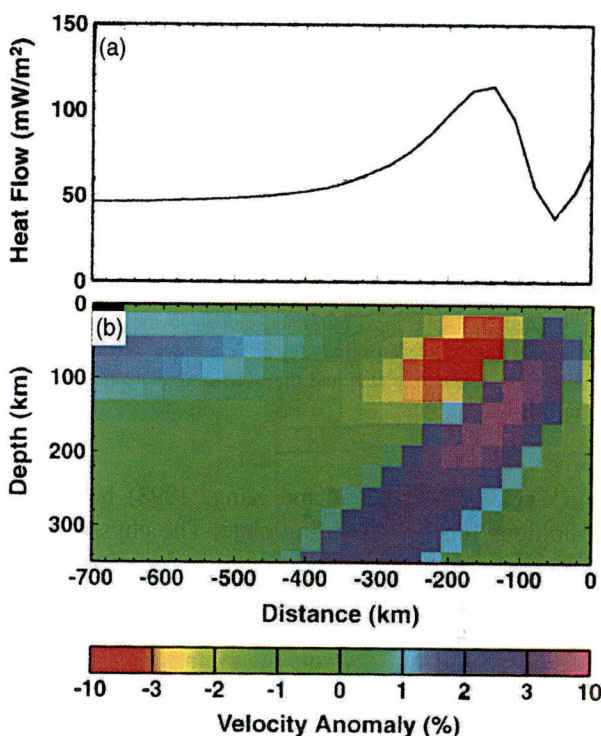

Plate 4. (a) Across-arc variation of surface heat flow and (b) vertical cross-section of P-wave velocity anomalies calculated from the thermal structure obtained by a numerical simulation of mechanically-induced flow in subduction zone [*Eberle et al.*, 2002]. P-wave velocity anomalies are shown as percentage deviations from PREM.

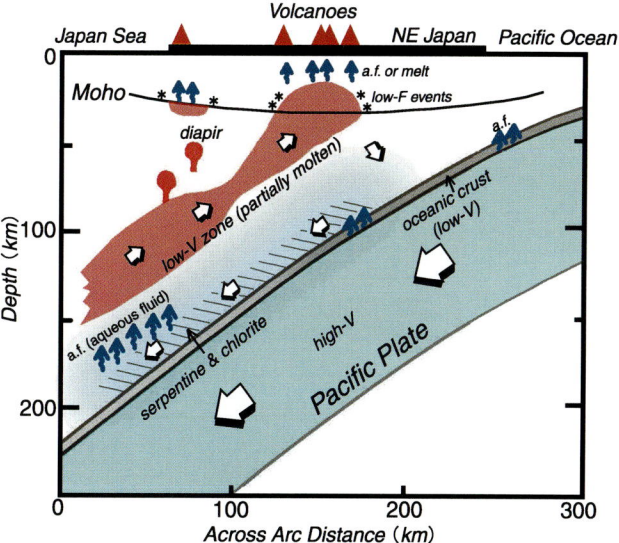

Plate 5. Schematic diagram of vertical cross-section of the crust and upper mantle beneath NE Japan showing inferred transportation paths of aqueous fluids.

resolution to image the seismic velocity anomalies in the back-arc deeper regions.

3.2 Seismic Attenuation Structure

Seismic attenuation structure provides additional information on the physical states of the earth's interior. Much of our current understanding of the upper mantle attenuation structure is based on inversions of long-period seismic data. However, recent high quality data also enable us to estimate the attenuation structure of subduction zones on local scales.

Tsumura et al. [2000] estimated three-dimensional P-wave attenuation structure beneath NE Japan by a joint inversion for source parameters, site response, and Qp values. Results are shown in Plate 6. Inclined high-attenuation (low-Qp) zones are distributed continuously in the mantle wedge, sub-parallel to the subducting slab. The high-attenuation zones are consistent with the inclined low-V zones in Plate 2. Thus the upwelling flow induced by the slab subduction is imaged as the inclined low-V and high-attenuation zones in the mantle wedge. Similar inclined high-attenuation zones have been revealed in the mantle wedge of Kanto-Tokai, central Japan [*Sekiguchi*, 1991], North Island, New Zealand [*Satake and Hashida*, 1989], Tonga-Fiji [*Roth et al.*, 1999], and northern Chile [*Haberland and Rietbrock*, 2001]. High-attenuation zones detected in the mantle wedge of Tonga-Fiji correspond to the low-V zones [*Zhao et al.*, 1997] as well.

Spatial variation of temperature in the mantle wedge can be constrained by seismic attenuation data, since the presence of melts have little effect on seismic attenuation [*e.g., Sato et al.*, 1989]. Seismic attenuation anomalies are primarily caused by thermal anomalies [*e.g., Roth et al.*, 2000]. *Nakajima and Hasegawa* [2003a] estimated the thermal structure in the wedge by applying experimental studies of olivine-dominated rocks [*e.g., Karato and Spetzler*, 1990; *Jackson et al.*, 1992] to the seismic attenuation structure reported by *Tsumura et al.* [2000]. The results suggest that the temperature in the mantle wedge is 1000–1200 °C. Inclined high temperature zones corresponding to high-attenuation zones can be seen in the mantle wedge, which is nearly agreement with the temperature distribution predicted by numerical simulations of the subduction-induced convection flow [*e.g., Eberle et al.*, 2002]. Temperatures in the high-attenuation zones of the mantle wedge beneath NE Japan are higher than the wet solidus of peridotite [*e.g., Kushiro et al.*, 1968; *Wood*, this volume] but lower than the dry solidus of peridotite [*e.g., Iwamori*, 1998], suggesting that the addition of aqueous fluids supplied from the subducting slab is required to generate arc magmas. Thus aqueous fluids expelled from the slab probably cause partial melting in the upwelling flow.

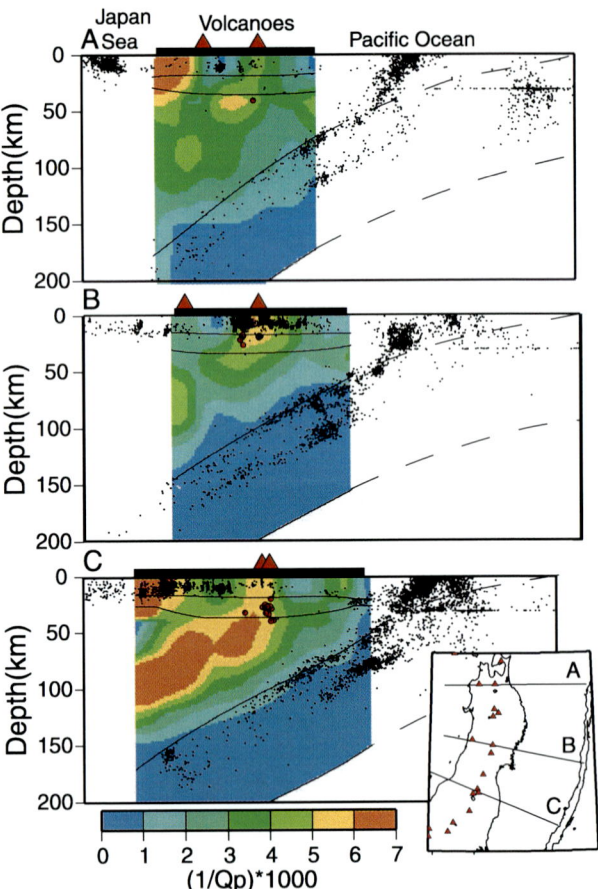

Plate 6. Vertical cross-sections of P-wave attenuation structure along lines in the inserted map of NE Japan [*Tsumura et al.*, 2000]. Red and blue colors represent high and low attenuations, respectively, according to the scale at the bottom. Other symbols are the same as in Plate 2.

3.3 Seismic Anisotropy Structure

Studies on seismic anisotropy structure have provided important information associated with tectonic deformation and dynamic processes of the mantle. Many researchers have investigated shear-wave polarization anisotropy in various subduction zones, and have discussed flow patterns in the mantle wedge [see *Savage*, 1999 and references therein].

We investigated shear-wave polarization anisotropy in the mantle wedge of NE Japan using waveform data of intermediate-depth earthquakes recorded at more than 70 stations above them. The direction of the fast shear-wave and averaged delay time between the leading and following shear-waves are plotted at each station, and are shown by oriented segments in Figure 1. The results clearly show a systematic spatial variation in directions of fast shear-waves. The fast directions in the back-arc side are nearly E-W, sub-parallel to the direction of relative plate motion. Most of stations with such trench-perpendicular fast directions seem to be located above the inclined low-V zones in the mantle wedge, corresponding to the hot upwelling flow. Lattice preferred orientation of minerals caused by flow-induced strain is a most likely candidate for the observed trench-perpendicular fast directions. This interpretation is consistent with numerical studies on lattice preferred orientation development in polycrystalline olivine or peridotite aggregates [e.g., Ribe, 1992; Tommasi, 1998] and with experimental results [e.g., Zhang and Karato, 1995].

On the other hand, it seems that N-S trending fast directions are not related directly to the upwelling flow. Perhaps other mechanisms are working to cause the trench-parallel fast directions in the fore-arc mantle wedge. The difference between the fore-arc and back-arc mantle wedge is also observed in seismic velocity and attenuation structures. The fore-arc side does not exhibit low velocity and high attenuation (Plates 2 and 6), suggesting that the upwelling flow is confined to the back-arc side and does not extend into the fore-arc mantle wedge beneath NE Japan.

Similar patterns of shear-wave polarization anisotropy to that of NE Japan have been observed in Tonga [Fischer and Wiens, 1996; Smith et al., 2001] and northern Chile [Polet et al., 2000]. Fast directions of shear waves obtained in the northern Chile subduction zone show trench-parallel directions in the fore-arc side and trench-perpendicular directions in the back-arc side. Fast directions of shear-waves obtained for the Tonga subduction zone show a slightly complex pattern. Smith et al. [2001] inferred that the complex pattern of shear-wave polarization anisotropy beneath Tonga is caused by the complex pattern of mantle flows. On the other hand, trench-parallel fast directions have been observed, for example, in the eastern Aleutians [Yang et al., 1995] and beneath Kamchatka [Peyton et al., 2001]. Numerical simulations by Hall et al. [2000] suggest that flow models incorporating plate motion boundary conditions produce splitting parameters consistent with trench-parallel fast directions.

4. DEEP, LOW-FREQUENCY TREMORS AND MICROEARTHQUAKES

The recent development of dense regional seismic networks has led to the detection of deep, low-frequency tremors just above the subducting slab. Obara [2002] detected non-volcanic deep tremors occurring above the slab surface of 35–45 km depth in southwestern Japan. Their focal depths are ~30 km, and they are distributed along the strike of the subducting Philippine Sea Plate over a length of ~600 km. He showed that these deep tremors are sometimes triggered by nearby relatively large earthquakes, and that the tremors do not always remain in one region in an episode but sometimes migrate at a velocity of ~10 km/day. These observations suggest that the tremors are caused by fluids supplied from the dehydration process of the subducting slab.

Deep, low-frequency tremors are often accompanied by impulsive body waves with a predominant frequency of about 1 to 2 Hz, which are similar to S wave parts of deep low-frequency microearthquakes detected at many locations in the Japan Islands [e.g., Katsumata and Kamaya, 2003]. This suggests that the low-frequency tremor is a continuous sequence of low-frequency microearthquakes [Obara, 2002]. Low-frequency microearthquakes in NE Japan occur in and around seismic low-velocity zones in the uppermost mantle (Plate 2), with implications that they are caused by deep magmatic activity [Hasegawa et al., 1991; Hasegawa and Yamamoto, 1994]. Recent investigations using seismic tomography [Nakajima and Hasegawa, 2003b; Nakajima et al., 2001b] show that these deep, low-frequency events are located just above low-Vp, low-Vs, and high-Vp/Vs areas as shown in Plate 7. Partially molten materials conveyed through the inclined upwelling flow reach the Moho beneath the volcanic front. Such materials accumulate and would spread below the Moho along the volcanic front since the buoyancy of ascending magmas should be decrease at the Moho. Seismic tomography studies imaged them as low-Vp, low-Vs and high Vp/Vs zones [Nakajima et al., 2001a, b]. Some of them intrude into the lower crust beneath active volcanoes (Plate 7). The tops of these partially molten zones are cooled by the shallow man-

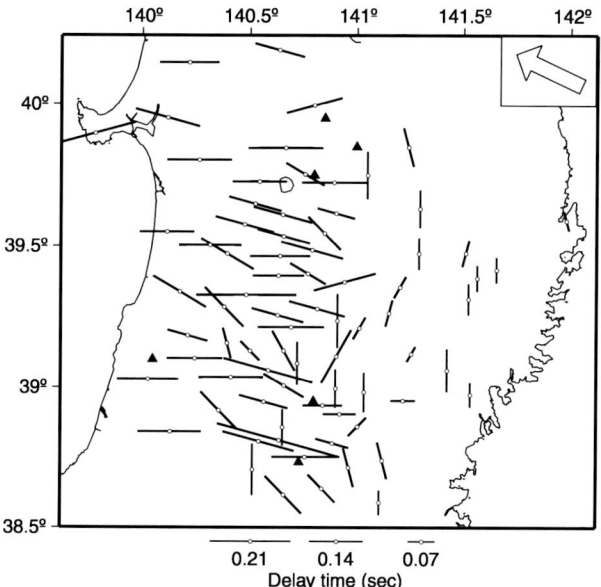

Figure 1. Direction of the fast shear-waves and delay time plotted at stations. Black lines show the direction of the fast shear-wave and the length is proportional to average delay times between the leading and following shear-waves. A white arrow shown at the upper right denotes the direction of relative plate motion. Solid triangles show active volcanoes.

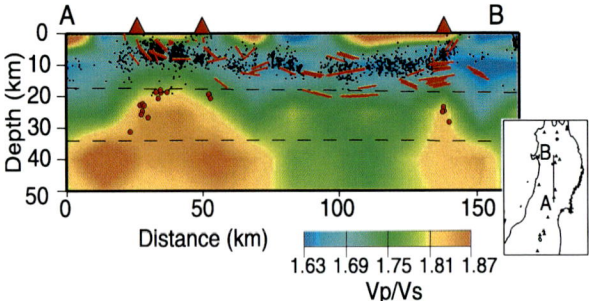

Plate 7. NS vertical cross-section of Vp/Vs structure in NE Japan along a line in the inserted map [*Nakajima et al.*, 2001b]. Red and blue colors represent high and low Vp/Vs, respectively, according to the scale at the bottom. Red circles and dots show low-frequency microearthquakes and shallow earthquakes, respectively. Red lines show S-wave reflectors (bright spots). Red triangles on the top denote active volcanoes.

tle and lower crust and these partial melts probably differentiate and crystallize, supplying water to the crust immediately above. We infer that deep, low-frequency microearthquakes are caused by water thus released by solidification of the partially molten materials. Occurrence of deep, low-frequency microearthquakes perhaps shows the ascent of hot mantle materials there, and suggests a series of upward migrations of fluids from the slab to the crust.

5. HEAT FLOW AND OTHER OBSERVATIONS

In the previous sections, we have focused on the fluid migration inferred from seismic observations. Interpretation of heat flow data provides a means of determining the thermal structure of the crust and mantle wedge. Figure 2 shows observed across-arc variations of heat flow for NE Japan [*Furukawa*, 1993b] and for the central Andes [*Springer and Förster*, 1998]. The overall patterns of the observed surface heat flow are similar to each other; it is very low near the trench, steeply increases and reaches the maximum near the volcanic front, then gradually decreases toward the back-arc side. Numerical studies on induced flows in the mantle wedge predict a similar across-arc variation of the surface heat flow. The heat flow pattern calculated by *Eberle et al.* [2002] is shown in Plate 4 (a). The pattern of the predicted surface heat flow is in substantial agreement with those of observations (Figure 2), thus supporting the induced-flow model with temperature-dependent viscosity.

The geoid is an equi-potential surface of the earth gravity field. Geoid anomalies can reflect the density anomalies that drive mantle convection and plate motion. Geoid variation could also provide constraints on the viscosity structure in the earth [*e.g., Hager*, 1984], one of the most important factors affecting the dynamics in subduction zones but is poorly known. *Billen and Gurnis* [2001] showed, using dynamic models of Stokes flow, that a low viscosity mantle wedge has a dramatic influence on the force balance in subduction zones and leads to an observable signal in the topography, gravity and geoid. Their regional dynamic model for the Tonga-Kermadic subduction zone shows that the viscosity of the mantle wedge is at least a factor of 10 lower than the surrounding mantle, and the existence of a low-viscosity wedge creates a stronger component of horizontal flow entering the wedge and upward flow in the wedge. This result also supports an induced flow model with inclined low-viscosity zones.

6. ALONG-ARC VARIATION OF ARC MAGMATISM

Some observers suggest that volcanoes in active plate margins are regularly spaced along the strike of the arc [*e.g., Marsh*, 1979], and this pattern has been used to derive constraints on the characteristics of deep sources of magma and on the mechanisms of magma ascent. However, *Bremond d'Ars et al.* [1995] investigated the distribution of volcanoes in 16 active plate margins and concluded that volcanoes are randomly distributed along the arc.

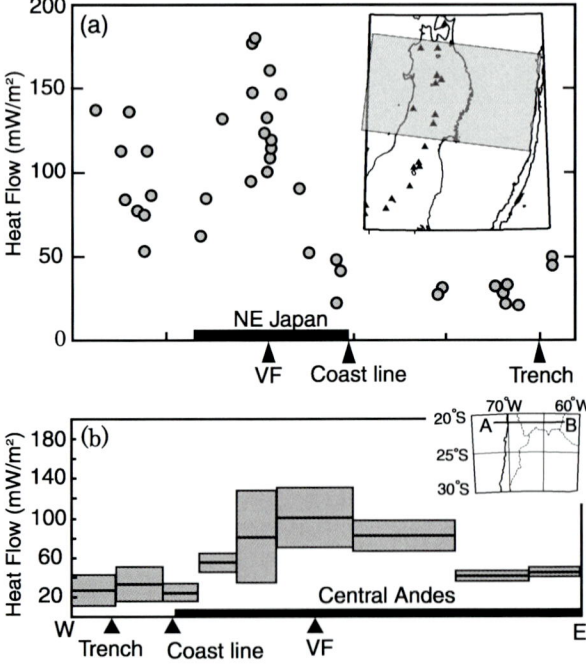

Figure 2. Across-arc variation of surface heat flow in (a) NE Japan [*Furukawa*, 1993b] and (b) central Andes along 21°S [*Springer and Förster*, 1998]. (a) Heat flow data in the hatched area in the inserted map are shown. (b) Mean heat flow values (black lines) with standard deviations (shaded ranges) are shown based on data from the area between 15°S and 30°S. VF: Volcanic front.

Recently, *Tamura et al.* [2002] investigated the spatial distribution of Quaternary volcanoes in the NE Japan arc. They confirmed randomly distributed volcanoes in this arc as suggested by *Bremond d'Ars et al.* [1995], but found additional clear patterns. They found that Quaternary volcanoes are grouped into 10 volcanic clusters striking transverse to the arc. They have an average width of 50 km, and are separated along arc by 30–75 km wide volcanic gaps. This clustering of volcanic centers is closely correlated with topographic profiles, low-velocity zones below the volcanic front, and local negative Bouguer gravity anomalies along the Japan Sea side of this volcanic arc. They proposed that these observations are surface manifestations of locally developed hot regions within the mantle wedge, having the form of inclined, ~50 km wide fingers extending from >150 km depth below the back-arc region to ~50 km depth below the volcanic front. The volcanic basement right above each finger has been uplifted by repeated injections of magmas into the crust, accompanied by Quaternary volcanic activity at the surface. The negative Bouguer anomalies below the rear of the arc could be caused by magmas supplied from the hot fingers and accumulated at the Moho.

To test this interpretation by *Tamura et al.* [2002], we conducted a specialized tomographic inversion of S-wave velocity structure of the mantle wedge beneath NE Japan. S-wave velocities of the mantle wedge portion alone were estimated by using the same data set of our previous study [*Nakajima et al.*, 2001a], but with those of the other portions (crust and slab) being fixed to the values obtained by the previous study. The images we thus obtained have better spatial resolution (~10 km) in both the horizontal and vertical directions, except in the southernmost area as explained below. Plate 8 (a) shows S-wave velocity perturbations along the core of the inclined low-V zones in the mantle wedge. We obtained this image by taking velocity perturbations along the slowest S-wave velocity portion of the mantle wedge. As predicted by *Tamura et al.* [2002], velocity decrease in the low-V zone has an along-arc variation and is locally large below the areas where Quaternary volcanoes are clustered. In other words, the hot mantle fingers, except the poorly-resolved southernmost one, have extremely low S-wave velocities as *Tamura et al.* [2002] predicted. Deep, low-frequency microearthquakes, shown by white circles, are also clustered above such localized low velocity areas. Spatial resolution in the back-arc side of southern NE Japan is locally low because of the sparse distribution of stations there, which caused the unclear image for the southernmost finger.

Comparison with the topography map (Plate 8b) shows that very low velocity areas within the inclined low-V zones of the mantle wedge, low-frequency microearthquakes in the lowermost crust, topographic highs, and Quaternary volcanoes at the surface in general closely correlate in space with each other. Low-frequency events occur immediately above the very low velocity areas, and Quaternary volcanoes and elevated basement are also built immediately above such areas.

Plate 9 schematically shows inferred mantle wedge structure and Quaternary volcanoes based on these observations. Low-V zones presently imaged in the mantle wedge show that they have the form of a single inclined sheet with varying thickness rather than the form of inclined and isolated fingers proposed by *Tamura et al.* [2002]. The volcanic front is formed at the surface where the upwelling flow meets the Moho. When materials in the upwelling flow reach the Moho, they evidently slow their ascent and spread along the volcanic front due to the decrease in density contrast and buoyancy. Some of them migrate upward and intrude into the lower crust beneath the volcanic areas, which are imaged as low-Vp, low-Vs, and high Vp/Vs areas in the lower crust (Plate 7). Thus magmas are supplied to active volcanoes at the surface along the volcanic front. These are consistent with petrological constraints that primary magmas supplied to volcanoes along the volcanic front are originally generated at pressures of ~1.1 GPa [*e.g., Tatsumi et al.*, 1983].

Clustering of volcanoes and low-frequency events, and the elevated basement toward the back-arc side that are distributed right above the very low velocity portions (or locally thick portions) of the single inclined sheet of the upwelling flow suggest that melts segregate from these portions of the upwelling flow and rise vertically into the overlying mantle wedge due to buoyancy (Plate 5). When melts reach the Moho, they would accumulate there. Then some of them intrude into the crust, accompanied by differentiation and crystallization. Such upward fluid migration would trigger deep low-frequency microearthquakes. Further upward migration of magmas forms volcanoes in the back-arc side. Repeated intrusions of such magmas, which have taken place simultaneously or just prior to the Quaternary volcanic activity, have uplifted the basement of these volcanoes [*Tamura et al.*, 2002]. This satisfies petrological constraints that magmas beneath the back-arc volcanoes are generated at deeper depths than those beneath the volcanic front [*e.g., Tatsumi et al.*, 1983].

We infer that the gaps of Quaternary volcanoes at the surface are due to insufficient melt contents in the upwelling flow for the segregation (Plate 9). This suggests that spacing of Quaternary volcanoes at the surface is determined by the along-arc variation in melt contents in the upwelling flow of the mantle wedge. It is not obvious what controls the spacing of very low velocity areas within the upwelling flow. Perhaps it is affected by irregular distribution of aqueous fluids released by dehydration reactions of slab minerals, gravity instability of buoyant aqueous fluids, the geometry of the slab subduction, and other factors. In any case, the present observations show the importance of three-dimensional modeling of arc magmatism.

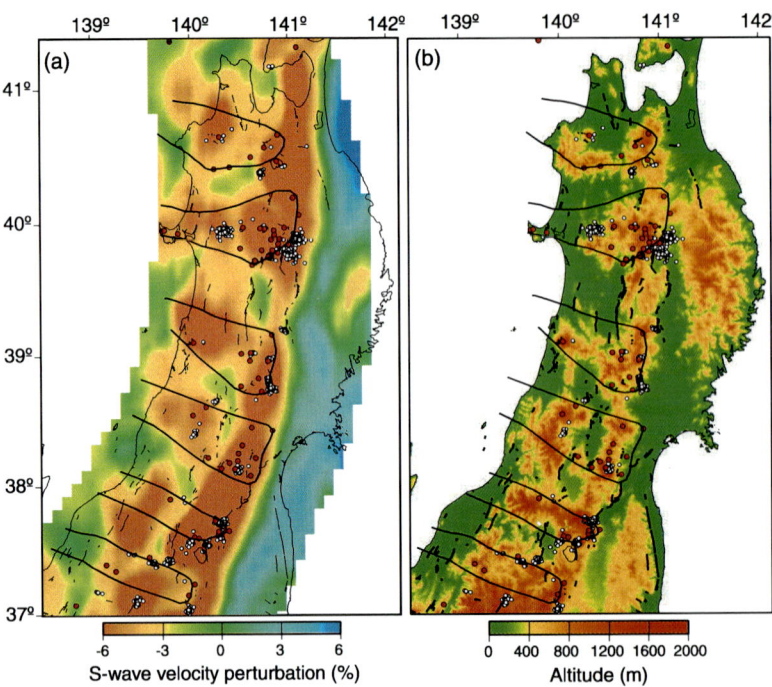

Plate 8. (a) S-wave velocity perturbations along the inclined low-velocity zone in the mantle wedge of NE Japan. Color scales are the same as in Plate 2. Red and white circles show Quaternary volcanoes and deep low-frequency microearthquakes located by JMA (Japan Meteorological Agency), respectively. Solid lines show the hot fingers by *Tamura et al.* [2002]. Thin lines show active faults. (b) Topography map of NE Japan. Others are the same as in (a).

7. SUMMARY

Intermediate-depth earthquakes occurring within subducted slab are probably caused by dehydration embrittlement of metamorphosed oceanic crust and serpentinized oceanic mantle [*Kirby et al.*, 1996]. Aqueous fluids thus released from hydrous minerals in the slab may migrate upward and may be hosted by serpentine and chlorite in the mantle wedge immediately above the slab. They are perhaps dragged by the subducting slab down to 150–200 km depth for fast, cold subduction zones such as NE Japan, and form fluid columns through which aqueous fluids are transported upward. Aqueous fluids transported from below would meet the inclined hot upwelling flow portion of subduction-induced convection, which lowers the solidus and perhaps generates partial melting. The dlnVs/dlnVp values obtained by seismic tomography studies suggest the existence of melt inclusions with volume fractions of 0.1–1 % in the upwelling flow portion for NE Japan. This upwelling flow finally meets the Moho, and magmas thus transported are perhaps stagnated directly below the Moho. Some of them may further migrate into the lower crust. These are also imaged by seismic tomography as low-Vp, low-Vs and high Vp/Vs areas in NE Japan. Further upward migration and repeated discharge to the surface form the volcanic front. The forgoing interpretations suggest that the location of the volcanic front is determined as the place where the upwelling flow meets the Moho. Observed across-arc variation of heat flow and trench-perpendicular fast shear-wave directions in the bark-arc also support the existence of this inclined upwelling flow in the mantle wedge.

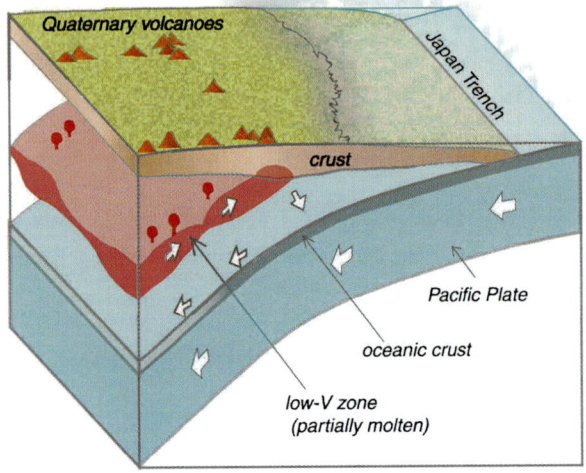

Plate 9. Schematic illustration of 3D structure of the crust and upper mantle of NE Japan showing the along-arc modulation of inclined zone of partial melting in the mantle wedge.

Our seismic-tomography study of the mantle wedge beneath NE Japan further revealed along-arc variations of the inclined low-V zone: very low S-wave velocity areas within the upwelling flow appear periodically every ~80 km along the strike of the arc in the mantle wedge beneath the arc and back-arc. This is consistent with the prediction by *Tamura et al.* [2002], who found a clear spatial relationship among clustering of Quaternary volcanoes, topographic highs, and low-V zones below the volcanic front. We also found that deep, low-frequency microearthquakes, probably caused by rapid movements of fluids, are generated right above the very low velocity areas in the mantle wedge. These observations suggest that melts could segregate from very low velocity areas in the inclined hot upwelling flow and rise vertically, triggering low-frequency microearthquakes in the lower crust and forming volcanoes at the surface. This means that the spacing of volcano clustering and topographic highs at the surface is determined by the along-arc variation in the temperature, ascent velocities and/or melt contents in the upwelling flow of the mantle wedge. Our images represent a present-day snapshot of these processes.

Acknowledgments. We thank Steve Sparks, Stephen Kirby, Dapeng Zhao and Aaron Sweeney for critical reviews that improved the manuscript. This research was supported in part by a grant from the Ministry of Education, Culture, Sports, Science and Technology of Japan.

REFERENCES

Abers, G. A., Relationship between shallow-depth and intermediate-depth seismicity in the eastern Aleutian subduction zone, *Geophys. Res. Lett.*, **19**, 2019–2022, 1992.

Abers, G. A., Three-dimensional inversion of regional P and S arrival times in the East Aleutians and source of subduction zone gravity highs, *J. Geophys. Res.*, **99**, 4395–4412, 1994.

Abers, G. A., Hydrated subducted crust at 100–250 km depth, *Earth Planet. Sci. Lett.*, **176**, 323–330, 2000.

Billen, M. I., and M. Gurnis, A low viscosity wedge in subduction zones, *Earth Planet. Sci. Lett.*, **193**, 227–236, 2001.

Bremond d'Ars, J., C. Jaupart, and R. S. J. Sparks, Distribution of volcanoes in active margins, *J. Geophys. Res.*, **100**, 20421–20432, 1995.

Cahill, T., and B. L., Isacks, Seismicity and Shape of the Subducted Nazca Plate, *J. Geophys. Res.*, **97**, 17503–17529, 1992.

Cassidy, J. F., and R. M. Eliiis, S wave velocity structure of the northern Cascadia subduction zone, *J. Geophys. Res.*, **98**, 4407–4421, 1993.

Comte, D., L. Dorbath, M. Pardo, T. Monfret, H. Haessler, L. Rivera, M. Frogneux, B. Glass, and C. Meneses, A double-layered seismic zone in Arica, northern Chile, *Geophys. Res. Lett.*, **26**, 1965–1968, 1999.

Conder, J. A., D. A. Wiens, and J. Morris, On the decompression melting structure at volcanic arcs and back-arc spreading centers, *Geophys. Res. Lett.*, **29**, doi:10.1029/2002GL015390, 2002.

Davies, J. H. and D. J. Stevenson, Physical model of source region of subduction zone volcanics, *J. Geophys. Res.*, **97**, 2037–2070, 1992.

Eberle, M. A., O. Grasset, and C. Sotin, A numerical study of the interaction between the mantle wedge, subducting slab, and overriding plate, *Phys. Earth Planet. Interiors*, **134**, 191–202, 2002.

Engdahl, E. R., and C. H. Scholtz, A double Benioff zone beneath the central Aleutians: An unbending of the lithosphere, *Geophys. Res. Lett.*, **4**, 473–476, 1977.

Fischer, K. M., and D. A. Wiens, The depth distribution of mantle anisotropy beneath the Tonga subduction zone, *Earth Planet. Sci. Lett.*, **142**, 253–260, 1996.

Frohlich, A., Geophysics—A break in the deep, *Nature*, **368**, 100–101, 1994.

Furukawa, Y., Magmatic processes under arcs and formation of the volcanic front, *J. Geophys. Res.*, **98**, 8309–8319, 1993a.

Furukawa, Y., Depth of the Decoupling Plate Interface and Thermal Structure Under Arcs, *J. Geophys. Res.*, **98**, 20005–20013, 1993b.

Gorbatov, A., G. Suerez, V. Kostoglodov, and E. Gordeev, A double-planed seismic zone in Kamchatka from local and teleseismic data, *Geophys. Res. Lett.*, **21**, 1675–1678, 1994.

Gorbatov, A., J. Dominguez, G. Suarez, V. Kostoglodov, D. Zhao, and E. Gordeev, Tomographic imaging of the P-wave velocity structure beneath the Kamchatka peninsula, *Geophys. J. Int.*, **137**, 269–279, 1999.

Haberland, C., and A. Rietbrock, Attenuation tomography in the western central Andes: A detailed insight into the structure of a magmatic arc, *J. Geophys. Res.*, **106**, 11151–11167, 2001.

Hacker, B. R., S. M. Peacock, G. A. Abers, and S. D. Holloway, Subduction factory, 2, Are intermediate-depth earthquakes in subducting slabs linked to metamorphic dehydration reactions?, *J. Geophys. Res.*, **108**, 2030, doi:10.1029/2001JB001129, 2003.

Hager, B. H., Subducted Slabs and the Geoid: Constraints on Mantle Rheology and Flow, *J. Geophys. Res.*, **89**, 6003–6015, 1984.

Hall, C. E., K. M. Fischer, E. M. Parmentier, and D. K. Blackman, The influence of plate motions on three-dimensional back arc mantle flow and shear wave splitting, *J. Geophys. Res.*, **105**, 28009–28033, 2000.

Hasegawa, A., and I. S. Sacks, Subduction of the Nazca plate beneath Peru as determined from seismic observations, *J. Geophys. Res.*, **86**, 4971–4980, 1981.

Hasegawa, A., and A. Yamamoto, Deep, low-frequency microearthquakes in or around seismic low-velocity zones beneath active volcanoes in northeastern Japan, *Tectonophysics*, **233**, 233–252, 1994.

Hasegawa, A., N. Umino, and A. Takagi, Double-planed structure of the deep seismic zone in the northeastern Japan arc, *Tectonophysics*, **47**, 43–58, 1978.

Hasegawa, A., D. Zhao, S. Hori, A. Yamamoto and S. Horiuchi, Deep structure of the northeastern Japan arc and its relationship to seismic and volcanic activity, *Nature*, **352**, 683–689, 1991.

Helffrich, G., Subducted Lithospheric Slab Velocity Structure: Observations and Mineralogical Inferences, in Subduction Top to Bottom, *Geophys. Monogr. Ser.*, vol. **96**, edited by G.E. Bebout et al. pp. 215–222, AGU, Washington, D.C., 1996.

Igarashi, T., T. Matsuzawa, N. Umino, and A. Hasegawa, Spatial distribution of focal mechanisms for interplate and intraplate earthquakes associated with the subducting Pacific plate beneath the northeastern Japan arc: A triple-planed deep seismic zone, *J. Geophys. Res.*, **106**, 2177–2191, 2001.

Iwamori, H., Transportation of H_2O and melting in subduction zones, *Earth Planet. Sci. Lett.*, **160**, 65–80, 1998.

Jackson, I., M. S. Paterson, and J. D. F. Gerald, Seismic wave dispersion and attenuation in Åheim dunite: an experimental study, *Geophys. J. Int.*, **108**, 517–534, 1992.

Kao, H., and L.G. Liu, A hypothesis for the seismogenesis of a double seismic zone, *Geophys. J. Int.*, **123**, 71–84, 1995.

Karato, S., and H.A. Spetzler, Defect microdynamics in minerals and solid-state mechanisms of seismic wave attenuation and velocity dispersion in the mantle, *Rev. Geophysics*, **28**, 399–421, 1990.

Katsumata, A., and N. Kamaya, Low-frequency continuous tremor around the Moho discontinuity away from volcanoes in the southwest Japan, *Geophys. Res. Lett.*, **30**, 1020, doi:10.1029/2002GL 015981, 2003.

Kawakatsu, H., Downdip tensional earthquakes beneath the Tonga arc- A double seismic zone, *J. Geophys. Res.*, **91**, 6432–6440, 1986.

Kirby, S., interslab earthquakes and phase changes in subducting lithosphere, U. S. Natl. Rep. Int. Union Geod. Geophys. 1991–1994, *Rev. Geophys.*, **33**, 287–297, 1995.

Kirby, S., E. R., Engdahl, and R. Denlinger, Intermediate-Depth Intraslab Earthquakes and Arc Volcanism as Physical Expressions of Crustal and Uppermost Mantle Metamorphism in Subducting Slabs, in *Subduction: Top to Bottom, Geophys. Monogr. Ser.*, **96**, edited by G. E. Bebout et al., pp. 195–214, AGU, Washington, D.C., 1996.

Kushiro, I., Y. Syono, and S. Akimoto, Melting of a Peridotite Nodule at High Pressures and High Water Pressures, *J. Geophys. Res.*, **73**, 6023–6029, 1968.

Marsh, B. D., Island arc development: Some observations, experiments, and speculations, *J. Geology.*, **87**, 687–713, 1979.

Matsuzawa, T., N. Umino, A. Hasegawa, and A. Takagi, Upper mantle velocity structure estimated from *PS*-converted wave beneath the north-eastern Japan Arc, *Geophys. J. R. astr. Soc.*, **86**, 767–787, 1986.

McGuire, J. J., and D. A. Wiens, A double seismic zone in New-Britain and the morphology of the Solomon plate at intermediate depths, *Geophys. Res. Lett.*, **22**, 1965–1968, AGU, 1995.

McKenzie, D. P., Speculations on the Consequences and Causes of Plate Motions, *Geophys. J. R. astr. Soc.*, **18**, 1–32, 1969.

Nakajima, J., and A. Hasegawa, Estimation of thermal structure in the mantle wedge of northeastern Japan from seismic attenuation data, *Geophys. Res. Lett.*, **30**, doi:10.1029/2003GL017185, 2003a.

Nakajima, J., and A. Hasegawa, Tomographic imaging of seismic velocity structure in and around the Onikobe volcanic area, northeastern Japan: implications for fluid distribution, *J. Volcanol. Geotherm. Res.*, **127**, 1–18, 2003b.

Nakajima, J., and A. Hasegawa, Fluid distribution in the mantle wedge of NE Japan inferred from seismic velocity and attenuation structures, XXIII General Assembly of the International Union of Geodesy and Geophysics, JSS06b/10P/D-020, 2003c.

Nakajima, J., T. Matsuzawa, A. Hasegawa, and D. Zhao, Three-dimensional structure of V_p, V_s, and V_p/V_s beneath northeastern Japan: Implications for arc magmatism and fluids, *J. Geophys. Res.*, **106**, 21843–21857, 2001a.

Nakajima, J., T. Matsuzawa, A. Hasegawa, and D. Zhao, Seismic imaging of arc magma and fluids under the central part of northeastern Japan, *Tectonophysics*, **341**, 1–17, 2001b.

Obara, K., Nonvolcanic Deep Tremor Associated with Subduction in Southwest Japan, *Science*, **296**, 1679–1681, 2002.

Peacock, S. M., Thermal and petrologic structure of subduction zones, in *Subduction: Top to Bottom, Geophys. Monogr. Ser.*, **96**, edited by G. E. Bebout et al., pp. 119–133, AGU, Washington, D.C., 1996.

Peacock, S. M., Are the lower planes of double seismic zones caused by serpentine dehydration in subducting oceanic mantle?, *Geology*, **29**, 299–302, 2001.

Peyton, V., V. Levin, J. Park, M. Brandon, J. Lees, E. Gordeev, A. Ozerov, Mantle flow at a slab edge: Seismic Anisotropy in the Kamchatka region, *Geophys. Res. Lett.*, **28**, 379–382, 2001.

Polet, J., P. G. Silver, S. Beck, T. Wallace, G. Zandt, S. Ruppert, R. Kind, and A. Rudloff, Shear wave anisotropy beneath the Andes from the BANJO, SEDA, and PISCO experiments, *J. Geophys. Res.*, **105**, 6287–6304, 2000.

Raleigh, C. B., and M. S. Paterson, Experimental deformation of serpentinite and its tectonic implications, *J. Geophys. Res.*, **70**, 3965–3985, 1965.

Ribe, N. M., On the Relation Between Seismic Anisotropy and Finite Strain, *J. Geophys. Res.*, **97**, 8737–8747, 1992.

Roth, E. G., D. A. Wiens, L. M. Dorman, J. Hildebrand, and S. C. Webb, Seismic attenuation tomography of the Tonga-Fiji region using phase pair methods, *J. Geophys. Res.*, **104**, 4795–4809, 1999.

Roth, E. G., D. A. Wines, and D. Zhao, An empirical relationship between seismic attenuation and velocity anomalies in the upper mantle, *Geophys. Res. Lett.*, **27**, 601–604, 2000.

Satake, K., and T. Hashida, Three-dimensional attenuation structure beneath North Island, New- Zealand, *Tectonophysics*, **159**, 181–194, 1989.

Savage, M. K., Seismic anisotropy and mantle deformation: What have we learned from shear wave splitting?, *Rev. Geophysics*, **37**, 65–106, 1999.

Sato, H., I. S. Sacks, T. Murase, G. Muncill, and H. Fukuyama, Qp-Melting Temperature Relation in Peridotite at High Pressure and Temperature: Attenuation Mechanism and Implications for the Mechanical Properties of the Upper Mantle, *J. Geophys. Res.*, **94**, 10647–10661, 1989.

Schmidt, M. W., and S. Poli, Experimentaly based water budgets for dehydrating slabs and consequences for arc magma generation, *Earth Planet. Sci. Lett.*, **163**, 361–379, 1998.

Sekiguchi, S., Three-dimensional Q structure beneath the Kanto-Tokai district, Japan, *Tectonophysics*, **195**, 83–104, 1991.

Seno, T., and Y. Yamanaka, Double Seismic Zones, Compressional Deep Trench-Outer Rise Events and Superplumes, in *Subduction: Top to Bottom, Geophys. Monogr. Ser.*, **96**, edited by G. E. Bebout et al., pp. 347–355, AGU, Washington, D.C., 1996.

Sisson, T. W. and S. Bronto, Evidence for pressure-release melting beneath magmatic arcs from basalt at Galunggung, Indonesia, *Nature*, **391**, 883–886, 1998.

Smith, G. P., D. A. Wiens, K. M. Fischer, L. M. Dorman, S. C. Webb, and J.A. Hildebrand, A Complex Pattern of Mantle Flow in the Lau Backarc, *Science*, **292**, 713–716, 2001.

Springer, M., and A. Forster, Heat-flow density across the Central Andean subduction zone, *Tectonophysics*, **291**, 123–139, 1998.

Tamura, Y., Y. Tatsumi, D. Zhao, Y. Kido, and H. Shukuno, Hot fingers in the mantle wedge: new insights into magma genesis in subduction zones, *Earth Planet. Sci. Lett.*, **197**, 105–116, 2002.

Tatsumi, Y., Formation of the volcanic front in subduction zones, *Geophys. Res. Lett.*, **13**, 717–720, 1986.

Tatsumi, Y., M. Sakuyama, H. Fukuyama, and I. Kushiro, Generation of Arc Basalt Magmas and Thermal Structure of the Mantle Wedge in Subduction Zones, *J. Geophys. Res.*, **88**, 5815–5825, 1983.

Tommasi, A., Forward modeling of the development of seismic anisotropy in the upper mantle, *Earth Planet. Sci. Lett.*, **160**, 1–13, 1998.

Tsumura, N., S. Matsumoto, S. Horiuchi, and A. Hasegawa, Three-dimensional attenuation structure beneath the northeastern Japan arc estimated from spectra of small earthquakes, *Tectonophysics*, **319**, 241–260, 2000.

Wood, B. J., Geochemical processes in the mantle wedge, this volume, 2004.

Yamasaki, T. and T. Seno, Double seismic zone and dehydration embrittlement of the subducting slab, *J. Geophys. Res.*, **108**, doi:10.1029/2002JB001918, 2003.

Yang, X., K. M. Fischer, and G. A. Abers, Seismic anisotropy beneath the Shumagin Island segment of the Aleutian-Alaska subduction zone, *J. Geophys. Res.*, **100**, 18165–18177, 1995.

Zhang, S. and S. Karato, Lattice preferred orientation of olivine aggregates deformation in simple shear, *Nature*, **375**, 774–777, 1995.

Zhao, D. and A. Hasegawa, P wave Tomographic Imaging of the Crust and Upper Mantle Beneath the Japan Islands, *J. Geophys. Res.*, **98**, 4333–4353, 1993.

Zhao, D., A. Hasegawa, and H. Kanamori, Deep structure of Japan subduction zones as derived from local, regional, and teleseismic events, *J. Geophys. Res.*, **99**, 22313–22329, 1994.

Zhao, D., D. Christensen, and H. Pulpan, Tomographic imaging of the Alaska subduction zone, *J. Geophys. Res.*, **100**, 6487–6504, 1995.

Zhao, D., Y. Xu, D.A. Wiens, L. Dorman, J. Hildebrand, and S. Webb, Depth Extent of the Lau Back-Arc Spreading Center and its Relation to Subduction Processes, *Science*, **278**, 254–257, 1997.

A. Hasegawa, and J. Nakajima, Research Center for Prediction of Earthquakes and Volcanic Eruptions, Graduate School of Science, Tohoku University, Sendai 980-8578, Miyagi, Japan (e-mail:hasegawa@aob.geophys.tohoku.ac.jp, nakajima@aob.geophys.tohoku.ac.jp)

Evolution of Arc Magmas and Their Volatiles

Richard J. Arculus

Department of Earth and Marine Sciences, Australian National University, Canberra, ACT 0200, Australia

Of the volumetrically significant magma types, those emplaced in arcs at plate convergence zones are typically richest in dissolved volatiles (H_2O, CO_2, and S species). These volatiles are mostly products of large-scale recycling, derived though multi-stage processes involving devolatilization of variably hydrated and carbonated, sediment-bearing, subducted lithosphere, and transported towards the surface by magmas generated in the mantle overlying the subducted plate. Volatile contents of parental arc basalts are globally variable, but mostly range from ~ 0.5 to 10 wt% H_2O, ≤ 1000 ppm CO_2, and ≤ 3500 ppm S (as H_2S and/or SO_2). Arc magmas are also generally more oxidized than those of ridges and hot-spots. These characteristics lead to distinctive differences in the course of magmatic crystallization compared with dry, reduced types: plagioclase saturation is delayed, and Ca-rich upon appearance; olivine persists in the crystallization sequence to higher SiO_2 contents; a spinel phase appears early and persists throughout crystallization. Resultant relatively voluminous Na-K-feldspar- and SiO_2-rich residual magmas dominate the bulk continental crust. The complementary SiO_2-poor olivine-clinopyroxene-dominated fraction is probably recycled into the upper mantle. Volatile fluxes through the subduction cycle are not straightforwardly determined; total magma volume flux estimates range from ~ 1.2 to 7 km^3/year, but are not well constrained for the full diversity of global arc systems. Even at the high end of this range, more C is subducted than returned via arcs, presumably with long-term effects on the global C cycle. At the lower end, deficiencies may exist in return fluxes of H_2O and S species.

INTRODUCTION

There are a number of reasons for strong geological interest in the quantities of volatiles dissolved in arc magmas: the first is the critical role played by volatiles in determining the sequence of crystallization and the specific compositions of the phases appearing in magma, the viscosity of the magma, and its eruption characteristics; the second concerns the overall character of global recycling of volatile compounds and elements through the Earth's interior via plate tectonic processes, and possible temporal variability in these cycles; the third involves the environmental impact of emitted aerosols and gases; and, fourth, is the development of some metallic ore deposits through degassing processes of arc magmas.

Ultimately, the distinctive inter-terrestrial planet character of Earth's granodioritic continental crust [*Hofmann*, 1988] results from the high concentrations of H_2O in island and continental arc magmas relative to those forming mid-ocean ridges (MOR) and intra-plate hot-spots, leading to enrichment of alkali feldspar and quartz components in residual arc magmas. The aphorism "no water, no granites – no oceans, no continents" [*Campbell and Taylor*, 1985] summarizes the planetary significance of the issues.

Fluxes of H_2O, CO_2, S species, halogens, and other volatiles emitted by arcs result primarily from immense global recycling processes. Cycling of some volatiles ultimately controls major chemical balances in the hydrosphere, atmosphere, and biosphere. While the noble gases are insignificant in terms of chemical reactions in the mantle, crust and exospheric reservoirs, ^{3}He in particular comprises a crucial tracer of volatile recycling more generally through subduction zone systems [*Hilton et al.*, 2002]. Although subducted fluxes are reasonably well constrained overall, albeit globally variable, return fluxes in arcs are much harder to estimate, and may be deficient by large factors indicating deep return and long-term storage in the mantle. Given the amount of H_2O currently subducted (~ $1.3 * 10^{13}$ kg/per km of arc strike/million years) [*Peacock*, 1990], the oceans would be drained in ~3,000 million years in the absence of a return flux. We have no evidence that this has occurred [*Wise*, 1972] and assume that the interior of the Earth and its exterior are more-or-less in steady state in terms of a global H_2O cycle, at least during the Phanerozoic. Given the complexities of the recycling processes however, it is not clear why steady state in terms of external hydrosphere volumes should prevail.

In this article, I summarize our knowledge of the major volatile (H_2O, CO_2, S species, and halogens) contents of arc magmas, the distinctive magmatic compositional evolution trends controlled by H_2O in particular, and address the mass balance issues of volatile recycling via the subduction of tectonic plates.

EVIDENCE FOR HIGH VOLATILE CONTENTS IN ARC MAGMAS

Our primary lines of evidence for the presence of high volatile contents in arc magmas are through direct analytical measurements of quenched glasses (as matrix and inclusions in crystals); the presence of volatile-bearing phases in some types of arc magma such as amphibole, mica, apatite, anhydrite, sulfides, and haüyne-sodalite; the explosive eruptive style and associated discharges of gas from arc volcanoes; and the experimental duplication of observed phenocryst assemblages with controlled pressure, temperature, and concentrations of volatiles.

Direct Measurements

A major interpretative difficulty for all direct measurements, however, is the ubiquitous degassing experienced by subaerially-erupted, variably crystalline magmas, driven primarily by the strong insolubility of CO_2 relative to H_2O and S compounds (Figure 1). In order to mitigate this problem, submarine-erupted glasses and phenocryst-hosted glass inclusions are preferred materials for study. The greater ambient pressure of eruption on the seafloor combined with rapid quenching in cold sea water means that a greater proportion of the volatile load remains dissolved in the magma. Likewise melt can be trapped as inclusions in phenocrysts at crustal pressures and retain dissolved volatiles. Chemical analysis involves standard gravimetric and spectroscopic techniques for bulk rocks, combined with small aperture, Fourier transform infrared and Raman spectroscopy [*Fine and Stolper*, 1985/86] plus electron microprobe (for S and halogens) and ion microprobe analysis of microscopic volumes. Countering the depletion of volatile compounds and elements through degassing is enrichment through the fractional crystallization of volatile-poor phases such as spinel, olivine, pyroxene, and feldspar.

There have been relatively few studies of volatile loads in the bulk matrix glasses of submarine-erupted arc lavas [*Newman et al.*, 2000]. There are more data available for backarc basin lavas, generally erupted at greater water depths and hence confining pressures than on submarine cones forming the arc volcanic front [*Newman et al.*, 2000]. A number of these backarc magmas have overall compositional variabili-

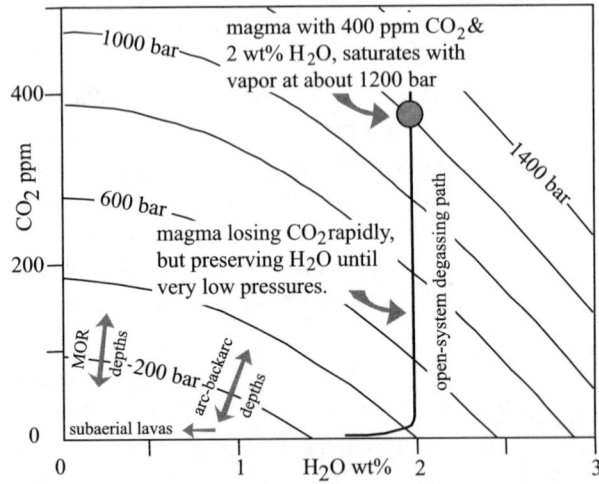

Figure 1. Solubilities of CO_2 and H_2O in vapor-saturated basaltic melts at 1200°C after *Dixon and Stolper* [1995]. The continuous curves are curves of constant pressure (isobars). A representative arc basalt magma with 400 ppm CO_2 and 2 wt% H_2O is saturated with a vapor phase at 1200 bars (indicated by filled circle). If this magma continuously loses vapor (open system) during ascent, the proportions of CO_2 and H_2O remaining in the melt are given by the line labeled "open-system degassing path". Note that drastic CO_2 loss occurs from the melt, but minimal H_2O loss takes place until very low pressures (labeled "subaerial lavas"). The depth ranges for MOR and (deep) arc-backarc eruption sites are indicated.

ties indicative of involvement of one end-member comprising a volatile-rich subduction component characteristic of the arc itself; this permits identification of the volatile characteristics of the arc-like end-member [*Stolper and Newman*, 1994]. A key point is that during fractional degassing of ascending magma, the partitioning between the escaping supercritical fluid phase and magma results in relatively little decrease in concentrations of H_2O and, to a lesser extent, S compounds in the magma (Figure 1). The CO_2 contents on the other hand, decline dramatically in the magma. The analyzed CO_2 concentrations in bulk backarc glasses are equivalent to the solubilities at the pressures of eruption on the seafloor, clearly indicative of pre-eruption CO_2 loss. Many glasses have CO_2 concentrations below detection limits. With higher initial H_2O contents, degassing of H_2O and S species could occur through much of the crust.

The volatile loads in a wider range of samples from both island and continental arcs have been analyzed in melt inclusions. Technically, this is more challenging than for bulk glasses, and requires painstaking mineral separation, selection of sufficiently large inclusions for analysis, and petrologic caution in interpreting the origins of any given inclusion [*Danyushevsky et al.*, 2002]. However, melt inclusions have the distinct advantage of potentially preserving a higher pressure, relatively un-degassed, dissolved volatile complement.

Representative results of direct measurements of H_2O contents in bulk and inclusion glasses in basaltic/basaltic andesite arc magmas are displayed in Figure 2. There is clearly a large range from the low H_2O concentrations (~ 0.2 wt%) reported in olivine-hosted melt inclusions in primitive, MgO-rich magma of Galunggung, Indonesia [*Sisson and Bronto*, 1998]

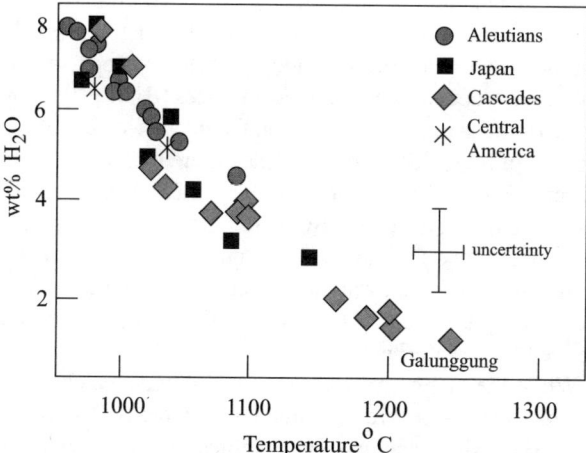

Figure 2. The range of H_2O contents and temperatures of phenocryst equilibration at 0.4 GPa, for a variety of arc basalts (high-Mg and high-alumina) from *Sisson and Grove* [1993b] and *Sisson and Bronto* [1998].

to ~ 6 to 8 wt% of Fuego magmas, Guatemala [*Sisson and Layne*, 1993] and Atka of the Aleutians (*Sisson and Grove*, 1993a) respectively. Even higher H_2O contents (~ 10 wt%) have been reported for mantle-derived, high-MgO andesites of Mt Shasta [*Grove et al.*, 2003]. For comparison, MOR basalts derived from the most depleted mantle sources have ~ 0.06 wt% H_2O [*Saal et al.*, 2002] and range up to ~ 0.3 wt% for those derived from somewhat less depleted sources [*Michael*, 1995].

Negligible contents of CO_2 have been detected in bulk glasses, but concentrations range up to 1000 ppm in melt inclusions from Cerro Negro, Nicaragua [*Roggensack et al.*, 1997]. A similar overall problem of determing primary, undegassed CO_2 contents in primary, mantle-derived basalts has been overcome for MOR through discovery of magmas erupted at depths greater than the saturation pressure for CO_2 [*Saal et al.*, 2002]. Through correlation of the geochemical behavior of CO_2 with involatile Nb, *Saal et al.* [2002] were able to demonstrate primary CO_2 concentrations in the most volatile-rich, MORB "popping-rocks" of <5,000 ppm. No equivalent arc samples have been recovered from ocean depths >3000m.

Most S concentrations in MORB range from 800 to 1000 ppm [*Wallace and Carmichael*, 1992]. As with H_2O and CO_2, the solubility of S is dependent on pressure and temperature, and bulk composition (particularly the FeO content). Most significantly however, the solubility of S is a strong function of the redox state of the magma (as controlled for example, by interactions between $Fe^{2+} \leftrightarrow Fe^{3+}$ establishing the effective concentration (or fugacity; f) of O_2 in a melt), and varies strongly within the common redox range of terrestrial basaltic magmas from sulfide at low- to sulfate at high-fO_2s [*O'Neill and Mavrogenes*, 2002](Figure 3). Complicating the situation further is that negative pressure dependence characterizes sulfide solubility, while positive pressure dependence appears to exist for sulfate [cf. *Carroll and Rutherford*, 1985; *Luhr*, 1990]. The presence of anhydrite ($CaSO_4$) and pyrrhotite ($Fe_{1-x}S$) in some arc magmas (e.g., Mt. Lamington, Papua New Guinea; El Chichon, Mexico; Mt. Pinatubo, Philippines) illustrates the importance of redox state on the partitioning of S between crystallizing magma and escaping gas. Many arc magmas contain negligible S presumably through degassing, but concentrations range up to levels equivalent to MORB at 1000 ppm [*Alt et al.*, 1993; *Nilsson and Peach*, 1993]. I have analyzed abundances of ≤3500 ppm S for melt inclusions in some arc basalts from Melanesia and the Kurile Islands.

While the abundance of F in arc magmas is generally in the same range (150 to 250 ppm) as MORB [*Jambon*, 1994], Cl may be markedly enriched in some arc magmas (~500 to 10,000 ppm in the Izu-Bonin arc; *Straub and Layne*, 2003) compared with 20 to 1000 ppm in MORB [*Jambon*, 1994]. The partitioning behavior of F and Cl between melt and exsolving

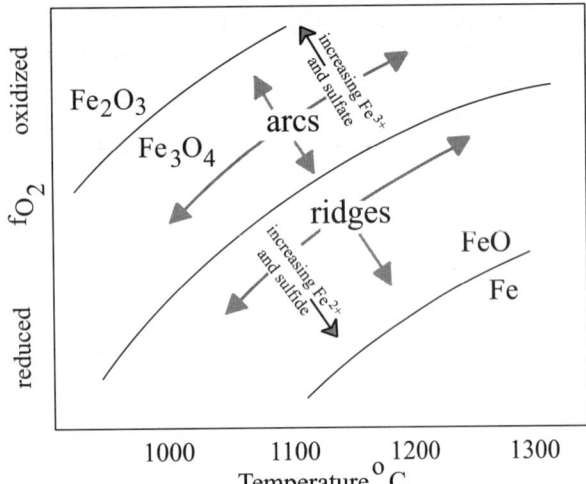

Figure 3. Schematic depiction of the range of common redox states reflected by dissolved Fe^{2+}/Fe^{3+} (FeO/Fe_2O_3) and SO_4^{2-}/S^{2-} (sulfate/sulfide) in mid-ocean ridge and island arc magmas, shown as the variation of relative oxygen concentration versus temperature. The bounding lines indicate stability of Fe_2O_3 and FeO alone; in the intermediate fO_2-T space, both FeO and Fe_2O_3 are stable. The dividing line between the arc and ridge fields corresponds in reality to a synthetic oxygen buffer ($3Fe_2SiO_4 + O_2 \leftrightarrow 2Fe_3O_4 + 3SiO_2$).

vapor is dissimilar with F being preferentially retained in the melt and Cl readily forming ligands with volatile metals in a supercritical fluid phase; both halogens are important however, in the formation of species partitioned between low density vapor and relatively high density brine generated through phase separation during decompression and/or cooling of supercritical fluids.

Experimental Duplication of Phenocryst Assemblages

Experimental petrologists established the important role of H_2O in the phase relations of silicate magmas in the first half of the 20th century [e.g., *Yoder and Tilley*, 1962]. Triggering of melting in the wedge of mantle overlying a subducted lithospheric plate by H_2O released from the plate was first suggested by *McBirney* [1969]. Investigations of the phase relations of arc-related magmas and quantification of the role of H_2O and other volatiles followed [*Eggler*, 1972; *Holloway and Burnham*, 1972]. Experimental studies provide the required pressure-temperature-fugacity of volatile species framework with which we can understand the chemical and physical evolution of arc magmas [*Merzbacher and Eggler*, 1984; *Sisson and Grove*, 1993a, b; *Johnson et al.*, 1994; *Gaetani and Grove*, 1998].

There is general concordance between estimates of volatile contents determined by experiment and the results of direct measurement when allowances are made for degassing. For example, *Gaetani et al.* [1993] showed that the characteristic early phenocryst assemblage of spinel+olivine+clinopyroxene in many primitive (high-MgO) arc basalts can result from crystallization at temperatures in the range 1000 to 1200°C, 0.2 to 0.4 GPa with 2–4 wt% H_2O (Figure 2), and suppression of early plagioclase crystallization unlike the case with MORB. Higher initial H_2O contents would enhance these effects. Once an arc magma reaches saturation with plagioclase however, a shift at given melt Ca/Na results in higher anorthite (i.e., Ca-rich) contents as a function of increased H_2O [*Sisson and Grove*, 1993a]. While the presence of phenocryst amphibole is unequivocal evidence for the presence of dissolved H_2O in a magma, stabilization requires at least 4 to 5 wt% H_2O [*Merzbacher and Eggler*, 1984] and relatively high Na contents (> 3 wt% Na_2O in the melt [*Sisson and Grove*, 1993a]. The absence of amphibole does not indicate an anhydrous magma. Experimental duplication of phenocryst assemblages in more evolved (SiO_2-rich) magmas, such as andesites and dacites characteristic of some of the most violently explosive eruptions of the last 30 years, have provided important constraints on the evolving conditions and depth locations of the tapped magma sources [e.g., Mt St Helens: *Rutherford et al.*, 1985; Mt. Pinatubo: *Rutherford and Devine*, 1996], either in isobaric magma chambers or via melt-crystal-vapor equilibria during polybaric acent paths. A number of these studies have demonstrated vapor-saturation must have taken place at intra-crustal depths (~ 2 to 8 km), providing a potent driving force for explosive eruption.

MAGMATIC COMPOSITIONAL EVOLUTION

The magmatic suites forming the igneous portions of island and continental arcs are compositionally distinctive when compared with spreading ridge, isolated hot-spot, or other terrestrial planet or meteoritic occurrences [*BVSP*, 1981]. My primary focus in this section is on the major element characteristics of SiO_2-rich products of the differentiation of parental basalt magmas. However, it is worth recalling that, even within the basalt range, the distinctive enrichments in arc magmas of trace alkalis, alkaline earths, U, Th, and Pb reflect the involvement during magma formation in the upper mantle of a H_2O-rich fluid phase [e.g., *Hofmann et al.*, 1986], presumably derived by dewatering of the subducted slab.

While the distinctive trace element abundance characteristics of arc basalts are inherited by the more SiO_2-rich rock types (andesites, dacites, and rhyolites), the primary distinguishing characteristic of evolved arc magmas compared with ridge or hot-spot magmas is the comparatively low total (FeO + Fe_2O_3, denoted FeO*)/MgO at a given SiO_2 content (Figure 4). The development of this compositional trend was

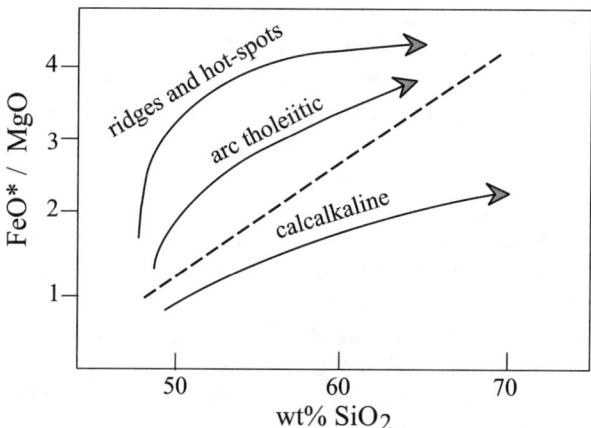

Figure 4. Schematic illustrating representative variation trends of wt% (FeO + Fe_2O_3) (denoted FeO*) / MgO vs. wt% SiO_2 for common subalkaline MOR, hot-spot and island arc volcanic suites. The dashed line is the discriminant boundary between so-called "tholeiitic" and "calcalkaline" suites from *Miyashiro* [1974].

the subject of much petrologic debate during the 20th century [*Young*, 1998]. The unfortunate term "calcalkaline" (or calcalkalic; both variably hyphenated) has been used to describe this "low FeO*" trend [*Arculus*, 2003], when neither calc nor alkali constitute individually or collectively the distinguishing characteristic.

Another empirical observation is that arc (and many backarc) magmas are oxidized (i.e., high Fe_2O_3/FeO) relative to MOR suites (Figure 3) [*Arculus*, 1985; *Carmichael*, 1991]. Arc-related peridotites are also oxidized relative to peridotites from non-subduction zone environments [*Parkinson and Arculus*, 1999], so this feature appears to be established in the source regions of arc magmas, and is not a secondary characteristic developed during contamination by crustal rock types. The origin of this oxidized character is controversial [*Frost and Ballhaus*, 1998], and cannot simply be the result of addition of slab-derived H_2O to the mantle source, but the relevance here is the stabilization earlier in the crystallization history of basaltic arc magmas than in MORB of a Fe^{3+}-rich spinel phase. In addition, the redox range of arc magmas extends from the solubility transition between sulfide and sulfate speciation to strongly sulfate-stable conditions.

The consequences of relatively high H_2O and Fe_2O_3/FeO in crystallizing arc magmas are displayed in Figure 5. The critical features are: 1. delayed appearance (saturation) relative to clinopyroxene of plagioclase feldspar (a solid solution of $CaAl_2Si_2O_8 - NaAlSi_3O_8$), and a shift towards more Ca-rich, Si-poor plagioclase in the wet arc compared with dry MORB case; 2. persistence of olivine crystallization (at the expense of plagioclase and orthopyroxene) to higher bulk magma SiO_2 concentrations; 3. early saturation with and persis-

tence throughout the crystallization interval of spinel (an oxide solid solution of general molecular formula $(Mg,Fe^{2+})(Cr,Al,Fe^{3+},Ti)_2O_4$) in the arc compared with MOR environment. The resulting compositional trends comprise enrichment in Al_2O_3 and Na_2O accompanying marked SiO_2 increase, and depletion in total Fe in evolved (andesite, dacite, and rhyolite) magmas. The distinctive phase assemblages forming the accumulative crystal fraction ("cumulates") complementary to the evolving arc melts are dunite (spinel and olivine), wehrlite (spinel, olivine, and clinopyroxene), and gabbro (spinel, clinopyroxene, Ca-rich plagioclase, ± olivine, ± orthopyroxene, ± amphibole). Examples outcrop in some exhumed lower crustal arc segments [*DeBari and Coleman*, 1989; *Khan et al.*, 1993; *Spandler et al.*, 2003], and are in contrast to cumulates forming the plutonic fraction of oceanic crust formed at spreading ridges: dunite, troctolite (olivine and plagioclase), and gabbro.

In addition to the andesite-dacite-rhyolite compositional arrays derived predominantly through fractional crystallization of hydrous parental basalt magmas, a number of additional processes are documented in arcs that are moderated by the presence of volatile components. These include: magma mixing of less and more SiO_2-rich types, where the viscosity contrasts between end-member magmas may be reduced several orders of magnitude especially during mixing of high temperature basalt with low temperature, hydrous felsic melts; assimilation of hydrated (hence with low solidus temperatures) rock types surrounding magma conduits and reservoirs, and re-melting of previously formed crustal lithologies by

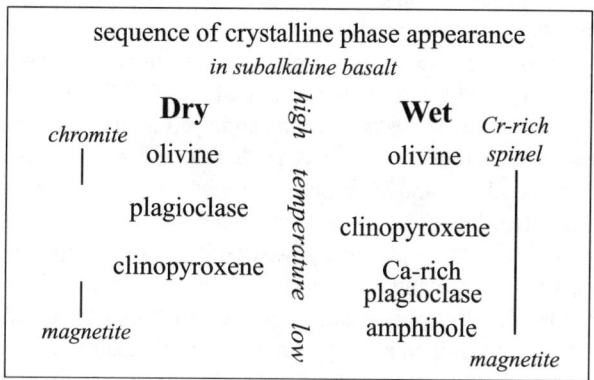

Figure 5. Diagrammatic representation of the major differences in progressive crystallization sequences of dry and wet, subalkaline basalt magmas. Basalts from MOR follow the dry crystallization sequence, whereas most arc magmas follow the wet crystallization path. The broken solid line on the left indicates early crystallization of low Fe^{2+}-Cr-rich spinel (chromite) in MORB, an absence of spinel and then later appearance of magnetite. The solid line on the right indicates continuous crystallization of a spinel phase in wet magmas.

invading magmas. Volumetrically minor arc magma types include boninites and adakites; the former are hydrous, high-MgO, intermediate-SiO$_2$ melts believed to form through subducted slab-derived, hydrous fluid-fluxing of extremely refractory harzburgite (i.e., olivine-orthopyroxene) assemblages [*e.g., Crawford*, 1989], and the latter are suggested to be direct partial melts of an eclogite or garnet-amphibolite source, either in subducted lithosphere [*Defant and Drummond*, 1990] or in the lower crust of an overriding plate [*Garrison and Davidson*, 2003].

Vapor-Mobile Metallic Trace Elements: Cu, Zn, Au, and Re

The orders of magnitude discrepancy between the amount of S released during explosive arc volcano eruptions and the amount of S dissolved in associated magma [*Andres et al.*, 1991] can be reconciled if the magma is already saturated with a vapor phase prior to eruption [*Wallace et al.*, 2003]. The presence of pre-eruption, vapor-saturated melt in basalt to dacite magmas of the eastern Manus Basin (Papua New Guinea) has been documented by *Kamenetsky et al.*, [2001] and *Yang and Scott* [2002]. These authors have also demonstrated the strong partitioning of metallic elements (e.g., Cu and Zn) into this vapor phase; escape of this phase from an evolving magma is likely important in the formation of epithermal and porphyry-type ore deposits [*Hedenquist and Lowenstern*, 1994].

For the Manus Basin suite [*Binns and Scott*, 1993], we have identified similar losses to a vapor phase of Re [*Sun et al.*, 2003a; 2003b], with consequences for understanding this element's geochemical cycle and the Re-Os isotopic system. Over the last 20 years, mass spectrometer instrumental developments have allowed the geochemical community to exploit the decay of ^{187}Re to ^{187}Os (half life of $42.3 * 10^9$ years), and the systematics of Os isotopes in, and cycling between, various terrestrial reservoirs. Re is much more incompatible than Os in the silicate and oxide phases forming crystalline residues during melting of the upper mantle, and is strongly enriched in the melt. Through the differentiation of the continental crust from the mantle during the evolution of the Earth, the Re/Os of melt-depleted mantle has become much lower than that of the continental crust, and the ^{187}Os/^{188}Os of the continental crust is greater than the mantle (^{188}Os is a non-radiogenic reference isotope).

A puzzling aspect that has emerged from studies of continental crust-building, subaerially-erupted arc rocks, is both their low abundances of Re and low Re/Os [*Righter et al.*, 2002]. Comparisons of Re abundances in submarine- and subaerially-erupted arc magmas appear however, to have resolved this problem [*Sun et al.*, 2003a, b]. The variation of Re with SiO$_2$ in basalt to dacite samples from the eastern Manus Basin, increases from about 0.6 to 3.5 ppb at 57 wt% SiO$_2$, and then decreases to about 1 ppb in the more SiO$_2$-rich magmas. All of these Re abundances are about 10* greater than those previously determined for subaerially-erupted arc volcanoes [e.g., *Woodland et al.*, 2002].

It is likely that Re (but not Os) has been lost in a volatile phase during degassing of the Manus Basin samples (erupted at ~ 2000m water depth); details of this process are not yet clear and presumably depend strongly on the specific nature of the behavior of the volatile-magma system and possible phase separation (low- and high-density vapour and brine respectively) [*Giggenbach*, 1992; *Lowenstern*, 2001] There is a strong positive correlation between Re and Cu in the Manus Basin suite; Cu loss in evolving vapor has previously been suggested for the Manus magmas by *Kamenetsky et al.* [2001] and *Yang and Scott* [2002]. The Re lost from degassing arc magmas may be fixed in hydrothermal mineral deposits or sediments; either way, the Re likely becomes fixed in rock types that eventually become part of the rock cycle forming the continental crust.

RECYCLING AT CONVERGENT MARGINS

The hydrologic cycle involving fluxes of H$_2$O between the hydrosphere, atmosphere, and crysophere is familiar to most of us as a dominating feature in the "weather". Few realize that spatially much larger cycles exist involving movement of H$_2$O between the Earth's exterior reservoirs and its deep interior. *If more H$_2$O is disappearing into the Earth than is currently reappearing in forearcs, arcs, backarcs, and eventually via hot-spot and ridge magmatism, the question arises: are we losing the oceans?* During deep recycling, fluids trigger melting in Earth's mantle, redistribute and transport trace elements from the interior to exterior. Some volatile-rich magmas and fluids become trapped in newly created crust, forming hydrated and fertile rock types that subsequently re-melt forming granites, while others are the prime cause of the highly explosive volcanism characteristic of arcs.

Oceanic lithospheric plates are variably hydrated and carbonated during exposure to the Earth's external hydrosphere and hydrothermal processes, and typically include H$_2$O- and carbonate-rich surficial sedimentary layers. Fluids are emitted from both accretionary [*Moore and Silver*, 2002] and non-accretionary forearc systems [*Fryer et al.*, 1995]. These are likely significant in terms of the mechanical (and earthquake) processes characteristic of these plate collision zones, but the total fluxes in these environments are not well known. Much sediment (currently estimated as ~ 1 km^3/year) [*von Huene and Scholl*, 1991] is apparently subducted into the Earth's mantle. Hydrous fluids (and possibly silicate melts) derived from the sediments, oceanic crust and mantle of the downgoing lithospheric plate, are released into and trigger melting of the mantle [*Pearce and Parkinson*, 1993; *Plank and Langmuir*, 1993]. Water can be retained in percent quantities in dense hydrous magnesium silicates (e.g., "Phase A: $Mg_7Si_2O_8(OH)_6$) of the mantle portion of the downgoing lithosphere under low dT/dP subduction trajectories [e.g., *Shieh et al.*, 1998; *Poli and Schmidt*, 2002], and at ~10^2 ppm levels in upper mantle olivine [*Ingrin and Skogby*, 2000].

While the subducted fluxes of potential volatiles are reasonably well constrained, albeit globally variable, the return fluxes in arcs and backarcs are not so well understood [*Jambon*, 1994; *Hilton et al.*, 2002]. The first order problem to be tackled then is the total mass flux in arcs, to be followed by our knowledge of individual volatile abundances to derive the fluxes of individual compounds and elements. The return volatile fluxes can be divided into two major categories: those that are highly insoluble at low pressures in arc magmas, and mostly degassed into the atmosphere and hydrosphere (e.g., He and CO_2); other volatiles such as H_2O and S species, which have some solubility in melts, may be partly retained in newly-formed arc lithosphere as amphiboles, micas, sulfides and sulfates.

Total Mass Flux in Arcs

In the MOR case, we can straightforwardly take the average rate of ocean floor spreading (e.g., 5 cm/year), the total ridge length (~$5.5*10^4$ km), and the average crustal thickness (7 km) to calculate a present-day production rate of ~ 20 km^3/year, equivalent to ~360 km^3 per km of ridge length per million years of spreading activity (Figure 6). By way of comparison, *Reymer and Schubert* [1984] estimate average production rates of oceanic crust production from the Mezozoic to Cenozoic of 25 km^3/year, equivalent to 450 km^3 per km of ridge length per million years. Over an equivalent time period, *Crisp* [1984] estimated a MOR crust flux of 19.5 to 25.5 km^3/year. Taking an average H_2O content of 0.1 wt% for MORB, a density of $2.8*10^3$ kg/m^3, and assuming all the H_2O is degassed, then the H_2O mass flux is $5.6*10^{10}$ kg/year. Even if "hot spot" magmas contain an order of magnitude more H_2O than MORB, the relatively low total volume flux (~ 1 km^3/year; *Crisp*, 1984) means a H_2O mass flux of only ~ $2.8*10^9$ kg/year.

For island and continental arcs, calculations of volume and mass fluxes are difficult because neither the nature (proportion and age) of the basement of an arc, nor the time interval over which any given volume of arc magmas has accumulated is well known. In addition, the nature and depth of the Moho are typically poorly constrained, and sometimes the arc crust is being deformed simultaneous with growth. Accordingly, it is difficult to determine what the crustal production rate is at present-day convergent margins.

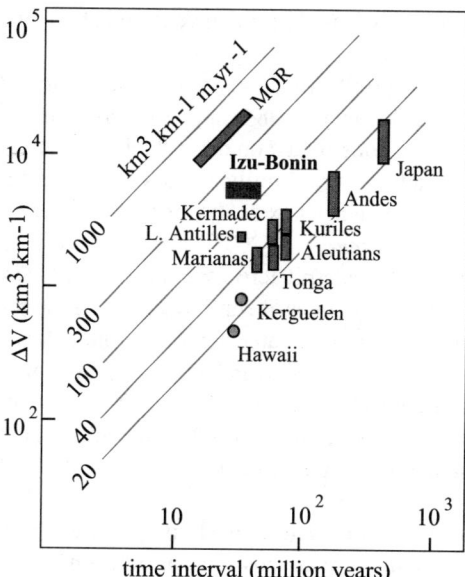

Figure 6. Growth rates of MOR, representative hot-spots (Hawaii and Kerguelen), and individual arc systems from *Reymer and Schubert* [1984] together with estimates for the Izu-Bonin system from *Arculus* [1999] and *Taira et al.* [1998].

Reymer and Schubert [1984] presented the results of a valiant attempt to surmount these difficulties: for intra-oceanic arcs, they calculated arc volumes from seismically-derived crustal profiles, subtracted the equivalent of 6 km of "oceanic basement" upon which the arcs were assumed to have been built, assumed durations of activity from estimates of dates of arc inception, and derived 30 km^3 per km of arc strike/million years giving a global mass flux of 1.1 km^3/year for an arc length of $3.7 * 10^4$ km (Figure 6). *Crisp* [1984] compiled global magma emplacement and output rates (for durations > 300 years) implicitly involving petrogenetic models, to derive a much higher global arc mass flux of ~5 to 7 km^3/year. These estimates do not include any volume of ultramafic cumulates located sub-Moho (*Fliedner and Klemperer*, 2000).

Recent seismic reflection/refraction studies of the intra-oceanic Izu-Bonin arc [*Suyehiro et al.*, 1996; *Takahashi et al.*, 1998] have permitted reexamination of this arc flux issue. By comparing crustal thicknesses (defined by compressional seismic wave velocities (V_p) < 7.8 km/s) of remnant and active arcs, and knowing the time interval over which incremental active arc growth had occurred relative to the magmatically-inactive remnant arc, *Arculus* [1999] calculated a flux of 200 km^3 per km of arc strike/million years. This estimate includes all sediments trapped within the arc crust from the Bonin trench to the margin of the backarc (Shikoku) basin with the active arc, and is equivalent to a global arc production rate of ~7 km^3/year; note this estimate may only be representative for the past 15 million years of activity in the Izu-Bonin system.

Taira et al. (1998) made a similar calculation using the Reymer and Schubert [1984] approach, assuming the Izu-Bonin arc was constructed on 6 km-thick oceanic crust, and was constructed over 45 million years (assumed age of arc inception) resulting in a flux of 80 km^3 per km of arc strike/million years. *Dimalanta et al.* [2002] have extended this approach to a number of western Pacific arcs to derive a mass flux of 100 km^3 per km of arc strike/million years. If the proto-Izu-Bonin arc was formed in an extensional (but still supra-subduction zone) environment, and not built on any pre-existing oceanic basement [as argued by *Taylor*, 1992], then the crustal growth rate to produce the entire 300 km wide by 20 km thick arc crust in ~ 50 million years [*Cosca et al.*, 1998] is 120 km^3 per km of arc strike/million years. *DeBari et al.* [1999] have identified some fragments of the Philippine Sea Plate trapped in the Izu-Bonin forearc, possibly forming a basement to the arc system, so that 120 km^3 per km of arc strike/million years would be an upper limit over 50 million years. The main points to note about these newer figures is they are a factor of 3 to 6 times greater than the global average estimate of *Reymer and Schubert* [1984] and closer to that of *Crisp* [1984], may only be representative of some of the western Pacific arcs, and represent limited time periods of arc construction.

Mass Flux of H_2O

Assuming steady state exists and arc magmatism is the only return flux for subducted H_2O, then with an estimated flux of subducted H_2O of $5 * 10^{11}$ kg/year (or $1.26 * 10^{13}$ kg per km strike/million years, given a 7.5 km thick ocean crust containing 2 wt% H_2O subducting at 3 cm/year; *Peacock,* 1990], an arc production rate of 30 km^3 per km of arc strike length/million years [*Reymer and Schubert*, 1984], and primitive arc magmas containing on average 2 wt% H_2O, only $1/_{6th}$ of the subducted H_2O flux can be returned directly to the Earth's surface via arc magmatism. If higher subduction rates (e.g., 5 cm/year as a current global average) prevail, then the discrepancy between subducted and return flux is increased. Note that the return fluxes of H_2O via MOR and hot-spot magmatism presented earlier, are respectively 1 to 2 orders of magnitude smaller than an arc return flux calculated with the *Reymer and Schubert* [1984] arc volume flux and an average of 2 wt% H_2O in parental arc magmas.

If parental arc magmas are wetter or the higher mass flux estimated for the Izu-Bonin system is typical, the discrepancy between subducted and return flux of H_2O could be halved or disappear. *Hilton et al.* [2002] calculate a global arc H_2O emission flux of $1.5 * 10^{12}$ kg/year, based on the assumption H_2O/CO_2 (molecular) = 50, and scaled to global estimates of SO_2 fluxes determined spectroscopically on sub-aerial volcanic plumes, implying steady state for H_2O through the subducted slab-arc magma generation-degassing cycle. However, both the suberial SO_2 flux and H_2O/CO_2 are currently poorly known [*Wallace et al.*, 2003].

I note further that assuming an average 2 wt% H_2O in parental arc magmas as a return flux is a preliminary estimate. Global heterogeneity of primitive arc magmas exists based on mineralogy and compositional characteristics [*Pearce and Parkinson,* 1993; *Arculus,* 1994], so varied H_2O contents are certain (Figure 2). For example, we know that some arc magmas are relatively dry with < 0.5 wt% H_2O [e.g., *Sisson and Bronto,* 1998] based on direct analytical evidence, and the early appearance of phenocryst plagioclase relative to clinopyroxene. Others however, contain up to 10 wt% H_2O [*Grove et al.*, 2003]. Heterogeneity derives from the range of independent and linked variables controlling magma production in arc systems. These include: 1. the potential fertility of a given mantle wedge source with respect to basalt production, and its temperature distribution. For example, there appears to be some linkage between the potential fertility of wedge sources below arcs and the presence of an adjacent backarc basin melting régime [*Woodhead et al.*, 1993]; 2. the amounts of subducted slab-derived volatiles (in the system C-O-H-S-halogens) and melts introduced into the wedge which may be independent of the previous melting history of the wedge; 3. the percentage of partial melting of the fluid-modified mantle wedge, which is a function both of fluid addition as well as the extent of decompression melting [*Pearce and Parkinson,* 1993]. At present, we do not have sufficient data to obtain a global arc average of parental magmatic H_2O contents. There may in fact be a trade-off between relatively low-temperature, volatile-rich, low mass flux systems and high-temperature, volatile-poor, high mass flux arc-backarc systems.

Mass Flux of S and C Compounds

If the S cycle is in steady state through subduction zones, and arc magmas are the only mechanism for return of S to the Earth's surface, then *Jambon* [1994] showed a S arc emission flux of ~ $5 * 10^{10}$ kg/year is required. This is based on the estimated flux of S into the oceans that becomes fixed in the oceanic crust in addition to any magmatic S already trapped within the crust during solidification at a MOR (~ $6*10^{10}$ kg/year; *Jambon,* 1994). An emission flux of $2 * 10^{10}$ kg/year of SO_2 was presented by *Hilton et al.* [2002] from a global assessment of data obtained via correlation spectrometer (COSPEC) studies of subaerial arc volcano gas plumes [e.g., *Stoiber et al.*, 1983]. The development of more accurate, remotely-sensed arc gas plumes with differential optical absorption spectrometers (DOAS) should result however, in improved estimates of this suberial arc S flux [e.g., *Galle et al.*, 2002].

With the *Reymer and Schubert* [1984] total mass flux of 30 km^3 per km of arc strike/million years, or about $3.4 * 10^{12}$ kg/year, the S concentration in arc magmas must be an implausible 1.5 wt%. Alternatively, if we set the S concentration in primitive arc magmas at ~ 0.3 wt% (3000 ppm) with a S flux of $5 * 10^{10}$ kg/year, then a total arc mass flux of ~ $1.7 * 10^{13}$ kg/year is required; this is 5 times the *Reymer and Schubert* [1984] estimate but consistent with the upper limits calculated for the Izu-Bonin system and the global estimate of *Crisp* [1984]. Given the relatively low abundances (<100 ppm) of S present in subaerially-erupted arc magmas, it is possible that most of the S is either degassed or trapped in the crust as magmatic and hydrothermal sulfide [*Alt et al.*, 1993].

The question of the C flux from arcs is clearly of special interest in terms of the long-term stability of the global C cycle. The relative insolubility of CO_2 at low pressures in arc magmas means that quantitative degassing occurs, commencing at crustal pressures (Figure 1). This is unlike the case for H_2O and S compounds which can be sequestered in part over a range of pressures (upper mantle to crust) in phases such as amphibole, mica, apatite, sulfides, and in some cases, anhydrite and haüyne.

In fact, more detailed analysis of the volatile fluxes from arcs would have to consider the details of residence times of volatiles within newly formed, intra- and underplated, amphibole-mica-apatite-sulfide-bearing crustal lithologies. These rock types are potentially fertile protoliths for the formation of various granitic magma types [e.g., *Chappell and White*, 2001], and the possible return to the atmosphere of H_2O and SO_2 via catastrophic, explosive rhyolitic volcanism. This type of igneous activity wherein the composition of the continental crust becomes modified through partial melting and vertical elemental redistributions is not restricted to convergent margins.

Given the relative stability of deeply recycled carbonate (as $CaCO_3$ in the subducted oceanic crust, and as $MgCO_3$ in the mantle) [*Kerrick and Connolly*, 2001], it might be anticipated that the output of C in arcs is less than the subducted flux. The temporal aspects of C recycling are also of interest in that the predominant component (calcic foraminifera) of pelagic carbonate ooze, typically forming the initial layer of the sedimentary blanket on newly-formed oceanic crust is a post-Jurassic development [*Lipps*, 1970]. Apart from potentially subductable carbonate of the post-Jurassic oceanic crust Layer 1, an additional flux of ~ $1 * 10^{11}$ kg/year of hydrothermally-generated carbonate trapped deeper in the oceanic crust likely exists [*Staudigel et al.*, 1990].

In the case of CO_2, it is worth recalling the steps by which the flux estimates at MOR have been made. Critical in the calculations for ridges has been the scaling between ^{3}He and heat flow [*Craig et al.*, 1975; *Elderfield and Schultz*, 1996]. For MOR, heat transported in hydrothermal fluids is the balance between the measured (conductive) heat flows and totals calculated on the basis of a cooling oceanic lithosphere. From these studies, an estimate of a ^{3}He flux of 1000 mol/year was derived, and knowing the abundance ratios of CO_2 to ^{3}He in quenched MORB glasses, a $CO_2/^3$He of $2.2\pm0.7 \times 10^9$ was established, giving a total MOR CO_2 flux of about $1 * 10^{12}$ kg/year. *Torgersen* [1989] assumed a total mass flux in arcs $1/5^{th}$ of the MOR flux [from *Crisp*, 1984], and consequently a ^{3}He flux in arcs of 200 mol/year.

Sano and Williams [1996] calculated a CO_2 emission flux from arcs of $3.1 * 10^{12}$ mol/year ($1.36 * 10^{11}$ kg/year) based on the ^{3}He flux of *Torgerson* [1989], $CO_2/^3$He, and isotopes of C and He. Assuming arc magmas contain 1000 ppm CO_2, then with the *Reymer and Schubert* [1984] mass flux, a maximum of $3.4 * 10^9$ kg/year CO_2 could be emitted. In a recent review with scaling (via direct gas sampling) of individual gas species such as CO_2 to COSPEC-determined, subaerial SO_2 emissions, *Hilton et al.*[2002] calculated an arc CO_2 flux of $1.63 * 10^{12}$ mol/year ($7.2 * 10^{10}$ kg/year), and a ^{3}He flux of 92 mol/year; from these figures, it appears that the amount of C being subducted exceeds the total return arc flux by an order of magnitude, but considerable uncertainty in many of the critical flux estimates persists.

Submarine Volcanic Plumes—A Complementary Approach

While we have learned much from spectral measurements of subaerial arc volcano gas plumes, a complementary (albeit expensive) strategy to determine gas fluxes directly is to take advantage of their entrapment in a water column, as has been done successfully for MOR settings [e.g., *Elderfield and Schultz*, 1996]. Furthermore, the dramatically enhanced quenching rates of lavas erupted at comparatively high pressures (equivalent to depths of 1000 to 4000m) in sea water rather than air means improved preservation of both magmatic glass and dissolved volatile loads.

The most significant results obtained to date using this approach have been summarized by *de Ronde et al.* [2001; 2003]. For the Tonga–Kermadec arc system, these authors note: 1. there is a major submarine arc volcano at ~ 30 km spacing along the volcanic front; 2. about 40% of these are emitting gas plumes (variably particle-rich); 3. These plumes are much richer in H_2S, CO_2, and Fe than MOR plumes. Stronger negative anomalies in pH for the arc (ΔpHmax = -0.35) compared with MOR plumes (ΔpHmax = -0.04) are interpreted by *Massoth et al.* [in press] to reflect mostly (70%) CO_2 degassing, coupled with a contribution from S species ($H_2S + SO_2$). The dimensions of these arc plumes are about ~

500m diameter and 250 m thick for 50% increases in ^{3}He/^{4}He above that of air. Ignoring the mega-hydrothermal plumes [e.g., *Baker et al.*, 1995] identified at MOR centres, possibly equivalent to the periods of intense gas emission associated with subaerial eruptions, confirmed plume sources along major ridges (e.g., along the East Pacific Rise at 15°S) tend to be larger (for the same elevations of ^{3}He/^{4}He above that of air) but less frequent (1 every 100 km) (*Lupton*, 1995).

An important point emerging from these submarine arc plume studies is the possibility that more than 25% of magmatic-derived hydrothermal fluid input to the ocean overlying the Pacific Plate may derive from submarine arc volcanoes (excluding contributions from intra-plate sources)[*Massoth et al.*, 2003]. Furthermore, the hydrothermal fluids are injected at depths of < 500m by a number of these arc volcanoes, and could be important in terms of high Fe contents for the productivity of photosynthetic organisms. At present however, we lack time-series measurements for these submarine arc plumes with which to accurately determine fluxes, and our scaling depends on ocean current models.

COMMENTS AND CONCLUSIONS

Global approaches are required to solve global issues such as volatile recycling through subduction zone systems. Nevertheless, a generalist stance has to be tempered with local details. For example, the specifics of erupted and intruded arc magma types coupled with the compositional characteristics of locally subducted lithosphere are clearly important [*Plank and Langmuir*, 1993]. All oceanic lithosphere does not bear the same (Layer 1) sediment types, the sediment load entering any given subduction zone varies with time, and not all of these sediments are subducted [*von Huene and Scholl*, 1991].

The eruptive and intrusive behavior, and compositional evolution of arc magma is also likely to be globally variable [*Arculus et al.*, 1995]. How significant these factors are at present is not known. Major issues surround our estimates of total mass and individual volatile species fluxes. For example, an intriguing feature of the well-constrained crustal structure of the Izu-Bonin arc is the large percentage (~40%) of relatively low V_p (6.1 to 6.3 km/s) rock types [*Suyehiro et al.*, 1996], consistent with a granodioritic composition (~60 – 70 wt% SiO_2). The bulk composition of the Izu-Bonin crust is not basaltic but likely andesitic (~58wt% SiO_2). Given that the arc magma flux from the mantle underlying the Izu-Bonin arc is most probably dominated by basalt, the question arises as to where the "missing complement" (i.e., mafic and ultramafic, relatively SiO_2-poor cumulates) of the more SiO_2-rich crustal types is located. These cumulates are likely to be sub-Moho, dominated by dunite and wehrlite lithologies with minor Ca-rich plagioclase feldspar (see section on Magmatic Compositional Evolution), with Vp's ranging from 7.8 to 8.3 km/s. These cumulates are hard to distinguish seismically from normal mantle, especially with elevated temperatures [*Fliedner and Klemperer*, 2000].

It has been suggested that "underplated" cumulate materials may be incorporated in the subjacent, advecting wedge, and recycled into the mantle [*Arculus*, 1999; *Davidson and Arculus*, 2004]. An important consequence in terms of flux calculations is that the total mass flux is likely greater (possibly double) than ~ 100 km^3 per km of arc strike/million years calculated from the supra-Moho thicknesses alone. With about 2 to 4 wt% H_2O dissolved in primary basalt magmas of the Izu-Bonin arc, a fractional crystallization history involving ~ 50% mass separation of anhydrous spinel-olivine-clinopyroxene assemblages will double the H_2O contents of crust-forming andesite magma. Such estimates of course radically alter the calculation of H_2O flux entering the arc crust.

The estimation of the subducted H_2O flux used here follows that of *Peacock* [1990], and is based on the degree of hydration of the igneous portions of the oceanic crust as determined by deep sea drilling and on-land exposures (ophiolites). The pore and structurally-bound H_2O in the sediments of Layer 1 has been ignored. If this was included, the total subducted flux is increased considerably. Pore water is likely expelled in the forearc, but the amount of H_2O bound in sheet silicates and amphiboles of Layer 1 that is transported into the mantle is not so clear. A number of possible H_2O–bearing hosts are stable along the low dT/dP trajectories of progressively dehydrating, subducting slabs [*Schmidt and Poli*, 1998], including phengite, epidote, lawsonite, zoisite, and amphibole. In their calculation of the H_2O cycle through subduction zones, *Hilton et al.* [2002] assume only ~ 1 wt% H_2O remains in the subducted slab in the regions of supposed arc magma generation. However, the magmatic (and dissolved volatile) conduit represented by an island arc likely acts in the manner of a collecting funnel in the sense of channeling melts and volatiles dispersed over some distance down-dip and normal to the subduction zone [*Spiegelman and McKenzie*, 1987]. In this framework, H_2O contents towards the upper end of the range 2 to 6 wt% are required in crust-forming, arc magmas to balance the H_2O cycle.

On the basis of isotopic and trace element systematics, some petrological models currently propose involvement of two volatile-rich fluid components in arc magmas; one derived by dehydration of the igneous portion of the oceanic crust, and the other possibly a H_2O-rich melt of Layer 1[e.g., *Elliott et al.*, 1997]. While these fluids are generated in different positions in the subducted lithosphere, both may be involved in triggering melting of the overlying mantle and become dissolved in the resultant magma.

Spreading ridges in backarcs represent an additional potential return path for recycled volatiles; those proximal to arcs are known to bear the imprint of a subduction component in the form of distinctive trace element abundance patterns and isotopic characteristics, and H_2O enrichments relative to MORB [*e.g., Stolper and Newman*, 1994; *Kamenetsky et al.*, 1997]. We currently lack the necessary remnant-active arc crustal structural details akin to those of the Izu Bonin system in regions of well-studied backarc-arc systems (e.g., West Mariana Ridge – Mariana Trough – Mariana Arc; Lau Ridge – Lau Basin – Tonga Arc) where known magmatic volatile loads could be placed in mass flux context. This is clearly an area of promising future research.

To summarize: of the volumetrically significant magma types, those emplaced in arcs at plate convergence zones are typically richest in dissolved volatiles (H_2O, CO_2, and S species). To a large extent, these volatiles are the products of large-scale recycling, although a significant fraction must also be derived from the mantle wedge. For example, *Hilton et al.* [2002] argue on the basis of the overlap of $^3He/^4He$ with MORB ($\sim 8 *$ atmospheric) and a global mean of $5.4 *$ atmospheric, that the majority of the ^3He emitted from arc magmas must be mantle wedge-derived. Similarly, about 20% of the CO_2 flux in arcs is suggested to be wedge-derived with the balance from the subducted lithosphere [*Sano and Williams*, 1996], but large uncertainties are attached to these estimates.

These volatiles follow a complex path though multi-stage and poorly understood processes, involving: devolatilization (and possibly dehydration melting) of variably hydrated and carbonated, sediment-bearing, subducted lithosphere [*Bebout*, 1995]; migration into the mantle wedge overlying the subducted plate, triggering melting and basalt generation in the wedge; transportation towards the surface dissolved in basalt magmas; achieving vapor saturation, initiated by highly insoluble CO_2, at crustal pressures; variably escaping to the atmosphere and hydrosphere or remaining trapped (hosted by volatile-rich crystalline phases) within newly formed arc crust. We have a preliminary global coverage of SO_2 fluxes in subaerial arc gas plumes, but the deployment of DOAS promises to improve these estimates considerably. Subsequently, we should be able to estimate the fluxes of other species such as H_2O and CO_2 in these plumes by scaling their abundances relative to SO_2 in gas samples [e.g., *Giggenbach*, 1992].

The volatile contents of parental arc basalts are globally variable, but mostly range from $\sim$ a fraction of a percent to 10 wt% H_2O, ≤ 1000 ppm CO_2, and ≤ 3500 ppm S (as H_2S and/or SO_2). For as yet unexplained reasons, arc magmas are generally more oxidized than those of ridges and hot-spots. These characteristics lead to distinctive differences in the course of crystallization compared with dry, reduced magma types: saturation with plagioclase is delayed, and is Ca-rich upon first appearance;

olivine persists for longer in the crystallization sequence; a spinel phase appears early in the crystallization sequence and persists throughout. The resultant relatively voluminous Na-K-feldspar- and SiO_2-rich residual magmas constitute the granodioritic building blocks of the Phanerozoic continental crust [*Taylor and McLennan*, 1985; *Davidson and Arculus*, 2004]. The complementary SiO_2-poor olivine-clinopyroxene-dominated fraction is probably recycled into the upper mantle.

The fluxes of volatiles through the subduction cycle are not straightforwardly determined; total magma volume flux estimates range from ~ 1.2 to 7 km^3/year, but are not well constrained for the full diversity of global arc systems. Even at the high end of this range, more C is being subducted than returned in arcs, presumably with long-term effects on the global C cycle. Similarly, we have little evidence for the amount of H_2O that remains hosted in deeply subducted lithosphere, both crust and mantle sections.

At the lower end of the total arc volume flux estimates, deficiencies exist in return fluxes of H_2O and S species. At the higher end, given the probability that some fraction of the H_2O and S remains locked in newly-formed arc crust rather than emitted to the atmosphere and hydrosphere, the subduction cycle may be in steady state, and the oceans are not draining. Overall, it appears from the very dry nature of MORB in which a recognizable component of recycled oceanic lithosphere is involved, that dehydration during subduction is very efficient [*Dixon et al.*, 2002]. How much of this H_2O is returned through arc systems however, has not yet been rigorously established.

Acknowledgments. My plume-sniffing, rock-dredging mariner collaborators (Ed Baker, Ray Binns, Dave Christie, Cornel de Ronde, Gary Massoth, Tim McConachy, Peter Michael, Chris Yeats, and Tim Worthington) together with ANU colleagues (David Ellis, John Mavrogenes, and S. Ross Taylor) have been educative and sources of much wisdom. Steve Sparks, Chris Hawkesworth, and Jon Davidson provided critical reviews. The Australian Research Council, the ANU and its Department of Earth and Marine Sciences have provided financial support, and I am grateful to the Steering Committee of the Australian National Marine Facility for use of research vessels (*RV Franklin* and *RV Southern Surveyor*).

REFERENCES

Alt, J. C., Shanks, W. C. and Jackson, M. C., Cycling of sulfur in subduction zones: the geochemistry of sulfur in the Mariana island arc and back-arc trough, *Earth Planet. Sci. Lett., 119*, 477–494, 1993.

Andres, R. J., Rose, W. I., Kyle, P. R., deSilva, S., Francis, P., Gardeweg, M. and Moreno Roa, H., Excessive sulfur dioxide emissions from Chilean volcanoes, *J. Volcanol. Geotherm. Res., 46*, 323–329, 1991.

Arculus, R. J., Oxidation status of the mantle: past and present, *Ann Rev. Earth Planet. Sci., 13*, 75–95, 1985.

Arculus, R. J., Aspects of magma genesis in arcs, *Lithos, 33,* 189–208, 1994.

Arculus, R. J., Origins of the continental crust, *Journal & Proc. Royal Soc. New South Wales*, *132,* 83–110, 1999.

Arculus, R. J., The use and abuse of the terms calcalkaline and calcalkalic, *J. Petrol., 44,* 929–935, 2003.

Arculus, R J., Gill, J. B., Cambray, H., Chen, W. and Stern, R.J., Geochemical evolution of arc systems in the western Pacific: the ash and turbidite record recovered by drilling. In (B. Taylor & J. Natland, eds.) *Active Margins and Marginal Basins of the Western Pacific,* American Geophysical Union, *Geophysical Monograph Series, 88,* 78–101, 1995.

Baker, E. T., German, C. R. and Elderfield, H., Hydrothermal plumes over spreading-center axes: global distribution and geological inferences, *Seafloor Hydrothermal Systems, AGU Geophysical Monograph, 91,* 47–71, 1995.

Basaltic Volcanism Study Project, *Basaltic Volcanism on the Terrestrial Planets.* Pergamon Press, Inc., New York. 1286 pp., 1981.

Bebout, G. E., The impact of subduction-zone metamorphism on mantle-ocean chemical cycling, *Chem. Geol., 126,* 191–218, 1995.

Binns, R. A. and Scott, S. D., Actively forming polymetallic sulfide deposits associated with felsic volcanic rocks in the eastern Manus backarc basin, Papua New Guinea, *Econ. Geol, 88,* 2226–2236, 1993.

Campbell, I. H. and Taylor, S.R., No water, no granites—No oceans, no continents, *Geophysical Research Letters, 10,* 1061–1064, 1985.

Carmichael, I. S. E., The redox states of basic and silicic magmas—a reflection of their source regions, *Contrib. Mineral. Petrol., 106,* 129–141, 1991.

Carroll, M. R. and Rutherford, M. J., Sulfide and sulfate saturation in hydrous silicate melts, *Proc. 15th Lunar Planet. Sci. Conf., J. Geophys. Res., 90,* C601–C612, 1985.

Chappell, B. W. and White, A. J. R., Two contrasting granite types: 25 years later, *Australian J. Earth Sci., 48,* 489–499, 2001.

Cosca, M. A., Arculus, R. J., Pearce, J. A. and Mitchell, J. G., $^{40}Ar/^{39}Ar$ and K-Ar geochronological age constraints for the inception and early evolution of the Izu-Bonin-Mariana arc system, *The Island Arc, 7,* 579–595, 1998.

Craig, H., Clarke, W. B. and Beg, M. A., Excess ^{3}He in the deep water on the East Pacific Rise, *Earth Planet. Sci. Lett., 26,* 125–132, 1975.

Crawford, A. J., *Boninites,* Unwin Hyman, London, 465pp, 1989.

Crisp, J. A., Rates of magma emplacement and volcanic output, *J. Volc. Geotherm. Res., 20,* 177–211, 1984.

Danyushevsky, L. V., McNeill, A. W. and Sobolev, A. V., Experimental and petrological studies of melt inclusions in phenocrysts from mantle-derived magmas: an overview of techniques, advantages and complications, *Chem. Geol., 183,* 5–14, 2002.

Davidson, J. P. and Arculus, R. J., The significance of Phanerozoic arc magamtism in generating continental crust, In (M. Brown and T. Rushmer, eds.), *Evolution and Differentiation of the Continental Crust,* Cambridge University Press, Cambridge, in press, 2004.

DeBari, S. M. Coleman, R. G., Examination of the deep levels of an island arc: evidence from the Tonsina ultramafic-mafic assemblage, Tonsina, Alaska, *J. Geophys. Res., 94(B),* 4373–4391, 1989.

DeBari, S. M., Taylor, B., Spencer, K. and Fujioka, K., A trapped Philippine Sea plate origin for MORB from the inner slope of the Izu-Bonin trench, *Earth Planet. Sci. Lett., 174,* 183–197, 1999.

Defant, M. J. and Drummond, M. S., Derivation of some modern arc magmas by melting of young subducted lithosphere, *Nature, 347,* 662–665, 1990.

de Ronde, C. E. J., Baker, E. T., Massoth, G. J., Lupton, J. E., Wright, I. C., Feely, R. A. and Greene, R., Intra-oceanic subduction-related hydrothermal venting, Kermadec volcanic arc, New Zealand, *Earth Planet. Sci. Lett., 193,* 359–369, 2001.

de Ronde, C. E. J., Massoth, G. J., Baker, E. T. and Lupton, J. E., Submarine hydrothermal venting related to volcanic arcs, *Econ. Geol., Special Pub., 10,* 91–110, 2003

Dimalanta, C., Taira, A., Yumul, G. P., Tokuyama, H., Mochizuki, K. New rates of western Pacific island arc magmatism from seismic and gravity data, *Earth Planet. Sci. Lett., 202,* 105–115, 2002.

Dixon, J. E., Leist, L., Langmuir, C. H. and Schilling, J-G., Recycled dehydrated lithosphere observed in plume-influenced mid-ocean-ridge basalt, *Nature, 420,* 385–389, 2002.

Dixon, J. E. and Stolper, E. M., An experimental study of water and carbon dioxide solubilities in mid-ocean ridge basaltic liquids. Part II: applications to degassing, *J. Petrol., 36,* 1633–1646, 1995.

Eggler, D. H., Water saturated and undersaturated melting relations in a Paracutin andesite and estimate of H_2O content in the natural magma, *Contrib. Mineral. Petrol., 34,* 261–271, 1972.

Elderfield, H. and Schultz, A., Mid-ocean ridge hydrothermal fluxes and the chemical composition of the ocean, *Ann. Rev. Earth Planet. Sci., 24,* 191–224, 1996.

Elliott, T., Plank, T., Zindlet, A., White, W. and Bourdon, B., Element transport from slab to volcanic front at the Mariana arc, *J. Geophys. Res., 102,* 14991–15019, 1997.

Fine, G. and Stolper, E., Dissolved carbon dioxide in basaltic glasses: concentrations and speciation, *Earth Planet. Sci. Lett., 76,* 263–278, 1985/1986.

Fliedner, M. M. and Klemperer, S. L., Crustal structure transition from oceanic arc to continental arc, eastern Aleutian Islands and Alaska Peninsula, *Earth Planet. Sci. Lett., 179,* 567–579, 2000.

Frost, B. R. and Ballhaus, C., Comment on "Constraints on the origin of the oxidation state of mantle overlying subduction zones: an example from Simcoe, Washington, USA" by A. D. Brandon and D. S. Draper, *Geochim. Cosmochim. Acta, 62,* 329–331, 1998.

Fryer, P., Mottl, M., Johnson, L., Haggerty, J., Phipps, S. and Maekawa, H., Serpentine bodies in the forearcs of western Pacific convergent margins: origin and associated fluids, *Active Margins,Marginal Basins Western Pacific, AGU Geophysical Monograph, 88,* 259–279, 1995.

Gaetani, G. A., Grove, T. L. and Bryan, W. B., The influence of water on the petrogenesis of subduction-related igneous rocks, *Nature, 365,* 332–334, 1993.

Gaetani, G. A. and Grove, T. L., The influence of water on melting of mantle peridotites, *Contrib. Mineral. Petrol., 131,* 323–346, 1998.

Galle, B., Oppenheimer, C., Geyer, A., McGonigle, A. J. S., Edmonds, M. and Horrocks, L., A miniaturized ultraviolet spectrometer for remote sensing of SO_2 fluxes: a new tool for volcano surveillance, *J. Volcanol. Geotherm. Res., 119,* 241–254, 2002.

Garrison, J. M. and Davidson, J. P., Dubious case for slab melting in the

Northern volcanic zone of the Andes, *Geology, 31,* 565–568, 2003.

Giggenbach, W. F., SEG Distinguished Lecture – Magma degassing and mineral deposition in hydrothermal systems aong convergent plate boundaries, *Econ. Geol.,87,* 1927–1944, 1992.

Gill, J. B., *Orogenic Andesites and Plate Tectonics,* Springer-Verlag, Berlin, 392 pp., 1981.

Grove, T. L., Elkins-Tanton, L. T., Parman, S. W., Chatterjee, N., Müntener. O. and Gaetani, G. A., Fractional crystallization and mantle-melting controls on calc-alkaline differentiation trends, *Contrib. Mineral. Petrol., 145,* 515–533, 2003.

Hedenquist, J. W. and Lowenstern, J. B., The role of magmas in the formation of hydrothermal ore deposits, *Nature, 370,* 519–527, 1994.

Hilton, D. R., Fischer, T. P. and Marty, B., Noble gases in subduction zones and volatile recycling, *Revs Mineralogy Geochemistry, 47,* 319–370, 2002.

Hofmann, A. W., Chemical differentiation of the Earth: the relationship between mantle, continental crust, and oceanic crust, *Earth Planet. Sci. Lett., 90,* 297–314, 1988.

Hofmann, A. W., Jochum, K. P., Seufert, M. and White, W. M., Nb and Pb inoceanic basalts: new constraints on mantle evolution, *Earth Planet. Sci. Lett., 79,* 33–45, 1986.

Holloway, J. R. and Burnham, C. W., Melting relations of basalt with equilibrium water pressure less than total pressure, *J. Petrol, 13,* 1–29, 1972.

Ingrin, J. and Skogby, H., Hydrogen in nominally anhydrous upper mantle minerals: concentration levels and implications, *European J. Mineralogy, 12,* 543–570, 2000.

Jambon, A., Earth degassing and large-scale geochemical cycling of volatile elements, *Volatiles in Magmas, Reviews in Mineralogy, Min. Soc. Amer., 30,* 479–517, 1994.

Johnson, M. C., Anderson, A. T., Jr. and Rutherford, M. J., Pre-eruptive volatile contents of magmas, *Volatiles in Magmas, Reviews in Mineralogy, Min. Soc. Amer., 30,* 281–330, 1994.

Kamenetsky, V., Crawford, A. J., Eggins, S. M. and Mühe, R., Phenocryst and melt inclusion chemistry of near-axis seamounts, Valu Fa Ridge, Lau Basin: insight into mantle wedge melting and the addition of subduction components. *Earth Planet. Sci. Lett., 151,* 205–223, 1997.

Kamenetsky, V. S., Binns, R. A., Gemmell, J. B., Crawford, A. J., Mernagh, T. P., Maas, R. and Steele, D., Parental basaltic melts and fluids in eastern Manus backarc Basin: implications for hydrothermal mineralisation, *Earth Planet. Sci. Lett., 184,* 685–702, 2001.

Kerrick, D. M. and Connolly, J. A. D., Metamorphic devolatilization of subducted marine sediments and the transport of volatiles into the Earth's mantle. *Nature, 411,* 293–296, 2001.

Khan, M. A., Jan, M. Q. and Weaver, B. L., Evolution of the lower arc crust in Kohistan, N. Pakistan: temporal arc magmatism through early, mature and intra-arc rift stages. In (P. J. Treloar & M. P. Searle, eds.) *Himalayan Tectonics,* Geol. Soc. London Spec. Pub., 74, 123–138, 1993.

Lipps, J. H., Plankton evolution, *Evolution 24,* 1–22, 1993.

Lowenstern, J. B., Carbon dioxide in magmas and implications for hydrothermal systems, *Mineralium Deposita, 36,* 490–502, 2001.

Luhr, J. F., Experimental phase relations of water- and sulphur-saturated arc magmas and the 1982 eruptions of El Chichon Volcano, *J. Petrology, 31,* 1071–1114, 1990.

Lupton, J. E., Hydrothermal plumes: near and far field, *Seafloor Hydrothermal Systems, AGU Geophysical Monograph, 91,* 317–346, 1995.

Massoth, G. J., de Ronde, C. E. J., Lupton, J. E., Feely, R. A., Baker, E. T., Lebon, G. T. and Maenner, S.M., Chemically rich and diverse submarine hydrothermal plumes of the southern Kermadec volcanic arc (New Zealand), In (Larter, R. and Leat, P., eds) *Intra-oceanic subduction systems: tectonic and magmatic processes,* Geological Society of London Special Publication, 2003, in press.

McBirney, A. R., Compositional variations in Cenozoic calc-alkaline suites of Central America, *Proc. Andesite Conf., Dept Geology and Mineral Industries, State of Oregon, Bull., 65,* 185–189, 1969.

Merzbacher, C. and Eggler, D. H., A magmatic geohygrometer: application to Mount St. Helens and other dacitic magmas, *Geology, 12,* 587–590, 1984.

Michael, P. J., Regionally distinctive sources of depleted MORB: evidence from trace elements and H_2O, *Earth Planet. Sci. Lett., 131,* 301–320, 1995

Miyashiro, A., Volcanic rock series in island arcs and active continental margins, *Am. J. Sci., 274,* 321–355, 1974.

Moore, C. and Silver, E., Fluid flow in accreting and eroding convergent margins, *JOIDES Journal, 28,* 91–96, 2002.

Newman, S., Stolper, E. and Stern R., H_2O and CO_2 in magmas from the Mariana arc and backarc systems, *Geochemistry Geophysics Geosystems, 1,* 1999GC000027, 2000.

Nilsson, K. and Peach, C. L., Sulfur speciation, oxidation state, and sulphur concentration in backarc magmas, *Geochim. Cosmochim. Acta, 57,* 3807–3813, 1993.

O'Neill, H. StC. and Mavrogenes, J. A., The sulphide capacity and sulphur content at sulphide saturation of silicate melts at 1400°C and 1 bar, *J. Petrology, 43,* 1049–1087, 2002.

Parkinson, I. J. and Arculus, R. J., The redox state of subduction zones: insights from arc peridotites, *Chem Geol., 160,* 409–423, 1999.

Peacock, S. M., Numerical simulation of metamorphic pressure-temperature-time paths and fluid production in subducting slabs, *Tectonics, 9,* 1197–1211, 1990.

Pearce, J. A. and Parkinson, I. J., Trace element models for mantle melting: application to volcanic arc petrogenesis, *Geological Society of London Special Pub., 76,* 373–403, 1993.

Plank, T. and Langmuir, C. H., Tracing trace elements from sediment input to volcanic output at subduction zones, *Nature, 362,* 739–743, 1993.

Poli, S. and Schmidt, M. W., Petrology of subducted slabs, *Annu. Rev. Earth Planet. Sci., 30,* 207–235, 2002.

Reymer, A. and Schubert, G., Phanerozoic addition rates to continental crust and crustal growth, *Tectonics, 3,* 63–77, 1984.

Righter, K., Chesley, J. T. and Ruiz, J., Genesis of primitive arc-type basalt: constraints from Re, Os, and Cl on the depth of melting and role of fluids, *Geology, 30,* 619–622, 2002

Roggensack, K., Hervig, R. L., McKnight, S. B. and Williams, S. N., Explosive basaltic volcanism from Cerro Negro volcano: influence of volatiles on eruptive style, *Science, 277,* 1639–1642, 1997.

Rutherford, M. J. and Devine, J. D., Preeruption pressure-temperature conditions and volatiles in the 1991 dacitic magma of Mount

Pinatubo, in *Fire and Mud* (eds. C. G. Newhall and R. S. Punongbayan), Phil. Inst. Volc. Seismol. And Univ. Washington Press, 751–766, 1996.

Rutherford, M. J., Sigurdsson, H. and Carey, S., The May 18, 1980 eruption of Mount St. Helens, 1. Melt compositions and experimental phase equilibria, *J. Geophys. Res., 90,* 2929–2947, 1985.

Saal, A. E., Hauri, E. H., Langmuir, C. H. and Perfit, M. R., Vapour undersaturation in primitive MORB and the volatile content of Earth's upper mantle, *Nature, 419,* 451–455, 2002.

Sano, Y. and Williams, S. N., Fluxes of mantle and subducted carbon along convergent plate boundaries, *Geophys. Res. Lett., 23,* 2749–2752, 1996.

Schmidt, M. W. and Poli, S., Experimentally based water budgets for dehydrating slabs and consequences for arc magma generation, *Earth Planet. Sci. Lett., 163,* 361–379, 1998.

Shieh, S. R., Mao, H-K., Hemley, R. J. and Ming, L. C., Decomposition of phase D in the lower mantle and the fate of dense hydrous silicates in subducting slabs, *Earth Planet. Sci. Lett., 159,* 13–23, 1998.

Sisson, T. W. and Bronto, S., Evidence for pressure-release melting beneath magmatic arcs from basalt at Galunggung, Indonesia, *Nature, 391,* 883–886, 1998.

Sisson, T. W. and Grove, T. L., Temperature and H_2O contents of low-MgO high-alumina basalts, *Contrib. Mineral. & Petrol. 113,* 167–184, 1993a.

Sisson, T. W. and Grove, T. L., Experimental investigations of the role of H_2O in calc-alkaline differentiation and subduction zone magmatism, *Contrib. Mineral. & Petrol. 113,* 143–166, 1993b.

Sisson, T. W. and Layne, G. D., H_2O in basalt and basaltic andesite glass inclusions from four subduction-related volcanoes, *Earth Planet. Sci. Lett., 117,* 619–635, 1993.

Spandler, C. J., Arculus, R. J., Eggins, S. M., Mavrogenes, J. A., Price, R. C. and Reay, A. J., Petrogenesis of the Greenhills Complex, Southland, New Zealand: magmatic differentiation and cumulate formation at the roots of a Permian island-arc volcano, *Contrib. Mineral. & Petrol., 144,* 703–721, 2003.

Spiegelman, M. and McKenzie, D., Simple 2-D models for melt extraction at mid-ocean ridges and island arcs, *Earth Planet. Sci. Lett., 83,* 137–152, 1987.

Staudigel, H., Hart, S. R., Schmincke, H. U. and Smith, B. M. Cretaceous ocean crust at DSDP Sites 417 and 418: carbon uptake from weathering versus loss by magmatic outgassing, *Geochim. Cosmochim Acta, 63,* 3091–3094, 1990.

Stoiber, R. E., Malinconico, L. L., J. and Williams, S. N., Use of the correlation spectrometer at volcanoes, In *Forecasting Volcanic Events* (H. Tazieff and J. Sabroux, eds.), Elsevier, New York, 425–444, 1983.

Stolper, E. and Newman, S., The role of water in the petrogenesis of Mariana Trough magmas, *Earth Planet. Sci. Lett., 121,* 293–325, 1994.

Straub, S. M. and Layne, G. D., Decoupling of fluids and fluid-mobile elements during shallow subduction: evidence from halogen-rich andesite melt inclusions from the Izu arc volcanic front, *Geochemistry Geophysics Geosystems, 4,* doi:10.1029/2002GC000349, 2003.

Sun, W., Bennett, V. C., Eggins, S. M., Kamenetsky, V. S. and Arculus, R. J., Enhanced mantle-to-crust rhenium transfer in undegassed arc magmas, *Nature, 422,* 294–297, 2003a.

Sun, W., Arculus, R. J., Bennett, V. C., Eggins, S. M. and Binns, R. A., Evidence for rhenium enrichment in the mantle wedge from submarine arc-like volcanic glasses (Papua New Guinea), *Geology, 31,* 845–848, 2003b.

Suyehiro, K., Takahashi, N. and Ariie, Y., Continental crust, crustal underplating, and low-Q upper mantle beneath an oceanic island arc, *Science, 271,* 390–392. 1996.

Taira, A., Saito, S., Aoike, K., Morita, S., Tokuyama, H., Suyehiro, K., Takahashi, N., Shinohara, M., Kiyokawa, S., Naka, J. and Klaus, A., Nature and growth rate of the northern Izu-Bonin (Ogasawara) arc crust and their implications for continental crust formation, *The Island Arc, 7,* 395–407, 1998.

Takahashi, N., Suyehiro, K. and Shinohara, M., Implications from the seismic crustal structure of the northern Izu-Bonin arc, *The Island Arc, 7,* 383–394, 1998.

Taylor, B., Rifting and the volcanic-tectonic evolution of the Izu-Bonin-Mariana arc, In (B.Taylor, and K. Fujioka et al., eds.) *Proceedings of the Ocean Drilling Program, Scientific Results, 126,* 625–651, 1992.

Taylor, S. R. and McLennan, S. M., *The Continental Crust: Its Composition and Evolution,* Blackwell, Oxford, 312 pp., 1985.

Torgersen, T., Terrestrial helium degassing fluxes and the atmospheric helium budget: implications with respect to the degassing processes of continental crust, *Chem. Geol., 79,* 1–14, 1989.

Wallace, P. and Carmichael, I. S. E., Sulfur in basaltic magmas, *Geochim. Cosmochim Acta, 56,* 1863–1874, 1992.

Wallace, P. J., Carn, S. A., Rose, W. I., Bluth, G. S. J. and Gerlach, T., Integrating petrologic and remote sensing perspectives on magmatic volatiles and volcanic degassing, *EOS, 84,* 441–447, 2003

von Huene, R. and Scholl, D. W., Observations at convergent margins concerning sediment subduction, subduction erosion, and the growth of the continental crust, *Rev. Geophysics* 29: 279–316, 1991.

Wise, D. U., Freeboard of continents through time, *Geol. Soc. Am. Mem., 132,* 87–96, 1972.

Woodhead, J. Eggins, S. and Gamble, J., High-field strength and transition element systematics in island-arc and back-arc basin basalts—evidence for multiphase extraction and depleted mantle wedge, *Earth Planet. Sci. Lett., 114,* 491–504, 1993.

Woodland, S. J., Pearson, D. G. and Thirlwall, M. F., A platinum group element and Re-Os isotope investigation of siderophile element recycling in subduction zones: comparison of Grenada, Lesser Antilles arc and the Izu-Bonin arc, *J. Petrol., 42,* 171–198, 2002.

Yang, K. and Scott, S. D., Magmatic degassing of volatiles and ore metals into a hydrothermal system on the modern sea floor of the eastern Manus back-arc basin, western Pacific, *Econ. Geol., 97,* 1079–1100, 2002.

Yoder, H. S. Jr. and Tilley, C. E., Origin of basaltic magmas—an experimental study of natural and synthetic rock systems, *J. Petrology, 3,* 342–532, 1962.

Young, D. A., *N. L. Bowen and Crystallization-Differentiation, the Evolution of a Theory,* Min. Soc. Amer., 276 pp., 1998.

R.J. Arculus, Department of Earth and Marine Sciences, Australian National University, Canberra, ACT, 0200, Australia. (Richard.Arculus@anu.edu.au)

Volatile Controls on Magma Ascent and Eruption

Katharine V. Cashman

Department of Geological Sciences, University of Oregon, Eugene, Oregon

"Gas is the active agent, and the magma is its vehicle" F. A. Perret

Volatiles provide the primary driving force for volcanic eruptions, thus understanding magma degassing is fundamental to understanding volcanic activity in volatile-rich arc environments. A complete picture of volatile behavior requires knowledge of not only (1) how, when and where volatiles saturate, exsolve, and accumulate in magma reservoirs and (2) how they nucleate, expand and coalesce during magma ascent; but also (3) how volatiles affect the phase relations, rheology, and fluid dynamics of the magma, and (4) how volatiles escape to and interact with surrounding wall rocks and hydrothermal systems. Together these interactions determine degassing conditions, rates of magma ascent to the Earth's surface, and, ultimately, the style and intensity of volcanic eruptions. For example, rapid closed-system degassing provides the explosivity of silicic plinian eruptions, while open-system gas escape permits passive effusion of lava domes. Variations in the details of magma decompression rates and paths create the rich variability in eruptive style that characterizes arc volcanism. Development of a fully integrated perspective on the role of volatiles in volcanic systems requires both better constraints on the time scales and dynamics of processes occurring within volcanic conduits and coupling of conduit processes to other parts of subvolcanic systems.

1. INTRODUCTION

Magma produced in arc environments is typically rich in volatile elements, particularly H_2O. Volatile elements are so-named because they dissolve in silicate melts at high pressures but form a gas at lower pressures. Formation of a free vapor phase creates overpressures in magma reservoirs and adds buoyancy that drives magma ascent; rapid near-surface gas expansion provides the kinetic energy of explosive eruptions. Thus although volatile processes have not been explicitly incorporated into common eruption classification schemes [e.g., *Lacroix*, 1904; *Mercalli* 1907; *Walker*, 1973; *Pyle*, 1989; Fig. 1], implicit assumptions about the role of volatiles underlie most interpretations of eruptive activity. These assumptions are most obvious in the Volcanic Explosivity Index [*Newhall and Self*, 1983], which assumes that eruption magnitude and intensity are correlated and together describe eruptive style. However, subsequent work has shown that these parameters are linked only for steady explosive eruptions, and must be decoupled for descriptions of unsteady and effusive activity [*Pyle*, 2000]. Here I show that much of this decoupling arises from complex feedbacks that develop as a result of competing time scales of decompression-driven vesiculation, crystallization, magma ascent, and volatile loss. Weaving these complexities into new descriptions of eruptive behavior and predictive models of volcanic activity presents the main challenge for the future.

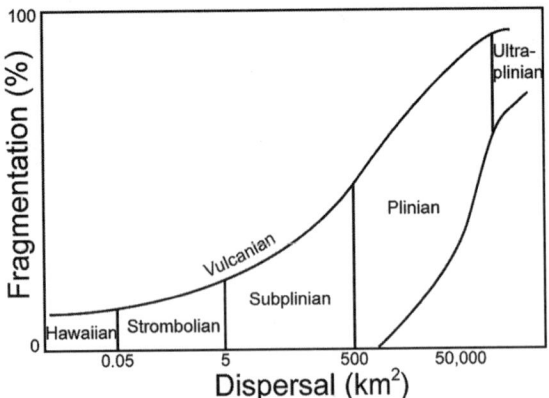

Figure 1. Walker's [1973] classification of volcanic eruptions based on deposit characteristics.

2. MAGMA STORAGE

Most magma resides in shallow (upper crustal) storage regions before beginning the final ascent toward the surface. Here it may saturate in volatiles by cooling and crystallizing, by adding volatiles from a deep source, or by decompression. In this section I review evidence from melt inclusions, volatiles released during eruption, and phase equilibria that many arc magmas are not only saturated in volatiles prior to eruption, but contain several wt% of an exsolved C-O-H-S vapor phase. Understanding the temporal development and spatial distribution of this vapor phase is central to questions of eruption triggering and transfer of magmatic volatiles to ore-forming systems.

2.1. Magmatic Volatiles

Reviews of magmatic volatiles are provided in Carroll and Holloway [1994] and Wallace and Anderson [2000] and will be only briefly summarized here. Water and carbon dioxide are the two most common volatile components in natural magmas, followed by sulfur, chlorine and fluorine. Water is much more soluble in silicate melts than is carbon dioxide, although the solubility of both species decreases with decreasing pressure. Solubility is also a function of melt composition, with both H_2O and CO_2 being somewhat less soluble in basaltic melts than in rhyolitic melts (Fig. 2a,b). In a vapor-saturated melt, the solubility of a given volatile species is determined by its equilibrium partial pressure in the gas phase. For this reason, addition of H_2O to a CO_2-bearing melt at constant pressure increases P_{H2O}, which decreases both P_{CO2} and dissolved CO_2 (Fig. 2c). The low abundance of S, Cl and F suggests that exsolution of these components exerts only a minor control on eruption style, although sulfur emissions play a critical role in determining the climactic effects of large eruptions.

Magmatic volatiles may be measured directly when they are quenched in melt pockets (inclusions) trapped in phenocrysts (Fig. 3a; inset). Melt inclusions are best preserved in crystals with poor cleavage, such as olivine and quartz, thus detailed melt inclusion studies are limited to end member basaltic or rhyolitic compositions [*Lowenstern*, 1994;1995; *Metrich et al.*, 2001; *Roggensack et al.*, 1997; *Schmitt*, 2001; *Sisson and Layne*, 1993; *Wallace et al.*, 1995;1999]. These studies show that both arc-related basalts and large rhyolitic magma bodies may initially contain > 6 wt% H_2O. However, melt inclusions in individual phenocrysts within a given eruptive deposit often contain variable amounts of H_2O and CO_2, reflecting crystallization over a wide range of magmatic pressures. This range is particularly striking in studies of large ignimbrite units such as the Bishop and Pine Grove Tuffs (Fig. 3a). Correlations between CO_2 content and incompatible trace element abundance in these inclusions indicate that crystallization occurred under gas-saturated conditions and created an exsolved volatile phase that would have comprised up to 6 wt% (30 vol%) of the magma reservoir. Measurements of SO_2 emissions associated with explosive eruptions support this interpretation. Volcanic plumes often contain "excess" S, that is, sulphur volumes in excess of the amount dissolved in the erupted magma. This excess S must have been exsolved in a C-O-H-S vapor phase prior to eruption [Fig. 3b]. Rec-

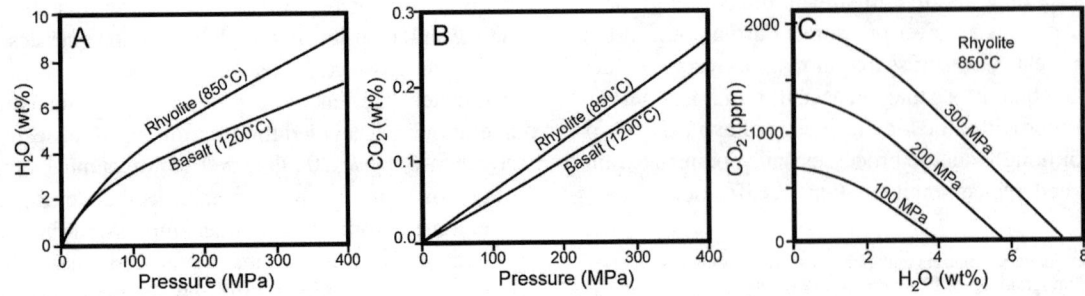

Figure 2. (a) H_2O solubility in rhyolite (850°C) and basalt (1200°C); (b) CO_2 solubility in rhyolite (850°C) and basalt (1200°C); H_2O and CO_2 solubility in volatile-saturated rhyolitic melt at 850°C. Redrafted from Wallace and Anderson [2000].

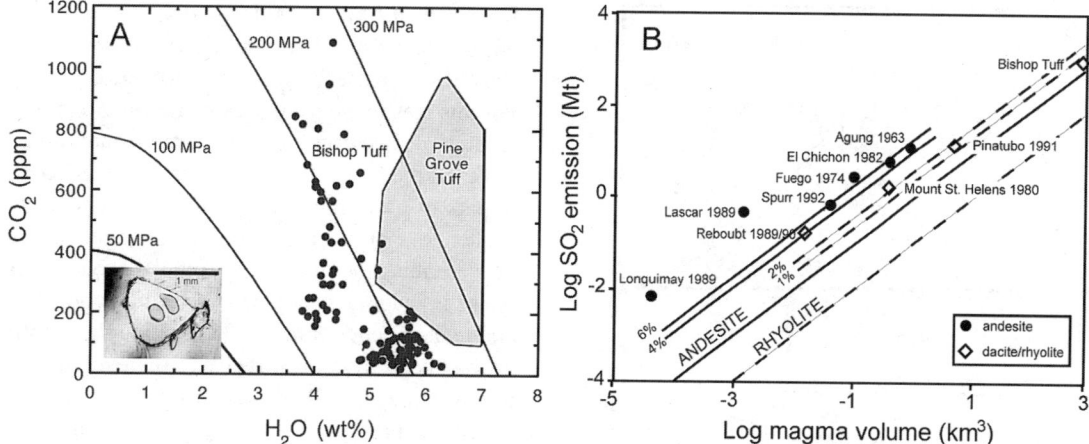

Figure 3. (a) H_2O and CO_2 concentrations in quartz melt inclusions from two large silicic systems: Pine Grove Tuff [outlined field, *Lowenstern, 1994*]; Bishop Tuff [points, *Wallace et al., 1995*]. (b) Anticipated (lines) and measured (points) relationship between erupted magma volume and SO_2 emissions for andesitic (filled circles; solid lines) and silicic (open diamonds, dashed lines) compositions. Calculated amounts of excess sulfur required to match observed emissions are labeled in wt% next to the appropriate line. Redrafted from Wallace [2003].

onciliation of measured and dissolved S abundances also requires 1–5 wt% vapor [*Gerlach and McGee*, 1994; *Luhr*, 1990; *Scaillet et al.*, 1998; *Wallace*, 2003].

2.2. Experimental Phase Equilibria

The volatile content of intermediate composition magmas is more difficult to constrain by melt inclusion analysis because the common phenocryst phases (plagioclase, pyroxene and hornblende) tend to leak volatiles. As a consequence, the volatile contents of these magmas are best determined through phase equilibria studies [e.g., *Costa and Scaillet*, 2003; *Moore and Carmichael*, 1998; *Rutherford and Devine*, 1988;1996], as illustrated in Figure 4a. The phases most indicative of volatile abundance are hornblende, which generally requires a minimum of 3–4 wt% H_2O in the melt, and plagioclase, the stability (liquidus temperature) of which is extremely sensitive to P_{H2O}. As plagioclase abundance is easily measured, it can be used in conjunction with experimental data to infer the water content and P_{H2O} at which magma last resided. For example, an isothermal (950°C) section through the phase diagram in Figure 4a shows that volatile-saturated decompression from 200 to 100 MPa would reduce the H_2O content of the liquid from 5.5 to 3.5 wt%, causing an increase in plagioclase abundance from 10 to 25% (Fig. 4b). Thus the phase assemblage of plagioclase, hornblende, pyroxene and Fe-Ti oxides common to intermediate composition calc-alkaline magmas provides an estimate of both the minimum equilibration pressures (from hornblende stability) and the extent of volatile-saturated decompression and crystallization [*Blundy and Cashman*, 2001; *Carmichael*, 2002; *Sisson and Grove*, 1993]. Evidence of extensive volatile-saturated crystallization in some systems supports SO_2-emission data that require a substantial (= 4 wt%) free volatile phase for most recent eruptions of andesitic arc volcanoes (Fig. 3b).

2.3. Porphyry Ore Deposits

Further evidence of volatile concentration in the upper parts of magma reservoirs lies in granite-hosted porphyry Cu and Mo ore deposits [*Burnham, 1997; Cloos*, 2001; *Lowenstern*, 1994; *Shinohara and Kazahaya*, 1995]. Porphyry ore deposits form in arc environments in association with shallow intrusions of intermediate to evolved compositions. Solidified granite porphyries have low bulk volatile contents that contrast with the high volatile content of most silicic melts. This mismatch requires that silicic magma chambers exsolve and discharge most of their volatile components during solidification, either to the atmosphere (via volcanic eruptions) or to hydrothermal systems. Evidence for high volatile concentrations above magma reservoirs can be found in the key stratigraphic components of porphyry systems: a volatile-enriched 'cupola' that typically overlies an elongate porphyry intrusion and is tapped by radial and concentric fluid-rich veins that feed the ore deposit. Although this physical picture is consistent with inferred volatile gradients in active volcanic systems, questions remain about the temporal and spatial relationship between ore formation and volcanism. Traditionally viewed as separate, growing evidence for high temperature magmatic fluids involved in ore formation and extensive volatile transfer accompanying volcanic eruptions suggests that links between these two environments need to be re-evaluated.

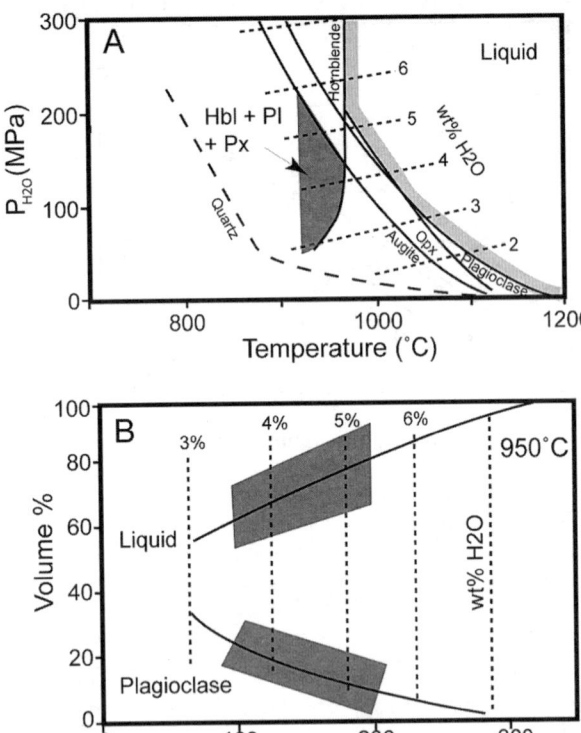

Figure 4. H$_2$O-saturated andesite phase relations [redrafted from *Moore and Carmichael*, 1998]. (a) Phase fields and equilibrium H$_2$O contents. Gray line is liquidus; gray field shows P$_{H_2O}$-T range of andesitic magmas; dashed line shows approximate quartz phase boundary [estimated from *Barclay et al.*, 1998; *Brugger et al.*, 2003; *Hammer and Rutherford*, 2002a; *Martel and Schmidt*, 2003]. (b) Isothermal section through (a) shows co-variation of melt and plagioclase phase proportions as a function of P$_{H_2O}$ (gray field).

2.4. Implications

The presence of a free volatile phase in regions of magma storage has enormous implications for magma ascent and eruption. Extensive decompression-induced crystallization explains the plagioclase-phyric nature of many arc magmas, and, perhaps, the common association of porphyry intrusives with magmatic hydrothermal ore deposits. Overpressures generated by exsolution of a free volatile phase may be sufficient to fracture wall rock and permit both gas escape and magma ascent toward the surface. Volatile accumulation and overpressure generation at the roofs of magma reservoirs may trigger rapidly propagating gas-driven fractures [*Carrigan*, 2000; *Lister*, 1990; *Rubin*, 1993] and transfer of volatiles into overlying hydrothermal systems. Such gas-driven fractures may precede magma ascent, causing phreatic (groundwater-driven) activity before the onset of a full-scale eruption [*Menard and Tait*, 2001]. Finally, free volatiles contained within the magma reservoir at the time of eruption will substantially increase the volatile contribution to the atmosphere. For this reason, petrologic estimates of volatile contributions from explosive eruptions are minima, and independent estimates of the free volatile phase are required to assess the full impact of past eruptions on global climate. Future research challenges include (1) identifying the origin of excess volatiles; (2) describing the spatial and temporal evolution of the volatile phase in magma reservoirs of varying depths, compositions, and tectonic environments; (3) linking the volatile evolution of magmatic systems directly to the evolution of ore-forming systems; and (4) determining potential trigger mechanisms for magma ascent and eruption provided by a free volatile phase.

3. MAGMA ASCENT

Prior to eruption, magma must be transferred from the storage reservoir to the surface, a process now recognized as critical to the study of eruptive processes [*Jaupart*, 2000]. Early models calculated the velocity of a homogeneous fluid moving steadily up a vertical pipe, with frictional losses dictated by wall shear stresses and volatile exsolution/expansion controlled by equilibrium solubility [*Wilson et al.*, 1980]. More recent models allow disequilibrium between the gas and liquid phases [*Jaupart and Allegre*, 1991; *Papale et al.*, 1998; *Woods and Koyaguchi*, 1994], compressible fluid flow [*Massol and Jaupart*, 1999; *Massol et al.*, 2001], coupling with thermodynamic models to calculate phase changes [*Mastin and Ghiorso*, 2000], open system degassing and crystallization [*Melnik and Sparks*, 1999;2002a;b] and variable bubble nucleation kinetics [*Mangan et al.*, 2004; *Papale*, 2001]. These models demonstrate the sensitivity of magma ascent to the kinetics of vesiculation and crystallization, and to related changes in the rheology of bubble- and crystal-bearing melt, the dynamics of magma flow through conduits, and the explosivity of resulting eruptions. Here I review field, experimental and theoretical constraints on both the kinetics of phase changes during decompression and the rheological changes induced by the addition of bubbles and/or crystals to silicic melts.

3.1. Phase Changes Resulting from Decompression

Ascent of magma through volcanic conduits decompresses the melt and, as a consequence, decreases the volatile solubility (Fig. 2) and changes the stability of crystalline phases (Fig. 4). In response, volatiles exsolve from the melt by either diffusion into existing gas bubbles or nucleation of new bubbles, hydrous minerals break down, and anhydrous minerals crystallize. The kinetics of these phase changes are controlled by melt composition, volatile content, viscosity and rate of magma decompression.

3.1.1. Bubble nucleation. Nucleation of bubbles within a melt may occur by one of two mechanisms: nucleation may be *heterogeneous* on existing crystal phases or *homogeneous* within the melt itself. Experiments on silicic melts show that heterogeneous nucleation occurs at low supersaturations ($\Delta P < 20$ MPa) and volatiles exsolve under near-equilibrium conditions. Heterogeneous nucleation produces a single population of bubbles whose number is controlled by the number of crystal sites [*Gardner et al.*, 1999; *Hurwitz and Navon*, 1994]. Further decompression causes these bubbles to grow, creating a Gaussian size distribution with a mean size that increases with decreasing pressure (Fig. 5a). In the absence of sufficient nucleation sites or at very high rates of decompression, nucleation is homogeneous and continuous [*Mangan et al.*, 2004; *Mangan and Sisson*, 2000; *Mourtada-Bonnefoi and Laporte*, 1999;2002], and bubble number decrease exponentially with increasing bubble size (Fig. 5b). Homogeneous nucleation requires high supersaturations ($\Delta P = \sim 100$ MPa) that increase with decreasing abundance of H_2O and CO_2 in the melt. Degassing paths are far from equilibrium, and continuous nucleation creates high bubble number densities ($> 10^8/cm^3$; Fig. 5c), particularly when the exsolving phase is a mixture of H_2O and CO_2.

In natural systems, bubble nucleation mechanisms may be inferred from the bubble populations preserved in eruptive products. Vesicular clasts of rhyolitic pumice produced by high energy Plinian eruptions have bubble number densities of 10^8–$10^{10}/cm^3$, equalling or exceeding those produced by homogeneous nucleation of H_2O–CO_2 vapor mixtures. Bubble numbers decrease exponentially with increasing bubble size [*Klug et al.*, 2002], and groundmass crystals that might serve as nucleation sites are rare or absent. These textural characteristics indicate that Plinian eruptions are driven by homogeneous bubble nucleation at high supersaturation. Initial bubble number densities, in turn, control (1) inter-bubble diffusion distances [*Gardner et al.*, 1999; *Mangan and Sisson*, 2000], (2) subsequent rates of bubble growth [*Proussevitch and Sahagian*, 1998], and (3) times required for local volatile depletion and brittle fragmentation [*Zhang*, 1999]. Conditions of bubble nucleation are more difficult to constrain for other eruption styles because of complications introduced by syn-eruptive crystallization and variable amounts of passive gas loss. However, experimental data suggest that heterogeneous, near-equilibrium, volatile exsolution should increase in importance as magma decompression rates decline.

3.1.2. Degassing-induced crystallization. First suggested early in the 20th century to explain the famous spine of Mt. Pelee [*Shepherd and Merwin*, 1927; *Williams*, 1932] and quantified during the 1980–1986 eruption of Mount St. Helens, WA [*Cashman*, 1992], decompression-induced crystallization has now been documented during effusive and intermittent explosive phases of numerous recent andesitic and dacitic eruptions [e.g., *Hammer et al.*, 1999;2000; *Martel et al.*, 2000; *Nakada et al.*, 1995;1999]. These observations have led to new models of cyclical dome growth and explosive activity controlled by complex inter-relations among processes of magma ascent, degassing, and crystallization [*Melnik and Sparks*, 1999; *Sparks*, 1997; *Voight et al.*, 1999].

Important constraints on crystallization rates are provided by H_2O-saturated decompression experiments on rhyolitic melts [*Couch et al.*, 2003; *Geschwind and Rutherford*, 1995; *Hammer and Rutherford*, 2002; *Martel and Schmidt*, 2003]. As anticipated from phase equilibria studies, plagioclase is the most abundant crystallizing phase. The effective undercooling (ΔT) driving plagioclase crystallization is determined by both the final pressure (P_f) and the decompression rate. The extent of crystallization generally increases with

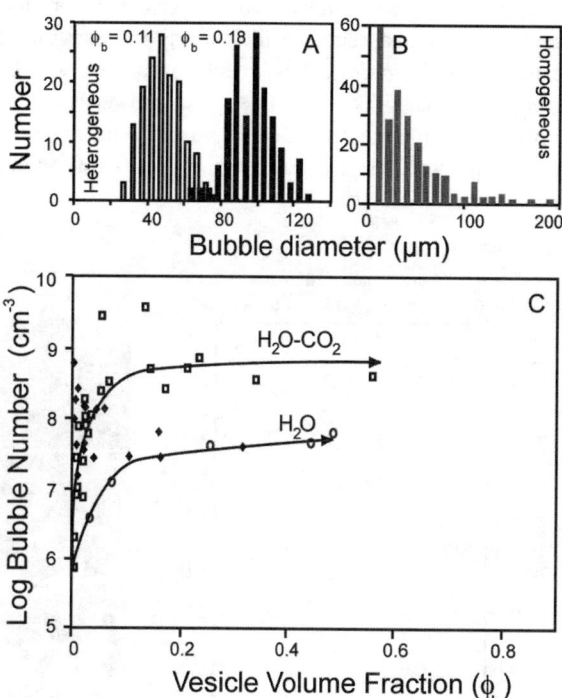

Figure 5. Results of bubble nucleation experiments: (a,b) bubble size distributions for heterogeneous and homogeneous nucleation; (c) log bubble number density (no./cm^3) vs. vesicle volume fraction (ϕ_b) for homogeneous (open symbols) and heterogeneous (solid symbols) nucleation [data from *Gardner et al.*, 1999; *Mourtada-Bonnefoi and Laporte*, 1999; 2002; *Mangan and Sisson*, 2000]. Arrows show general trend for homogeneous nucleation of H_2O- and H_2O-CO_2-bearing melts.

increasing ΔT, as does the mechanism of phase change, with crystal nucleation dominant at high ΔTs (P_f = 5–15 MPa) and crystal growth important at intermediate ΔTs (P_f ~ 100 MPa). Long times (slow decompression) permit extensive crystallization while short times (rapid decompression) limit plagioclase precipitation. In the extreme, very rapid magma transit to the Earth's surface will inhibit crystallization altogether.

Groundmass textures preserved in erupted samples show evidence of variable decompression rates and magma equilibration depths. When magma rises quickly but stalls at shallow levels, as is common in pulsatory (subplinian) eruptive sequences, crystallization occurs primarily by nucleation, and resulting pyroclasts contain abundant small skeletal or acicular plagioclase crystals (Fig. 6b). When magma ascent is slow (as in many effusive eruptions) or when magma stalls at intermediate depths, crystal growth dominates and the groundmass contains abundant plagioclase microphenocrysts (Fig. 6c). When magma ascent from magma storage depths is sufficiently rapid to permit vesiculation but not crystallization, the result is vesicular pumice (Fig. 6d). The same textural features may be seen in the anhydrous breakdown products of hornblende: rapid ascent to shallow levels produces thin, fine-grained breakdown rims; slow ascent permits extensive breakdown and growth of coarse-grained rims; and very rapid ascent prevents hornblende breakdown Thus both hornblende and groundmass textures provide important constraints on magma decompression histories.

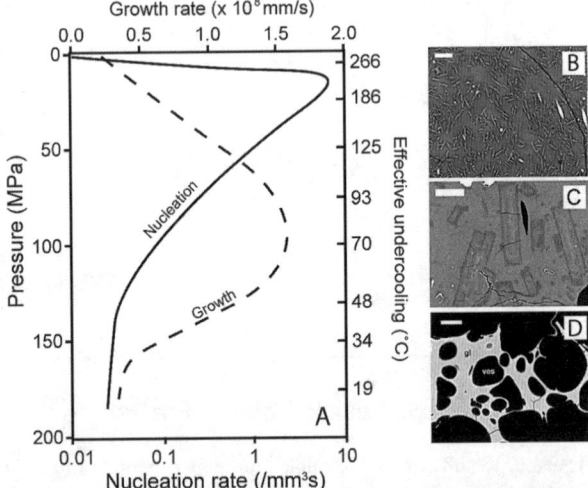

Figure 6. (a) Experimentally determined rates of plagioclase growth and nucleation in rhyolitic melt [data from *Hammer and Rutherford*, 2002]. (b–d) Groundmass textures representative of melts last equilibrated at (b) shallow (~ 10 MPa; Pichincha, Ecuador, breadcrust bomb), (c) intermediate (~ 100 MPa; MSH dome), and (d) deep (> ~ 150 MPa, MSH pumice) pressures.

3.2. Degassing-induced Changes in Magma Rheology

All of the physical changes experienced by magma during ascent and degassing translate into dramatic changes in the bulk rheology of bubble- and/or crystal-bearing melts, rheological changes that modulate the timing and explosivity of ensuing eruptions. There are three main controls on magma rheology in volcanic conduits: (1) exsolution of H_2O from the melt causes an increase in the melt viscosity, particularly at low water contents (low pressures); (2) addition of bubbles to a fluid creates shear-thinning rheologies; and (3) addition of crystals increases the magma viscosity and, when crystals interact, generates non-Newtonian responses to imposed shear stresses. Together these effects create complex feedback mechanisms that control rates of magma ascent and eruption.

Loss of H_2O has little effect on melt viscosity over much of the anticipated H_2O range of silicic melts: viscosity increases by only an order of magnitude as the water content of a rhyolite melt drops from 7 to 3 wt% (Fig. 7a). However, melt viscosity changes rapidly as H_2O decreases to low levels, increasing four orders of magnitude from 1.5 to 0.1 wt% H_2O (a pressure change of < 50 MPa). Thus shallow degassing is an effective means of producing large rheological contrasts over small pressure drops.

The effect of bubbles on magma rheology depends not only on bubble volume but also on bubble size (radius r), melt viscosity (μ) and shear rate (G; Fig. 7b). Shear-rate sensitivity is expressed by the capillary number $Ca = rG\mu/\Gamma$, a ratio of competing stresses imposed by shearing that deforms and surface tension (Γ) that restores a bubble to a spherical shape. At high shear rates, bubbles deform and reduce the shear viscosity; at low shear rates bubbles remain close to spherical and increase the shear viscosity [*Manga et al.*, 1998; *Pal*, 2003; *Rust and Manga*, 2002]. Although the effect of bubbles on magma viscosity is relatively small (less than a factor of 10 at either Ca limit), the shear-thinning behavior of bubbly magma concentrates shear near conduit walls to create large horizontal pressure and velocity gradients [*Massol and Jaupart*, 1999]. High shear rates may stretch bubbles to form tube pumice [*Klug et al.* 2002; *Polacci et al.* 2001], enhance coalescence and gas escape to produce dense obsidian [*Rust et al.*, 2003] or facilitate shear-induced fragmentation [*Papale*, 2001].

Adding crystals to a melt increases the suspension viscosity and, at moderate crystal concentrations, changes its rheology. The relationship between viscosity and particle volume fraction is controlled by the maximum packing fraction ϕ_m (Fig. 7c), commonly taken as 0.6 [*Marsh*, 1981]. However, the particle concentration at maximum packing changes with variations in both crystal shape and imposed stress [*Zhou et al.*, 1995]. Additionally, well before maximum packing is

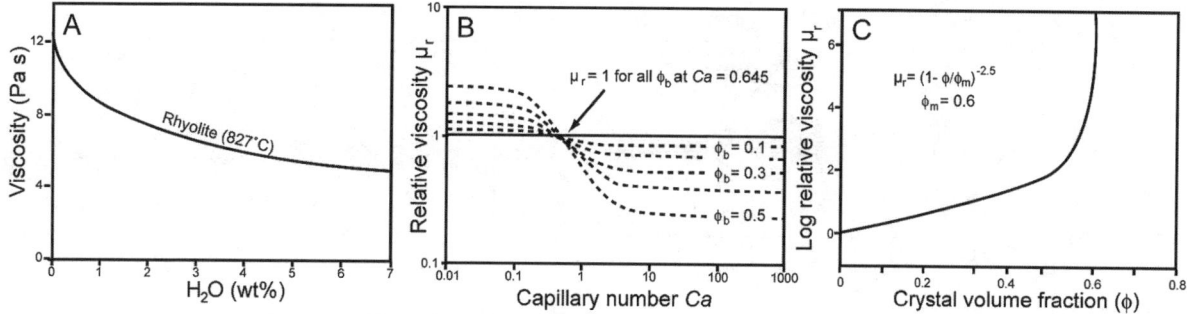

Figure 7. Viscosity variations in bubble- and crystal-bearing magmas. (a) Melt viscosity as a function of dissolved H_2O [*Hess and Dingwell*, 1996]. (b) Variation in relative viscosity (μ_r) with capillary number Ca for different bubble volume fractions (ϕ_b; from *Rust and Manga* [2002] and *Pal* [2003]). (c) Variation in log μ_r with increasing crystal volume fraction (ϕ) according to the simple Roscoe-Einstein relation for $\phi_m = 0.6$ [e.g., *Marsh*, 1981].

achieved, touching crystal networks create complex (non-Newtonian) responses to shear [*Kerr and Lister*, 1991]. Near the packing limit, large imposed shear stresses at conduit walls causes individual crystals to break; when accompanied by viscous heating this process may help to lubricate the ascent of crystal-rich magmas through volcanic conduits [*Polacci et al.*, 2001; *Rosi et al.*, 2003]. When the crystal concentration exceeds the packing limit, magma may dilate under shear [*Smith*, 2000]. As high crystallinities are common in very shallow magma instrusions [e.g., *Cashman and Hoblitt*, 2004], shear-dilatancy may contribute to both deformation and seismic precursors to some eruptions.

3.3. Modeling Volatiles and Magma Ascent

Rates of magma ascent thus depend on complex feedbacks determined largely by decompression rate. Rapid homogeneous bubble nucleation induced by very high rates of decompression causes rapid magma expansion and large horizontal velocity gradients because of shear-thinning rheologies. At more modest decompression rates, ascent rates will be controlled by the relative rates of bubble and crystal nucleation. This coupled problem of vesiculation and crystallization during decompression is poorly constrained. Thus future challenges for improving models of magma ascent include (1) improving our understanding of the kinetics of decompressing magmatic systems sufficiently to predict both the bubble- and crystal- content of the magma as a function of location in the conduit, (2) determining the effect of changing phase proportions on magma rheology, and (3) predicting the effect of changing rheologies on continued magma flow. To date, experimental and textural studies have been largely confined to rhyolitic melts, thus need to be extended to more mafic compositions. Additionally, while the effects of water content, temperature and bubble concentration on melt rheology are fairly well constrained, existing rheological models of crystal-melt suspensions are fairly crude, a reflection of difficulties posed by experimental and numerical studies.

4. DEGASSING AND ERUPTION

The physical changes that accompany magma ascent and decompression translate into a wide range of eruptive styles. For example, high rates of gas expansion that accompany rapid decompression and vesiculation cause explosive disruption of magma, generating large and sustained Plinian eruption columns. Alternatively, slow magma ascent accompanied by extensive crystallization and gas loss produces effusion of viscous lava domes. Between these end members lies a continuous range of eruptive styles characterized by variable decompression and degassing trajectories. Here I briefly review degassing styles, physical mechanisms by which gas may be lost during magma ascent, and the translation of degassing style into eruptive behavior.

4.1. Styles of Degassing

Volatile exsolution may occur under equilibrium or non-equilibrium conditions, may precede or accompany volcanic eruptions, and may involve variable amounts of degassing (gas escape). When bubbles remain within the melt during magma ascent, the mixture expands as a 'closed-system' and the gas volume fraction is a direct measure of the extent of degassing. In contrast, when volatiles segregate from the melt prior to eruption, the system is considered 'open' (Fig. 8a). Here vesiculation and eruption may be decoupled, and the gas volume fraction of the erupted material no longer indicates the extent or final equilibration pressure of degassing. Taking a broader perspective, an entire eruptive sequence may be viewed as closed when the rise and decompression of a single magma batch causes both magma and gas emission rates to decrease exponentially with time [*Gerlach and McGee*, 1994;

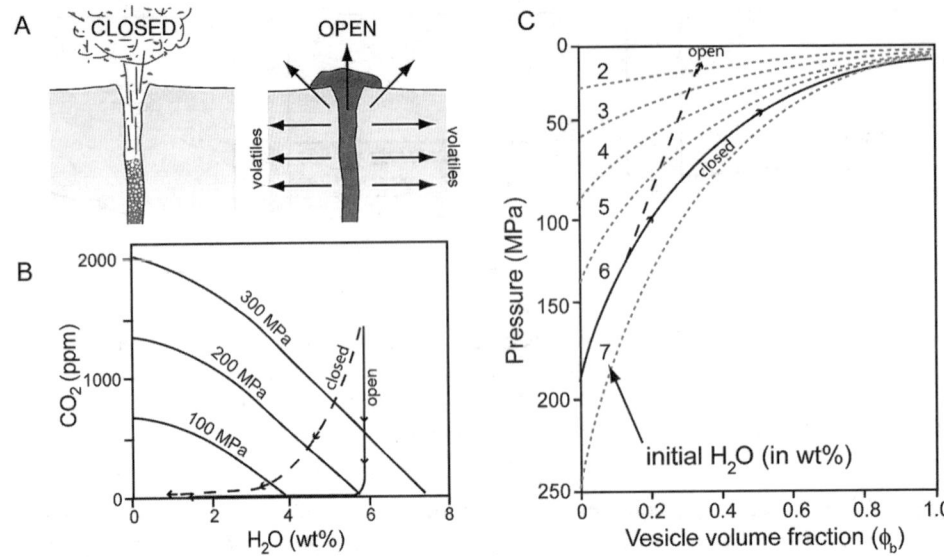

Figure 8. (a) Schematic illustration of open- and closed-system degassing. (b) CO_2-H_2O solubility diagram illustrating schematic open- and closed-system degassing paths. (c) Calculated change in vesicle volume fraction with pressure under closed-system conditions for a range of initial volatile contents (labeled on curves in wt%). Highlighted are example open- and closed-system degassing paths for an initial H_2O content of 5 wt%.

Zapata et al., 1997]. Volcanic systems may be considered open when gases are either lost from the conduit prior to eruption, or when volatiles fed from a deep source flux through the magmatic system. In the latter case, eruptive activity is modulated by the ability of the system to act as a conduit for the volatile phase [*Edmonds et al.*, 2003], and ultimately controlled by deep inputs to the magmatic reservoir [*Sparks and Young*, 2002].

Volatile exsolution under closed-system conditions causes a decrease in H_2O and CO_2 over much of the decompression range [Fig. 8b; e.g., *Newman and Lowenstern*, 2002]. Equilibrium calculations show that even modest amounts of H_2O produce final gas volume fractions > 0.99 if closed-system conditions are maintained to atmospheric pressure (Fig. 8c). In contrast, early gas loss during open-system degassing causes rapid CO_2 depletion of the melt at near-constant H_2O values (Fig. 8b). Once CO_2 is lost, continued decompression depletes H_2O. The gas volume fraction of the magma will be less (by an amount equal to the gas loss from the system) than predicted by equilibrium models.

4.2. Physical Controls on Open-system Degassing

Gas segregation in open systems may cause violent explosions if those gases are contained (and pressurized) within the conduit, or effusion of bubble-poor magma if gases are removed from the conduit [e.g., *Jaupart*, 1998]. Degassing conditions depend on both magma ascent rate and the relative permeabilities of magma and wall rock. Magma permeability may develop by brittle fracture of viscous magma or development of connected bubble networks within expanding bubbly melt. Magma will fracture when either the melt viscosity or the strain rate is sufficiently high, a condition most likely to be met along conduit walls [*Gonnerman and Manga*, 2003]. Evidence for syn-eruptive magma fracture includes brecciated obsidian clasts in subplinian pyroclastic deposits [*Rust et al.*, 2004], tuffisite veins found within exposed magma conduits [*Tuffen et al.*, 2003] and the occurrence of hybrid earthquakes interpreted to result from some combination of fracture and fluid flow [*Chouet*, 1996; *White et al.*, 1998].

Formation of connected bubble networks requires both a sufficiently high volume fraction of the gas phase and rupture of adjacent bubble walls (coalescence). Most models of permeability development in bubble-melt suspensions rely on percolation theory, which predicts development of touching networks of spherical bubbles at a bubble volume fraction ~ 30% [*Sahimi*, 1994; *Garboczi et al.*, 1995; *Saar and Manga*, 1999]. Coalescence is controlled by rates of fluid drainage and capillary pressures in stationary low viscosity melts. In rapidly decompressing melts, bubble expansion is likely to play a major role in film thinning, as are local pressure differences resulting from adjacent bubbles of different sizes [*Klug and Cashman*, 1996]. Silicic pyroclasts have high permeabilities over a range of porosities, and show no evidence of a threshold in gas volume fraction for the development of permeable networks (Fig. 9). Early and rapid expansion-driven coales-

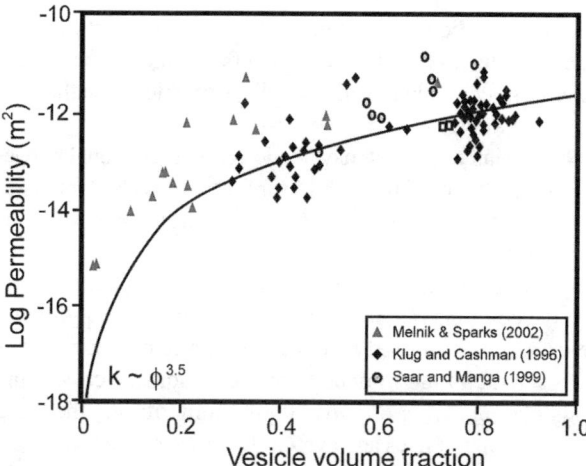

Figure 9. Permeability-ϕ_b relations measured in volcanic rocks; line is empirical fit to data of Klug and Cashman [1996].

cence observed in recent vesiculation experiments [*Mourtada-Bonnefoi and Laporte,* 2002], and high SO_2 fluxes from Soufriere Hills Volcano immediately following dome collapse [*Edmonds et al.,* 2003], provide additional evidence for rapid and extensive development of permeable bubble networks.

4.3. Degassing and Explosive Eruptions

By definition, explosive eruptions involve fragmentation of bubbly magma, where fragmentation refers to the transformation of a liquid or solid with dispersed gas bubbles to a gas with dispersed liquid drops or isolated solid particles. Fragmentation provides the explosive force of magmatic eruptions by converting the potential energy of expanding magma to the kinetic energy of individual fragments [*Cashman et al.,* 2000], and may be viewed as either dynamic or static. Dynamic fragmentation occurs when rapidly expanding bubbly melt disintegrates as the result of fluid instabilities when melt viscosity is low [e.g., *Mader,* 1998], or by exceedance of the tensile strength (brittle failure) when melt viscosity is high (Fig. 10a). In contrast, fragmentation is static when rapid decompression of previously vesiculated melt causes successive brittle fracture of the porous material [e.g., *Dingwell,* 1998; Fig. 10b].

4.3.1. Explosive eruptions involving rapid closed-system degassing. For magma to degas as a closed system, rates of bubble nucleation, expansion, and fragmentation must exceed those of bubble coalescence and gas migration. Eruptions driven by closed-system degassing are typically explosive, with eruption intensity and magnitude correlated, volatile emissions proportional to erupted volume, and highly expanded pyroclasts. Of the range of eruptive activity exhibited by andesitic to rhyolitic volcanoes, plinian eruptions most closely meet these criteria.

Plinian eruptions produce large volumes of magma that are emitted rapidly to produce sustained eruption columns and thick deposits of pumice and ash [*Cioni et al.,* 2000]. Eruption intensity is typically $> 10^7$ kg/s and correlates approximately with erupted volume [*Pyle,* 2000]. Pumice clasts are

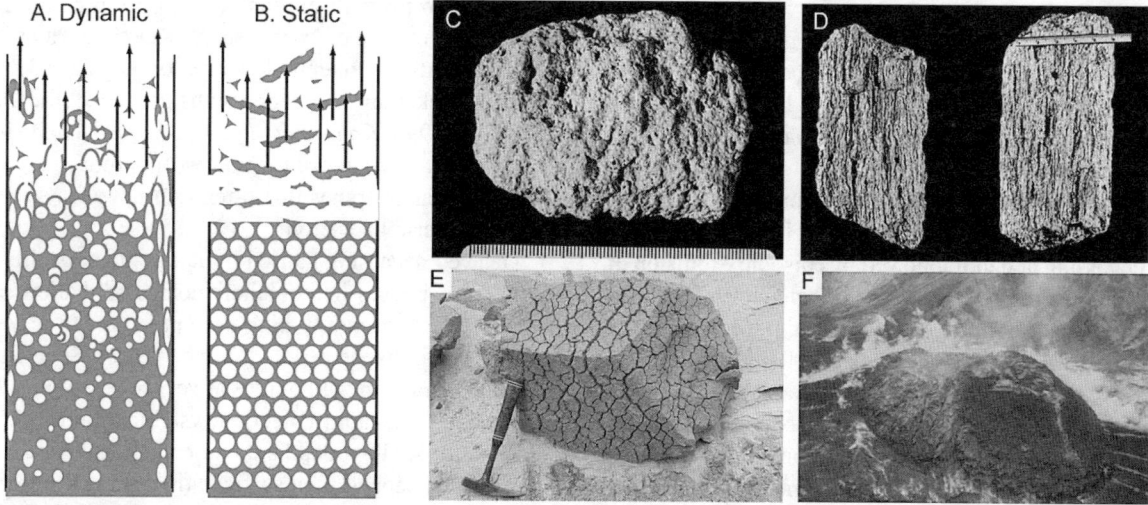

Figure 10. (a, b) Cartoons illustrating dynamic (syn-eruptive vesiculation) and "static" (pre-eruptive vesiculation) fragmentation mechanisms. (c) "Normal" and (d) "Tube" pumice samples from the 1991 eruption of Pinatubo Volcano, Philippines [from *Polacci et al.,* 2001]. (e) Breadcrust bomb from the 1999 eruption of Guagua Pichincha Volcano, Ecuador; note the angularity of the original fracture surfaces. (f) Smooth and scoriaceous surface textures on the 1980 Mount St. Helens lava dome.

vesicular, with high bubble number densities indicating homogeneous nucleation, rapid decompression and degassing under largely closed-system conditions. In detail, mean clast vesicularities of ~ 75% indicate fragmentation after near-spherical bubbles achieve close packing, which inhibits continued expansion [*Sparks*, 1978]. Fragmentation is dynamic, with isolated bubbles rupturing individually to form ash and permeable melt regions preserved as pumice [*Klug and Cashman*, 1996]. The extent of permeability development prior to fragmentation thus controls the volumetric proportion of pumice, while the extent of bubble deformation, expansion and collapse preserved in individual pyroclasts records variations in strain rate within the conduit (Fig. 10c,d).

4.3.2. Explosive eruptions involving pre-eruptive degassing. Magma that ascends slowly maintains near-equilibrium melt volatile concentrations and may experience variable amounts of degassing prior to eruption. Eruption intensity is proportional to the pre-eruptive overpressure achieved in the conduit, fragmentation is (initially) static, and emitted gas volumes may be either greater or less than amounts initially dissolved in the erupted magma. Pre-eruptive degassing and partial open-system behavior are responsible for the range of explosive activity classified in andesitic-to-rhyolitic volcanoes as vulcanian, pelean, or subplinian.

Vulcanian explosions are defined as small to moderate volcanic outbursts that last seconds to minutes [*Morrissey and Mastin*, 2000]. Vulcanian eruptions are powered by overpressures ≤ 10–15 MPa and produce discrete violent explosions, ballistic blocks, and atmospheric shock waves. The juvenile products of vulcanian eruptions range from dense blocks to vesicular pumice and include ubiquitous breadcrusted bombs. Prismatic breadcrusted blocks (Fig. 10e) and angular pumice clasts indicate extensive brittle fragmentation of previously vesiculated and variably degassed material from the uppermost conduit or capping dome. Suggested source models for magmatic vulcanian eruptions include pressurization of a lava cap or dome at shallow levels, and closed-system degassing at depth followed by episodic expulsion of gas into the conduit. The first type of behavior is exemplified by eruptions of Soufriere Hills Volcano, Montserrat, in 1997 [*Clarke et al.*, 2002; *Druitt and al.*, 2002; *Formenti et al.*, 2003], where volatile exsolution precedes eruption, the extent of degassing varies with position in the conduit, and fragmentation occurs by brittle fracture in response to a suddenly imposed decompression wave. Pelean (dome explosion) eruptions probably occur by a similar mechanism, but with a pressure source that is within the dome rather than within the conduit. The second model arises from seismic and deformation signals from volcanoes such as Popocatepetl, Mexico, that indicate gas accumulation at depths of > 2–3 km prior to episodic release [e.g., *Arcinega-Ceballos et al.*, 2003]. The mechanism by which this occurs and the structure of the conduit system that permits rapid gas transfer to shallow levels is not known.

Subplinian eruptions may be viewed as transitional between vulcanian and plinian, with lower magnitudes and intensities than plinian eruptions but longer durations than vulcanian explosions. Like vulcanian eruptions, however, subplinian eruptions are often characterized by unsteady (pulsatory) convective columns [*Cioni et al.*, 2000].. Subplinian pyroclastic deposits are dominated by vesicular pumice but include clasts with a wide range in density and groundmass crystallinity [e.g., *Cashman and Blundy*, 2000]. Similarities in the instantaneous eruption rates and pyroclast characteristics (vesicularity and texture) of plinian and subplinian eruptions suggest similar fragmentation mechanisms. However, the wide density range of subplinian deposits indicates a component of pre-eruptive volatile exsolution and gas escape during temporary magma arrest in shallow conduits. Thus it is not surprising that many subplinian eruptions are preceded by vulcanian explosions.

4.4. Degassing and Effusion of Lava Flows and Domes

Effusive eruptions produce lava flows or domes, with eruptive activity that may be continuous or episodic and may occur by extrusion of lava flow lobes (exogenous growth; Fig. 10f) or by intrusion of magma into an existing dome (endogenous growth). Continuous exogenous growth occurs when eruption rates are high and the system is sufficiently open to volatiles to prevent explosive disruption of the magma. Decreasing the rate of magma supply causes activity to become episodic until sufficiently low rates of magma ascent permit crustal thickening and endogenous growth styles [*Fink and Griffiths*, 1998; *Kaneko et al.*, 2002; *Swanson and Holcomb*, 1990]. Very low rates of magma ascent produce highly crystalline volcanic spines that often (but not always) represent the end of eruptive activity.

Delayed degassing caused by late-stage crystallization of anhydrous phases within silicic domes or shallow conduits may create sufficiently high pore pressures to drive explosive dome-collapse events [*Sparks*, 1997]. Specifically, as quartz appears on the liquidus at very low pressures (~ 5–10 MPa; Fig. 4a), rapid cotectic crystallization of quartz and feldspar at shallow depths could create explosive overpressures when capped by a sufficiently rigid and impermeable plug [*Blundy and Cashman*, 2001]. Evidence for high internal pressures within growing domes includes impulsive seismic signals, emitted jets of high-pressure gas, and high SO_2 emissions from domes immediately after collapse events [*Sparks and Young*, 2002].

4.5. Degassing Determined by Rates of Magma Ascent

As illustrated above, the rate of magma decompression controls the kinetics of vesiculation and crystallization, which in turn determine a magma's degassing history. This history is preserved (imperfectly) in the textures (vesicle and crystal content) of the erupted material (Fig. 11a). In general, closed-system degassing during plinian eruptions generates abundant vesicular juvenile clasts, while partial gas loss from conduits prior to subplinian and vulcanian eruptions produces variably degassed and crystallized juvenile material along with low-density pumice. More extensive degassing prior to effusive eruption of lava domes and flows creates crystal-rich and vesicle-poor lava. This comparison suggests that variations in the relative abundance of different clast density populations should provide a measure of pre-eruptive gas loss from the system. Using this interpretation, Figure 11a presents a general picture of volcanic systems that become increasingly open to volatiles as eruption styles become less energetic (change from explosive to effusive). As one goal of volcano monitoring is prediction of transitions in eruptive styles, it would be useful to quantify this general observation.

A measure of eruptive energy is the intensity, or mass eruption rate (MER, in kg/s). In sustained eruptions, MER is a direct measure of magma supply rate (MSR) from depth. In pulsatory (subplinian, vulcanian) eruptions, however, the instantaneous MER is a more a measure of overpressure and fragmentation conditions than of MSR. In fact, the condition MSR < MER may define pulsatory activity [*Scandone and Malone*, 1985]. Under these conditions, processes of magma ascent and degassing can be considered separately from those of fragmentation and eruption. Separation of these processes suggests a classification framework that both incorporates variable degassing scenarios and provides a direct link to recent decompression experiments.

The relationship between decompression rate (MSR) and degassing history is shown schematically in Figure 11b. The vertical axis uses clast density ranges shown in Figure 11a to approximate the extent of pre-eruptive degassing. Limiting magma supply rates are estimated at >10^7 kg/s for plinian eruptions, 10^5 to 10^7 kg/s for subplinian eruptions, 10^4 to 10^5 kg/s for vulcanian eruptions, and < ~10^4 for dome effusion, based on direct observations of recent activity [e.g., *Druitt et al.*, 2002; *Geschwind and Rutherford*, 1995; *Nakada and Motomura*, 1999; *Pyle*, 2000; *Scandone and Malone*, 1985]. Exact values will depend on the material properties of both the erupting magma and the wall rock. To estimate decompression rates corresponding to bounding MSRs, I assume a minimum conduit diameter of 10m (Fig. 11b). This yields maximum decompression rates of ~ 1 MPa/s and 0.01 MPa/s for the subplinian/plinian and subplinian/vulcanian boundaries, respectively. Similarly, magma ascent rates > 0.02 m/s required to preserve pristine hornblende phenocrysts [*Rutherford and Gardner*, 2000] correspond to MSR > 4000 kg/s (0.0005 MPa/s).

Although approximate, these boundaries illustrate the utility of using MSR as a framework for linking eruption style to

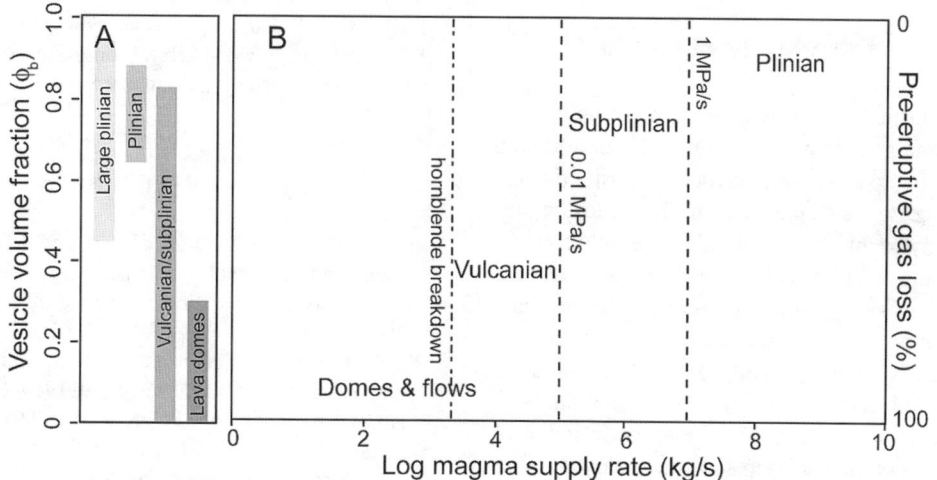

Figure 11. (a) Clast vesicularity range of large Plinian eruptions [*Houghton and Wilson*, 1989; *Polacci et al.* 2003], moderate Plinian eruptions [*Gardner et al.*, 1996; *Houghton and Wilson*, 1989; *Polacci et al.*, 2001], vulcanian and subplinian eruptions [*Gardner et al.*, 1998; *Formenti et al.*, 2003], and dome effusion [*Eichelberger et al.*, 1986; *Fink et al*, 1992]. (b) Alternative classification scheme based on magma supply rate and exsolution/degassing style. Limiting MSR values are provided by estimated eruption intensities for sustained eruptions [*Pyle*, 2000], subplinian and vulcanian activity [*Formenti et al.*, 2003; *Scandone and Malone*, 1985], and rates of dome growth [*Geschwind and Rutherford*, 1995; *Nakada et al.*, 1995; *Pyle*, 2000].

degassing conditions. Decompression rates in excess of ~ 1 MPa/s inhibit crystal nucleation [*Martel and Schmidt*, 2003] and promote rapid, homogeneous bubble nucleation [*Mangan and Sisson*, 2004]. These rates correspond to MSR > 10^7 kg/s, consistent with the estimated minimum MSR for plinian activity. When nucleation is heterogeneous (microlites are present) decompression rates < 0.025 MPa/s are required for maintenance of melt-vapor equilibrium over a large pressure range [*Gardner et al.*, 1999]. This value is approximately equivalent to an MSR of ~ 10^5 kg/s, and may define limiting conditions for vulcanian eruptions (which require extensive pre-eruptive degassing). For 10^7 > MSR > 10^5 kg/s, decompression rates are sufficiently slow to allow limited crystallization and degassing, but sufficiently fast that eruptions are probably driven by syn-, rather than pre-eruptive vesiculation. Finally, decompression rates of 0.005 MPa/s required for hornblende breakdown lie below the limiting MSR for lava dome effusion (~ 10^4 kg/s), consistent with observations of dome lavas with and without hornblende breakdown rims [*Rutherford and Gardner*, 2000].

In summary, Figure 11b, although schematic, provides a framework that (1) links explosive and effusive eruptive styles, (2) describes the control exerted by volatiles in determining eruptive style, and (3) can be calibrated using a combination of decompression experiments, numerical models, and observations of active volcanoes. This approach could easily be extended to include MSR limits to eruptive styles characteristic of hydrous mafic volcanism, which is also an important component of arc environments.

5. SUMMARY

Volatiles play a critical role in all aspects of volcanic activity in arc environments, from exsolution and accumulation in magma reservoirs, to degassing that occurs during magma ascent to the surface, to the eruptions themselves. Most importantly, the rates, styles and timing of volatile exsolution and degassing determine conditions of eruption. Recent advances in methods of measuring volatiles in melts, fumaroles and eruptive plumes provide excellent data on the behavior of the gases themselves. New experimental work constrains the degassing and crystallization behavior of silicic melts, although these studies have yet to be extended to more mafic and alkalic compositions. Of particular interest are the potential interactions between degassing and crystallization in melts where crystals nucleate easily, and may thus directly influence the evolution of the gas phase. Also important are the implications of strain-rate dependent rheology of bubble- and crystal-rich melts for both permeability development and for the flow of magma through volcanic conduits.

A recent focus on processes occurring along conduit margins is exciting, as it marks the first step toward filling a large gap in volatile studies, that is, coupling of magmatic degassing processes to conditions of volatile loss through conduit walls [e.g., *Jaupart and Allegre*, 1991]. Newly recognized links between gas emissions, seismic signals and magmatic processes within conduits provide the tantalizing vision that real-time monitoring of gas migration within subvolcanic systems will soon be possible. Also promising are new research avenues arising from rapid developments in remote sensing. Volatile emissions can now be measured directly during large eruptions, greatly enhancing our view of syn- and post-eruptive volatile behavior. Fully integrated volatile studies, however, will require extension of this linked chemical and physical analysis in both space and time. Spatial scales of observation are improving with the use of InSAR (radar interferometry) to document periodic deep replenishments to magma reservoirs, migration of hydrothermal waters in response to changing stress fields, and intrusions of magma into the upper crust. Improved temporal scales of observation will come from integration of different monitoring techniques (to improve resolution of short time scales) with studies conducted on active and exhumed magmatic-hydrothermal systems.

Acknowledgments. I would like to thank Maggie Mangan, Bernard Chouet, Hugh Tuffen, Paul Wallace and Helge Gonnerman for preprints of their papers. Stimulating discussions in both the field and the laboratory with Rafaello Cioni, Lucia Gurioli, Laura Pioli, Mauro Rosi, Alison Rust, Paul Wallace and Heather Wright helped to develop many of the ideas presented here. Mauro Rosi, Tony Fowler, Steve Sparks, Claude Jaupart, Chris Hawkesworth, Heather Wright and Rob Nicholson provided helpful comments on the manuscript. This work was supported by the National Science Foundation grant EAR-0207362.

REFERENCES

Arcinega-Ceballos, A., B. A. Chouet, and P. Dawson, Long-period events and tremor at Popocatepetl volcano (1994–2000) and their broadband characteristics, *Bulletin of Volcanology*, 65, 124–135, 2003.

Blundy, J., and K. V. Cashman, Ascent-driven crystallisation of dacite magmas at Mount St. Helens, 1980–1986, *Contributions to Mineralogy and Petrology*, 140, 631–651, 2001.

Burnham, C. W. (1997) Magmas and hydrothermal fluids. in *Geochemistry of Hydrothermal Ore Deposits*, ed. H. L. Barnes, 63–125.

Carmichael, I. S. E., The andesite aqueduct: perspectives on the evolution of intermediate magmatism in west-central (105–99°W) Mexico, *Contributions to Mineralogy and Petrology*, 143, 641–663, 2002.

Carrigan, C. R., Plumbing systems, in *Encyclopedia of Volcanoes*, edited by H. Sigurdsson, pp. 149–170, Academic Press, San Diego, 2000.

Carroll, M. R., and J. R. Holloway, *Volatiles in Magmas*, Mineralogical Society of America, 1994.

Cashman, K. V., Groundmass crystallization of Mount St. Helens dacite, 1980–1986: A tool for interpreting shallow magmatic processes, *Contributions to Mineralogy and Petrology*, *109*, 431–449, 1992.

Cashman, K. V., and J. Blundy, Degassing and crystallization of ascending andesite and dacite, *Phil. Trans. R. Soc. Lond.*, *358*, 1487–1513, 2000.

Cashman, K. V. and R. P. Hoblitt, Magmatic precursors to the 18 May 1980 eruption of Mount St. Helens, USA, *Geology, 32*, 141–144, 2004.

Cashman, K. V., B. Sturtevant, P. Papale, and O. Navon, Magmatic fragmentation, in *Encyclopedia of Volcanoes*, edited by H. Sigurdsson, pp. 421–430, Academic Press, New York, 2000.

Chouet, B. A., Long-period seismicity: its source and use in eruption forecasting, *Nature*, *380*, 309–316, 1996.

Cioni, R., P. Marianelli, R. Santacroce, and A. Sbrana, Plinian and subplinian eruptions, in *Encyclopedia of Volcanoes*, edited by H. Sigurdsson, pp. 477–494, Academic Press, San Diego, 2000.

Clarke, A. B., B. Voight, A. Neri, and G. Macedonio, Transient dynamics of vulcanian explosions and column collapse, *Nature*, *415*, 897–901, 2002.

Cloos, M., Bubbling magma chambers, cupolas, and porphyry copper deposits, *International Geology Review*, *43*, 285–311, 2001.

Costa, F., and B. Scaillet, Massive atmospheric sulfur loading of the AD 1600 Huaynaputina eruption and implications for petrologic sulfur estimates, *Geophysical Research Letters*, *30*, doi:10.1029/2002GL016402, 2003.

Couch, S., R. S. J. Sparks, and M. R. Carroll, The kinetics of degassing-induced crystallization at Soufriere Hills Volcano, Montserrat, *Journal of Petrology*, *44*, 1477–1502, 2003.

Dingwell, D. B., Recent experimental progress in the physical description of silicic magma relevant to explosive volcanism, *Geological Society of London Special Publication*, *145*, 9–26, 1998.

Druitt, T. H., et al., Episodes of cyclic Vulcanian explosive activity with fountain collapse at Soufriere Hills Volcano, Montserrat, *Geological Society of London Memoirs*, *21*, 281–304, 2002.

Edmonds, M. C., C. Oppenheimer, D. M. Pyle, R. A. Herd, and G. Thompson, SO2 emissions from Soufriere Hills Volcanoe and their relationship to conduit permeability, hydrothermal interaction and degassing regime, *Journal of Volcanology and Geothermal Research*, *124*, 23–43, 2003.

Eichelberger, J. C., C. R. Carrigan, H. R. Westrich, and R. H. Price, Non-explosive silicic volcanism, *Nature*, *323*, 598–602, 1986.

Fink, J. H., S.vW. Anderson, and C. R. Manley, Textural constraints on effusive silicic volcanism: Beyond the permeable foam model, Journal of Geophysical Research, 97, 9073–9083, 1992.

Fink, J. H., and R. W. Griffiths, Morphology, eruption rates, and rheology of lava domes: Insights from laboratory models, *Journal of Geophysical Research*, *103*, 527–545, 1998.

Formenti, Y., T. H. Druitt, and K. Kelfoun, Characterisation of the 1997 Vulcanian explosions of Soufriere Hills Volcano, Montserrat, by video analysis, *Bulletin of Volcanology*, *65*, 587–605, 2003.

Garboczi, E. J., K. A. Snyder, J. F. Douglas, and M. F. Thorpe, Geometrical percolation threshold of overlapping ellipsoids, *Physical Review E*, *52*, 819–828, 1995.

Gardner, C. A., K. V. Cashman, and C. A. Neal, Tephra-fall deposits from the 1992 eruption of Crater Peak, Alaska: implicatinos of clast textures for eruptive processes, Bulletin of Volcanology, 59, 537–555, 1998.

Gardner, J., M. Hilton, and M. R. Carroll, Experimental constraints on degassing of magma: isothermal bubble growth during continuous decompression from high pressure, *Earth and Planetary Science Letters*, *168*, 201–218, 1999.

Gardner, J. E., R. M. E. Thomas, C. Jaupart, and S. Tait, Fragmentation of magma during Plinian volcanic eruptions, Bulletin of Volcanology, 58, 144–162, 1996.

Gerlach, T. M., and K. A. McGee, Total sulfur dioxide emissions and pre-eruption vapor-saturated magma at Mount St. Helens, 1980–88, *Geophysical Research Letters*, 1994.

Geschwind, C.-H., and M. J. Rutherford, Crystallization of microlites during magma ascent: the fluid mechanics of 1980–1986 eruptions at Mount St. Helens, *Bulletin of Volcanology*, *57*, 356–370, 1995.

Gonnerman, H., and M. Manga, Explosive volcanism may not be an inevitable consequence of magma fragmentation, *Nature*, *426*, 432–435, 2003.

Hammer, J., and M. Rutherford, An experimental study of the kinetics of decompression-induced crystallization in silicic melt, *Journal of Geophysical Research*, *107*, doi:10.1029/2001JB000281, 2002.

Hammer, J. E., K. V. Cashman, R. Hoblitt, and S. Newman, Degassing and microlite crystallization during the pre-climactic events of the 1991 eruption of the Mt. Pinatubo, Phillipines, *Bulletin of Volcanology*, *60*, 355–380, 1999.

Hammer, J. E., K. V. Cashman, and B. Voight, Magmatic processes revealed by textural and compositional trends in Merapi dome lavas, *Journal of Volcanology and Geothermal Research*, *100*, 165–192, 2000.

Hess, K.-U. and D. B. Dingwell, Viscosity of hydrous leucogranitic melts: A non-Arrhenian model, *American Mineralogist, 81*, 1297–1300, 1996.

Houghton, B. F., and C. J. N. Wilson, A vesicularity index for pyroclastic deposits, Bulletin of Volcanology, 51, 451–462, 1989.

Hurwitz, S., and O. Navon, Bubble nucleation in rhyolitic melts: Experiments at high pressure, temperature, and water content, *Earth and Planetary Science Letters*, *122*, 267–280, 1994.

Jaupart, C., Gas loss from magmas through conduit walls during eruption, *Geological Society of London Special Publication*, *145*, 73–90, 1998.

Jaupart, C., Magma ascent at shallow levels, in *Encyclopedia of Volcanoes*, edited by H. Sigurdsson, pp. 237–245, Academic Press, New York, 2000.

Jaupart, C., and C. J. Allegre, Gas content, eruption rate and instabilities of eruption regime in silicic volcanoes, *Earth and Planetary Science Letters*, *102*, 413–429, 1991.

Kaneko, T., M. J. Wooster, and S. Nakada, Exogenous and endogenous growth of the Unzen lava dome examined by satellite infrared image analysis, *Journal of Volcanology and Geothermal Research*, *116*, 151–160, 2002.

Kerr, R. C., and J. R. Lister, The effects of shape on crystal settling and the rheology of magmas, *Journal of Geology*, *99*, 457–467, 1991.

Klug, C., and K. V. Cashman, Permeability development in vesiculating magma, *Bulletin of Volcanology, 58, 87–100*, 1996.

Klug, C., K. V. Cashman, and C. R. Bacon, Structure and physical characteristics of pumice from the climactic eruption of Mt. Mazama (Crater Lake), Oregon, *Bulletin of Volcanology, 64,* 486–501, 2002.

Lacroix, A., *La Montagne Pelee et ses eruptions*, 62 pp., Masson et Cie, Paris, 1904.

Lister, J. R., Buoyancy-driven fluid fracture: the effects of material toughness and of low-viscosity precursors, *Nature*, *411*, 678–680, 1990.

Lowenstern, J. B., Dissolved volatile concentrations in an ore-forming magma, *Geology*, *22*, 893–896, 1994.

Lowenstern, J. B., Applications of silicate melt inclusions to the study of magmatic volatiles, in *Magmas, Fluids, and Ore Deposits*, edited by J. F. H. Thompson, pp. 71–99, Mineralogical Society of Canada, 1995.

Luhr, J. F., Experimental phase relations of water- and sulfur- saturated arc magmas and the 1982 eruptions of El Chichon Volcano, *Journal of Petrology*, *31*, 1071–1114, 1990.

Mader, H. M., Conduit flow and fragmentation, *Geological Society of London Special Publication*, *145*, 51–71, 1998.

Manga, M., J. Castro, K. V. Cashman, and M. Loewenberg, Rheology of bubble-bearing magmas, *Journal of Volcanology and Geothermal Research*, *87, 15–28,* 1998.

Mangan, M. T., L. G. Mastin, and T. Sisson, Gas evolution in eruptive conduits: combining insights from high temperature ad pressure decompression experiments with steady-state flow modeling, *Journal of Volcanology and Geothermal Research*, *129*, 23–36, 2004.

Mangan, M. T., and T. Sisson, Delayed, disequilibrium degassing in rhyolite magma: decompression experiments and implications for explosive volcanism, *Earth and Planetary Science Letters*, *183*, 441–455, 2000.

Marsh, B. D., On the crystallinity, probability of occurrence, and rheology of lava and magma, *Contributions to Mineralogy and Petrology*, *78*, 85–98, 1981.

Martel, C., J.-L. Bourdier, M. Pichavant, and H. Traineau, Textures, water content and degassing of silicic andesites from recent plinian and dome-forming eruptions at Mont Pelee volcano (Martinique, Lesser Antilles arc), *Journal of Volcanology and Geothermal Research*, *96*, 191–206, 2000.

Martel, C., and B. C. Schmidt, Decompression experiments as an insight into ascent rates of silicic magmas, *Contibutions to Mineralogy and Petrology*, *144*, 397–415, 2003.

Massol, H., and C. Jaupart, The generation of gas overpressure in volcanic eruptions, *Earth and Planetary Science Letters*, *166*, 57–70, 1999.

Massol, H., C. Jaupart, and D. W. Pepper, Ascent and decompression of viscous vesicular magma in a volcanic conduit, *Journal of Geophysical Research*, *106*, 16223–16240, 2001.

Mastin, L. G., and M. S. Ghiorso, A numerical program for steady-state flow of magma-gas mixtures through vertical eruptive conduits, *U.S. Geological Survey Open-File Report*, *00–209*, 1–59, 2000.

Melnik, O., and R. Sparks, Nonlinear dynamics of lava dome extrusion, *Nature*, *402*, 37–41, 1999.

Melnik, O., and R. S. J. Sparks, Dynamics of magma ascent and lava extrusion at Soufriere Hills Volcano, Montserrat, *Geological Society of London Memoirs*, *21*, 153–170, 2002a.

Melnik, O., and R. S. J. Sparks, Modeling conduit flow dynamics during explosive eruptions at Soufriere Hills Volcano, Montserrat, *Geological Society of London Memoirs*, *21*, 307–317, 2002b.

Menard, T., and S. R. Tait, A phenomenological model for precursor volcanic eruptions, *Nature*, *411*, 678–680, 2001.

Mercalli, G., *I vulcani attivi della Terra*, 421 pp., Ulrico Hoepli, Milan, 1907.

Metrich, N., A. Bertagnini, P. Landi, and M. Rosi, Crystallization driven by decompression and water loss at Stromboli Volcano (Aeolian Islands, Italy), *Journal of Petrology*, *42*, 1471–1490, 2001.

Moore, G., and I. S. E. Carmichael, The hydrous phase equilibria (to 3 kbar) of an andesite and basaltic andesite from western Mexico: constraints on water content and conditions of phenocryst growth, *Contributions to Mineralogy and Petrology*, *130*, 304–319, 1998.

Morrissey, M. M., and L. G. Mastin, Vulcanian eruptions, in *Encyclopedia of Volcanoes*, edited by H. Sigurdsson, pp. 463–475, Academic Press, San Diego, 2000.

Mourtada-Bonnefoi, C. C., and D. Laporte, Experimental study of homogeneous bubble nucleation in rhyolitic magmas, *Geophysical Research Letters*, *26*, 3505–3508, 1999.

Mourtada-Bonnefoi, C.C., and D. Laporte, Homogeneous bubble nucleation in rhyolitic magmas: An experimental study of the effect of H_2O and CO_2, *Journal of Geophysical Research*, *107*, doi:10.1029/2001JB000290, 2002.

Nakada, S., and Y. Motomura, Petrology of the 1991–1995 eruption at Unzen: effusion pulsation and groundmass crystallization, *Journal of Vocanology and Geothermal Research*, *89*, 173–196, 1999.

Nakada, S., Y. Motomura, and H. Shimizu, Manner of magma ascent at Unzen Volcano (Japan), *Geophysical Research Letters*, *22*, 567–570, 1995.

Newhall, C. G., and W. G. Melson, Explosive activity associated with the growth of volcanic domes, Journal of Volcanology and Geothermal Research, 17, 111–131, 1983.

Newman, S. and J. B. Lowenstern, VolatileCalc: a silicate melt-H_2O-CO2 solution model written in Visual Basic for Excel, *Computers and Geosciences, 28,* 597–604, 2002.

Pal, R., Rheological behavior of bubble-bearing magmas, *Earth and Planetary Science Letters*, *207*, 165–179, 2003.

Papale, P., Dynamics of magma flow in volcanic conduits with variable fragmentation efficiency and nonequilibrium pumice degassing, *Journal of Geophysical Research*, *106*, 11043–11066, 2001.

Papale, P., A. Neri, and G. Macedonio, The role of magma composition and water content in explosive eruptions I. Conduit ascent dynamics, *Journal of Volcanology and Geothermal Research*, 87, 75–93, 1998.

Polacci, M., P. Papale, and M. Rosi, Textural heterogeneities in pumices from the climactic eruption of Mount Pinatubo, 15 June 1991, and implications for magma ascent dynamics, *Bulletin of Volcanology*, 63, 83–97, 2001.

Polacci, M., L. Pioli, and M. Rosi, The Plinian phase of the Campanian Ignimbrite eruption (Phlegrean Fields, Italy): evidence from density measurements and textural characterization of pumice, Bulletin of Volcanology, doi 10.1007/s00445-002-0268-4, 2003.

Proussevitch, A. A., and D. L. Sahagian, Dynamics and energetics of bubble growth in magmas: analytical formulation and numerical modeling, *Journal of Geophysical Research*, 103, 18223–18251, 1998.

Pyle, D. M. (1989) The thickness, volume and grain size of tephra fall deposits. *Bulletin of Volcanology* 51, 1–15.

Pyle, D. M., Sizes of Volcanic Eruptions, in *Encyclopedia of Volcanoes,* edited by H. Sigurdsson, pp. 263–269, Academic Press, New York, 2000.

Robertson, R.e.a., The explosive eruption of Soufriere Hills Volcano, Montserrat, West Indies, 17 September, 1996, Geophysical Research Letters, 25, 3429–3432, 1998.

Roggensack, K., R. L. Hervig, S. B. McKnight, and S. N. Williams, Explosive basaltic volcanism from Cerro Negro Volcano: Influences of volatiles on eruptive style, *Science*, 277, 1639–1642, 1997.

Rosi, M., P. Landi, M. Polacci, A. DiMuro, and D. Zandomeneghi, Role of conduit shear on ascent of the crystal-rich magma feeding the 800-year-BP Plinian eruption of Quilotoa Volcano (Ecuador), *Bulletin of Volcanology*, DOI:10.1007/s00445-03-0312-z, 2003.

Rubin, A. M., On the thermal viability of dikes leaving magma chambers, *Geophysical Research Letters*, 20, 257–260, 1993.

Rust, A. C., K. V. Cashman, and P. J. Wallace, Magma degassing buffered by vapor flow through brecciated conduit margins, *Geology, (in press)*, 2004.

Rust, A. C., and M. Manga, Effects of bubble deformation on the viscosity of dilute suspensions, *Journal of nonNewtonian Fluid Mechanics*, 104, 53–63, 2002.

Rust, A. C., M. Manga, and K. V. Cashman, Determining flow type, shear rate and shear stress in magmas from bubble shapes and orientations, *Journal of Volcanology and Geothermal Research*, 122, 111–132, 2003.

Rutherford, M. D., and J. D. Devine, The May 18, 1980 eruption of Mount St. Helens 3. Stability and chemistry of amphibole in the magma chamber, *Journal of Geophysical Research*, 93, 11949–11959, 1988.

Rutherford, M. J., and J. D. Devine, Pre-eruption pressure-temperature conditions and volatiles in the 1991 dacitic magma of Mount Pinatubo, in *Fire and Mud: Eruptions and Lahars of Mount Pinatubo, Phillipines*, edited by C. Newhall, and R. Punongbayan, pp. 751–766, University of Washington Press, Seattle, 1996.

Rutherford, M. R., and Gardner, J. E., Rates of magma ascent, in *Encyclopedia of Volcanoes*, edited by H. Sigurdsson, pp. 207–218, Academic Press, New York, 2000.

Saar, M. O., and M. Manga, Permeability-porosity relationship in vesicular basalts, *Geophysical Research Letters*, 26, 111–114, 1999.

Sahimi, M., *Applications of Percolation Theory*, 258 pp., Taylor and Francis, London, 1994.

Scaillet, B., B. Clemente, B. W. Evans, and M. Pichavant, Redox control of sulfur degassing in silicic magmas, *Journal of Geophysical Research*, 103, 23937–23949, 1998.

Scandone, R., and S. D. Malone, Magma supply, magma discharge and readjustment of the feeding system of Mount St. Helens during 1980, *Journal of Volcanology and Geothermal Research*, 23, 239–262, 1985.

Schmitt, A. X., Gas-saturated crystallization and degassing in large-volume, crystal-rich dacitic magmas from the Altiplano-Puna, northern Chile, *Journal of Geophysical Research*, 106, 30561–30578, 2001.

Shepherd, E. S., and H. E. Merwin, Gases of the Mt. Pelee lavas of 1902, *Journal of Geology*, 35, 97–116, 1927.

Shinohara, H., and K. Kazahaya, Degassing processes related to magma chamber crystallization, edited by J. Thompson, GAC-MAC, 1995.

Sisson, T., and G. D. Layne, H_2O in basalt and basaltic andesite glass inclusions from four subduction-related volcanoes, *Earth and Planetary Science Letters*, 117, 167–184, 1993.

Sisson, T. W., and T. L. Grove, Temperatures and H_2O contents of low MgO high-alumina basalts, *Contributions to Mineralogy and Petrology*, 113, 167–184, 1993.

Smith, J. V., Textural evidence for dilatant (shear thickening) rheology of mama at high crystal concentrations, *Journal of Volcanology and Geothermal Research*, 99, 1–7, 2000.

Sparks, R. J. S., The dynamics of bubble formation and growth in magmas: a review and analysis, *Journal of Vocanology and Geothermal Research*, 3, 1–37, 1978.

Sparks, R. S. J., Causes and consequences of pressurization in lava dome eruptions, *Earth and Planetary Science Letters*, 150, 177–189, 1997.

Sparks, R. S. J., and S. R. Young, The eruption of Soufriere hills Volcano, Montserrat (1995–1999): overview of scientific results, *Geological Society of London Memoirs*, 21, 45–69, 2002.

Swanson, D. A., and R. T. Holcomb, Regularities in the growth of the Mount St. Helens dacite dome, 1980–1986, in *Lava Flows and Domes*, edited by J. H. Fink, pp. 3–24, Springer-Verlag, Berlin, 1990.

Tuffen, H., D. B. Dingwell, and H. Pinkerton, Repeated fracture and healing of silicic magma generates flow banding and earthquakes?, *Geology*, 31, 1089–1092, 2003.

Voight, B., et al., Magma flow and cyclic instability at Soufriere Hills volcano, Montserrat, British West Indies, *Science*, 283, 1138–1142, 1999.

Walker, G. P. L., Explosive volcanic eruptions—a new classification scheme, *Geologische Rundschau*, 62, 431–446, 1973.

Wallace P. J. (2003) From mantle to atmosphere: Magma degassing, explosive eruptions, and volcanic volatile budgets. In: Bodnar BJ, DeVivo B (Eds.) *Developments in Volcanology*, Elsevier Science, pp. 105–127.

Wallace, P. J., and A. T. Anderson, Volatiles in Magmas, in *Encylopedia of Volcanoes*, edited by H. Sigurdsson, pp. 149–170, Academic Press, San Diego, 2000.

Wallace, P J., A. T. Anderson, and A. M. Davis, Gradients in H2O, CO2, and exsolved gas in a large-volume silicic magma system: Interpreting the record preserved in melt inclusions from the Bishop Tuff, *Journal of Geophysical Research, 104*, 20097–20122, 1999.

Wallace, P. J., A. T. Anderson Jr., and A. M. Davis, Quantification of pre-eruptive exsolved gas contents in silicic magmas, *Nature, 377*, 612–616, 1995.

White, R. A., A. D. Miller, L. Lynch, and J. A. Power, Observations of hybrid seismic events at Soufriere Hills Volcano, Montserrat: July 1995 to September 1996, *Geophysical Research Letter, 25*, 3657–3660, 1998.

Williams, H., The history and character of volcanic domes, *University of California, Bulletin of the Department of Geological Sciences, 21*, 51–146, 1932.

Wilson, L., R. S. J. Sparks, and G. P. L. Walker, Explosive volcanic eruptions—IV. The control of magma properties and conduit geometry on eruption column behavior, *Geophys J R Astr Soc, 63*, 117–148, 1980.

Zapata, J. A.et al., SO_2 fluxes from Galeras Volcano, Columbia, 1989–1995: Progressive degassing and conduit obstruction of a Decade Volcano, *Journal of Volcanology and Geothermal Research, 77, 195–208*, 1997.

Zhang, Y., A criterion for the fragmentation of bubbly magma based on brittle failure theory, *Nature, 402*, 648–650, 1999.

Zhou, J. Z.Q., T. Fang, G. Luo, and P. H. T. Uhlerr, Yield stress and maximum packing fraction of concentrated suspensions, *Rheologica Acta, 34*, 544–561, 1995.

Katharine V. Cashman, Department of Geological Sciences, University of Oregon, Eugene, Oregon

Climatic Impact of Volcanic Emissions

Alan Robock

Department of Environmental Sciences, Rutgers University, New Brunswick, New Jersey

Studying the impacts of volcanic eruptions on climate is important because it helps us improve climate models, it allows us to make seasonal and interannual climate forecasts following large eruptions, it provides support for nuclear winter theory, and it allows us to separate the natural causes of interdecadal climate change from anthropogenic effects, giving us greater confidence in the attribution of recent global warming to anthropogenic causes. While much has been learned since the large 1991 eruption of Mt. Pinatubo in the Philippines, there are still quite a few outstanding research problems, which are discussed here. These questions include: What exactly goes into the atmosphere during an explosive eruption? How can we better quantify the record of past climatically-significant volcanism? Can we design an improved system for measuring and monitoring the atmospheric gases and aerosols resulting from future eruptions? How can we better model the climatic impact of eruptions, including microphysics, chemistry, transport, radiation, and dynamical responses? How do high-latitude eruptions affect climate? How important are indirect effects of volcanic emissions on clouds? Where are the important potential sites for future eruptions?

1. INTRODUCTION

Volcanism has long been implicated as a possible cause of weather and climate variations. *Franklin* [1784], *Humphreys* [1913, 1940] and *Mitchell* [1961] were pioneers in their association of volcanic eruptions with climate change. More recently, *Lamb* [1970, 1977, 1983], *Toon and Pollack* [1980], *Toon* [1982], and *Robock* [2000, 2002a, 2003a, 2003b] presented reviews of these effects. An AGU monograph [*Robock and Oppenheimer*, 2003] contains many articles on all aspects of this subject. There are however many outstanding research problems and this paper focuses on frontiers and challenges in this area. First, a brief summary of the known effects serves to set the stage for a focus on these outstanding research questions.

a. How volcanic eruptions affect weather and climate

Volcanic eruptions can inject into the stratosphere tens of teragrams of chemically and microphysically active gases and solid aerosol particles, which affect the Earth's radiative balance and climate, and disturb the stratospheric chemical equilibrium (Figure 1). The volcanic cloud forms in several weeks by SO_2 conversion to sulfate aerosol, and its subsequent microphysical transformations. The resulting cloud of sulfate aerosol particles, with an e-folding decay time of approximately 1 year, has important impacts on both shortwave and longwave radiation. The resulting disturbance to the Earth's radiation balance affects surface temperatures through direct radiative effects as well as through indirect effects on the atmospheric circulation. In cold regions of the stratosphere, these aerosol particles also serve as surfaces for heterogeneous chemical reactions that liberate chlorine to destroy ozone in the same way that water and nitric acid aerosols in polar stratospheric clouds produce the seasonal Antarctic ozone hole (Figure 1).

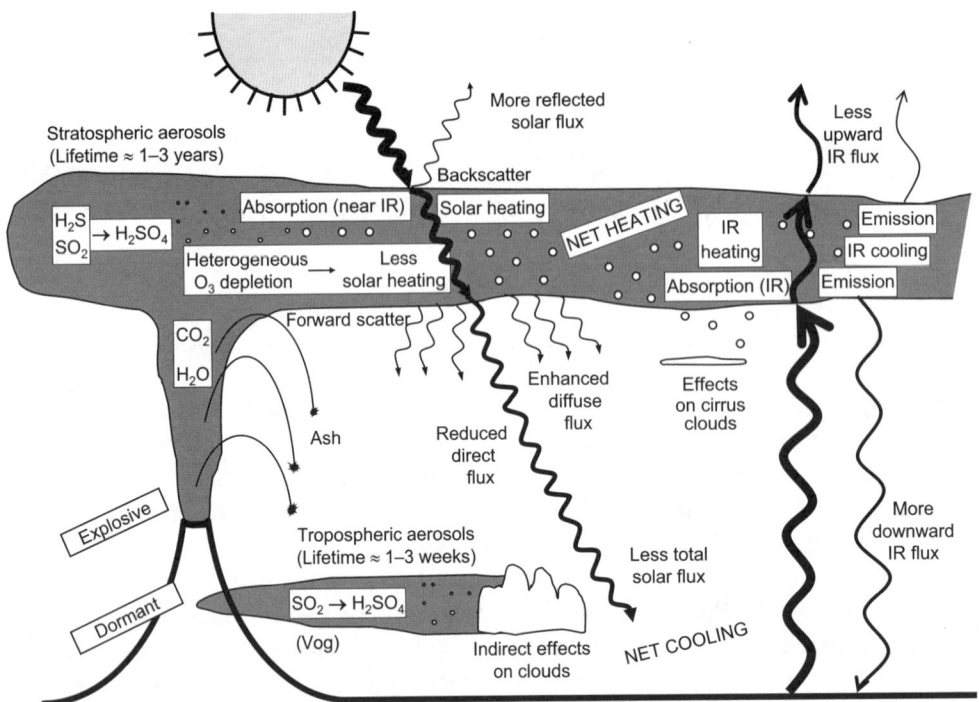

Figure 1. Schematic diagram of volcanic inputs to the atmosphere and their effects. [Adapted from Plate 1 of Robock (2000), © Copyright, American Geophysical Union.]

The major component of volcanic eruptions is magmatic material, which emerges as solid, lithic material or solidifies into large particles, which are referred to as ash or tephra. These particles fall out of the atmosphere very rapidly, on time scales of minutes to a few weeks in the troposphere. Figure 2 illustrates the dramatic effects of the ash from the 1991 Mt. Pinatubo eruption in the Philippines. Small amounts can last for a few months in the stratosphere, but have very small climatic impacts. *Symons* [1888], after the 1883 Krakatau eruption, and *Robock and Mass* [1982], after the 1980 Mt. St. Helens eruption, showed that this temporary large atmospheric loading reduced the amplitude of the diurnal cycle of surface air temperature in the region of the tropospheric cloud. But these effects disappear as soon as the particles settle to the ground. Continuous volcanic emissions, including fumaroles and small episodic eruptions, add sulfates to the troposphere, but their lifetimes there are much shorter than those of stratospheric aerosols (Figure 1). Therefore they are not important for climate change, but could be if there is a sudden change or a long-term trend in them develops. Global sulfur emission of volcanoes to the troposphere is about 14% of the total natural and anthropogenic emission [*Graf et al.*, 1997], but has a much larger relative contribution to radiative effects. Many volcanic emissions are from the sides of mountains, above the atmospheric boundary layer, and thus they have longer lifetimes than anthropogenic aerosols.

Large volcanic eruptions inject sulfur gases into the stratosphere, which convert to sulfate aerosols with an e-folding residence time of about 1 yr (Figure 1). Large ash particles fall out much quicker. The radiative and chemical effects of this aerosol cloud produce responses in the climate system. By scattering some solar radiation back to space, the aerosols cool the surface, but by absorbing both solar and terrestrial radiation, the aerosol layer heats the stratosphere (Figure 1). Because the sulfate aerosol particles have an effective radius of about 0.5 μm, equivalent to the wavelength of visible light, they interact more strongly with the shortwave solar radiation than the longwave (~10 μm) radiation being emitted by the surface and atmosphere. The shortwave interaction includes some absorption in near infrared wavelengths, producing some heating of the top of the aerosol cloud, but mostly scattering. The light which is backscattered is essentially reflected from the Earth-atmosphere system, cooling the planet. Much of it is forward scattered, resulting in more downward diffuse radiation, and less direct downward radiation. In the longwave, the aerosol cloud absorbs upward radiation from the surface and atmosphere, heating the aerosol cloud. It emits downward, producing some compensation for the reduction in down-

Figure 2. A World Airways DC-10 at Cubi Point Naval Air Station, 40 km from Mt. Pinatubo, tipped backwards from the volcanic ash on the stabilizer. What appears to be snow is actually volcanic ash on the ground. These large particles, while producing devastating local effects and short-term weather effects, fall out of the atmosphere too quickly to affect climate. [U.S. Navy photograph by R. L. Rieger]

ward solar radiation, but this longwave effect is an order of magnitude smaller than the shortwave effect, so the net effect is surface cooling, except in the polar night where there is no sunlight. The aerosol cloud also emits upward, but much less radiation than the surface. Thus, to satellites trying to measure sea surface temperatures or vegetation indices, the volcanic aerosol cloud produces an interference, which must be addressed. These effects are illustrated in Figure 1.

For a tropical eruption, the stratospheric heating is larger in the tropics than in the high latitudes, producing an enhanced pole-to-equator temperature gradient, especially in winter. In the Northern Hemisphere winter, this enhanced gradient produces a stronger polar vortex, and this stronger jet stream produces a characteristic stationary wave pattern of tropospheric circulation resulting in winter warming of Northern Hemisphere continents [*Perlwitz and Graf*, 1995; *Kodera et al.*, 1996; *Thompson and Wallace*, 1998, 1999a, b]. This indirect advective effect on temperature is stronger than the radiative cooling effect that dominates at lower latitudes and in the summer. After the Pinatubo eruption, the observed ozone depletion was largest in the high latitudes of the Northern Hemisphere in the second year after the eruption, enhancing the strengthening of the polar vortex and enhancing winter warming in the second year after the eruption [*Stenchikov et al.*, 2002]. The quasi-biennial oscillation of stratospheric winds also modulates these effects [*Stenchikov et al.*, 2004].

b. Reasons to study the volcanic impacts on climate

Studying the effects of volcanic eruptions on climate is important for several reasons. First, it helps us to improve climate models. Volcanic eruptions are an important natural cause of climate change on many time scales. Studying the responses of climate to volcanic eruptions helps us to better understand important radiative and dynamical processes in the climate system. Large volcanic eruptions, like the 1991 Mt. Pinatubo eruption in the Philippines, produce short (2–3 yr) but large perturbations to the climate system. While we cannot use them to test long-term processes, such as changes in the thermohaline circulation, we can take advantage of them to examine some short time-scale feedback processes and impacts. If a climate model responds correctly to a volcanic eruption, it gives us more confidence that the timing and amplitude of future global warming will be well simulated by the model. For example, *Soden et al.* [2002] showed that without the most important positive feedback in the climate system, the water vapor feedback, a state of the art climate model is unable to reproduce the observed global cooling following the Pinatubo eruption. These results provide quantitative evidence of the reliability of water vapor feedback in current climate models, which is crucial to their use for global warming projections.

Second, our current understanding of the effects of large volcanic eruptions provides us a capability to make seasonal and interannual climate forecasts of Northern Hemisphere winter warming and summer cooling following large tropical eruptions [*Robock*, 2000]. Thus we can be confident in predictions of global cooling in the summer following large tropical eruptions, unless there is a large simultaneous El Niño, as in 1982 following the El Chichón eruption. We can also now predict that there will be a circulation response in the first one or two Northern Hemisphere winters following a large tropical eruption, resulting in winter warming of continents. We can account for the effects of the modulation of this response by the phase of the quasi-biennial oscillation.

Third, volcanic eruptions provide an analog for some parts of nuclear winter theory. The theory of nuclear winter, the climatic effects of a massive injection of soot aerosols into the atmosphere from fires following a global nuclear holocaust [*Turco et al.*, 1983, 1990; *Robock*, 1984, 1996], includes upward injection of the aerosols to the stratosphere, rapid global dispersal of stratospheric aerosols, heating of the stratosphere, and cooling at the surface under this cloud. As we cannot perform this experiment in the real world to test the theory, the similar climate responses to volcanic eruptions provide an analog for these phenomena and give us more confidence in climate model simulations of this worst of all possible anthropogenic impacts on climate. In fact, *Rampino and Self* [1992] explicitly acknowledge this when they refer to a "volcanic winter" as the effect of a massive volcanic eruption.

Fourth, the study of volcanic effects on climate allows us to separate the natural causes of interdecadal climate change from anthropogenic effects. The major potential causes of climate change on an interdecadal to centennial time scale are solar variations, volcanic eruptions, and anthropogenic greenhouse gases. Anthropogenic land surface changes and aerosols are less important globally, but may be more important regionally. To understand how important anthropogenic effects are on climate, and to attribute the 0.6°C ± 0.2°C warming of the past century [*Jones et al.*, 1999; *Folland et al.*, 2001] to its causes, we must understand the strength of the natural causes of climate change. *Free and Robock* [1999] showed that the major causes of climate change from 1600 to 1850 were natural, with volcanic eruptions being the most important. However, they found that natural causes could not explain the warming of the past century and a half, but that it could be well simulated when anthropogenic forcing was included. These and other results led *Houghton et al.* [2001] to conclude that most of the warming of the second half of the 20th century was caused by human pollution of the atmosphere. In addition, sulfur emissions into the troposphere above the boundary layer from volcanoes produce a substantial amount of tropospheric aerosols [*Graf et al.*, 1997]. Resolving uncertainty in the amount and timing of these slow volcanic emissions into the troposphere will produce a better understanding of the record of tropospheric sulfate aerosols, their contribution to radiative and cloud microphysical changes, and thus their relative role in climate change.

Fifth, the response to volcanic eruptions allows us to better understand the impacts of anthropogenic climate change on life [*Robock*, 2003b]. The response of the biosphere to the Pinatubo eruption illustrated its sensitivity to climate change and clarified portions of the carbon cycle. For example, the unusually large number of polar bear cubs born in the summer of 1992 was due to the characteristic winter and summer temperature responses of the climate system. The largest cooling in the summer of 1992 was in the center of North America, and as a result, the ice on Hudson Bay melted almost a month later than normal that year. Polar bears, who feed and have babies on the ice, were much heavier and had more healthy cubs that summer. Biologists call them the "Pinatubo Bears" [*Stirling*, 1997]. While the cool conditions in the summer after the Pinatubo eruption were very beneficial for the Hudson Bay polar bears, and there were many more bears born that year than the year before or after, the long-term concern for these bears, and many other plants and animals in the Arctic, however, is the opposite impact from global warming. This is an example of how the benefits of Pinatubo from global cooling teach us about the negative impacts of anthropogenic global warming. In addition, enhanced vegetation growth from more diffuse and less direct solar radiation took more carbon dioxide out of the atmosphere than normal, temporarily reducing the observed long-term increase in carbon dioxide [*Gu et al.*, 2002, 2003; *Farquhar and Roderick*, 2003], thus adding to our understanding of the carbon cycle.

2. OUTSTANDING RESEARCH PROBLEMS

Robock [2002b] discussed outstanding problems in understanding the effects of volcanic eruptions on the atmosphere and on climate, but did not have room to go into detail and completely explain why these issues were important and what current ideas have been proposed. An enumeration of these scientific questions is important as guidance for funding agencies and to give ideas to graduate students and other researchers for dissertation topics and project ideas. Here I expand on that article and discuss several of these issues in greater depth.

What exactly goes into the atmosphere during an explosive eruption? The impacts of volcanic eruptions on weather, climate, and atmospheric chemistry depend on what materials eruptions put into the atmosphere. Climatically significant inputs include sulfur species (especially SO_2), halogens, H_2O, and fine silicate particles. *Textor et al.* [2004] discussed our current understanding of the input to the atmosphere from both tropospheric slow release of gases by fumaroles and steam plumes from dormant volcanoes and the eruption of lava with weak degassing and also episodic, explosive eruptions which put material into the stratosphere, but many questions remain. What are the magmatic controls on how much sulfur is emitted from eruptions? For example, the partitioning of sulfur between solid igneous mineral phases (sulfides and sulfates) that remain in the crust and gaseous phases (H_2S and SO_2) is critical to understanding volcanic sulfur budgets and emissions, but is far from being well understood [e.g., *Wallace*, 2001]. Volcanoes appear to erupt far more SO_2 than they should on the basis of solubility laws and this excess sulfur problem remains unresolved.

When eruptions take place in wet environments, how much of the water in the plume is primary magmatic water, as compared to entrained water from the atmosphere or from lakes or the ocean? What are the detailed chemical and microphysical transformations that occur in the eruption column and downwind plume, and how do they affect the composition of the stratospheric injection? In particular, if the eruption is rich in Cl, would all of it be washed out in the plume or would some be injected into the stratosphere, with consequences for ozone chemistry [*Tabazadeh and Turco*, 1993]? Does the injection of water vapor from the volcanic input plus that entrained from the atmosphere, especially in a moist, tropical environment, provide an important source of stratospheric water vapor, or is the amount negligible? Would subsequent heating of the lower stratosphere by radiative absorption by the aerosol cloud produce a warmer tropopause in the tropics, reducing the cold trap and producing an increased flux of water vapor into the stratosphere for a year or two? This seems to have happened after the 1982 El Chichón eruption, but not after the 1991 Mt. Pinatubo eruption [*Shepherd*, 2002]. Why? Does this mechanism explain part of the interdecadal upward trend in stratospheric water vapor, as suggested by *Joshi and Shine* [2003]?

How can we better quantify the record of past climatically-significant volcanism? To measure the natural climatic forcing from volcanic eruptions for the past, so that we may place anthropogenic climate change in context, we need a better record of the frequency and magnitude and sulfur-producing potential of past eruptions [e.g., *Halmer et al.*, 2002]. Previous volcanic indices, the dust veil index [DVI; *Lamb*, 1970, 1977, 1983], the volcanic explosivity index [VEI; *Newhall and Self*, 1982; *Simkin et al.*, 1981, *Simkin and Siebert*, 1994], and *Mitchell*'s [1970], all have limitations [*Rampino and Self*, 1982], as discussed in detail by *Robock and Free* [1995]. *Robock and Free* [1995] concluded that an ice-core-based index would potentially be superior, as it would include actual physical observations of the remnants of the climatically significant component (sulfate) of each eruption, if sufficient cores were available to overcome the natural noise in each individual core.

Although *Rind* [1996] estimates that the volcanic forcing is potentially easier to capture than the solar forcing (in particular the low frequency part), the initial volcanic sulfate loading in the atmosphere is uncertain if estimated using ice core sulfate flux data from only one or a few cores. *Robock and Free* [1995] showed that any pair of ice core records of volcanic sulfate is poorly correlated, but when a larger number are combined, the volcanic signal becomes clearer. *Zielinski et al.* [1997] pointed out that there still is no complete understanding of variability from core to core resulting in only a 75% chance that an El Chichón sized event is even recorded in any particular core. This is even the case for large eruptions as discussed by *Cole-Dai et al.* [2000] for the 1259 eruption [see also *Mosley-Thompson, et al.*, 2003]. Events can be missing and biases are potentially large using single or a small number of cores.

Now the scientific community has the opportunity to take advantage of a significantly larger number of ice core observations, and combine them with a new understanding of stratospheric transport, to produce an improved index of past volcanism to force climate models. *Robock and Free* [1995, 1996] and *Robertson et al.* [2001] collected data from a number of cores and generated hemispherically averaged volcanic sulfate fluxes. However, since the mid-1990s, a number of new high-resolution cores with very accurate age control have been drilled and new data have been published. New techniques have been developed to analyze cores for the volcanic signal [*Naveau et al.*, 2003] and to prescribe the latitudinal and seasonal distribution of stratospheric aerosols in the stratosphere [*Ammann et al.*, 2003]. Among the new cores becoming available are those of *Budner et al.* [2003] and *Mosley-Thompson et al.* [2003].

A major advance to allow better interpretation of the location of eruptions that produce ice core signatures would be better atmospheric models of transport and deposition that could trace sulfate aerosols from the vent to the ice. Such transport and deposition models are now being developed.

Ice cores will not give detailed information about tropospheric emissions of volcanoes, which could be very important for regional weather and climate. However, real-time observations and modeling are now being conducted by the nine Volcanic Ash Advisory Centers around the world, which issue warnings for aircraft, and their output could be archived over time. In addition, the new "A-Train" of satellites (Aqua, Cloudsat, Calipso, Parasol, and Aura) [*Stephens et al.*, 2002] will provide enhanced measurement capability for tropospheric aerosols.

Can we design an improved system for measuring and monitoring the atmospheric gases and aerosols resulting from future eruptions? In spite of current technology, without better planning and an investment in equipment, there will be significant gaps in observations of the next major volcanic eruption. To be ready for the next major eruption, given the lack of a global satellite monitoring system, we should have a fleet of stratospheric balloons, lidar-equipped airplanes, and stratospheric airplanes with the capability for in situ observations ready to be deployed within weeks of the eruption. While there are many lidar observatories in the Northern Hemisphere midlatitudes, and several in the Southern Hemisphere midlatitudes, there are no lidars in the Tropics designed for measuring stratospheric aerosols, with the exception of the one in Bandung, Indonesia (6.9°S, 107.6°E), which is plagued by bad weather. It would be relatively cheap and quick to fill in this gap [Robock and Antuña, 2001]. While the A-Train [Stephens et al., 2002] will include the first space-based lidars, they will not be as sensitive as ground-based ones, and will need ancillary observations for calibration and validation. However, a new technique to measure SO_2 using observations from the TIROS Operational Vertical Sounder (TOVS) instrument, which has been in orbit on different satellites for the past 25 yr, will prove useful [Prata et al., 2003].

Near-vent observations of volcanic gases and aerosol emissions, unless the eruption is forecast in advance as were Mount St. Helens in 1980 and Mount Pinatubo in 1991, will depend on work with local observers. As many volcanoes are in developing countries, a program to train, work with, and support local observers would significantly enhance our ability to monitor small and medium size eruptions.

Because of the diversity of observations available for eruptions, a data assimilation system using atmospheric models must be developed, which will be the only way to produce a stratospheric aerosol data set that can be used for atmospheric chemistry and climate calculations. The need for a stratospheric aerosol data set for general circulation model (GCM) studies covering an extended time interval has long been recognized [Robock, 2000, 2001]. Sato et al. [1993] developed a data set of zonally-averaged stratospheric aerosol optical depth at $\lambda = 0.55$ μm for the period 1850-1990. However, for GCM experiments, information about the spectral dependence of aerosol radiative properties and their vertical distribution is necessary. Andronova et al. [2000] calculated volcanic radiative forcing at the tropopause level using optical depths from Sato et al. [1993] and employed a regression relation estimated for the post-Pinatubo period using aerosol characteristics from Stenchikov et al. [1998], but did not produce the vertical distribution of aerosols.

The Stratospheric Aerosol and Gas Experiment (SAGE) and Stratospheric Aerosol Measurement projects [McCormick et al., 1979; Mauldin et al., 1985; McCormick, 1987; Thomason, 1991; Veiga, 1993] have provided more than 20 years of three-dimensional data of stratospheric aerosol spectral extinction, the longest such record. Hitchman et al. [1994] and Stevermer et al. [2000] used these data to study the zonal mean aerosol climatology. However, SAGE II only samples each latitude band every 40 days, does not cover polar regions, and has significant gaps in regions of heavy aerosol loading, where the aerosol cloud causes so much extinction of the solar signal that no retrievals are possible. SAGE III [Thomason and Taha, 2003] is now providing high-latitude coverage but sketchy low latitude observations. The eruption of El Chichón in 1982 (the second most important in terms of atmospheric impact in the second half of the 20th century after Mt. Pinatubo) was not covered by SAGE observations because the SAGE I instrument failed in 1981, and SAGE II was only launched in 1984. The Stratosphere Mesosphere Explorer satellite did take some observations during the El Chichón period, but its observations will have to be combined with lidar [Antuña et al., 2002, 2003] and other data in the context of a data assimilation system to produce a consistent three-dimensional global stratospheric aerosol data set that can be used to study the climatic response during the last 20 years of the 20th century, which included the El Chichón and Pinatubo eruptions.

How can we better model the climatic impact of eruptions, including microphysics, chemistry, transport, radiation, and dynamical responses? A few GCMs have simulated the general climatic response to the 1991 Pinatubo eruption using a specified distribution of aerosols [Stenchikov et al., 1998]. Remaining problems include adequately accounting for the effects of the Quasi-Biennial Oscillation, microphysical evolution and transport of the aerosols, effects on ozone, the amount and impacts of water vapor injection into the stratosphere, and the regional response, although some progress is being made along those lines [Stenchikov et al., 2003, 2003]. Data assimilation experiments and model intercomparison programs, like the Pinatubo Model Intercomparison Project (PINMIP) now being carried out under the GCM-Reality Intercomparison Project for SPARC (GRIPS) [Pawson et al., 2000], will help to improve the models. The ultimate goal would be to couple conduit models of magma, plume models, and microphysical and transport models in the stratosphere to climate models to predict the impact of the next large eruption as soon as it occurs.

How do high-latitude eruptions affect climate? Most research on the impacts of volcanic eruptions on climate has focused on tropical explosive eruptions, such as the recent 1963 Agung, 1982 El Chichón, and 1991 Pinatubo eruptions. However, two of the largest eruptions of the past five centuries occurred at high latitudes of the Northern Hemisphere, the 1783–1784 Laki eruption in Iceland [Franklin, 1784; Thordarson and Self, 2003], and the 1912 Katmai eruption in Alaska. Mt. Hudson in Chile erupted in 1991, but its S emissions were much smaller than those of Pinatubo, so it was not possible to isolate its effects. The 8-month-long Laki eruption from a 27-km long fissure (Figure 3) affected air quality and climate for most of the Northern Hemisphere and if it occurred today could halt air traffic there for six months [*Thordarson and Self*, 2003]. The summer of 1783 had extreme and unusual weather. In the summer, the Arctic was unusually cold, but Western Europe was unusually hot. The people of Japan suffered through the most severe famine in their history. The winter of 1783–84, on the other hand, was one of the most severe in Europe and North America in the last 250 years.

The mechanisms by which this eruption produced these climate changes are not well understood. How could volcanic aerosols, which we normally think of as cooling the surface in the summer [*Robock*, 2000], have produced the record warm temperatures in Europe? Was the severe cold in the winter of 1783–84 a negative mode of the Arctic Oscillation, and if so how was it produced? The 1783 Asama eruption in Japan may have contributed to these effects, and needs to be considered when unraveling the climatic response. Questions that still need answers include whether high latitude eruptions can affect the climate in the other hemisphere, and what the effects would be of eruptions from high latitude Southern Hemisphere volcanoes.

How important are indirect effects of volcanic emissions on clouds? The indirect effect of tropospheric sulfate emissions on clouds [*Penner et al.*, 2001] is an area of intensive research. Sulfate aerosols serve as cloud condensation nuclei, and adding sulfate aerosols to clouds produces more, but smaller could droplets. This increases the albedo of the cloud, producing additional net cooling of the climate system, and is referred to as the "first indirect effect" of sulfate aerosols. In addition, the cloud lifetime and structure can change, the "second indirect effect." Thus, sulfate aerosols can affect the atmospheric thermal structure, surface temperatures, and precipitation.

Can volcanic examples be used to improve current models? While climate scientists are concerned with anthropogenic sulfate emissions, they tend to be diffuse and mixed with other pollutants. Could the cloud response to intense volcanic emissions, such as in Hawaii or Sicily, be used to isolate and study these processes? Another, more difficult question to address,

Figure 3. The Laki cone row, looking southwest from the Laki mountain. Note bus and cars parked at bottom for scale. Cones in distance are obscured by rain shower. Photograph by Alan Robock, August 30, 2002.

is whether volcanic aerosols from the stratosphere seed cirrus clouds and affect their optical properties and lifetimes?

Where are the important potential sites for future eruptions? For monitoring, emergency response, warning aircraft, and real-time prediction of climatic response, it would be helpful to know which volcanoes would be most likely to erupt. This will involve production of improved risk maps and catalogs of hazards. It will again require working with and supporting local observers.

3. DISCUSSION

Some research aimed at answering the above questions is already under way, nurtured by the interactions of the volcanology and climate communities at the 2002 "Volcanism and the Earth's Atmosphere" Chapman Conference [*Robock and Oppenheimer*, 2003], and subsequent conference sessions, such as those at the 2003 IUGG meeting in Sapporo. Future progress in understanding this important aspect of planetary interactions of the lithosphere with the climate system will result from well-funded projects that address the above questions. I look forward to seeing graduate students and other researchers tackle these interesting issues in the next few years.

Acknowledgments. I thank Steve Sparks, Chris Hawkesworth, and an anonymous reviewer for valuable suggestions, which resulted in important improvements in the paper. Supported by NASA grant NAG 5-9792 and NSF grants ATM-9988419 and ATM-0313592.

REFERENCES

Ammann, C. M., G. A. Meehl, W. M. Washington, and C. S. Zender, A monthly and latitudinally varying volcanic forcing dataset in simulations of the 20th century climate, *Geophys. Res. Lett.*, *30(12)*, 1657, doi:10.1029/2003GL016875, 2003a.

Andronova, N., E. Rozanov, F. Yang, M. Schlesinger, and G. Stenchikov, Radiative forcing by volcanic aerosols from 1850 through 1994, *J. Geophys. Res.*, *104*, 16,807-16,826, 1999.

Antuña, J. C., A. Robock, G. L. Stenchikov, L. W. Thomason, and J. E. Barnes, Lidar validation of SAGE II aerosol measurements after the 1991 Mount Pinatubo eruption. *J. Geophys. Res.*, *107 (D14)*, 4194, doi:110.1029/2001JD001441, 2002.

Antuña, J. C., A. Robock, G. Stenchikov, J. Zhou, C. David, J. Barnes, and L. Thomason, Spatial and temporal variability of the stratospheric aerosol cloud produced by the 1991 Mount Pinatubo eruption. *J. Geophys. Res.*, *108 (D20)*, 4624, doi:10.1029/2003JD003722, 2003.

Budner, D., and J. Cole-Dai, The number and magnitude of large explosive volcanic eruptions between 904 and 1865 A.D.: Quantitative evidence from a new South Pole ice core, in *Volcanism and the Earth's Atmosphere*, edited by A. Robock and C. Oppenheimer, (American Geophysical Union, Washington, DC), 165-176, 2003.

Cole-Dai, J., E. Mosley-Thompson, S. Wight, and L. Thompson, A 4100-year record of explosive volcanism from an East Antarctica ice core, *J. Geophys. Res.*, *105*, 24,431-24,441, 2000.

Farquhar, G. D., and M. L. Roderick, Pinatubo, diffuse light, and the carbon cycle, *Science*, *299*, 1997-1998, 2003.

Folland, C. K., T. R. Karl, J. R. Christy, R. A. Clarke, G. V. Gruza, J. Jouzel, M. E. Mann, J. Oerlemans, M. J. Salinger and S.-W. Wang, Observed Climate Variability and Change, Chapter 2 of *Climate Change 2001: The Scientific Basis*, edited by J. T. Houghton et al., Cambridge Univ. Press, Cambridge, UK, 99-181, 2001.

Franklin, B., Meteorological imaginations and conjectures, *Manchester Literary and Philosophical Society Memoirs and Proceedings*, *2*, 122, 1784. [Reprinted in *Weatherwise*, 35, 262, 1982.].

Free, M., and A. Robock, Global warming in the context of the Little Ice Age, *J. Geophys. Res.*, *104*, 19,057-19,070, 1999.

Graf, H.-F., J. Feichter, and B. Langhmann, Volcanic sulfur emission: Estimates of source strength and its contribution to the global sulfate distribution, *J. Geophys. Res.*, *102*, 10,727-10,738, 1997.

Gu, L., D. Baldocchi, S. B. Verma, T. A. Black, T. Vesala, E. M. Falge, and P. R. Dowty, Advantages of diffuse radiation for terrestrial ecosystem productivity, *J. Geophys. Res.*, *107(D6)*, 10.1029/2001JD001242, 2002.

Gu, L., D. Baldocchi, S. C. Wofsy, J. W. Munger, J. J. Michalsky, S. P. Urbanski, and T. As. Boden, Response of a deciduous forest to the Mount Pinatubo eruption: Enhanced photosynthesis, *Science*, *299*, 2035-2038, 2003.

Halmer M. M., Schmincke H.-U., Graf H.-F., The annual volcanic gas input into the atmosphere, in particular into the stratosphere: A global data set for the past 100 years. *J. Volc. Geotherm. Res.*, *115*, 511-528, 2002.

Hitchman, M. H., M. McKay, and C. R. Trepte, 1994: A climatology of stratospheric aerosols, *J. Geophys. Res.*, *99*, 20,689-20,700.

Houghton, J. T., Y. Ding, D. J. Griggs, M. Noguer, P. J. van der Linden, X. Dai, K. Maskell, and C. A. Johnson, Eds., *Climate Change 2001: The Scientific Basis, Contribution of Working Group I to the Third Assessment Report of the Intergovernmental Panel on Climate Change*, Cambridge Univ. Press, Cambridge, UK, 881 pp., 2001.

Humphreys, W. J., Volcanic dust and other factors in the production of climatic changes, and their possible relation to ice ages, *J. Franklin Institute*, *August*, 131-172, 1913.

Humphreys, W. J., *Physics of the Air*, Dover, New York, 676 pp., 1940.

Jones, P. D., M. New, D. E. Parker, S. Martin, and I. G. Rigor, Surface air temperature and its changes over the past 150 years, *Rev. Geophys.*, *37*, 173-199, 1999.

Joshi, M., and K. Shine, A GCM study of volcanic eruptions as a cause of increased stratospheric water vapor, *J. Climate*, *16*, 3525–3534, 2003.

Kodera, K., M. Chiba, H. Koide, A. Kitoh, and Y. Nikaidou, Interannual variability of the winter stratosphere and troposphere in the northern hemisphere, *J. Meteorol. Soc. Japan*, *74*, 365-382, 1996.

Lamb, H. H., Volcanic dust in the atmosphere; with a chronology and assessment of its meteorological significance, *Philos. Trans. Royal Soc. London, A266*, 425-533, 1970.

Lamb, H. H., Supplementary volcanic dust veil index assessments, *Climate Monitor, 6*, 57-67, 1977.

Lamb, H. H., Update of the chronology of assessments of the volcanic dust veil index, *Climate Monitor, 12*, 79-90, 1983.

Mauldin III, L. E., N. H. Zaun, M. P. McCormick, J. H. Guy, and W. R. Vaughn, Stratospheric Aerosol and Gas Experiment II instrument: A function description, *Opt. Eng., 24*, 307-312, 1985.

McCormick, M. P., P. Hamill, T. J. Pepin, W. P. Chu, T. J. Swissler, and L. R. Master, Satellite studies of the stratospheric aerosol, *Bull. Am. Meteorol. Soc., 60*, 1038-1046, 1979.

McCormick, M. P., SAGE II: An overview, *Adv. Space Res., 7*, 219-226, 1987.

Mitchell, J. M., Jr., A preliminary evaluation of atmospheric pollution as a cause of the global temperature fluctuation of the past century, in *Global Effects of Environmental Pollution*, edited by S. F. Singer, Reidel, Dordrecht, 139-155, 1970.

Mosley-Thompson, E., T. Mashiotta, and L. Thompson, Ice core records of late Holocene volcanism: Current and future contributions from the Greenland PARCA cores, in *Volcanism and the Earth's Atmosphere*, edited by A. Robock and C. Oppenheimer, (American Geophysical Union, Washington, DC), 153-164, 2003.

Naveau, P., C. M. Ammann, H.-S. Oh, and W. Guo, A statistical methodology to extract the volcanic signal in climatic time series, in *Volcanism and the Earth's Atmosphere*, edited by A. Robock and C. Oppenheimer, (American Geophysical Union, Washington, DC), 177-186, 2003.

Newhall, C. G., and S. Self, The Volcanic Explosivity Index (VEI): An estimate of explosive magnitude for historical volcanism, *J. Geophys. Res., 87*, 1231-1238, 1982.

Pawson, S., et al., The GCM-Reality Intercomparison Project for SPARC (GRIPS): Scientific issues and initial results, *Bull. Am. Meteorol. Soc., 81*, 781-796, 2000.

Perlwitz, J., and H.-F. Graf, The statistical connection between tropospheric and stratospheric circulation of the northern hemisphere in winter, *J. Climate, 8*, 2281-2295, 1995.

Penner, J. E., et al., Aerosols, their direct and indirect effects, Chapter 5 of *Climate Change 2001: The Scientific Basis, Contribution of Working Group I to the Third Assessment Report of the Intergovernmental Panel on Climate Change*, Cambridge Univ. Press, Cambridge, UK, edited by Houghton et al., 289-348, 2001.

Prata, A. J., D. M. O'Brien, W. I. Rose, and S. Self, Global, long-term sulphur dioxide measurements from TOVS data: A new tool for studying explosive volcanism and climate, in *Volcanism and the Earth's Atmosphere*, A. Robock and C. Oppenheimer, Eds. (American Geophysical Union, Washington, DC), 75-92, 2003.

Rampino, M. R. and S. Self, Sulphur-rich volcanic eruptions and stratospheric aerosols, *Nature, 310*, 677-679, 1984.

Rampino, M. R. and S. Self, Volcanic winter and accelerated glaciation following the Toba super-eruption, *Nature, 359*, 50-52, 1992.

Rind, D., The potential for modeling the effects of different forcing factors on climate during the past 2000 years, in *Climatic Variations and Forcing Mechanisms of the Last 2000 Years*, P. D. Jones, R. S. Bradley, and J. Jouzel, Eds., (Springer-Verlag, Berlin), 563-581, 1996.

Robertson, A., et al., Hypothesized climate forcing time series for the last 500 years, *J. Geophys. Res., 106*, 14,783-14,803, 2001.

Robock, A., Snow and ice feedbacks prolong effects of nuclear winter, *Nature, 310*, 667-670, 1984.

Robock, A., Nuclear winter, in *Encyclopedia of Weather and Climate, 2*, edited by S. H. Schneider, Oxford Univ. Press, New York, 534-536, 1996.

Robock, A., Volcanic eruptions and climate, *Rev. Geophys., 38*, 191-219, 2000.

Robock, A., Stratospheric forcing needed for dynamical seasonal prediction, *Bull. Amer. Meteor. Soc., 82*, 2189-2192, 2001.

Robock, A., Pinatubo eruption: The climatic aftermath, *Science, 295*, 1242-1244, 2002a.

Robock, A., Blowin' in the wind: Research priorities for climate effects of volcanic eruptions. *EOS, 83*, 472, 2002b.

Robock, A., Volcanoes: Role in climate, in *Encyclopedia of Atmospheric Sciences*, J. Holton, J. A. Curry, and J. Pyle, Eds., (Academic Press, London), 10.1006/rwas.2002.0169, 2494-2500, 2003a.

Robock, A., Introduction: Mount Pinatubo as a test of climate feedback mechanisms, in *Volcanism and the Earth's Atmosphere*, A. Robock and C. Oppenheimer, Eds. (American Geophysical Union, Washington, DC), 1-8, 2003b.

Robock, A., and J. C. Antuña, Support for a tropical lidar in Latin America, *EOS, 82*, 285, 289, 2001.

Robock, A., and M. P. Free, Ice cores as an index of global volcanism from 1850 to the present, *J. Geophys. Res., 100*, 11,549-11,567, 1995.

Robock, A., and M. P. Free, The volcanic record in ice cores for the past 2000 years, *Climatic Variations and Forcing Mechanisms of the Last 2000 Years*, edited by P. D. Jones, R. S. Bradley, and J. Jouzel, Springer-Verlag, Berlin, 533-546, 1996.

Robock, A., and C. Mass, The Mount St. Helens volcanic eruption of 18 May 1980: Large short-term surface temperature effects, *Science, 216*, 628-630, 1982.

Robock, A., and C. Oppenheimer, Eds., *Volcanism and the Earth's Atmosphere*, Geophysical Monograph 139, (American Geophysical Union, Washington, DC), 360 pp., 2003.

Sato, M., J. E. Hansen, M. P. McCormick, and J. B. Pollack, Stratospheric aerosol optical depths, 1850-1990, *J. Geophys. Res., 98*, 22,987-22,994, 1993.

Shepherd, T. G., Issues in stratosphere-troposphere coupling, *J. Meteorol. Soc. Japan, 80*, 769-792, 2002.

Simkin, T., L. Siebert, L. McClelland, D. Bridge, C. G. Newhall, and J. H. Latter, *Volcanoes of the World*, Hutchinson Ross, Stroudsburg, Pa., 232 pp., 1981.

Simkin, T., and L. Siebert, *Volcanoes of the World, Second Ed.*, Geoscience Press, Tucson, Az., 349 pp., 1994.

Soden, B. J., R. T. Wetherald, G. L. Stenchikov, and A. Robock, Global cooling following the eruption of Mt. Pinatubo: A test of climate feedback by water vapor, *Science, 296*, 727-730, 2002.

Stenchikov, G. L., I. Kirchner, A. Robock, H.-F. Graf, J. Carlos Antuña, R. G. Grainger, A. Lambert, and L. Thomason, Radiative forcing from the 1991 Mount Pinatubo volcanic eruption, *J. Geophys. Res., 103*, 13,837-13,857, 1998.

Stenchikov, G., A. Robock, V. Ramaswamy, M. D. Schwarzkopf, K.

Hamilton, and S. Ramachandran, Arctic Oscillation response to the 1991 Mount Pinatubo eruption: Effects of volcanic aerosols and ozone depletion, *J. Geophys. Res.*, *107 (D24)*, 4803, doi:10.1029/2002JD002090, 2002.

Stenchikov, G., K. Hamilton, A. Robock, V. Ramaswamy, and M. D. Schwarzkopf, Arctic Oscillation response to the 1991 Pinatubo eruption in the SKYHI GCM with a realistic Quasi-Biennial Oscillation, *J. Geophys. Res.*, 109, D03112, doi:10.1029/2003JD003699, 2004.

Stephens, G. L., et al., The CLOUDSAT mission and the A-Train, *Bull. Am. Meteorol. Soc.*, *83*, 1771-1790, 2002.

Stevermer, A., I. Petropavlovskikh, J. Rosen, and J. DeLuisi, Development of global stratospheric aerosol climatology: Optical properties and applications for UV, *J. Geophys. Res.*, *105*, 22,763-22,776, 2001.

Stirling, I., The importance of polynyas, ice edges, and leads to marine mammals and birds, *J. Marine Systems*, *10*, 9-21, 1997.

Symons, G. J., Ed., *The Eruption of Krakatoa, and Subsequent Phenomena*, Trübner, London, England, 494 pp., 1888.

Tabazadeh, A., and R. P. Turco, Stratospheric chlorine injection by volcanic eruptions: HCl scavenging and implication for ozone, *Science*, *260*, 1082-1086, 1993.

Textor, C., H.-F. Graf, C. Timmreck, and A. Robock, Emissions from volcanoes, Chapter 7 of *Emissions of Atmospheric Trace Compounds*, C. Granier, P. Artaxo, and C. Reeves, Eds., (Kluwer, Dordrecht), 269–303, 2004.

Thomason, L. W., A diagnostic aerosol size distribution inferred from SAGE II measurements, *J. Geophys. Res.*, *96*, 22,501-22,508, 1991.

Thomason, L. W., and G. Taha, SAGE III aerosol extinction measurements: Initial results, *Geophys. Res. Lett.*, *30(12)*, 1631, doi:10.1029/2003GL017317, 2003.

Thordarson, T., and S. Self, Atmospheric and environmental effects of the 1783-84 Laki eruption: A review and reassessment, *J. Geophys. Res.*, *108 (D1)*, 4011, doi:10.1029/2001JD002042, 2003.

Toon, O. B., Volcanoes and climate, in *Atmospheric Effects and Potential Climatic Impact of the 1980 Eruptions of Mount St. Helens*, edited by Adarsh Deepak, pp. 15-36, NASA Conference Publication 2240, NASA, Washington, DC, 1982.

Toon, O. B., and J. B. Pollack, Atmospheric aerosols and climate, *American Scientist*, *68*, 268-278, 1980.

Turco, R. P., O. B. Toon, T. P. Ackerman, J. B. Pollack, and C. Sagan, Nuclear winter: Global consequences of multiple nuclear explosions, *Science*, *222*, 1283-1292, 1983.

Turco, R. P., O. B. Toon, T. P. Ackerman, J. B. Pollack, and C. Sagan, Nuclear winter: Climate and smoke: An appraisal of nuclear winter, *Science*, *247*, 166-176, 1990.

Thompson, D. W. J., and J. M. Wallace, The Arctic Oscillation signature in the wintertime geopotential height and temperature fields, *Geophys. Res. Lett.*, *25*, 1297-1300, 1998.

Thompson, D. W. J., and J. M. Wallace, Annular modes in the extratropical circulation, I, Month-to-month variability, *J. Climate*, *13*, 1000-1016, 2000.

Thompson, D. W. J., J. M. Wallace, and G. C. Hegerl, Annular modes in the extratropical circulation, II, Trends, *J. Climate*, *13*, 1017-1036, 2000.

Veiga, R. E., SAGE II measurements of volcanic aerosols, *1993 Technical Digest Series*, *5*, 467-470, Optical Society of America, Washington, D.C., 1993.

Wallace P. J., Volcanic SO_2 emissions and the abundance and distribution of exsolved gas in magma bodies, *J. Volcanol. Geotherm. Res.*, *108*, 85-106, 2001.

Zielinski, G. A., J. E. Dibb, Q. Yang, P. A. Mayewski, S. Whitlow and M. S. Twickler, Assessment of the record of the 1982 El Chichón eruption as preserved in Greenland snow. *J. Geophys. Res.*, *102*, 30,031-30,045, 1997.

Alan Robock, Department of Environmental Sciences, 14 College Farm Road, Rutgers University, New Brunswick, New Jersey, 08901. (robock@envsci.rutgers.edu)

High-Resolution Gravity Mapping: the Next Generation of Sensors

Christopher Jekeli

Laboratory for Space Geodesy and Remote Sensing, Ohio State University, Columbus, Ohio

Gravity sensors have long been used to infer Earth's mass density variations at all spatial scales. Instruments on airborne platforms are now commonly employed for oil exploration and geoid modeling. Satellite systems (CHAMP and GRACE) have recently been launched and new missions are in development (GOCE) to map the gravity field globally up to resolutions better than 100 km. In addition, the precision of these space-borne sensors is sufficient to discern the monthly (and longer-term) temporal variations of the field, thus enabling improved modeling of the time variations in continental hydrology, sea level, polar ice mass, and ocean bottom pressure. Improvements in terrestrial instrumentation, primarily in gravity gradiometry, portend similar capabilities at much finer spatial scales. Regional temporal variations in the continental water storage, and the crust's vertical motion associated with earthquakes, volcanic eruptions, and extractions of water and oil deposits, can all potentially be detected with new gradiometers currently under development. One of the leading new technologies is based on the cold-atom interferometer that uses the wave-like properties of laser-cooled atoms to infer their acceleration in a gravitational field. This paper gives a statistical analysis of gravitational signals and errors of current and future moving-base sensors and provides an outline of capabilities, limitations, and requirements for applications in geodesy and geophysics. It is shown that gravity gradiometry, with a near-term goal of 0.1 E/$\sqrt{Hz}$ sensitivity, easily detects mass changes due to major earthquakes and borders on sensing regional significant hydrological changes.

1. INTRODUCTION

Results from the study of Earth dynamics form an integral part in an assessment of the State of the Planet, a theme of the 2003 General Assembly of the International Union of Geodesy and Geophysics (I.U.G.G.). Earth dynamics include changes at all temporal scales in the various regimes that constitute the planet, from the magnetosphere, atmosphere (ionosphere and troposphere), cryosphere, biosphere, and hydrosphere (including the oceans and continental water) to the lithosphere, and mantle and core of the Earth. These regimes change either directly or indirectly in response to internal and external forces that cause a redistribution of mass. The atmosphere, hydrosphere, cryosphere, and even the interior of the Earth may be thought of as fluid (at different time scales) and thus they are subject to mass transport. The secular and long-period changes in these regimes affect the Earth's climate directly in obvious ways, as well as through their mutual interactions. In addition, mass transport of these "fluids" affects the occurrence of episodic natural hazards, such as hurricanes, earthquakes, even insect infestations (with associated health and agricultural hazards), and long-term hazards, such as sea-level rise and desertification, all of which have a direct, immediate and long-term impact on the economics and quality of life for humans on the planet.

As *Chao* [2003] recently noted, transport of Earth's masses affects four directly measurable terrestrial features: Earth's center of mass, Earth's rotation rate, deformation of its surface, and the gravitational field. The latter has been the subject of geodetic study for over one hundred fifty years, though only in the last few decades with respect to temporal change detection, and only in this decade at temporal–spatial wavelengths shorter than several thousand kilometers. The dedicated satellite mission, GRACE (Gravity Recovery and Climate Experiment; [*Tapley* and *Reigber*, 1998]), launched in March 2002, specifically targets the determination of changes in the gravitational field at wavelengths longer than several hundred kilometers and with monthly temporal resolution. It is notable that the other three measurable features (center of mass, rotation rate, and surface deformation) are detected and analyzed with essentially geometric observations, namely range and time interval, that yield essentially direct evidence of change in the particular feature being considered. The measurement of the gravitational field, on the other hand, encompasses the response to all mass distributions and transports, and therein lies one of two fundamental separability issues with this type of measurement. That is, the principle of superposition, being a basic tenet of Newtonian gravitational theory, implies that measurements of the field, in whatever form, cannot necessarily discriminate between different sources since they all add to the total generated field. The geophysical inverse problem (interpret underground structure from above-ground field measurements) is mathematically ill-posed as a consequence of this principle. The other separability issue concerns Einstein's Principle of Equivalence and the inertial nature of the measurement sensors themselves, and is discussed later.

When considering the measurement of Earth's gravitational field, there are, today, essentially three types of platforms: the static measurement using relative and absolute gravimeters; terrestrial (or near-surface) vehicles, such as aircraft, ships, etc.; and satellites, usually in low Earth orbit, say, less than 500 km in altitude. Static measurements, dating from the eighteenth century are sensitive to all spatial and temporal scales of the gravity field, up to the accuracy of the instrument, but cannot be used to distinguish different spectral constituents without extensive networks in space (on Earth's surface) and in time. Such networks are expensive to establish and maintain, which motivates the preferred airborne and satellite-borne instrumentation. Satellite gravity systems, in particular, are suited for both spatial and temporal analysis due to the global coverage afforded with polar orbits and their longevity (e.g., many years) obtained with some form of orbital maintenance (countering orbital decay due to atmospheric drag). Such systems, long envisaged, are typified by the current mission, GRACE, and the mission GOCE (Gravity Field and Steady-State Ocean Circulation Mission; [*Sünkel*, 2000]) to be launched in 2006.

On the other hand, the great speed of the satellite through the field inherently constrains satellite-borne systems in recovering the spectrum of the gravitational field. Assuming a polar, circular orbit and using simple Keplerian motion, the cross-track resolution, Δy, of the ground-track (projection of the orbit onto the Earth's surface) at the equator, assuming no ground-track repetition, is given by

$$\Delta y = \frac{(2\pi)^2 R}{86400 d} \sqrt{\frac{(R+h)^3}{GM}}, \quad (1)$$

where d is the mission duration in units of days, $R = 6371$ km is Earth's mean radius, $GM = 3.986 \times 10^{14}$ m³/s² is Newton's gravitational constant times Earth's mass, and h is the satellite altitude in units of km. The along-track resolution, Δx, is given simply by the integration time, Δt, of the measurement (which is non-zero for any practical system) times the velocity:

$$\Delta x = \Delta t \sqrt{\frac{GM}{R+h} \frac{R}{R+h}}, \quad (2)$$

which includes a projection onto the ground. The total resolution is given by

$$\Delta \rho = \sqrt{\Delta x^2 + \Delta y^2}, \quad (3)$$

and is plotted in Figure 1 for various values of d and Δt (and a nominal $h = 450$ km). Clearly, the recovered spatial and temporal resolutions increase with mission duration, but there is a limit in spatial resolution imposed by the integration time, irrespective of mission duration. Integration times of 5 s to 10 s are typical for satellite-borne instrumentation and are necessary to filter the high-frequency noise in the sensors.

To obtain higher resolution requires either much shorter integration time (while maintaining accuracy) or slower speed, as with an aircraft. Airborne gravimetry and gravity gradiometry are also limited in resolution by the integration time, but with a typical aircraft speed of 100 m/s, even 60 s of integration time yields 6 km along-track resolution. The need for such high resolution in gravimetry is well illustrated, e.g., in [*Dickey* et al., 1997]. Figure 2, adapted from this reference, depicts the various geophysical signals in terms of their wavelength constituents and corresponding gravitational accuracy requirements. Also indicated is the spatial predicted gravitational accuracy (cumulative in wavelength) of GRACE and GOCE. Airborne gravimetry and gradiometry are capa-

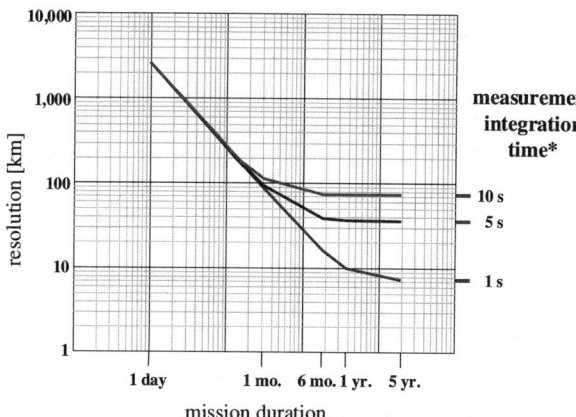

Figure 1. Spatial resolution of satellite measurements versus mission duration and integration time. * Satellite missions GRACE and GOCE have integration times of 5 s and 10 s (anticipated), respectively.

ble of detecting the signals with spatial resolution (half-wavelength) better than a few hundred kilometers, while satellite systems, generally, are limited to resolutions lower than about 50–70 km (see also Figure 1). Clearly, the two types of gravimetry platforms overlap in capability, but they also possess their own niche in the spectral domain of geophysical signals. Much has been analyzed and predicted regarding dedicated satellite gravity missions, such as CHAMP (Challenging Minisatellite Payload [*Reigber* et al., 2002]), GRACE, and GOCE [e.g., *Rummel* et al., 2002], as well as follow-on missions with ultra-high accuracy laser tracking systems [*Bender* et al., 2002]. However, the focus here is on airborne systems, especially in view of the new cold-atom-interferometer technology that presages accelerometry with 10^{-10} g bias stability [*Kasevich*, 2002]. This technology takes advantage of the quantum-mechanical, particle-wave duality of atoms, treating them as proof masses that respond classically to inertial accelerations, and using their wave properties in an interferometer to sense these inertial responses (including rotations). The sensitivity of the matter-wave interferometer compared to the optical interferometer is orders of magnitude higher due to the much shorter wavelengths associated with slowly moving (cold) atoms.

Airborne gravity gradiometry benefits significantly from such advancements in inertial sensor technology. It is the aim of this paper to contrast airborne gravimetry and gradiometry, to introduce the atom interferometer technology and its application to gradiometry, and to illustrate the capabilities of airborne gradiometry in the detection of regional geophysical signals. Only a brief outline of the interferometer technology is possible here and reference is made to [*Berman*, 1997] and [*Storey* and *Cohen-Tannoudji*, 1994], as well as the papers mentioned in Section 3 for further details.

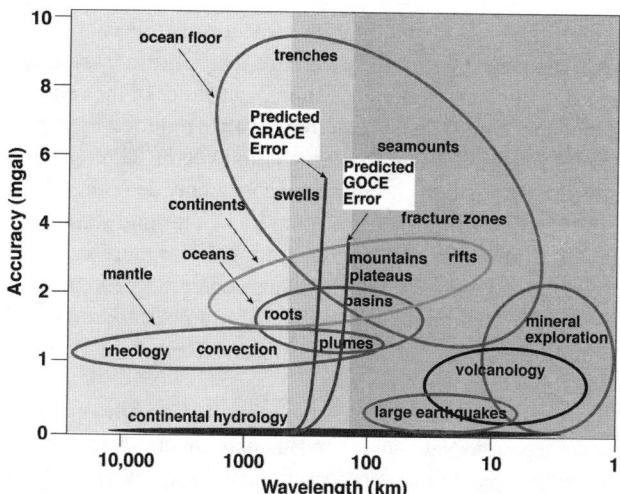

Figure 2. Wavelength (2 × resolution) versus cumulative spectral accuracy requirements for various gravitational signals identified by source. Shaded areas of the graph refer to the resolution capabilities of airborne gravity systems (center and right) and satellite gravity systems (center and left), with overlap in the central region. The indicated cumulative accuracy predicted for GRACE and GOCE gives the lower bounds of resolution (half-wavelength) for these satellite gravity missions.

2. AIRBORNE GRAVIMETRY AND GRADIOMETRY

The theoretical foundation for airborne gravimetry, based on measurements of accelerations, is Newton's second law of motion. Provided that all accelerations are properly transformed into the inertial frame (where this law only holds), one has:

$$g^n = C_i^n \left(\ddot{x}^i - C_b^i a^b \right), \qquad (4)$$

where (from right to left) a^b is the specific force or inertial acceleration in the body-frame (b-frame); C_b^i is the rotation matrix that transforms this into the inertial frame (i-frame); $\ddot{x}^i$ is the kinematic acceleration of the vehicle in the i-frame; C_i^n transforms the difference in these accelerations to some desired frame, such as the local north-east-down (NED) frame (n-frame); and g^n is the gravitation vector in that frame. (Superscripts on vectors denote the frame of coordinates; and, on the rotation matrices, they denote the target frame, while the subscript denotes the source frame.) The two types of accelerations, specific force and kinematic acceleration, must be measured by two independent systems in order to recover the gravitational acceleration. Accelerometers sense specific force, and the kinematic acceleration can be derived from a time series of position coordinates, obtained, e.g., from the Global Positioning System (GPS). Determining the rotation transformation from the body frame to the inertial frame can be done with gyroscopes (gyros), which measure the angular rate of the body (vehicle) with respect to the inertial frame. (Alternatively, they actively stabilize the platform by isolating it, using gimbals, from the rotational dynamics of the vehicle.) Three gyros and three accelerometers are required to resolve the total three-dimensional gravitation vector. Three-dimensional GPS position coordinates are generally determined precisely with respect to one or more known ground stations on the basis of mutual range observations to 6 or more satellites. Computing kinematic accelerations accurate to 1 mgal (10^{-5} m/s^2 ≈ 10^{-6} g) requires, as a rule of thumb, an along-track sampling interval not longer than 1 s (1 Hz data rate) and an averaging time of about 60 s.

A gravity gradiometer, on the other hand, while operating under the same Newtonian fundamentals, in principle, requires only one type of gradient measurement. Essentially, most gradiometers are differential accelerometers, where an observation comes from the difference in accelerations of two accelerometers divided by the known geometric baseline length between them. The common linear accelerations of the platform cancel. Yet the rotational velocity and rotational acceleration still couple indistinguishably into the gradient measurement if the platform is rotating with respect to inertial space. The tensor of gravitational gradients, Γ^n, say, in the NED frame, is related to the differential acceleration measurements according to

$$\Gamma^n = -C_b^n \left(\frac{\delta a^b}{\delta x^b} - \Omega_{ib}^b \Omega_{ib}^b - \dot{\Omega}_{ib}^b \right) C_n^b, \qquad (5)$$

where $\delta a^b / \delta x^b$ represents the differential acceleration tensor (3×3 matrix) in the body frame and Ω_{ib}^b is the skew-symmetric matrix of angular rates, ω_j, of the body frame with respect to the inertial frame (as indicated by the subscripts):

$$\Omega_{ib}^b = \begin{pmatrix} 0 & -\omega_3 & \omega_2 \\ \omega_3 & 0 & -\omega_1 \\ -\omega_2 & \omega_1 & 0 \end{pmatrix}, \qquad (6)$$

with components given in the b-frame (superscript).

The gravitational gradient tensor, Γ^n, is symmetric (because its components are the second-order partial derivatives of the gravitational potential and the order of differentiation is immaterial) and it is traceless in free space (because of Laplace's equation), irrespective of the n-frame:

$$\Gamma^n = \begin{pmatrix} \Gamma_{1,1} & \Gamma_{1,2} & \Gamma_{1,3} \\ \Gamma_{2,1} & \Gamma_{2,2} & \Gamma_{2,3} \\ \Gamma_{3,1} & \Gamma_{3,2} & \Gamma_{3,3} \end{pmatrix}, \qquad (7)$$

$$\Gamma_{j,k} = \Gamma_{k,j}, \quad \Gamma_{1,1} + \Gamma_{2,2} + \Gamma_{3,3} = 0.$$

The diagonal and off-diagonal components are called the in-line and cross gradients, respectively. The differential acceleration tensor, in contrast, is neither symmetric nor traceless in the general case. If all elements are measured ("full-tensor" gradiometer) then the symmetry of Γ^n can be used to eliminate the angular acceleration term:

$$\Gamma^n = -C_b^n \left(\frac{1}{2}\left(\frac{\delta a^b}{\delta x^b} + \left(\frac{\delta a^b}{\delta x^b}\right)^T \right) - \Omega_{ib}^b \Omega_{ib}^b \right) C_n^b. \qquad (8)$$

Similarly, the angular acceleration term is irrelevant if only in-line gradients are measured and the platform is stabilized such that $C_b^n = I$.

Turning now to errors, a differential perturbation of the observation model for gravimetry, equation (4), yields the relationship between the gravitation error and the contributing sensor errors. For the sake of simplicity, it is assumed that the n-frame is identical to the i-frame (which is reasonable for short duration). Thus:

$$\delta g^i + \frac{\partial g^i}{\partial x^i} \delta x^i = \delta \ddot{x}^i - \delta C_b^i a^b - C_b^i \delta a^b, \qquad (9)$$

where the second term on the left is the *registration error* due to unknown position. The error in the transformation matrix, δC_b^i, can be expressed [*Jekeli*, 2000] in terms of the orientation errors, $\psi = (\psi_1, \psi_2, \psi_3)^T$, of the coordinate axes of the b-frame with respect to the i-frame:

$$\delta C_b^i = -\begin{pmatrix} 0 & -\psi_3 & \psi_2 \\ \psi_3 & 0 & -\psi_1 \\ -\psi_2 & \psi_1 & 0 \end{pmatrix} C_b^i = -\Psi C_b^i, \qquad (10)$$

$$\delta C_i^b = \left(\delta C_b^i\right)^T.$$

With a further simplification that $C_b^i \approx I$, equation now becomes

$$\delta g^i = \delta \ddot{x}^i + \Psi a^i - \Gamma^i \delta x^i. \qquad (11)$$

Similarly, from the observation model for gradients, equation (8) (see [*Jekeli*, 2003] for details), it is easy to find an expression for the gradient error in terms of contributing sensor errors:

$$\delta \Gamma^i = \Gamma^i \Psi - \Psi \Gamma^i - \left(\delta B - \delta \Omega_{ib}^i \Omega_{ib}^i - \Omega_{ib}^i \delta \Omega_{ib}^i \right) - \sum_j \Xi_j^i \delta x^j \qquad (12)$$

where Ξ_j^i is the spatial gradient of Γ^i with respect to the coordinate x_j^i, and

$$\delta B = \frac{1}{2}\left(\delta\left(\frac{\delta a^b}{\delta x^b}\right) + \delta\left(\frac{\delta a^b}{\delta x^b}\right)^T\right). \quad (13)$$

Inspection of the error equations (11) and (12) reveals that in the case of gravimetry the orientation error couples with the *specific forces* on the vehicle, a^i (including the reactions to gravity and centrifugal accelerations, etc.); whereas, in the case of gradiometry, it couples with the gravitational gradients, Γ^i. Thus, since the *acceleration* dynamics of the aircraft cannot be strictly controlled and are much larger than the anomalous gravitational signal, the orientation error may be expected to cause greater problems in airborne gravimetry. On the other hand, errors in the gravitational gradients are affected by the dynamics of the angular rate, Ω_{ib}^i, through its coupling to the angular rate error.

A rudimentary error analysis may be performed in the spectral domain using these equations. Table 1 lists nominal values for the parameters associated with the sensor errors, here assumed to consist of only white noise for the accelerometer, the differential accelerometer, and the positioning system, and including also a random bias in addition to white noise for the gyros. These values are typical of conventional high-accuracy inertial measurement units. Also, the uncalibrated orientation error is assumed to be a random bias. Aircraft dynamics are defined in terms of power spectral densities (psd's) typical of a small twin-engine aircraft flying essentially straight and level at 1000 m altitude and 250 km/hr. The corresponding psd's, Φ, of the error in the vertical gravitation component, as well as in the in-line and cross gradients are shown in Figures 3, 4, and 5, respectively. For details the reader is referred to [*Jekeli*, 2003].

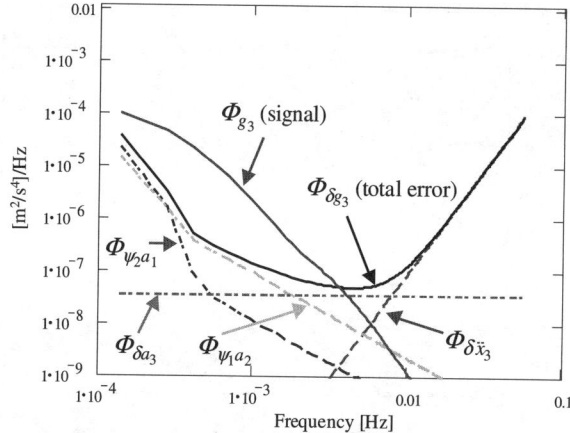

Figure 3. Along-track power-spectral densities of the vertical gravitational component, Φ_{g_3}, of its total error, $\Phi_{\delta g_3}$, and of its constituent errors (dashed lines), due to orientation error and sensor errors, identified by the subscript notation (see equation (11)).

Table 1. Nominal values* for airborne gravimetry and gradiometry error parameters

differential accelerometer white noise	psd($\delta(\delta a/\delta x)$) = (30 E/$\sqrt{Hz}$)^2
accelerometer noise	var(δa) = (25 mgal)^2
position noise	var(δx) = (0.1 m)^2
gyro white noise	psd($\delta\omega$) = (0.06 °/hr/$\sqrt{Hz}$)^2
gyro rate bias (uncalibrated part)	var($\delta\omega_0$) = (0.003 °/hr)^2
orientation bias error (uncalibrated part)	var(ψ_0) = (0.005 °)^2

* 1 mgal = 10^{-5} m/s^2, 1E = 10^{-9} s^{-2}

The essential feature of the gravimetry error psd (Figure 3) is the "window" within which the gravitational signal can be detected. This well-known spectral window [*Schwarz et al.*, 1992] is framed by the long-wavelength errors associated with the orientation errors and by the short-wavelength errors arising in the kinematic acceleration obtained by differentiating position coordinates. Greater aircraft velocity shifts the window to the right relative to the psd of the gravitational signal. Positioning error is particularly critical as it defines the accuracy of the derived kinematic acceleration; it is much less critical (and not shown in Figure 3) for the geospatial registration of the measurement. In gradiometry, such types of restricting bounds in the spectral domain, if any, are generally due mostly to the gradiometer, itself, rather than to the auxiliary sensors, with the exception that the orientation error has some long-wavelength influence on the cross gradients. Another critical component in this case is the angular rate noise of the gyro that couples with the angular rate dynamics of the aircraft. In order to reduce this effect to the level of 0.1 E/$\sqrt{Hz}$ requires a reduction in the assumed nominal value in the gyro noise by almost an order of magnitude, to about 0.01°/hr/$\sqrt{Hz}$. At this level the orientation error also has a definite effect at the longer wavelengths. Positioning error is hardly a concern in airborne gradiometry since the geospatial variation of gradients within several tens of meters is negligible compared to existing (and predicted) typical instrument resolution (100 m) and accuracy (0.1 E).

In terms of required sensors, the difference between airborne gravimetry and gradiometry lies in the reliance of the former on two completely different types, a geometric type (GPS) and an inertial type (the accelerometer (and gyro)), while the latter requires essentially only one type, the iner-

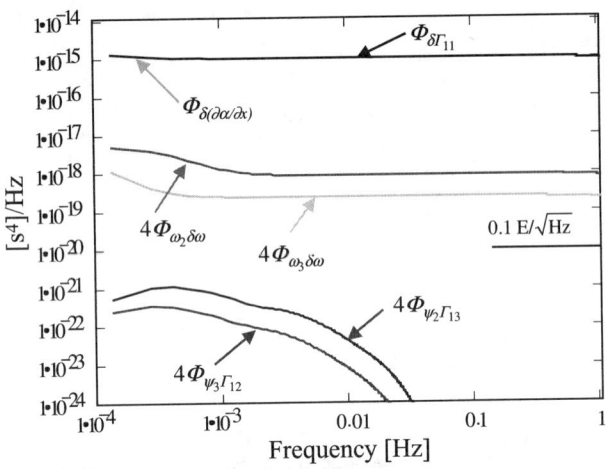

Figure 4. Along-track power-spectral densities of the error in the in-line gradient, $\delta\Gamma_{1,1}$, and of its constituent errors, due to orientation error and sensor errors, identified by the subscript notation (see also equation (12)). Also indicated is the goal sensitivity of 0.1 E/√Hz.

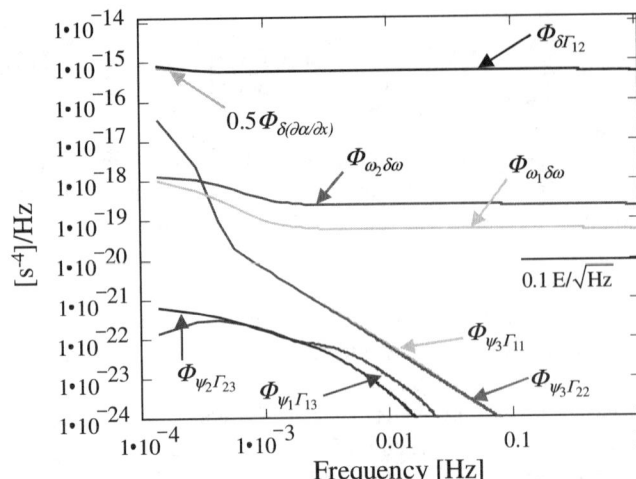

Figure 5. Along-track power-spectral densities of the error in the cross gradient, $\delta\Gamma_{1,2}$, and of its constituent errors, due to orientation error and sensor errors, identified by the subscript notation (see also equation (12)). Also indicated is the goal sensitivity of 0.1 E/√Hz.

tial sensor. In fact, both linear and angular inertial accelerations can be measured with an appropriate configuration of simple linear accelerometers [e.g., *Zorn*, 2002]. This dichotomy in sensor types for gravimetry and gradiometry has profound implications for the future of airborne gravity systems. It is doubtful, at present, that the geometric type (such as GPS) will realize significant (orders of magnitude) improvement in accuracy in the foreseeable future, thus limiting the overall potential of airborne gravimetry. Airborne gradiometry, on the other hand, benefits from relying on only one type of sensor; and, the inertial sensor is currently undergoing a literal quantum leap in technological advancement.

3. NEW INERTIAL SENSOR TECHNOLOGY

This development makes use of atoms as proof masses in an inertial sensor that is based on the cold-atom interferometer [*Kasevich* and *Chu*, 1992]. According to the quantum mechanical, wave-particle duality of atoms, their inertial mass is sensitive to acceleration (and rotation), while their wave properties can be utilized to extract exquisite sensitivity from a measurement of interfering beams. In quantum mechanics, atoms may be treated as de Broglie waves with wavelength

$$\lambda = \frac{\hbar}{mv}, \qquad (14)$$

where $\hbar$ is Planck's constant and mv is the linear momentum of the atom. The sensitivity of the interferometer output (fringe pattern) to changes in the path length of interfering beams of atoms depends inversely on their wavelength. Light beams have wavelengths limited to the order of micrometers ($3\times10^{-7} - 3\times10^{-6}$ m), but by cooling atoms (of cesium, e.g.) in an otherwise evacuated chamber to μK, thus reducing the velocity to the order of 1 m/s, the equivalent wavelength is of the order of 3×10^{-9} m. Consequently, already from this simple analysis, the sensitivity of the cold-atom interferometer is orders of magnitude greater than that of the optical interferometer. *Clauser* [1988] shows that the potential sensitivity is up to 10^{10} greater.

The recent applications of the cold-atom interferometer as inertial sensor have been oriented primarily to measuring the gravitational acceleration and its gradient in the static mode. However, the concept hardly differs for accelerometers, in general, as would be used for inertial navigation systems or for gradiometers on moving platforms. Such progress is currently underway at Stanford University by the Kasevich group [*McGuirk* et al., 2002]. In addition, the concepts can be applied to sense rotation in the form of Coriolis acceleration, and in fact, the atom interferometer is sensitive to both angular rate and specific force. The following describes the technology as applied to static gravimetry, but is analogous to inertial acceleration measurements.

In this case, ultra-cold atoms are launched and then allowed to fall in the gravitational field, much like the corner-cube of the FG5 absolute gravimeter (which is only dropped, not launched [*Carter* et al., 1994]). First, the atoms are collected and loaded into a magneto-optic trap (MOT) where they are cooled by a set of three mutually orthogonal pairs of counter-propagating lasers that create a light field whose interaction with the atoms slows them to near zero velocity [*Metcalf* and *van der Straten*, 1999; *Dalibard* and *Cohen-Tannoudji*, 1989].

The magnetic field is used to keep the atoms from falling in the gravitational field. With the field turned off, after cooling, the cloud of atoms is launched and allowed to follow a free-fall trajectory, during which another set of lasers interact with them to create interfering sets of atoms.

These interactions are optical stimulations, produced with opposing laser pulses, that cause the atoms to absorb and emit photons, thus changing their internal states (their phase) and also inducing a net mechanical recoil, changing their momentum and their physical trajectory in the gravitational field [*Kasevich* and *Chu*, 1991]. These so-called Raman state transitions simulate splitters and mirrors of a conventional interferometer in terms of transition probability. That is, an initial pulse (at $t = t_1 = 0$) is adjusted to transition (probabilistically) the states of half the atoms (analogue of a splitter), thus adding a quantum mechanical phase, $\phi(t_1)$. The two resulting atomic wave packets travel different paths (A and B) due to the recoil of one, and both are affected equally by the gravitational acceleration (if uniform). Subsequently, at time, $t = t_2 = T$, an optical pulse transitions the states of both wave packets (adding or subtracting a phase, $\phi(t_2)$, depending on their existing state) and deflects them (analogue of mirrors), in essence returning the originally transitioned set to the initial state and changing the state of the originally non-transitioned set. As shown in Figure 6, the trajectories of the two atomic beams are not symmetric due to the gravitational acceleration. When the two beams recombine, at $t = t_3 = 2T$, half of each is again stimulated (analogue of a splitter) with an added phase. Each change in phase can be expressed as [*Kasevich* and *Chu*, 1991; *Gustavson*, 2000, p.38]

$$\phi(t) = kz(t) + \phi_0(t), \qquad (15)$$

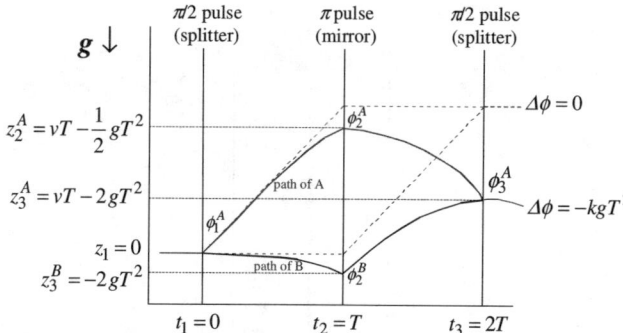

Figure 6. Paths, as functions of time, of atoms stimulated by Raman lasers with (solid lines) and without (dashed lines) gravitational acceleration (after Peters et al., 2001). Vertical positions on the left axis are derived from: $z(t) = z_0 + v_0(t - t_0) + 0.5\, g(t - t_0)^2$. Phase differences detected at time, $t_3 = 2T$, by the interferometer are indicated next to the right axis.

where k is the resultant wave number of the two counter-propagating Raman lasers, z is the vertical position of the atom, and ϕ_0 depends on the difference in frequencies of the lasers.

Combining the phase shifts of the two atomic wave packets as they pass through the interferometer, we have

$$\Delta\phi = \left(\phi^A(t_1) - \phi^A(t_2) + \phi^A(t_3)\right) - \phi^B(t_2). \qquad (16)$$

Substituting equation (15) into equation (16) and using expressions for the positions along the free-fall trajectories, $z(t)$, given in Figure 6, yields

$$g = -\frac{1}{kT^2}(\Delta\phi - \Delta\phi_0). \qquad (17)$$

On a vibrationally quiet platform, $\Delta\phi_0 = 0$. Furthermore, in the absence of an acceleration, $\Delta\phi = 0$. The cycle time in laboratory tests for one measurement (loading and cooling the atoms in the MOT, launching them, stimulating them with the Raman lasers, and detecting the phase change) is of the order of one second, but cycle times shorter than a tenth of a second are in development. Details of the atom-interferometric gravimeter may be found in [*Kasevich* and *Chu*, 1992] and [*Peters* et al., 2001], who also report a measurement precision of 2×10^{-7} m/s^2 (1.3 s integration time), comparable to conventional relative gravimeters, and 2×10^{-9} m/s^2 (2 day integration time), comparable to cryogenic absolute gravimeters.

The phase associated with the Raman coupling, according to equation (15), locates the atom with respect to the frame defined by the optical field of the lasers. Thus, we have the analogy with the free-fall absolute gravimeter where the falling corner-cube is located by a laser interferometer relative to the frame of the laser. More importantly, this means that the atom-interferometric gravimeter (like any gravimeter) more accurately is an accelerometer that senses action forces (in the static case, the reaction force of Earth's surface that keeps the instrument from falling to Earth's center). The input axis is along the Raman laser axis that may be oriented arbitrarily, not only in the vertical direction. Thus, this accelerometer could serve as an inertial measurement unit in precision inertial navigation systems. On a moving platform, it is also sensitive to Coriolis accelerations due to rotations. These could be cancelled by adding the output of two such accelerometers whose atomic beams travel in opposite directions. The atom interferometer as an inertial sensor is currently under intense development at Stanford University.

Moving-base gravimetry, as already noted, suffers from the inseparability of kinematic and inertial accelerations; and, the well-known alternative to this problem is the airborne (and satellite) gravity gradiometer. A particular problem with conventional gradiometers (being differential accelerometers) is the non-cancellation of systematic errors in the accelerometers. The error characteristics (depending significantly on the acceleration environment) of the accelerometers must be exquisitely matched and the input axes must be precisely aligned in order to achieve accuracy of the order of 1 E. In the airborne case, special filters that rely on rotating platforms to modulate the gradients and errors at different frequencies [*Jekeli*, 1988] have achieved an accuracy approaching better than 10 E [*Talwani*].

However, a gradiometer based on the atom-interferometric accelerometer inherently has no alignment errors and the proof masses, e.g., cesium atoms, of both accelerometers are identical (perfectly matched). As Figure 7 shows, a single pair of opposing Raman lasers can stimulate the sets of atoms in both accelerometers thus defining an unambiguous (in-line) gradient input axis (cross-gradients would still require parallel, non-colinear Raman lasers). Furthermore, in principle, the gradiometer operation is not constrained by the baseline length between accelerometers, although practically it must fit into the moving vehicle. Increasing the baseline length increases the sensitivity, but also potentially requires accounting for higher-order gradients.

Gradiometer technology has a long history [*Jekeli*, 1988] and extremely precise sensors have been developed by others, e.g., by *H.J. Paik* at the University of Maryland for potential space applications [*Moody* et al., 1986] and airborne deployment [*Paik*, et al., 1997], where the latter has a predicted sensitivity of 0.1 E/√Hz to 1 E/√Hz. These sensors take advantage of the extremely low noise associated with cryogenic, superconducting devices, however, they still rely on macroscopic proof masses, where alignment and matching of other systematic errors are also critical.

Although tested primarily in the laboratory (field tests are planned for 2004, private communication, *M. Kasevich*, 2003), the atom-interferometric gradient sensor may be deployed on a moving platform. From equation (5) it is clear that all common-mode accelerations cancel in gradiometry ($\delta a_{common\,mode} = 0$); and, from equation (17) it is equally clear that the gradient measurement is insensitive to the vibrational noise that appears in $\Delta\phi_0$ but is common to both sets of atoms if a single optical reference is used. Laboratory success of the atom-interferometric gradiometer has been reported by *Snadden* et al. [1998], *McGuirk* [2001], and *McGuirk* et al. [2002], the latter demonstrating a sensitivity of 4 E/√Hz. Current efforts at Stanford University are designed to achieve a sensitivity of at least 0.1 E/√Hz on a moving platform (*M. Kasevich*, private communication, 2003).

4. GRAVITATIONAL SIGNALS

While the new sensor technology for airborne gradiometry promises orders of magnitude improved sensitivity over existing systems, it is worthwhile to consider the consequent capabilities of detecting mass transport at regional spatial scales. Figure 2 already gives some indication; the two examples below put some bounds on the prospects. In the first case, consider the monthly change in the mean water content in the upper 2 m of soil with respect to a long period of time (e.g. 6 years). Such data are available in the form of 1°×1° mean values over the continents from the IERS Global Geophysical Fluids Center (NCEP/NCAR CDAS-1 data set). The water content values are given in equivalent columns of water height, Δh_w; and, these generate a gravitational potential (Newtonian potential of a surface density layer) given by

$$\delta V_w(\theta,\lambda) = G\rho_w R^2 \iint_\sigma \frac{\Delta h_w(\theta',\lambda')}{\ell} d\sigma, \qquad (18)$$

where $\rho_w = 1000$ kg/m³ is the density of water; σ is the unit sphere; and ℓ is the distance between the integration point, with spherical polar coordinates (θ',λ'), and the evaluation point, (θ,λ), of the potential, δV_w. The mean value, standard deviation, and extreme (maximum or minimum) value of the water heights for the U.S. and tropical South America are shown in Table 2. Expressing the water height function as a series of spherical harmonics, $\overline{Y}_{n,m}(\theta,\lambda)$:

$$\Delta h_w(\theta,\lambda) = \sum_{n=0}^{\infty}\sum_{m=-n}^{n} \Delta H_{n,m}\overline{Y}_{n,m}(\theta,\lambda), \qquad (19)$$

the coefficients, $\Delta H_{n,m}$, lead directly to corresponding degree variances (power spectral density accumulated over orders, m):

$$\sigma^2_{\Delta h_w}(n) = \sum_{m=-n}^{n} (\Delta H_{n,m})^2. \qquad (20)$$

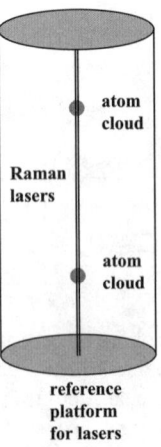

Figure 7. In-line gradiometer (Kasevich, 2002), showing single Raman laser interacting with two atom wave packets.

From equation (18), and

$$\frac{1}{\ell} = \frac{1}{R}\sum_{n=0}^{\infty}\frac{1}{2n+1}\sum_{m=-n}^{n}\bar{Y}_{n,m}(\theta',\lambda')\bar{Y}_{n,m}(\theta,\lambda), \quad (21)$$

the corresponding degree variances of the potential are given by [see also *Wahr* et al., 1998]:

$$\sigma^2_{\delta V_w}(n) = \left(\frac{4\pi RG\rho_w}{2n+1}\right)^2 \sigma^2_{\Delta h_w}(n), \quad (22)$$

where the secondary crustal loading effect on the potential can safely be neglected for the higher degrees, n (shorter wavelengths).

Wahr et al.'s [1998] global harmonic spectral analysis of the gravitational potential due to these water heights was used to elucidate the sensitivity of satellite gravity missions, such as GRACE, to the continental hydrological cycles. However, for regional analyses, regional spectral transforms are more appropriate. A relationship analogous to equation (22) holds in planar approximation for the isotropic (averaged over frequency direction) power spectral density:

$$\Phi_{\delta V_w}(\mu) = \left(\frac{G\rho_w}{\mu}\right)^2 \Phi_{\Delta h_w}(\mu), \quad (23)$$

where the spatial cyclical frequency, μ, is related approximately to harmonic degree, n, by

$$\mu \approx \frac{n}{2\pi R}. \quad (24)$$

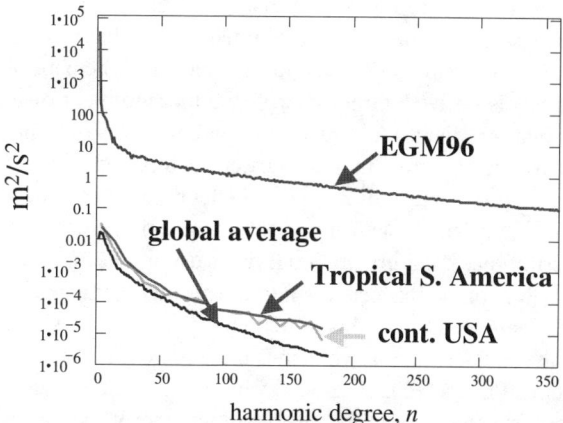

Figure 8. Root-mean-square gravitational potential per spherical harmonic degree, $\sigma_{\delta V}(n)$, due to monthly soil water change with respect to a 6-year mean. The global average and the regional signals for the U.S. and for tropical South America are compared to the Earth's gravitational potential represented by the model EGM96.

The planar, isotropic water height psd's for the continental U.S. and for tropical South America were computed using the classical two-dimensional periodograms [*Marple*, 1987]. Converting these to degree variances according to the approximate relationship:

$$\sigma^2(n) \approx \frac{n}{2\pi R^2}\Phi(\mu), \quad (25)$$

yields the comparison in Figure 8, that also includes the Earth's total gravitational potential as modeled by the global model EGM96 [*Lemoine* et. al, 1998].

The much lower average of the global hydrological signal is due to the analysis of zero-values over the oceans, which cover 70% of the planet. Resolution of the signal higher than that limited by the 1° data can be modeled under the assumption that its spectral decay, like that of the total gravitational signal, obeys a simple power law (a straight line in logarithmic scale). The following type of model is used to approximate that decay in the spectral domain [*Jekeli*, 2003]:

$$\tilde{\Phi}_{\delta V_w}(\mu) = \frac{1}{\mu}\sum_{j=1}^{J}\frac{\sigma_j^2}{\alpha_j}e^{-2\pi\frac{\mu}{\alpha_j}}, \quad (26)$$

with parameters, J, σ_j^2, and α_j. Figure 9 shows this (somewhat conservative) model, now in terms of the vertical-vertical gradient psd, obtained from the psd of the gravitational potential according to:

$$\Phi_{\delta\Gamma}(\mu) = (2\pi\mu)^4 \Phi_{\delta V}(\mu). \quad (27)$$

Also shown in Figure 9 are the gradiometer noise levels corresponding to gradient surveys comprising tracks spaced at $\Delta y = 1$ km and with aircraft velocity of $v = 100$ m/s (altitude attenuation of the signal is neglected). The relationship between the temporal and spatial error psd's is given by

$$\bar{\Phi}_{\delta\Gamma}(\mu) = v \cdot \Delta y \cdot \Phi_{\delta\Gamma}(f). \quad (28)$$

The second example concerns the vertical crustal displacement due to a large earthquake, such as the 1960 Chilean earthquake centered at $(\theta,\lambda) = (128.26°, 287.85°)$, with magnitude $M = 9.5$. The displacements, provided on a $0.25° \times 0.25°$ grid (A. Braun, private communication, 2003), were *modeled* based on earthquake source parameters associated with the main fault plane. The displacements are viewed as changes in mass from a nominal distribution and thus may be analyzed like the hydrological change signals considered above. The displacement statistics (mean value, standard deviation, and extreme value) over the region bounded by $\{120° \leq \theta \leq 134°,$

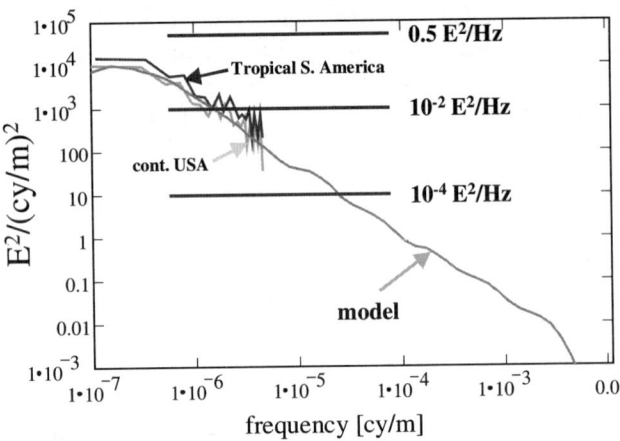

Figure 9. Power spectral densities, $\Phi_{\delta\Gamma}(\mu)$, of the vertical-vertical gravitational gradient of hydrological signals and a model extension to higher resolution. Gradiometer noise psd's are indicated assuming 1 km track spacing and 100 m/s velocity.

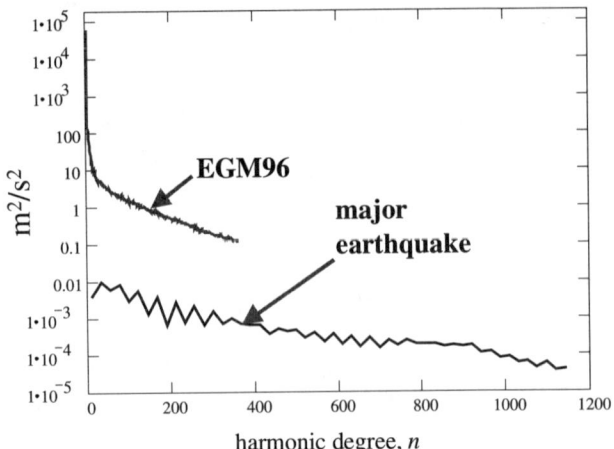

Figure 10. Root-mean-square gravitational potential per spherical harmonic degree, $\sigma_{\delta V}(n)$, due to *modeled* vertical crustal displacement caused by the Chilean 1960 earthquake. The gravitational potential model, EGM96, is shown for comparison.

$280° \leq \lambda \leq 291°\}$ are also given for comparison in Table 2. Figure 10 shows the analyzed signal from the earthquake-induced displacement model in terms of gravitational potential per spherical harmonic degree (periodogram of $\Delta h \to \Phi_{\Delta h} \to \Phi_{\delta V}$ by equation (23) $\to \sigma^2_{\delta V}$ by equation (25)), showing approximate agreement with a similar analysis by *Gross* and *Chao* [2000] of this and other earthquakes. The same procedures were applied as before to compute a corresponding gradient psd, using a power-law decay model for the higher spatial frequencies. Figure 11 shows the gradient psd, the extended model, and a gradiometer noise psd under the same assumptions of track spacing and aircraft velocity.

Figures 9 and 11 show that a gradiometer with sensitivity of 0.1 E/√Hz will barely detect regional hydrological change but will easily sense mass changes due to major earthquakes (even those with an order of magnitude displacement less than the extremely powerful Chilean quake).

Table 2: Statistics for hydrological and earthquake signals in terms of height of mass change

	mean	st. dev.	extremum
continental U.S. hydrology (1°×1° grid)	10.4 cm	8.9 cm	36.7 cm
tropical S. American hydrology (1°×1° grid)	7.1 cm	12.6 cm	31.2 cm
Chilean earthquake model (0.25°×0.25° grid)	2.0 cm	88.6 cm	688 cm

5. SUMMARY

The detection of gravitational signatures due to mass transport associated with Earth dynamics is ultimately limited using satellite techniques to spatial resolution of about 50 km to 70 km because of the integration time of the measurement, typically 5 s to 10 s. Significantly higher resolution can be achieved only with airborne (or other terrestrial) systems whose speed is much slower than the satellite ground speed of about 7.7 km/s (for satellites at 400 km altitude). The two fundamental types of gravitational field measurements on aircraft (or other terrestrial moving platforms) are gravimetry and gradiometry. Gravimetry (the measurement of the gravitational acceleration) requires two essentially different types of sensors, the inertial accelerometer (also the gyro) and a geometric positioning system from which the kinematic acceleration is derived. Gradiometry (the measurement of gravitational gradients), on the other hand, requires only inertial measurement units, that is, accelerometers and gyros. Both gravimetry and gradiometry are encumbered by Einstein's equivalence principle that prevents the discrimination between inertial mass and gravitational mass using measurements of acceleration or of acceleration gradients on rotating platforms. The manifestation of this principle in gradiometry is in terms of the squared angular velocities that mimic gravitational gradients. Recent technological advances in inertial sensors, particularly the cold-atom interferometer, promise orders of magnitude improvement in measurement accuracy. This new technology can be used to construct extremely sensitive accelerometers (10^{-10} g/√Hz), gradiometers (0.1 E/√Hz), and gyroscopes (10^{-6} °/hr/√Hz), *M. Kasevich,* private communication, 2003). Gravity gradiometry, relying in principle only

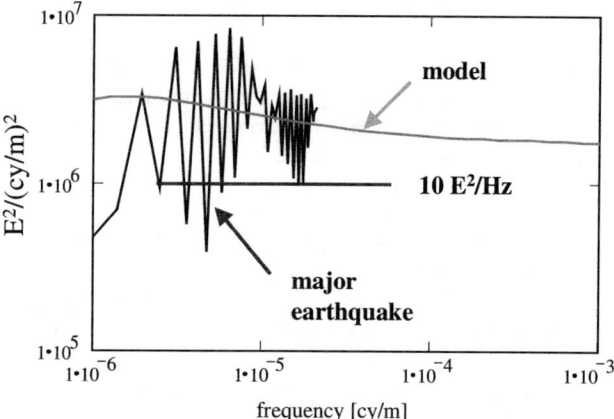

Figure 11. Gravitational gradient power spectral density, $\Phi_{\delta\Gamma}(\mu)$, of the modeled, earthquake-induced, vertical crustal displacement and a modeled extension to higher resolution. Gradiometer noise psd of 10 E²/Hz is indicated assuming 1 km track spacing and 100 m/s velocity.

on inertial sensors, will benefit significantly from these technological advancements. Analyses shown here demonstrate the ability of these new sensors to detect low-amplitude gravitational signals associated with regional mass transport, such as vertical crustal deformation (e.g., due to earthquakes and volcanism) and perhaps even regional hydrological changes.

Acknowledgments. This work was supported by the National Imagery and Mapping Agency, contracts NMA202-98-1-1110 and NMA401-02-1-2005. The author is grateful to *Mark Kasevich* for information provided on the cold-atom interferometer development at Stanford University, and to *Steve Sparks* and two anonymous reviewers for helping to smooth out the manuscript.

REFERENCES

Bender, P. L., J. L. Hall, J. Ye, and W. M. Klipstein (2002): Satellite-satellite laser links for future gravity missions. *Space Science Review*, in press.

Berman, P. R. (ed.) (1997): *Atom Interferometry*. Academic Press, San Diego.

Carter, W. E., G. Peter, G. S. Sasagawa, F. J. Klopping, K. A. Berstis, R. L. Hilt, P. Nelson, G. L. Christy, T. M. Niebauer, W. Hollander, H. Seeger, B. Richter, H. Wilmes, and A. Lothammer (1994): New gravity meter improves measurements. *EOS, Trans. Am. Geophys. Union*, 75 (8).

Chao, B. F. (2003): Geodesy is not just for static measurements any more. *EOS, Trans. Am. Geophys. Union*, 84(16), 145.

Chu, S. and M. Kasevich (1992): Measurement of the gravitational acceleration of an atom with a light-pulse atom interferometer. *Appl. Phys. B*, 54, 521.

Dalibard, J. and C. Cohen-Tannoudji (1989): Laser cooling below the Doppler limit by polarization gradients: simple theoretical models. *J. Opt. Soc. Am. B*, 6(11), 2023.

Dickey, J. O. (ed.) (1997): Satellite gravity and the geosphere – contributions to the study of the solid Earth and its fluid envelope. Report of the National Research Council, Washington, D.C.

Gross, R. S. and B. F. Chao (2000): The gravitational signatures of earthquakes. In: Gravity, Geoid and Geodynamics 2000, M. G. Sideris (ed.), IAG Vol.123, pp. 205–209, Springer-Verlag, Berlin.

Gustavson, T. L. (2000): Precision rotation sensing using atom interferometry. Dissertation, Department of Physics, Stanford University.

Jekeli, C. (1988): The Gravity Gradiometer Survey System. *EOS, Trans. Am. Geophys. Union*, 69 (8), 105, 116–117.

Jekeli, C. (2000): *Inertial Navigation Systems with Geodetic Applications*. W. deGruyter, Berlin.

Jekeli, C. (2003): Statistical analysis of moving-base gravimetry and gravity gradiometry. Report No.466, Laboratory for Space Geodesy and Remote Sensing Research, Geodetic Science, Ohio State University, Columbus, Ohio.

Kasevich, M. (2002): Moving platform gravity gradiometry with ultra-cold atoms. Proceedings of the Weikko A. Heiskanen Symposium in Geodesy, 1–4 October 2000, Ohio State University, Columbus, Ohio.

Kasevich, M. and S. Chu (1991): Atomic interferometry using stimulated Raman transitions. *Phys. Rev. Lett.*, 67(2), 181–184.

Kasevich, M. and S. Chu (1992): Measurement of the gravitational acceleration of an atom with a light-pulse atom interferometer. *Appl. Phys. B*, 54, 321–332.

Lemoine F. G., Kenyon S. C., Factor J. K., Trimmer R. G., Pavlis N. K., Chinn D. S., Cox C. M., Klosko S. M., Luthcke S. B., Torrence M. H., Wang Y. M., Williamson, R. G., Pavlis E. C., Rapp R. H., Olson T. R. (1998): The development of the joint NASA GSFC and the National Imagery and Mapping Agency (NIMA) geopotential model EGM96, NASA Technical Paper NASA/TP-1998–206861, Goddard Space Flight Center, Greenbelt.

Marple, S. L. (1987): *Digital Spectral Analysis*. Prentice-Hall, Inc., Englewood Cliffs, New Jersey.

McGuirk (2001): High precision absolute gravity gradiometry with atom interferometry. Dissertation, Department of Physics, Stanford University.

McGuirk, J. M., G. T. Foster, J. B. Fixler, M. J. Snadden, and M. A. Kasevich (2002): Sensitive absolute gravity gradiometry using atom interferometry. *Phys. Rev. A*, 65, 033608.

Metcalf, H. J. and P. van der Straten (1999): *Laser cooling and trapping*. Springer-Verlag, New York.

Moody, M. V., H. A. Chan, and H. J. Paik (1986): Superconducting gravity gradiometer for space and terrestrial applications. *J. Appl. Phys.*, 60(12), 4308–4315.

Paik, H. J., E. R. Canavan, and M. V. Moody (1997): Airborne/shipborne SGG survey system. Proceedings of the International Symposium on Kinematic Systems in Geodesy, Geomatics, and Navigation, 3–6 June 1997, Banff, Canada, pp.565–570.

Peters, A., K. Y. Chung, and S. Chu (2001): High-precision gravity measurements using atom interferometry. *Metrologia*, 38, 25–61.

Reigber, C., G. Balmino, P. Schwintzer, R. Biancale, A. Bode, J. M. Lemoine, R. Koenig, S. Loyer, H. Neumayer, J. C. Marty, F. Barthelmes, F. Perosanz, and S. Y. Zhu (2002): A high quality global gravity field model from CHAMP GPS tracking data and

Accelerometry (EIGEN-1S). *Geophysical Research Letters*, 29(14), 10.1029/2002GL015064.

Rummel, R, G. Balmino, J. Johannessen, P. Visser, and P. Woodworth (2002): Dedicated gravity field missions – principles and aims. *Journal of Geodynamics*, 33, 3–20.

Schwarz, K. P., O. L. Colombo, G. Hein, and E. T. Kmickmeyer (1992): Requirements for airborne vector gravimetry. Proc. IAG Symp., From Mars to Greenland: Charting Gravity with Space and Airborne Instruments, General Assembly of the IUGG, Vienna, 1991, Springer Verlag, New York, pp.273–283.

Snadden, M. J., J. M. McGuirk, P. Bouyer, K. G. Haritos, and M. A. Kasevich (1998): Measurement of the Earth's gravity gradient with an atom interferometric-based gravity gradiometer. *Phys. Rev. Lett.*, 81(3), 971–974.

Storey, P. and C. Cohen-Tannoudji (1994): The Feynman path integral approach to atomic interferometry—a tutorial. *Journal de Physique II France*, 4, 1999–2027.

Sünkel, H. (ed.) (2000): From Eötvös to Milligal, ESA Final Report, ESA/ESTEC, contract no.13392/98/NL/GD, Graz, Austria.

Talwani, M.: http://www.geophysics.rice.edu/department/faculty/talwani/sanandreas.html.

Tapley, B. and C. Reigber (1998): GRACE: A satellite-to-satellite tracking geopotential mapping mission, Proceedings of Second Joint Meeting of the Int. Gravity Commission and the Int. Geoid Commission, Trieste, Sep. 7–12, 1998.

Wahr, J., M. Molenaar, and F. Bryan (1998): Time variability of the Earth's gravity field – hydrological and oceanic effects and their possible detection by GRACE. *Journal of Geophysical Research*, 103(B12), 30,205–30,229.

Zorn, A. H. (2002): A merging of system technologies—all-accelerometer inertial navigation and gravity gradiometry. Presented at IEEE Position Location and Navigation Symposium (PLANS) 2002, Palm Springs, California, 15–18, April 2002.

Christopher Jekeli, Laboratory for Space Geodesy and Remote Sensing, 470 Hitchcock Hall, Ohio State University, 2070 Neil Ave., Columbus, Ohio 43210

Satellite Magnetic Field Measurements: Applications in Studying the Deep Earth

Catherine G. Constable and Steven C. Constable

Institute of Geophysics and Planetary Physics, Scripps Institution of Oceanography, University of California at San Diego, La Jolla, California

Following a 20 years hiatus, there are several magnetometry satellites in near-Earth orbit providing a global view of the geomagnetic field and how it changes. The measured magnetic field is an admixture of all field sources, among which one must identify the contributions of interest, namely (1) the field generated in Earth's core, and (2) the fields induced in Earth's mantle by external magnetic variations used in studies of electrical conductivity. Models of the core field can be downward continued to the core surface under the assumption that Earth's mantle is a source free region with zero electrical conductivity. Additional assumptions are invoked to estimate the fluid flow at the core surface. New satellite measurements provide an unprecedented view of changes in the core over the past 20 years; further measurements will clarify the temporal spectrum of the secular variation. Secular changes are coupled to changes in length of day, and recent modeling of torsional oscillations in the core can provide an explanation for the abrupt changes in the field known as geomagnetic jerks. Mantle induction studies require a comprehensive approach to magnetic field modeling. Unwanted internal field contributions are removed to yield time series of external variations and their induced counterparts: improved modeling, combined with the increased data accuracy, and longer term magnetic measurements make conductivity studies feasible. One-dimensional global conductivity responses have been estimated under strong assumptions about the structure of the source field. Ongoing improvements to this work will take account of more complicated source-field structure, three-dimensional Earth structure, and spatio-temporal aliasing due to satellite motion. Modeling of three-dimensional near surface conductivity structure, and the use of time-domain rather than frequency-domain techniques to estimate the 3-D Earth response are needed. Progress could be furthered by future magnetometer missions that involve multiple satellite configurations.

EARTH'S MAGNETIC ENVIRONMENT

Earth has its own internally generated magnetic field, the bulk of which arises from a self-sustaining dynamo that operates in the liquid outer part of the core. The magnetic field is a dynamic entity and varies significantly on all temporal and spatial scales. The most dramatic changes are field reversals, which take several thousand years to complete and occur at irregular intervals, typically a few times per million years, although the reversal rate has varied over Earth's history. Time variations arising from the geodynamo induce magnetic fields in Earth's electrically conducting mantle, and the internal field is also responsible for both remanent and induced magnetic field anomalies found in Earth's lithosphere. The internal geomagnetic field plays a protective role in shielding the environment from cosmic rays: the magnetic pressure also holds off the solar wind and prevents stripping of Earth's atmosphere. The cold plasma that forms the solar wind contributes to Earth's complicated magnetic environment: the solar wind interacts with the internal field and is the cause of temporal variations on a broad range of time scales in both the external magnetospheric and ionospheric fields and their induced counterparts in Earth's lithosphere and mantle [see *e.g., Campbell*, 1997].

This chapter is concerned with magnetic field measurements made from satellites, what they contribute to our knowledge of the geodynamo, and how they can be used to probe the electrical conductivity of Earth. Compared with the relatively sparse array of a few hundred land-based magnetic observatories, magnetic satellites provide essentially complete coverage of the entire globe. For example, the International Geomagnetic Reference Field (IGRF), the definitive model of Earth's magnetic field, is limited by observatory distribution to spherical harmonic degree and order 13, or about a 3,000 km resolution. Satellite measurements, however, are limited only by the flight altitude, or about 500 km, and models up to degree 60 are possible. Satellites provide continuous measurements during the mission, and although untangling the spatial and temporal variations in the field presents novel challenges, the long period changes in the field that are generated by secular variation in the geodynamo and the shorter period variations that are useful for probing mantle conductivity are both resolved in a unique way. Vector magnetic satellite missions are the best way to study Earth's magnetic field.

The magnetic field has its origin in a number of distinct locations and processes which are defined in Plate 1. The measured field is an admixture of fields from external sources in the magnetosphere and ionosphere, internal sources in the crust, and core, and secondarily induced fields in Earth's electrically conducting ocean, lithosphere, and mantle which arise from primary time varying fields from within the core and external to the Earth. Separating the individual contributions to the magnetic field remains an active area of research. We are concerned with isolating (1) the field generated in Earth's liquid outer core and associated secular variations that provide a view of the workings of the geodynamo, and (2) fields induced in Earth's mantle by large scale external magnetospheric variations; these arise from the interaction of Earth's internal magnetic field with the solar wind and are used in studies of electrical conductivity. Crustal magnetization, itself a major topic of geophysical interest [*Langel and Hinze*, 1998], contributes significantly to the measured field, and must be considered in attempts to isolate the fields in both core and mantle. The rotation of the earth beneath the sun providing a daily heating cycle is evident in the diurnal variation of ionospheric Sq currents which can also be used in electrical conductivity studies using ground-based observatory data: however, since they lie below the satellite measuring region we do not consider that application here, but treat them as an unwanted source of noise.

Two simplifying strategies used in geomagnetic studies exploit the fact that individual sources contribute magnetic fields that vary on distinct temporal or spatial scales. It may be possible to isolate a particular part of the magnetic field purely on the basis of how it changes with time. Figure 1 presents an amplitude spectrum of geomagnetic variations as a function of frequency, with annotations indicating the predominant physical processes that contribute at the various timescales. It is clear that the largest variations occur at the very long time scales associated with geomagnetic reversals. It is less than 50 years since the OGO satellites (1965–1971) provided the first high precision global geomagnetic surveys, and the temporal variations that can be measured directly are rather small in comparison with the overall signal. Nevertheless, satellite measurements sample a number of significant processes in the geomagnetic spectrum shown in Figure 1, with frequencies ranging from about 25 Hz to periods of several years. This covers a range of external field phenomena and extends well into the regime of internal field secular variations. One complication is that both solar controlled magnetospheric and core processes contribute to field variations on timescales ranging from several months to at least decades (solar variability certainly occurs at much longer timescales), obviously a unique separation cannot be accomplished just using time variations in the field for a single location.

The second strategy we bring to bear is a spatial separation of contributions to the magnetic field: firstly, according to whether the source lies above or below the region in which measurements are made, and secondly according to spatial scale. To achieve this the usual assumption is to regard the measurement region as being free of magnetic field sources: this is an excellent approximation for ground-based measurements, as the atmosphere beneath the ionosphere is essentially an insulator. In a source free region the static magnetic field, $\vec{B}$, can be represented as the gradient of a scalar potential, Ψ, satisfying Laplace's equation ($\nabla^2 \Psi = 0$). The general solution to Laplace's equation yields a spherical harmonic representation for the geomagnetic field that provides a formal separation between internal and external sources, and within each source region decomposes the field according to spherical harmonic degree, l, and order m. The degree and order of a term within the spherical harmonic expansion determines the spatial scale of that contribution to the magnetic field: low degrees correspond to the largest spatial scales, with $l = 1$ being the largest scale dipole part of the field. This representation was used by Gauss in the 19th century to generate a least squares fit to magnetic observatory data and show that by far the largest part of the geo-

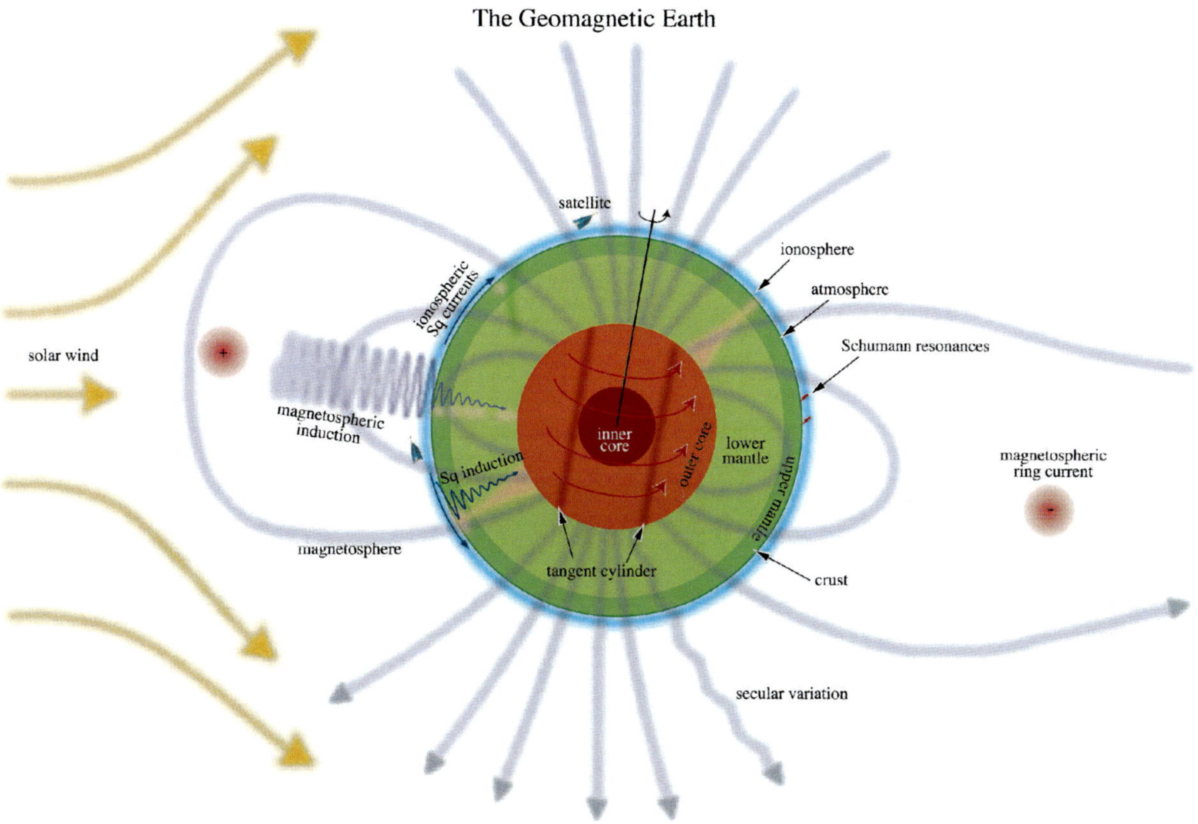

Plate 1. Earth's magnetic field is generated in the liquid **outer core**, where fluid flow is influenced by Earth rotation and the **inner core** geometry (which defines the **tangent cylinder**). Core fluid flow produces a **secular variation** in the magnetic field, which propagates upward through the relatively electrically insulating **mantle** and **crust**. The crust makes a small static contribution to the overall field. Above the insulating **atmosphere** is the electrically conductive **ionosphere**, which supports **Sq currents** as a result of dayside solar heating. Lightning generates high frequency **Schumann resonances** in the Earth/ionsophere cavity. Outside the solid Earth the **magnetosphere**, the manifestation of the core dynamo, is deformed and modulated by the solar wind, compressed on the sunside and elongated on the nightside. At a distance of about 3 earth radii, the **magnetospheric ring current** acts to oppose the main field and is also modulated by solar activity. Magnetic fields generated in the magnetosphere and ionosphere propagate by **induction** into the conductive Earth, providing information on electrical conductivity variations in the crust and mantle. **Magnetic satellites** fly above the ionosphere, but below the magnetospheric induction sources.

magnetic field is of internal origin. Solution of Laplace's equation by least squares estimation or regularized inversion [see for example, *Parker*, 1994] forms the backbone of many magnetic field analyses conducted today. *Langel* [1987] provides a review of the methods used.

The spherical harmonic representation can be used to provide a spatial analog of the amplitude/frequency spectrum of Figure 1. The spatial power spectrum for the magnetic field is usually defined as the average power in $\bar{B}$ at Earth's surface as a function of spherical harmonic degree: spherical harmonic degree plays the role of spatial wavelength, and an approximate length scale is given by $\pi a/l$ with a the radius of the earth. Plate 2 shows the spatial power spectrum for a range of recent models derived from both satellite [CO2, *Holme et al.*, 2003; MF, *Maus et al.*, 2002, 2003; OSVM *Olsen*, 2002; LPPC, *Lowe et al.*, 2001) and aeromagnetic data [KCP, *Korte et al.*, 2002]. Symbols represent spectra calculated from spherical harmonic models as an intermediate step. Curves (KCP and LPPC spectra) are derived from along and cross track power and cross spectra estimated directly from satellite passes or very long-track, high-altitude, aeromagnetic surveys using a technique described by *O'Brien et al.* [1999]. The range of spatial scales represented here is from some 10's of km for the aeromagnetic models to the scale of the Earth for the dipole part of the field. The spatial spectrum makes clear that the degree 1 dipole part of the field is

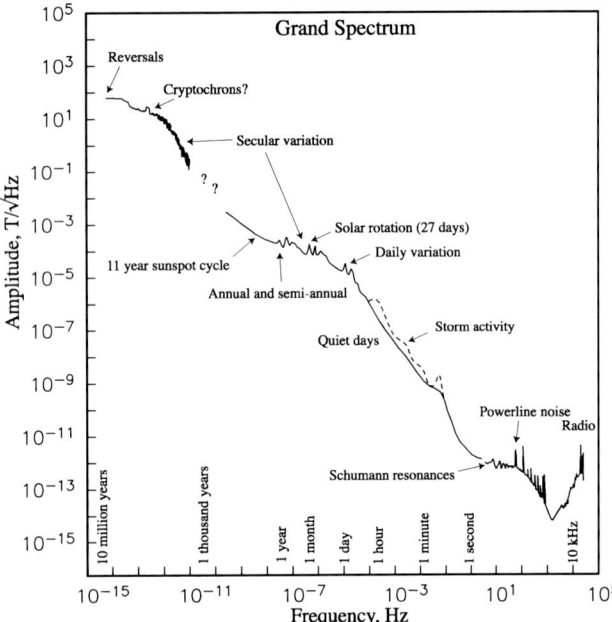

Figure 1. Amplitude spectrum of Earth's geomagnetic field. From 10^{-15} to 10^{-10} Hz, data are from *Constable, Tauxe & Parker* [1998]. Between 10^{-10} and 1 Hz, we have redrawn Figure 3 of *Filloux* [1987]. Above 1 Hz, we use the results of Nichols et al. [1988]. Internal variations associated with motions of the fluid core dominate at periods longer than a few months, culminating in whole reversals of the dipolar part of the field on 10^5 to 10^6 year time scales. The eleven year sunspot cycle, solar rotation, and Earth's orbit modulate the distortions of the field associated with geomagnetic storms, which themselves have energy in the several hour to several second band. Energy at the daily variation and harmonics comes from diurnal heating of the ionosphere. Lightning creates high frequency energy in the Earth/ionosphere cavity, which resonates at 7–8 seconds and harmonics. At the highest frequencies man-made sources dominate—it is unlikely that the natural spectrum abandons its red nature as shown in this figure.

The external fields due to the magnetospheric ring current (often referred to as the disturbance storm time, or *Dst*, because large magnetic disturbances are caused by magnetic storms) and *Sq* currents in the ionosphere are the major contributors to short term variations (see Figure 1) although the static part of these fields is not large. Magnetic storms caused by changes in the solar wind can generate rapid changes in these external fields. Short term variations originating in the core are attenuated by their passage through the (slightly) electrically conducting mantle. However, there is overlap in the temporal spectrum of internal and external variations. The shortest term internal variations that have been identified are the sudden changes in $d\vec{B}/dt$, known as geomagnetic jerks [*Courtillot et al.*, 1978], detected in monthly or annual mean observatory records. No jerk has yet been captured in satellite measurements, although there is evidence that jerks recur at approximately decadal intervals [see for example *Mandea et al.*, 2000]. At periods ranging from months to tens of years and longer there are significant changes in both external and internal parts of the field. The 11-year solar cycle variations are a well known example of long period modulations of external field variations. Of course, these external variations induce corresponding internal variations in addition to the changes arising in Earth's core. There are ongoing efforts to model all significant internal and external magnetic field contributions and their time variations using both satellite and observatory data. In the geomagnetic community this approach is known as Comprehensive Magnetic Field Modeling [*Sabaka et al.*, 2002].

MAGNETIC SATELLITE MISSIONS

Magnetic measurements have routinely been carried out on the ground and over the oceans since the 16th century [see *Jackson et al.*, 2000], but it is only since the OGO missions that we have acquired the global view of the geomagnetic field from space. Early magnetometer missions carried scalar magnetometers, but *Backus* [1970] showed that field models derived from fixed altitude intensity data alone are intrinsically non-unique. This led to the development of the first vector magnetic field satellite, known as Magsat, which was active from November 1979 to May 1980. Following Magsat there was a 20 year hiatus in satellite magnetometer missions until the launch of the Danish satellite Ørsted in February 1999. Renewed interest in geomagnetic measurements has led to the promotion of an International Decade of Geopotential Research, and 2 additional satellites are currently mapping the field (CHAMP and Ørsted-2/SAC-C). In contrast to Magsat, which was confined to an 06:00/18:00 local time orbit, Ørsted and CHAMP will sample all local times. Ørsted-2/SAC-C is in a 10:30/22:30 local time orbit. Altitudes range from 400 km (circular) for CHAMP to 650–850 km for Ørsted.

dominant, but there are substantial higher degree contributions. Somewhere between degrees 11 and 15 the contribution of the internal core field is overwhelmed by that of the lithosphere, whose contribution remains relatively flat out to about degree 1000. The slight increase in power at highest degrees for each of the satellite spherical harmonic crustal models is probably an artifact due to measurement and external field noise and the truncation level chosen in least squares spherical harmonic modeling. The KCP and LCCP spectra have had the core field (below degree 13) removed, prior to estimating the crustal power spectra. Also shown on Plate 2 are the power contributions from the large scale degree 1 and 2 static part of the external field, and the power in the core field secular variation in $(nT/yr)^2$ for the year 2000.0 as a function of spherical harmonic degree for the OSVM [*Olsen*, 2002].

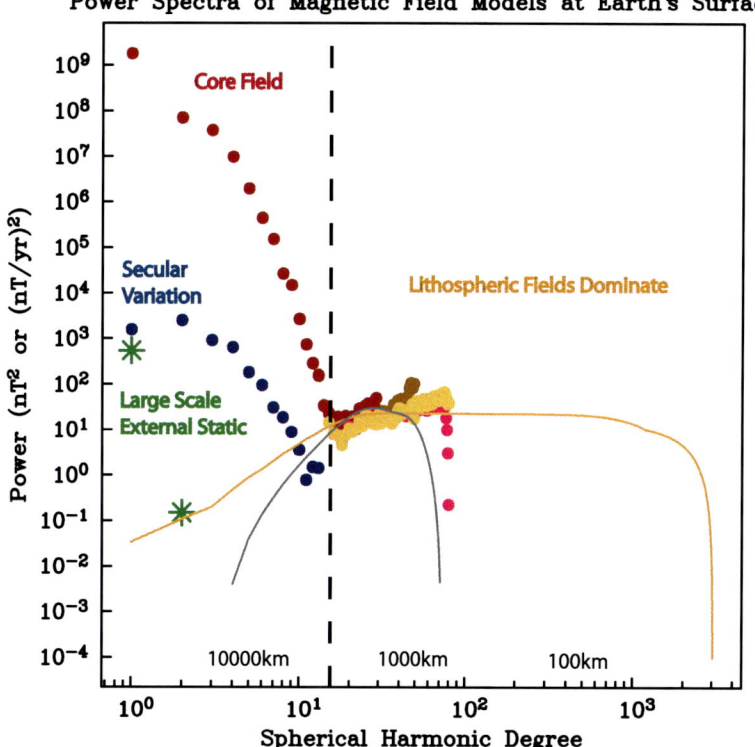

Plate 2. The spherical harmonic power spectrum for the geomagnetic field. For each degree l the power is the average of the square of the magnetic field intensity (or for secular variation, $(d\bar{B}/dt)^2$) over Earth's surface calculated for the designated model. The spherical harmonic degree is a measure of the lengthscale under consideration. CO2+ (brown main field, green external static) is described by *Holme et al.* [2003], OSVM (red main field, blue secular variation) by *Olsen* [2002], MF1 (yellow, crustal) and MF2 (purple, crustal) by *Maus et al.* [2002, 2003], KCP (orange) by *Korte et al.* [2003] and LPPC (gray) by *Lowe et al.*, [2001].

Vector field missions typically consist of a fluxgate magnetometer combined with one or more star cameras to determine attitude of the spacecraft, GPS for positioning, and a scalar magnetometer for calibration of the vector field instrument. Details of the Ørsted satellite are given by *Stauning* [2003] and the instrument calibration is described by *Olsen et al.* [2003a], and for CHAMP by *Reigber et al.* [2002]. The fact that individual magnetic missions now provide several years of continuous field mapping is facilitating exploration of the part of the geomagnetic temporal spectrum where internal and external field variations overlap (Figure 1).

Satellite observations have a number of obvious advantages over measurements made on the ground by observatories or surveys. The number of observatories is limited and they are irregularly spaced over Earth's surface with a bias to coastlines. Observatory measurements are affected by small scale heterogeneity of the lithospheric field resulting in what are known as *observatory crustal biases* [*Langel*, 1987, p.314]; these can contribute several hundreds of nT to the measured signal. In addition to these very localized effects the many observatories located on or near the coast are particularly sensitive to the *coast effect*, a large scale induction influence due to the conductivity contrast between the ocean and the continental crust (see Plate 4(b)). Satellites provide excellent global coverage and the crustal field attenuates rapidly with increasing distance from the source.

On the downside, satellites fly several hundred kilometers above Earth's surface so that the details of the crustal field are difficult to identify. When mapping the core field the crustal contribution is generally treated as noise: however, the spatial averaging inherent in upward continuation to satellite altitude results in crustal errors that are spatially coherent [*Jackson*, 1990;1994], and this must be considered in the data selection and inversion [*Rygaard-Hjalsted et al.*, 1997]. A further complication arises from the fact that the ionospheric field, and the associated Sq variations, lie beneath the satellite measurement region, and the measurement region itself is not entirely free of magnetic sources. The Sq variations may be dealt with in core modeling by choosing data from magnetically quiet times of day and/or modeling them using observatory data for which they are an external field. The comprehensive field modeling effort [*Sabaka et al.*, 2002] includes the influence of sources in the measurement region using a technique developed by *Olsen* [1996,1997; *Engels & Olsen*, 1998].

CORE FIELD MODELING USING SATELLITE OBSERVATIONS

As already indicated, core field modeling is usually accomplished by selecting data with minimal external field contributions and solving Laplace's equation for the scalar potential under the assumption that the measurement region is field free. Various data selection criteria are used, ranging from using only night-time data (or, more stringently, data that fall entirely on the shadow side of Earth) to choosing data on the basis of the magnetic indices [*Campbell*, 1997], such as the *Kp* index (usually $Kp \leq 1+$) and the *Dst* index ($\leq \pm 10\ nT$), both of which indicate the strength of external geomagnetic field activity. At high latitudes using only scalar field measurements minimizes the effect of field aligned currents found in the satellite measurement region. An additional selection criterion based on the strength of the dawn-dusk component, $|B_y|$, of the interplanetary magnetic field may also be used to minimize the effect of polar cap ionospheric currents. Recent improvements in internal field modeling have focused on improved corrections for the external field, and the inclusion of higher degree terms for the crustal part: it is impossible to separate the core and crustal contributions unambiguously, but general practice is to attribute at least to degree 11 predominantly to the core. Non-Gaussian data errors are accommodated by using iteratively reweighted least squares with Huber weights to achieve a robust fit [*Olsen*, 2002]. External fields are often modeled using the *Dst* index as a proxy for large scale external fields, and a fixed ratio of 0.27 for the internally induced variations, combined with appropriate terms to accommodate seasonal variations. However, as is well known from conductivity studies, this ratio is actually frequency dependent.

Details for two recent models (OSVM and CO2+) based on Ørsted and CHAMP satellites combined with observatory measurements are supplied by their respective authors [*Olsen*, 2002; *Holme et al.*, 2003). The data used in these models span a significant time interval making it possible to derive a secular variation model in addition to the static main field and low degree external field variations. OSVM is a model for epoch 2000.0 with linear secular variation (SV) in spherical harmonic coefficients up to degree 13, and main field coefficients, representing the core and crust, up to degree 29. CO2+ for epoch 2001.0 uses more data and extends the main field further into the crustal dominated regime, reaching to degree 49. The power spectra for these two models are shown in Plate 2 along with crustal models MF1 and MF2 derived from CHAMP data [Maus et al., 2002, 2003]. The spectrum of secular variations for OSVM indicates that the dominant secular changes are large scale when viewed at Earth's surface, but that the SV spectrum falls off less rapidly with increasing *l* than that for the main field. The extension of secular variation models to smaller spatial scales with spherical harmonic degree as high as $l = 13$ (probably reliable to $l = 11$ or 12) is one result of these new highly accurate satellite missions: previously SV models have not been computed past degree 8 or 10. The importance of such estimates for our understanding of field behavior is apparent from Figure 1: although the high degree static fields generated in Earth's core are masked by the crustal contribution this is not true for the secular variation (only the induced parts of the crustal field will change in a minor way as a result of SV, the remanent parts remain static). Thus there are greatly improved prospects for studying small scale field variations as more data and better parameterized SV models are developed.

The spherical harmonic representation of the field allows it to be downward or upward continued throughout the region where there are no magnetic sources. The relatively large scale and slow time variations in the core field, combined with the approximation of the mantle as an electrical insulator, make this a good candidate for downward continuation of satellite based models to either Earth's surface or the core-mantle boundary. The neglect of mantle conductivity results in a temporal filtering of the core signal [*Benton & Whaler* 1983; *Backus*, 1983] so that short period field variations will be preferentially attenuated, while the presence of sources in Earth's lithosphere masks the short wavelength time-invariant core field. Downward continuation to the surface of Earth's core provides a completely different view from that at Earth's surface, where higher degree spatial variations are greatly attenuated. This is illustrated in Plate 3 (a) and (b) where the radial component of the core field (up to degree 11) from CO2+ is plotted at Earth's surface, $r = a$, and at the core-mantle boundary, $r = c$. Smaller scale structures dominate the field at $r = c$: the generally low field at Earth's surface over South America and the South Atlantic is resolved into a complex region of magnetic flux with opposite polarity to its surroundings; similarly the north polar region, which has strong negative radial fields at $r = a$, has weak positive flux at the core-mantle boundary. The small-scale details vary somewhat among different models at $r = c$ (downward continuation amplifies the small scale differences in models that are essentially the same at Earth's surface), but the general appearance of such models is similar, with pairs of flux lobes at high latitudes in both northern and southern hemispheres, and a number of significant (and some less significant) regions of magnetic flux that are of opposite polarity to their immediate surroundings.

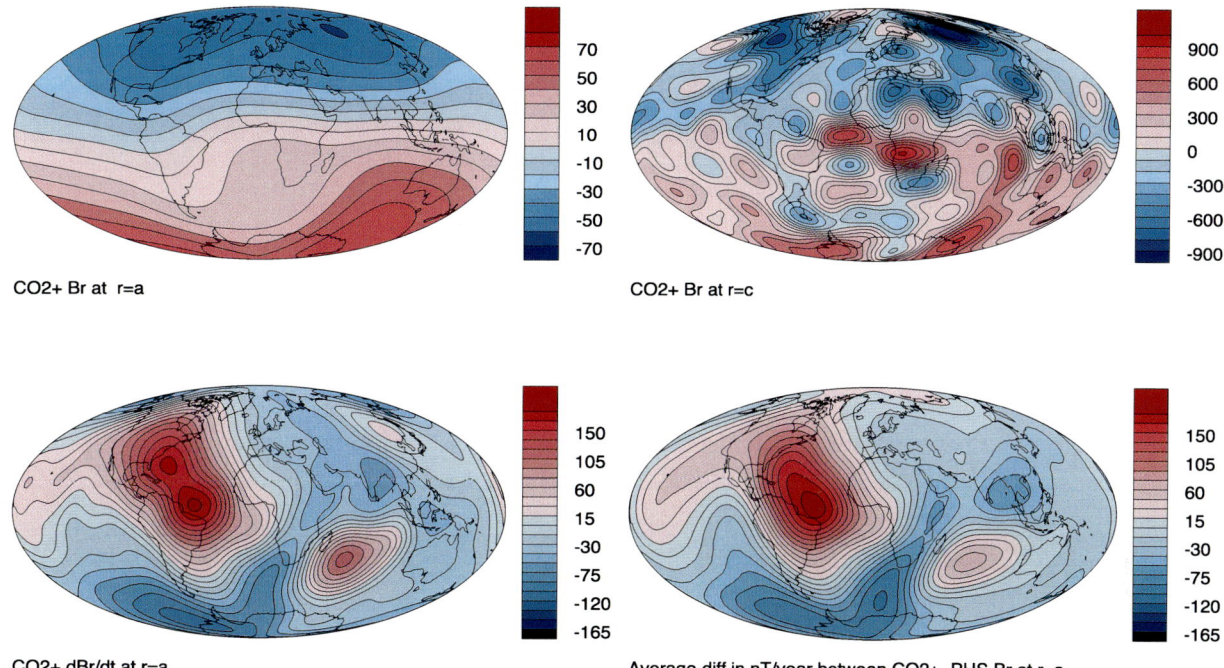

Plate 3. The radial magnetic field for model CO2+ at (a) Earth's surface in 2001.0 and, (b) the core mantle boundary truncated at l = 13. (c) dB_r/dt at Earth's surface for CO2+, the model extends to degree 13 (d) average annual change between CO2+ and PHS, the 1980 field model of *Shure, Parker and Langel* [1985].

Maps of the field at the CMB have been generated from historical data spanning the past 400 years [*Jackson et al.*, 2000], and show the same large scale features, so the important issues concern changes in the field and how these can be interpreted. The lower panels in Plate 3 show dB_r/dt in nT/year at $r = a$, for the CO2+ model in (c), and for comparison the average change in B_r per year calculated from the raw difference between B_r for epochs 1980 and 2001. As might be expected, the CO2+ instantaneous SV differs slightly from the 21 year difference, especially at small scales, reflecting the need for high resolution magnetic satellite data to resolve short temporal variations. On large scales however, there is good agreement between Figs 4(c) and (d). We see that secular variation is largest in the Atlantic hemisphere, especially over the central and southernmost Atlantic, and relatively weak in the Pacific hemisphere. These changes reflect the well-known recent decrease in the dipole moment (at a rate of a few percent/century), but the largest changes are currently in the quadrupole part of the field as seen in the SV spectrum for OSVM in Plate 2 (higher degree variations also contribute). The analog for Plate 3(d) at the CMB would have a great deal of small scale structure that has been amplified by the downward continuation. *Hulot et al.* [2002] computed these changes for an epoch 2000 model primarily based on Ørsted data and note that the changes remain smaller beneath the Pacific, but can be as large as several tens of percent of the field at polar latitudes and below southern Africa. Some of these changes are certainly robust, but there remains the possibility of residual contamination by external fields in the polar regions. The relatively low magnetic field strength beneath the South Atlantic is of considerable interest for two reasons: satellites flying in this region are exposed to increased risk of radiation damage because of the diminished protection associated with weak fields; the rapid field changes in this region suggest the possibility of relatively short term secular variation processes that may involve flux expulsion and ohmic dissipation in the core.

Magnetic field models can also provide dynamical information, and are used to infer the fluid flow at the surface of the core, generally under the additional assumption that the frozen-flux approximation holds: this neglects the effects of electrical diffusion in core field motions and implies that magnetic field lines are locked to the fluid motion. Two components of velocity must be deduced, making such inversions fundamentally non-unique; this dilemma is resolved by imposing further physical constraints on the flow. *Bloxham & Jackson* [1991] provide a review of various options, which include supposing that the flow is steady over time, that it is toroidal (no upwelling or downwelling) or tangentially geostrophic (the horizontal component of the Coriolis force is mainly balanced by the dynamical pressure gradient). The various means of resolving the non-uniqueness result in very similar flows, which may be a consequence of the fact that they all implicitly assume that the flow is large scale. Independent evidence of the validity of such inversions is supplied by the fact that core surface flows can be used to predict length of day variations on decadal time scales when angular momentum exchanges between core and mantle are taken into account [*Jault et al.*, 1988; *Jackson et al.*, 1993]. The core angular momentum can be calculated from the top of core flow, because the flow in the outer fluid core on decadal time scales is dominated by torsional oscillations. These are simple oscillatory flows that describe differential rigid rotations about the rotation axis of coaxial cylinders (like the inner core tangent cylinder in Plate 1) and for which the magnetic field supplies the restoring force. Torsional oscillations are expected from theory and found in simulated dynamos and real observations. *Bloxham et al.* [2002] have shown that the combination of steady core flow with torsional oscillations can explain the features occurring at roughly decadal intervals (in 1969, 1978, 1991, and 1999) that have been interpreted as geomagnetic jerks.

A recent example of such a core flow is given by *Hulot et al.* [2002], who have calculated the tangentially geostrophic flow that can account for the main field changes they observe between 1980 and 2000. They find a mainly westward axisymmetric flow with some vortices embedded in it. The axisymmetric flow is also symmetric about the equator. Strong polar vortices show westward flow inside the tangent cylinder, an imaginary region outlined by a cylinder parallel to the rotation axis and tangent to the solid inner core. Prograde vortices do exist in some regions. *Hulot et al.* [2002] note that in numerical dynamo models retrograde vortices are associated with upwelling flows and prograde vortices with downwelling. Upwelling flows tend to expel field of reverse polarity (compared with the dipole field) from within the core. However, considerable caution is needed in transferring this interpretation to Earth's field because of the way in which the inherent non-uniqueness is resolved in flow modeling. Under the tangential geostrophic assumption, upwelling or down-welling must occur at the edge of vortices, not the center as seen in numerical simulations. Toroidal flows suffer from a similar lack of realism in that there can be no upward or downward motions in the fluid: nevertheless either kind of flow model can fit the observations. The fact that these kinds of flows satisfy the geomagnetic observations is a start, but it does not guarantee that they accurately reflect the details of what is going on in the core.

INDUCTION STUDIES USING SATELLITE OBSERVATIONS

Induction studies using satellite magnetic measurements rely on the technique known as geomagnetic depth sounding or GDS [*Parkinson*, 1983]. The magnetic field response induced by externally generated field variations follows Faraday's law, and depends on the electrical conductivity in the crust and mantle which in turn reflects the composition, the presence of phase transitions, mantle temperature, and the influence of volatiles and trace materials [*Xu et al.*, 2000]. The depth of penetration into the mantle depends on the frequency content of the time-varying fields and the conductivity of the medium and is characterized by the electromagnetic skin depth, the length scale of exponential decay into Earth. Estimates of the frequency domain transfer function between external and internal fields are the usual data which are interpreted to provide estimates of electrical conductivity as a function of depth. This has proved quite successful in interpreting observations on Earth's surface, which allows the collection of time series of observations at fixed locations [*e.g., Banks*, 1969; *Schultz & Larsen*, 1987; *Roberts*, 1984; *Olsen*, 1998, 1999a; *Constable*, 1993], from which the GDS method is usually used to provide locally one-dimensional (1-D) transfer functions. In principle, such transfer functions can be extended to interpret 3-D variations in electrical conductivity with depth, as well as to accommodate arbitrarily complicated source field structure. In practice these extensions are rarely attempted at a global scale, although differences in the 1-D structure inferred at different locations have been remarked on [*Schultz & Larsen*, 1990; *Weiss & Everett*, 1998]. Recently, substantial progress has been made in forward modeling of 3-D conductivity variations [*Kuvshinov et al.* 1999, 2002b; *Everett & Schultz*, 1996; *Martinec*, 1999; *Uyeshima & Schultz*, 2000]. However, the situation is substantially more complicated for satellite observations than for those on the ground, because the satellite motion results in aliasing of spatial and temporal field variations. In this section we discuss recent progress in using satellite observations to infer 1-D electrical conductivity structure in the mantle and prospects for extending this effort to recover 3-D structures.

1-Dimensional Conductivity Studies

Unlike the situation described in modeling the corefield, in induction studies one is interested in only a small fraction of the signal, typically some tens of nT compared with an average surface field of 45 μT. The first task is to separate some part of the external field variations along with its induced counterpart from the much larger signal generated in Earth's core and the smaller crustal contributions. In ground-based GDS studies the *Sq* ionospheric signal is an external source field and can be used for induction studies, but the ionospheric currents lie below the satellite orbit: thus they are an additional source of noise that must considered in isolating the Dst field variations associated with the magnetospheric ring current. *Tarits & Grammatica* [2000] have shown that the ionospheric contributions at 400 km can range from 2–4 nT depending on the local time. *Olsen* [1999b] reviews early attempts to use satellite observations for induction studies, which mainly relied on the removal of the core field contribution and its secular variation before analysing the residual signal. More recent induction studies [*Constable & Constable*, 2003; *Korte et al.*, 2003; *Olsen et al.*, 2003] make use of the Comprehensive Magnetic Field Model developed by *Sabaka et al.* [2002], which allows the removal of core and crustal contributions along with an estimate of the quiet time ionospheric *Sq* variations. This provides a significantly better estimate of the signal due to Dst variations in the ring current, although residual *Sq* contamination remains a problem at periods close to 1 day and its harmonics. The spatial structure of the ring current variations is usually approximated by an axial external dipole field configuration in geomagnetic coordinates: these are geocentric coordinates with the z-axis aligned with that of the dipolar part of the internal field: for the CO2+ model the northern end of this axis cuts the Earth's surface at colatitude 10.4°, longitude 288.4°. The simple large scale structure provides a reasonable approximation at geomagnetic latitudes lower than 50°, above this there is substantial contamination by currents in the auroral regions. The size of the auroral signal and attempts to model the current systems that generate it are also discussed by *Fujii & Schultz* [2002] in the context of analysing observatory data.

Once the *Dst* signal has been separated from the bulk of the magnetic signal, one is left with an almost continuous series of vector field measurements in time and space. For each satellite pass (between ± 50° geomagnetic latitude) one can estimate the internal, i_1^0, and external, e_1^0, axial dipole field contributions, thereby producing time series of $i_1^0(t)$ and $e_1^0(t)$ sampled at approximately hourly intervals, the exact interval depending on the satellite orbit. The sum of these times series can be compared directly with the *Dst* index computed from ground observatories. Good agreement between the two [*Constable & Constable*, 2003] indicates that the methodology is robust.

Once the time series $i_1^0(t)$ and $e_1^0(t)$ are obtained the geomagnetic deep sounding method is used to find an electrical impedance response for Earth as a function of frequency. The transfer function,

$$Q_1^0(f) = i_1^0(f)/e_1^0(f)$$

or a transformed variant of it to a complex admittance function [*Weidelt*, 1972], can then be inverted to determine a one-dimensional conductivity profile in the crust and mantle. Figure 2 (inset) shows the admittance function estimates for the Magsat data derived by *Constable & Constable* [2004]. The solid and dashed curves in the inset show the predictions from the 1-D conductivity models plotted in the main part of the figure. As is generally the case in such inversion problems the best-fitting 1-D conductivity profile (derived using *Parker & Whaler's* [1981] D+ inversion algorithm) is unphysical, consisting of 4 delta-functions of conductivity in an insulating half-space. A more realistic conductivity profile with rms misfit of 1.15 (cf 0.95 for the D+ model) is found using regularized inversion, and indicated by the dashed curves: the smoothest model in the sense of minimum first derivative in log conductivity is obtained using a spherical earth variant of the Occam algorithm of *Constable et al.* [1987]. As in the real Earth the model is assumed to have a highly conducting core at a depth of 2886 km. The response functions in Figure 2 are in agreement with estimates obtained by others for Magsat (see also *Olsen et al.* [2003] who attempt the first such analysis using Ørsted data), but have been extended in frequency range at both the high and low ends relative to earlier Magsat analyses by *Olsen* [1999]. This is possible because of improvements in ionospheric field modeling and data processing at high frequencies, and the use of multitaper cross-spectral analyses [*Riedel & Siderenko*, 1995] as opposed to spectral techniques that use section averaging at low frequency.

Inversions of the data in Figure 2 demonstrate that the electromagnetic responses are sensitive to conductivities as shallow as the oceans. The D+ algorithm recovers a surface conductance of 8300 S corresponding nicely with an average ocean conductivity of 3 S/m and 2770 m depth. The regularized model also requires this enhanced near-surface conductivity, a feature not usually recovered from observatory analyses because they are all on land. The upper mantle is relatively uniform in conductivity at around 0.01 S/m which corresponds to that of dry olivine at about 1500° C [*Constable et al.*, 1992]. There is little evidence for a large conductivity increase in the transition zone, but a jump to about 2 S/m occurs at about 700 km depth. This feature has been reported in many conductivity studies and is believed to be associated with the phase transition from garnet and olivine spinel above 670 km to magnesiowüstite and silicate pervoskite at greater depth [e.g., *Xu et al.*, 2000]. A feature in Figure 2 not widely reported in other studies is the further increase at about 1300 km. It is generated by the small imaginary component in the admittance function at long period. This is also observed in a small number of observatory admittance functions (Honolulu is one example, *Schultz & Larsen*, [1987]), which are less representative in their sampling than the satellite. It will be of some interest to see whether this result is confirmed by the newer satellite studies which can deliver longer times series than Magsat. One difficulty that arises is in reliably separating the long period externally induced signal from that induced in the mantle by the shortest period variations arising in the core.

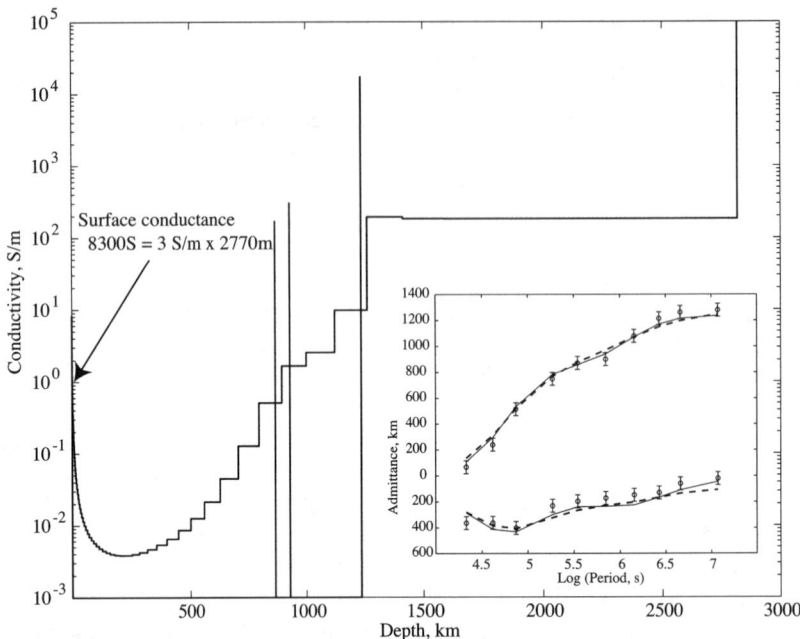

Figure 2. Radially averaged electrical conductivity structure of Earth, from *Constable & Constable's* [2004] treatment of magnetic satellite data. Two models are shown: The vertical bars are a representation of the least squares best fitting model, which is composed of delta functions of conductivity in an otherwise insulating sphere, and includes a surface conductance which nicely agrees with a globally averaged ocean. The inset shows the data (symbols) with the response of this model (solid lines). The stepped conductivity model shows individual layers in a regularized inversion which does not fit the data quite as well (broken lines in inset) but is more realistic. The model is dominated by the conductivity of the surface oceans and deep mantle.

Towards 3-Dimensional Conductivity Studies

The surface conductance recovered from the 1-D conductivity modeling indicates the powerful influence of the oceans and near surface layer in the induction problem. This may mask any interesting 3-D conductivity structures in the mantle, unless its influence can be adequately accounted for and removed. There are at present two approaches to studying the importance of 3-D structures in satellite observations; both rely on time-domain rather than frequency domain analyses, because of the spatio-temporal aliasing that is introduced in the 3-D problem by the satellite's motion. The first approach, adopted by Constable & Constable [2003], is purely qualitative. They suppose that the ratio of $i_1^0(t)/e_1^0(t)$ represents a kind of global time domain response. The normalization of the induced field by the primary field accounts for variations in magnetospheric activity, but since this response can be calculated for a very large number of times and locations the average values in spatial bins of dimension a few degrees can be used to study geographic variations in Earth's electrical response. A further normalization by $\sqrt{1+3\cos^2\theta}$ removes the dependence of the induced dipole field on magnetic colatitude. Plate 4(a) shows the induced magnetic field in Earth as inferred from a stack of over 5000 passes of Magsat data. As might be anticipated the induced fields are systematically lower over continental areas and higher over the more conductive oceanic regions, although this is not the only signal present.

The second approach is illustrated in Plate 4(b) which shows a global model of near-surface conductance constructed from *a priori* knowledge of seawater, and sediment conductivity and thickness [after *Everett et al.*, 2003]. Such models provide a means of undertaking a more systematic analysis of the effects of near-surface heterogeneity in conductivity on the satellite signal. Several forward modeling algorithms now exist [*Kuvshinov et al.*, 2002a,b; *Hamano*, 2002; *Velimsky et al.*, 2003] that allow the prediction of the time domain response of a 3-D earth to a specified primary forcing field. If the appropriate forcing functions can be supplied then in principle these provide a means of stripping out near-surface effects so that underlying variations in mantle conductivity can be studied. One clear result of these studies is that there are major problems associated with interpreting data from discrete observatory locations that are often situated near coastlines [*Kuvshinov et al.*, 2002a].

CONCLUSIONS

We conclude by noting that the burgeoning number of satellite observations are fostering new studies of Earth's core and

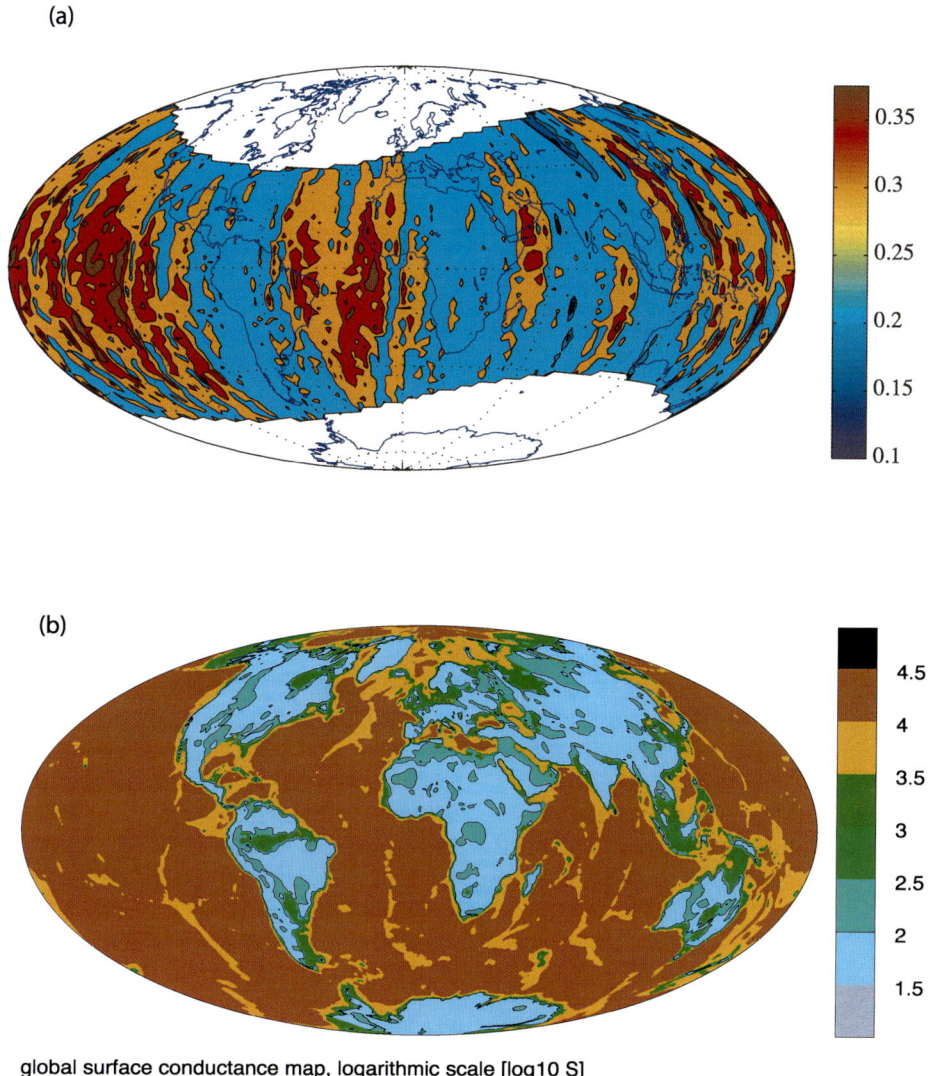

Plate 4. a) The induced internal field normalized by $e_1^0 \sqrt{1+3cos^2\theta}$ and averaged in $2 \times 3°$ bins. See text for details (b) Global surface conductance model of *Everett et al.* [2003].

mantle. Observatory estimates of Earth's electrical impedance have been extended spatially and in frequency, and there are good prospects for continued improvement. Inversions demonstrate that the responses are sensitive to conductivities as shallow as the oceans, and we can recover the ocean conductance. Upper mantle conductivity is consistent with hot, dry olivine, and there is little evidence for a large increase in conductivity in the transition zone. The longest period response function estimates probe the lowermost mantle, where satellite responses suggest a substantially more conductive region than is obtained from the average of observatory responses. Electrical conductivity measurements have the potential to play an important role in constraining phase, composition, and temperature variations in the deep mantle, complement-

ing the information garnered from studies that rely on seismic tomography, mineral physics and geochemical arguments. Further investigations of deep mantle structure require longer satellite records and better processing of internal/external field separations.

The 3-D induction problem is most easily tackled in the time domain for satellite data. By stacking the induced field we can get an image of 3-D structure, revealing increased induction over the oceans and smaller induced fields beneath the continents. Forward modeling algorithms exist that allow predictions of the signals expected at the satellite; current limitations are in understanding the spatio-temporal character of the external field variations. More progress can be expected in this area as better analysis tools are developed and long time series of data become available from multiple satellite configurations like the proposed European SWARM mission.

In the area of core-field modeling the greatest improvements have been in the area of secular variation modeling and the detection of decadal scale field changes. Continued monitoring of the field for a number of years is likely to capture global records of a magnetic jerk as it is occurring, with the possibility of substantial insight into the physical origin of this phenomenon. Other applications of magnetic satellite data include detection of tidal signals from motions in the ocean and using magnetic field measurements to map ocean currents [*Tyler et al.*, 2003].

Acknowledgments. This work was supported by NSF grants EAR-0112290 and EAR-0087391

REFERENCES

Backus, G. E., 1970. Non-uniqueness of the external geomagnetic field determined by surface intensity data. *J. Geophys. Res.*, **A75**, 6339–6341.

Backus, G. E., 1983. Application of mantle filter theory to the magnetic jerk of 1969. *Geophys. J.R. astr. Soc.*, **74**, 713–746.

Banks, R. J., 1969. Geomagnetic variations and the conductivity of the upper mantle. *Geophys. J.R. astr. Soc.*, **17**, 457–487.

Benton, E R & Whaler, K A, 1983. Rapid diffusion of the poloidal geomagnetic field through the weakly conducting mantle: A perturbation solution. *Geophys. J. R. astr. Soc.*, **75**, 77–100.

Bloxham, J, & A. Jackson, 1991. Fluid flow near the surface of Earth's outer core. *Rev. Geophys.*, **29**, 97–120.

Bloxham, J., S. Zatman, & M. Dumberry, 2002. The origin of geomagnetic jerks. *Nature*, **420**, 65–68.

Campbell, W. H., 1997. *Introduction to Geomagnetic Fields.* Cambridge University Press.

Constable, C. G., L. Tauxe, & R. L. Parker, 1998. Analysis of 11 Myr of geomagnetic intensity variation. *J. Geophys. Res.*, **103**, 17,735–17,748.

Constable, S. C., 1993. Constraints on mantle electrical conductivity from field and laboratory measurements. *J. Geomagn. Geoelectr.*, **45**, 707–728.

Constable, S. C. and C. G. Constable, 2004. Observing geomagnetic induction in magnetic satellite measurements & associated implications for mantle conductivity. *Geochem. Geophys. Geosyst.*, **5(1)**, doi:10.1029/2003GC000634.

Constable, S. C., Parker, R. L., and Constable, C. G., 1987. Occam's Inversion: a practical algorithm for generating smooth models from EM sounding data. *Geophysics*, **52**, 289–300.

Constable, S. C., Shankland, T. J. and Duba, A., 1992. The electrical conductivity of an isotropic olivine mantle. *J. Geophys. Res.*, **97**, 3397–3404.

Courtillot, V., J. Ducruix, & J.-L. Le Mouël, 1978. Sur une accélération récente de la variation séculaire du champ magnétique terrestre. *C. R. Acad. Sci D*, **287**, 1095–1098.

Engels, Uta & Nils Olsen, 1998. Computation of magnetic fields within source regions of ionospheric and magnetospheric currents. *J. Atm. Solar-Terr. Phys.*, **60**, 1585–1592.

Everett, M. E., S. C. Constable, & C. G. Constable, 2003. Effects of near-surface conductance on global satellite induction responses. *Geophys. J. Int.*, **153**, 277–286.

Everett, M. E. & A. Schultz, 1996. Geomagnetic induction in a heterogeneous sphere: azimuthally symmetric test computations and the response of an undulating 660 km discontinuity. *J. Geophys. Res.*, **101**, 2765–2783.

Filloux, J. H., 1987. Instrumentation and experimental methods for oceanic studies. In *"Geomagnetism"*, ed. J. A. Jacobs, Academic Press, London, pp. 143–248.

Fujii, I., & A. Schultz, 2002. The 3-D electromagnetic response of the Earth to ring current and auroral excitation. *Geophys. J. Int.*, **151**, 689–709.

Hamano, Y., 2002. A new time-domain approach for the electromagnetic induction problem in a three-dimensional heterogeneous earth. *Geophys. J. Int.*, **150**, 753–769.

Holme, R., N. Olsen, M. Rother, & H. Lühr, 2003. CO2—A CHAMP magnetic field model. In *"First CHAMP Mission Results for Gravity, Magnetic and Atmospheric Studies"*, ed. Ch. Reigber, H. Luehr and P. Schwintzer, Springer-Verlag Berlin, pp. 220–225.

Hulot, G., C. Eymin, B. Langlais, M. Mandea & N. Olsen, 2002. Small-scale structure of the geodynamo inferred from Ørsted and Magsat satellite data. *Nature*, **416**, 620–623.

Jackson, A., 1990. Accounting for crustal magnetization in models of the core magnetic field. *Geophys. J. Int.*, **103**, 657–674.

Jackson, A., 1994. Statistical treatment of crustal magnetization. *Geophys. J. Int.*, **119**, 991–998.

Jackson, A. *et al.*, 1993. Time-dependent flow at the core surface and conservation of angular momentum in the coupled core-mantle system. *Dynamics of Earth's Deep Interior and Earth Rotation*, **Vol. 72**, 97–108.

Jackson, A., A. R. T. Jonkers, M. Walker, 2000. Four centuries of geomagnetic secular variation from historical. *Phil. Trans. R. Soc. Lond.*, **A359**, 957–990.

Jault D., Gire C., and Le Moul J.-L., 1988. Westward drift, core motions and exchanges of angular momentum between core and mantle. *Nature*, **333**, 353–356.

Korte, M., C. Constable, R. L. Parker, 2002. Revised magnetic power spectrum of the oceanic crust. *J. Geophys. Res.*, **107**, 2205, doi:10.1029/2001JB001389.

Korte, M., Constable, S., & Constable, C., 2003. Separation of external magnetic signal for induction studies. In *"First CHAMP Mission Results for Gravity, Magnetic and Atmospheric Studies"*, ed. Ch. Reigber, H. Luehr and P. Schwintzer, Springer-Verlag Berlin, pp. 315–320.

Kuvshinov A.V., N. Olsen, D. B. Avdeev & O. V. Pankratov, 2002a. Electromagnetic induction in the oceans and the anomalous behaviour of coastal C-responses for periods up to 20 days. *Geophys. Res. Lett.*, **29**, 2001GL0144.

Kuvshinov A.V., D. B. Avdeev, O. V. Pankratov, S. A. Golyshev, & N. Olsen, 2002b. Modelling electromagnetic fields in 3-D spherical earth using fast integral equation approach. In *" 3-D Electromag-*

netics", ed. M. S. Zhdanov and P. E. Wannamaker, Elsevier, Holland, pp. .

Kuvshinov A.V., D. B. Avdeev, & O. V. Pankratov, 1999. Global induction by *Sq* and *Dst* sources in the presence of oceans: bimodal solutions for non-uniform spherical surface shells above radially symmetric earth models in comparison to observations. *Geophys. J. Int.*, **137**, 630–650.

Langel, R. A., 1987. The Main Field. In *"Geomagnetism"*, ed. Jacobs, J. A., Academic Press, Orlando, Fla., pp. 249–512.

Langel, R. A., and Hinze, W. J., 1998. *The Magnetic Field of the Earth's Lithosphere: the Satellite Perspective.* Cambridge University Press.

Lowe, D. A. J., R. L. Parker, M. E. Purucker, & C. G. Constable, 2001. Estimating the crustal power spectrum from vector Magsat data. *J. Geophys. Res.*, **106**, 8589–8598.

Mandea, M., E. Bellanger & Jean-Louis Le Mouël, 2002. A geomagnetic jerk for the end of the 20th century?. *Earth Planet. Sci. Lett.*, **183**, 369–373.

Martinec, Z., 1999. Spectral finite-element approach to three-dimensional electromagnetic induction in a spherical Earth. *Geophys. J. Int.*, 136, 229–250.

Maus, S., K. M. Rother, H. Lühr, N. Olsen & V. Haak, 2002. First scalar magnetic anomaly map from CHAMP satellite data indicates weak lithospheric field. *Geophys. Res. Lett.*, **29**, 10.1029/2001GL013685.

Maus, S., K. Hemant, M. Rother, & H. Lühr, 2003. Mapping the lithospheric magnetic field from CHAMP scalar and vector magnetic data. In *"First CHAMP Mission Results for Gravity, Magnetic and Atmospheric Studies"*, ed. C. Reigber, H. Lühr, P. Schwinzer, Springer-Verlag, Berlin, pp. 269–274.

Nichols, E. A., Morrison, H. F., Clarke, J., 1988. Signals and noise in measurements of low-frequency geomagnetic fields. *J. Geophys. Res.*, **93**, 13743–13754.

O'Brien, M. S., R. L. Parker, & C. G. Constable, 1999. The magnetic power spectrum of the ocean crust on large scales. *J. Geophys. Res.*, **104**, 29,189–29,202.

Olsen, N., 1997. Ionospheric F-Region Currents at Middle and Low Latitudes Estimated From MAGSAT Data. *J. Geophys. Res.*, **102 (A3)**, 4563–4576.

Olsen, N., 1996. A new tool for determining ionospheric currents from magnetic satellite data. *Geophys. Res. Lett.*, **23**, 3635–3638.

Olsen, N., 1998. The electrical conductivity of the mantle beneath Europe derived from C-Responses from 3 h to 720 h. *Geophys. J. Int.*, **133**, 298-308.

Olsen, N., 1999a. Long period (30 days–1 year) electromagnetic sounding and the electrical conductivity of the Mantle beneath Europe. *Geophys. J. Int.*, **138**, 179–187.

Olsen, N., 1999b. Induction Studies With Satellite Data. *Surveys in Geophysics*, **20**, 309–340.

Olsen, N., L. Tøffner-Clausen, T. J. Sabaka, P. Brauer, J. M. G. Merayo, J. L. Jørgensen, J.-M. Lger, O. V. Nielsen, F. Primdahl & T. Risbo, 2003a. Calibration of the Ørsted Vector Magnetometer. *Earth, Planets and Space*, **55**, 11–18.

Olsen, N., S. Vennerstrøm and E. Friis-Christensen, 2003b. Monitoring magnetospheric contributions using ground-based and satellite magnetic data. In *"First CHAMP Mission Results for Gravity, Magnetic and Atmospheric Studies"*, ed. Ch. Reigber, H. Luehr and P. Schwintzer, Springer-Verlag Berlin, pp. 245–250.

Parker, R. L., 1994. *Geophysical Inverse Theory.* Princeton University Press, 386pp.

Parker, R. L., & Whaler, K. A. W., 1981. *Numerical methods for establishing solutions to the inverse problem of electromagnetic induction,* J. Geophys. Res., **86**, 9574–9584.

Parkinson, W. D., 1983. *Introduction to Geomagnetism.* Scottish Academic Press,. Edinburgh.

Reigber, Ch.; Lühr, H.; Schwintzer, P., 2002. CHAMP mission status. *Advances in Space Research*, **30(2)**, 129–134.

Riedel, K., and A. Sidorenko, 1995. Minimum bias multiple taper spectral estimation. *IEEE Trans. on Signal Process.*, **43**, 188–195.

Roberts, R. G., 1984. The long period electromagnetic response of the Earth. *Geophys. J.R. astr. Soc.*, **78**, 547–572.

Rygaard-Hjalsted, C., C. G. Constable, and R. L. Parker, 1997. The influence of correlated crustal noise in modeling the main geomagnetic field. *Geophys. J. Int.*, **130**, 717–726.

Sabaka, T., N. Olsen & R.A. Langel, 2002. A comprehensive model of the near-Earth magnetic field: Phase 3. *Geophys. J. Int.*, **151**, 32–68.

Schultz, A. and J. C. Larsen, 1987. On the electrical conductivity of the mid-mantle: 1. Calculation of equivalent scalar magnetotelluric response functions. *Geophys. J.R. astr. Soc.*, **88**, 733–761.

Schultz, A. and J. C. Larsen, 1990. On the electrical conductivity of the mid-mantle: 2. Delineation of heterogeneity by application of extremal inverse solutions. *Geophys. J.R. astr. Soc.*, **101**, 565–580.

Shure, L., R. L. Parker and R. A. Langel, 1985. A preliminary harmonic spline model from Magsat data. *J. Geophys. Res.*, **90**, 11,505–11,512.

Tarits, P. & N. Grammatica, 2000. Electromagnetic induction effects by the solar quiet magnetic field at satellite altitude. *Geophys. Res. Lett.*, **27**, 4009–4012.

Tyler, R. H., S. Maus, & H. Lühr, 2003. Satellite Observations of magnetic fields due to ocean tidal flow. *Science*, **299**, 239–241.

Uyeshima, M., & A. Schultz, 2000. Geoelectromagnetic induction in a heterogeneous sphere: A new three-dimensional forward solver using a conservative staggered-grid finite difference method. *Geophys. J. Int.*, **140**, 635–650.

Velimsky, J., M. E. Everett, & Z. Martinec, 2003. The transient Dst induction signal at satellite altitudes for a realistic 3-D distribution of electrical conductivity in the crust and mantle. *Geophys. Res. Lett.*, **30**, art. no. 2002GL016671.

Weiss, C. J., & M. E. Everett, 1998. Geomagnetic induction in a heterogenous sphere: fully three-dimensional test computations and the response of a realistic distribution of oceans and continents. *Geophys. J. Int.*, **135**, 650–662.

Xu, Y. S., Shankland, T. J., Poe, B. T., 2000. Laboratory-based electrical conductivity in the Earth's mantle. *J. Geophys. Res.*, **105**, 27865–27875.

C. G. Constable and S. C. Constable Institute of Geophysics and Planetary Physics, Scripps Institution of Oceanography, University of California at San Diego, La Jolla, Ca 92093-0225, USA. (e-mail: cconstable@ucsd.edu; sconstable@ucsd.edu)

Global Navigation Satellite Sounding of the Atmosphere and GNSS Altimetry: Prospects for Geosciences

T. P. Yunck and G. A. Hajj

Jet Propulsion Laboratory, California Institute of Technology, Pasadena, California

The vast illuminating power of the Global Positioning System (GPS), which transformed space geodesy in the 1990s, is now serving to probe the earth's fluid envelope in unique ways. Three distinct techniques have emerged: ground-based sensing of the integrated atmospheric moisture; space-based profiling of atmospheric refractivity, pressure, temperature, moisture, and other properties by active limb sounding; and surface (ocean and ice) altimetry and scatterometry with reflected signals detected from space. Ground-based GPS moisture sensing is already in provisional use for numerical weather prediction. Limb sounding, while less mature, offers a bevy of attractions, including high accuracy, stability, and vertical resolution; all-weather operation; and exceptionally low cost. GPS bistatic radar, or "reflectometry," is the least advanced but shows promise for a number of niche applications.

INTRODUCTION

Atmospheric sounding with the Global Positioning System arose in the 1980s from GPS geodesy. It had long been expected that signal delays through the neutral atmosphere would ultimately limit the accuracy of all forms of space geodesy [*Abshire and Gardner*, 1985; *Davis et al.*, 1985]. The major concern was the variable effects of water vapor, which represented the greatest model uncertainty. Early investigators explored a variety of techniques for reducing that critical error. A particularly effective strategy involved modeling the atmospheric delay as a random walk and estimating a delay correction directly from the GPS data at every time step – typically every few minutes [*Lichten and Border*, 1987; *Lichten and Bertiger*, 1989].

Although atmospheric delay estimates were at first treated as "nuisance" parameters, analysts soon found that by using surface pressure data to calibrate the delay due to dry air they could recover the delay due to moisture with exceptional fidelity [*Elgered et al.*, 2003]. This "wet" delay can be readily converted to an estimate of precipitable water (PW) for use in weather and climate modeling. Alternatively, one can assimilate the delay estimates themselves directly into numerical models [*Ha et al.*, 2003; *Ridal and Gustafsson*, 2003]. The GPS geodetic ground networks now expanding rapidly around the world are thus yielding a windfall of atmospheric moisture data, one of the most critical quantities for numerical weather prediction (NWP).

Over the past decade, a number of weather organizations have been investigating ground-based GPS sounding (see, for example, the papers from the first International Workshop on GPS Meteorology in Tsukuba, Japan, Jan 2003, and the Second CHAMP Science Meeting in Potsdam, Sep 2003). Plate 1 shows a sequence of PW contour maps derived from GPS data by the Japanese Meteorological Agency [*Hatanaka*, 2003] at the sites of Japan's vast GEONET array. The maps span a

162 GNSS ATMOSPHERIC SOUNDING AND ALTIMETRY

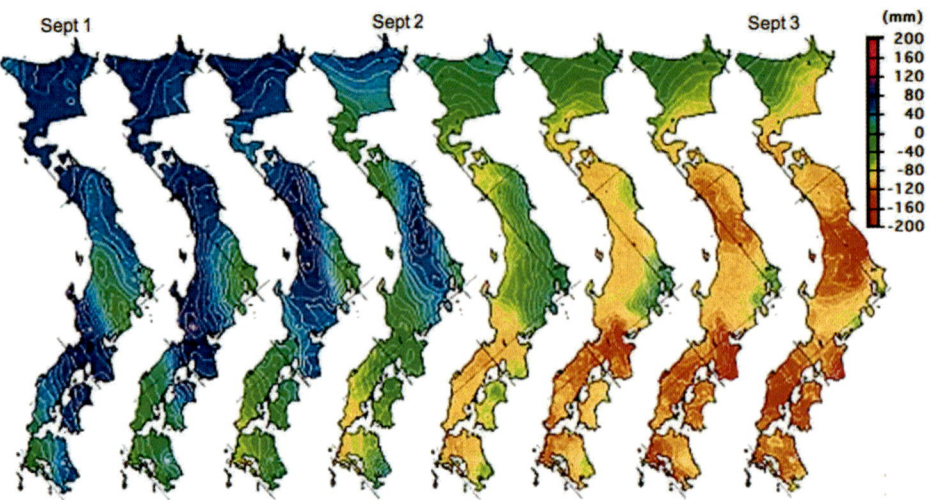

Plate 1. GEONET PW maps over a 3-day period in September 2001.

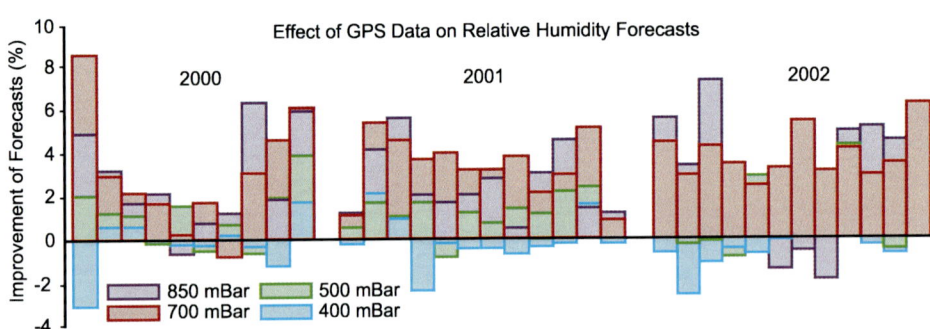

Plate 2. Impact of GPS data on 12-hr predictions of relative humidity in the US at 4 pressure levels over 3 years. Positive value indicates improvement (from *Gutman et al.* 2003).

Plate 3. Observing geometry for atmospheric limb sounding by GPS radio occultation.

3-day period in September 2001, as a weather system moved across the islands.

GROUND-BASED GPS AND WEATHER PREDICTION

The keen interest in GPS PW data stems from its potential for improving weather prediction, from short-term "now-casting" out to 10 days and beyond. For several years the Forecast Systems Laboratory (FSL) of the US National Oceanic and Atmospheric Administration (NOAA) has produced daily experimental predictions across the continental US of such quantities as relative humidity (RH) and precipitation, with and without GPS PW assimilated along with other standard data sources [*Gutman et al.*, 2003]. Plate 2 summarizes the 12-hr RH predictions from 2000 through 2002, with statistics collected into 1-month bins. The scale on the left gives the average effect of the GPS data, in percent accuracy change, for the full month. Values are given for four different altitudes (or, more precisely, pressure levels) for each month.

The GPS data have a consistently beneficial effect, except at the highest altitude, where ground-based GPS offers no useful discrimination. While the overall impact may appear small, these improvements are significant given the large volume of other data already assimilated. Furthermore, the average values are deceptive. Most days are relatively benign, without marked weather activity. The challenge for weather prediction is not the majority of easy days when forecasts are reliable, but those rarer days where major transitions and serious weather events are in progress or pending, and which are often inaccurately forecast. The value of the GPS data on those days tends to be higher than is suggested by the monthly averages.

A limitation of these early assimilation efforts is their restriction to a scalar PW value at each time step, representing the integrated zenith moisture for each observing site. This value is then related to other elevations by means of a mapping function, a method that is only approximate and that necessarily misses direction-dependent irregularities. Such irregularities can be pronounced at moist sites, particularly at times when weather systems are active. Because the mapping function does not fit these irregularities, they will show up in the post-fit residuals, combined with such other errors as instrumental noise, signal multipath, and higher order ionospheric effects. It is reasonable to presume that in many conditions the moisture irregularities will dominate the post-fit residuals and, therefore, that by adding back the residual to the mapped zenith delay solution one might obtain a value closer to the true line-of-sight moisture between the receiver and transmitter. That is the approach now being taken by several research groups seeking to improve ground-based GPS meteorology [*Braun and Rocken*, 2003; *Shoji et al.*, 2003].

There remain, however, important limitations to ground-based GPS sounding: It yields just one principal product, integrated PW; it offers little useful vertical resolution; and it is restricted to land areas fitted with extensive GPS networks. It is not, in short, a global solution. Space-based atmospheric limb sounding offers a valuable complement that addresses all of these issues.

ATMOSPHERIC SOUNDING FROM SPACE

In the late 1980s a few groups began to consider the possibility of atmospheric limb sounding from Earth orbit by radio occultation, exploiting GNSS signals as sources. This was inspired by the great success of planetary radio occultation over the previous 25 years, beginning with the 1964 Mariner IV mission to Mars [*Fjeldbo*, 1964; *Kliore et al.*, 1964]. Although Earth radio occultation had been proposed in various forms since the 1960s, the first account of a practical approach exploiting GPS did not appear until *Yunck et al.* [1988]. Progress has since been rapid. The first successful space-based demonstration of GPS occultation was the 1994 GPS/MET experiment, led by the University Corporation for Atmospheric Research (UCAR) in Boulder [*Rocken et al.*, 1997]. For a more complete history of radio occultation see *Melbourne et al.* [1994] and *Yunck et al.* [2000].

Plate 3 illustrates the GNSS occultation geometry. A low Earth orbiter (LEO) at a typical altitude of 400–800 km, equipped with limb-directed antennas, tracks GNSS signals as they rise and set through the atmosphere. The receiver continuously measures the changing carrier phase at both L-band frequencies. The atmosphere acts as a lens, bending and retarding the signals, inducing an excess path delay. This delay appears in the measurements as additional accumulated phase and a Doppler shift. Depending on the observing geometry, it may take a minute or more for the observed signal to pass from the top of the neutral atmosphere (~100 km altitude) to the surface, or vice versa. Since the changing geometry is dominated by the rapid orbital motion of the LEO, most soundings occur within ±45° of the forward and reverse velocity directions.

Plate 4 shows the geographical distribution of soundings over 24 hours from a single high-inclination receiver for a 24-satellite GPS constellation. There are roughly 700 in all. With a future GNSS constellation of 60 satellites, a 12-element LEO occultation array will darken the coverage map, acquiring more than 15,000 soundings daily. Both Galileo and the "Block 3" GPS satellites, to be deployed within the next decade, will boost signal strength by 6 dB above today's levels, improving both retrieval quality and consistency of penetration to the surface.

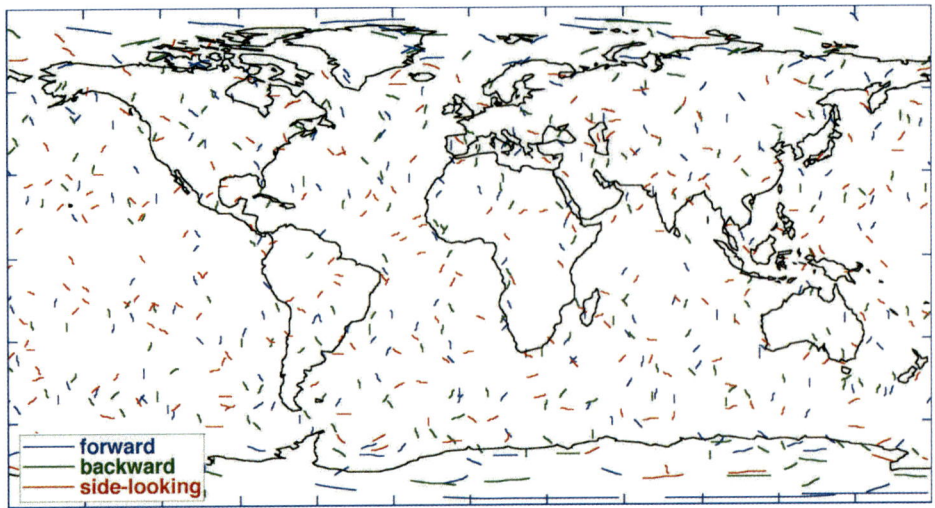

Plate 4. Distribution of soundings in 1 day with 1 receiver and 24 GPS satellites.

Sounding the atmosphere by radio occultation differs fundamentally from ground-based geodetic delay estimation. Because we are looking at (usually) a single ray traversing the atmosphere horizontally, we obtain extremely high vertical resolution — of the order of 100 m from the surface to the stratosphere. Moreover, radio occultation delivers a diverse assortment of atmospheric parameters distributed almost uniformly around the globe, enabling a broader variety of weather and climate applications [*Anthes et al.*, 2000]. In addition, from the precisely known observing geometry we can compute the absolute height of each measurement with respect to the geoid (or other reference surface) to within a few meters, limited largely by horizontal structure. This allows construction of precise 2D contour maps, pressure gradients, and such derived products as geostrophic wind fields in the troposphere and stratosphere [*Leroy*, 1997]. This ability to extract precise "geopotential heights" is unique among spaceborne sensors. The basic retrieval method, yielding initial profiles of atmospheric refractivity, density, pressure, and temperature (or moisture), is summarized below.

Profile Retrieval

To begin, we must first isolate the atmospheric excess delay from the measured total phase by precise geometric modeling and dual-frequency ionospheric correction, and then compute the atmosphere-induced Doppler shift. From this one can derive a precise estimate of the changing bending angle along the full vertical profile. The bending angle depends directly on the atmospheric refractivity, which one can compute by means of an Abel integral inversion. (For these steps we assume local spherical symmetry in the atmosphere, a usually adequate approximation). The water molecule's permanent dipole moment causes moisture to affect atmospheric refractivity more strongly than dry air. In the dryer regions, above about 5 km altitude, we can ignore the effect of moisture on signal refraction. The refractivity then depends only on the number density of atmospheric molecules and their types. Since we know the molecular weights of the different species and their relative abundance we can compute the atmospheric density accurately along the profile. And since the pressure field is in near hydrostatic equilibrium, we can sum up the total weight of a column of air from the top downwards to determine atmospheric pressure along the profile. With pressure and density known, we can apply the standard gas laws to derive the temperature profile.

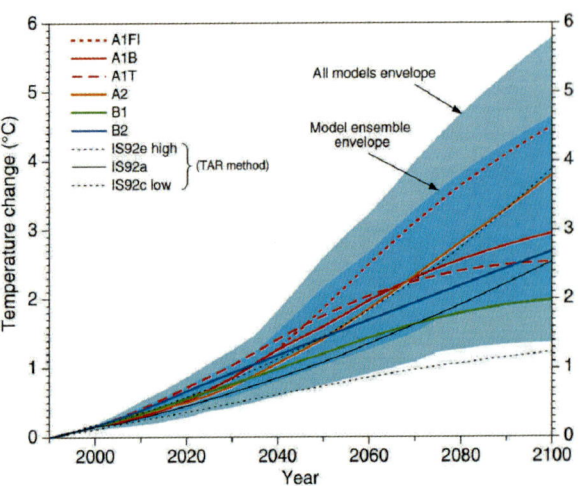

Plate 5. Projected change in average global surface temperatures over the next century according to several climate models (from *Cubasch et al.*, 2001).

Moisture in the lower troposphere can increase the path delay markedly, particularly in the tropics, where it can contribute up to 30% of the refractivity. Because of the little variability of temperature in that region, one can apply climatology (or other external information) to extrapolate the temperature profile downward and estimate moisture. (For a more thorough discussion see *Hajj et al.* [2002].) The processing sequence thus proceeds as follows:

- Isolate the atmospheric excess phase and compute the observed Doppler shift.
- Compute a profile of the signal bending angle from the changing Doppler shift.
- Compute a refractivity profile from the bending angle by an Abel integral inversion.
- Compute a molecular density profile from the observed refractivity and known atmospheric constituents.
- Compute a pressure profile from molecular density and known molecular weights, summing from top down.
- Compute a temperature profile from standard gas laws, or...
- In the lower troposphere, model the temperature and estimate moisture density.

There are several sophisticated variations on this approach, most notably the one-dimensional variational (1DVAR) method, which derives temperature and humidity jointly by use of atmospheric models with assigned a priori values and variances, along with the measured refractivity [*Kursinski et al.*, 2000; *Palmer et al.*, 2000]. Complications often arise in deriving signal bending in the lower troposphere as a result of atmospheric multipathing and signal loss from super-refraction, or "ducting." At present only about 70% of retrievals reach the lowest 1 km. "Open-loop" signal acquisition and advanced retrieval techniques will soon largely remove the multipath issue, while other specialized algorithms now under study promise to improve recovery below ducting layers [*Gorbunov et al.*, 1996; *Gorbunov*, 2001 & 2002; *Ao et al.*, 2003; *Sokolovskiy*, 2003]. A new generation of sensors and retrieval algorithms, to debut operationally on COSMIC in 2005 or 2006 [*Lee et al.*, 2000], should reach the lowest 1 km in more than 90% of profiles globally.

Performance

It may be surprising that this sequence of steps with its various approximations can yield useful accuracy. The key is in the precision with which we can measure the GPS carrier phase. A geodetic receiver with an antenna gain of 6–10 dB can measure the instantaneous phase with a precision of a few millimeters in a fraction of a second. With the power of modern GNSS-based orbit determination we can recover the atmospheric Doppler shift observed from low orbit to about 0.1 mm/s. At 70 or 80 km altitude the atmospheric path delay is negligible, even at the level of millimeters. That delay grows exponentially as the ray descends, reaching up to 2.5 km at the surface within about 60 s, with phase rates exceeding 100 m/s, or nearly a million times the precision of the measurement. The standard analysis assumptions—local spherical symmetry, hydrostatic equilibrium, mixing ratios of atmospheric constituents—while imperfect, are extremely good in most conditions and result in relatively small errors.

In the first performance analysis for GPS limb sounding, conducted in March 1988, the JPL team estimated that temperature accuracies of a few tenths of a Kelvin might be achieved over altitudes ranging from about 5 to 30 km [*Yunck et al.*, 1988]. More detailed studies in later years [*Kursinski et al.*, 1997; *Hajj et al.*, 2003] supported this conclusion. These studies suggest that because the major errors tend to be random or quasi-random, averaging of multiple profiles for long-term climate studies could reduce the effective temperature error below 0.1 K. That is, the averaged profiles may be accurate, independent of the instrument used, to better than 0.1 K over an extended altitude range. This would represent an improvement by a factor of anywhere from 10 to 40 over current global techniques, depending on just where the occultation error bottoms out (0.05 K?) and what one believes to be the long term stability of other techniques (of order 1–2K).

Climate Signal Detection

Arguably, the greatest concern in Earth science today is global climate change. That is the focus of NASA's Earth science program, exemplified by the flagship spacecraft of the Earth Observing System—Terra, Aqua, and Aura—and the dozen or so specialized probes, such as GRACE and ICESat, now in operation. Many more international missions, in flight or in preparation, address related themes.

A central question is whether or not Earth has entered a period of accelerated global warming (as seems likely) and, if so, to what extent this may result from stresses imposed by human activity. While the evidence from both ground and space is compelling [*Mann et al.*, 2003; *Vinnikov and Grody*, 2003; *Cubasch et al.*, 2001], to discern the effects of distinct drivers and trace the operative climate processes we require accurate data at all levels of the atmosphere. Despite vast investments in sensors we are at present severely limited in our ability to gather such data globally.

The United Nations Intergovernmental Panel on Climate Change (IPCC) has estimated that human-induced global

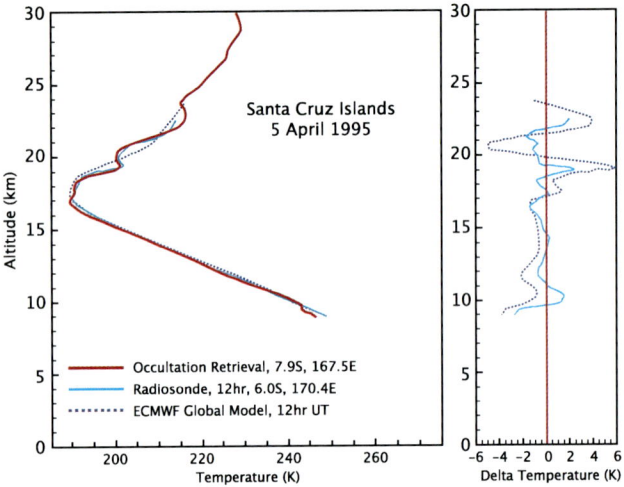

Plate 6. GPS/MET profile compared against a radiosonde and ECMWF analysis. Discrepancies are consistent with the expected accuracies of the comparison standards (from *Kursinski et al.*, 2000).

warming may be in the range of 0.1–0.3 K per decade at the surface over the next 20 years, and 0.2–0.5 K per decade over the next century. The effects may be subtler (or even reversed) in upper strata. Plate 5 shows global temperature projections for a variety of climate models [*Cubasch et al.*, 2001]. Detecting such signals requires great measurement precision and long-term stability: 0.1 K/decade or better. Current spaceborne instruments fall well short of this. The sensor suite is dominated by passive infrared and microwave radiometers, including MISR, MODIS, MLS, AIRS/AMSU, and AVIRIS. Although great ingenuity is employed to maintain calibration, over a decade or more this is elusive. Sensor aging and replacement with new designs undermine stability. Even external calibration against in situ radiosondes is limited by the variable quality of the radiosonde record. Over a decade or more the best that can reasonably be achieved is 1–2 K absolute stability. Thus, current sensors lack by at least an order of magnitude the stability needed to address some of the most critical questions facing Earth science.

In a 2002 research solicitation NASA declared: "Perhaps the greatest roadblock to our understanding of climate variability and change is the lack of robust and unbiased long-term global observations." They went on to say that for climate monitoring "the focus is on...construction of consistent datasets from multi-instrument [and] multi-year observations with careful attention to calibration and validation over the lifetime of the measurement" [*NASA*, 2002].

GNSS occultation exploits principles entirely different from those of radiometers, depending primarily on the timing of delay variations over about one minute—something we can accomplish with extreme accuracy. Each profile is largely self-calibrating and virtually unbiased, independent of the particular instrument used or when in its lifetime the profile is acquired. Individual and averaged profiles will therefore maintain their accuracies through future instrument evolution, indefinitely. A profile taken today will be directly comparable, at the level of 0.1 K or better, with a profile taken 50 years hence, no matter the specific designs of the instruments or the frequencies and structures of the signals employed.

The consequences for research in climate change could be profound. For the first time we will have the means to detect atmospheric temperature trends (above ~5 km) at the level of 0.1 K within a matter of years. Moreover, GNSS occultation can provide a continuing source of calibration data to normalize and stabilize atmospheric products from sensors now in place, transforming them into effective climate sensors. In short, if its current promise is validated, GNSS radio occultation may offer a key to removing the

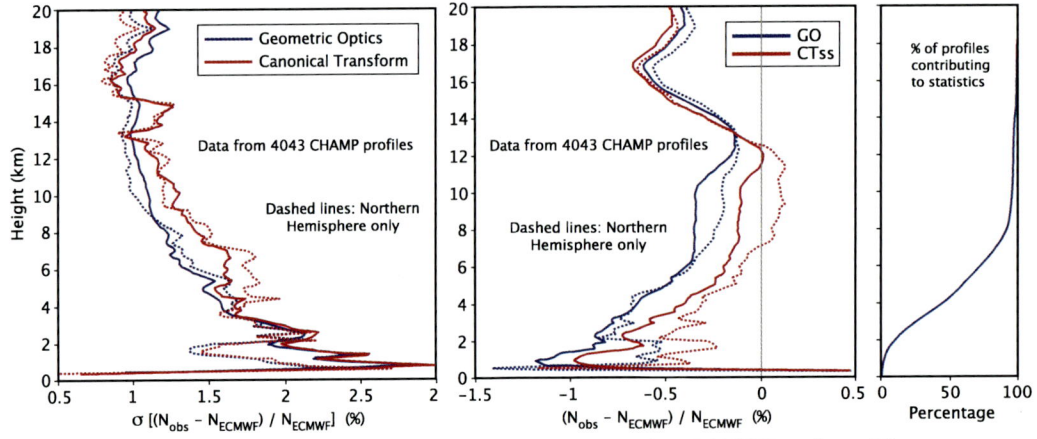

Plate 7. Average of several hundred GPS/MET profiles differenced against ECMWF analyses. The mean agreement (dark line) is at the level of ~1 K (from *Hajj et al.*, 2002).

"greatest roadblock to our understanding of climate variability and change."

Early Experimental Results

Establishing average temperature accuracies of 0.1 K or better over a broad altitude range presents difficulties. The challenge is to find adequate standards of comparison. To date, the primary standards have included radiosonde data and climate analyses generated by the US National Centers for Environmental Predictions (NCEP) and by the European Centre for Medium-Range Weather Forecasts (ECMWF), derived by assimilating radiosonde and space data into numerical models. For the former we must find radiosondes that are close in space and time to the profiles to be validated. While we can map the analyses to any time and location, they involve a good deal of interpolation and spatial smoothing. In neither case can we expect better than 1–2 K accuracy from the comparison data.

Plate 6 shows a temperature profile produced at JPL in 1995 from GPS/MET data, compared against both a radiosonde and the ECMWF analysis. Agreement is consistent with the expected 1–2 K accuracy of the analysis. The lack of fine vertical detail in the analysis suggests that its inherent smoothing has compromised vertical resolution. From such tests we cannot assess the accuracy of the GPS data beyond the evident level of agreement. Figure 1 shows the average agreement with ECMWF climate analyses for several hundred GPS/MET profiles. The standard deviation (shaded area) over an altitude range of 5–25 km is roughly 2 K. The mean difference (solid line) over the same range is reduced to 0.5–1.0 K, well off the 0.1 K or better predicted for the GPS occultations alone.

A new generation of occultation receiver, JPL's BlackJack, first flew on the German CHAMP mission in July of 2000. Plate 7 shows a statistical summary of more than 4000 CHAMP refractivity profiles from May and June 2001, compared with ECMWF daily analyses [*Beyerle et al.*, 2003]. The GFZ analysis team applied two different techniques, the standard "geometric optics" technique described above (blue), and a "canonical transform" technique (red) [*Gorbunov*, 2002], which offers the advantage of separating multiple tones in the lower troposphere and enhancing vertical resolution. The solid curves are for the whole earth, while the dashed curves are for the northern hemisphere, for which, owing to more abundant radiosonde data, the analyses are more reliable. Mean offsets (right) and standard deviations (left) correspond roughly to temperature mean offsets of 0.5 K and sigmas of 1–2 K, above 2–3 km. That level of temperature agreement is shown explicitly in *Hajj et al.* [2003]. These values do not differ appreciably from the earlier GPS/MET results and leave open the question of actual GPS occultation accuracy.

CHAMP / SAC-C Comparisons

Analysis prospects improved markedly when the Argentine SAC-C spacecraft, carrying a second BlackJack receiver, reached orbit in November of 2000. For the first time we could compare data directly from two advanced occultation instruments, provided they observed the same location at about the same time. It happens that such coincidences are rare but not unknown. Of more than 60,000 soundings acquired from 10 July 2001 to 9 June 2002, sixty pairs occurred within 200 km and 30 min of one another. Of those, fewer than thirty pairs occurred within 100 km and 30 min. As these coincidences are not exact there are real differences between pairs, estimated to be typically several tenths of a Kelvin.

Plate 8 compares near-coincident CHAMP and SAC-C occultations. The left panel shows a typical coincident pair (solid red and green) plotted against standard NCEP and ECMWF analyses computed for the two profiles. The observing directions differed by 137°. What stands out is the closeness of the two occultation profiles despite their differences in space and time. Both differ from the analyses, which are similar to one another but again lack fine vertical detail. This provides stronger evidence of the essential correctness of the occultation result. The NCEP and ECMWF analyses involve a fair degree of smoothing and cannot reliably capture such sharp features as

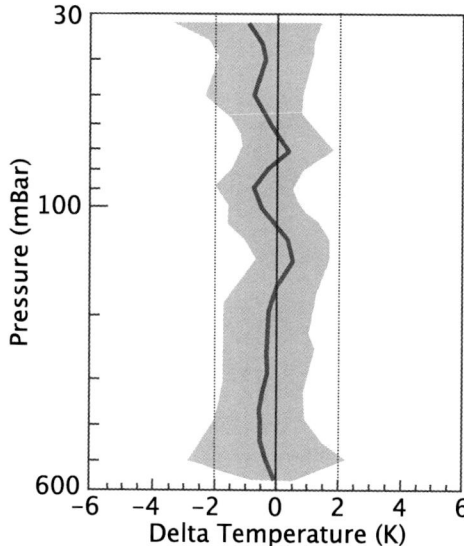

Figure 1. Summary of 4,043 CHAMP profiles made with geometric optics (blue) and canonical transform retrieval techniques, differenced with ECMWF analyses. Standard deviations (left) and mean offsets (center) are similar to those of GPS/MET (from *Beyerle et al.*, 2003).

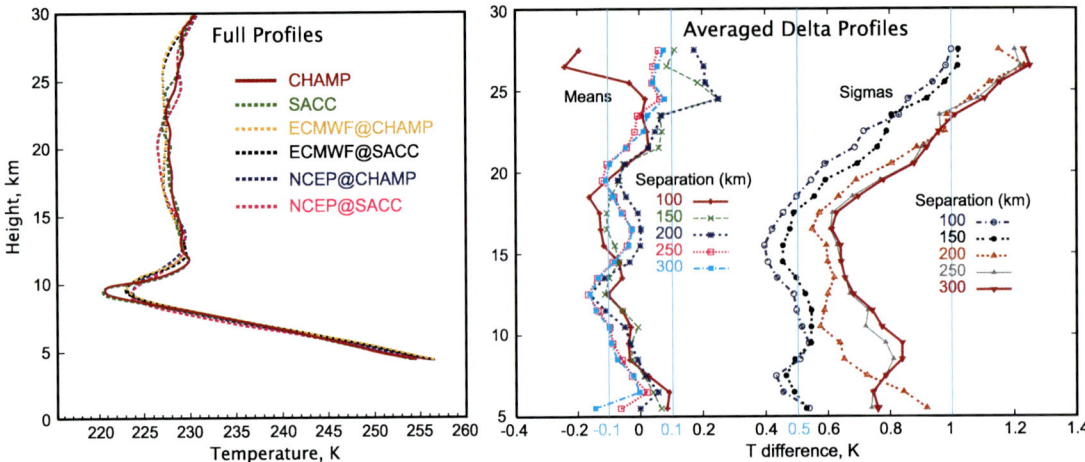

Plate 8. Coincident profile pair from CHAMP and SAC-C plotted against NCEP and ECMWF analyses at each profile location (left); and the average differences of several dozen coincident pairs at five different maximum separations (right) (from *Hajj et al.*, 2003.)

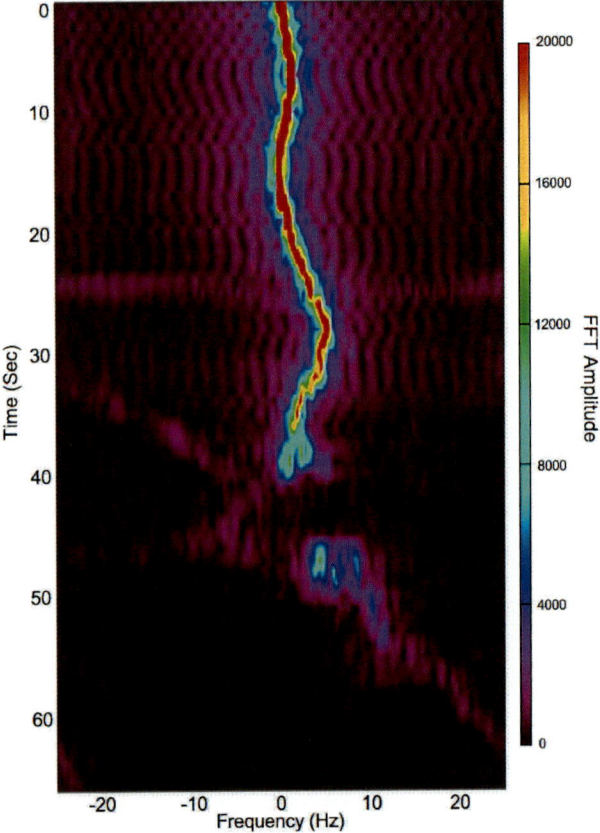

Plate 9. Spectrogram of an occultation from CHAMP after known Doppler shifts have been removed (bright vertical feature). Also visible is a faint surface reflection moving from the far left (about halfway up) to the lower right. Direct modeling would enhance the reflection (from *Hajj et al.*, 2003).

the tropopause at ~9.5 km shown in the occultations. That two profiles acquired from different instruments and directions agree so exactly in this detail is compelling.

Plate 8 also shows the mean differences and sigmas for all occultation pairs coincident within 30 min, for maximum separations ranging from 100 to 300 km. Between 5 and 25 km altitude the average differences fall within or near 0.1 K. Compare this with Figure 1 showing differences with the weather analyses, where the mean offsets are about 5 times greater. For the closest pairs, the sigmas (when reduced by root-2 to account for the combining of two profiles) are near 0.5 K at 20 km and below. These increase steadily for the more widely separated pairs, at least partly because of actual differences between the locations. For a more thorough presentation see *Hajj et al.* [2003].

While these results are consistent with an occultation bias of 0.1 K or less, they do not prove it. There could be, for example, common biases or trends in the occultation profiles not revealed in these comparisons. We know that at higher altitudes where the retrieval is initialized (above 30 km), and near the surface where the observations can become corrupted by complex atmospheric effects, biases can creep in. Their causes, however, are well understood [*Sokolovskiy*, 2001a; *Beyerle et al.*, 2003a; *Ao et al.*, 2003; *Sokolovskiy*, 2003]. Refined techniques now in development may all but eliminate refractivity biases near the surface and extend the upper range of profile accuracy.

Retrieving Moisture Profiles

GNSS occultation is often faulted for its inability to distinguish the refractive effects of moisture and dry air, leading to uncertainties in the structures and behaviors of the lower troposphere. While this is true in the strictest sense—the pure measurement does not allow a deterministic separation—as with nearly all science data there is a rich and informative context: physical constraints, spatial and temporal signatures, external and meta data to supply further clues. Sharp gradients, for example, cannot be sustained in dry air. In moist regions refraction from water begins to dominate below 5 km altitude. A simple tactic is to use temperature from a weather analysis to estimate moisture near the surface from the refractivity signature. *Kursinski and Hajj* [2001] show that this can yield moisture with a precision of 0.2 g/kg, surpassing other remote sensing instruments. More promising is the direct assimilation of bending angles or refractivity profiles into numerical models for a more nuanced separation of effects. Some scientists have begun to examine the additional information in glancing surface reflections (discussed below) and in polarization effects. As these cues become better understood and the models correspondingly refined, such approaches are likely to provide further gains.

Early Assimilation Studies

Perhaps the greatest appeal of GNSS occultation is its potential for improving weather forecasts. As yet, with the still meager supply of high quality occultation data, little has been accomplished in the way of data assimilation and impact studies. But what has been done shows much promise. Preliminary assimilation studies have been reported by *Zou et al.* [2003], *Healy et al.* [2003], *Kuo et al.* [2003], and *Aoyama et al.* [2003]. In the first of these, Zou et al. assimilated CHAMP bending angle data taken over a two-week period in July 2002. This approach is attractive because the effect of refractivity along the ray path can be modeled, ionospheric effects are more readily removed, and problems unique to refractivity retrieval (e.g., the upper boundary condition, various limiting assumptions, and the ill-posed nature of the inversion under super-refraction) can be avoided. By contrast, the Healy and Kuo teams chose the more tractable refractivity profiles for assimilation, and Aoyama et al. assimilated the final temperature profiles. Despite the extreme sparseness of the data sets, all reported slight improvements in forecasts of various kinds—some going out 4 or 5 days—when occultation data were included. Healy et al. note that the "results are very encouraging [and] support the case for assimilating RO measurements operationally."

Despite these successes GNSS occultation science is still in its infancy. The gulf between the tens of profiles used in assimilations today and a future of tens of thousands is beyond our power to bridge by intuition. Table 1 tabulates seven key attractions—the Seven Cardinal Virtues—of GNSS limb sounding. Soon this simple trick of radio metrology may offer a key to one of the most vexing problems in Earth science: discerning the faint signatures of global climate change. The potential for monitoring of the global troposphere has equally profound implications for weather prediction.

GNSS SURFACE REFLECTIONS

Traditional radar altimetry (e.g., TOPEX/Poseidon and Jason-1) observes vertically, obtaining one nadir height value at a time. GNSS reflectometry, by contrast, can track a dozen or more surface returns from many angles at once with a single LEO receiver. This offers the prospect of higher temporal and spatial resolution for resolving finer scale, short-lived features, such as mesoscale eddies [*Treuhaft et al.*, 2003], which play an important role in the transport of momentum, heat, salt, nutrients, and chemicals within the ocean. Mesoscale eddies are analogous to atmospheric storms and create mean sea-height changes of about 10 cm over distances of 10–100 km, persisting typically for 1 week to 1 month. Ocean coverage from a single orbiting receiver collecting reflections from 24 GPS satellites is illustrated in Figure 2. *Hajj and Zuffada* [2003] show that an array of eight orbiters acquiring GPS and Galileo reflections could provide global 3-cm ocean heights in one day with 200-km spatial resolution, or sub-decimeter heights in four days with 25–50 km resolution.

Another attractive application is the monitoring of barotropic waves that move across ocean basins too quickly to be captured within the Jason-1 10-day repeat cycle [*Stammer et al.*, 2000]. While the technical feasibility of bistatic GNSS is clear, it remains to be seen whether it will offer practical advantages over alternative approaches, such as wide swath altimetry

Table 1. Seven Attractions of GNSS Occultation

- Offers order-of-magnitude improvement in accuracy and long-term stability of atmospheric temperature profiles.
- Operates undiminished in all weather, day or night.
- Offers near-uniform global coverage from the upper stratosphere to the surface.
- Provides vertical resolution to better than 100 m throughout the troposphere.
- Yields absolute heights of measured quantities to better than 10 m, permitting derivation of global pressure contours and non-equatorial geostrophic wind fields.
- Exploits different physical principles from other atmospheric sensors, remote or in situ, presenting an independent comparison and calibration standard.
- Requires only a palm-sized module costing <1 % of the tens to hundreds of M$ of many of today's passive sensors.

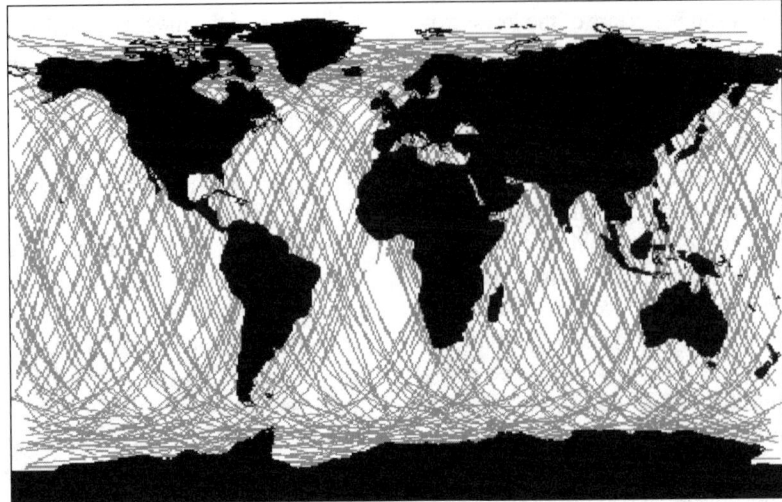

Figure 2. Ocean coverage achieved in one day from one receiver in LEO collecting reflections from 24 GPS satellites (from *Hajj and Zuffada*, 2003)

[*Rodriguez and Pollard*, 2001] or flying multiple nadir altimeters [*Raney*, 2001a,b; *AVISO*, 2003]. Here we consider some basic requirements and configurations for a practical bistatic GNSS system.

An early concept for GNSS-based ocean altimetry was proposed by *Martin-Neira* [1993]. Since then a good deal of theoretical system performance analysis has appeared [*Picardi et al.*, 1998; *Zavorotny and Voronovich*, 2000; *Fung et al.*, 2001; *Hajj and Zuffada*, 2003]. The first known GPS ocean reflection observed from space was collected fortuitously by the SIR-C L-band radar carried aboard the space shuttle (STS-68) in 1995, and detected in an intensive search by a team at JPL several years later [*Lowe et al.*, 2002].

A Few Principles of GNSS Reflectometry

The ocean reflects incident signals to an orbiting receiver from an extended portion of the surface. A given transmitted wavefront will arrive first from the specular reflection point, representing the shortest reflection path. Steadily decreasing energy is received from surrounding points, incurring greater path delays. By cross-correlating the received signal against a model signal at many different closely spaced delays one can map the returned energy vs delay. Figure 3 shows the nominal shape of the cross-correlation function for the direct GPS signal, assuming no multipath (left), along with typical cross-correlation functions for a signal reflected from the rough ocean surface, for different surface wind speeds. Because the greatest reflected energy arrives on the shortest path, the function rises abruptly, then decays gradually as the weaker outlying reflections trail in. The detailed shape of this function contains a good deal of information about the ocean surface. To extract this information, the receiver must provide hundreds of individual correlators to map the correlation functions from a dozen or more concurrent reflections.

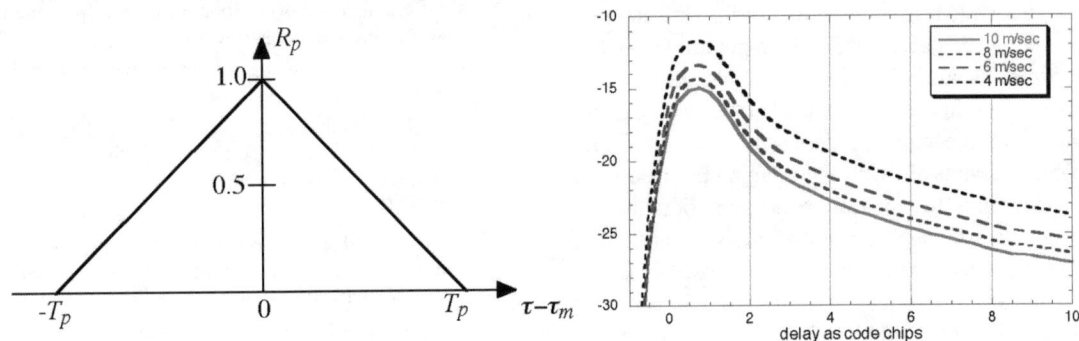

Figure 3. Plots of the ideal GPS auto-correlation function with no multipath (left) and of a reflection from the ocean surface cross-correlated with a model signal at many different lags, for four different surface wind speeds.

The sharp leading edge of the correlation function permits precise determination of the arrival time—and hence total delay—of the reflection from the specular point; from that delay, along with other geometrical modeling information, one can derive an estimate of the ocean height at that point. Analysis of the correlation tail can reveal information about the surface roughness, significant wave height, and wind speed. Thus, from the same reflections one can, in principle, perform both altimetry and scatterometry.

Practical Considerations

Many technical challenges arise in making such measurements from Earth orbit (for a succinct discussion see *Hajj and Zuffada*, 2003). Among these are the weakness of the reflected signals, the wide angular extent over which reflections may be seen (ranging from nadir to the Earth horizon) and the rapid signal de-coherence due to ocean roughness, which is typically large in comparison to the ~20 cm GNSS wavelengths. The last of these implies that under most ocean surface conditions and viewing geometries we cannot acquire the continuous phase observable that is so essential in many precise GNSS applications. Instead we must rely on heavily averaged pseudorange measurements to recover signal delay and ocean height. That task is made more difficult by the weak signal, necessitating antenna gains of 20 dBIC or greater. High gains imply large collecting areas, in this case at least 2 m². Moreover, to attain the precision and spatial resolution needed to study ocean eddies, we must seek to acquire nearly all detectable reflections, of which (with GPS and Galileo) there may be a dozen or more at once. Thus, we require a bouquet of beams that can be steered broadly and independently. This is further complicated by the varying polarizations of signals reflected from different angles. A good deal of creative design is needed to devise efficient systems that can compete with more traditional techniques.

There are, however, reasons for optimism. Among the less obvious properties of bistatic GNSS is the complex dependence of observation quality on viewing angle. For a given delay measurement precision, the greatest sensitivity to ocean height occurs at nadir. This sensitivity degrades roughly as 1/cos of the angle off nadir. However, reflected signal quality is poorest at nadir, owing both to a lower returned signal strength and greater decoherence. The delay accuracy can improve by a factor of two or more at 60° off nadir. Figure 4, from *Hajj and Zuffada* [2003], shows the expected L1 reflected range measurement error as a function of the angle off nadir for different antenna gains. In addition, for a given observation beamwidth the number of reflections to be seen is least at nadir. As the beam moves off nadir it subtends increasingly large surface areas, and those areas exhibit greater curvature (hence more angles that reflect to the sensor). Figure 5 quantifies this tradeoff for a receiver at 700 km. At the horizon the average number of reflected signals

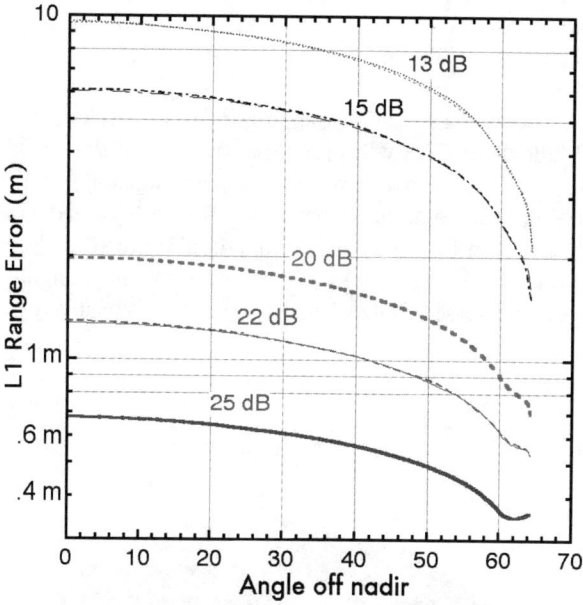

Figure 4. Estimated net reflected range errors for a 4-sec averaged measurement as a function of the angle off nadir, for 5 different receiver antenna gains.

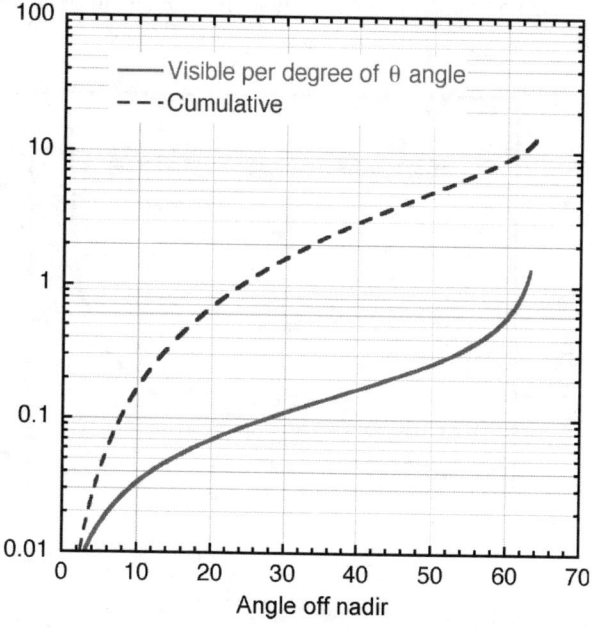

Figure 5. Average number of visible reflections per degree of off-nadir angle (red) and cumulative from nadir outward, as a function of the off-nadir viewing angle (from *Hajj and Zuffada*, 2003).

within a unit solid angle is fully two orders of magnitude greater than at nadir.

OCCULTATION / REFLECTION SYNERGY

New realms of investigation often bring surprises and GNSS reflectometry is no exception. We've noted that the coherence time of the reflected signal tends to increase with increasing angle off nadir. The Rayleigh criterion is traditionally used to define the onset of incoherent scattering and is given by

$$h = \lambda / 8 \sin \varepsilon,$$

where h is the wave height, λ is the signal wavelength, and ε is the observing angle with respect to a tangent line at the reflection point. This says simply that when ocean wave heights exceed h, delays from the crest and trough differ by more than $\lambda/4$, leading to loss of coherence. In Figure 6 we plot h as a function of the angle off nadir for a LEO at 700 km. For normal incidence (nadir reflection) the ocean scatters coherently only for h <2 cm; near the horizon (~64° off nadir) this occurs for h <1 m. The former condition is almost never satisfied while the latter is satisfied much of the time.

GNSS satellites broadcast their signals with right-hand circular polarization (RCP). For a reflection directly at nadir this is fully reversed, requiring an LCP antenna. For reflections off nadir the polarization reversal is only partial; hence they contain both RCP and LCP components. Glancing reflections near the horizon remain predominantly RCP and thus can be detected with a standard RCP GNSS antenna. We see then that reflections visible well off nadir offer a suite of attractions: they are typically stronger, having reflected from a considerably larger surface area; they have longer coherence times, approaching continuous coherence (and thus offering the power of the centimeter-quality phase observable) near the horizon; and they can be acquired with a standard RCP antenna. Evidently, we have not yet exhausted the virtues of GNSS limb sounding. To the seven listed earlier we can add the prospect of acquiring high-precision glancing surface reflections. Figure 7 illustrates the geometries for nadir and near-limb reflections.

Recent Observations

This indeed is more than just a prospect. Plate 9 shows a spectrogram of a typical GPS occultation acquired by CHAMP. The horizontal axis is frequency and the vertical axis is time (advancing downward), which can also be thought of as altitude, decreasing with time as the occultation descends. The changing frequency of the occulting signal, including the gross effect of the atmosphere, has been removed, so we see the occultation as the bright, nearly constant-frequency vertical feature. Small frequency variations, including occasional splitting of the signal into two or more tones in the lower troposphere, result from atmospheric structure and provide the basic information of the measurement. Also evident in Plate 9 is a more curious feature: a faint line slanting in from the left edge, intersecting the occultation near the surface. A similar feature appears in a high percentage of CHAMP and SAC-C occultations. While these were at first puzzling, analysts at JPL and GFZ quickly discerned them to be glancing reflections of the occulting signal—essentially a form of multipath.

Students of GPS will know that because of the pseudonoise (PN) code modulation on the GPS signals, when the receiver tracks a direct signal, any reflection that is delayed by more than 1.5 code chips (about 45 m for the P-code and 450 m for the C/A-code) with respect to the direct signal is largely suppressed. Only when the occultation is relatively near

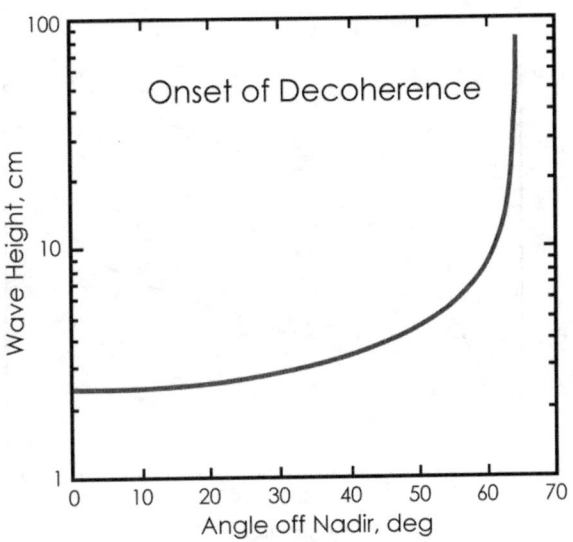

Figure 6. The nominal wave height at which the reflected carrier phase loses coherence, based on the Rayleigh criterion (from *Hajj*, 1998).

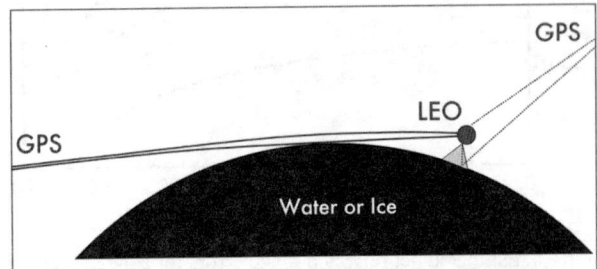

Figure 7. The differing geometries of reflections acquired near nadir and at the horizon.

the surface and the excess reflected delay is sufficiently small can we expect to see the glancing reflection, and then it will be attenuated owing to its offset from the modeled code. Future upgrades to the receiver software will anticipate these reflections and model them directly in order to recover them at full strength.

Analysis at GFZ

The CHAMP science team at GeoforschungZentrum (GFZ) in Potsdam has done a thorough study of these glancing reflections, with promising results [*Beyerle et al*, 2002]. Figure 8 shows all occultations acquired by CHAMP over a 4-week period in the spring of 2001 (dots) together with all detected reflections (circles). Clearly evident is the relative scarcity of reflections in the tropics. This is presumed to owe to the combination of dense tropical moisture, attenuation from the PN code offset, and the less frequent penetration of tropical occultations sufficiently close to the surface to yield the required small delay offsets. With coming refinements in "open-loop" signal acquisition [*Sokolovskiy*, 2001b, 2002] and direct onboard modeling of the reflections, this shortfall should be considerably alleviated.

Also evident in Figure 8 is the presence at high latitudes of reflections in almost equal densities from water and ice, offering the possibility of cryosphere as well as ocean sensing. In fact, at high latitudes nearly all occultations occurring over water or ice are accompanied by a detectable reflection. These uninvited guest signals may effectively double the science value of occultation sensors, a reward we can realize with relatively simple flight software updates. We are thus seeing an unexpected convergence of GNSS occultation science and reflectometry.

In their scrupulous analysis, Beyerle et al. chose as their fundamental observable the anomalous Doppler shift (evident in Plate 9) of the reflected signal. They show that this depends on a number of factors, including the height of the reflecting surface, its slope (when ice), and the moisture density of the atmosphere. They argue that in different circumstances the Doppler signal may be used to estimate each of these. For example, for ocean reflections, where the height and slope are known, one can estimate moisture in the lower troposphere. In frigid regions, where atmospheric moisture is negligible, one can recover topography to investigate ice mass balance and the evolution of ice sheets—central issues in global climate change. More recently, *Cardellah et al.* [2003] showed how one can analyze the direct and reflected signals interferometrically to infer ice height with possible sub-decimeter accuracy.

THE BLACKJACK OCCULTATION RECEIVER

Plate 10 shows the 48-channel BlackJack receiver now flying on CHAMP and SAC-C. That model is based on c. 1997

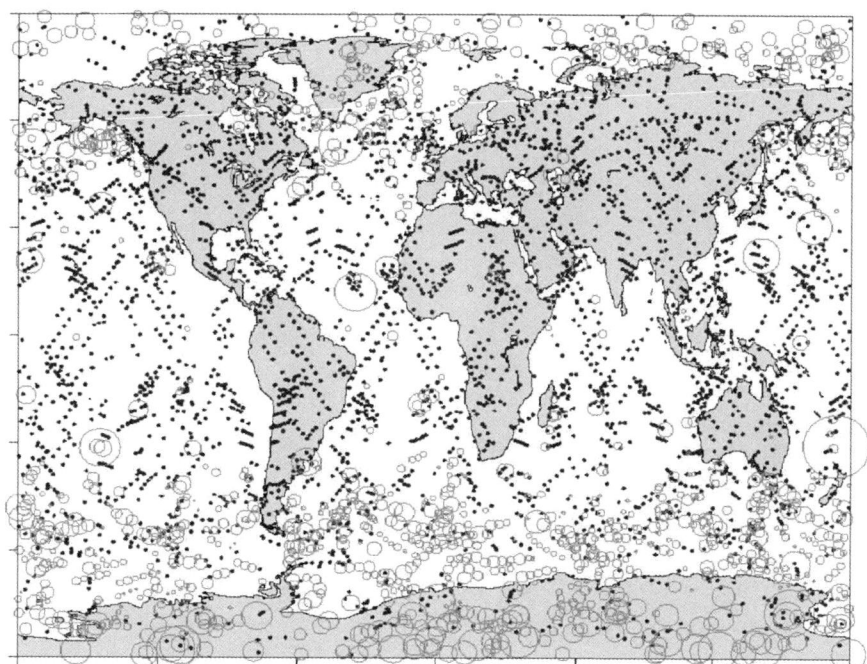

Figure 8. Occultations acquired by CHAMP between 14 May and 10 June of 2001 (dots) and accompanying reflections (circles). Circle size indicates relative strength (from *Beyerle et al.*, 2002).

technology and by no means exploits the level of integration possible even then. Also shown is a postage-stamp-size 12-channel consumer GPS receiver of the kind now found in cell phones and wristwatches. While far less capable than the BlackJack, it illustrates what we have come to expect in consumer electronics and hints at a possible future for GNSS flight sensors. We note that consumer electronics—laptop computers, PDAs, DVD players—are now flown routinely on the space shuttle and space station in just the sorts of low orbits favored for occultation and reflection sensing.

To achieve its high performance, the BlackJack itself is made almost entirely of commercial (rather than flight rated) electronic parts. The BlackJack and its predecessor, the TurboRogue Space Receiver, have now accumulated well over 60 years in space without an on-orbit failure. The occasional radiation-induced single-event upset (typically one every week or two on SAC-C) is cleared within minutes by an automatic software reset. Plate 11 shows a concept for a more highly integrated GNSS science receiver based on current digital technology, an area that continues to advance swiftly. Such a device could be reproduced even in limited quantities (ignoring development costs) for less than $50,000 apiece.

In the context of commercial electronics, $50,000 for the module in Plate 11 may seem high. In the world of spaceborne sensors, however, that is a trifle. Typical acquisition costs for sensors like MODIS, MISR, AIRS/AMSU, ATMS, and so on, are many tens of millions of dollars. Once built, such a module could be carried by a great variety of LEO spacecraft. What is more, most satellites today carry GPS for navigation and timing. Reduced to a single card, the BlackJack could serve both science and utility functions while

Plate 11. Concept for future occultation instrument exploiting modern LSI technology.

consuming no more power, space, or dollars than a typical navigation instrument today.

FUTURE PROSPECTS

The question arises, how many sensors are enough? Much is expected scientifically from the emerging occultation array to consist of COSMIC (Taiwan, 6 satellites, 2005), EQUARS (Brazil–Japan, 2005), TerraSAR-X (Germany, 2006), ACE+ (European Space Agency, 2007?), and others to follow. For ocean reflections we've seen that a dozen could be of value. Perhaps the best hint of an upper limit on useful numbers comes from the recommendations of a national science panel convened by NASA in Easton, PA, in 1997. Their charter was to define the foreseeable scope of Earth remote sensing data needed in the next 10–15 years to address NASA's overall Earth science agenda. The Easton panel embraced GPS occultation and recommended that an operational constellation be deployed. A key requirement articulated for that system was to provide global atmospheric profiles with an average horizontal spacing of 50 km, twice per day, to initialize operational weather models. That translates to roughly 400,000 profiles per day. Achieving such numbers with a 60-satellite GNSS constellation will require 250–300 LEO sensors.

An array of eight or ten occultation sensors is, to be sure, incomparably superior to none. It will represent a milestone as consequential as the early GPS geodetic networks. The marginal return for the next dozen will surely be less. But surveying the rich history of GNSS-based science, we have little doubt that the dividends will grow steadily with larger numbers, even into the hundreds. Global numerical weather

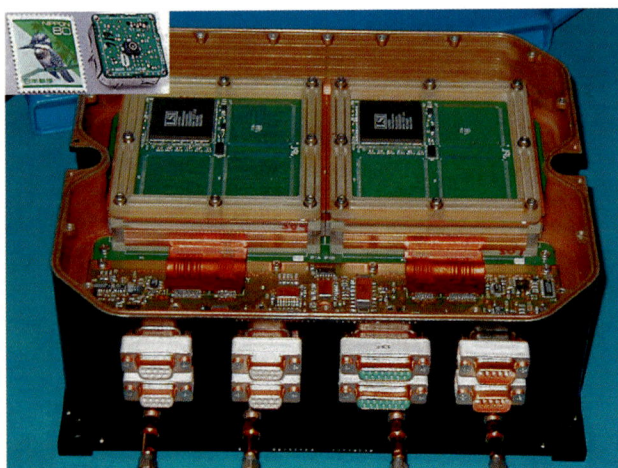

Plate 10. The BlackJack GPS occultation receiver now flying on CHAMP and SAC-C. Inset shows a commercial 12-channel GPS receiver found in some cell phones and watches.

prediction and snapshot 3D ionospheric imaging will be among the obvious beneficiaries. Whatever the value of the first ten sensors, it will be far surpassed by the value of 50.

Any effort we make now to catalogue the science to emerge from such an array is necessarily constrained by current concepts and paradigms. Sufficient changes in quantity, as Engels noted, at length become changes in quality. A global occultation array is a new breed of instrument for which have little experience or insight. Uncovering the riches within GNSS geodesy has taken two decades and continues still. That process has transformed our understanding of what is possible in measuring the Earth. A similar conceptual retooling will likely be required for Earth remote sensing with spaceborne GNSS. When all is said and done, the power and compass of this new tool will surely surpass what we can imagine now.

Acknowledgment. Work described in this paper was performed in part by the Jet Propulsion Laboratory under contract with the National Aeronautics and Space Administration.

REFERENCES

Abshire, J. B. and S. Gardner (1985), Atmospheric refractivity corrections in satellite laser ranging, *IEEE Trans. Geosci. & Remote Sensing, 23,* 414–425.

Anthes, R. A, C. Rocken and Y.-H. Kuo (2000), Applications of COSMIC to meteorology and climate, in Lee et al. (2000), 115–156.

Ao, C. O., T. K. Meehan, G. A. H ajj, A. J. Mannucci, and G. Beyerle (2003), Lower-troposphere refractivity bias in GPS occultation retrievals, *J. Geophys. Res.,* 108(D18), 4577, doi:10.1029/2002JD003216.

Aoyama, Y., E. Ozawa and H. Tada (2003), Data assimilation experiment using temperature retrieved from CHAMP occultation data, Proc. Int. Workshop on GPS Meteorology, Tsukuba, Japan.

AVISO (2003), High-precision altimetry with satellites working together, http://www-aviso.cls.fr/html/alti/multi_sat_ uk.html.

Beyerle, G. et al. (2003), The radio occultation experiment aboard CHAMP (Part II): Advanced retrieval techniques in atmospheric sounding and GPS reflectometry, Proc. Int. Workshop on GPS Meteorology, Tsukuba, Japan.

Beyerle, G. et al. (2002), GPS radio occultations with CHAMP: A radio holographic analysis of GPS signal propagation in the troposphere and surface reflections, *J. Geophys. Res.,* 107(D24), 4802, doi:10.1029/2001JD001402.

Beyerle, G; M. E. Gorbunov and C. O. Ao (2003a), Simulation studies of GPS radio occultation measurements, *Radio Sci.,* 39(5), 1084, doi:10.1029/2002RS002800.

Braun, J. and C. Rocken (2003), Slant water vapor in the United States southern Great Plains, Proc. Int. Workshop on GPS Meteorology, Tsukuba, Japan.

Cardellach, E., C. O. Ao., M. de la Torre Juarez and G. A. Hajj (2004), Carrier-phase delay altimetry with GPS-reflection/ occultation interferometry from low earth orbiters, submitted to *Geophys. Res. Lett.*

Cubasch et al. (2001), Projections of future climate change, In: *Climate Change 2001: The Scientific Basis,* Houghton, J. T. et al. (eds.), Cambridge Univ. Press, p. 555.

Davis, J. L., T. A. Herring, I. I. Shapiro, A. E. E. Rogers and G. Elgered (1985), Geodesy by radio interferometry: Effects of atmospheric modeling errors on estimates of baseline length, *Radio Sci., 20,* 1593–1607.

Elgered, G., B. Stoew, L. Gradinarsky and H. Bouma (2003), Analysis of atmospheric parameters derived from ground-based GPS observations, Proc. Int. Workshop on GPS Meteorology, Tsukuba, Japan.

Fjeldbo, G. (1964), Bistatic-radar methods for studying planetary ionospheres and surfaces, *Sci. Rpt. #2,* NsG-377, SU-SEL-64-025, Stanford Electronics Labs, Stanford, CA.

Fung, A. K, C. Zuffada and C.Y. Hsieh (2001), Incoherent bistatic scattering from the sea surface at L-band, *IEEE Trans. Geosci. and Rem. Sensing,* 39(5): 1006–1012.

Gorbunov, M. E. (2002), Canonical transform method for processing radio occultation data in the lower troposphere, *Radio Sci.* 37(5), doi:10.1029/2000RS002592.

Gorbunov M. E. (2001), Radioholographic methods for processing radio occultation data in multipath regions, *Danish Meteorological Inst. Scientific Report, 01–02,* Copenhagen.

Gorbunov, M. E., A. S. Gurvich, and L. Bengtsson (1996), Advanced algorithms of inversion of GPS/MET satellite data and their application to reconstruction of temperature and humidity, *Rep. 211,* Max Planck-Inst. Meteorol., Hamburg.

Gutman, S. I. et al. (2003), Rapid retrieval and assimilation of ground based GPS-Met observations at the NOAA Forecast Systems Laboratory: Impact on weather forecasts, Proc. Int. Workshop on GPS Meteorology, Tsukuba, Japan.

Ha, S.-Y., Y.-H. Kuo, G.-H. Lim (2003), Assimilation of GPS slant wet delay data and its impact on the short-range NWP, Proc. Int. Workshop on GPS Meteorology, Tsukuba, Japan.

Hajj, G. A. (1998), Reflections on GPS signal surface reflection, *JPL Memorandum,* 335–98–003.

Hajj, G. A. et al. (2004), CHAMP and SAC-C atmospheric occultation results and intercomparisons, *J. Geophys. Res.,* in press.

Hajj, G. A., E. R. Kursinski, L. J. Romans, W. I. Bertiger, S. S. Leroy (2002), A technical description of atmospheric sounding by GPS occultation, *J. Atmos. and Solar–Terrestrial Physics,* 64, 451–469.

Hajj, G. A. and C. Zuffada (2003), Theoretical description of a bistatic system for ocean altimetry using the GPS signal, *Radio Sci.,* 38(5), 1089, doi:10.1029/2002RS002787.

Hatanaka, Yuki (2003), Estimation of troposphere delay and accuracy of GEONET solutions, Proc. Int. Workshop on GPS Meteorology, Tsukuba, Japan.

Healy, S., A. Jupp and C. Marquardt (2003), A forecast impact trial with CHAMP radio occultation measurements, presented at EGS annual meeting, Nice, France, http://web.dmi.dk/pub/GRAS_SAF/presentations/egs-sh2-apr-2003.pdf.

Kliore, A. J., T. W. Hamilton, and D. L. Cain (1964), Determination of some physical properties of the atmosphere of Mars from changes in the Doppler signal of a spacecraft on an earth occultation trajectory, *Technical Report 32-674*, JPL.

Kuo, Y-H, T.-K. Wee and S. Sokolovskiy (2003), Analysis and assimilation of radio occultation data, Proc. Int. Workshop on GPS Meteorology, Tsukuba, Japan.

Kursinski E. R., Healy S. B., Romans L. J. (2000), Initial results of combining GPS occultations with ECMWF global analyses within a 1DVar framework, *Earth Planets and Space,* 52 (11): 885–892.

Kursinski, E. R. and G. A. Hajj (2001), A comparison of water vapor derived from GPS occultations and global weather analyses, *J. Geophys. Res.,* 106(D1): 1113–1138.

Kursinski, E. R., G. A. Hajj, K. R. Hardy, J. T. Schofield, and R. Linfield (1997), Observing Earth's atmosphere with radio occultation measurements, *J. Geophys. Res.,* 102(D19): 23429–23465.

Lee, L., C. Rocken and E. Kursinski, eds. (2000), *Applications of Constellation Observing System for Meteorology, Ionosphere & Climate,* Springer-Verlag, ISBN 962-430-135-2.

Leroy, S. S. (1997), Measurement of geopotential heights by GPS radio occultation, *J. Geophys. Res.,* 102, 6971–6986.

Lichten, S. M. and W. I. Bertiger (1989), Demonstration of submeter GPS orbit determination and 1.5 parts in 10^8 three-dimensional baseline accuracy, *Bull. Geod.,* 63, 167–189.

Lichten, S. M. and J. S. Border (1987), Strategies for high-precision Global Positioning System orbit determination, *J. Geophys. Res.,* 92, 12751–12762.

Lowe, S. T. et al. (2002), First spaceborne observation of an Earth-reflected GPS signal, *Radio Sci.,* 37(1), article 1007.

Mann, M. et al. (2003), On past temperatures and anomalous late-20th century warmth, *EOS, Trans. AGU* 84(27): 256–257.

Martin-Neira, M. (1993), A passive reflectometry and interferometry system (PARIS): application to ocean altimetry, *ESA Journal,* 17, 331–355.

Melbourne, W.G. et al. (1994), *The Application of Spaceborne GPS to Atmospheric Limb Sounding and Global Change Monitoring,* JPL Publication 94–18.

NASA (2002), Cooperative Agreement Notice CAN-02-OES-01, Earth Science Research, Education and Applications Solutions Network (REASoN).

Palmer, P. I., J. J. Barnet, J. R. Eyre, S. B. Healy (2000), A nonlinear optimal, estimation inverse method for radio occultation measurements of temperature, humidity, and surface pressure, *J. Geophys. Res.—Atmos.,* 105(D13): 17513–17526.

Picardi, G., R. Seu, S. G. Sorge and M. Martin-Neira (1998), Bistatic model of ocean scattering", *IEEE Trans. Antennas and Propagation,* 46(10): 1531–1541.

Raney, R.K. (2001a), Bistatic WITTEX Altimetry, *SRO-01-05,* Johns Hopkins University, Applied Physics Laboratory.

Raney, R. K. and D. L. Porter (2001b), WITTEX: An innovative multi-satellite radar altimeter constellation: A summary statement for the high-resolution ocean topography science WG, http://www.deos.tudelft.nl/gamble/docs/wittex.pdf.

Ridal, M. and N. Gustafsson (2003), Assimilation of ground-based GPS data within the European COST-716 action, Proc. Int. Workshop on GPS Meteorology, Tsukuba, Japan.

Rocken, C. et al. (1997), Verification of GPS/MET data in the neutral atmosphere, *J. Geophys. Res.,* 102, 29,849–29,866.

Rodriguez, E. and B.D. Pollard (2002), The measurement capabilities of wide-swath ocean altimeters, GAMBLE web site: http://www.deos.tudelft.nl/gamble/documents.shtml

Shoji, Y. et al. (2003), Improvement of GPS analysis of slant path delay by stacking one-way postfit phase residuals, Proc. Int. Workshop on GPS Meteorology, Tsukuba, Japan.

Sokolovskiy, S. (2002), Fundamentals of open-loop tracking of radio occultation signals in the lower troposphere, Proc. OPAC-1 International Workshop, Graz, Austria.

Sokolovskiy, S. V. (2001a), Modeling and inverting radio occultation signals in the moist troposphere, *Radio Sci.,* 36(3): 441–458,

Sokolovskiy, S. (2001b), Tracking tropospheric radio occultation signals from low Earth orbit, *Radio Sci.,* 36(3), doi:10.1029/1999RS002305.

Sokolovskiy, S (2003), Effect of superrefraction on inversions of radio occultation signals in the lower troposphere, *Radio Sci.,* 38(3), 1058, doi:10.1029/2002RS002728.

Stammer D., C. Wunsch and R. M. Ponte (2000), De-aliasing of global high frequency barotropic motions in altimeter observations *Geophys. Res. Lett.,* 27 (8): 1175–1178.

Treuhaft, R. N., Y. Chao, S. T. Lowe, L. E. Young, C. Zuffada and E. Cardellach (2003), Monitoring coastal eddy evolution with GPS altimetry, Proc. Int. Workshop on GPS Meteorology, Tsukuba, Japan.

Vinnikov and Grody (2003), Global warming trend of mean tropospheric temperature observed by satellites, *Science 302* (5643): 269–272.

Yunck, T. P., C.-H. Liu and R. Ware (2000), A history of GPS sounding, in Lee et al. (2000), pp. 1–20.

Yunck, T. P., G. F. Lindal and C.-H. Liu (1988), The role of GPS in precise Earth observation, *Proc. IEEE Position Loc. and Nav. Symp. (PLANS 88),* Orlando, FL, 251–258.

Zavorotny, V.U. and A. G. Voronovich (2000), Scattering of GPS signals from the ocean with wind remote sensing application, *IEEE Trans Geosci. and Rem. Sensing,* 8(2): 951–964.

Zou, X., H. Liu and H. Shao (2003), Further investigations on the assimilation of GPS occultation measurements, Proc. Int. Workshop on GPS Meteorology, Tsukuba, Japan.

Thomas P. Yunck, M/S 126–347, Jet Propulsion Laboratory, 4800 Oak Grove Drive, Pasadena, CA 91109.

George A. Hajj, M/S 238–600, Jet Propulsion Laboratory, 4800 Oak Grove Drive, Pasadena, CA 91109.

Dense GPS Array as a New Sensor of Seasonal Changes of Surface Loads

Kosuke Heki[1]

Division of Earth Rotation, National Astronomical Observatory, Mizusawa, Iwate, Japan

Global Positioning System (GPS) receivers have been deployed worldwide to study inter- and intraplate crustal motions and local deformations associated with earthquakes and volcanic activities. A dense array of GPS is also useful for studying seasonally changing load through periodic components in crustal movements. This article reviews observed and predicted seasonal crustal movements in the Japanese Islands, where both nationwide dense GPS array and meteorological sensor network are available. From comprehensive evaluation of various sources contributing to seasonal signals, the largest factor in Japan is found to be snow, weighing over 1000 kg per square meter in some regions. This is followed by various kinds of loads on the land area, such as atmosphere, soil moisture and water impoundment in reservoirs, and non-tidal ocean loads also cause certain seasonal signatures. Seasonal crustal deformations are calculated by synthesizing all these seasonal load changes, some of which are directly measured meteorologically and others are inferred through models. They are compared with real data observed by the dense GPS array in Japan, and their agreement was examined. The seasonal signals observed by GPS also include artifacts, such as those caused by atmospheric delay gradients and scale changes due to atmospheric refraction. We often discuss subtle crustal deformation signals, e.g. those associated with silent earthquakes, isolating them by removing secular and periodic components. Understanding seasonal signals and their interannual variability is crucial in removing these unwanted signals. The article discusses the Japanese case, but the methods proposed here will be useful worldwide to study seasonal mass redistributions. Dense GPS arrays may play a complementary role to satellite gravity missions in studying seasonal mass redistribution on the Earth in a regional scale.

1. INTRODUCTION

Ground stations (receiver and antenna) of Global Positioning System (GPS) are less expensive than other space geodetic techniques like Very Long Baseline Interfereometry (VLBI) and Satellite Laser Ranging (SLR), and suitable for dense deployment near plate boundaries. Relatively small amount of GPS raw data enables data transfer through internet or public telephone line for rapid data analysis in a central station. Crustal deformation data provided by Interferometric Synthetic Aperture Radar (InSAR) are dense in space, but sparse in time because they need almost identical orbits over studied areas. A high sampling rate is possible for GPS, which makes GPS and InSAR complementary techniques. Considering these benefits, a nationwide array of continuous GPS stations, GEONET

---

[1] Division of Earth and Planetary Science, Hokkaido University, Sapporo, Japan.

The State of the Planet: Frontiers and Challenges in Geophysics
Geophysical Monograph 150, IUGG Volume 19
Copyright 2004 by the International Union of Geodesy and Geophysics and the American Geophysical Union.
10.1029/150GM15

(GPS Earth Observation Network), has been established in Japan by the Geographical Survey Institute (GSI).

Comprehensive study of global-scale seasonal mass redistribution, including the Earth's response to it, has been done first to explain the observed annual polar motion with geophysical models of mass transports [*Munk and MacDonald*, 1960; *Chao and Au*, 1991]. Recent progress in space geodesy has enabled us to detect such changes in several different ways. Seasonal mass redistribution induces motion of the Earth's center of mass [*Dong et al.*, 1997], which can be estimated by analyzing satellite geodesy data [*Bouille et al.*, 2000; *Crétaux et al.*, 2002]. It also induces deformation of the Earth, which can be measured with worldwide deployed receivers of GPS. *Blewitt et al.* [2001] found seasonal "degree one" deformation (i.e. long-wavelength deformation that can be expressed by the degree one spherical harmonic functions) due to the mass exchange between the northern and southern hemispheres. Higher degree seasonal deformation signals have been detected by *Wu et al.* [2003]. Such mass exchanges can also be observed as the seasonally changing part of the lower order gravity potential coefficients [*Cazenave*, 1999]. Detection of the seasonal changes in higher degree/order gravity coefficients due to mass redistribution is the main target of recent and future satellite gravity missions [*Wahr et al.*, 1998; *Reigber*, 2003].

Comprehensive studies of potential seasonal displacement signal sources were carried out on a global scale by *Dong et al.* [2002]. They evaluated joint contributions of relatively well-known sources, and compared the predicted seasonal coordinate changes with real GPS data to investigate less well-known sources. In order to identify what caused differences between observed and calculated seasonal changes, GPS receivers must be densely deployed, which is not (and probably will not be) the case on a global scale. The Japanese regional network GEONET covers the whole country with inter-site distances of 20–30 km and is ideal for such mass redistribution studies. Seasonal signals have been found in the GEONET GPS site coordinate time series by *Murakami and Miyazaki* [2001]. They investigated their signal structures in detail, but failed to identify the mechanisms responsible for the variation. Later, seasonal signals in northeastern Honshu were interpreted to be largely due to snow loads along the western flanks of the backbone range by *Heki* [2001]. *Hatanaka* [2003a] pointed out the existence of arbitrary seasonal signals irrelevant to crustal deformation in GEONET, namely seasonal scale changes and spurious displacement signals due to atmospheric gradients. Comprehensive evaluation of seasonal loads in Japan is also important to establish a standard method to study seasonal mass redistribution with a dense GPS network in other parts of the world.

Modern satellite geodesy is based on precise determination of satellite orbits. For GPS, this task has been done by the IGS (International GPS Service) for a decade by tracking GPS satellites at fiducial stations whose positions and velocities are given in a terrestrial reference frame. Investigation of seasonal signals is important in assessing the stability of the terrestrial reference frame (and its influence on the errors in the determined orbits) at seasonal time scales. In active plate boundary regions, investigation of seasonal crustal movements is important for crustal deformation studies. *Ozawa et al.* [2002] tried to reveal moment release history of the Tokai slow event, a slow earthquake that started in the middle of 2000. The method they employed to isolate its transient signal is the removal of seasonal and secular components inferred from the time series before the onset of the event. However, such extrapolation is dangerous since it is not guaranteed that these components, especially seasonal changes, behave similarly from year to year. Many such silent events, including silent fault slip after normal earthquakes, are observed in Japan [*Heki et al.*, 1997; *Hirose et al.*, 1999; *Ozawa et al.*, 2003]. Understanding and predicting load-induced crustal deformation would be necessary to extract signals of such events correctly in near real time.

Activities of some kinds of earthquakes in Japan are known to be higher in certain seasons; e.g. interplate thrust events due to the subduction of the Philippine Sea Plate occur mainly in autumn and winter [*Ohtake and Nakahara*, 1999] and inland earthquakes in snowy areas occur more in spring and summer [*Okada*, 1982]. *Heki* [2003] investigated the causal relationship between the snow load and seasonal variation of inland seismicity in Japan by quantitatively estimating disturbances of the Coulomb failure function due to seasonally changing snow loads. He suggested that the removal of snow load associated with the spring thaw may encourage faulting. However, this discussion is based only on the snow load estimated earlier by *Heki* [2001], and lacks evaluation of other kinds of seasonally changing loads. Therefore it is important to evaluate all such loads in terms of their influence upon seismicity.

2. SEASONAL CHANGES IN LOADS

The study by *Heki* [2001] had three shortcomings: (1) only the snow was considered as the seasonally changing load, (2) only the northeastern Honshu was studied, and (3) only peak-to-peak amplitudes of the load were discussed. Although it is likely that snow load is the largest seasonally changing load in the northeastern Japan, here I perform a comprehensive study of five different kinds of loads on the entire Japanese islands. This will expand the scope of the study, i.e. the same approach may be applied for other countries where non-snow seasonal loads are important. I also consider the waveforms of the seasonal load changes to enable better correlation with

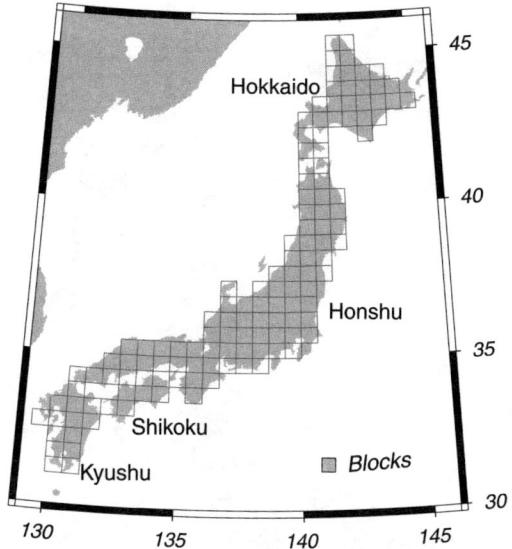

Figure 1. The Japanese Islands were approximated with 130 blocks as large as 0.5° in latitude and 0.67° in longitude. Monthly (atmosphere, soil moisture, and dam) or bimonthly (snow) values of loads were obtained for these blocks to calculate elastic response of the Earth (Figures 5–8). Ocean loading was calculated for blocks 1° x 1° in the oceanic area (Figure 9).

the observed signals. For these purposes, I divide the Japanese Islands into 130 blocks as large as 0.5° (latitude) and 0.67° (longitude) (Figure 1) and obtain the sum of loads at epochs set up every month (every half month for snow loads) for each kind of loads for each of the blocks. Displacement at each of the GEONET GPS stations by such loads can be calculated in a standard way as the sum of the Green's function describing the elastic response of the Earth to a point load given in *Farrell* [1972].

I did not consider viscous or viscoelastic effects because of the relatively short timescale of the crustal movement discussed in this article. Apart from load-induced displacements, *Hatanaka* [2003a] found that relatively large (and perhaps largely artificial) seasonal scale change of about five parts per billion exists in the GEONET data. I do not try to identify fully the mechanism giving rise to scale changes, but, later in this paper, I model its behavior empirically so that we can remove its contribution from GPS signals. I also exclude discussions on periodic crustal movements whose behaviors are well known and are already corrected in standard GPS data analysis software packages. They include solid earth tide, pole tides (readjustments of the Earth's equatorial bulge to the polar motion), and ocean tidal loading (the Earth's elastic response to ocean tides). Treatments of these corrections in the software package are given in *Hugentobler et al.* [2001], and those for the GEONET routine analysis are given in *Hatanaka* [2003b].

2.1. Snow Load

Snow depth data are obtained daily throughout the snowy area of the country by unmanned snow depth meter as a part of the Automated Meteorological Data Acquisition System (AMeDAS), run by Japan Meteorological Agency (JMA). Figure 2 shows the snow depth curves at three AMeDAS points and averages of the maximum snow depths at AMeDAS stations of the last seven winters. Although the AMeDAS stations are densely deployed, they tend to be located along inhabited valleys rather than in remote mountain areas (they need public electricity and telephone lines). Hence, the elevations of the sites are negatively biased from the national average (the average elevations of snow depth meters is ~200 m, about a half of the Japanese national average). Since snow depths are highly dependent on the elevation in Japan [e.g. *Kazama and Sawamoto*, 1995], this causes serious under estimation of snow depths.

To circumvent this problem, I performed the "altitude correction" for the snow depths by modeling the relationship between the measured snow depths and altitudes of the AMeDAS stations. In Figure 3, I plot snow depths against elevations of the individual AMeDAS sites in different geographical districts. Altitudes of most AMeDAS sites are lower than the district averages. For such districts, I basically used prefectures (about fifty in total) since they correspond to areas with relatively uniform meteorological setting surrounded by natural boundaries such as mountain ranges and seashore (from this viewpoint, blocks shown in Figure 1 are not suitable for this purpose). Average altitudes of Japanese prefectures are available at http://iide.hp.infoseek.co.jp/zatu/zatu6.htm. Some small neighboring prefectures with similar climates were amalgamated as one large district (e.g. Toyama Pref. and Ishikawa Pref.) while large prefectures composed of snowy highlands and snow-free lowlands were commonly divided into two districts (e.g. Fukushima Pref.).

Within such a district, snow depths increase approximately linearly with altitudes suggesting that the altitude is the main factor to control the amount of snow on a local scale. Slopes are slightly different between districts reflecting different geographical and meteorological settings. In this study, I performed linear regression for each district, and then the altitude correction for every AMeDAS site was obtained by multiplying the slope and the difference in altitudes between the AMeDAS site and the prefectural average. This correction was added to the raw AMeDAS snow depth value to make it more realistic. Then the daily snow depth data at the AMeDAS sites after district-wise altitude corrections were compiled to calculate block-wise bimonthly average snow depths. Scatter in Figure 3 reflect within-district differences of factors controlling snow, e.g. orientation of topography relative to

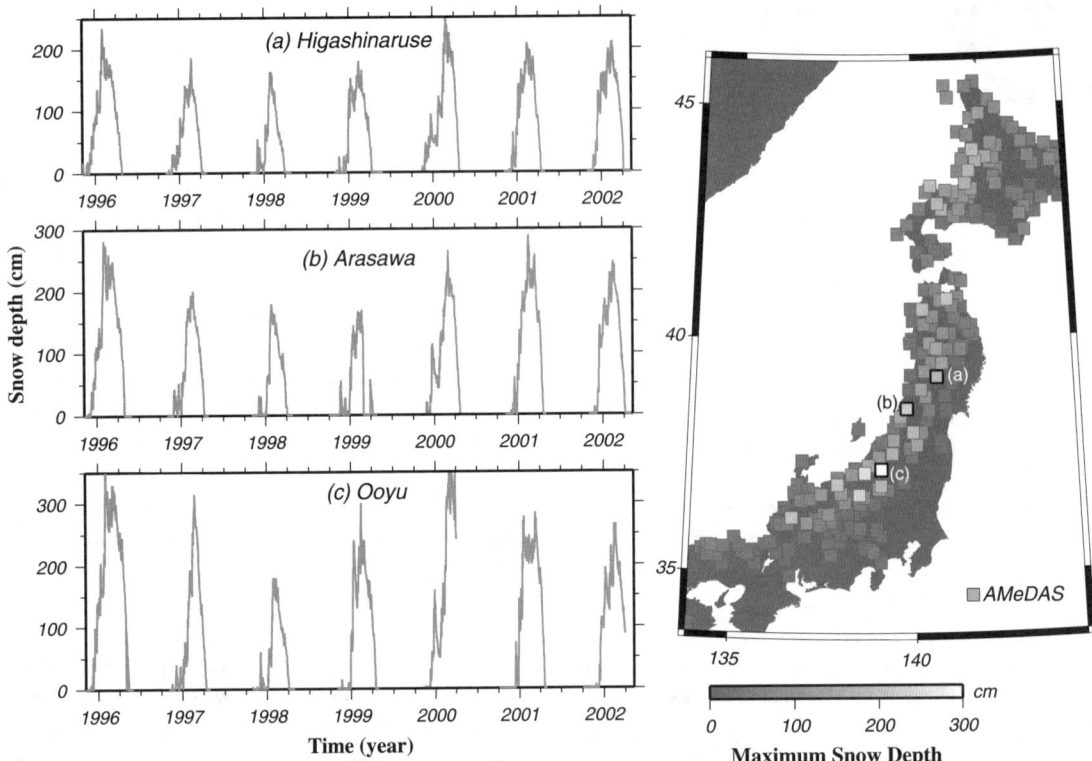

Figure 2. Daily snow depth measurements at three AMeDAS sites, (a) Higashinaruse (Akita Pref.), (b) Arasawa (Yamagata Pref.) and (c) Ooyu (Niigata Pref.), indicated as squares with bold frames in the map, from north to south, over the seven winter seasons 1996–2002 (left graphs), and distribution of the average of the maximum snow depths of the seven winters at the AMeDAS sites equipped with snow depth meters (right map).

main weather systems, overall size of mountain region and local topographic effects. This makes the uncertainties of the corrected snow depths 10–30% of the estimated values depending on the amount of scatter.

To convert snow depths to loads, we need the average density of the snow cover, which generally increases toward the end of winter due to compaction. *Heki* [2001] adopted 400 kg/m^3 as the instantaneous snow density at the time when snow loads are the largest (assumed as middle of March), but here I need its continuous temporal change from December to May to model bimonthly loads. I compiled snow density data collected at four points in Japan (Figure 4), namely Sapporo (snow cross section data were available annually from the Low Temperature Res. Inst., Hokkaido Univ., Sapporo, Hokkaido, e.g. *Hachikubo et al.* [1997]), Shinjo (Yamagata Pref.), Nagaoka (Niigata Pref.) (both observatories are run by Nat. Inst. Disaster Prevention and Earth Sci., and data are available at http://romalia.bosai.go.jp/seppyo/), and the top of Mt. Zao (boundary between Yamagata and Miyagi Prefs.) [*Yamaya et al.*, 1999].

Despite the difference in winter temperatures and altitudes, the average density values at these points are fairly uniform, starting with ~200 kg/m^3 and reaching ~400 kg/m^3 when the snows are deepest. In this study, I use the unified model of the snow density; it starts with 250 kg/m^3 at the beginning of December, increasing with time by ~40 kg/m^3/month, attaining 400 kg/m^3 at the end of March. This model is shown as a line in each panel of Figure 4. By multiplying the snow depth by such time-dependent density, I obtained bimonthly blockwise snow loads. Figure 4 shows that the observed snow densities scatter around the model by a few tens of kg/m^3, which will cause additional error of ~10% to the estimated snow loads. Figure 5 shows differences between the maximum and minimum (zero in this case) snow loads for the blocks. Large snow loads are distributed along the northwestern flanks of the backbone range, while they are almost absent along the Pacific coast and in the southwestern Japan. As shown in the insets of Figure 5, snow loads are characterize by strong peak in winter and a relatively flat period during the rest of the year.

2.2. Atmospheric Loads

Changes in atmospheric pressure and the corresponding displacements of space geodetic stations have been studied for

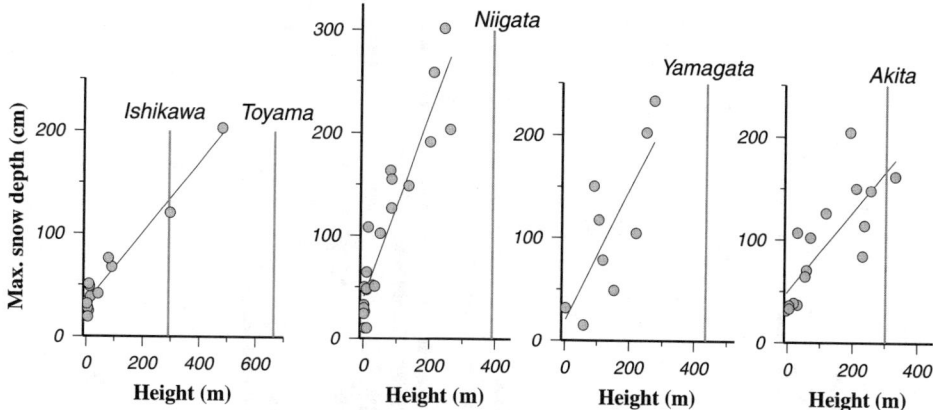

Figure 3. Relationship between the average maximum snow depths (same as Figure 2 right) and elevations of the AMeDAS snow depth meter sites (best-fit lines are shown in black). They were plotted for snowy prefectures along the Japan Sea coast of Honshu (Akita, Yamagata, Niigata, Toyama and Ishikawa Prefectures, from northeast to southwest). Ishikawa and Toyama are plotted together in one diagram. Vertical gray lines show average elevation for these prefectures, and dots denote AMeDAS stations. Most AMeDAS sites are located below average altitudes, and realistic snow depths of the entire prefectures have to be inferred considering their altitude dependence.

many years [*van Dam et al.*, 1994]. Response of the solid earth to the atmospheric load obeys the same theory as the snow. Atmospheric load, unlike snow, acts both on the land and ocean. However, relatively long period (e.g. annual frequency) pressure changes do not exert on the sea floor since they are effectively compensated by natural adjustment of sea surface height (inverted barometer hypothesis). The coefficient to convert pressure changes to displacements hence depends on the shoreline geometry around the site. In this study I assume that atmospheric load works only on the land area. I model monthly mean pressures for individual blocks (Figure 1), using average meteorological data over the last few years at 80 sites available in *Naional Astronomical Observatory* [2002]. Pressure values in blocks without data points

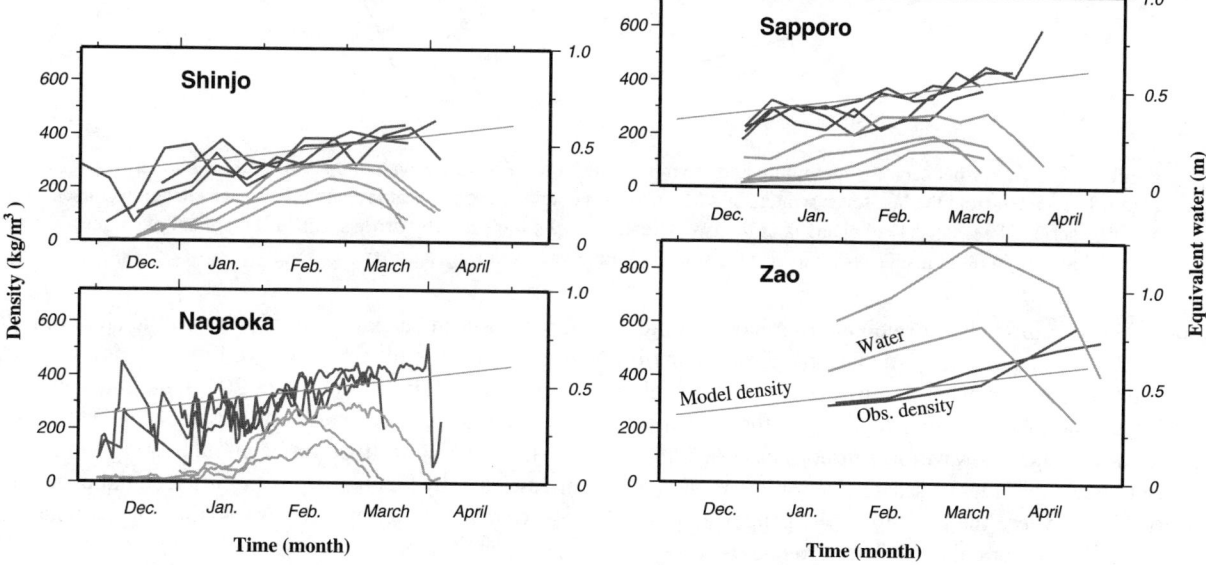

Figure 4. Light and dark gray lines indicate the time series of the total amount of snow in terms of equivalent water depth (scale to the right) and average density of snow (scale to the left) over winter months. The data were taken at four sites, namely Sapporo (Hokkaido), Zao (Miyagi-Yamagata Pref. boundary), Shinjo (Yamagata Pref.) and Nagaoka (Niigata Pref.). Years when the data were obtained are different between the sites. It is shown that snow mass is the largest in late February or early March, and snow density increases monotonously with time. Model density shown by gray straight lines is assumed as the common standard.

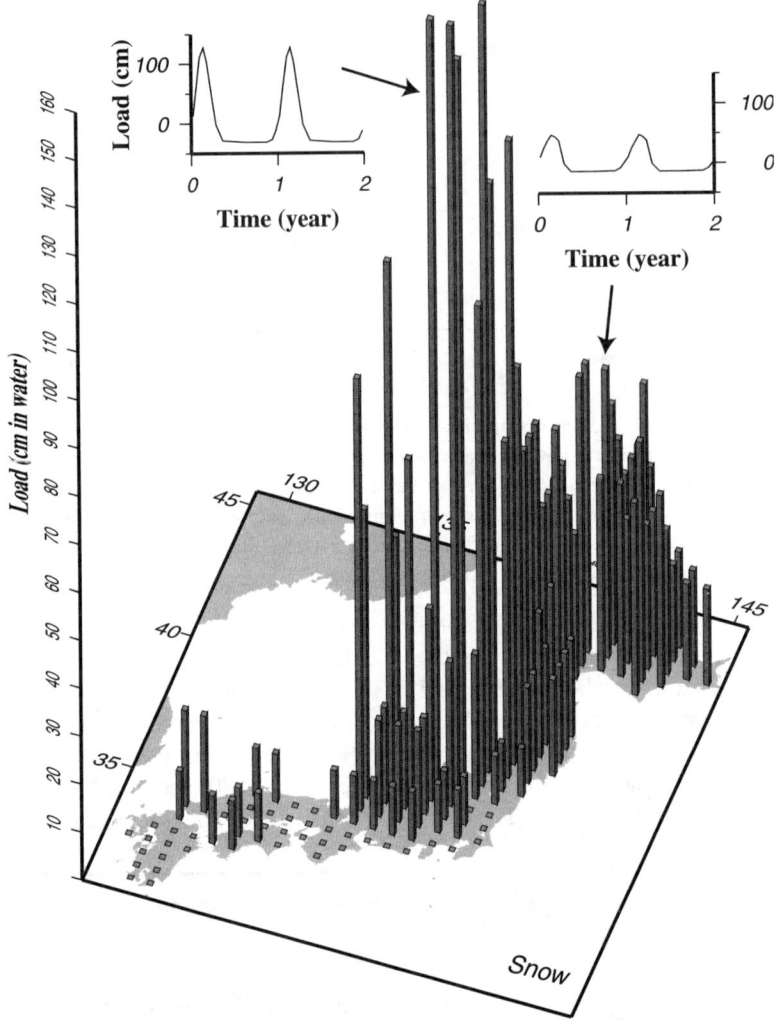

Figure 5. Difference between the maximum and the minimum (zero in this case) value of snow load for blocks shown in Figure 1, obtained from the last seven winters of AMeDAS data converted to those at prefectural average elevations. Insets are time series of seasonal load changes at the two selected blocks indicated by arrows. Since the annual averages are adjusted to zero in these inset graphs, loads take small negative values during summer.

were interpolated using values of neighboring blocks. I did not use AMeDAS data because of smaller spatial variation of the pressure than snow. In general, pressure tends to be high in winter and low in summer in Japan, and the seasonal change has a complicated waveform in northeastern Japan (Figure 6 inset). Amplitudes of seasonal variation are larger in western Japan, where the peak-to-peak changes amount to 1.5 kPa (~15 cm water column). They are less than 1 kPa in northeastern Japan. Figure 6 shows distribution of peak-to-peak atmospheric load changes.

Short-term intra-seasonal pressure variations often exceed annual variations in amplitudes. For example, the 21st typhoon of 2002 hit the Kanto Area on 1st October, and brought several kPa pressure dip lasting for a few hours.

Crustal strain changes caused by such transient removal of atmospheric load were observed by borehole-type three-component strainmeter [*NIED*, 2003], but the signature was not obvious in coordinate time series of GPS stations made of daily averages. It would be an interesting future issue to investigate, by kinematic positioning technique, how far the inverted barometer hypothesis remains valid for such short-term disturbances.

2.3. Soil Moisture

Seasonal variation of continental water storage, including snowpack, soil moisture, and groundwater, causes mm to cm level vertical crustal movements [*van Dam et al.*, 2001]. Here

I model the soil moisture changes in individual blocks (Figure 1), with a simple model constructed using the following three quantities, i.e., (1) potential evapotranspiration (PET), (2) precipitation, (3) water-holding capacity in soil. Following a simple model by *Milly* [1994], soil water is assumed to increase by precipitation and decrease by evapotranspiration. When the water exceeds the storage capacity of soil, the excess amount runs off keeping soil water at its maximum level. When the PET is larger than the available soil water, all soil water is lost due to evapotranspiration and the water level becomes zero.

Such calculations were performed at 80 points where meteorological data are available with monthly time steps. The monthly PET was calculated as a function of temperature and length of daytime following the formula given in *Thornthwaite* [1948]. Monthly mean temperature and precipitation were taken from *National Astronomical Observatory* [2002]. Here I did not use the AMeDAS data, either, considering rather smooth spatial variation of the meteorological data. Water storage capacity depends on the type and thickness of the soil, and is basically unknown. Precipitation exceeds PET almost all time in most places in Japan, and the soil water level is maintained at levels close to the water storage capacity. Thus, unless the storage capacity is unrealistically small, the soil water never runs out and we could calculate seasonal soil water changes without information on the capacity.

Simulation of monthly soil moisture values for individual blocks showed that the load changes are smaller than those related to the atmosphere in most regions (Figure 7). The largest variation occurs in the inland area of the southwestern Japan and Hokkaido, where summer precipitation falls below PET and soil water levels temporarily become lower than maxima for a few months. In cold regions, input of solid precipitation (e.g. snow) into soil could delay until the spring thaw. Such a delay is neglected in this study since it causes delay in runoff rather than in soil moisture change, because soils are nearly saturated with water during winter and spring. Change in groundwater (water in aquifer below soil layer) may locally cause vertical crustal movements in Japan [*Munekane et al.*, 2003], but this was not modeled in this study because its seasonal change is not well known. Impoundments during summer months of rice fields (usually as deep as ten centimeters or so), which occupy about ten percent of the land, are not considered, either. Its influence would not exceed those considered in this article.

2.4. Water Impoundment in Reservoirs

Water mass reserved in increasing number of reservoirs is large enough to change the low degree gravity coefficients [*Chao*, 1995]. Impoundment of several largest dams, e.g. the Three Gorges Dam, China, causes crustal deformation large enough to be detected by space geodesy [*Wang et al.*, 2002]. Japanese dams are smaller; even the largest one (the Okutadami dam, Fukushima Pref.) has the capacity of only 600 million tons, two orders of magnitudes smaller than the world largest

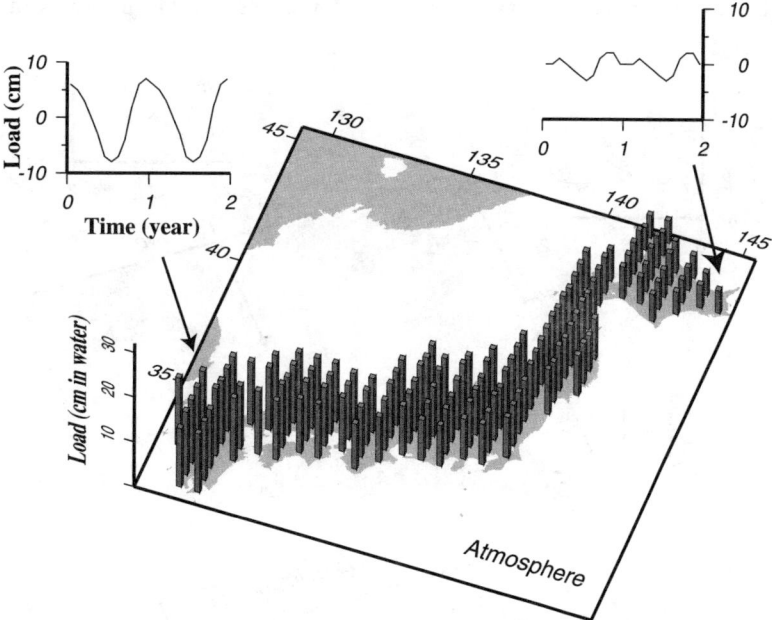

Figure 6. Peak-to-peak amplitudes of seasonal atmospheric pressure changes at the blocks, with two insets showing variations at two selected blocks.

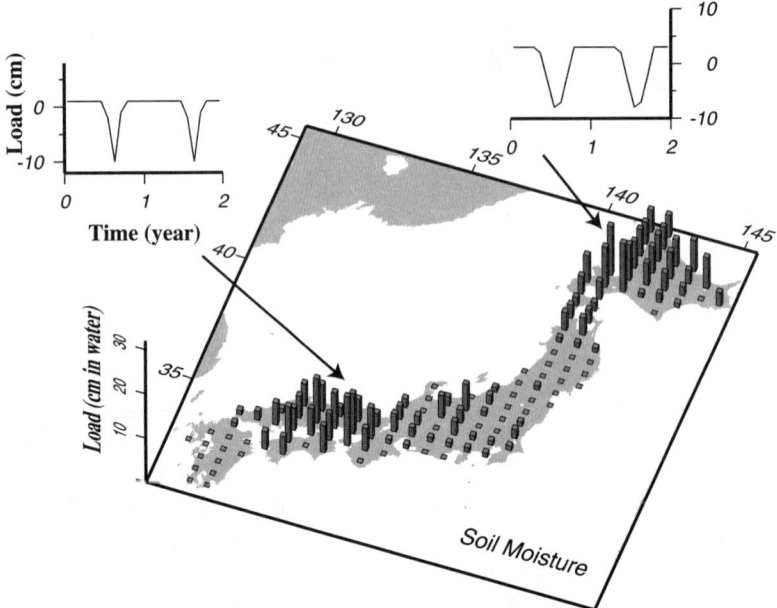

Figure 7. Peak-to-peak amplitudes of seasonal soil moisture load changes at the blocks.

dams. However, dams are densely distributed in Japan, and their total water storage may be worth considering. I compiled water capacities, coordinates of 546 dams with capacities larger than five million tons (information available from the Japan Dam Foundation, http://wwwsoc.nii.ac.jp/jdf/Dambinran/binran/TopIndex.html).

Pattern of seasonal change of water level varies from dam to dam as well as from year to year. Roughly speaking, there are two types of patterns, i.e. those of reservoirs in snowy areas and in snow-free areas. The latter is characterized by one peak, i.e. high water level in winter and low level in summer (Figure 8 left inset). The former shows more complicated patterns. Water levels decrease during snow-covered period, and increase in spring thaw attaining the first peak of a year. Then, they fall again in summer and rise during autumn toward the second peak around the onset of snowy season (Figure 8 right inset). I adopted the 9 years average seasonal water level change curve of the reservoirs in the Kanto District available in

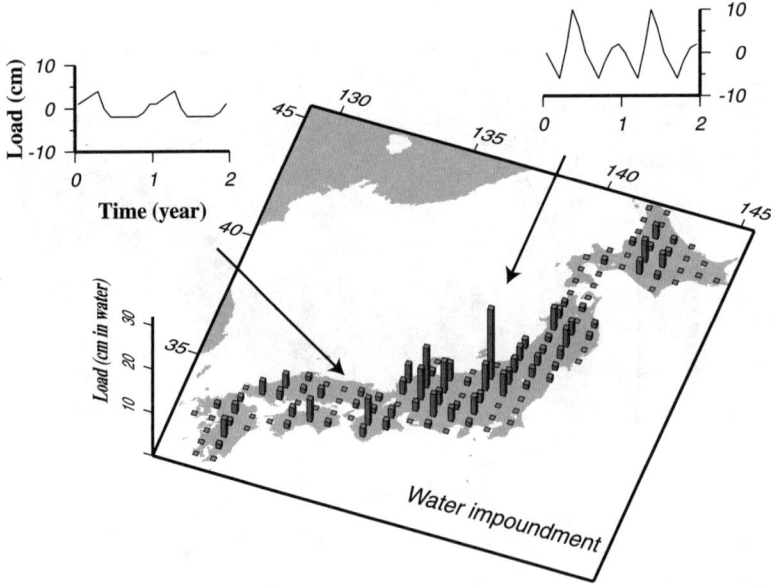

Figure 8. Peak-to-peak amplitudes of seasonal load changes by water impoundment of reservoirs at the blocks.

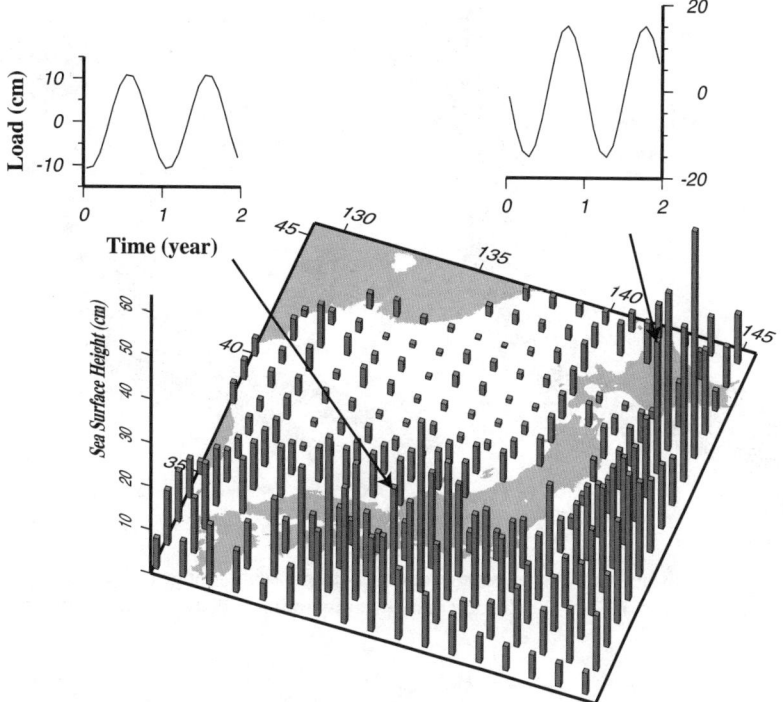

Figure 9. Peak-to-peak amplitudes of seasonal ocean load changes at the 1° x 1° blocks in the oceanic area around Japan, with two insets showing variations at two selected blocks.

http://www.ktr.mlit.go.jp/kyoku/1_topics/16_mzshi/mizu_02.htm. The peak-to-peak amplitude of water level variation of a reservoir was adjusted to a half of its total capacity. I classified Japanese reservoirs into the two types, and calculated monthly sums of the water loads at these reservoirs for individual blocks. As seen in Figure 8, the variation is large in some mountainous blocks in central Japan, often exceeding 10 cm in within-block averages. However, they are much less than the two major contributors, i.e., snow and atmosphere.

2.5. Ocean Loading

Tidal ocean loading causes half-daily crustal movement above the detection level of GEONET [*Hatanaka et al.*, 2001] but is corrected in the GEONET solutions [*Hatanaka*, 2003b]. Sea surface heights (SSH) change seasonally such that the sea surface in September is higher than in March by 20 cm or more [*Okada*, 1982; *Kato*, 1983]. This includes thermal expansion irrelevant to load changes, which must be removed to consider seasonal changes of pressure at the ocean bottom. Annual changes in gravity are considered to be the sum of the contributions of polar motion, solid earth tide, and the effect of sea water mass changes [*Sato et al.*, 2001]. They corrected the SSH seasonal variation in the Parallel Ocean Climate Model (POCM) [*Stammer et al.*, 1996] for thermal expansion assuming various thermal steric coefficients. The observed annual gravity changes best coincide with predictions when the coefficient is 6.0 mm/degree. They removed thermal expansion contribution using this coefficient and the seasonal variation of sea surface temperature in POCM, and found residual annual variation of SSH with amplitudes exceeding 10 cm around Japan (Figure 3 of *Sato et al.*, [2001]). Peak-to-peak seasonal variations of sea water mass at the 1° x 1° squares around Japan show a large increase in ocean loads in the Pacific Ocean in summer while they are almost stationary throughout a year in the Japan Sea (Figure 9). This gives rise to seasonal translation of the whole Japanese Islands by about 1 mm, but crustal deformation within Japan does not exceed those by soil moisture and water impoundment in reservoirs. Ocean loading models still have large uncertainties and we would need their further refinement, e.g. by satellite measurements of seasonal gravity changes [*Wahr et al.*, 1998] and by deploying ocean bottom pressure gauges [*Fujimoto et al.*, 2003].

3. SYNTHESIS

3.1. Seasonal Scale Change

Geodetic positioning often suffers from the scale problem, namely an ambiguity in uniform expansion or contraction of the whole network. It is typically seen in classical triangulation surveys; measured length of a certain baseline determines

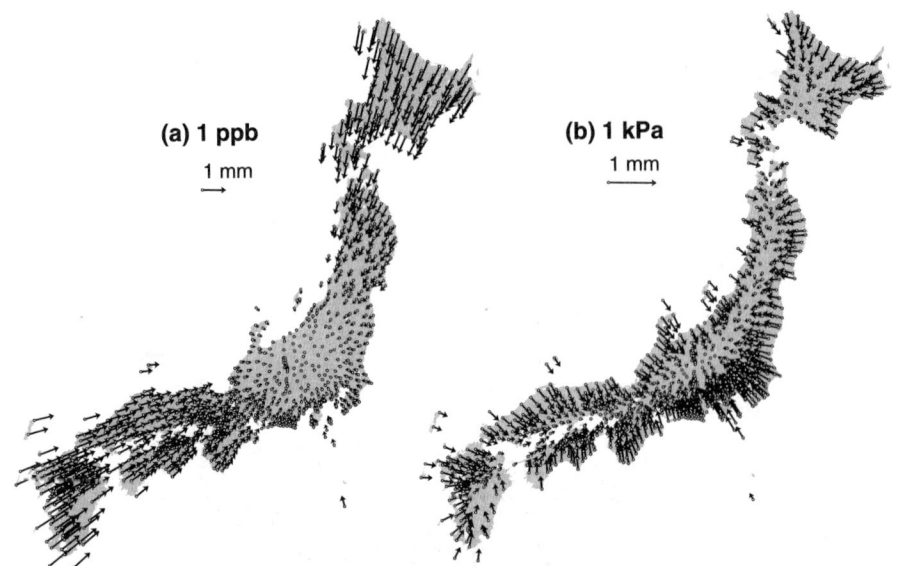

Figure 10. Comparison of seasonal coordinate changes due to isotropic scale change (1 ppb contraction) (a), and load (1 kPa uniform load on land area) (b). They are characterized by arc-parallel and arc-normal contraction, respectively, and can be easily discriminated.

the scale of the whole network. Even in space geodesy, the scale is controlled by several quantities, e.g. the speed of light, the Earth's GM (product of the universal gravity constant and the Earth's mass), error in GPS center-of-mass to antenna phase-center offsets, and many other factors [*Beutler et al.*, 1988]. In most cases, such scale changes are considered to be artifacts irrelevant to real crustal deformation, and scales have been readjusted in combining positioning results obtained with different techniques to establish global terrestrial reference frames [e.g. *Altamimi et al*, 2002].

In Japan, *Hatanaka* [2003a] found that the whole GPS network repeats annual cycles of uniform expansion in summer and contraction in winter (seasonal scale change) of about 5–6 part per billion (ppb) in addition to real crustal deformation due to seasonally changing loads. Isotropic contraction causes substantial shortening of arc-parallel baselines as shown in Figure 10a. Because Japan is an island arc elongated in northeast–southwest, load uniformly distributed over land mainly gives rise to arc-normal contraction (Figure 10b). In order that surface loads cause such a uniform contraction, there should be an equal amount of seasonal changes of loads over the land and seafloor. This is quite unlikely. The seasonal degree one deformation [*Blewitt et al,* 2001] causes almost uniform regional expansion (contraction) in summer (winter), but this is no more than 1 ppb in Japan. Therefore substantial part of the 5–6 ppb scale change in GEONET would not be seasonal deformation of the real Earth, but an artifact.

GEONET site coordinates are estimated using the Tsukuba station, Ibaraki, as the fixed reference. *Munekane et al.* [2003] reported that this station repeats seasonal vertical movement with peak-to-peak amplitude of ~2 cm due to extraction of groundwater for agricultural purposes in spring. *Beutler et al.* [1988] suggests that such a vertical coordinate error of the fixed site causes a ~0.6 ppb seasonal scale change (contraction in winter). This effect, together with the seasonal degree one deformation [*Blewitt et al*, 2001], accounts for ~20% of the observed scale changes. In addition the pole tide could give rise to annual and Chandler period height change with amplitude of ~1 cm. However, it does not cause scale changes because they are corrected beforehand in the GEONET solution [*Hatanaka*, 2003b], following the IERS standards [*McCarthy*, 1996]. *Beutler et al.* [1988] also suggests that an error in absolute atmospheric delay causes scale changes. *Hatanaka* [2003a] considered that this possibly causes the unexplained part of the scale change. He plotted the difference between wet atmospheric zenith delay obtained by GPS and by water vapor profile obtained by radiosonde, and found that the seasonal changes in the differences are almost consistent with the observed scale changes. However, there is no a-priori reason to consider only the GPS-based water vapor data wrong, and it is premature to conclude that the scale problem is solved.

In the present study, instead of trying to clarify the mechanisms for the scale change, I apply an empirical model according to the following procedure. I take out ten stations which are well distributed over the country and free from transient signatures related to recent seismic or volcanic activities. I remove secular components from their daily coor-

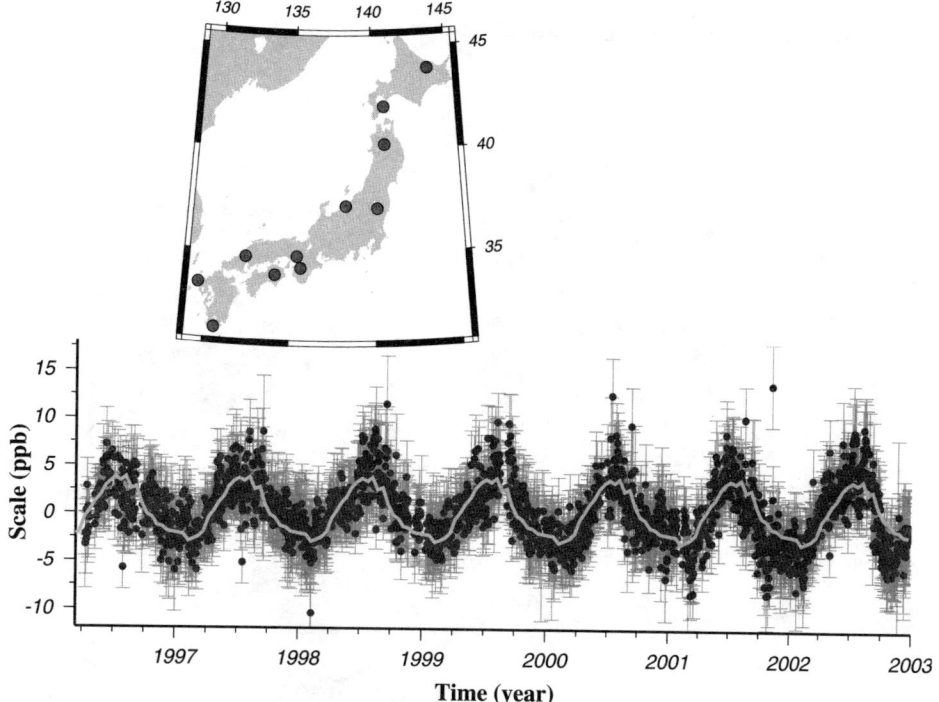

Figure 11. Daily scale values were estimated using time series of site coordinates at ten sites in the inset map after removing secular trends. Average seasonal scale variation composed of bimonthly values (thick gray curve) was estimated by stacking these scale changes over seven years.

dinate time series. Then the residual coordinate time series 1996–2003 are used to estimate daily values of scales (Figure 11). Seasonal behavior of scale is fairly uniform year after year, and from them I obtained the average seasonal changes composed of bi-monthly scale values. This is possible because the scale change signature is very "different" from load signature as shown in Figure 10. Such a "difference" comes from the elongated shape of the Japanese Islands, and so the discrimination between scale and load signatures could be difficult in other countries.

In addition to seasonal scale changes, *Hatanaka* [2003a] pointed out atmospheric delay gradients [*MacMillan*, 1995; *Miyazaki et al.*, 2003] are significant in the north–south direction at stations along the Pacific coast of the central and southwestern Japan. They cause apparent northward movement of southern stations in winter relative to the Japan Sea side of the arc. This contributes to arc-normal contraction in southwestern Japan in winter, and enhances signals of seasonally changing loads. By repeating data analyses with and without estimation of atmospheric gradient parameters, *Hatanaka* [2003a] found that this contribution may amount up to a few mm. Because GSI does not estimate atmospheric gradients in their official GEONET solution at the moment [*Hatanaka*, 2003b], the current solution includes apparent seasonal crustal movement due to this effect.

3.2. Sum of Contributions

In Figure 12, I show the sum of the displacements of GPS sites in winter relative to summer of the five seasonal loads considered in this study (Figures 5–9). In this figure, a station on the Japan Sea coast (the Oogata station) is held fixed. Arc-normal contraction in winter of up to 4–5 mm is seen in the northeastern Japan. Since the fixed station subsides in winter due to snow load, GPS sites in snow-free southwestern Japan apparently go up in winter. These features are consistent with the basic pattern of seasonal crustal deformation in northeastern Honshu discussed in *Heki* [2001] although he considered only snow loads. Snow-induced signals are thus predominant as long as relatively small regions (up to a few hundreds of kilometers level) in the northeastern Japan are concerned. Actual GPS data include components coming from the seasonal scale changes, whose contribution becomes predominant as the studied area expands.

Examples showing relative importance of various kinds of loads and scale are shown in Figure 13. Contribution of scale change simply depends on baseline lengths; those longer than 1000 km vary by 5–6 mm by seasonal scale changes, which is larger than any real load signals. For example, major source of seasonal changes for a long baseline connecting Hokkaido and Kyushu (Figure 13a) is the scale change, with additional

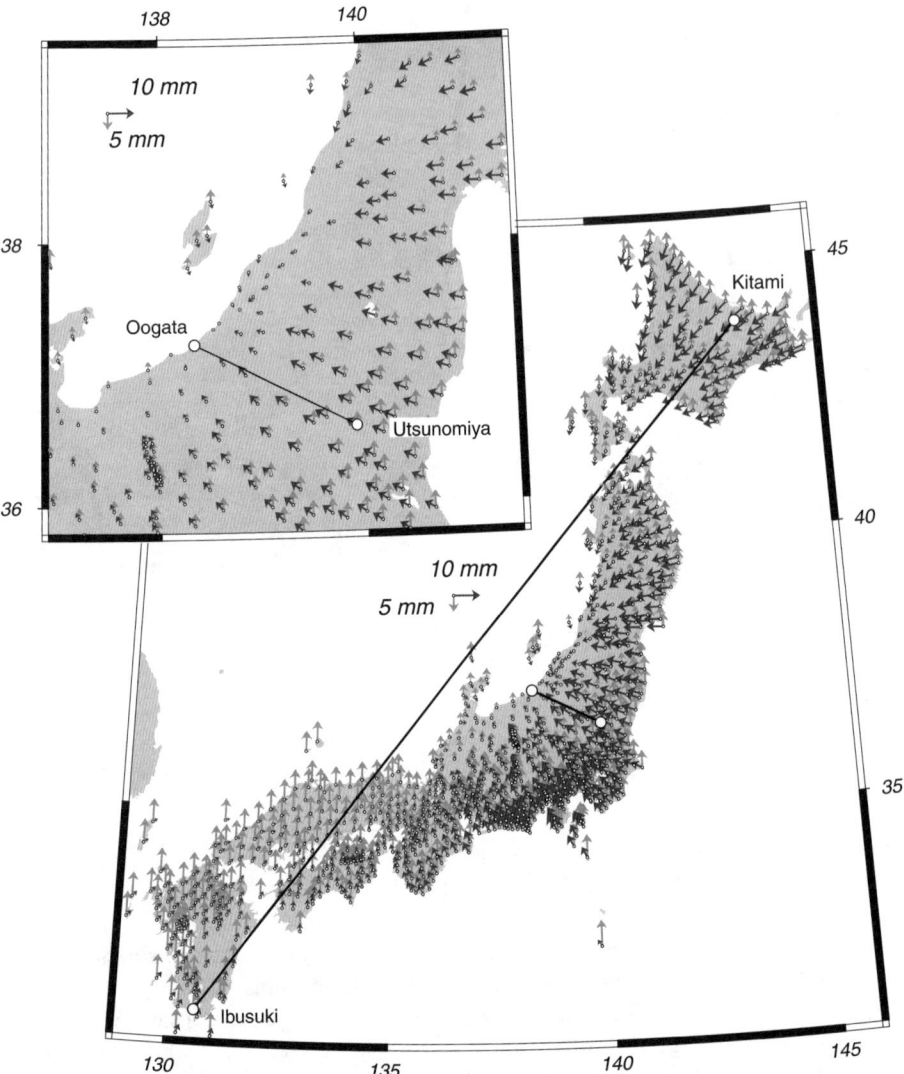

Figure 12. Seasonal displacement (early February position with respect to the early August position) due to joint contribution of snow, atmosphere, soil moisture, water impoundment in dam and ocean loads. The Oogata station is held fixed. Contributions from individual sources are compared in Figure 13 for lengths of the two baselines shown in the map.

contributions of snow, ocean loading and atmosphere. In a relatively short baseline in central Japan (Figure 13b) contribution of snow is the largest, being followed by scale change and atmosphere. Contribution of snow loads is characterized by sharp negative peaks in winter and relatively flat period lasting from spring to autumn. Other components, mainly scale and atmospheric load, bring broad positive peak in summer making the total waveform more like a sine curve.

Soil moisture, water impoundment in reservoirs, and nontidal ocean loading were relatively minor in Japan, but these factors could be major contributors in other parts of the world. For example, in relatively dry areas with precipitations concentrating on a certain time of the year, soil moisture may change by several tens of centimeters as a equivalent water column. In an oceanic island, seasonal changes in ocean circulation could cause significant displacement of the island. In this article I have tried to explain the observed seasonal signals in Japan. However, the same approach to study seasonally changing loads could be used in other parts of the world if there are networks of geophysical (meteorological) sensors with quality and density similar to the present case.

3.3. Comparison With GPS Observations

Here I compare synthesized seasonal signals with those observed by GEONET. I employ the newest solution based on the improved analysis strategy [*Hatanaka*, 2003b]. Figure 14 compares seasonal signals in the lengths of five arc-normal baselines. They are relatively short, and real crustal deformation signals are predominant like in Figure 13b. Several features seen in Figure 13b, e.g. strong negative peak in winter and broad positive peak in summer, are reproduced to some extent in Figure 14. The southernmost baseline (baseline E, Figure 14) show somewhat larger seasonal variations than predicted. Part of such difference may be an artifact coming from the north–south atmospheric gradient of the southern station as discussed in section 3-1. Arc-parallel baselines are shown in Figure 15. They are longer and are more affected by scale changes than those in Figure 14, although snow load has significant contributions in northern Japan. The observed seasonal signals are also fairly consistent with predicted curves; northern baselines have sharper winter peaks reflecting the snow signature, and such peaks get broader toward the south reflecting the dominance of the scale change.

Such features are confirmed by the behavior of the root-mean-squares (rms) for various cases shown to the right of Figures 14 and 15. There I show the two extremes, i.e. rms for the case of best-fit seasonal (annual plus semiannual) change curves (black column), and rms for the case without seasonal signals (rightmost gray column). Cases with a-priori seasonal curves considering all or part of the six contributors (scale, snow, atmosphere, soil moisture, water impoundment in reservoirs, and ocean loads) are shown between these two extremes.

In spite of the large uncertainties in the snow load as explained in section 2-1, the a-priori corrections give almost as small rms as the a-posteriori best-fit seasonal changes, suggesting that the current approach has been acceptable.

While baseline length changes in Figures 14 and 15 mainly reflect horizontal coordinates and include significant amount of scale change contributions, vertical components are less affected by scale changes and respond mainly to loads. Figure 16 shows seasonal vertical movements at four sites in snow covered area and one in snow-free area, relative to a fixed station in a snow-free area. Although the signals are noisier than the baseline lengths shown in Figures 14 and 15, winter subsidence of about 1 cm can be seen at snow-covered stations. We also notice unexpected strong summer peaks in observed signals, which makes the amplitudes larger than predicted. Such tendency is not clear in horizontal components as seen in Figures 14 and 15, and is possibly of atmospheric origin. Relative atmospheric delay errors tend to give rise to height errors [*Beutler et al.*, 1988]. The four sites in a snowy area in Figure 16 are at much higher elevations than the fixed reference. Large height difference, i.e. large contribution of tropospheric delay to the site-to-site phase differences, makes the estimated vertical coordinates more susceptible to systematic errors. In addition to the height difference, existence of large atmospheric delay gradients in summer may have increased the vertical coordinate errors [*Iwabuchi et al.*, 2003; *Miyazaki et al.*, 2003].

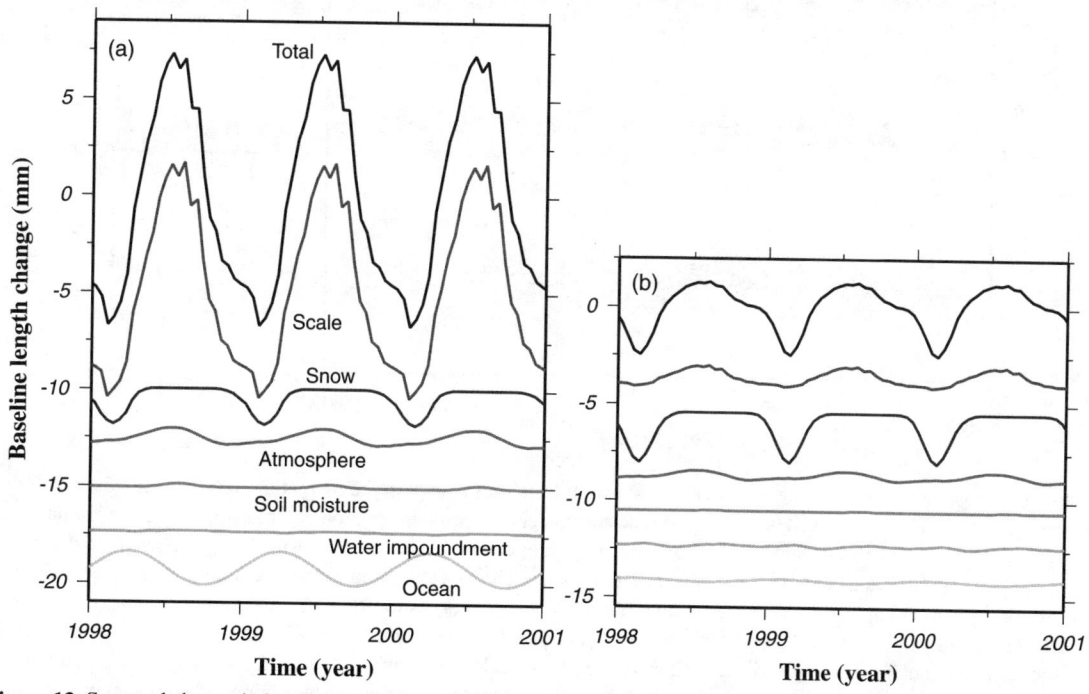

Figure 13. Seasonal change in lengths for two baselines shown in Figure 12, Kitami-Ibusuki (a) and Oogata-Utsunomiya (b). Contributions from scale, snow, atmosphere, soil moisture, water impoundment and ocean are shown from top to bottom.

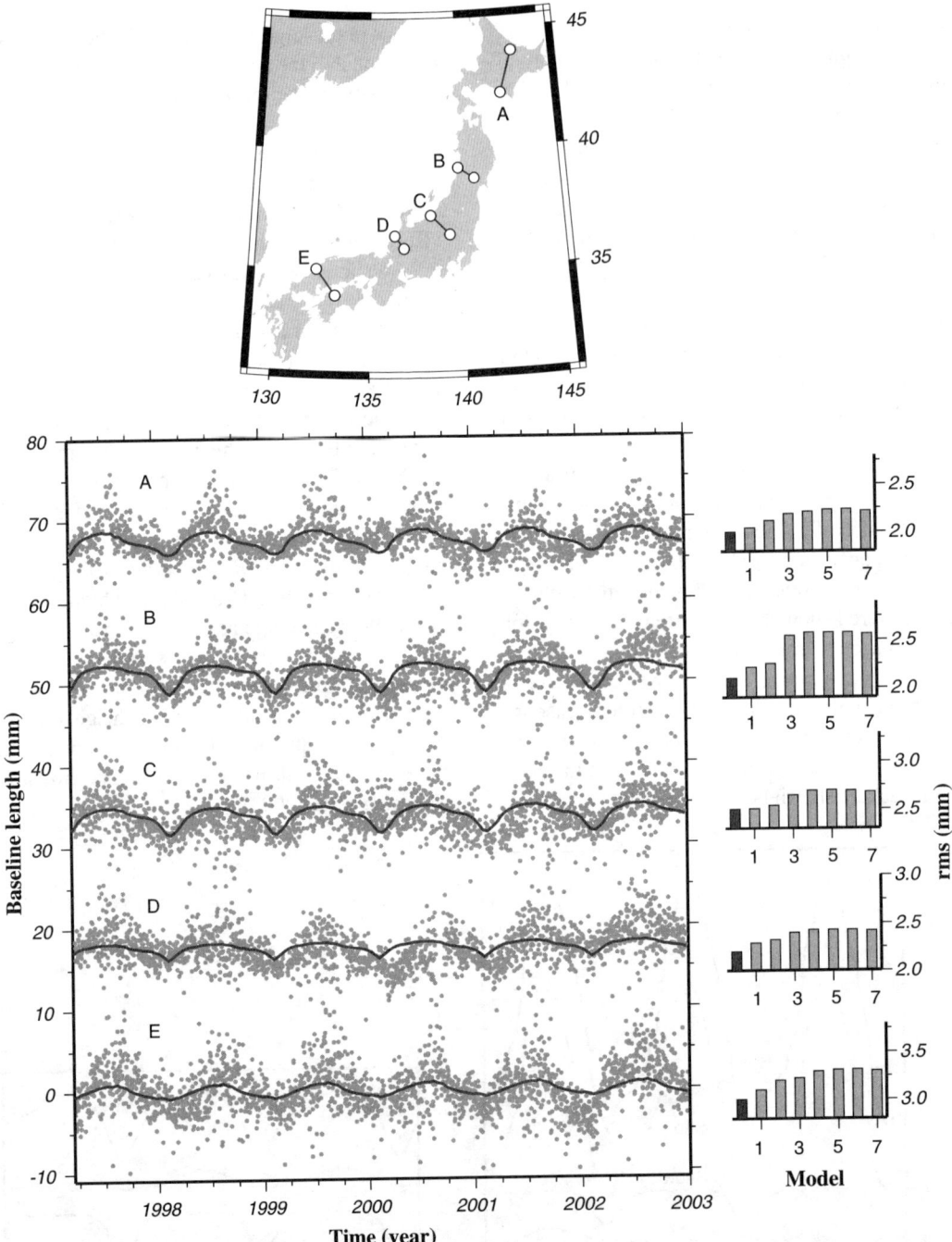

Figure 14. Seasonal change in lengths for five arc-normal baselines shown in the map at the top. Dots are daily solution of GPS (secular trends are removed beforehand) while curves denote seasonal variation curves synthesized in a similar manner to Figure 13. Diagrams indicated to the right for individual baseline data show root-mean-square (rms) of the differences between the GPS data and the synthesized model values. The black column to the left denotes rms values obtained a-posteriori by estimating best-fit seasonal (annual plus semiannual) curves for the GPS data. Gray columns (model number 1 to 7) are rms values obtained a-priori by assuming (1) all of the seven contributors, (2) without scale, (3) without scale and snow, (4) without scale, snow and atmosphere, (5) without scale, snow, atmosphere and soil moisture (i.e. only water impoundment and ocean), (6) without scale, snow, atmosphere, soil moisture, and water impoundment (i.e. only ocean), (7) without scale, snow, atmosphere, soil moisture, water impoundment and ocean (i.e. no seasonal variation).

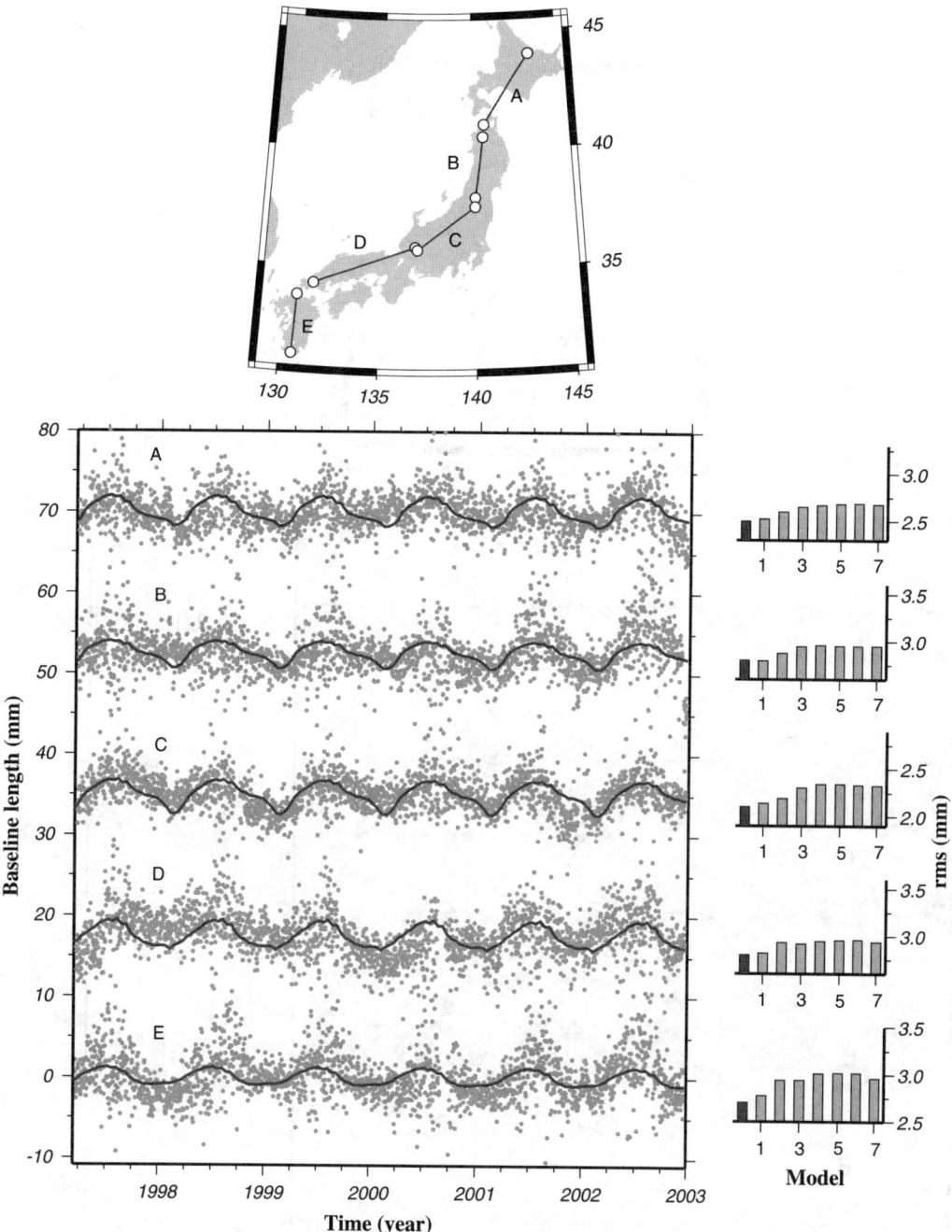

Figure 15. Seasonal change in lengths for five arc-parallel baselines shown in the map at the top. See the caption of Figure 14 for detail.

4. DISCUSSION

4.1. Interannual Changes and Isolation of Transient Signals

The depths of snow cover, the largest seasonally changing load in Japan, vary from year to year by tens of percents. The AMeDAS record (Plate 1) indicates that the snow depths were relatively small during the three winters from the end of 1996 to the early 1999, and has been relatively large for the next three winters. Such tendency might be present in amplitudes of wintry negative peaks in Figures 14–16 but its significance seems marginal. Change in amplitudes of annual signals in

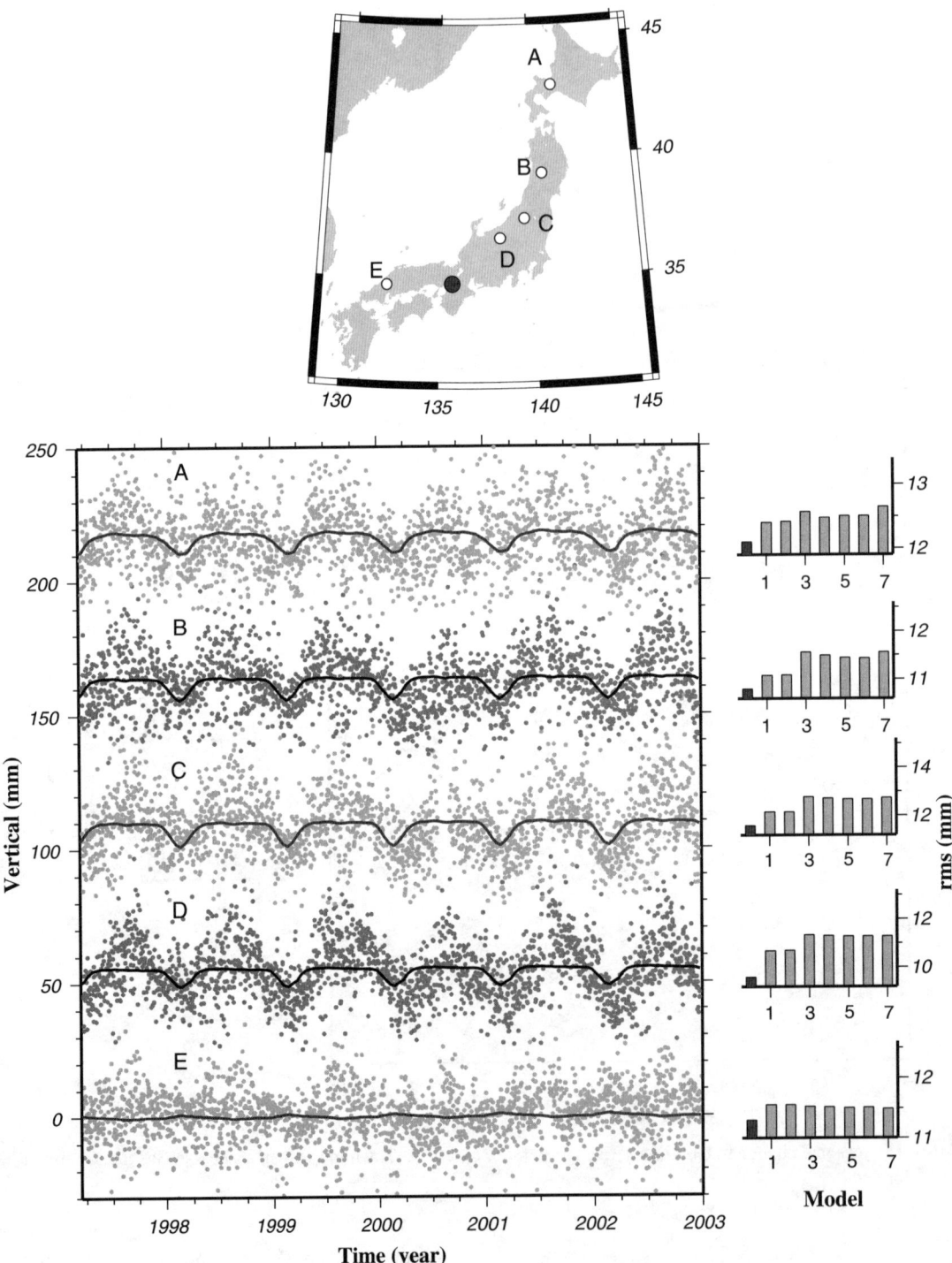

Figure 16. Seasonal change in vertical positions for five GPS stations (white circles in the map at the top) relative to a reference station (black circle in the map). Dots are GPS daily solutions in which secular components are removed (A,C,E and B,D are shown with different darkness for visual clarity). Curves denote synthesized seasonal variations. See the caption of Figure 14 for detail.

GPS data is reported by *Yamamoto* [2003]; he suggested that the amplitudes increased at the beginning of 2000. *Kedar et al.* [2003] discussed the effect of second order ionospheric delay, which is not considered yet by standard GPS data analysis software packages, on GPS positioning. The 11 years solar activity period has been increasing the Total Electron Content (TEC) during the time period shown in Figures 14–16. This might be related to the increasing amplitudes of seasonal signals.

As mentioned before, investigation of seasonal signals is important to isolate transient crustal deformation signals out of the mixture of secular, transient and periodic crustal movements. Figure 17 shows the time series of east–southeastward component of the Hamamatsu station coordinate relative to Oogata. Hamamatsu is just above the focal region of the ongoing Tokai slow event [*Ozawa et al., 2002*], and Oogata is fairly much affected by seasonal crustal movement due to snow load. They result in complicated time series made up of multiple components. Sudden departure in the middle of 2000 from the normal trend corresponds to the onset of the Tokai event. Such a transient signal is better perceived by removing secular components (Figure 17 left, middle), or both secular and seasonal components (Figure 17 left, below).

The "isolated" transient signal seems to indicate that the slow fault slip decelerated sometime in 2001 autumn as recognized by a small break of the curve. There I modeled the secular and seasonal components (in this case, annual plus semiannual) empirically using the three-year time series before the start of this event. Such an extrapolation is based on the implicit expectation that seasonal variations behave similarly every year. As suggested by the snow load variation shown in Figure 2, this may not be the case, and the slowing down of the Tokai transient signal in Figure 17 might only reflect inappropriate removal of seasonal signals. Slow fault slips in transient depths of subduction zones are often associated with non-volcanic tremor activities [*Rogers and Dragert*, 2003]. Such tremor activities are also present in the Tokai area [*Obara*, 2002], and their temporal changes may provide independent information on the change in the slow slip rate. At the moment, the seasonal coordinate change models are not sufficiently accurate (especially vertical components in summer, behaviors of interannual variations). The ultimate goal of the future seasonal crustal deformation studies is to make it possible to perform a-priori correction of seasonal signals in near real time, e.g. by using the current

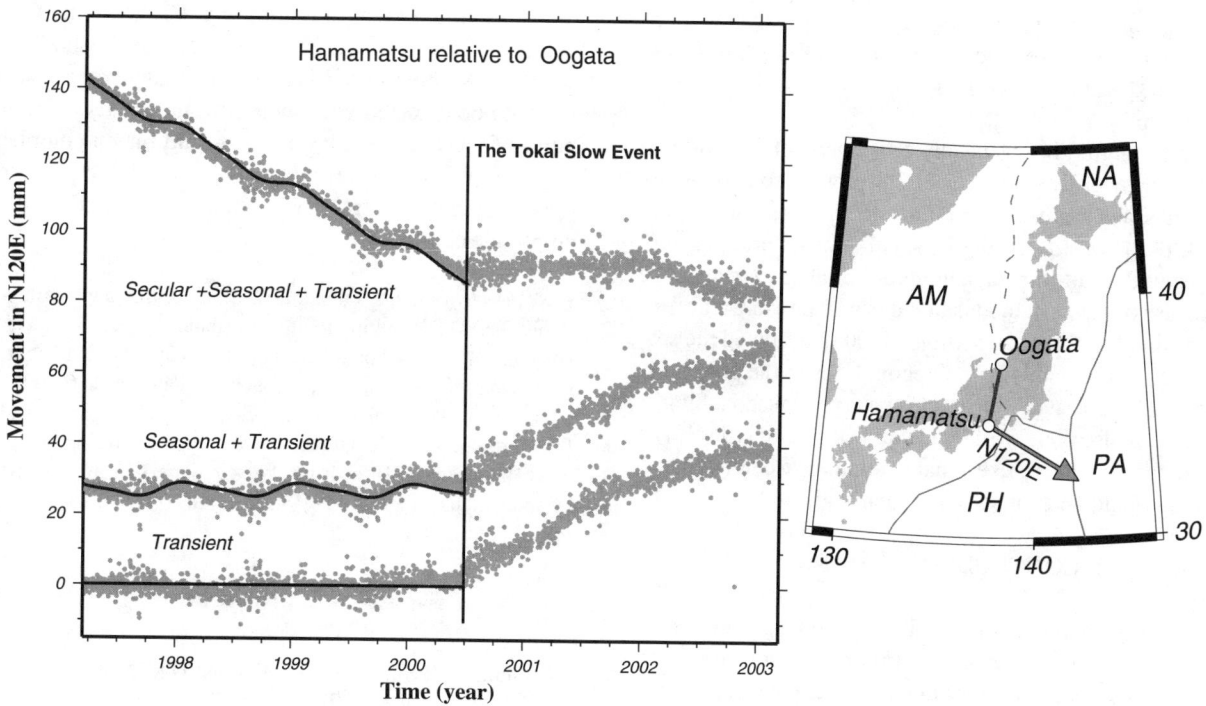

Figure 17. Movement in the ESE direction of the Hamamatsu GPS station, in the Tokai area, with respect to the Oogata station (see the map to the right for locations). The raw data (left, up) includes secular, transient, and seasonal components. The left middle curve is de-trended, and in the left lower curve seasonal changes are also removed. Seasonal and secular movements are inferred using the time series before the onset of the 2000 Tokai Slow Event [*Ozawa et al., 2002*]. Solid curves indicate modeled secular and seasonal components. Abbreviations in the map denote tectonic plates, i.e. the North American (NA), Pacific (PA), Amurian (AM) and Philippine Sea (PH).

snow depth, and other loading data inferred by meteorological sensors and geophysical models.

4.2. Load Changes and Triggering of Inland Earthquakes

Seismic activity in some regions in Japan is known to change seasonally as mentioned in the introduction. Using the newest earthquake catalog, *Heki* [2003] recognized the tendency that inland earthquakes in snow-covered regions in Japan occur more in spring and summer as originally suggested by *Okada* [1982], and concluded that it may be related to the removal of snow load. In the present study, other kinds of loads have been evaluated (Figures 5–8), and the snow was found to be by far the largest seasonally changing load in Japan. This thus justifies the approach by *Heki* [2003]. Interplate thrust event along the Nankai, Suruga and Sagami Troughs are known to occur mainly in autumn and winter [*Ohtake and Nakahara*, 1999]. This is qualitatively consistent with the seasonal change of non-tidal oceanic loads; the SSH shows maximum in August and September (Figure 9), and the highest seismicity in December coincide with the period of the rapid load removal (this period comes 1/4 year after the SSH peak). However, thermal steric correction may be still imperfect, and the deployment of ocean bottom pressure gauges around Japan to monitor in-situ water mass changes above the sensor [e.g. *Fujimoto et al.*, 2003] would derive final conclusion for their causal relationship.

It is meaningful, in seismically active parts of the world, to understand the relative importance of all possible sources of seasonal loads, and evaluate their influences upon seismicity. *Gao et al.* [2000] considered the seasonal atmospheric pressure changes with peak-to-peak amplitudes of about 2 kPa as the main cause of seasonal modulation of seismicity in California. On the other hand, *Saar and Manga* [2003] attributed the seasonal change in seismicity in Oregon, USA to the pore pressure changes associated with the change in precipitation. Similar discussion is also done world-wide by *Matsumura* [1986]. The method described here to evaluate various seasonally changing loads would be applicable for such studies.

5. CONCLUDING REMARKS

I performed comprehensive studies of changing loads within the Japanese Islands, i.e. snow, atmosphere, soil moisture, water impoundment of reservoirs, and over the oceanic area around Japan, and calculated crustal deformation caused by them. Then synthesized seasonal coordinate change signals are compared with real data observed by the dense GPS array. The following conclusions have been obtained.

(1) In Japan, by far the largest contribution comes from snow, and it causes a few mm arc-normal contractions and subsidence of up to 1 cm in northeastern Japan in winter. To evaluate snow contributions quantitatively, special attentions are needed for the altitude correction of the snow depth data, and for temporal changes of average snow densities.

(2) Other loads are of minor importance in Japan, being atmosphere the second largest source to snow. However, the order of importance would change in other parts of the world.

(3) Seasonal change in scales of unknown origin exists in the GEONET solution.

(4) Seasonal change in atmospheric gradients also contributes to seasonal signals in some areas [*Hatanaka*, 2003a], and it is preferable to estimate gradient parameters in the GEONET official solutions in the future.

(5) Amplitudes and waveforms predicted by the changing loads and scales are mostly in accordance with real GEONET data, although small inconsistency remains in summer behaviors.

(6) Understanding of seasonal signals including their inter-annual changes is important to isolate transient crustal deformation signals.

(7) Seasonal change in the occurrences of inland earthquakes is possibly influenced by snow loads. Seasonality in submarine interplate earthquakes needs further investigation.

Study of mass redistribution is one of the main targets of recent satellite gravity missions, e.g. CHAMP and GRACE [*Reigber*, 2003]. However, those in relatively small spatial scales, as studied in the Japanese Islands here, are still too small to be detected with their current accuracy. From this point of view, dense GPS arrays would play a complementary role to these satellite missions by detecting surface mass changes with relatively short wavelength and large amplitudes.

Acknowledgements. I thank C.K. Shum (OSU) and Clark Wilson (Univ. Texas) for inviting me to contribute a paper to the U6 symposium. I also thank Soh Kazama (Tohoku Univ.) for sending reprints related to estimation of mountain snows, Tadahiro Sato (NAO) for his advice on the seasonal ocean load changes, Koji Matsumoto (NAO) for subroutines to calculate the Earth's response to a point load, and Yuki Hatanaka (GSI) for fruitful discussions. Constructive reviews by anonymous referees and by Steve Sparks (Univ. Bristol) are gratefully acknowledged.

REFERENCES

Altamimi, Z., P. Sillard, and C. Boucher, ITRF2000: A new release of the International Terrestrial Reference Frame for earth science applications, *J. of Geophys. Res.*, 107, 2214, doi:10.1029/2001JB000561, 2002.

Beutler, G., I. Bauersima, W. Gurtner, M. Rothacher, T. Schildknecht, and A. Geiger, Atmospheric refraction and other important biases in GPS carrier phase observations, *Monograph 12*, School of Surveying, Univ. New South Wales, Kensington, pp.26, 1988.

Blewitt, G., D. Lavallée, P. Clarke, and K. Nurutdinov, A new global mode of earth deformation: seasonal cycle detected, *Science,* 294, 2342–2345, 2001.

Bouille, F., A. Cazenave, J. M. Lemoine, and J. F. Crétaux, Geocenter motion from the DORIS space system and laser data on Lageos satellites: Comparison with surface loading data, *Geophys. J. Int.,* 143, 71–82, 2000.

Cazenave, A., Global-scale interactions between the solid earth and its fluid envelopes at the seasonal time scale, *Earth Planet. Sci. Lett.,* 171, 549–559, 1999.

Chao, B. F., Anthropogenic impact on global geodynamics due to reservoir water impoundment, *Geophys. Res. Lett,* 22., 3529–3532, 1995.

Chao, B. F., and A. Y. Au, Atmospheric excitation of the Earth's annual wobble, 1980–1988, *J. of Geophys. Res.,* 96, 6577–6582, 1991.

Crétaux, J.-F., L. Soudarin, F. J. M. Davidson, M.-C. Gennero, M.Bergé-Nguyen, and A. Cazenave, Seasonal and interannual geocenter motion from SLR and DORIS measurements: Comparison with surface loading data, *J. of Geophys. Res.,* 107, 2374, doi:10.1029/2002JB001820, 2002.

Dong, D., J.O. Dickey, Y. Chao, and M.K. Cheng, Geocenter variations caused by atmosphere, ocean and surface ground water, *Geophys. Res. Lett.,* 24, 1867–1870, 1997.

Dong, D., P. Fang, Y. Bock, M. K. Cheng, and S. Miyazaki, Anatomy of apparent seasonal variations from GPS-derived site position time series, *J. of Geophys. Res.,* 107, 10.1029/2001JB000573, 2002.

Farrell, W. E., Deformation of the Earth by surface loads, *Rev. Geophys. Space Phys.,* 10, 761–797, 1972.

Fujimoto, H., M. Mochizuki, K. Mitsuzawa, T. Tamaki, and T. Sato, Ocean bottom pressure variations in the southeastern Pacific following the 1997–98 El Niño event, *Geophys. Res. Lett.,* 30, 1456, doi:10.1029/2002GL016677, 2003.

Gao, S. S., P. G. Silver, A. T. Linde, and I. S. Sacks, Annual modulation of triggered seismicity following the 1992 Landers earthquake in California, *Nature,* 406, 500–504, 2000.

Hachikubo, A., T. Kaihara, and Y. Ito, Report of pit-wall observations of snow cover in Sapporo 1996–97, *Low Temperature Sci., Ser. A., Data Report,* 56, 1–8, 1997 (in Japanese with English abstract).

Hatanaka, Y., A. Sengoku, T. Sato, J. M. Johnson, C. Rocken, and C. Meertens, Detection of tidal loading signals from GPS permanent array of GSI Japan, *J. of Geod. Soc. Japan,* 47, 187–192, 2001.

Hatanaka, Y., Seasonal variation of scale of GEONET network and ZTD biases, *paper presented at the Symposium JSG01,23rd IUGG General Assembly, Sapporo, Japan,* July 7, 2003a.

Hatanaka, Y., Improvement of the analysis strategy of GEONET, *Bull. of Geograph. Survey Inst.,* 49, 11–37, 2003b.

Heki, K., Seasonal modulation of interseismic strain buildup in Northeastern Japan driven by snow loads, *Science,* 293, 89–92, 2001.

Heki, K., Snow load and seasonal variation of earthquake occurrence in Japan, *Earth Planet. Sci. Lett.,* 207, 159–164, 2003.

Heki, K., S. Miyazaki, and H. Tsuji, Silent fault slip following an interplate thrust earthquake at the Japan Trench, *Nature,* 386, 595–598, 1997.

Hirose, H., K. Hirahara, F. Kimata, N. Fujii, and S. Miyazaki. A slow thrust slip event following the two 1996 Hyuganada earthquakes beneath the Bungo Channel, Southwest Japan, *Geophys. Res. Lett.,* 26, 3237–3240. 1999.

Hugentobler, U., S. Schaer, and P. Fridez (eds.), *Bernese GPS Software Version 4.2,* Astronomical Institute, University of Berne, pp 515, Bern, 2001.

Iwabuchi, T., S. Miyazaki, K. Heki, I. Naito and Y. Hatanaka, An impact of estimating tropospheric delay gradients on tropospheric delay estimations in the summer using the Japanese nationwide GPS array, *J. of Geophys. Res.,* 108, 4315, doi:10.1029/2002JD002214, 2003.

Kato, T., Secular and earthquake-related vertical crustal movements in Japan as deduced from tidal records (1951–1981), *Tectonophys.,* 97, 183–200, 1983.

Kazama, S., and M. Sawamoto, Estimation of the snow depth distribution and water resources volume using NOAA/AVHRR, *J. Japan Soc. Hydrol. Water Resour.,* 8, 477–483, 1995 (in Japanese with English abstracts).

Kedar, S., G.A. Hajj, B. D. Wilson, and M. B. Heflin, The effect of the second order GPS ionospheric correction on receiver positions, *Geophys. Res. Lett.,* 30, 1829, doi:10.1029/2003GL017639, 2003.

MacMillan, D. S., Atmospheric gradients from very long baseline interferometry observations, *Geophys. Res. Lett.,* 22, 1041–1044, 1995.

Matsumura, K., On regional characteristics of seasonal variation of shallow earthquake activities in the world, *Bull. of Disas. Prev. Res. Inst., Kyoto Univ.,* 36, 43–98, 1986.

McCarthy, D. D. (Ed.), IERS Conventions 1996, *IERS Tech. Note 21,* Int. Earth Rotation Serv., Obs. de Paris, Paris, Jul. 1996.

Milly, P. C. D., Climate, soil water storage, and the average annual water balance, *Water Resour. Res.,*30, 2143–2156, 1994.

Miyazaki, S., T. Iwabuchi, K. Heki, and I. Naito, An impact of estimating tropospheric gradient on precise positioning in summer using the Japanese nationwide GPS array, *J. of Geophys. Res.,* 108, 2335, doi:10.1029/2000JB000113, 2003.

Munekane, H., M. Tobita, K. Takashima, S. Matsuzaka, Y. Kuroishi, and Y. Masaki, Groundwater-driven vertical movement in Tsukuba detected by GPS, *paper presented at the American Geophys. Union, 2003 Fall Meeting,* San Francisco, USA, Dec. 12, 2003.

Munk, W. H., and G. J. F. MacDonald, *The Rotation of the Earth, a Geophysical Discussion,* Cambridge University Press, 323pp, 1960.

Murakami, M., and S. Miyazaki, Periodicity of strain accumulation detected by permanent GPS array: possible relationship to seasonality of major earthquakes' occurrence, *Geophys. Res. Lett.,* 28, 2983–2986, 2001.

National Astronomical Observatory (ed.), *Chronological Scientific Tables,* 76, 942pp, Maruzen, 2002..

National Res. Inst. For Earth Sci. Disaster Prevention (NIED), Interesting phenomena detected by Sakata-type three-component strainmeters, *Rept. Coord. Comm. Earthq. Pred.,* 69, 205–211, 2003 (in Japanese)

Obara, K., Nonvolcanic deep tremor associated with subduction in southwest Japan, *Science,* 296, 1579–1681, 2002.

Ohtake, M., and H. Nakahara, Seasonality of great earthquake occurrence at the northwestern margin of the Philippine Sea Plate, *Pure Appl. Geophys.*, 155, 689–700, 1999

Okada, M., Seasonal variation in the occurrence rate of large earthquakes in and near Japan and its regional differences, *Zisin 2*, 35, 53–64, 1982 (in Japanese with English abstracts).

Ozawa, S., M. Murakami, M. Kaidzu, T. Tada, T. Sagiya, Y. Hatanaka, H. Yarai, and T. Nishimura, Detection and monitoring of ongoing aseismic slip in the Tokai Region, Central Japan, *Science,* 298, 1009–1012, 2002.

Ozawa, S, S. Miyazaki, Y. Hatanaka, T. Imakiire, M. Kaidzu, and M. Murakami, Characteristic silent earthquakes in the eastern part of the Boso Peninsula, Central Japan, *Geophys. Res. Lett.,* 30, doi: 2002GL016665, 2003.

Reigber, C., The gravity field of the Earth determined by new satellite missions, *paper presented at the Symposium U6, 23rd IUGG General Assembly,* Sapporo, Japan, July 9, 2003.

Rogers, G. and H. Dragert, Episodic tremor and slip on the Cascadia subduction zone: the chatter of silent slip, *Science,* 300, 1942–1943, 2003.

Saar, M. O., and M. Manga, Seismicity induced by season groundwater recharge at Mt. Hood, Oregon, *Earth Planet. Sci. Lett.*, 214, 605–618, 2003.

Sato, T., Y. Fukuda, Y. Aoyama, H. McQueen, K. Shibuya, Y. Tamura, K. Asari, and M. Ooe, On the observed annual gravity variation and the effect of sea surface height variations, *Phys. Earth Planet. Inter.,* 123, 45–63, 2001.

Stammer, D., R. Tokmakian, A. Semter, and C. Wunsh, How well does a 1/4° global circulation model simulate large-scale oceanic observations? *J. of Geophys. Res.,* 101, 25779–25811, 1996.

Thornthwaite, C. W., An approach toward a rational classification of climate, *Geogr. Rev.,* 38, 55–94, 1948.

van Dam, T. M., G. Blewitt, and M. Heflin, Detection of atmospheric pressure loading using the Global Positioning System, *J. of Geophys. Res.* 99, 23,939–23,950, 1994.

van Dam, T. M., J. Wahr, P. C. D. Milly, A. B. Shmakin, G. Blewitt, D. Lavallée, and K. M. Larson, Detection of atmospheric pressure loading using the Global Positioning System, *Geophys. Res. Lett.,* 28, 651–654, 2001.

Wahr, J. M., M. Molenaar, and F. Bryan, Time variability of the Earth's gravity field: Hydrological and oceanic effects and their possible detection using GRACE, *J. of Geophys. Res.,* 103, 30,205–30,229, 1998.

Wang, H., H. T. Hsu, and Y. Z. Zhu, Prediction of surface horizontal displacements, and gravity and tilt changes caused by filling the Three Gorges Reservoir, *J. of Geodesy,* 76, 105–114, 2002.

Wu, X.-P., M. B. Heflin, E. R. Ivins, D. F. Argus, and F. H. Webb, Large-scale global surface mass variations inferred from GPS measurements of load-induced deformation, *Geophys. Res. Lett.,* 30, 1742, doi:10.1029/2003GL017546, 2003.

Yamamoto, T., Change in the annual variation patterns of the GEONET site coordinate time series and its implication for the analysis of the Tokai Slow Event, *The Earth Monthly*, 41, 71–76, 2003 (in Japanese).

Yamaya, M., K. Numazawa, K. Yano, A. Kumagai, H. Abiko, A. Sato, H. Abe, Observation of icing and snow accretion on Jyuhyo (Ice Monsters) at Mt.Zao, *Tohoku J. Natural Disaster Sci.,* 36, 171–176, 2000 (in Japanese).

Kosuke Heki, Division of Earth and Planetary Science, Hokkaido University, Kita-ku, N10 W8, Sapporo 060-0810, Japan. (heki@ep.sci.hokudai.ac.jp)

Remote Sensing of Terrestrial Water Storage, Soil Moisture and Surface Waters

James S. Famiglietti

Department of Earth System Science, University of California, Irvine, California, USA

Comprehensive monitoring of terrestrial water is critical for characterizing changes in water availability, hydrologic extremes, to determine human impacts on the water cycle, and more generally, for enhanced predictive understanding of regional and global water cycles and their interactions within the Earth system. In this paper, the current and near-future capabilities of remote sensing of terrestrial water are assessed, with a focus on liquid water. The potential for GRACE observations of time-variable gravity to monitor monthly and longer changes in total water storage for regions greater than 200,000 km^2 is discussed. Near-future AMSR observations of surface soil moisture at 60 km resolution with 2-day repeat are described, as is the future HYDROS mission. Current and future capabilities of altimetric observations of terrestrial surface waters are reviewed. An important perspective of this paper is that the current and near-future sensors described in this paper will offer unprecedented opportunities for monitoring terrestrial hydrology, and that their joint use will enable new, simultaneous views of both the lateral and vertical distribution of water on land that have not been previously possible.

1. INTRODUCTION

The lateral and vertical distribution of terrestrial water is a fundamental Earth system state that interacts with climate on a range of temporal and spatial scales. For example, surface waters, including rivers, inland water bodies, wetlands and inundated floodplain, represent a renewable source of freshwater, respond to flood and drought conditions, and are an important link to biogeochemical cycles. Soil moisture, defined here as water stored in the upper few centimeters of soil, is an important control on partitioning rainfall into runoff and infiltration and solar radiation into latent and sensible heat, and can provide a significant source of moisture for precipitation on land [*Brubaker et al.*, 1993; *Eltahir and Bras*, 1996; *Bosilovich and Schubert*, 2002]. Terrestrial, or total water storage, which includes all of the surface waters, surface and deeper soil moisture and groundwater, as well as snow and ice, is an integrated measure of water availability and the terrestrial freshwater stock, and its variations have implications well beyond water supply, including impacts on earth rotation [*Guttierez and Wilson*, 1987; *Chao and O'Connor*, 1988; *Chao*, 1995] and rates of sea level change [*Chen et al.*, 1998; *Chambers et al.*, 2000; *Church et al.*, 2001]. Hence comprehensive monitoring of terrestrial water is critical for characterizing changes in water availability, hydrologic extremes, to determine human impacts on the water cycle, and more generally, for enhanced predictive understanding of regional and global water cycles and their interactions within the Earth system.

In this paper, the current state of remote monitoring of terrestrial water is assessed, with a view towards present and near-future capabilities. The focus is on liquid water, and will specifically address satellite observations of total water storage, surface soil moisture, and surface waters. Key sensors and techniques discussed include satellite observations of the time-variable gravity field for estimating changes in total water storage, e.g. using the Gravity Recovery and Climate

Experiment [GRACE, *Wahr et al.*, 1998; *Rodell and Famiglietti*, 1999]; microwave soil moisture remote sensing, e.g. using the Advanced Microwave Scanning Radiometer [AMSR, *Njoku et al.*, 2003], and altimetric observations of surface waters [*Alsdorf et al.*, 2003; *Alsdorf and Lettenmaier*, 2003]. An important perspective of this paper is that the current and near-future sensors described here offer unprecedented opportunities for monitoring terrestrial hydrology: in particular, their joint use will enable new, simultaneous views of both the lateral and vertical distribution of water on land that have not been previously possible. Further, ingesting the hydrologic products of these sensors into advanced land modeling-assimilation systems such as the Global Land Data Assimilation System [GLDAS, *Rodell et al.*, 2003], will result in well-constrained estimates of global-scale terrestrial water storage variations that are very likely the best available to date. As such, we are entering a new era of information on the distribution of water on land, one with the potential to greatly enhance our understanding of the storage and movement of water in the terrestrial branch of the hydrologic cycle and its interactions within the Earth system.

2. ESTIMATING CHANGES IN TOTAL WATER STORAGE USING GRACE

2.1. Basic Principles and Feasibility

The underlying principle in deriving changes in total water storage (ΔTWS) from GRACE is that temporal variations in Earth's gravity field on short time scales (months to years), which the satellite will be able to detect with unprecedented accuracy, are due to air and water mass redistribution [*Guttierez and Wilson*, 1987]. In an atmosphere-land column, this mass redistribution or gravity change is due primarily to the redistribution of water. Hence, given a surface load change estimate from GRACE, the atmospheric water mass change can be removed (e.g. using reanalysis products from the National Center for Environmental Prediction / National Center for Atmospheric Research (NCEP/NCAR) [*Kalnay et al.*, 1996] or the European Center for Medium Range Weather Forecasts (ECMWF) [*Gibson et al.*, 1997]), leaving ΔTWS as the residual.

Rodell and Famiglietti [1999] explored the feasibility of this approach for the drainage basins shown in Plate 1 by comparing uncertainty in potential GRACE mass-change estimates to the magnitude of modeled ΔTWS using time series from the Global Soil Wetness Project [*Dirmeyer et al.*, 1999]. The GRACE ΔTWS error budget included uncertainty in the atmospheric mass signal (estimated from comparison of the NCEP/NCAR and ECMWF reanalysis products), errors in the estimate of the postglacial rebound (PGR) contribution (determined from a PGR model [*Peltier*, 1994, 1995]), and GRACE instrument error (provided by the GRACE science team as a function of area and time period). Note that GRACE instrument errors decrease with increasing spatial scale and with increasing observation time.

Selected results for four basins of varying area are shown in Plate 2. Overall, great potential for detectability (signal greater than noise) was demonstrated in basins larger than 200,000 km^2 for monthly and longer time periods. At these spatial-temporal scales, instrument errors were relatively small; uncertainty in the atmospheric mass change component dominated the error budget. Because GRACE instrument error increases dramatically with decreasing spatial scale (note the increase in error with decreasing basin area apparent in Figure 2), potential detectability was characterized as poor for the basins below the 200,000 km^2 threshold in the study.

For the 15 largest basins in *Rodell and Famiglietti* [1999], average monthly error was roughly 5 mm, which decreased at longer seasonal and annual timescales. Relative error is a function of the magnitude of ΔTWS: even in a smaller basin like the 201,000 km^2 Chao Phraya, a scale at which instrument error begins to increase significantly, relative error for seasonal estimates of ΔTWS was 25% owing to the greater magnitude of the seasonal change in water storage. This work demonstrated that the limiting aspect of GRACE ΔTWS estimation is basin size, and that a threshold of 200,000 km^2 is appropriate for first products. *Wahr et al.* [1998] reached a similar conclusion regarding a spatial threshold for detectability.

2.2. Production of Basin-Scale ΔTWS Time Series

There are several important steps required in order to produce estimates of watershed-scale ΔTWS from GRACE products (i.e. spherical harmonic (Stokes) coefficient time series representing change in mass). Much of this work is now underway in the author's research group with collaborators at the University of Texas at Austin, the University of Colorado, NASA Goddard Space Flight Center, and NASA Jet Propulsion Laboratory. First, as discussed above, the contribution of mass variations other than those due to water redistribution on land must be quantified and removed. On monthly to annual timescales, important contributions result from atmospheric and oceanic mass redistribution. These must be quantified and removed, e.g. using atmospheric reanalyses, ocean model predictions, satellite altimeters (e.g. TOPEX, JASON) and tide gauge records.

In order to convert monthly mean Stokes coefficients to a basin-scale water mass change (ΔTWS), a set of filter coefficients specific to that basin must be developed that will give an estimate as a linear combination of the GRACE products. It is convenient to consider these "basin functions" as the spherical harmonic representation of 0–1 masks that filter

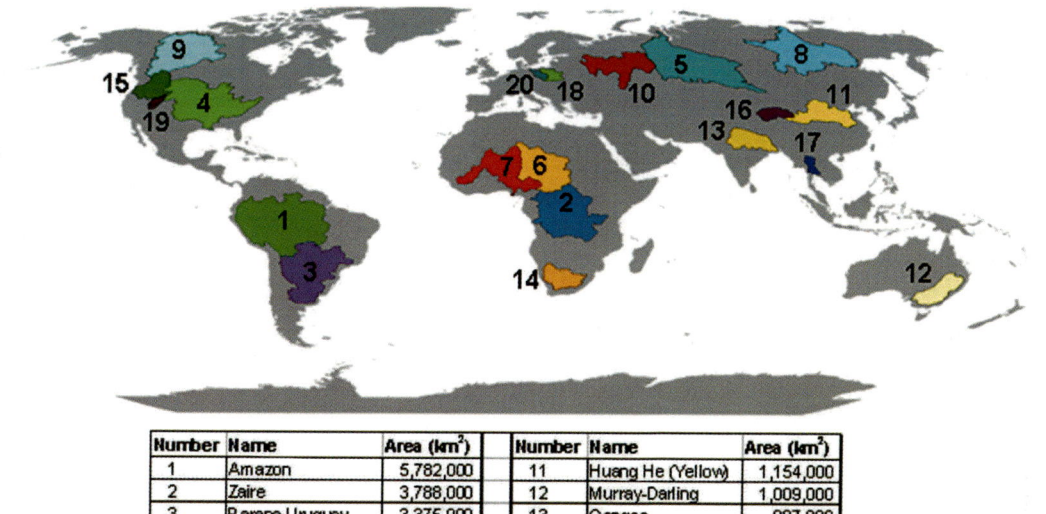

Plate 1. Twenty continental scale drainage basins where feasibility of GRACE-derived ΔTWS was explored by *Rodell and Famiglietti* [1999].

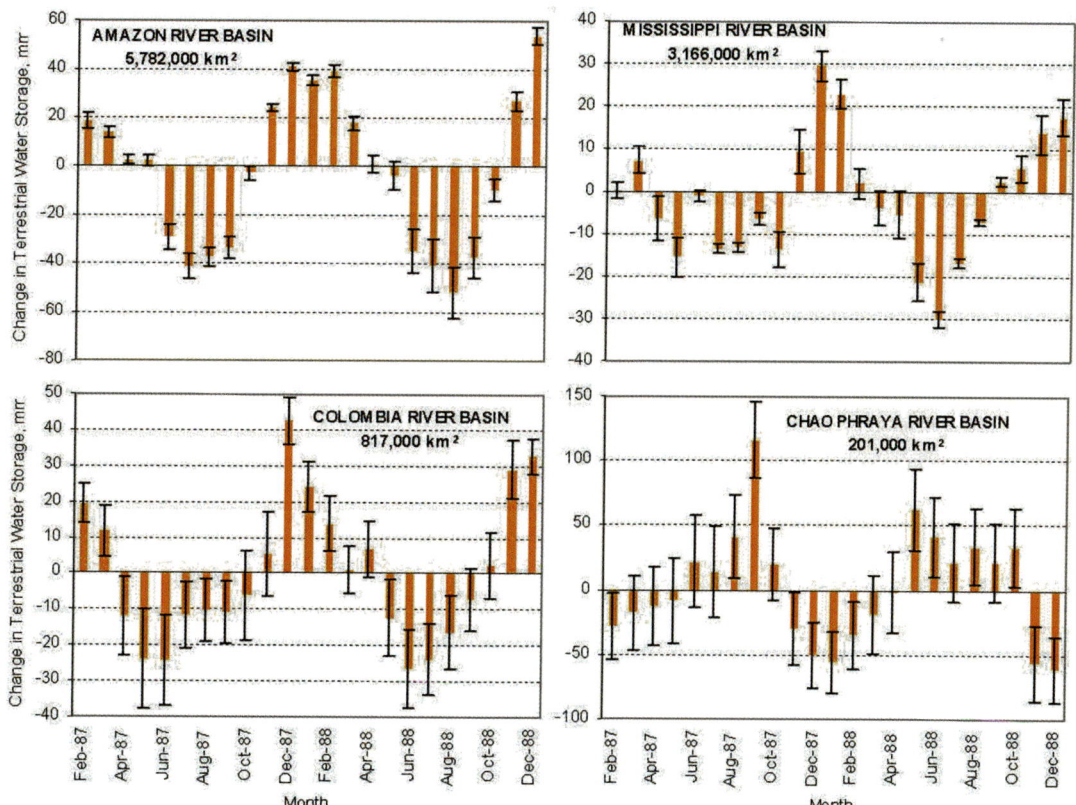

Plate 2. Orange bars are modeled monthly changes in total water storage. Error bars represent the total uncertainty in GRACE-derived estimates; error increases with decreasing basin area. From *Rodell and Famiglietti* [1999].

out mass change variations outside of the basin of interest. Such functions have been derived by *Swenson and Wahr* [2002] and *Seo et al.* [2003].

A third step is to quantify uncertainty in each ΔTWS estimate. Error budgets should include the uncertainties previously identified, namely atmospheric, oceanic, and PGR mass errors and instrument errors. However, GRACE ΔTWS errors should also reflect uncertainties due to the "leakage" from mass variations outside of the basin into a basin-scale estimate, which results from the incomplete masking by the basin function; and the uncertainty due to "temporal aliasing," or the fact that the hydrologic mass change signal is changing at higher frequencies (sub-monthly) than the one month period required to complete a global gravity field measurement.

A final important step is validating GRACE ΔTWS estimates. Current plans for GRACE validation include computing independent, basin-scale terrestrial and coupled land-atmosphere water balances and comparing GRACE estimates to observed ΔTWS. Comparison of GRACE ΔTWS estimates to those simulated by land model-assimilation systems such as GLDAS will also be an important component in the validation strategy. Finally, in situ validation sites should be established where the terrestrial water balance, ΔTWS and related changes in gravity can be monitored directly. Logical sites may include some of those identified for remotely-sensed soil moisture validation (see below), where *in situ* monitoring of soil water is currently being enhanced [*Njoku*, 2003].

Given the vast potential of GRACE for hydrologic monitoring, what can the scientific community reasonably expect in terms of routinely-available hydrologic products and the new science and applications they will enable? GRACE launched in March, 2002, and the first gravity field, GRACE Gravity Model 01, was released in July, 2003. Subsequent gravity field releases enabled the preliminary calculation of ΔTWS which show good correlation with global assimilating model outputs for the period November 2002 to April, 2003 [*Famiglietti et al.*, 2003; *Wahr et al.*, 2003]. These first results are extremely encouraging given that the GRACE mission is in its early stages.

Current expectations are that the release of monthly GRACE products, from which monthly basin-scale ΔTWS can be determined, will begin in Spring, 2004. Hence monthly ΔTWS time series and their uncertainty for selected large basins in Plate 1 should follow shortly after this release. An important, longer-term goal that would greatly enhance the utility of GRACE ΔTWS products is to work towards continuous coverage of Earth's land surface partitioned into large drainage basins or regions. Plate 3 shows this partitioning for North America. Such a dataset will greatly complement expected soil moisture and snow products and will provide an important contribution towards understanding the lateral and vertical distribution of water on land.

2.3. New Applications

Monthly time series of ΔTWS for large watersheds or other hydrologic units (e.g. aquifers) will support a range of new research and application activities. For example, continental-scale GRACE ΔTWS time series will provide new information on interannual variations in continental water storage, with implications for the terrestrial response to changing climate, as well as for sea level change. In a study of the High Plains aquifer in the United States, *Rodell and Famiglietti* [2002] showed that GRACE has the potential to monitor changing water levels in large groundwater systems, which can provide important information for water resources managers. *Famiglietti et al.* [2003] explored how GRACE ΔTWS time series can be used to solve the basin scale water balance for evapotranspiration, ET, as $P - Q - \Delta TWS$ (P is precipitation; Q is streamflow), and provide an important complement to other remote-sensing-based approaches that require significant, ancillary land surface and meteorological data [e.g. *Jiang and Islam*, 2001; *Batiaanssen et al.*, 2002; *Liu et al.*, 2003]. Given the potential of GRACE to monitor water mass changes, it is also likely that we will soon see new applications, for example, in monitoring snow load variations or seasonal floodplain inundation, where these hydroclimatological features dominate ΔTWS over large regions. In summary, GRACE ΔTWS time series will represent an important new source of information on changes in terrestrial water storage that will complement and significantly enhance the capabilities of existing and near-future satellite missions for remote monitoring of the terrestrial water cycle.

3. SOIL MOISTURE REMOTE SENSING

3.1. Optimal Sensors

Main Considerable research over the past several decades, including numerous plot-to-regional scale field and aircraft experiments [e.g. *Jackson et al.*, 1995; *Jackson et al.*, 1999], have clearly demonstrated that the optimal sensor for remote observation of soil moisture (i.e. water in the upper few centimeters of the soil) is an L-band, 21 cm wavelength passive microwave radiometer. In the L-band part of the electromagnetic spectrum, thermal emissions from the land surface, which are recorded as microwave brightness temperatures, are a strong function of the dielectric constant, which itself is strongly dependent on water content in the upper 5 cm of the soil. The brightness temperature is also sensitive to other land surface characteristics, including vege-

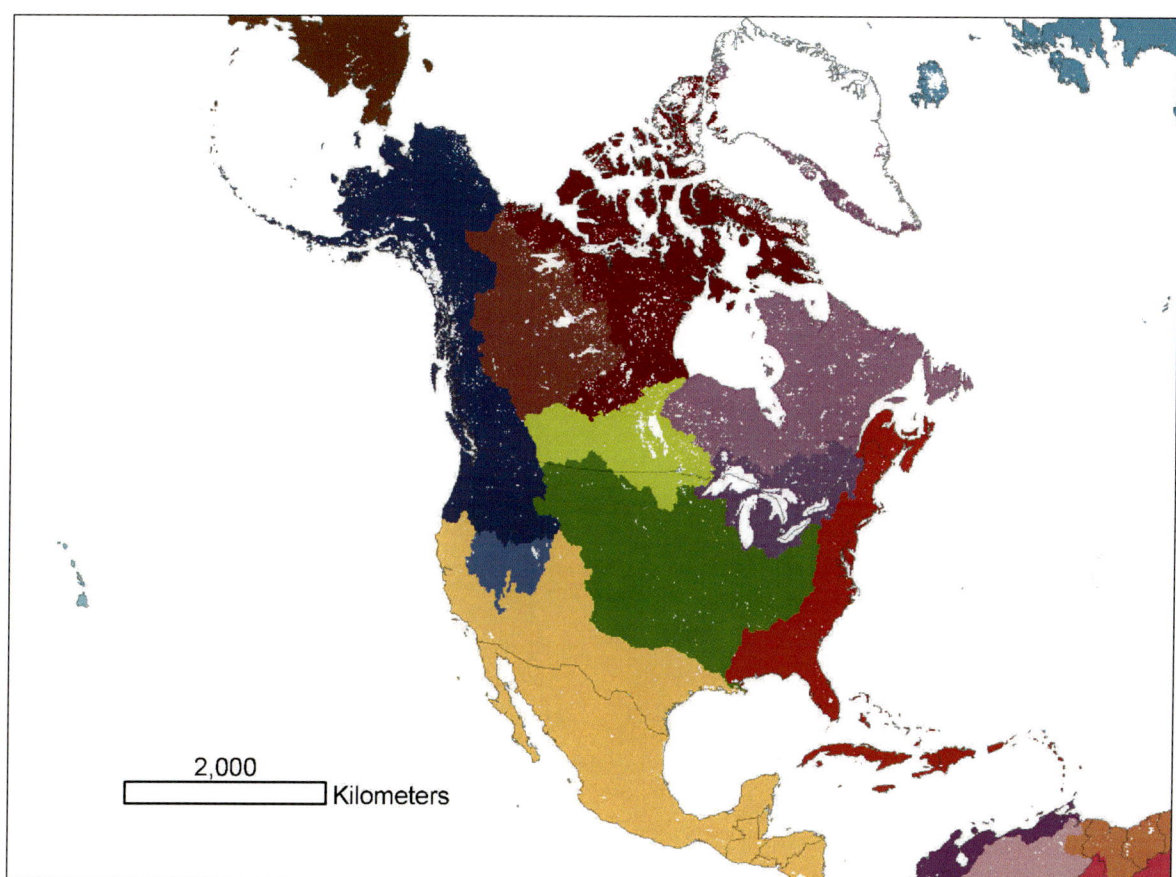

Plate 3. Large drainage basins and drainage regions of North America as candidates for routine production of monthly GRACE ΔTWS estimates.

tation water content, surface temperature and roughness, and soil texture, so that inversion algorithms and ancillary datasets are required to extract soil moisture content (cm^3/cm^3) [see e.g. *Jackson et al.*, 1995]. Plate 4 shows an example of aircraft L-band passive microwave sensor performance from the Southern Great Plains 1997 Hydrology Experiment (SGP97), a field experiment held in central Oklahoma during the summer of 1997 (from *Jackson et al.*[1999]).

These results, and others like them, have led the international hydrologic community to propose two passive microwave satellite soil moisture missions, the ESA Soil Moisture and Ocean Salinity (SMOS; *Silvestrin* et al., 2001) and the NASA Hydrosphere State (HYDROS; http://essp.gsfc.nasa.gov/hydros) satellites. The SMOS instrument is an L-band passive microwave-interferometer that will provide 50-km resolution soil moisture with a 3-day global revisit. At present, SMOS is expected to launch in early 2007. The HYDROS instrument is an integrated passive and active L-band sensor that will deliver soil moisture at two resolutions, 40 km and 10 km, for hydroclimatolog-

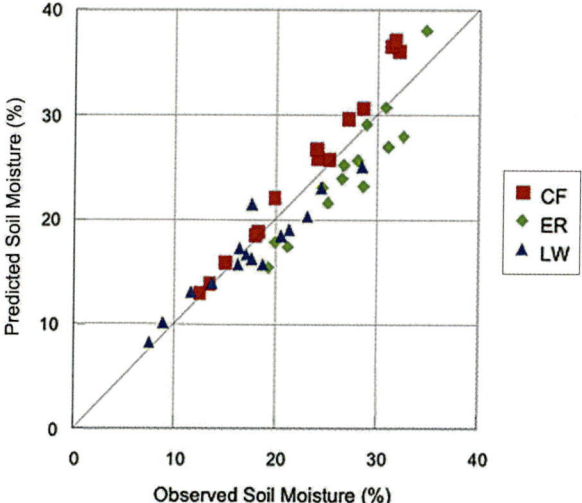

Plate 4. Soil moisture measured by an airborne L-band radiometer versus observations during the SGP97 experiment. Symbols represent location of ground sampling sites (ARM/CART Central Facility (CF), El Reno (ER) and Little Washita (LW). Standard error is 3% cm^3/cm^3. From *Jackson et al.* [1999].

ical and hydrometeorological applications respectively, with a 3-day global revisit. HYDROS will also monitor the freeze/thaw transition at 3 km spatial resolution and 1–2 day global revisit. HYDROS was recently approved for launch in 2009 as a NASA Earth System Science Pathfinder satellite. While it will be several years before either the SMOS or HYDROS L-band soil moisture missions are operational, the hydrologic community looks forward to their launches with great anticipation.

3.2. Current Missions

While a satellite L-band soil moisture mission will not be operational in the next few years, there is great anticipation for C-band soil moisture products from the AMSR-E sensor, which is on the NASA EOS Aqua platform (launched in May, 2002), and AMSR, on the Japanese ADEOS-II (launched in December, 2002). *Wigneron et al.* [1995], *Jackson et al.* [2002] and *Njoku et al.* [2003] have all demonstrated the sensitivity of C-band sensors to soil moisture.

There are some drawbacks to C-band sensors which will constrain the utility of their products relative to those from L-band sensors. C-band brightness temperatures are representative of the moisture content in roughly only the upper 2 cm of the soil (compared to the upper 5 cm for L-band), and thermal emissions from the surface are attenuated more by overlying biomass and the associated vegetation water content, relative to L-band [*Njoku et al.*, 2003]. In addition to the depth of monitoring difference, the differences in signal attenuation by overlying biomass limits soil moisture retrievals to regions where vegetation water content is less than 5 kg/m2 for L-band sensors (roughly 75% of the global land surface), compared to 1.5 kg/m2 for C-band (roughly 55% of the global land surface). Note that these percentages include ice-covered lands, so that the actual soil moisture coverage will be somewhat less. More recently, problems with contamination of the 6.9 GHz brightness temperatures over urban areas have been encountered, which have required inversion algorithm modifications [*Njoku et al.*, 2003].

Despite these differences, AMSR-E soil moisture products represent a critical first step on the road towards future L-band missions. At the time of writing, regular release of 60 km spatial resolution (mapped to a 25 km EASE grid), global 2-day repeat visit soil moisture maps are anticipated by the end of 2003 [E. Njoku, NASA Jet Propulsion Laboratory, personal communication, 2003]. These products will likely see wide usage, from initialization and assimilation in climate and weather prediction models, to flood and drought prediction, fire hazard assessment, precision agriculture, and across disciplines into biogeochemistry and ecology.

3.3. Field Experiments and Research Opportunities

Clearly an important component of evaluating the utility of remotely-sensed soil moisture products is validation. *Njoku et al.* [2003] have proposed several AMSR validation sites in the U. S., including enhanced *in situ* instrumentation for long-term monitoring. To complement these sites with more dense, footprint-scale measurements, NASA has sponsored two recent series of AMSR validation field campaigns, including the Soil Moisture Experiments in 2002 and 2003 (SMEX02 [*Jackson*, 2002] and SMEX03 [*Jackson*, 2003]). The SMEX02 experiment was held in a 50 km by 100 km region surrounding Ames, Iowa; while SMEX03 was actually a series of footprint-scale validation campaigns held near Tifton, Georgia, Huntsville, Alabama, and central Oklahoma. These experiments provide an important opportunity to explore sensor and inversion algorithm performance across a range of terrains and land cover types, as well as for simultaneous investigation of land-atmosphere interaction processes, and the behavior of soil moisture variability across scales, up to the scale of satellite footprints [*Cosh and Brutsaert*, 1999; *Famiglietti et al.*, 1999; *Kim and Barros*, 2002, *Oldak et al.*, 2002; *Famiglietti et al.*, 2003]. Such information is critical in order to assess the uncertainty of ground-based validation data from the field campaigns, for the design of future, footprint-scale in situ networks, and for the development of more realistic parameterizations of soil moisture dynamics in land surface models. As mentioned earlier, these validation sites and campaigns may also provide important opportunities to expand the scope of the measurements in order to accommodate GRACE validation as well.

4. REMOTE SENSING OF TERRESTRIAL SURFACE WATERS

4.1. The Need for Global Surface Water Measurements

The importance of terrestrial surface waters to life on Earth cannot be overstated, and was discussed briefly in the introduction. Here we define surface waters as river discharge, seasonal floodplain inundation, wetlands, and inland water bodies such as lakes and reservoir impoundments. In spite of the importance of surface waters to Earth system function, no comprehensive global monitoring system exists to fully characterize rates of river discharge and the amount of surface storage for a specified time period. While stream gauges have long been operable in many of the world's watersheds, global coverage is far from complete, and the number of stream gauges is in decline [*Stokstad*, 1999; *IAHS*, 2001; *Shiklomanov* et al., 2002]. Further, significant flows may occur in braided channels or overbank, as inundated floodplain; in either case, an important quantity of discharge may go ungauged. Additionally, streamflow discharge records in certain countries are often withheld for national security reasons. Even with efforts such as that of the Global Runoff Data Centre [*GRDC*, 2003] to archive world streamflow data, a consistent global picture of river discharge remains elusive, and in near-real time, does not currently exist.

4.2. Current Capabilities

Remote sensing can play a critical role in characterizing global river discharge as well as surface water storage within floodplains, wetlands, lakes and reservoirs [*Smith*, 1997; *Vörösmarty et al.*, 1999]. In particular, satellite altimetry can measure the elevations of large water bodies and rivers, which are directly related to changes in storage volumes. For example, several recent studies have demonstrated the potential of satellite altimetry to monitor variations in lake levels [*Birkett*, 1995; *Mercier et al.*, 2002] and stream heights [*Koblinsky et al.* 1993; *Birkett*, 1998, 2002; *Meheu et al.*, 2003]. In addition, *Smith et al.* [1996] showed the utility of synthetic aperture radar (SAR) for monitoring braided river discharge, and *Alsdorf et al.* [2000] demonstrated the role of interferometry in observing changes in floodplain water levels. The January, 2003 launch of the Ice, Cloud and Land Elevation Satellite (ICESat) Geoscience Laser Altimeter System (GLAS) [http://www.csr.utexas.edu/glas/] will provide new opportunities for continued demonstration of the role of altimetry in surface water monitoring.

4.3. A Potential Future Surface Water Mission

With the need for enhanced global monitoring of terrestrial surface waters and the role of altimetry clearly demonstrated, the NASA Surface Water Working Group [*Alsdorf et al*, 2003; *Alsdorf and Lettenmaier*, 2003] is currently working towards the development of a surface water-based altimetry mission for monitoring stream discharge and wetland, floodplain, and inland water body heights. Key mission requirements and technological design issues are now being explored. An altimetry-based surface water mission could provide the first-ever near-simultaneous view of continental river discharges as well as the spatial extent of inundation due to floodplain, wetland, lake and reservoir storage. Snapshots such as these could reveal an integral component of the terrestrial water cycle, yet, one that has not been well characterized but has tremendous importance for man, weather and climate, biogeochemical cycles and ecosystem dynamics.

Beyond monitoring, enhanced knowledge of surface water dynamics and global river discharge is essential for improved modeling and prediction of terrestrial hydrology and climate.

River discharge is an integrated measure of Earth system processes acting within a watershed, and the ability to constrain model predictions or to assimilate surface water observations into land models is a critical step towards improved prediction. Further, freshwater outflows from the continents are important inputs into global ocean models owing to their impact on sea surface salinities and temperatures. The full implications of these outflows on global ocean processes and circulation are only now being explored, but considerable links between global river outflows and the thermohaline circulation are clear [e.g. *Branstetter*, 2001]. Hence land, ocean, atmospheric and coupled Earth system modeling would greatly benefit from the observations provided by a terrestrial surface water monitoring mission.

5. SUMMARY AND FUTURE PROSPECTS

The goal of this paper was to review the current state of remote sensing the soil moisture, surface water and total water storage components of terrestrial hydrology. The status of the GRACE mission for monitoring changes in total water storage was reviewed, and its potential and new applications were discussed. Soil moisture remote sensing using the C-band AMSR sensor was reviewed, and the future of optimal L-band missions such as SMOS and HYDROS were discussed. Finally, the potential for global, altimetric observations of terrestrial surface waters was highlighted, and the activities of the NASA Surface Water Working Group towards designing a hydrology-specific mission were presented.

The point was made that a new era has arrived in which remote sensing will be providing new insights into terrestrial hydrology that were not available to previous generations of scientist. For example, GRACE ΔTWS and AMSR soil moisture products will be available in the very near term; and altimetry (e.g. from TOPEX or ICESat/GLAS) can cur-

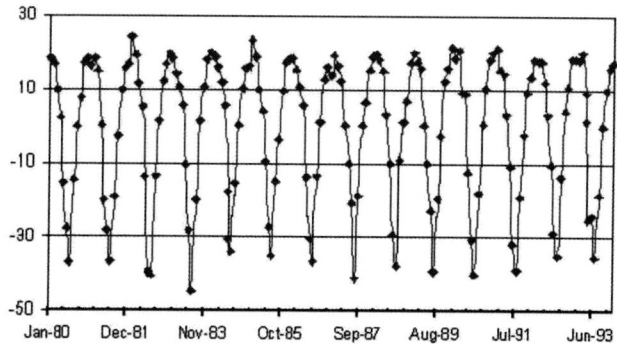

Plate 5. Time series of global, monthly terrestrial water storage change, in mm. Model simulations using the Mosaic land model [*Koster and Suarez*, 1992] driven by forcing data of *Berg et al.* [2003] data in the GLDAS framework.

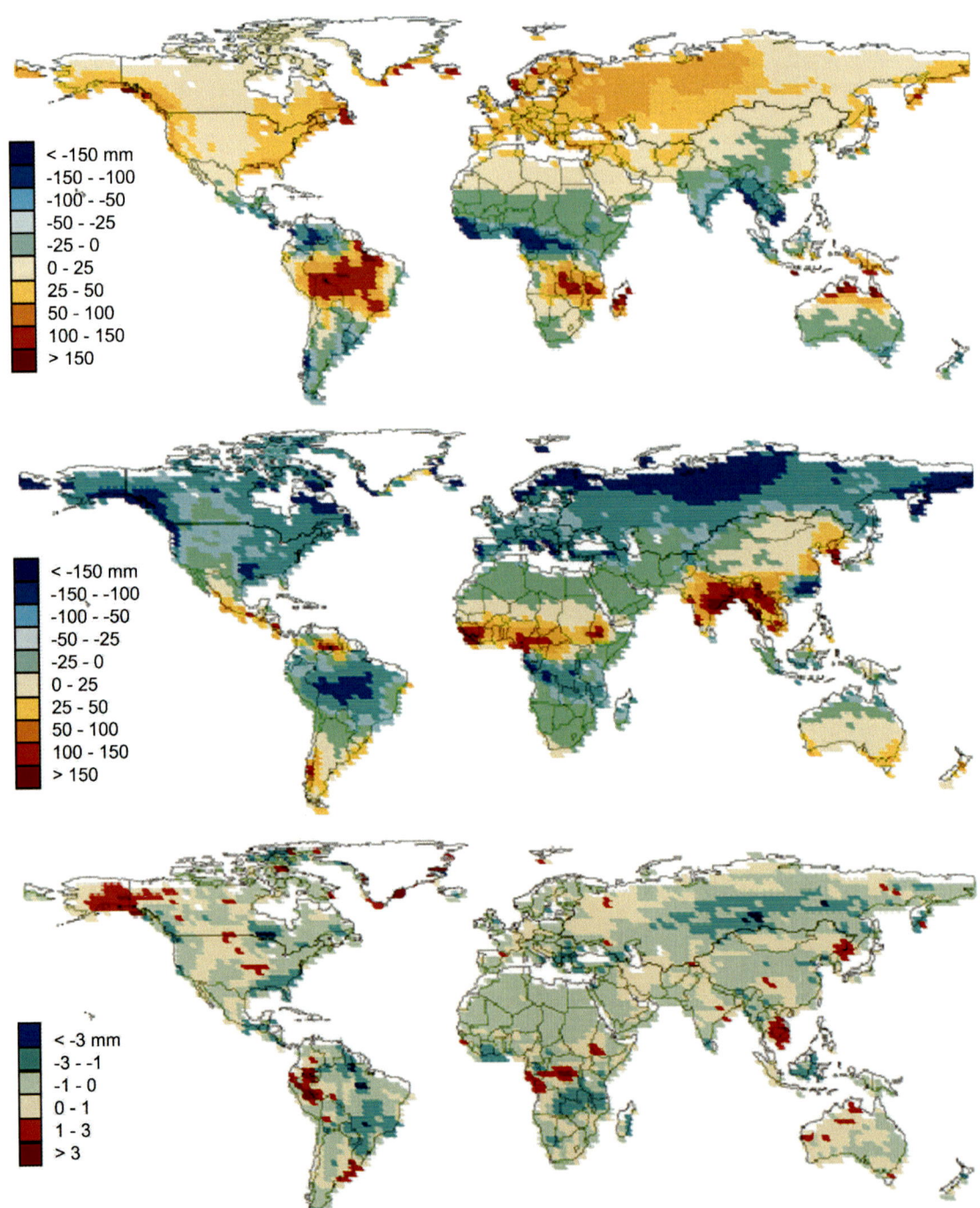

Plate 6. Average monthly water storage changes, 1979-1993, for January (top), July (middle), and for the entire simulation period. Units are mm. Simulation details as in Plate 5.

rently provide water heights for several river, floodplain and inland water body targets, albeit not in the comprehensive manner of a potential dedicated mission. L-band soil moisture, with its greater penetration depth and global coverage, should be available in 2007. As such, hydrologists are urged to think beyond traditional problems and to consider creative applications that these new views of the terrestrial water cycle will enable.

For example, the joint use of GRACE ΔTWS, AMSR soil moisture and altimetric surface water observations could be assimilated into land model-assimilation systems like GLDAS to begin comprehensive monitoring of water storage on land, including interannual variations in both the lateral and vertical distribution of water. Such an exercise would result in a clear view of how the storage of water is changing in response to changes in climate and to human activities, and could lead to important improvements in predictive understanding of land-atmosphere interaction.

Another new scientific endeavor for hydrologists that these new observations will enable is characterizing the contribution of continental water storage to sea level variations. Changes in continental water storage are a key term in the sea level rise budget, equal in magnitude to the contributions of thermal expansion and glaciers and ice caps. However, uncertainty in continental water storage contributions is the largest of the major sources, by a factor of three [Church et al., 2001]. The advent of these new sensor products, combined with recent advances in land surface modeling and data assimilation, opens new doors for hydrologists to produce very-well-constrained estimates of global-scale terrestrial water storage variations for understanding the contribution of continental water storage to sea level rise, while greatly minimizing previous uncertainties. Plates 5 and 6 show time series and spatial patterns of global water storage changes produced using the GLDAS framework, but without assimilation of the sensor products discussed here. Assimilation of GRACE ΔTWS, AMSR or SMOS soil moisture, and altimetric measurements of surface waters will greatly minimize the uncertainty in these current estimates.

When one considers other types of hydrological, biophysical and Earth remote sensing, beyond those described here, a very bright picture emerges. In particular, the potential for tremendous advances in monitoring and characterizing the terrestrial branch of the water cycle is growing greater with each passing month. We should plan now to capitalize on the relative near-term availability of the new hydrologic data, and on the current and near-term availability of complementary remote observations, in anticipation of new opportunities for elucidating the connections between terrestrial hydrology, the global water cycle, and their role in a broader Earth system context.

Acknowledgments. This work was supported by NASA grants NGT5-30443 and NAG5-11718, 12344, and 11753. Thanks to Tom Jackson, Eni Njoku, Dara Entekhabi and Paul Houser for discussions on microwave soil moisture remote sensing; to Matt Rodell, Ki-Weon Seo, Clark Wilson, John Wahr, Steve Nerem, Victor Zlotnicki, Mike Watkins and Jean Dickey for continued collaboration, discussion and support of GRACE applications in hydrology; and to Doug Alsdorf for his leadership of the NASA Surface Water Working Group and discussions regarding GRACE and altimetry.

REFERENCES

Alsdorf, D. C. Birkett, T. Dunne, J. Melack, John, L. Hess, 2001, Water level changes in a large Amazon lake measured with spaceborne radar interferometry and altimetry, Geophys. Res. Lett. 28(14), 2671–2674

Alsdorf, D., D. Lettenmaier, C. Vorosmarty and the NASA Surface Water Working Group, 2003, The need for global, satellite-based observations of terrestrial surface waters, EOS, Transactions, American Geophysical Union, 84(29), 269.

Alsdorf, D. and D. Lettenmaier, 2003, Tracking fresh water from space, Science, 301, 1491–1494.

Alsdorf, D., J. Melack, T. Dunne, L. Mertes, L. Hess and L. Smith, 2000, Interferometric radar measurements of water level changes on the Amazon floodplain, Nature, 404, 164–177.

Bastiaanssen, W., M. Ahmad and Y. Chemin, 2002, Satellite surveillance of evaporative depletion across the Indus Basin, Wat. Resour. Res., 38(12), 1273, doi:10.1029/2001WR000386.

Berg, A. A., J. S. Famiglietti, J. Walker and P. R. Houser, 2003, The Impact of Bias Correction to Reanalysis Products on Simulation of North American Soil Moisture and Hydrologic Fluxes, to appear, J. Geophys. Res.-Atm.

Birkett, C., 1995, The contribution of TOPEX/POSEIDON to the global monitoring of climatically sensitive lakes, J. Geophys. Res., 100(C12), 25179–25204.

Birkett, C., 1998, Contribution of the TOPEX NASA radar altimeter to the global monitoring of large rivers and wetlands, Wat. Resour. Res., 34(5), 1223–1240.

Birkett, C. M., L. A. K. Mertes, T. Dunne, M.H. Costa, and M.J. Jasinski, 2002, Surface water dynamics in the Amazon Basin: Application of satellite radar altimetry, J. Geophysical Res., doi: 10.1029/2001JD000609, vol. 107, D20, 8059, pages 26.1 – 26.21.

Bosilovich, M. G. and S. D. Schubert, 2002, Water vapor tracers as diagnostics of the regional hydrologic cycle, J. Hydrometeor., 3(2), 149–165.

Branstetter, M., 2001, Development of a parallel river transport algorithm and applications to climate studies, Dissertation thesis, The University of Texas at Austin, J. Famiglietti, supervisor.

Brubaker, K. L., D. Entekhabi and P. S. Eagleson, 1993, Estimation of continental precipitation recycling, J. Clim. 6, 1077–1089.

Chambers, D. P., J. Chen, R. S. Nerem and B. D. Tapley, 2000, Interannual mean sea level change and the Earth's water mass budget, Geophys. Res. Lett., 27(19), 3073–3076.

Chao, B. F., 1995, Anthropogenic impact on global geodynamics due to water impoundment in major reservoirs, Geophys. Res. Lett., 22, 3533–3536.

Chao, B. F. and W. P. O'Connor, 1988, Global surface-water induced seasonal variations in Earth's rotation and gravitational field, Geophys. J. R. Astron. Soc., 94, 263–270.

Chen, J. L., C. R. Wilson, D. P. Chanbers, R. S. Nerem and B. D. Tapley, 1998, Seasonal global water mass budget and mean sea level variations, Geophys. Res. Lett., 25(19), 3555–3558.

Church, J. A., J. M. Gregory, P. Huybrechts, M. Kuhn, K. Lambeck, M. T. Nhuan, D. Qin, P. L. Woodworth, A. O. Anisimov, F. O. Bryan, A. Cazenave, K. W. Dixon, B. B. Fitzharris, G. M. Flato,

A. Ganopolski, V. Gornitz, J. A. Lowe, A. Noda, J. M. Oberhuber, S. P. O'Farrel, A. Ohmura, M. Oppenheimer, W. R. Peltier, S. C. B. Raper, C. Ritz, G. L. Russell, E. Schlosse, C. K. Shum, T. F. Stocker, R. J. Stouffer, R. S. W. van de Wal, R. Voss, E. C. Wiebe, M. Wild, D. J. Winghma and H. J. Zwally, 2001, Changes in sea level, . In: Climate Change 2001: The Scientific Basis. Contribution of Working Group I to the Third Assessment Report of the Intergovernmental Panel on Climate Change [Houghton, J. T., Y. Ding, D. J. Griggs, M. Noguer, P. J. van der Linden, X. Dai, K. Maskell and C. A. Johnson (eds.)] Cambridge University Press, Cambridge, United Kingdom and New York, NY, USA, 881pp.

Cosh, M. H. and W. Brutsaert, 1999, Aspects of soil moisture variability in the Washita '92 study region, J. of Geophys. Res.-Atmospheres 104(D16): 19751–19757.

Dirmeyer, P. A., A. J. Dolman, and N. Sato, 1999, The Global Soil Wetness Project: A pilot project for global land surface modeling and validation, Bull. Am. Meteorol. Soc.,80, 851–878.

Eltahir, E. A. B. and R. Bras, 1996, Precipitation recycling, Rev. Geophys., 34(3), 367–378.

Famiglietti, J. S., A. A. Berg, S. L. Holl, M. Rodell, K-W. Seo and C. R. Wilson, 2003, Potential for basin-scale water balance evapotranspiration estimation using GRACE, in preparation for Geophys. Res. Let.

Famiglietti, J. S., A. A. Berg, M. Rodell, R. Bindlish, M. Cosh and T. J. Jackson, 2003, Soil moisture variability from the plot to the footprint scale: Field observations from SGP97, SGP99 and SMEX02, submitted, Remote Sens. Environ.

Famiglietti, J., J. Chen, M. Rodell, K. Seo and C. Wilson, 2003, Terrestrial water storage variations from GRACE: Estimation, validation and uncertainty, EOS, Trans. AGU, 84(46), Fall Meet. Suppl., Abstract G32A-0724

Famiglietti, J. S., J. A. Devereaux, C. Laymon, T. Tsegaye, P. R. Houser, T. J. Jackson, S. T.Graham, M. Rodell and P. J. van Oevelen, 1999, Ground-based investigation of spatial-temporal soil moisture variability within remote sensing footprints during SGP97, Wat. Resour. Res., 35(6), 1839–1851.

Gibson, J.K., P. Kållberg, S. Uppala, A. Nomura, A. Hernandez, E. Serrano, ERA Description, ECMWF Reanal. Proj. Rep. Ser.1 72 pp., Eur. Cent. For Medium-Range Weather Forecasts, Reading, England, 1997.

GRDC, 2003, GRDC Status Report 2002, Report No. 29, Federal Institute of Hydrology, Koblenz, Germany, 60 pp.

Guttierez, R. and C. R. Wilson, 1987, Seasonal air and water mass redistribution effects on LAGEOS and Starlette, Geophys. Res. Lett., 14(9), 929–932.

IAHS Ad Hoc Group on Global Water Data Sets, 2001, Global water data: A newly endangered species, EOS Trans., AGU, 82, 54–58.

Kalnay, E., and co-authors, The NCEP/NCAR 40-year reanalysis project, Bull. Amer. Meteorol. Soc., 77, 437–471, 1996.

Kim, G. and A. P. Barros, 2002, Space-time characterization of soil moisture from passive microwave remotely sensed imagery and ancillary data, Remote Sens. Environ., 81, 393–403.

Koblinsky, C. J., R. T. Clarke, A. C. Brenner and H. Frey, 1993, Measurement of river level variations with satellite altimetry, Wat. Resour. Res., 29, 1839–1848.

Koster, R.D., and M.J. Suarez, 1992, Modeling the land surface boundary in climate models as a composite of independent vegetation stands, J. Geophys. Res., 97, 2697–2715.

Jackson, T. J., 2003, Soil Moisture Experiments in 2003 (SMEX03), Experiment Plan, available at http://hydrolab.arsusda.gov/smex03/.

Jackson, T. J., 2002, Soil Moisture Experiments in 2002 (SMEX02), Experiment Plan, available at http://hydrolab.arsusda.gov/smex02/.

Jackson, T. J., A. J. Gasiewski, A. Oldak, M. Klein, E. G. Njoku, A. Yevgrafov, S. Christiani and R. Bindlish, 2002, Soil moisture retrieval using the C-band polarimetric scanning radiometer during the Southern Great Plains 1999 Experiment, IEEE Trans. Geosci. Remote Sensing, 2151–2161.

Jackson, T. J., D. M. LeVine, C. T. Swift, T. J. Schmugge and F. R. Schiebe, 1995, Large area mapping of soil moisture using the ESTAR passive microwave radiometer in Washita '92, Remote Sens. Environ., 53, 27–37

Jackson, T. J., D. M. LeVine, A. Y. Hsu, A. Oldak, P. J. Starks, C. T. Swift, J. D. Isham and M. Haken, 1999, Soil moisture mapping at regional scales using microwave radiometry: The Southern Great Plains Hydrology Experiment, IEEE Trans. Geosci. Remote Sensing, 37(5), 2136–2151.

Liu, J., J. M. Chen and J, Cihlar, 2003, Mapping evapotranspiration based on remote sensing: An application to Canada's landmass, 39(7), DOI 10.1029/2002WR001680.

Maheu, C., A. Cazenave and C. R. Mechoso, 2003, Water level fluctuations in the Plata Basin (South America) from TOPEX/Poseidon satellite altimetry, Geophys. Res. Lett, 30(3), 1143, doi:10.1029/2002GL016033.

Mercier, F., A. Cazenave and C. Maheu, 2002, Interannual lake level fluctuations (1993–1999) in Africa from TOPEX/Poseidon: connections with ocean-atmosphere interactions over the Indian Ocean, Global Planet. Change, 32, 141–163.

Mohanty, B. P., J. S. Famiglietti and T. H. Skaggs, 2000a, Evaluation of soil moisture spatial structure in a mixed-vegetation pixel during the SGP97 Hydrology Experiment, Wat. Resour. Res., 36(12), 3675–3686.

Mohanty, B. P., T. H. Skaggs and J. S. Famiglietti, 2000b, Analysis and mapping of field-scale soil moisture variability using high resolution ground-based data during the Southern Great Plains 1997 (SGP97) Hydrology Experiment, Wat. Resour. Res., 1023–1031.

Njoku, E. G., T. J. Jackson, V. Lakshmi, T. K. Chan, and S. V. Nghiem, 2003, Soil moisture retrieval from AMSR-E, IEEE Trans. Geoscii. Remote Sensing, 41(2) 215–229.

Oldak, A., Y. Pachepsky, et al., 2002, Statistical properties of soil moisture images revisited, J. Hydrology , 255(1–4), 12–24.

Peltier, W. R., 1994, Ice-Age Paleotopography, Science, 265, 195–201.

Peltier, W. R., 1995, Paleotopography of Ice Age ice sheets, Science, 267, 536–538.

Rodell, M., P. R. Houser, U. Jambor, J. Gottschalck, K. Mitchell, C.-J. Meng, K. Arsenault, B. Cosgrove, J. Radakovich, M. Bosilovich, J. K. Entin, J. P. Walker, D. Lohmann, and D. Toll, 2003, The Global Land Data Assimilation System, Bull. Amer. Meteor. Soc., to appear

Rodell, M. and J. S. Famiglietti, 2002, The potential for satellite-based monitoring of groundwater storage changes using GRACE: The High Plains Aquifer, Central U. S., J. Hydrol., 263, 245–256.

Rodell, M. and J. S. Famiglietti, 1999, Detectability of variations in continental water storage from satellite observations of the time-variable gravity field, Wat. Resour. Res., 35(9), 2705–2723.

Seo, K-W., C. R. Wilson, J. Chen, J. S. Famiglietti, and M. Rodell, 2003, Filters to estimate water storage variations from GRACE, J. Geodesy, in review.

Silvestrin, P., M., Berger, Y. Kerr and J. Font, 2001, ESA's second Earth Explorer Opportunity Mission: The Soil Moisture and Ocean Salinity Mission – SMOS, IEEE Geosci. and Remote Sensing Newsletter, 118, 11–14.

Shiklomanov, A., R. Lammers and C. J. Vörösmarty, 2002, Widespread decline in hydrological monitoring threatens Pan-Arctic research, Eos. Trans. AGU, 83, 13–16.

Smith, L. C., 1997, Satellite remote sensing of river inundation area, stage and discharge: A review, Hydrol. Processes, 11, 1427–1439.

Smith, L., B. Isacks, A. Bloom and A. Murray, 1996, Estimation of discharge from three braided rivers using synthetic aperture radar (SAR) satellite imagery: Potential application to ungaged basins, Wat. Resour. Res., 32, 2021–2034.

Stokstad, E., 1999, Scarcity of rain, stream gages threatens forecasts, Science, 285, 1199.

Swenson, S. and J. Wahr, 2002, Methods for inferring regional surface-mass anomalies from Gravity Recovery and Climate Experiment (GRACE) measurements of time-variable gravity, J. Geophys. Res., 107 (B9), ETG 3-1 to ETG 3-13, 0148-0227, 2193, doi:10.1029/2001JB000576.

Vörösmarty, C., C. Birkett, L. Dingman, D. Lettenmaier, Y. Kim, E. Rodriguez and G. Emmit, 1999, NASA Post-2002 land surface hydrology mission component for surface water monitoring: HYDRA-SAT HYDRological Altimetry SATellite, A report from the NASA post-2002 land surface hydrology planning workshop, Irvine, CA, 12–14 April, 1999.

Wahr, J., M. Molenaar, and F. Bryan, 1998, Time variability of the Earth's gravity field: Hydrological and oceanic effects and their possible detection using GRACE, J. Geophys. Res., 103(B12), 30,205–30,230.

Wahr, J., S. Swenson and I. Velicogna, 2003, An initial look at GRACE time-variable results, EOS, Trans. AGU, 84(46), Fall Meet. Suppl., Abstract G31A-05.

Wigneron, J-P., A. Chanzy, J.-C. Calvet and N. Bruguier, 1995, A simple algorithm to retrieve soil moisture and vegetation biomass using passive microwave measurements over crop fields, Remote Sens. Environ., 51, 331–341.

James S. Famiglietti, Department of Earth System Science, University of California, Irvine, Irvine, California, 92692-3100.

High-Resolution Measurement of Ocean Surface Topography by Radar Interferometry for Oceanographic and Geophysical Applications

Lee-Lueng Fu and Ernesto Rodriguez

Jet Propulsion Laboratory, California Institute of Technology, Pasadena, California

This paper describes a new approach to making altimeter measurements from off-nadir radar signal returns and its oceanographic and geophysical applications. The approach is based on the technique of radar interferomety and the new instrument is called the Wide-Swath Ocean Altimeter (WSOA). WSOA is designed to be flown with a Jason class conventional dual-frequency altimeter system, including a multi-frequency radiometer for the correction of the effects of water vapor in the troposphere. WSOA will extend the measurement from a line along the nadir to a swath of 200 km centered on the nadir track. The most important application of WSOA is to provide the first synoptic maps of the global oceanic eddy field with a spatial resolution of 15 km x 15 km. The strong currents and water property anomalies (in temperature, salinity, oxygen, etc.) associated with ocean eddies are a major factor affecting the oceanic general circulation and the biogeochemical cycles of the ocean. WSOA will also provide measurements that allow the monitoring and study of coastal currents and tides that affect the lives of half of the world's population. The intrinsic resolution of WSOA in the look direction of the interferometric radar is about 1 km, allowing the estimation of sea surface slope with an accuracy of 1 micro radian down to a wavelength of 20 km for a 2-year mission (15 km for a 5-year mission). This capability will make contributions to the mapping of the details of the sea floor topography.

1. INTRODUCTION

The height of sea surface, or the ocean surface topography (denoted as OST hereafter), reveals a great deal of information about the physical state of the ocean as well as the earth's gravity field. To a first approximation, OST is determined by the earth's gravity field, which conforms to the sea surface close to an equi-geopotential surface called the geoid. A variety of oceanic processes may cause OST to deviate from the geoid, including waves, tides, and currents. The knowledge of OST thus has many applications to oceanography and geophysics. The idea of making measurement of OST using a radar altimeter from space was developed in the 1970's when the technology of artificial satellite and microwave remote sensing became available. A comprehensive review of the scientific applications of satellite altimetric measurement can be found in Fu and Cazenave (2001).

The past decade has seen the most intensive observations of the global oceans by satellite altimeters. The Joint U.S./France TOPEX/Poseidon (T/P) Mission has become the longest single radar mission ever flown in space, providing the most accurate measurements for the study of large-scale ocean circulation and variability beginning in October, 1992. Jason, the follow-on to T/P was launched in December, 2001. The European Space Agency's ERS-1 and –2 missions and their follow-on, the ENVISAT mission, have also provided continuous altimetric observations since 1991. Furthermore, the U.S. Navy launched Geosat Follow-on (GFO), which became operational in 2000. The combined data from these missions have higher spatial resolution and greater coverage than the data from individual missions. However, the turbulent ocean is rich in spatial scales not resolvable by a single conventional altimeter or the combination of the few existing ones. The unresolved scales play important roles in ocean general circulation and transport of heat and water properties affecting global climate.

Satellite altimetry has also revolutionized geophysics by providing the first globally uniform and detailed description of the marine gravity field. Perhaps the most important results are the confirmation of plate tectonics and the findings of new features that were not anticipated by the theory. Another important application is the estimation of sea floor topography from the altimetrically-determined gravity anomalies. However, there is yet more to be learned from the small-scale variability of OST beyond the spatial resolution provided by the existing data record.

Measurements of OST by radar altimetry inevitably involve a tradeoff between spatial resolution and temporal resolution (Figure 1). Improvement of one results in degradation of the other. For example, the sampling pattern for the early Geodetic Phase of the ERS-1 mission had a very fine longitudinal ground track spacing of 0.15°. This small track spacing is very useful for geophysical studies, but not for oceanographic studies because of temporal aliasing of mesoscale variability owing to the long 168-day orbit repeat period. At the other extreme, the short 3-day exact repeat during the Ice Phases of the ERS-1 mission and during the last few weeks of the Seasat mission resulted in a coarse longitudinal ground track spacing of 8.37°. The tradeoff between spatial resolution and temporal resolution thus forces a compromise in mission design

The T/P orbit configuration was chosen specifically for measuring the large-scale OST field for studies of ocean variability on monthly and longer time scales. This dictates an orbit with approximately a 10-day repeat period in order to minimize temporal aliasing of mesoscale variability at the measurement locations along the exactly repeating ground track. The precise repeat period of 9.9156 solar days for T/P was chosen to minimize the effects of tidal aliasing (Parke et al., 1987). The corresponding 2.834° longitudinal spacing of ground tracks (Figure 1) significantly limits the spatial scales of OST variability that can be resolved by the T/P sampling pattern. Processes with spatial scales shorter than approxi-

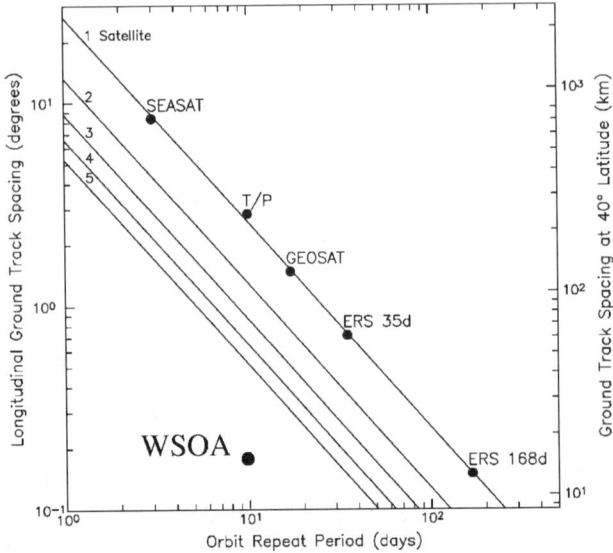

Figure 1. The relationship between orbit repeat period and the longitudinal separation of neighboring ground tracks for altimeters in exact repeat orbit configurations. The ground track spacing is displayed in degrees along the left axis and in kilometers at 40° latitude along the right axis. In log-log space, the choices of exact repeat period and ground track spacing for a single satellite fall approximately along the top straight line in the figure. The repeat periods and ground track separations of past and present altimeter missions are shown by the solid circles. The two sampling patterns shown for the ERS altimeter correspond to the Multi-Disciplinary Phase (35-day repeat) and the Geodetic Phase (168-day repeat) of the ERS-1 satellite. The improvements in the resolution that would be obtained from multiple satellites in coordinated orbit configurations with evenly spaced ground tracks are shown for constellations of 2, 3, 4 and 5 satellites. Also shown is the resolution expected to be achieved by WSOA. (The figure without the WSOA information is adapted from Chelton, 2001.)

mately 500 km are not fully resolved in the T/P data (Chelton and Schlax, 2003).

OST varies over a continuum of space and time scales in association with a wide range of physical processes that interact in very complicated ways. A schematic diagram of the spatial and temporal scales of various oceanic phenomena is shown in Plate 1. The box bordered by dashed lines indicates the range of the spatial and temporal scales that can be fully resolved by a single altimeter in the T/P orbit configuration. Variability with spatial scales that fall outside of the box cannot be adequately addressed from T/P data. Note that the right-side border of the box represents the largest scales of the ocean basins. Different orbit configurations result in slightly different resolutions, but the fundamental limitation is imposed by the tradeoff between spatial and temporal resolution shown in Figure 1. While very useful for studies of large-scale phenomena such

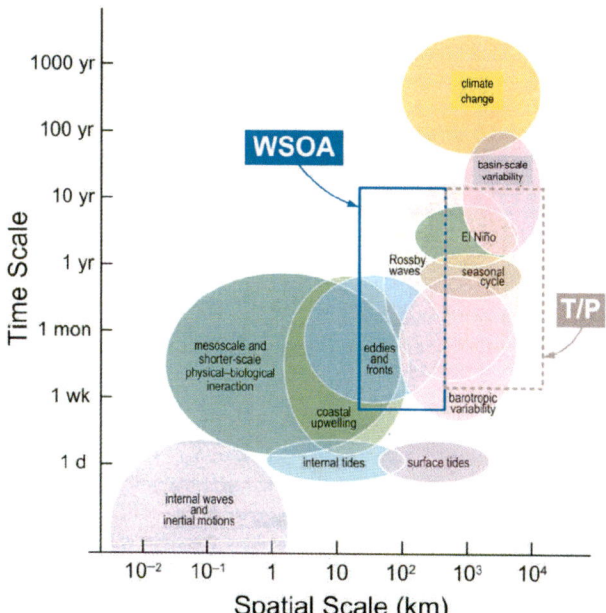

Plate 1. The approximate space and time scales of phenomena of interest. The box bordered by dashed lines represents the domain covered by the TOPEX/Poseidon Mission. The box bordered by solid lines represents the additional coverage provided by WSOA flown in the T/P orbit. (adapted from Chelton, 2001)

as seasonal variability of OST, interannual variability associated with the El Niño–Southern Oscillation phenomenon, Rossby wave dynamics and basin-scale variability (Fu and Chelton, 2001), it is evident from Plate 1 that there are many processes with spatial and temporal scales that are too short to be resolved by the sampling pattern of a single altimeter.

An important spatial scale in determining the structure of ocean eddies is the Rossby radius of deformation, which ranges from 15 km at high latitudes to 230 km in the tropics (Chelton et al., 1998). Although this scale is well sampled along track by a single altimeter, it is not resolved in the cross-track direction owing to the wide spacing between the tracks. It is well known that the kinetic energy of ocean circulation in the open ocean is dominated by eddies (Wyrtki et al., 1976). The kinetic energy associated with the mean flow in the open ocean away from the boundary currents is less than that associated with eddies by an order of magnitude. The sampling by a single conventional altimeter is thus not able to resolve the synoptic structures of the most energetic component of ocean circulation in the open ocean. It is shown in Section 3.1 that the effects of eddies are crucial in transporting heat from the tropics to high latitudes. High resolution measurement is thus required to improve the understanding of the roles of ocean in climate variability.

Regarding geophysical applications, the main interests are focused on wavelengths less than 3000 km (Cazenave and Royer, 2001), especially less than 100 km where the sea floor topography can be predicted from OST (Sandwell and Smith, 2001). However, there is a physical limit on the resolution of gravity anomalies reflected in OST. This limit is twice the regional ocean depth (Sandwell and Smith, 2001), or about 8 km in the deep ocean. The data record compiled from the Geosat and ERS altimeters has a spatial resolution of 25–45 km (Sandwell et al., 2001). There is significant amount of important information in the unresolved scales from 8–25 km.

In recognition of the importance of the phenomena with scales shorter than can be resolved in past and present altimeter datasets, NASA Earth Science Enterprise established the High-Resolution Ocean Topography Science Working Group (HOTSWG) to evaluate the scientific and operational rationales and requirements for high-resolution measurements of ocean topography, as well as to review available and developing technologies and make recommendations on strategies of implementation (Chelton, 2001). As summarized in Figure 1, one approach to obtaining high-resolution measurements of ocean topography is from a constellation of altimeters in coordinated orbit configurations. For the 10-day exact repeat orbit of T/P, for example, the 2.834° longitudinal spacing of ground tracks for a single altimeter would be reduced to 0.945° with a triplet altimeter mission with evenly spaced ground tracks. The ground track spacing would decrease to 0.567° (or 48 km at 40 degree latitude) with a pentad altimeter mission with 10-day exact repeat. However, such resolution is not sufficient for resolving the first-mode Rossby radius of deformation (the scale of ocean eddies), which is about 40 km at 40 degrees latitude (Chelton et al., 1998). It would take 10 conventional altimeters in formation flight to create a 20 km ground track spacing at 40 degree latitude to resolve the 40 km Rossby radius of deformation.

The requirement on spatial resolution for geophysical studies is even more stringent, but it does not demand exactly repeating ground tracks. The geodetic phases of Geosat and ERS have led to major advances in geophysics but they are not useful for oceanographic studies, which require repeat observations to address the temporal variability of ocean circulation. Therefore there is a fundamental conflict in the requirements between oceanography and geophysics. This conflict is difficult to resolve by the approach of multiple conventional altimeters.

An alternate approach is to explore new technologies that have recently been developed for high-resolution measurements of ocean topography over broad swaths using methods different from traditional nadir-looking radar altimetry. Described in this paper are the measurement principles and applications of the Wide-Swath Ocean Altimeter (WSOA) proposed to be flown on the Ocean Surface Topography Mission (OSTM) for feasibility demonstration (Fu, 2003). OSTM is a follow-on mission to Jason and is currently planned for launch in 2007.

WSOA is able to provide 15 km x 15 km spatial resolution for oceanographic applications (Figure 1). The crossover coverage of the wide swath also shortens the revisit time and thus enhances the temporal resolution. WSOA is thus able to extend the resolved spatial and temporal scales to the box bordered by solid lines in Plate 1. The increased resolution will allow the study of ocean eddies and their roles in ocean general circulation, climate, and biogeochemical processes. Knowledge of coastal circulation and tides will also be improved. The improvement of the corrections for coastal tides will retrospectively make the decade-long T/P observations near the coastal regions of the world's oceans a great asset for coastal oceanography.

For geophysical applications, WSOA's intrinsic resolution will be exploited. With averaging of repeat observations, accurate estimate (error less than 1 micro radian) of the cross-track sea surface slope can be obtained down to a wavelength of 20 km. Such measurements will improve the current poor knowledge of the east–west component of sea surface slopes for geophysical applications.

2. WIDE-SWATH ALTIMETRY WITH RADAR INTERFEROMETRY

2.1 Measurement Principle

The goal of WSOA is to extend the coverage of OST measured at the nadir point, as provided by conventional altimeters, to a large swath on either side of the satellite track. Conventional altimetry measures only one variable related to OST: the round-trip travel time to the nearest point on the surface. Because geometry constrains the nearest surface point to be the nadir point, it is straightforward to determine the location of the OST measurement if the satellite position is known.

In order to determine the location of the radar returns that are not from the nearest point on the surface, WSOA has been developed from the technique of radar interferometry (Rodriguez and Martin, 1992). This technique exploits the relative delay between the target-reflected signals received by two antennas separated by a baseline distance. The range measurements from the two antennas and the baseline form a triangle which can be used for determining the location of the target in the plane of the observation (Figure 2). The measurement triangle is made up of the baseline B, and the ranges from the target to the two antennas, r_1 and r_2. The baseline is known by construction and knowledge of the spacecraft attitude. The range r_1 is determined by system timing measurements. The range difference between r_1 and r_2 is determined by measuring the relative phase shift Φ between the two sig-

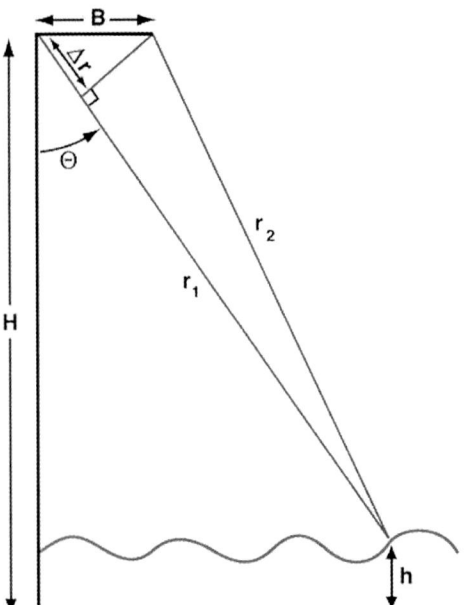

Figure 2. Geometric concept used for radar interferometry.

nals. The phase shift is related to the range difference Δr by the equation $\Phi = 2\pi \Delta r/\lambda$, where λ is the radar wavelength. The additional information required for determining the target location—the incidence angle θ, can be obtained from the range difference by means of the relationship $\Phi = 2\pi B \sin(\theta)/\lambda$. Given these measurements, the height h above a reference plane can be obtained using the equation $h = H - r_1 \cos(\theta)$, where H is the altitude of the satellite determined from orbit ephemeris.

A major difference between conventional altimetry and interferometry is that the interferometric measurement of the range relies on the complex phase information, which is available for each imaged pixel in the scene. In contrast, the altimeter measurement relies on the power and the specific shape of the leading edge of the return waveform, which is only available for the nadir point. Thus, the interferometric measurement of the range is intrinsically more accurate than the altimeter measurement (since it relies on the phase, which can be determined to a fraction of a wavelength), and is available for all imaged points in the scene.

Shown in Figure 3 is the proposed configuration of WSOA as part of a Jason-class altimeter mission, which is OSTM in the present case. With a deployable 6.4 m baseline (length limited by the Jason spacecraft dimension), the Ku-band interferometry system will be integrated with the standard Jason instrument package: Ku and C band altimeters, a three-frequency radiometer, a DORIS receiver and a GPS receiver. The interferometric antennas have beams illuminating both sides of the nadir track to produce a total swath width of 200

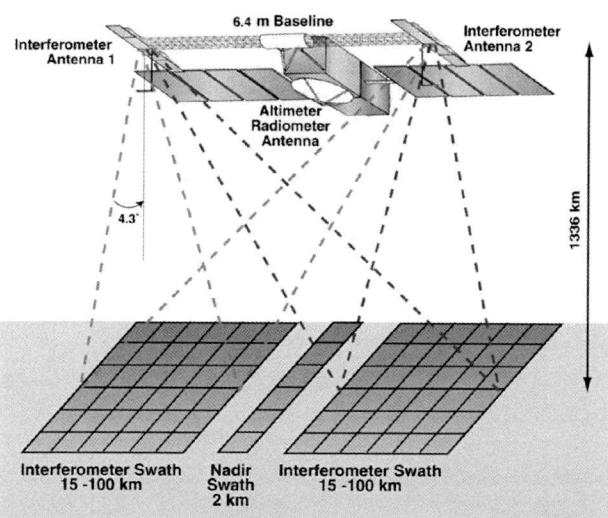

Figure 3. Conceptual operation of the WSOA instrument to measure a 200 km swath.

km, when optimally oriented. However, for the flight demonstration of WSOA on OSTM, some reduction in coverage will occur due to the steering requirements of the platform.

The instrument spatial resolution characteristics are determined by the size of the antenna beam in the direction perpendicular to the baseline direction, and by the system bandwidth (or intrinsic range resolution) in the direction parallel to the baseline direction. For the WSOA system, the first resolution is 11 km, while the intrinsic range resolution varies from about 500 m in the inner swath to about 100 m in the outer swath. The height estimates measured at these resolutions are noisy, so noise is reduced by averaging all measurements within a box. The topographic results are posted on a 15 km grid. The improvement in coverage which can be obtained by a single WSOA is compared against the coverage obtained by two conventional altimeters flown in interleaving T/P-type orbits in Plate 2. An additional advantage of wide-swath coverage is that the revisit time between observations is reduced, on average, to half an orbit cycle, since most points are imaged by both ascending and descending satellite passes (see Plate 2c).

2.2 Random and Systematic Measurement Errors

Random errors in the interferometric measurements are introduced through measurement noise of the interferometric phase difference. Spatial averaging will reduce the random errors by the square root of the number of pixels averaged. There are three different kinds of systematic errors: range errors; roll errors; and systematic phase errors. Range errors are caused by the same sources that are responsible for the altimeter range errors due to the reduction of the speed of light in the atmosphere and ionosphere depending on the medium condition. The electromagnetic bias caused by ocean waves also produces a range error, as in the nadir altimeter measurement (Chelton et al., 2001). The only additional error source for the WSOA is the fact that the range corrections across the swath are assumed to be identical to the nadir corrections, which are available from the nadir altimeter and used for the corrections across the WSOA swath. In order to assess the spatial variability of the media and the electromagnetic bias corrections, we examined the TOPEX/Poseidon data and computed the average change in the correction for varying cross-track distances. This error ranges from 1 cm in the near-nadir edge of the swath to 2 cm in the outer swath edge.

The roll errors are induced by the lack of knowledge in the roll angle of the interferometric baseline. The effect of an error in the roll knowledge of $\delta\theta$ is to introduce a tilt to the measured surface of $x\delta\theta$, where x is the distance from the nadir track. As an order of magnitude, a roll error of 0.1 arcsec will result in a height error of 4.5 cm at the edge of the swath. The main source of roll error is uncertainty in the knowledge of the spacecraft attitude and small distortion of the antenna system. Current attitude measurement systems have the precision appropriate for measuring the roll to the desired tolerance, but they suffer from long-term (fractions of a revolution) drifts, which must be corrected by the calibration process described later.

The phase errors are due to changes in the delays in the electronic system. For the small look angles of WSOA (~4°), their effect is identical to the roll error: they induce linear tilts over the swath. The changes in the system delay are driven by temperature changes, which change slowly over an orbit, so the systematic phase errors can be calibrated in the same fashion as the roll errors, as described in the next section.

2.3 WSOA Performance and Data Products

The accuracy of the WSOA instrument, prior to calibration, is governed by the systematic errors. These errors vary slowly over the orbit and have known geometric signatures. In order to reduce these errors significantly, we have developed a calibration process which uses the fact that the nadir altimeter accompanying the WSOA is not affected by the roll in the same way as the WSOA. If the ocean did not move, the systematic errors could be estimated by taking height differences between the heights measured by the nadir altimeter in ascending track and the heights measured by the interferometer in the descending track (or conversely). Since the altimeter measurement is unaffected by the roll, the height differences will be proportional to the roll error, and can be estimated using least squares inversion. In addition to these differences, the

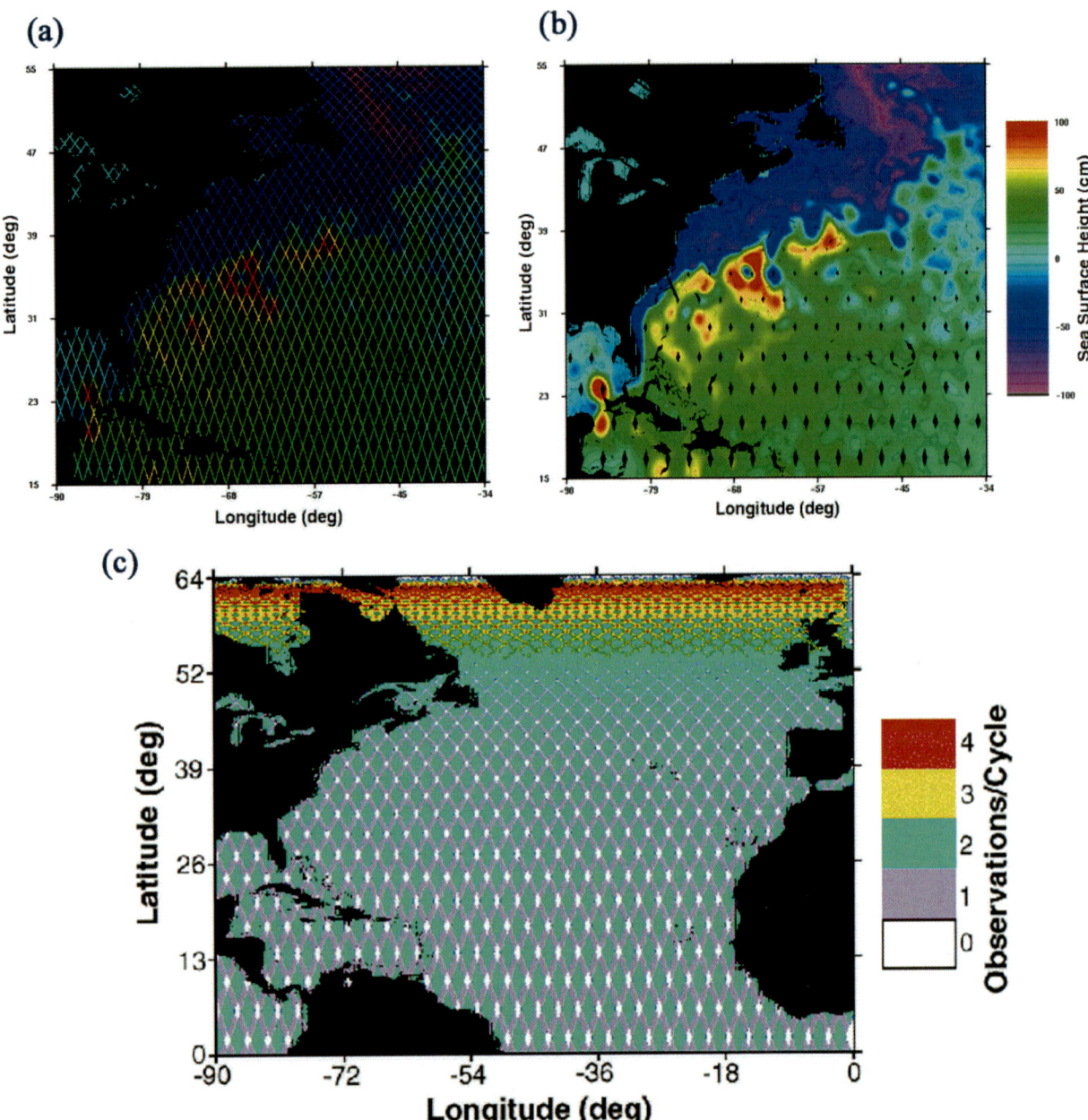

Plate 2. (a) Coverage of the Gulf Stream and the North Atlantic Ocean by two coordinated TOPEX/Poseidon-class altimeters and (b) by a single WSOA altimeter. (c) Coverage map for the WSOA altimeter showing the number of observations per ten-day cycle for a 200 km swath.

interferometer–interferometer ascending/descending height differences can be used to further constrain the difference between the errors at the cross-over diamonds. This is similar to the standard cross-over adjustment for reducing orbit errors in nadir altimetry applications (e.g., Fu and Vazquez, 1988).

In order to validate the cross-over calibration scheme, and to study the impact of the oceanic motions on the calibration accuracy, we used the Los Alamos 0.1 degree eddy-resolving model (Smith et al., 2000) as an input to a measurement and calibration simulator that includes simulated random and systematic errors of the measurement system. The resulting RMS residual error estimated from an experiment in the North Atlantic basin is shown in Plate 3. The error field has a geographic pattern with higher values near the Gulf Stream where the effects of oceanic motions are the greatest. However, the magnitude of the errors is generally much smaller than the signals.

Based on the results from the experiment and the height errors described above, we estimate the expected error budget for a single pass of the WSOA instrument as given by Table 1. Note that orbit errors are not included in the budget. Because the measurement objective of WSOA is to improve the resolution of small-scale signals, orbit errors whose spatial scales are thousands of kilometers are not relevant for discussing the WSOA measurement performance.

One of the primary advantages of the WSOA system is that it produces OST maps in two dimensions, not just along the nadir track. This implies that both the surface gradient (which can be converted to the geostrophic vector velocity) and the surface Laplacian (which can be converted to the relative geostrophic vorticity) can be computed at every point. The vector velocity has been computed using standard altimetry and been discussed extensively in the literature Schlax and Chelton, 2003). The most accurate computation is obtained at the crossover points, where the direction of the velocity can be estimated directly. However, the accuracy of the computed vector velocity degrades with latitude, because the angle between the crossing tracks approaches zero. Furthermore, the ascending and descending data are collected at two different times, and the vector velocity will be in error for regions with short correlation

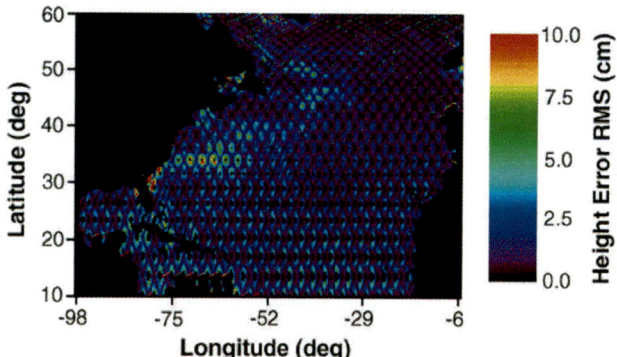

Plate 3. Residual height errors from the simulation using the Los Alamos circulation model as input. Errors increase near the Gulf Stream, but are much smaller than the dynamic topography signals in the region (see Plate 2).

times compared to the crossover period. It is not possible to obtain an accurate estimate of the relative geostrophic vorticity using nadir altimeter tracks. Plate 4 shows examples of the velocity magnitude and relative vorticity fields retrieved from the simulated data discussed previously. The possibility of calculating vector velocities at every point within the WSOA swath opens up the possibility of calculating flow-related quantities, such as the Reynolds' stresses, which can presently be computed only at the crossover points. Plate 4 shows an example of the potential data product from WSOA.

Table 1. WSOA height error budget (in cm) compared to the Jason height error budget. The WSOA errors vary as a function of cross-track distance, so four representative cross-track distances are presented (26, 47, 68, and 89 km). The values quoted here represent single pass values, and do not include performance gains which can be made by combining ascending and descending or overlapping passes.

Error Source	26 km	47 km	68 km	89 km	Jason
Interferometric Precision	1.9	1.7	1.9	2.5	1.7
Residual Systematic Error	1.6	1.9	2.4	2.9	1.4
EM Bias	2.0	2.0	2.0	2.0	2.0
Wet Troposphere	1.2	1.2	1.2	1.2	1.2
Dry Troposphere	0.7	0.7	0.7	0.7	0.7
Ionosphere	0.5	0.5	0.5	0.5	0.5
Cross-track Media Decorrelation	1.0	1.3	1.5	1.8	
Height Error	3.7	3.8	4.2	4.9	3.3

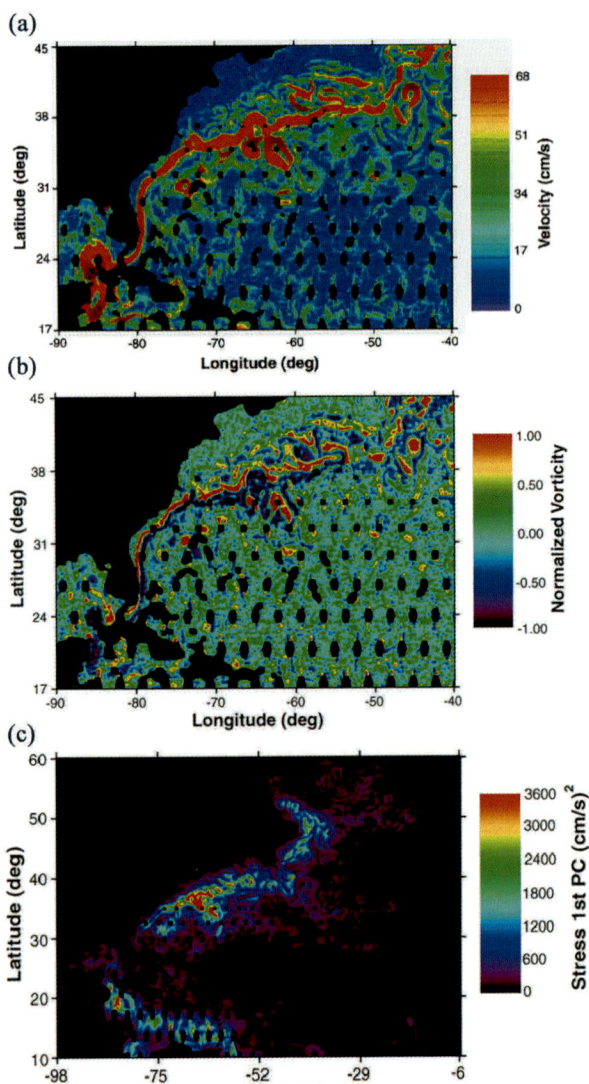

Plate 4. Sample geophysical data products from the WSOA instrument: (a) the geostrophic velocity; (b) the relative geostrophic vorticity; (c) the Reynolds stress tensor. These quantities may be derived at all points in the WSOA swath, not just at the altimeter crossovers.

An additional feature of the WSOA instrument is that, unlike conventional altimeters, no dynamic tracking of the range gate is required on board the satellite. Without the need of range-tracking which is interrupted at land-sea boundaries, WSOA is always able to make valid measurement within a resolution cell next to the coast. The proximity to land where valid measurement can be made depends on the geometry of the track relative to the coastline. The best case is when the satellite track is parallel to the coast. The observation closest to the coast is within one cross-track resolution cell, which is 500 m (This resolution may be degraded to 1 km due to the constraint of data downlink). The worst case is when the satellite track is perpendicular to the coast. Then the closet observation is offshore by one along-track resolution cell, which is 11 km.

3. SCIENTIFIC APPLICATIONS

3.1 Mesoscale Eddies

Satellite altimetry has made significant contributions to observing and understanding the oceanic mesoscale eddy variability (see Le Traon and Morrow, 2001 for a recent review). Altimeter data analyses provide, for the first time, global description of the eddy energy and its seasonal and interannual variations. The time and space scales of the mesoscale circulation are assessed. The eddy/mean flow interactions are mapped and the results provide important contributions towards understanding the dynamics of the western boundary currents and the Antarctic Circumpolar Current.

Most of these studies are based on Geosat, TOPEX/Poseidon, or ERS altimeter data separately. The contributions of the merging of T/P and ERS are well illustrated by Ducet et al. (2000) and Ducet and Le Traon (2001). The variance, Reynolds stress, and eddy momentum flux associated with the velocity field have been mapped globally, providing description of the eddy kinetic energy with a resolution never achieved before. Despite the progress in the knowledge of the statistics and dynamics of the eddy field, it is still difficult to map the synoptic evolution of individual eddy motions because of insufficient sampling capability of two or even three simultaneously flying altimeters. Yet the eddies are important in the momentum and energy budgets of the oceanic general circulation as well as in transporting heat and nutrients that affect climate and biogeochemical cycles. Ocean eddies also affect maritime operations such as offshore oil drilling, ship routing, fisheries, marine debris dispersion, etc.

Recent modeling studies (e.g., Bryan and Smith, 1998; Smith et al., 2000) have demonstrated the sensitivity of the simulated oceanic general circulation to the extent of the resolution of mesoscale eddies. Using the Parallel Ocean Program (POP) ocean model (Dukowicz and Smith, 1994), Bryan and Smith (1998) conducted three experiments that differ only in horizontal resolution and the associated dissipation parameters. A Mercator grid was used with equatorial resolution of 0.1°, 0.2° or 0.4°. The domain included the Atlantic Ocean from 20°S to 73°N. Clearly shown in Figure 4 is that the improvement in model resolution is responsible for the nearly 50% increase in the simulation of the meridional heat transport and brings the model simulation closer to the observations, especially at mid latitudes. However, the models are not able to simulate the heat transport at low latitudes. In order to improve ocean models, we need to make eddy-resolving observations of the ocean circulation to fully test the model's performance.

Another important application of WSOA is to remove the effects of eddies in the analysis of the Argo float array. Argo is an international program to launch thousands of profiling floats to measure the temperature, salinity, and circulation of the upper 2000 m of the global oceans (Wilson, 2000; also see http://www-argo.ucsd.edu/designdoc.html). With the goal of reaching 3000 floats globally, the ultimate spatial resolution of Argo is about 4° x 4°. Such resolution is not able to resolve the energetic ocean eddies and the analysis of the data will suffer from serious aliasing effects of eddies. The ability of resolving eddies makes WSOA highly complementary to the Argo program. The two systems are a powerful combination for a global ocean observing system.

3.2 Biogeochemical Processes

Measurements of ocean color from space have revealed complex patterns at mesoscales (Doney et al., 2003). At these scales, ocean color variability is thought to be modulated by physical processes either directly by turbulent advection and stirring (Mackas *et al.*, 1985) or indirectly via impacts on phytoplankton growth rates and trophic interactions (Garçon *et al.*, 2001). From a global perspective, satellites provide the only feasible approach for mapping ocean surface chlorophyll at the required time and space scales to characterize mesoscale variability.

To illustrate the scales of biological variability, the semivariogram approach from geostatistics was used by Doney et al. (2003) to calculate the mesoscale variance and spatial decorrelation length scales globally for a full year of SeaWiFS ocean color data (McClain *et al.*, 1998). The resulting spatial structure function provides a useful measure for quantifying biological–physical interactions and discriminating among theoretical models. Of interest here are the spatial scales of the biological activity as revealed by the semivariogram analysis. The pattern of the zonal spatial scale (Plate 5) exhibits a strong, banded latitudinal pattern with maximal values of 250–300 km along the Equator. The scale decreases poleward to less than 50 km near the poles. The latitudinal decrease in biological spatial scales is similar to the latitudinal variation of the scales of mesoscale eddy motions, suggesting that the mesoscale eddies play an important roles in determining the biological variability. The capability of WSOA in making eddy-resolving synoptic observations thus has important applications to the study of biogeochemical processes.

3.3 Coastal Circulation

Transport processes in the coastal ocean are of interest to both scientific research and operational applications. Scientific issues include: coastal upwelling and cross-shelf transport

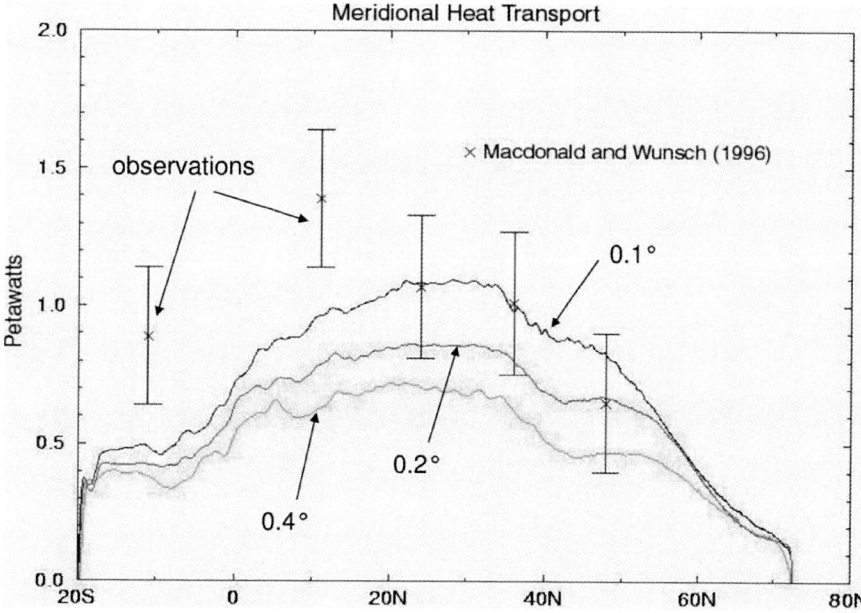

Figure 4. Meridional heat transport in the Atlantic Ocean simulated by an ocean circulation model (see the text for details) with a horizontal resolution of 0.1°, 0.2° and 0.4°, respectively. Also shown is the inverse model estimate of Macdonald and Wunsch (1996). (from Bryan and Smith, 1998)

of mass, salt, and nutrients; the degree of topographic control over the shelf (widths of 20–300 km); the effect of the coastal circulation on marine organisms; changes to coastal ocean ecosystems caused by climate variability, etc. The need for operational nowcasts and forecasts of the coastal ocean circulation arises from the heavy human utilization of the coastal ocean—fishing, transportation, recreation, waste disposal, habitation next to the water's edge, and the search and rescue services needed by those engaged in the above activities.

As the space and time scales for some coastal processes decrease rapidly as one approaches the coast, the sampling requirement for coastal observation is quite a challenge to satellite altimetry. However, the processes that create significant signatures in OST generally have spatial scales close the Rossby radius of deformation and temporal scales of a few days. Shown in Plate 6 are model simulations of the surface density and currents offshore from the U.S. northwest coast (Strub, 2001). There are significant changes over a time span of 4 days. To a large extent, such variability can be sampled by the 15-km and 5-day resolution provided by WSOA at mid latitudes (see Plate 2).

Another issue is how close to the coast WSOA can make valid measurement. As described earlier, the cross-track resolution of WSOA is adjustable, allowing the measurement to be made at about 500 m from the coast if the coast is parallel to the ground tracks. For oblique orientation between the coast and ground tracks, the distance between the coast and the closest observations varies from 500m–11 km depending on the coast orientation.

3.4 Shallow-Water Tides

Besides the scientific significance of the ocean tides in its own right, the OST variability caused by tides, if not corrected for, often overwhelms the signals of ocean circulation. In the deep ocean, the major constituents of tides have been mapped from satellite altimetry with an accuracy of 3 cm (Le Provost, 2001). Most of the state-of-the-art global ocean tide models derived from altimetry are generated by applying long-wavelength corrections to *a priori* hydrodynamic model results. The short wavelengths in these solutions are not corrected because of the limitation of the spatial resolution of the T/P Mission, which has been the primary source of observations for developing the deep ocean tide models.

The wavelengths of tidal motions are governed by the shallow-water gravity wave dynamics and can be expressed as $T(gh)^{1/2}$, where T is the tidal period, g is the gravitational constant, and h is the water depth. For semi-diurnal tides, the wavelength decreases from 4430 km at a depth of 1000 m to 625 km at a depth of 20 m. Such scales are not adequately resolved by the coarse cross-track spacing of the T/P ground tracks. Furthermore, the amplitudes of tides over sallow waters are much larger than those in the deep ocean, posing a significant problem for studying the coastal ocean circulation using altimetry data. Displayed in Plate 7 is the mean kinetic energy of the M_2 tides, clearly revealing the locations of the short-wavelength tides that are associated with high velocities.

In coastal areas, further complexity is caused by non-linear dynamical processes, which distort the tidal waves (Le Provost, 1991). They are associated with the sensitivity of wave propagation to the variation in depth between high and low tides, spatial acceleration of the flows around capes, and non-linear bottom friction. In harmonic analysis of the tides, these non-linear effects lead to the generation of high-order harmonics in all tidal constituents. These nonlinear constituents can reach several tens of centimeters. The complexity of their patterns increases with the tidal frequency. The short wavelengths of these higher harmonic tidal waves are also beyond the cross-track resolution of the T/P Mission.

The high-resolution data from WSOA in the T/P orbit are expected to resolve the two-dimensional characteristics of the shallow water tides as well as their non-linear complications. The improved tidal knowledge is not only important to coastal navigation and other offshore operations, but also to the sci-

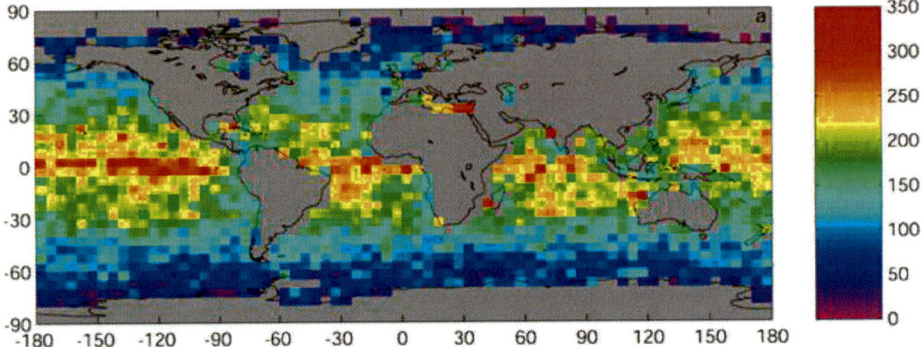

Plate 5. Spatial scales of the variability of the ocean color observations from SeaWiFS (from Doney et al., 2003). The color scales are in km.

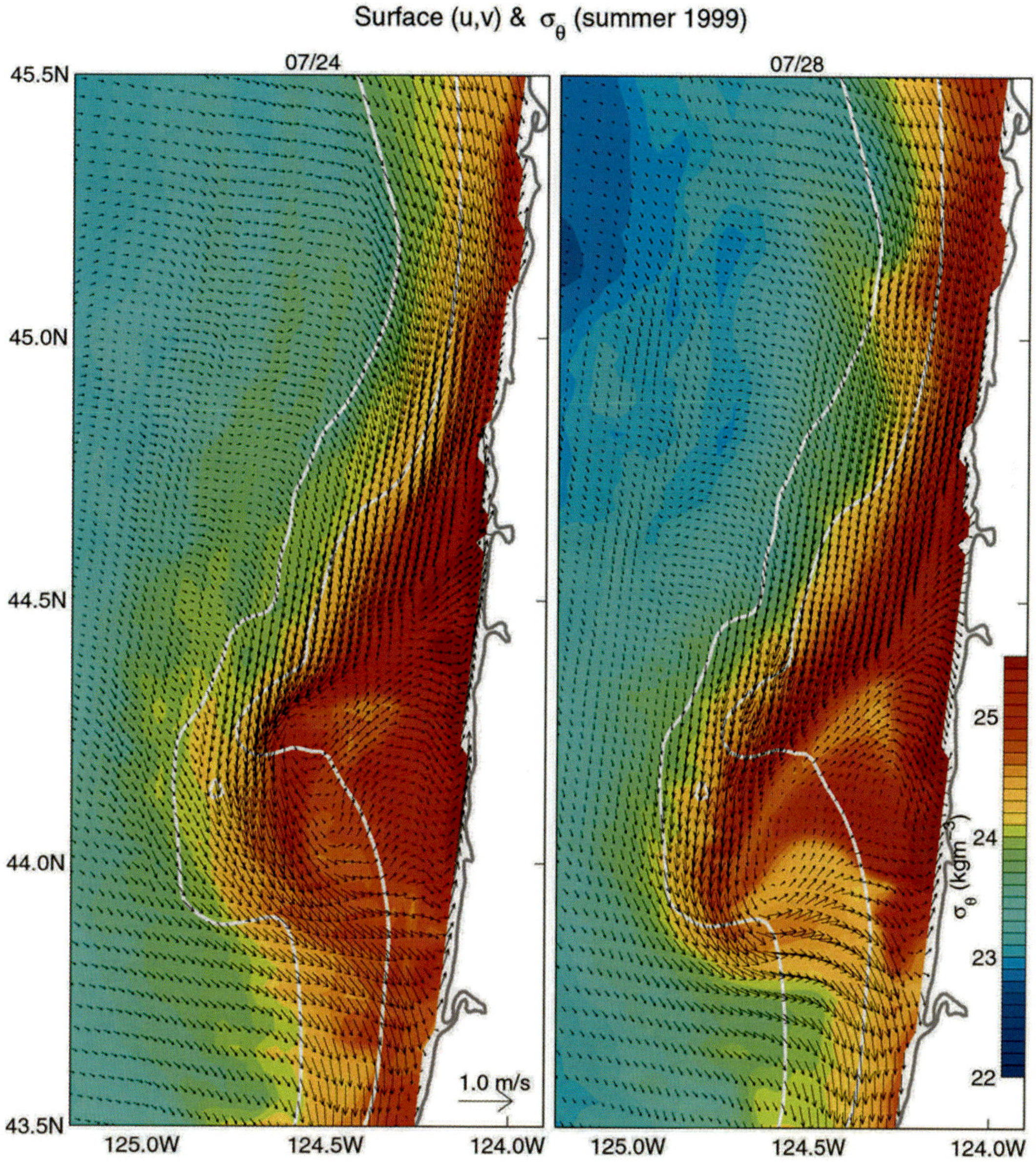

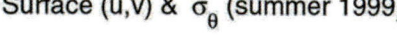

Plate 6. Surface velocities (vectors) and densities from a numerical model forced by wind fields from July 1999 (from Strub, 2001).

entific investigation of coastal processes, especially to the retroactive tidal corrections of the decade-long altimetry record in the world's coastal oceans.

3.5 Geophysical Applications

As noted in Cazenave and Royer (2001) and Sandwell and Smith (2001), small-scale seamounts and sea-floor roughness hold a great deal of information about the physics of the oceanic lithosphere and upper mantle. In the wavelength band of 10–100 km, the details of sea-floor topography can be estimated from the slope of OST, providing the fine-scale ocean bathymetry that is important to the modeling of ocean currents, eddies and mixing, in addition to geophysical and practical applications. Although the geophysical community prefers the flight of an advanced nadir-looking altimeter in a sweeping (non-repeating) orbit for obtaining the small-scale OST information (Sandwell et al., 2001), the sampling characteristics of WSOA should also be considered for extracting information for geophysical applications.

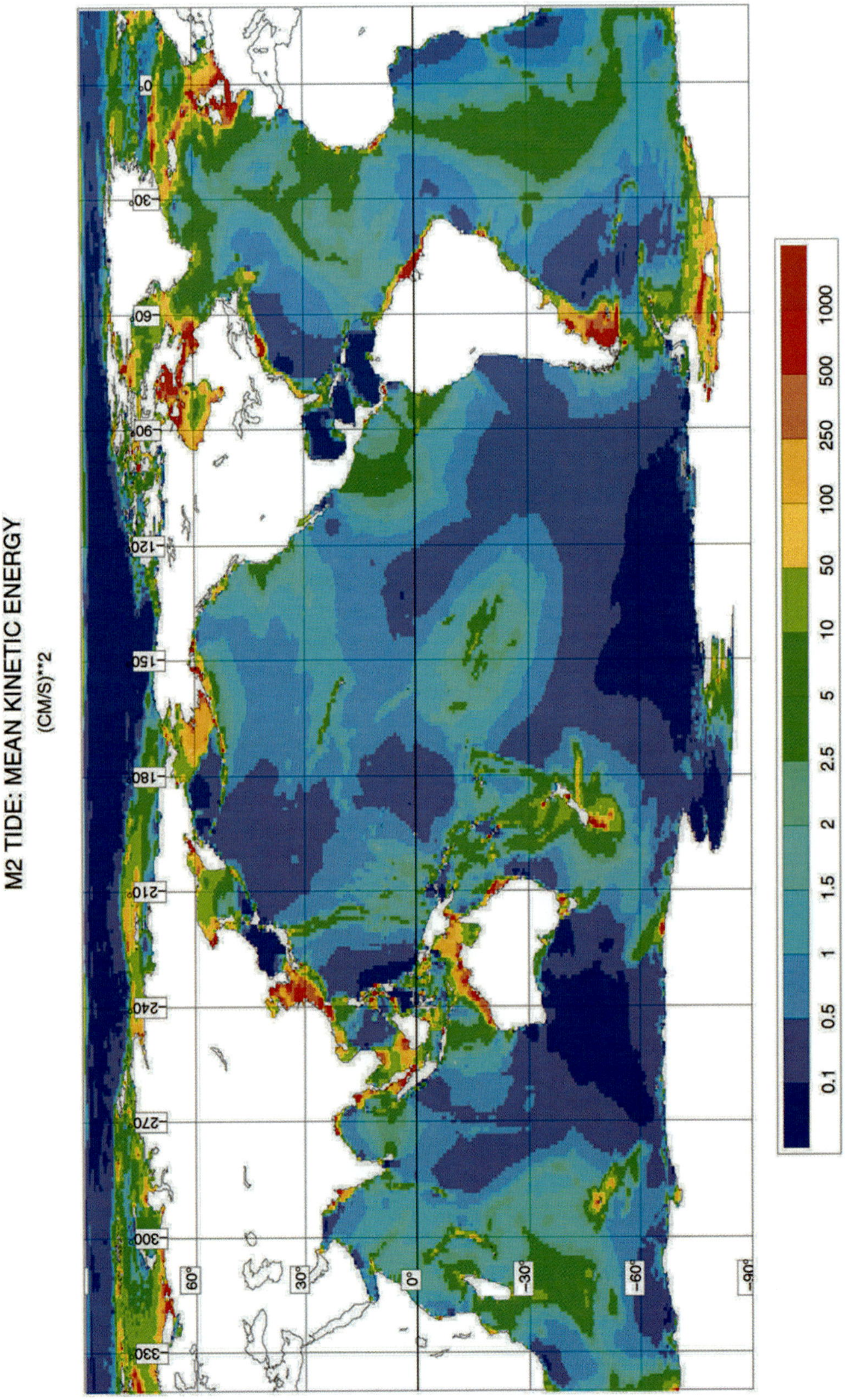

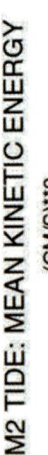

Plate 7. Mean kinetic energy of the M_2 tide—short wavelength signatures are over areas where the kinetic energy is in the range 5 to 50 $(cm/s)^{**}2$ (green to yellow). (from Le Provost, 2001)

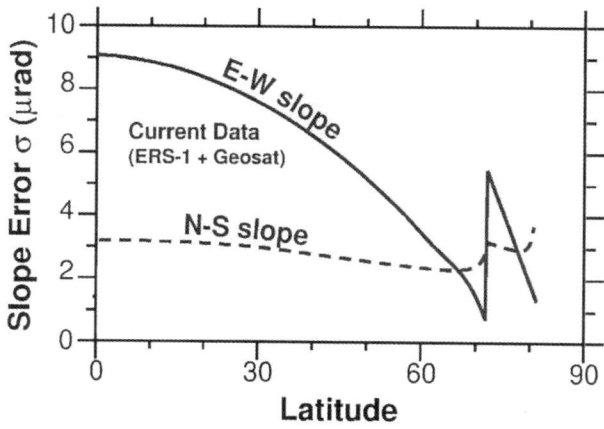

Figure 5: Propagation of along-track slope error from 1.5 years of dense Geosat coverage and 1 year of dense ERS-1 coverage into east (solid) and north (dashed) components of sea surface slope recovery versus latitude. At the equator, the Geosat and ERS tracks mainly run N–S so the N–S component is well determined while the E–W component of sea surface slope is poorly determined. (adapted from Sandwell et al., 2001)

Shown in Figure 5 is the accuracy of slope estimates currently achievable from existing altimeter data records. Owing to the high inclination of altimetry missions, the north–south slopes are much better known (~ 3 micro radians) than the east–west slopes (~ 3–9 micro radians). Note that 1 micro radian corresponds to the slope of 1 mm over 1 km. Sandwell et al. (2001) advocated that future OST slope measurement ought to meet an accuracy of 1 micro radian with a wavelength resolution of 15 km. We would like to examine the potential performance of WSOA in meeting this objective.

The raw measurements of WSOA have along-track (cross-radar-look) resolution of 12 km (limited by the interferometric antenna size) and cross-track (along-radar-look) resolution of 1 km (limited by the transmission bandwidth and data downlink capacity). For most oceanographic applications, the cross-track measurements are averaged spatially to reduce measurement errors to create more accurate observations at coarser resolution.

In order to retain the high spatial resolution for geophysical applications, one could retain the intrinsic instrument resolution of 1 km in the cross-track direction to reduce errors for estimating the time-invariant surface slopes. The resulting data have a resolution of 1 km in the cross-track direction and 15 km in the along-track direction. The resulting height precision is degraded relative to the 15 km x 15 km WSOA ocean data, and is presented in Figure 6 as a function of distance from the satellite nadir point.

Additional measurement error sources are introduced by estimation errors for the platform roll and media effects. The roll errors will typically produce tilts in the look direction on the order of 2 cm/200 km, or 0.1 micro radians and can be neglected. A similar argument applies to the media corrections, and is quantified by Sandwell and Smith (2001). The largest additional error contribution is due to mesoscale eddies and boundary currents, but these contaminate the data only in a relatively small fraction of the ocean (Sandwell and Smith, 2001).

If single passes of the WSOA observations were the only information available, the resolution in the cross-look (nominally along-track) direction would be rather coarse due to the limited antenna size. However, owing to the wide-swath coverage of WSOA, ascending and descending data can be combined for most locations in the ocean to obtain a higher resolution estimate in both directions. This problem is equivalent to estimating an image given two versions of the image, blurred in different directions. Solutions to the problem are well known in the inverse problem literature, but estimating the performance of an optimal algorithm, which will depend on the track geometry and signal and noise characteristics, is beyond the scope of this initial assessment. In general, the accuracy of the slope will be greatest in the cross-track direction, in contrast to conventional altimetry, where the accuracy is greatest in the along-track direction, due to the track geometry. This means that, at the very least, the WSOA data will provide slope estimates complimentary to those shown in Figure 5.

In order to provide an order of magnitude assessment, we take estimated slope error in the cross-track direction as a

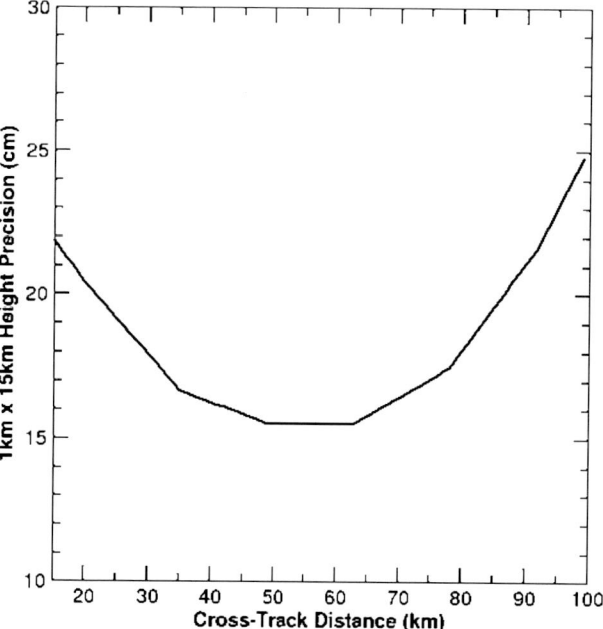

Figure 6: WSOA height precision for a spatial resolution of 1 km in the cross-track direction and 15 km in the along-track direction.

best case indicator of performance. We estimate the cross-track slope by performing a local linear least squares fit to the topography data. Reduction in the slope noise is obtained by averaging the estimated slopes over time. Assuming that all data can be used in the estimation, the slope error will reduce by a factor of the square root of the number of altimeter repeat pass cycles used in the average. Plate 8 (a) shows the expected slope accuracy as a function of the spatial scale used for the estimation and averaging time. The purple regions correspond to regions where the slope requirements given by Sandwell et al. (2001) are met. The blue regions correspond to regions where the data would represent an improvement over currently available data. Notice that for a minimal mission duration of 2 years, the smallest resolvable wavelength which meets the accuracy requirement is ~20 km, but data improvements are observed down to wavelengths of ~13 km. If the mission lifetime is 5 years, the smallest resolvable wavelength which meets the requirements is ~15 km, while improvements in performance are observed down to wavelengths of ~8 km.

Plate 8 (a) assumes that all data will be available and useful for the estimation. In reality, due to satellite yaw steering which will be present in the OSTM mission, it is likely that the number of useful points might be reduced. (Notice that the effect of yaw steering is twofold: it decreases the coverage and it changes the track geometry so that the ascending and descending track angle is different than that of the nadir track. However, slope estimates in two dimensions are still obtained even when yaw-steering occurs.) As a pessimistic estimate, we assume that only 1/4 of all points are useful for estimation (this corresponds to the worst case yaw steering scenario where only the data without yaw steering are useful for the slope estimation). In that case the estimated slope error is presented in Plate 8 (b). The performance after 4 years for the degraded case is equivalent to the performance after 2 years using all the data. Nevertheless, a performance of 2 micro radians at a 20 km wavelength, which would still represent a significant improvement relative to the current results shown in Figure 5, can still be obtained.

4. CONCLUSIONS

The technique of radar interferometry provides an effective approach to wide-swath measurement of OST. The effects of the uncertainties in the satellite's roll angle and the phase of the transmitted signals, the major sources for measurement errors, can be reduced significantly by the application of a crossover calibration technique. By flying with a Jason class altimeter, WSOA will be able to measure the OST over a swath of 200 km in width with rms accuracy ranging from 3.7 cm (inner swath) to 4.9 cm (outer swath) at a spatial resolution of 15 km. Orbit errors are not included in the accuracy

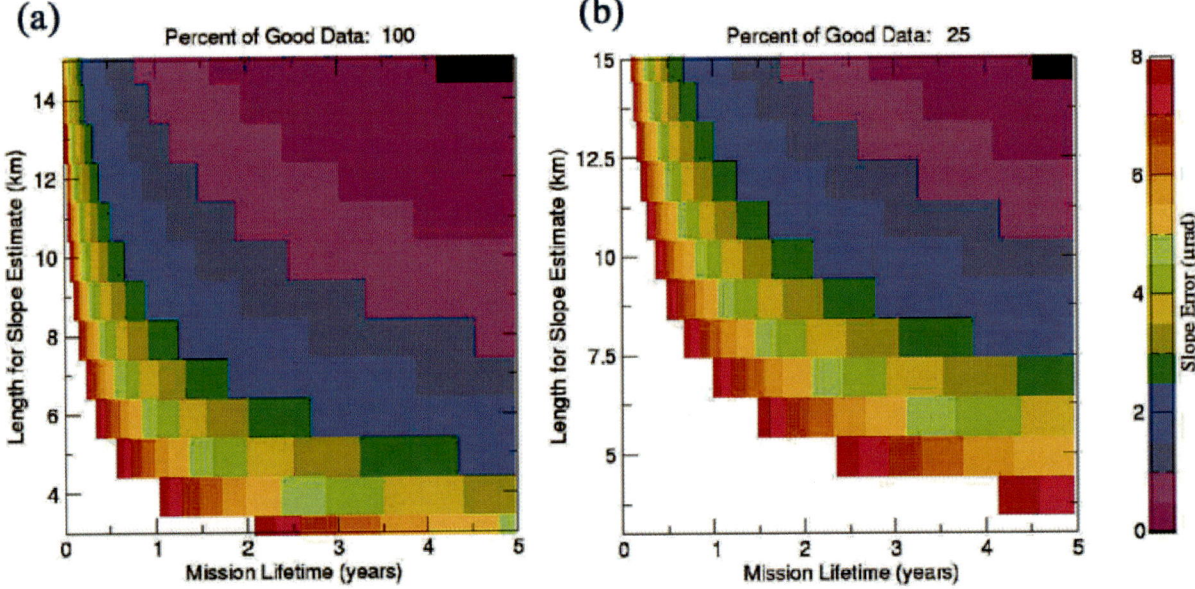

Plate 8. (a) Cross-track slope accuracy as a function of averaging time and distance used to estimate the slope. The parameter space which meets the Sandwell et al. (2001) requirements is colored in shades of purple. The parameter space which improves current measurements is roughly given by the area shaded in blue. This figure assumes that 100% of the data are available for estimating the slopes. (b) The same as (a), but this figure assumes that only 25% of the data are available for slope estimation.

estimates because their scales are much larger than the scales of the phenomena to be addressed by WSOA.

Within a 10-day orbit repeat period as that of TOPEX/Poseidon and Jason, WSOA will make at least two measurements at a given resolution cell at most mid- and high-latitude locations. One can make use of these multiple observations to either enhance the temporal resolution or reducing the measurement errors by averaging them. The capability of WSOA is equivalent to that of a formation flight of more than 5 nadir-looking conventional altimeters.

The most important application of WSOA is to provide the first synoptic maps of the global oceanic eddy field. Although many of the statistical and dynamical properties of ocean eddies have been documented from the analysis of conventional altimeter data, global synoptic observations of the evolution of individual eddies are not possible with the existing altimeters flown in space. The knowledge of eddies carries a significant impact on the general circulation, heat transport, and biogeochemical processes of the ocean. It is shown that only eddy-resolving models can simulate the right amount of the meridional heat transport in the ocean, a key component in the earth's climate system. Such high-resolution ocean models need to be constrained by high-resolution observations from WSOA in order to produce realistic simulations of the ocean for both scientific and practical applications.

The coastal ocean, which has many critical impacts on the lives of half of the world's population, is rich in processes of small scales demanding high-resolution measurements. WSOA will provide all-weather measurements that allow the monitoring and study of coastal currents and tides. For example, the current coastal tide models have an uncertainty of 20 cm in many places, preventing altimeter data from being useful for studying the coastal circulation. WSOA will have sufficient resolution and accuracy for reducing this uncertainty to a few centimeters, the level of accuracy of open ocean tide models.

The strong currents and water property anomalies (in temperature, salinity, oxygen, etc.) associated with ocean eddies are a major factor in affecting maritime operations such as offshore oil drilling, ship routing, fisheries, marine debris dispersion, etc. WSOA is expected to be an essential part of the future ocean observing system for addressing these applications.

The intrinsic resolution in the look direction of the interferometric radar of WSOA is about 1 km, allowing the estimation of sea surface slope with an accuracy of 1 micro radian over a wavelength of 20 km for a 2-year mission (15 km for a 5-year mission). Because the interferometric radar looks in the cross-track direction, this capability provides improvement in the knowledge of the OST slope in the east–west direction, a direction in which the current knowledge of the OST slope is poorest.

Acknowledgements. The research presented in the paper was carried out at the Jet Propulsion Laboratory, California Institute of Technology, under contract with the National Aeronautic and Space Administration. The writing of the paper benefits a great deal from *Report of the High-Resolution Ocean Topography Science Working Group Meeting* edited by Dudley Chelton of the Oregon State University.

REFERENCES

Bryan, F. O. and R. D. Smith, Modelling the North Atlantic circulation: From eddy-permitting to eddy-resolving, *International WOCE Newsletter. 33*, 12–14, WOCE International Project Office, Southampton, U.K., 1998.

Cazenave, A., and J. F. Royer, Applications to marine geophysics, in *Satellite Altimetry and Earth Sciences: A Handbook for Techniques and Applications"* edited by L.-L. Fu and A. Cazenave, 407–439, Academic Press, San Diego, 423 pp., 2001.

Chelton, D. B., editor, *Report of the High-Resolution Ocean topography Science Working Group Meeting,* Ref. 2001-4, Oregon State University, Corvallis, OR, 224 pp, 2001.

Chelton, D. B., J. Ries, B. Haines, L.-L. Fu, and P. Callahan, Satellite altimetry, in *Satellite Altimetry and Earth Sciences: A Handbook for Techniques and Applications,* edited by L.-L. Fu and A. Cazenave, 1–131, Academic Press, San Diego, 423 pp., 2001.

Chelton, D. B., and M. G. Schlax, The accuracies of smoothed sea surface height fields constructed from tandem altimeter datasets, *J. Atmos. Oceanic Tech., 20,* 1276–1302, 2003.

Chelton, D. B., R. A. deSzoeke, M. G. Schlax, K. E. Naggar, and N. Siwertz, Geographical variability of the first baroclinic Rossby radius of deformation, *J. Phys. Oceanogr., 28,* 433–460, 1998.

Doney, S. C., D. M. Glover, S. J. McCue and M. Fuentes, Mesoscale Variability of SeaWiFS Satellite Ocean Color: Global Patterns and Spatial Scales, *J. Geophys. Res., 108,* 3024, doi:10.1029/2001JC000843, 2003.

Ducet, N., P.-Y. Le Traon, and G. Reverdin, Global high resolution mapping of ocean circulation from TOPEX/Poseidon and ERS-1/2, *J. Geophys. Res., 105,* 19477–19498, 2000.

Ducet, N. and P.-Y. Le Traon, A comparison of surface eddy kinetic energy and Reynolds stresses in the Gulf Stream and the Kuroshio Current systems from merged TOPEX/Poseidon and ERS-1/2 altimetric data, *J. Geophys. Res., 106,* 16603–16622, 2001.

Dukowicz, J. K., and R. D. Smith, Implicit free-surface method for the Bryan-Cox-Semtner ocean model, *J. Geophys. Res., 99,* 7991–8014, 1994.

Fu, L.-L. editor, *Wide-Swath Altimetric Measurement of Ocean Surface Topography,* JPL Publication 03–002. Jet Propulsion Laboratory, Pasadena, CA, 67 pp, 2003.

Fu, L.-L., and A. Cazenave, editors, *Satellite Altimetry and Earth Sciences: A Handbook of Techniques and Applications.* Academic Press, San Diego, 463 pp, 2001.

Fu, L.-L., and D. B. Chelton, Large-scale ocean circulation, in *Satellite Altimetry and Earth Sciences: A Handbook for Techniques and Applications,* edited by L.-L. Fu and A. Cazenave, 133–169, Academic Press, San Diego, 423 pp., 2001.

Fu, L-L., and J. Vazquez, On correcting radial orbit errors for altimetric satellites using crossover analysis, *J. Atmos. Oceanic Tech.*, 5, 466–471, 1988.

Garçon, V. C., A. Oschlies, S. C. Doney, D. McGillicuddy, J. Waniek, The role of mesoscale variability on plankton dynamics, *Deep-Sea Res. II, 48*, 2199–2226, 2001.

Le Provost, C., Generation of overtidess and compound tides (review), in *Tidal Hydrodynamics*, B. Parker, ed., John Wiley and Sons, New York, 269–296, 1991.

Le Provost, C. Tides over ridges, shelves, and near the coasts, in *Report of the High-Resolution Ocean topography Science workinbg Group Meeting,* Ref. 2001-4, edited by D.B. Chelton, Oregon State University, Corvallis, OR, 224 pp., 2001.

Le Traon, P.-Y. and R. A. Morrow, Ocean currents and mesoscale eddies, in *Satellite Altimetry and Earth Sciences. A Handbook of Techniques and Applications*, Academic Press, L.-L. Fu and A. Cazenave, Eds., Academic Press, 171–215, 2001.

MacDonald, A. M., and C. Wunsch, An estimate of global ocean circulation and heat fluxes, *Nature, 382*, 436–439, 1996.

Mackas, D. L., K. L. Denman, and M. R. Abbott, Plankton patchiness: biology in the physical venacular, *Bull. Mar. Sci., 37(2)*, 652–674, 1985.

McClain, C. R., M. L. Cleave, G. C. Feldman, W. W. Gregg, S. B. Hooker, and N. Kuring, Science quality SeaWiFS data for global biosphere research, *Sea Technology, 39*, 10–14, 1998.

Parke, M. E., R. L. Stewart, D. L. Farless and D. E. Cartwright, On the choice of orbits for an altimetric satellite to study ocean circulation and tides, *J. Geophys. Res., 92*, 11,693–11,707, 1987.

Rodriguez, E., and J. Martin, Theory and design of interferometric synthetic aperture radars, *IEEE Proceedings F, 139, 2*, 147–159, 1992.

Sandwell, D. T., and W. H. F. Smith, Bathymetric estimation, in *Satellite Altimetry and Earth Sciences: A Handbook for Techniques and Applications,* edited by L.-L. Fu and A. Cazenave, 441–457, Academic Press, San Diego, 423 pp., 2001.

Sandwell, D. T., W. H. F. Smith, S. Gille, S. Jayne, K. Soofi, and B. Coakley, Bathymetry from space, in *Report of the High-resolution Ocean topography Science working Group Meeting,*, edited by D. B. Chelton, Ref. 2001-4, College of Oceanic and Atmospheric Sciences, Oregon State University, Corvallis, OR, 224 pp., 2001.

Sandwell, D. T., W. H. F. Smith, and S. Gille, editors, Bathymetry from Space: Oceanography, Geophysics, and Climate, Geoscience Professional Service, Bethesda, MD, 24 pp., www.igpp.ucsd.edu/bathymetry_workshop, 2003.

Schlax, M. G., and D. B. Chelton, The accuracies of crossover and parallel-track estimates of geostrophic velocity from TOPEX/Poseidon and Jason-1 altimeter data, *J. Atmos. Oceanic Tech., 20*, 1196–1211, 2003.

Smith, R. D., M. E. Maltrud, F. O. Bryan, and M. W. Hecht, Numerical simulation of the North Atlantic Ocean at 1/10°, *J. Phys. Oceanogr., 30*, 1532–1561, 2000.

Strub, T., High-resolution ocean topography science requirements for coastal studies, in *Report of the High-Resolution Ocean topography Science Working Group Meeting,* Ref. 2001–4, edited by D.B. Chelton, Oregon State University, Corvallis, OR, 224 pp., 2001.

Wilson, S, Launching the Argo armada: taking the ocean's pulse with 3000 free-ranging floats, *Oceanus, 42*, 17–19, 2000.

Wyrtki, K., L. Magaard, and J. Hager, Eddy energy in the ocean, *J. Geophys. Res., 81*, 2641–2646, 1976.

Lee-Lueng Fu, MS 300–323, Jet Propulsion Laboratory, 4800 Oak Grove Drive, Pasadena, CA 91109

Ernesto Rodriguez, MS 300–319, Jet Propulsion Laboratory, 4800 Oak Grove Drive, Pasadena, CA 91109

The Global Water Cycle

Taikan Oki

Institute of Industrial Science, University of Tokyo, Meguro, Tokyo, Japan

Dara Entekhabi

Massachusetts Institute of Technology, Cambridge, Massachusetts

Timothy Ives Harrold

Research Institute for Humanity and Nature, Kamigyo, Kyoto, Japan

The global water cycle consists of the oceans, water in the atmosphere, and water in the landscape. The cycle is closed by the fluxes between these reservoirs. Although the amounts of water in the atmosphere and river channels are relatively small, the fluxes are high, and this water plays a critical role in society, which is dependent on water as a renewable resource. On a global scale, the meridional component of river runoff is shown to be about 10% of the corresponding atmospheric and oceanic meridional fluxes. Artificial storages and water withdrawals for irrigation have significant impacts on river runoff and hence on the overall global water cycle. Fully coupled atmosphere-land-river-ocean models of the world's climate are essential to assess the future water resources and scarcities in relation to climate change. An assessment of future water scarcity suggests that water shortages will worsen, with a very significant increase in water stress in Africa. The impact of population growth on water stress is shown to be higher than that of climate change. The virtual water trade, which should be taken into account when discussing the global water cycle and water scarcity, is also considered. The movement of virtual water from North America, Oceania, and Europe to the Middle East, North West Africa, and East Asia represents significant global savings of water. The anticipated world water crisis widens the opportunities for the study of the global water cycle to contribute to the development of sustainability within society and to the solution of practical social problems.

1. INTRODUCTION

Compared with other planets, the Earth system is unique in that water exists in all three phases, *i.e.*, water vapor, liquid water, and solid ice. Although the amount of water vapor is relatively small, large amounts of latent heat are released during the phase change to liquid water; therefore the water cycle is closely linked to the energy cycle [Oki, 1999]. The water cycle is also closely related to the Earth's biogeochemical cycles, particularly of carbon and nitrogen. Water is the key that holds and ties these interacting and closely coupled cycles together. Water also interacts with topography, and transports sediment to oceans.

In this paper, the existence of water on Earth and the components of the global water cycle are briefly described, with an emphasis on the role of rivers, and the depiction of rivers within global models. An assessment of future water resources is presented, focusing on changes in runoff due to climate change, and changes in water stress due to changes in population and per capita demand. The virtual water trade, which significantly eases the limitations of water scarcity in many arid regions, is also considered.

1.1 Water on the Earth

The total volume of water on or near the Earth's surface is estimated as approximately 1.4×10^{18} m^3, corresponding to a mass of 1.4×10^{21} kg. Compared with the total mass of the Earth (5.974×10^{24} kg), water constitutes only 0.02% of the planet, but it is critical for the survival of life, and the Earth is consequently called "the Blue Planet" and "the Living Planet." Approximately 70% of its surface is covered with the salty water of the oceans. Some of the remaining surface is covered by fresh water (lakes and rivers), solid water (ice and snow), and vegetation (which implies the existence of water). Even though the water content of the atmosphere is comparatively small (approximately 0.3% by mass and 0.5% by volume of the atmosphere), approximately 60% of the Earth is always covered by cloud [Rossow et al., 1993].

The reserves of water in the Earth's water cycle are shown in Table 1 (based on Table 9 in Korzun, 1978). The proportion in the ocean is dominant (96.5%). Other major reserves are glaciers and permanent snow cover, and ground water. The proportion of water stored in the atmosphere, soil, and in river channels is very small, and the residence times are short, but, of course, this transient water plays a critical role in the global hydrological cycle.

1.2 Components of the Global Water Cycle

Figure 1 (from Oki 1999) schematically illustrates the components of the global water cycle, including storages (with volumes taken from Table 1) and approximate fluxes (for 1989–1992). The fluxes are calculated from precipitable water, water vapor transport, and convergence estimated using the European Centre for Medium-Range Weather Forecasts objective analyses, obtained as a 4-year mean [Hoskins 1989], and from the precipitation estimates of Xie and Arkin [1996]. The runoff flux is estimated at the land/sea boundary and includes a subsurface component of approximately 10% of the flow. These fluxes are only approximations and they will be improved by more precise measurements. However, on a global

Table 1: World Water Reserves

Form of Water	Covering Area (km^2)	Total Volume (km^3)	Mean Depth (m)	Share of Volume (%)	Mean Residence Time
World oceans	361 300 000	1 338 000 000	3 700	96.539	2 500 years
Glaciers and permanent snow cover	16 227 500	24 064 100	1 463	1.736	56 years
Ground water[a]	134 800 000	23 400 000	174	1.688	8 years
Gound ice in zones of permafrost strata	21 000 000	300 000	14	0.0216	
Water in lakes	2 058 700	176 400	85.7	0.0127	
Soil moisture	82 000 000	16 500	0.2	0.0012	
Atmospheric water	510 000 000	12 900	0.025	0.0009	9 days
Marsh water	2 682 600	11 470	4.28	0.0008	
Water in rivers	148 800 000	2120	0.014	0.0002	18 days
Biological water	510 000 000	1 120	0.002	0.0001	
Total water reserves	510 000 000	1 385 984 610	2 718	100.00	

[a] excluding Antarctic groundwater (approximately 2 000 000 km^3).

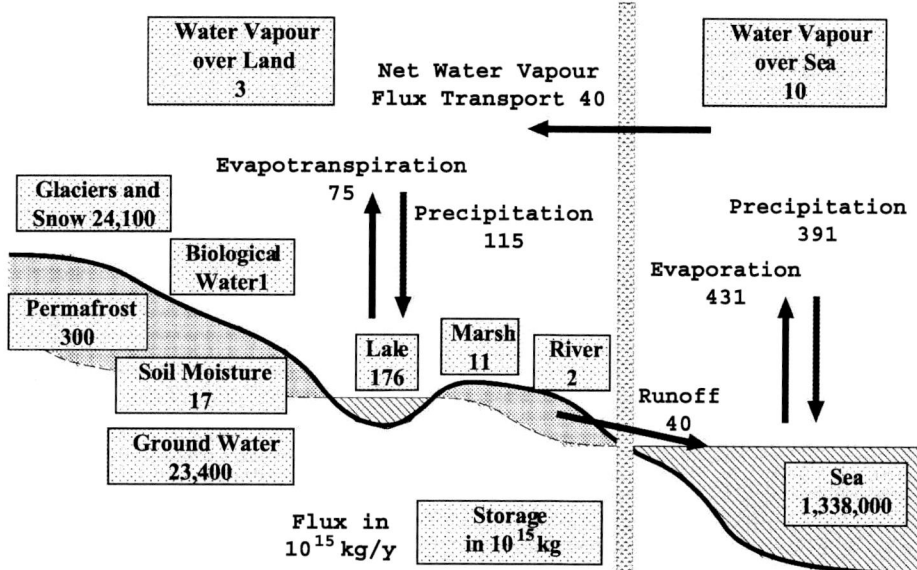

Figure 1. Schematic illustration of the water cycle on the Earth. Water storages (from Table 1) are indicated by boxes. Approximate fluxes (for 1989–1992) are represented by arrows.

scale, river runoff is about 8% of the atmospheric flux due to evaporation and precipitation.

The fluxes shown in Figure 1 are important for water resources assessment, because most social applications ultimately depend on water as a renewable and sustainable resource. Further, the current challenge is to estimate the variability of these fluxes at fine spatial and temporal scales. The regional impacts of variability can be severe, and include extreme weather events, floods, and droughts. The prediction of such variations in the global water cycle is one of the most urgent issues in modern hydrology.

Some important components of the global water cycle are now briefly introduced.

• *Water vapor* is the major absorber in the atmosphere of both short-wave and long-wave radiation. Condensation of water vapor releases a large amount of latent heat (2.5×10^6 J/kg), warming the atmosphere, and affecting atmospheric circulation.

• *Liquid water in the atmosphere* is the result of condensation. *Clouds* significantly affect radiation fluxes in the atmosphere and at the Earth's surface. *Precipitation* is highly variable in both space and time. It drives the hydrological cycle on the land surface and changes surface salinity in the ocean.

• *Evapotranspiration* is the return flow of water from the Earth's surface to the atmosphere and provides latent heat flux from the surface. Evapotransiration is affected by atmospheric and soil conditions, and by vegetation.

• *Soil moisture* influences the energy balance at the land surface, as a lack of available water suppresses evapotranspiration and changes surface albedo. Soil moisture also affects runoff and infiltration.

• *Vegetation* can transpire water from deep soil layers, and affect the diurnal and seasonal cycle of evapotranspiration. Vegetation also modifies the surface energy and water balance by altering the surface albedo and by intercepting precipitation.

• *Snow cover* has special characteristics: snow may be accumulated, its albedo is high, and the surface temperature will not rise above 0°C until the completion of snow melting. Consequently the existence of snow cover significantly changes the surface energy budget. A snow surface typically reduces the aerodynamic roughness, so that it may also have a dynamical effect on the atmospheric circulation.

• *Ground water* contributes to runoff, especially during dry periods. Deep ground water may also reflect the long term climate.

• *Runoff* returns water to the ocean. The amount of water carried by rivers is smaller than that carried by the atmosphere and oceans, yet it is not negligible. Runoff into oceans is important for the fresh water balance and the salinity of the oceans.

• *The Oceans* are a major sub-system of the global water cycle. Even though classical hydrology has traditionally excluded ocean processes, the global hydrological cycle is never closed without them. The ocean circulation carries huge amounts of energy and water. Surface ocean currents are driven by surface winds, and the atmosphere itself is sensitive to sea surface temperatures. Temperature and salinity determine the density of ocean water, which contributes to overturning and the deep ocean general circulation.

The last column of Table 1 presents some values of the global mean residence time of water, which vary from a few days to thousands of years. The global mean residence times are a direct function of storage volumes and flux rates. There can be large regional variations in residence times, but they provide a broad indication of the susceptibility of each reserve to pollution, and to measures to improve water quality.

Precipitation, as stated above, drives the hydrological cycle on the land surface. Plate 1 illustrates the global distribution of mean annual precipitation for 1986–1995 from the forcing data for the Global Soil Wetness Project (GSWP) Phase II [IGPO, 2002]. The high spatial variability of precipitation can be clearly seen. The spatial variability of the runoff that results from this precipitation is also high. Plate 2 illustrates the annual mean runoff ratio for 1986–1995 estimated by a simple land surface model with forcing data from GSWP Phase II. The spatial distribution of the runoff ratio is partially influenced by the distribution of precipitation shown in Plate 1, but it is also influenced by the seasonal pattern of rainfall and potential evaporation, and by the characteristics of the land surface, such as topography and land cover. The runoff shown in Plate 2 does not disappear; it is conveyed to the oceans via rivers. The role of rivers within the global water cycle is discussed in the next section.

2 RIVERS IN THE GLOBAL WATER CYCLE

Rivers and their management are critical to the supply of fresh water in many parts of the world. Rivers are ecosystems of global importance, and much of hydrology is focused on understanding and modeling these systems. Rivers also have a key role in the global water cycle. Rivers carry water, sediment, chemicals, and various nutrients from continents to seas. The fresh water supply to the ocean has an important effect on the thermohaline circulation, because it changes the salinity and thus the density of the oceans, and it also influences the formation of sea ice.

Annual fresh water transport to and from the oceans is shown in Table 2. The net atmospheric flux is estimated using four dimensional data assimilation data [Hoskins 1989]. River discharge is estimated by the bucket model [Manabe, 1969] with forcing data provided by GSWP Phase II, except the estimates for Antarctica use water vapor flux as a proxy for runoff. The total annual river discharge balances with the atmospheric water vapor flux convergence.

The annual water flux in the meridional direction for 1989–1992 has also been estimated, with the results shown in Plate 3 (from Oki et al., 1995). Transports by the atmosphere and by the ocean have almost the same absolute values at each latitude but with different signs. The transport by rivers is about 10% of these other global fluxes. The negative (southward) peak by rivers at 30° S is mainly due to the Parana River in South America, and the peaks at the equator and 10° N are due to rivers in South America, such as the Magdalena and Orinoco. Large Russian rivers, such as the Ob, Yenisey, and Lena, carry freshwater towards the north between 50–70° N. These results indicate that hydrological processes over land play a significant role in the overall global circulation, not only by the exchange of energy and water at the land surface, but also through affecting the water balance of the oceans.

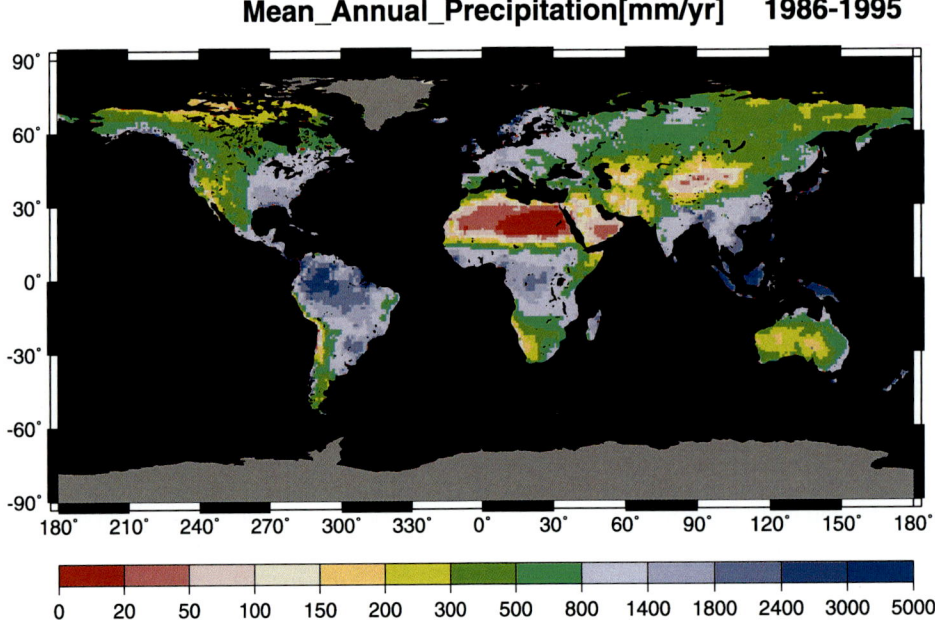

Plate 1. Mean annual precipitation (mm/y) for 1986-95 from the forcing data for the Global Soil Wetness Project Phase II.

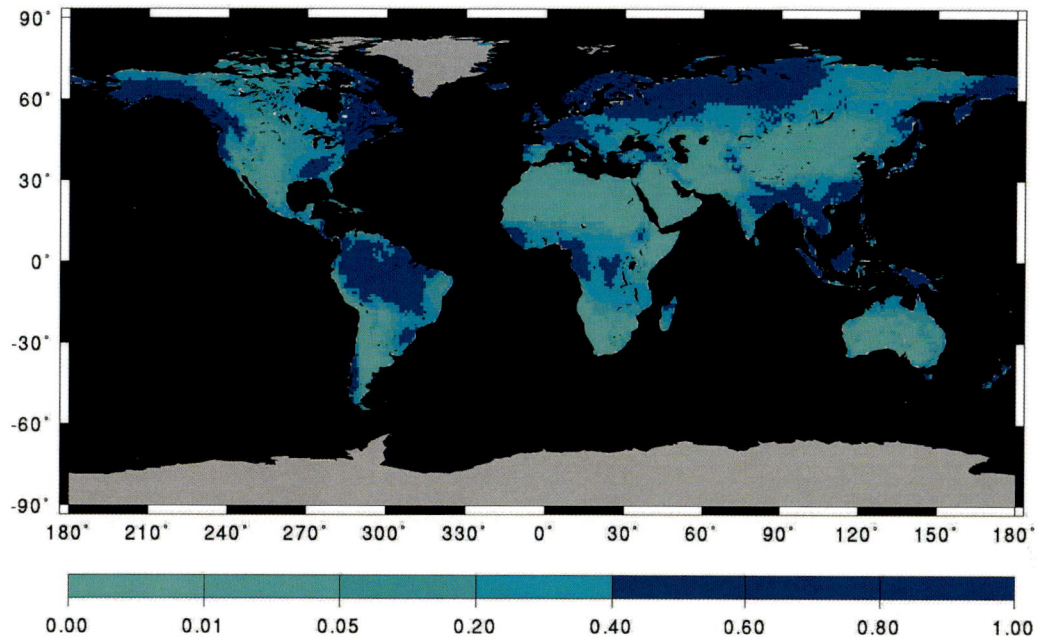

Plate 2. Annual mean runoff ratio for 1986-95 estimated by a simple land surface model with forcing data from the Global Soil Wetness Project Phase II.

2.1 Classification of River Representation in GCMs

Based on the arguments presented above, it is obvious that rivers should be included in modeling of the climate system.

The role of rivers within a General Circulation Model (GCM) is to receive runoff from precipitation, to transport the runoff downstream, and to discharge the runoff back into the ocean, thus closing the lower section of the water cycle that is depicted

Table 2: Annual Fresh Water Transport (10^{15} kg/y) To and From the Oceans

		North Pacific	South Pacific	North Atlantic	South Atlantic	Indian Ocean	Arctic Ocean	Internal Drainage	Total
River discharge	Asia	4.36	0.70	0.16		3.17	2.18	0.36	10.94
	Europe			3.51			1.20	0.70	5.41
	Africa			1.38	2.41	0.77		0.27	4.83
	N.America	2.29		3.06			0.28	0.58	6.21
	S.America	0.12	0.32	8.19	2.58			0.08	11.30
	Oceania		1.53			0.37		0.03	1.94
	Antarctica		(1.0)		(0.1)	(0.8)			(1.9)
Total River Discharge		6.78	2.56	16.30	4.99	4.32	3.66	2.02	40.62
(with Antarctica)		6.78	3.56	16.30	5.09	5.12	3.66	2.02	42.52
Net Atmospheric Flux		9.9	-11.1	-12.7	-14.0	-14.0	2.2	-2.02*	-41.72
Net Fresh Water Transport		16.68	-7.54	3.6	-8.91	-8.88	5.86		0.8[a]

*Assumed to match internal drainage, for mass balance purposes.
[a] In reality, this should be zero.

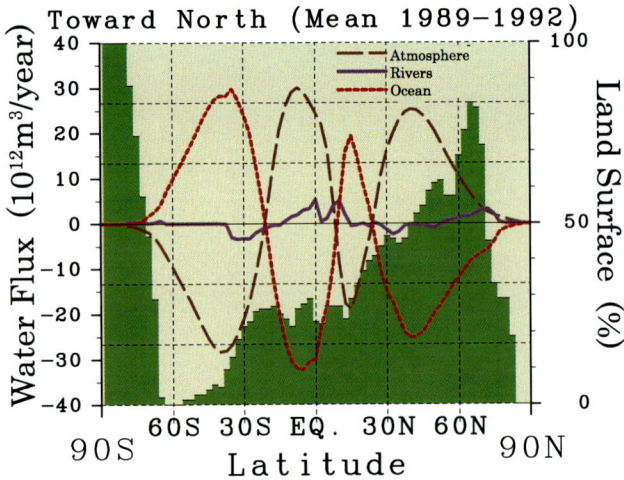

Plate 3. Annual water flux in the meridional direction by atmosphere, ocean, and rivers, 1989-1992. The background shading shows the percentage of the Earth's surface that is land, at each latitude.

in Figure 1. Table 3 (from Oki et al., 1999) shows a classification of the level of representation of rivers within GCM simulations, based on the destination of the river discharge, whether river routing is used, and whether interactions at downstream grid boxes occur. Level −1 through 0.5 are used for short term weather forecasting. For a coupled ocean-atmosphere model, level 1 representation of rivers may be adopted in order to close the mass balance of water in the model. A few GCMs use level 2 coupling [Miller et al., 1994; Sausen et al., 1994; Kanae et al., 1995].

Level 3 representation of rivers considers the effect of both artificial and natural water re-distribution from a river channel to the surrounding land surface. Runoff from an upstream grid box can evaporate at a downstream grid box by such an approach.

Even though the effect may be regional, the process should change the water and energy balance of the neighboring land surface and have some effect on the climate system. Efforts to model these processes using fully coupled (level 3) atmosphere-land-river-ocean models are needed, and improvements to Land Surface Models and river routing schemes are required in order to achieve this objective.

The challenge to accomplish the level 3 coupling is mainly due to the uncertainties associated with modeling anthropogenic activities, such as water withdrawals for irrigation and the regulation of river flow by the operation of artificial reservoirs. The total capacity of artificial reservoirs is estimated as 8,000 km^3 [Vörösmarty et al., 1997], which is large enough to affect the global river discharge of approximately 40,000 km^3/y globally.

2.2 Digital Rivers in GCMs

At least three components are required in order to describe digital rivers in GCMs.
These are;
- a river routing scheme,
- information on the direction of the lateral water movement (a global river channel network), and
- river discharge data.

River routing schemes are commonly based on the one-dimensional Navier-Stokes equation. Because of the limitations both in obtaining all the necessary coefficients and in the computational burden, simplified equations, consisting only of mass conservation and the balance of friction with gravitational forcing, are widely used.

Global river channel networks are reported in Miller et al. [1994], Sausen et al. [1994], Kanae et al. [1995], and Vörösmarty et al. [1997]. More recently, Oki and Sud [1998] prepared a global river channel network in 1° x 1° grid boxes, named the Total Runoff Integrating Pathways (TRIP). River basin areas are represented in TRIP with root mean square error of ±10%, and the ratio of the river length in TRIP com-

Table 3: Representation of River Processes in General Circulation Models

Level of Representation	Destination of Discharge	Timing
−1	Nowhere (it disappears)	Immediately
0	All ocean grids	Immediately
0.5	Nearest ocean grid	Immediately
1	Designated river mouth	Immediately
2	Designated river mouth	With time lag (river routing)
3	Level 2 +interactions at downstream grid boxes	

pared to the actual length is considered. TRIP is available to the public and has been used in various research studies, such as Chapelon et al. [2002] and Oki et al. [2003a], and included in general circulation models at the Hadley Centre (UK), Centre National de Recherches Meteorologiques, Meteo-France (France), and the Center for Climate System Research (CCSR) in the University of Tokyo (Japan). A part of TRIP in the Asian region is shown in Plate 4 (from Oki and Sud, 1998).

Another requirement for digital rivers is observed discharge data for the validation of the model. Observed river runoff represents integrated water fluxes over large areas. Therefore river runoff data should be used for the validation of GCM simulations and for assessments of variations of hydrologic cycles on a global scale.

Plate 5 shows the mean annual runoff for 1961–90 (from Oki et al., 1999). The plate is derived purely from observational data at river discharge gauging stations, with templates based on 1 degree mesh TRIP. The East-West transition of the runoff in North America can be seen; high runoff values are found in South America and Southeast Asia.

The negative runoff in some rivers, such as the Indus, Colorado, and other desert rivers, attract attention. Negative runoff occurs when downstream river discharge is less than upstream river discharge. Physically, negative runoff indicates that evapotranspiration exceeds precipitation in that river basin pixel. This situation often occurs artificially due to the diversion of river water to surrounding areas for irrigation, etc. Such anthropogenic effects are very important in most of the world's river basins, and cannot be considered without level 3 coupling of an atmosphere-land-river-ocean model. Modeling of socially relevant issues, such as the termination of the river flow at the lower reach of the Yellow River in China, will not be possible without level 3 coupling. Such coupling will also increase confidence in our ability to properly assess current and future world water resources.

3 FUTURE WORLD WATER RESOURCES

Water resources are under serious pressure in some regions of the Earth, and the situation is likely to worsen [Cosgrove and Rijsberman, 2000]. At the June 2003 G8 summit in Evian, France, the water crisis was addressed as a problem of the utmost gravity. "Water: a G8 Action Plan" is a summit consensus statement that appears immediately after the "Chair's Summary".

The estimation of the current level of water stress is important for reliable projections of the severity of the water crisis into the future. Global analyses of water scarcity have been carried out by Takahashi et al. [2000] and Vörösmarty et al. [2000]. Oki et al. [2001] presented water scarcities on a 0.5 degree by 0.5 degree grid for the globe.

Assessment of future water resources involves assessment of climate change, population growth, and changes in demand. The supply side of water resources for both present and future conditions can be directly taken from runoff simulated by a GCM. However, GCMs suffer biases in calculating water

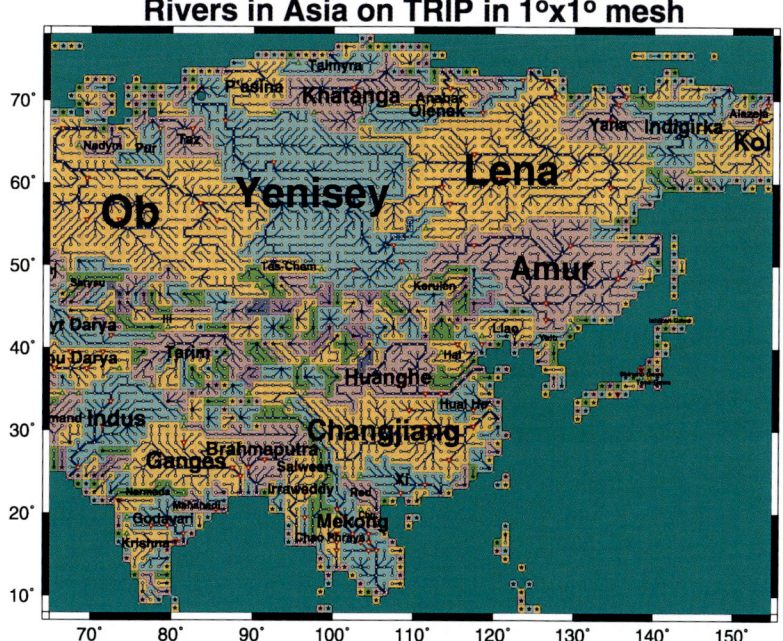

Plate 4. A part of the 1° grid Total Runoff Integrating Pathways (TRIP), in the Asian region.

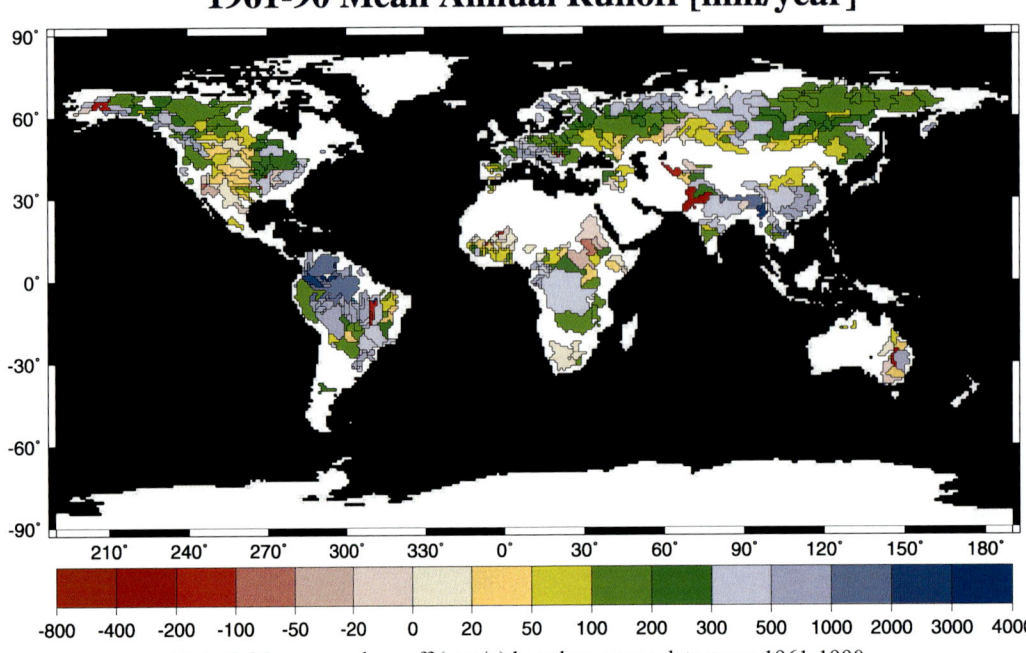

Plate 5. Mean annual runoff (mm/y) based on gauge data mean 1961-1990.

cycles, and direct values from GCMs are not frequently used. Instead, the change between current and future conditions from GCM studies can be applied to observed or best estimated current precipitation and temperature to produce future scenarios (see, for example, Arnell, 2004; Alcamo et al., 2003; Wallace, 2002). The modified precipitation and temperature are then used as future climatic inputs to a macro scale hydrological model, and future river runoff is calculated. However, such a procedure may have some inconsistencies with the original runoff simulation by GCMs.

In Oki et al. [2003a], projection of future world water resources was carried out by using current runoff estimated by land surface models (LSMs) and adding or subtracting the change in runoff estimated by a general circulation model (GCM), and considering future water withdrawals estimated by trend analysis. Results from Oki et al. [2003a] are discussed in sections 3.1 and 3.2.

3.1 Future Water Availability

To estimate the current runoff distribution on a global scale, offline runs by 11 different land surface models (LSMs) were used [Oki et al., 1999]. Offline simulation means LSMs are not coupled with the atmospheric component of GCMs but are instead driven by "forcing" data, with no feedback to the atmospheric circulation. Forcing data from ISLSCP were used (International Satellite Land Surface Climatology Project; Meeson et al., 1995). The data are based on observations of precipitation and radiation at the surface and from space, with data assimilation techniques applied to cover the regions with sparse monitoring networks. TRIP [Oki and Sud, 1998] was used for the river routing calculations to convert runoff from LSMs into river discharge. The estimated current annual discharge corresponded fairly well with river runoff observations [Oki et al. 2001], but was smaller compared to previous estimates by approximately 20%, mainly due to reduced bias in the forcing precipitation.

A CCSR/NIES atmospheric general circulation model (AGCM) [Numaguti et al., 1997] was run with boundary conditions of sea surface temperature in both 1990 and 2060, with sea surface temperatures for 2060 obtained from a coupled ocean-atmospheric general circulation model (AOGCM) run for doubled carbon dioxide conditions. The horizontal resolution of the AGCM was approximately 100km*100km globally. The simulated daily runoff from the AGCM was then routed using TRIP. The differences in annual mean runoff between 1990 and 2060 are shown in Plate 6 (from Oki et al. 2003a).

For this particular simulation, the Asian monsoon is enhanced due to the enhancement of the temperature difference between the Indian Ocean and Eurasia, and runoff is increased in the Indian sub-continent, the western part of the Indo-China Peninsula, northern China, and central and western Africa. Decreases in runoff are projected in central China and Europe. However Plate 6 is based on one possible climate scenario, based on modeling from a single GCM. The

Annual Change in River Discharge
[10⁶ m³/0.5°grid cell]

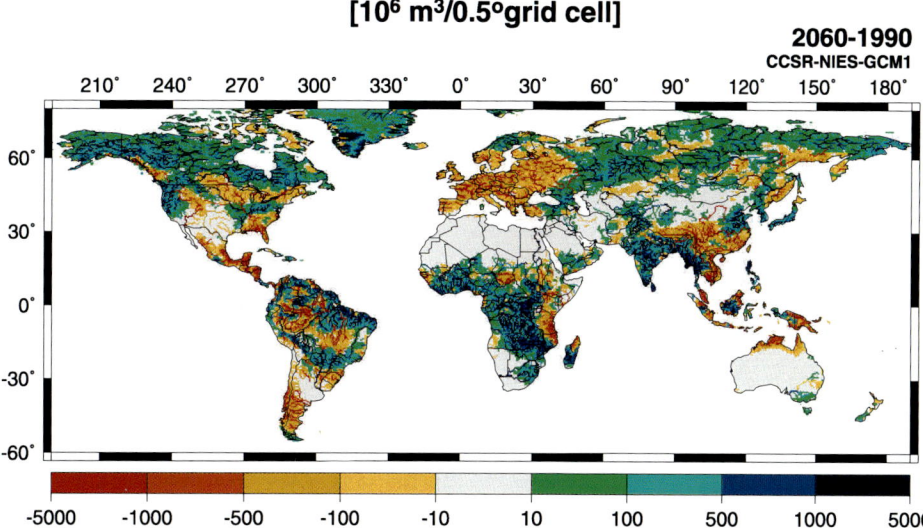

Plate 6. The difference in runoff ($10^6 m^3$ in each 0.5 x0.5° grid box) between 1990 and 2060, based on one possible climate scenario for 2060.

uncertainties involved with future climate scenarios are very high, and different GCMs will probably produce different patterns to the one presented here.

3.2 Future Water Scarcity

The assessment of water scarcity in Oki et al. [2003a] adopted the water scarcity index R_{ws} of Falkenmark et al. [1989], which is the ratio of the annual water withdrawal W to the available annual water Q, as used by United Nations [1997] and Vörösmarty et al. [2000]. The current water withdrawals estimated in Oki et al. [2001] were used as the baseline for the future prediction of the water withdrawals. The future withdraws were estimated separately for municipal, industrial, and agricultural water usages. The increase in irrigation withdrawal was assumed proportional to population growth. The resulting global distribution of projected R_{ws} in 2050 is shown in Plate 7 (from Oki et al., 2003a).

Annual Withdrawal-to-Availability Ratio

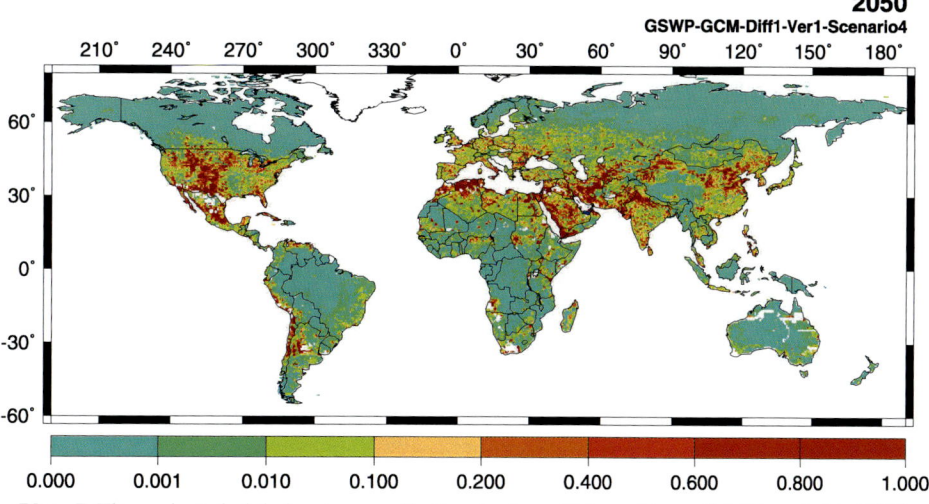

Plate 7. The projected global water scarcity R_{WS} (water withdrawal to availability ratio) in 2050.

Generally the severity of water scarcity is judged as:

$R_{ws} < 0.1$ no water stress
$0.1 = R_{ws} < 0.2$ low water stress
$0.2 = R_{ws} < 0.4$ moderate water stress
$0.4 = R_{ws}$ high water stress

Following this criteria, it is evident from Plate 7 that water scarcity will be severe in the river basins of Yellow, Indus, Ganges, and Amu-Darya, and in the middle west of the United States. These scarcities are generally similar to the current situation. However the scarcity anticipated in Africa is a change from the current situation. Rapid increases in African water withdrawals are projected according to the population growth estimated by the United Nations. Such sudden increases in R_{ws} should demand the development of infrastructure and management systems, and the situation will be serious if these objectives cannot be achieved.

The world population living in regions of no water stress, low water stress, moderate water stress, and high water stress in 1995 and 2050 is shown in Table 4 (from Oki et al., 2003a). The contributions of population growth, climate change, and per capita increases in water demand to the figures for 2050 are also shown in Table 4. Population growth has the largest impact on the increase in people under high water stress levels, with the impact of climate change being of lesser importance. Based on the GCM simulation used in this study, the water stress level is somewhat eased by climate change effects. This is opposite to the result by Vörösmarty et al. [2000] in which the population under high water stress increases slightly if the impact of climate change is considered. The difference between the two studies may be due to the large uncertainties associated with GCM projections of climate change; nonetheless both studies agree that the impact of population growth on water stress is higher than the impact of climate change.

4 THE VIRTUAL WATER TRADE

The concept of "Virtual Water" has been developed to explain how physical water scarcity in countries in arid regions is relaxed by importing water-intensive commodities [Allan, 1997]. The real water cycle is somewhat different from the natural water cycle, even on the global scale, in the Anthropocene [Crutzen, 2002], and the virtual water trade should be taken into account when discussing the global water cycle and water scarcity.

The unit requirement of water (UW) to produce each commodity should be estimated so that the virtual water trade can be quantified [Wichelns, 2001]. The database of UW can be utilized for assessment of water demand in the future [Yang and Zehnder, 2002]. Estimation of UW is not difficult:

$$UW = \frac{TWA}{TWE}$$

where TWA is total water use to produce the goods and TWE is the weight (or the value) of the goods

However, there is no consensus on the definition of "total water use" or "the weight of the goods". Some researchers only consider on-farm evapotranspiration, but others count water losses such as irrigation canal seepage and return flow. The weight of the goods can be considered as either gross weight or edible weight.

UW is highly dependent on the crop yield per area and is different in each country and changes with time. Since the original idea of virtual water is how much water can be saved in the importing country, the UW of the importing country should be used to estimate the how much virtual water is imported. In this aspect, it is obvious that "virtual water" is "virtually required water" in its original sense, and we may call the water really used in the exporting country "really required water" or "real water" in the same way.

From this point of view, "real water" in exporting countries and "virtual water" in importing countries generally do not correspond quantitatively. For example, one kilogram of soy bean is produced using 1.7 tons of water in the United States; however, it requires 2.5 tons of water if the same amount of soy bean is grown in Japan. In this case, 1.7 ton of "real water" in

Table 4: World Population (Billion People) Under Each Water Stress Level Projected in 2050

Stress Level	Current (1995)	Population Growth Only	Future +Climate Change	+Unit Demand Increase
High Water Stress	1.7	2.66	2.43	2.51
Moderate Water Stress	0.6	0.91	0.93	0.97
Low Water Stress	0.6	0.94	0.96	1.00
No Water Stress	2.8	4.31	4.45	4.28

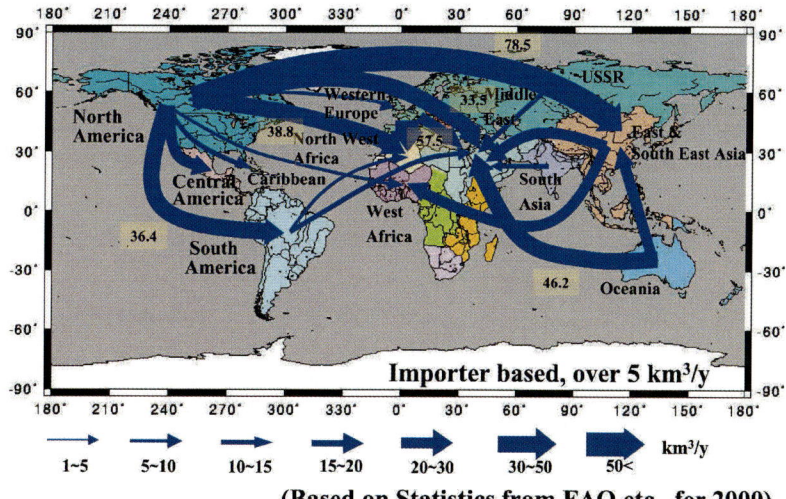

Plate 8. Annual virtually required water (virtual water) trade (km³/y) in year 2000 estimated for major crops of wheat, rice, maize, and barley.

the United States will be 2.5 ton of virtual water in Japan, associated with the import of 1 kilogram of soy bean.

Global estimation of the virtual water trade is now considered (from Oki et al., 2003b).

4.1 Global Virtual Water Flow

Global "virtual" and "real" water flows associated with major cereal (wheat, rice, maize, and barley) trade was estimated for each country where statistics were available for year 2000, and summarized into 16 regions. The annual virtual water flows are presented in Plate 8 (from Oki et al., 2003b). The plate shows that the Middle East, North West Africa, and East & South East Asia are gathering virtual water. The sources of virtual water are North America, Oceania, and Europe.

4.2 Does Virtual Water Save Water Globally?

Generally crop yields and water efficiencies in exporting countries are higher than in importing countries. Consequently, "real water" in exporting countries tends to be smaller than "virtual water" in importing countries. For example, as mentioned above, 1 kilogram of soy bean corresponds 1.7 ton of "real water" in the United States and 2.5 ton of "virtual water" in Japan. In this sense, the virtual water trade of 1 kilogram of soy bean from the United States to Japan saves 0.8 ton of global water resources.

Table 5 (from Oki et al. 2003b) shows total global "virtual water" and "real water" international transfers for year 2000 for maize, wheat, rice, and barley, soy bean, chicken, pork, and beef. Hoekstra and Hung's [2002] estimates of "virtual water" for the 1995–1999 cereal trade are also shown in Table 5, labeled as "VW". Hoekstra and Hung used global mean crop yield data to calculate VW (in contrast to Oki et al., who used country crop yield data wherever possible).

The total virtual water trade (imported virtual water) for the commodities in Table 5 is approximately 1,140 km³/y, however this corresponds to only 680 km³/y of real water, representing a water saving of 460 km³/y. The water savings are highest for crops and soybeans. It is interesting to see that the saving is less for chicken and pork, and is not significant for beef. This analysis provides a supporting argument for the globalization of trade; Table 5 claims that transferring commodities from water efficient regions to water inefficient regions will save global water resources. While the virtual water trade will not increase the total water resource, "saved" water in the importing country can be allocated to other purposes, such as municipal use and environmental use. However, one should be careful about interpreting these results since the idea of virtual water does not consider social, cultural, and environmental implications or limiting factors other than water.

5 CONCLUDING REMARKS

The global water cycle is essential in the Earth System. The ultimate goal of hydrological science is to increase understanding of the global water cycle through monitoring and modeling. The outcomes of hydrological research should be able to be used as tools to estimate, understand, and assess

Table 5: Global Virtual and Real Water Transfer (km3/y) Associated With Crop and Meat Trade

	Estimates by IIS, University of Tokyo				IHE[a] VW (km³/y)
	Virtual Water (km³/y)	Real Water (km³/y)	Saved (km³/y)	(%)	
Maize	127.0	51.7	75.3	59	61.4
Wheat	464.2	270.9	193.3	42	209.8
Rice	185.6	110.7	74.9	40	106.8
Barley	91.5	38.4	53.2	58	34.0
Cereal Total	868.4	471.7	396.6	46	(412.0)
Soy bean	118.1	84.0	34.1	29	
Chicken	37.4	25.3	12.0	32	
Pork	28.3	19.6	8.7	31	
Beef	86.2	82.4	3.9	5	
Meat total	151.9	127.3	24.6	16	
Total	1138.3	683.0	455.3	41	

[a]IHE: "Virtual Water Trade" [Hoekstra and Hung, 2002]

the water cycles on the Earth on various temporal and spatial scales, and these tools should be accessible to other scientific disciplines, the general public, and decision makers.

New developments and continued operation of existing global water cycle monitoring systems are essential to promote science and contribute to society. Innovative satellite observing missions, such as the Global Precipitation Mission (GPM) and the Hydrosphere Satellite Mission (HYDROS), are planned and should be pursued in addition to the maintenance and integration of surface observational networks of hydrological stations.

According to the paradigm shift of research in natural sciences after the wide recognition of global environmental problems, it is the era for geosciences to study the real situation of the Earth, including the impacts of anthropogenic activities. Water cycles are one of the most exposed natural cycles, vulnerable to human impacts. Therefore hydrological science should study water cycles, their impact on human society, and anthropogenic impact on water cycles. Regional characteristics and historical circumstances should be specially considered.

Scientifically excellent research and socially relevant research are not necessarily exclusive, and the anticipated world water crisis widens the opportunities for excellent scientific research to contribute to the solution of practical problems.

International and interdisciplinary frame works are required for the promotion of the hydrological sciences. More collaboration within hydrological sciences and with other disciplines is crucial, and philosophical and institutional development of hydrological sciences is ongoing.

The term "sustainable development" is often mistakenly interpreted as meaning that development must continue; hence, the term has come in for some criticism. In fact, the term's true meaning can be expressed as "the development of sustainability within society." At the present juncture, the most important requirement for a sustainable global water cycle is that governments, NGOs, universities, and private enterprises pool their wisdom and build sustainability into society through a fusion of policies, systems, scientific knowledge, and technical support for water.

Acknowledgements. The authors would like to thank all the data providing agencies and bodies. This research is partially supported by the Core Research for Evolutional Science and Technology of the Japan Science and Technology Corporation, and the No. 5 project of the Research Institute for Humanity and Nature, "Integrated management system for water issues of global environmental information library and world water model," and the Japan Society for the Promotion of Science.

REFERENCES

Alcamo, J., M. Märker, M. Flörke and S. Vassolo, 2003: Water and Climate: A global perspective. *Kassel World Water Series* **6**, Center for Environmental Systems Research, University of Kassel,

Kurt Wolters Strasse 3, 34109 Kassel, Germany.

Allan, J. A., 1997: 'Virtual Water': A long term solution for water short Middle Eastern economies?, Paper presented at the 1997 British Association Festival of Science, Roger Stevens Lecture Theatre, University of Leeds, Water and Development Session, TUE.51, 14.45.

Arnell, N. W., 2004: Climate change and global water resources: SRES emissions and socio-economic scenarios. *Global Environmental Change*, **14**, 31–52.

Chapelon, N., H. Douville, P. Kosuth and T. Oki, 2002: Off-line simulation of the Amazon water balance: a sensitivity study with implications for GSWP. *Climate Dynamics*, **19**, 141–154.

Cosgrove, W. J. and F. R. Rijsberman, 2000: World Water Vision. Earthscan Publications Ltd, London.

Crutzen, P. J., 2002: Geology of mankind—The Anthropocene. *Nature*, **415**, 23.

Falkenmark, M., J. Lundqvist and C. Widstrand, 1989: Macro-scale water scarcity requires micro-scale approaches; Aspects of vulnerability in semi-arid development. *Natural Resources Forum*, **14**, 258–267.

Hoekstra, A. Y. and P. Q. Hung, 2002: Virtual Water Trade, A Quantification of virtual water flows between nations in relation to international crop trade. *Technical Report Virtual Water Research Report Series* No.**11**, IHE Delft.

Hoskins, B. J., 1989: Diagnostics of the global atmospheric circulation based on ECMWF analyes 1979–1989, *Tech. Rep. WCRP-27, WMO/TD-No.326*, World Meteorological Organization.

IGPO, 2002: GSWP-2: The Second Global Soil Wetness Project Science and Implementation. Technical report, International GEWEX Project Office, Silver Spring, MD 20910.

Kanae, S., K. Nishio, T. Oki and K. Musiake, 1995: Hydrograph estimations by flow routing modelling from AGCM output in major basins of the world. *Annual Journal of Hydraulic Engineering*, JSCE, **39**, 97–102.

Korzun, V. I., 1978: World Water Balance and Water Resources of the Earth. Vol. **25** of *Studies and Reports in Hydrology*, UNESCO.

Manabe, S., 1969: Climate and the ocean circulation, I. The atmospheric circulation and the hydrology of the Earth's surface. *Mon. Wea. Rev.*, **97**, 739–774.

Meeson, B. W., F. E. Corprew, J. M. P. McManus, D. M. Myers, J. W. Closs, K.-J. Sun, D. J. Sunday and P. J. Sellers, 1995: ISLSCP Initiative I_Global Data Sets for Land-Atmosphere Models, 1987–1988., Published on CD-ROM by NASA (USA_NASA_ GDAAC_ISLSCP_001-USA_NASA_GDAAC_ ISLSCP_005).

Miller, J. R., G. L. Russell and G. Caliri, 1994: Continental-Scale River Flow in Climate Models. *J. Climate*, **7**, 914–928.

Numaguti, A., S.Sugata, M. Takahashi, T. Nakajima and A. Sumi, 1997: Study on the climate system and mass transport by a climate model. *CGER's Supercomputer Monograph Report*, **3**, National Institute for Environmental Studies, Environment Agency of Japan (Eds.).

Oki, T., 1999: The global water cycle. in Browning, K. and R. Gurney eds., *Global Energy and Water Cycles*, Cambridge University Press, pp.10–27.

Oki, T. and Y. C. Sud, 1998: Design of Total Runoff Integrating Pathways (TRIP)—A global river channel network. *Earth Interactions*, **2**.

Oki, T., K. Musiake, H. Matsuyama and K. Masuda, 1995: Global atmospheric water balance and runoff from large river basins. *Hydrol. Process.*, **9**, 655–678.

Oki, T., T. Nishimura and P. Dirmeyer, 1999: Assessment of annual runoff from land surface models using Total Runoff Integrating Pathways (TRIP). *J. Meteor. Soc. Japan*, **77**, 235–255.

Oki, T., Y. Agata, S. Kanae, T. Saruhashi, D. Yang and K. Musiake, 2001: Global Assessment of Current Water Resources using the Total Runoff Integrating Pathways. *Hydrol. Sci. J.*, **46**, 1159–1171.

Oki, T., Y. Agata, S. Kanae, T. Saruhashi and K. Musiake, 2003a: Global Water Resources Assessment under Climatic Change in 2050 using TRIP. No. **280** in *IAHS Publ.*, IAHS, 124–133.

Oki, T., M. Sato, A. Kawamura, M. Miyaka, S. Kanae and K. Musiake, 2003b: Virtual water trade to Japan and in the world. No. **12** in *Value of Water Research Report Series*, IHE Delft, 221–235.

Rossow, W. B., A. W. Walker and L. C. Garder, 1993: Comparison of ISCCP and Other Cloud Amounts. *J. Climate*, **6**, 2394–2418.

Sausen, R., S. Schubert and L. Dümenil, 1994: A model of river runoff for use in coupled atmosphere-ocean models. *J. Hydrol.*, **155**, 337–352.

Takahashi, K., Y. Matsuoka, Y. Shimada and R. Shimamura, 2000: Development of the model to assess water resources problems under climate change. *Proc. 8th Symposium on Global Environment*, **8**, 175–180.

UN, UNDP, UNEP, FAO, UNESCO, WMO, W. Bank, WHO, UNIDO and SEI, 1997: Comprehensive Assessment of the Freshwater Resources of the World. World Meteorological Organization, pp.33.

Vörösmarty, C. J., K. Sharma, B. Fekete, A. H. Copeland, J. Holden, J. Marble and J. Lough, 1997: The storage and aging of continental runoff in large reservoir systems of the world. *Ambio*, **26**, 210–219.

Vörösmarty, C. J., P. Green, J. Salisbury and R. B. Lammers, 2000: Global Water Resources: Vulnerability from Climate Change and Population Growth. *Science*, **289**, 284–288.

Wallace, J. S., 2002: Water resources and their use in food production systems. *Aquat. Sci.*, **64**, 363–375.

Wichelns, D., 2001: The role of 'virtual water' in efforts to achieve food security and other national goals, with an example from Egypt. *Agric. Water Manage.*, **49**, 131–151.

Xie, P. and P. A. Arkin, 1996: Analyses of global monthly precipitation using gauge observations, satellite estimates, and numerical model predictions. *J. Climate*, **9**, 840–858.

Yang, H. and A. J. B. Zehnder, 2002: Water scarcity and food import' A case study for Southern Mediterranean Countries. *World Development*, **30**, 1413–1430.

Dara Entekhabi, Massachusetts Institute of Technology, Room 48-331, Cambridge, MA 02139, USA.

Timothy Ives Harrold, Research Institute for Humanity and Nature, 335 Takashima-cho, Kamigyo-ku, Kyoto 602-0878, Japan.

Taikan Oki, Institute of Industrial Science, University of Tokyo, 4-6-1 Komaba, Meguro-ku, Tokyo 153-8505 Japan.

Cryosphere During the Twentieth Century

Atsumu Ohmura

Institute for Atmospheric and Climate Science, ETH Zürich, Winterthurerstrasse 190, CH-8057 Zürich, Switzerland

During the 20th century earth's ice bodies underwent substantial changes. This article reviews the present state of the cryosphere and investigates important variations observed in the seasonal snow cover, sea ice and glaciers. The seasonal snow cover area increased slowly until about 1970 when the area rapidly started to recede. The fastest recession was observed from the 1970s to the 1980s with a mean loss of 1.45×10^6 km^2. The most significant retreat occurred in late spring, amounting to 3×10^6 km^2 in area and shifting the end of the snow cover season by 3 weeks. While the sea ice extent decreased by 2.8%/decade in the north polar region, it increased by 1.3%/decade in the Antarctic during the last quarter century. The sea ice in the Arctic lost 40% of its thickness during the same time. The retreat of the sea ice limit was most significant in the warmer season. Most mountain glaciers and small ice caps showed significant mass loss during the second half of the last century. The mean increase in the equilibrium line altitude of 36 glaciers was 200 m for the second half of the 20th century. The small number of glaciers with a positive balance also showed increasing rates of melting. The global average specific mean mass balance for the last three decades was -265 mm y^{-1} or a loss of 135 km^3 y^{-1} corresponding to a 0.4 mm y^{-1} sea level rise. The rate of mass loss accelerated at 12 mm y^{-2}. The last decade showed an especially strong loss amounting to -475 mm y^{-1}, -242 km^3 y^{-1} or + 0.7 mm y^{-1} sea level equivalent. The mass balance of Greenland is negative: -50 to -90 km^3 y^{-1}, or -34 to -52 mm y^{-1} in mean specific balance, corresponding to a sea level contribution of 0.15 to 0.25 mm y^{-1}. The range of uncertainty for the Antarctic Ice sheet is still large. There is a report giving much less annual loss than the last IPCC report. The present work proposes 260 km^3 y^{-1} as the annual total discharge of the terrestrial cryosphere for the second half of 20th century, corresponding to +0.7 mm y^{-1} sea level equivalent. The most significant change in the Antarctic is the loss of ice shelves. Since the disappearance of an ice shelf tends to accelerate the flow of the background glaciers directly into the ocean, this may become a dominant mechanism in the future by which the Antarctic ice sheet can influence the sea level.

INTRODUCTION

This article is intended to summarise the present state of the cryosphere and to present important changes observed in the snow cover, sea ice and glaciers. The emphasis is on the mass balance of glaciers. A similar evaluation of the change in permafrost must wait until adequate information becomes available to provide a global picture.

PRESENT STATE OF CRYOPSHERE

At present, seasonal snow cover, sea ice, glaciers and permafrost collectively cover 70×10^6 km^2 or 14% of the earth's surface. In terms of the surface area, the seasonal snow cover and sea ice dominate the entire cryosphere, while glaciers are the major component with respect to volume. The surface areas of snow cover and sea ice show a large seasonal fluctuation, attaining the maximum in middle January and late October, respectively (Zwally et al., 1983; Parkinson et al., 1987; Rikiishi et al., 2003). As a result the maximum area of the cryosphere is found in January, covering 78×10^6 km^2 (15% of the earth's surface), while the minimum is in August with 59×10^6 km^2 (12 %) coverage. The area and volume of the cryosphere and its components near the end of the 20th century are summarised in Table 1.

Table 1 also shows the area and volume of the other two components of the cryosphere with much longer time scales, glaciers and permafrost. Permafrost occupies 4.5% of the earth's surface and 15% of the land surface. Permafrost is the least known of the four major cryosphere components. Based on Danilov (1996) and Zhang et al. (2000) the author estimates the present area and ice volume of permafrost at 23×10^6 km^2 and 0.4×10^6 km^3, respectively. It is noteworthy that the volume of frozen water in permafrost is almost 10 times that of mountain glaciers and small ice caps. The potential influence of permafrost on sea level should therefore be considered.

Although the glacier surface represents the smallest area of the four major components of the cryosphere, it contains by far the largest ice volume, estimated to be 33.1×10^6 km^3 glacier ice volume or 30.4×10^6 km^3 water equivalent (w.e.). This constitutes about 2.4% of the total water in the hydro- and atmosphere and corresponds to 75 m in the sea level equivalent. This value is about 10% less than the height expected from a sheer arithmetic division of the volume with the sea surface area. The ice volume which is presently in hydrostatic equilibrium with the sea level is excluded. Accurate estimates of glacier surface area and ice volume have been established during the last 20 years, especially due to the completion of the World Glacier Inventory (Müller, 1970) and the increasing number of the ice depth soundings by radars (Drewry, 1983). The regional distributions of the glacier area and volume are presented in Table 2. The striking feature of the glacier distribution is the dominance by Antarctica and Greenland. Antarctica and Greenland together harbour 97% of the glacier surface and 99.8% of the total glacier ice volume. Mountain glaciers distributed in the major mountain ranges such as the Andes, the Rocky Mountains, the Alaskan Range, the Himalayas and smaller mountain chains such as in the Pyrenees, account for only 0.2% of all glaciers in volume. It is, however, these small glaciers that have close contact to human society and whose volumes have become accurately known in recent years. A mean depth of 100 m might give an impression of a rather shallow glacier. It should, however, be noted that the average depth of all glaciers in the Alps is only 50 m. Although the Caucasus and the Alaskan Range have much larger glaciers, it is unlikely that the average depth of glaciers

Table 1: Global Cryosphere Area and Volume

	Area in [10^6 km^2]	Volume
Seasonal snow cover	4 (19) ~ 46 (58)	0.5 (1) ~ 5 (6) x 10^3 km^3
Sea ice	15.0 ~ 21.5	19 ~ 25 x 10^3 km^3
Glaciers	15.9	33.1 x 10^6 km^3
Permafrost	23	0.4 x 10^6 km^3 (11-35x10^3 km^3 segregated ice)
Total	59 ~ 78 x 10^6 km^2	33.2 x 10^6 km^3
Glacier ice distribution	[10^6 km^2]	[10^6 km^3]
Antarctic	13.59 (86 %)	30.11 (91.0 %)
Greenland	1.75 (11 %)	2.93 (8.9%)
Others (Mountain glaciers and small ice caps)	0.51 (3%)	0.05 (0.2%)
Total	15.85 x 10^6 km^2	33.09 x 10^6 km^3

Numbers in ( ) for seasonal snow cover are areas and volumes when the snow cover on sea ice is considered. The snow cover on glaciers is excluded in this calculation.
The sign ~ indicates the range of the seasonal fluctuation., while - means the range of uncertainty.

Sources: Mercer (1967), Ommanney (1972), Field (1975), Drewry (1983), Zwally et al. (1983), Hastenrath (1984), Björnsson (1986), Parkinson et al. (1987), Ohmura (1987), Østrem et al. (1988), Chen and Ohmura (1990), Haeberli et al. (1989), Weidick (1995), Zhang et al. (2000), Bamber et al. (2001), Rikiishi et al.(2003), A. Blochkov, M. Fukuda, A. Roesch (personal communication)

Table 2: Area and Volume of Glacier Ice

Region	Surface Area 10^3 km²	Ice Volume 10^3 km³	Sources
Greenland	1,748	2,931	Ohmura (1987), Weidick (1995), Bamber et al. (2001)
Iceland	11	1	Björnsson (1986)
Canadian Arctic Islands	152	15	Ommanney (1972)
Jan Mayen	10^{-1}	10^{-2}	Field (1975)
Svalbard	37	4	Haeberli et al. (1989)
Zemlya Franza Yosifa	14	1	Haeberli et al. (1989)
Novaya Zellya	23	2	Haeberli et al. (1989)
Severnaya Zemlya & Ostrov Ushakova	19	2	Haeberli et al. (1989)
Ostrava de Langa, Ostrov Vrangelya	10^{-1}	10^{-2}	Haeberli et al. (1989)
North America (continental)	124	12	Haeberli et al. (1989)
Scandinavia	3	10^{-1}	Østrem and others (1988)
Alps	3	0.14	Chen and Ohmura (1990)
Pyrenees and Cordillera Cantabrica	10^{-2}	10^{-3}	Haeberli et al. (1989)
Caucasus	1	10^{-1}	Haeberli et al. (1989)
Ural	10^{-2}	10^{-3}	Haeberli et al. (1989)
Asia	92	9	Haeberli et al. (1989)
Africa	10^{-2}	10^{-3}	Hastenrath (1984)
New Zealand	1	10^{-1}	Haeberli et al. (1989)
South America	25	3	Haeberli et al. (1989)
Sub-Antarctic islands	7	1	Mercer (1967)
Antarctica	13,586	30,110	Drewry (1983)
Total	15,855	33,092	

in these ranges exceeds 200 m. The present evaluation makes the sea level equivalent of all these glaciers outside Antarctica and Greenland only 15 cm. This value was previously grossly overestimated (Church and Gregory, 2001).

CHANGES IN THE 20TH CENTURY

Seasonal Snow Cover

The area of the seasonal snow cover is often expressed in terms of the snow cover extent. The snow cover extent is defined as the area of the surface with more than 50% snow cover. This definition corresponds to the snow cover for the surface characteristics in the synoptic meteorological observation. The more recent satellite definition of the snow cover extent is adjusted to the classic synoptic meteorological definition. Therefore, the actual area of the snow-covered surface is somewhat smaller than the snow cover extent.

The seasonal snow cover occupies a very unique place in the climate system. First of all it covers almost half of the cryospheric surfaces in its peak time. Even with respect to the annual mean, the snow cover accounts for about 1/3 of the entire cryosphere surface. Because the snow cover is present also at lower latitudes, it plays the most active role in the albedo/temperature positive feed-back process. As most snow cover is in the Northern Hemisphere, the albedo/temperature feedback process enhances the effect of the snow cover conditions of the Northern Hemisphere within the global climate system.

The characteristics of the snow cover have been observed since the 19th century. Since the geographic coverage with this type of observation was limited in earlier days, it became possible to discuss the global magnitude of the snow cover extent only in the second decade of the 20th century. Brown (2000) integrated the ground observations of the snow cover going back to 1915 and combined these manual observations with satellite observations for the 1972 to 1997 period, to form a relatively homogenous time series of snow cover extent, covering the major part of the 20th century. Brown (2000) found a slightly increasing trend to the 1970s and then a rapidly decreasing trend during the last 30 years on a hemispheric scale, with a mean retreat rate of 3.1×10^6 km² $(100\ y)^{-1}$. An especially strong retreat was found for the spring with a rate of 8.5×10^6 km² $(100\ y)^{-1}$. This tendency is probably mainly due to the temperature increase in the mid- to high latitudes of the Northern Hemisphere. The temperature sensitivity of the snow cover extent is calculated as -2.04×10^6 km² K^{-1}.

Rikiishi et al. (2003) compiled the Northern Hemisphere snow cover extent based on the information at the National Snow and Ice Data Center, and presented the hemispheric snow cover extent for three decades, 1970s, 1980s and 1990s. They concluded that over the last twenty years, the onset of the snow cover in autumn was delayed by one week, but there was a slight advance of early winter, while the disappearance of the snow cover in spring occurred two to three weeks earlier. The most significant change happened from the 1970s to 1980s. The mean retreat of the snow cover from 1970s to 1980s is 1.45×10^6 km^2. The largest retreat of 3×10^6 km^2 was observed in the late melt period of May to June. Globally this difference means about a 140 km^3 decrease of the snow w.e. on land.

Sea Ice

One of the most comprehensive reports on this subject was made by Folland and Karl (2001). This section is organised partly on their report, with information from more recent publications and the author's own material. The changes of sea ice are usually examined with respect to its surface coverage and thickness. The surface coverage is often measured in terms of the sea ice extent or sea ice area. The sea ice extent denotes the area within the sea ice limit which is covered by sea ice with more than a certain concentration. Therefore, the sea ice extent depends on the selection of the concentration that defines the sea ice limit. Normally, a concentration of 15% is used for the sea ice limit. The sea ice area, on the other hand, is more genuinely the surface area of the sea ice that is the product of the sea ice extent and the concentration. The hemispheric and global grasp of the surface coverage only became possible after high quality satellite remote sensing techniques were developed in early 1970s (Walsh, 1978), giving about a quarter century of information. There have been attempts to evaluate the sea ice area in older times using the ground-based observations (Vinje et al., 1998; Vinnikov et al., 1999). Although these are a valuable attempt to characterise the local ice conditions, their global significance is doubtful, as the terrestrial observations do not provide reliable information on sea surface conditions beyond 5 to 8 km from the shore. The visual obstructions caused by hummocks and ridges and optical refraction limit the area of accurate observation. Data on thickness are available dating back to the late 1950s. They are mostly submarine sonar soundings, and the thickness data are necessarily confined to the routes of navigation. Since it is extremely rare that the same track was traversed more than once, interpretation of the thickness changes should take account of the sampling errors. Bearing these limitations in mind, it can be concluded that important changes happened in the sea ice during the second half of 20th century which can be quantitatively discussed.

One peculiar circumstance is the seemingly opposite directions in the changes that took place in the sea ice extent in the Arctic and the Antarctic during the last 25 years. While the sea-ice extent is reported to have decreased in the Arctic, it has increased slightly in the Antarctic (Cavalieri et al., 1997; Parkinson et al., 1999; Zwally et al., 2002). Further in the case of the Arctic, Serreze et al. (2003) reported that the sea ice extent in September 2002 fell below the line of the previous record minimum of 5.47×10^6 km^2 observed in 1995 by about 2×10^5 km^2. This situation involves the decrease in the multiyear ice area from 4.4×10^6 to 3.5×10^6 km^2 over 22 years from 1978 to 2000 (M. R. Drinkwater, personal communication).

The thickness of the sea ice was reported to have thinned by 42% averaged for the six circum-Arctic seas over 40 years from 1958 to 1997 (Rothrock et al., 1999). For the cross-section at 0° longitude, Wadhams and Davis (2000) found a 43% thinning from 1976 to 1996. More recently Rothrock et al. (2003) reported a rapid decrease in thickness of about 1 m or 40% over the 11 year span from 1987 to 1997 for a distance of about 2000 km along 150° W in the Beaufort Sea. A decrease in the sea-ice season and an increase in the summer melt season are reported by Parkinson (2000). The air temperature near the shores of the Arctic Ocean increased during the same period. All these phenomena are compatible with the general trend of the sea ice retreat in the Arctic. The decreasing rate of $2.8 \pm 0.3\%$ /decade is well established (Parkinson et al., 1999). On the other hand, the sea ice was reported to have advanced at a rate of $1.3 \pm 0.2\%$ /decade in the Antarctic (Cavalieri et al., 1997). A report on the thickness change in the Antarctic sea ice over the equivalent period is not available.

The recent rapid retreat of the Arctic sea ice extent is attributed not only to the temperature rise but also to the change in the atmospheric circulation which is caused by the decrease in sea-level atmospheric pressure in the Arctic Basin (Serreze et al., 2003). This phase is sometimes referred to as positive in the Arctic Oscillation Index. A similar pressure deepening in the Antarctic region was also reported by Folland and Karl (2001). These trends in the sea level pressure distribution may be related to recent deepening of the polar vortex (J. Overland, personal communication). The same deepening trend of the surface pressure around the poles may cause different consequences in the Arctic and Antarctic, however. A direct result of the intensification of the pressure gradient towards the pole is the strengthening of the westerlies. Because of the Coriolis force, the eastward directed stress on the sea ice results in the divergent movement of the sea ice towards lower latitudes in both hemispheres. In the region of lower temperature like the Antarctic Ocean, this movement may result in expansion of the sea ice field. However, in the Arctic where temperature is higher and increasing in recent years, the loose pack caused by the divergence may promote melt owing to a decrease of the regional albedo. An observation-based explanation of the opposite trends in the sea ice cover is urgently needed.

Another important observation concerning the change in sea ice in the Arctic region is the seasonally variable shift of the sea ice limit. It is now well established that the shift in autumn and winter is less conspicuous in comparison with that in spring and summer. This means that the seasonal range of the sea ice area, and hence volume, has increased. The present author considers that these processes result in an increase in the annual production rate of the sea ice, and hence the intensification of the thermo-haline circulation in the North Atlantic. The actual strength of the thermohaline circulation depends on additional processes, such as the fresh water supply from precipitation and continental discharge. Apart from the possibility of an actual acceleration of the circulation, the present observation suggests that a summer recession of the ice limit does not *a priori* mean a slowing of the thermo-haline circulation. This finding is in contrast to some previous categorical speculations which envisaged a weakening of the thermohaline circulation in the warming climate(Stocker and Schmittner, 1997). This last conclusion is probably the most important result from the satellite-based sea ice observations.

Glaciers

Glaciers are the best documented component of the cryosphere. Geometrical characteristics (area, length, and depth) of glaciers are determined by the mass balance and dynamics of glaciers. As the mass balance of a glacier is regulated by the mass gain through accumulation and the mass loss through ablation, it is closely related to precipitation and temperature. This is the reason why glaciers react sensitively to climate changes. Furthermore glaciers are believed to influence the climate, as they are in a positive feedback relationship with temperature through their high albedo and large latent heat of melt. These characteristics make glaciers an important internal climate factor and also a useful object for climate monitoring.

Mountain Glaciers and Small Ice Caps

The variation in the geometry of glaciers was already documented in the 18th century. Systematic measurements of the changes in the length and surface area were made on some glaciers during the First International Polar Year (1882–83). They produced not only the measured mass balance for the 19th century but also high quality glacier maps which contain useful information on mass balance. The number of glaciers with annual mass balance records increased sharply on the occasions of the International Geophysical Year (IGY, 1957–58) and the International Hydrological Decade (IHD, 1965–74). Therefore, it is not possible to analyse the mass balance for the entire 20th century only with observed values. This section will present first the overview of mass balance for the entire 20th century which is based on observations and computations.

Then, the major part of the section will be dedicated to the status of the mass balance for the second half of the century for which observational data of reasonable quality and continuity can be expected.

Meier (1984; 1993) evaluated the long-term mass balance on a global scale, and obtained -380 mm y^{-1} for the period between 1900 and 1961. The method is a combination of the mass balance observations, an approximation with meteorological data and calculation of the volume change from geometrical variations of the glaciers. He was the first to point out the importance of the mass balance of mountain glaciers and small ice caps for influencing sea level. The main cause for the large mass loss was attributed to the warm period between 1920 and 1950. The Rhonegletscher, in the Central Swiss Alps belongs to a group of glaciers with rich historical and modern scientific observations. This glacier has one of the oldest mass balance observations which started in 1884, but terminated in 1909 until it was restarted in 1979. Chen and Funk (1990) bridged the missing observations between 1909 and 1979 with estimated mass balance calculated using meteorological data. The main result of this work is the mean specific mass balance of -250 mm w.e. y^{-1} over the 104 years from 1883 to 1987. This generally negative mass balance was interrupted for 20 years by a positive period between 1907 and 1927, mainly owing to heavier precipitation. This period was followed by about 30 years of strong negative balance from 1927 to 1957, which was caused by higher temperatures. The mass balance series of the four glaciers in the French Alps, Saint Sorlin, Gébroulaz, Argentière and Mer de Glace were constructed by Vincent (2002), based on short-term mass balance, meteorological observations and cartographic data for the period from 1907 to 1999. The mean specific mass balance for these glaciers was found to be -150 mm y^{-1}. The twentieth century for these glaciers was characterised by a rather long balanced period which was interrupted by two periods of major losses from 1940 to 1950 and 1980 to 1999. Therefore, the glaciers in the Alps during the 20th century experienced a strong mass loss centred around the 1940s and after 1980. More detailed analyses especially concerning the causes of the mass balance changes are not possible due to the limitation of the data.

A New Data Base for Seasonal Mass Balance

In order to analyse current conditions a new data base for annual and seasonal (winter and summer) mass balances had to be made. The reason for this rather laborious approach rests on the fact that the existing global data sets of the mass balance with winter and summer balances are insufficient. Further, the present analysis must be made solely based on the measured seasonal mass balances, without involving any meteorological estimations. When considering the causes of the mass balance change, it is meaningless just to analyse the annual balance, or to use the estimations usually based on meteorological data. The sources of the winter and summer mass balance data are Kasser (1967, 1973), Müller (1977), Haeberli (1985), Haeberli and Müller (1988), Haeberli and Hoelzle (1993), Haeberli et al. (1998), Kjøllmoen (1998, 1999, 2000, 2001, 2003), Dyurgerov (2002), R. J. Braithwaite (personal communication) L. N. Braun (personal communication) and C. Vincent (personal communication). There are about 40 glaciers on which the annual mass balance observations are currently conducted with at least 30 years of continuous records. When historical data are considered, there are more than 100 glaciers where the annual mass balance was measured for shorter periods. The author selected 75 glaciers which are considered to provide mean mass balance conditions and trends for the second half of the 20th century for representative regions and also for the globe.

The world is divided into 14 regions: the Andes (5 glaciers), the Rocky/ Coastal/Cascade/Olympic Mts. (8), Canadian Arctic (5), Alaska (2), Kamchatka (1), Altai (4), Himalayas (3), Tenshan/ Dhungariya (13), Pamir (1), Caucasus (6), Svalbard (3), Scandinavia (9), Alps (14) and Equatorial/Tropics (1). The regional mass balances were obtained by averaging the annual balances of the selected glaciers whose numbers appear in the brackets in the preceding sentence. They are also listed in Table 3. Several glaciers have mass balance records before 1952/53. Although these are important data on local conditions, they are not used in the present work, as their global meaning is limited. The global mean in Table 3 is for the period where reasonable data for at least three regions were available. The second mean (Mean 2) was calculated by excluding three regions with limited observational periods, the Andes, the Himalayas and Equatorial/Tropics. The 30 years from 1967 to 1996 was chosen for which the other 11 regions have continuous time series. Further, the area-weighted mean was calculated. The total area considered in this work is 350×10^3 km^2, or 68% of the total surface area of the glaciers in this category. A comparison of the first two arithmetic means in the second and third last columns shows that the global mean values for the period prior to 1967 can be relied upon as their arithmetic means are similar. The area-weighted mass balance (the last column) shows certain differences from the arithmetic means, but the decadal means are very similar. In fact the physical meaning of the area-weighted mass balance becomes more important when winter and summer balances are discussed. The annual balance of the 14 regions and the two mean values are illustrated in Plate 1.

The most important result of this analysis is that the global mean trend in mass balance of glaciers was negative in all regions for the second half of the 20th century. It appears that the negative tendency of mass balance started globally in the early 1960s. The global mean mass balance for the period of 1952 to 2000 is -279 mm y^{-1}. The area-weighted mean mass balance for the 30 years of the best observations (1967–1996) excluding the Andes, the Himalayas and the Equatorial/Tropics is -265 mm y^{-1}. During the same period the global area-weighted mean equilibrium line altitude (ELA) rose at a rate of 4 m y^{-1}. Therefore, it can be concluded that, globally glaciers lost mass during the second half of 20th century at a mean rate of -250 to -280 mm y^{-1}. Considering that the total area of mountain glaciers and small ice caps is 510×10^3 km^2, a rough estimate of the total discharge from the glaciers of this category for the corresponding period is 135 km^3 y^{-1} or 0.4 mm y^{-1} in sea-level equivalent. The present evaluation together with that of Meier (1984) gives 43 mm as the total sea level rise caused by glaciers outside Greenland and Antarctica during the 20th century. The present analysis further shows that the global mass balance is not only negative, but its trend is also negative, that is, the loss of mass is accelerating. The linear trends for the two means (arithmetic, and area-weighted means) give very similar trend-gradients, about -12 mm y^{-2}. This rather clear trend is due to a great extent to large losses in the 1990s. The beginning of the melt-dominated mode of the mass balance in the 1990s was reported by Hodge et al. (1998) with respect to glaciers in the Pacific Northwest region. This last decade of the 20th century saw a global mean mass balance of -475 mm y^{-1}. This is more than the sum of the preceding two decades. Examining each region, the largest loss was observed on Equatorial glaciers, represented by a single glacier, the Lewis Glacier where it was -903 mm y^{-1}. The smallest loss was found on Scandinavian glaciers where the mean balance was -53 mm y^{-1}. In fact the mass balances of many Scandinavian glaciers were slightly positive. Some glaciers even showed a tendency of increasing mass balance. Since this is the only region with such unusual features, it will be discussed later in more detail.

The investigation of the causes of changes in mass balance requires analysis of both winter and summer mass balances, as the annual balance is determined by these two seasonal balances which reflect rather different processes. The winter balance is to a great extent due to the winter solid precipitation. Summer balance is mostly a matter of melting for which air temperature is the single most influential cause. There are presently only about 30 glaciers where the seasonal mass balances are observed. The result of the seasonal mass balances for the 30 glaciers in the 11 regions is presented in Table 4. The periods of coverage for some glaciers are extended for as long as high quality data are available. The 11 regions cover wide ranges of climate, from the extremely maritime region of Kamchatka to a very dry zone in the Canadian Arctic. The area-weighted global mass balance for the 30 year period is

Table 3: Regional and Global Mean Mass Balance (Unit: mm y^{-1})

	Andes	Rockies, Cascade, Olympic	Coastal	Canadian Arctic	Alaska	Kamchatka	Altai	Tenshan,	Himalayas Dzungriya	Pamir	Caucasus	Svalbard	Scandinavia	Alps	Equatorial, Tropical	Mean	Mean 2	Area-Weighted Mean
1952/53		-580											-825	-471		-625		
1953/54		-180											-870	13		-346		
1954/55		710											-325	541		309		
1955/56		317											-325	-179		-63		
1956/57		-100											-140	-162		-96		
1957/58		-1847											-365	-670		-641		
1958/59		-87											-1125	-958		-584		
1959/60		-453											-1350	116		-335		
1960/61		-683								265			-805	-38		-408		
1961/62		-13								-625			970	-525		-234		
1962/63		-490					-400			-675			-827	-277		-155		
1963/64		1037					-340			700			289	-1099		79		
1964/65		-248					-280	20		-5			510	993		102		
1965/66		-117			281	570	-560	318		-425			-753	672		-215		
1966/67		-566			46	-780	-380	-13		40			911	139		-25		-163
1967/68		56		-577	-95	-520	290	-168		420			207	510		-154	-154	-286
1968/69		-81		-26	44	-300	-20	-13		-172		-227	-1497	-12		-221	-221	-87
1969/70		-999		-703	-177	-280	-666	-272		-384		-885	-911	-142		-273	-273	-208
1970/71		66		-109	3	-1900	186	-292		1349		-535	358	-813		19	19	-138
1971/72		-291		281	-62	580	250	260		312		-520	-333	-19		-167	-167	-138
1972/73		131			109	-985	120	333		793		-265	733	-855		-53	-53	-22
1973/74		-91			48	365	70	-70		328		-50	400	-25		-201	-201	-357
1974/75		834			-103	600	100	-56		-818		-905	369	270		-158	-158	-128
1975/76		28			86	1640	-1470	140		-311		-260	368	-678		4	4	-131
1976/77		835			123	-220	400	-666		-327		400	-328	818		-327	-327	-150
1977/78		-1002			-243	1330	680	186		-1034		95	-341	784		-354	-354	-195
1978/79		-333			-149	-1050	270	31		-1022		-490	-685	-115		-466	-466	-596
1979/80	625	-1450			-69	-940	-323	-154		-1459		-75	163	615		433		16
1980/81	-689				-87	-200	-663	-1116	-713	-1313	-299	-268	-1071	80	-70	295	-247	-85
1981/82	-220	-795			-241	-170	55	-297	-362	-387	-450	-685	195	-598	-1750	-377	-285	-85
1982/83	-1710	-22			-174	-1950	-243	-425	-169	-1081	-635	-475	-505	-438	-1210	-385	-273	-171
1983/84	2625	-174			-120	-247	-423	53	-262	129	120	-505	-10	-598	-320	-34	-201	-124
1984/85	-720	-93			-10	-295	230	-417	589	-774	-970	-10	-165	105	-720	-304	-216	-194
1985/86	270	30			-38	-420	313	-317	-644	-507	275	-220	659	-277	-900	-72	-85	-233
1986/87	-300	-790			-405	-340	223	-757	-108	-971	-240	-705	115	-916	-941	-534	-530	-185
1987/88	525	-144			495	2020	93	-475	-679	-855	-570	-515	-374	-617	-696	41	80	-35
1988/89	1290	-795			-155	-1660	195	-363	-266	-1010	975	-265	-245	-611	-721	-397	-353	-181
1989/90	-1055	-380			30	-300	328	-219	-580	240	395	230	844	-678	-2282	-169	-164	-634
1990/91	-844	-341			-209	-1940	135	-424		10	35	505	-1244	-992	770	-579	-496	-644
1991/92	-925	-99			-325	-740	98	-192		-230	215	-345	1683	-1115	953	-417	-335	-201
1992/93	-69	-950			-325	-1280	-459	-556		-530	-170	585	1074	-1032	-810	-317	-209	-414
1993/94	-1330	-760			-214	460	98	-769		-430	5	115	44	974	-1750	-233	-228	-469
1994/95	-75	-138			-138	410	-119	-202		400	-635	-170	974	-927	-440	-693	-534	-513
1995/96	-857	-1433			-525	-350	236	-442		300	-750	955	1076	-109	-2287	400	-315	-196
1996/97	-1275	-1322			-213	-670	16	-262		-780	705	-90	120	-928	-450	-306	-216	-196
1997/98	-1118	-912			-278	-230	-240	-389		-354	15	-785	544	444	-490	-508	-470	-526
1998/99	-1201	-225			-58	-915	83	-262		-101	215	-75	-669	-143	-242	-842		
1999/00	-901	-655			-147	870	-140	-989		-1737	-970	-570	480	-1554	-2165	-18		
2000/01	-1474	-2166			-316	-1280	-123	-503		204	-910	-725	217	-1115	-545	-780		
		705			-110	1645	-1110	-511				-350	-186					
		-1570			380		-113	-330				-25	1001					
					494	-995	-113					-405	-846					

| Mean | -384 | -422 | -150 | -365 | -224 | -78 | -269 | -319 | -409 | -179 | -383 | -53 | -292 | -903 | -279 | -248 | -265 |

Andes: Antizuma, Chacaltaya, Enchaurren Norte, Piloto East, Zongo
Rockies, Coastal, Cascade, Olympic: Blue, Helm, Lemon Creek, Peyto, Place, Ram, Sentinel, South Cascade
Canadian Arctic: Devon, Drambuie, Meighen, Melville South, White
Alaska: Gulkana, Wolverine
Kamchatka: Kozelskiy
Altai: Leviy Aktru, Maliy Aktru, No. 125, Praviy Aktru
Himalayas: Changmekhangpu, Dunagiri, Shaune Garang
Tenshan/Dhungariya:Golubin, Igly Tuyuksu, Kara Batkak, Kosmodemya, Manetovoy, Mayakovsko, Molodezhniy, Ordzhonikidze, Partizan, Shumskiy, Tsentralniy Tuyuksuyskiy, Urmuqihe S. No.1, Urmuqihe E. B.
Pamir: Abramov
Caucasus: Bezingi, Djankuat, Garabashi, Marukhskiy, Tbilisa, Zeiskiy
Svalbard: Austre Broeggerbreen, Midtre Lovénbreen, Finsterwalder
Scandinavia: Alfotbreen, Engabreen, Grásubreen, Hardangerjoekulen, Helistungubreen, Nigardsbreen, Rabots, Storbreen, Storglaciären
Alps: Aletsch, Careser, Gries, Hintereis, Jamtal, Kesselwand, Limmern, Plattalva, Saint Sorlin, Sarennes, Silvretta, Sonnblick, Vergagt, Wurten
Equatorial Region: Lewis

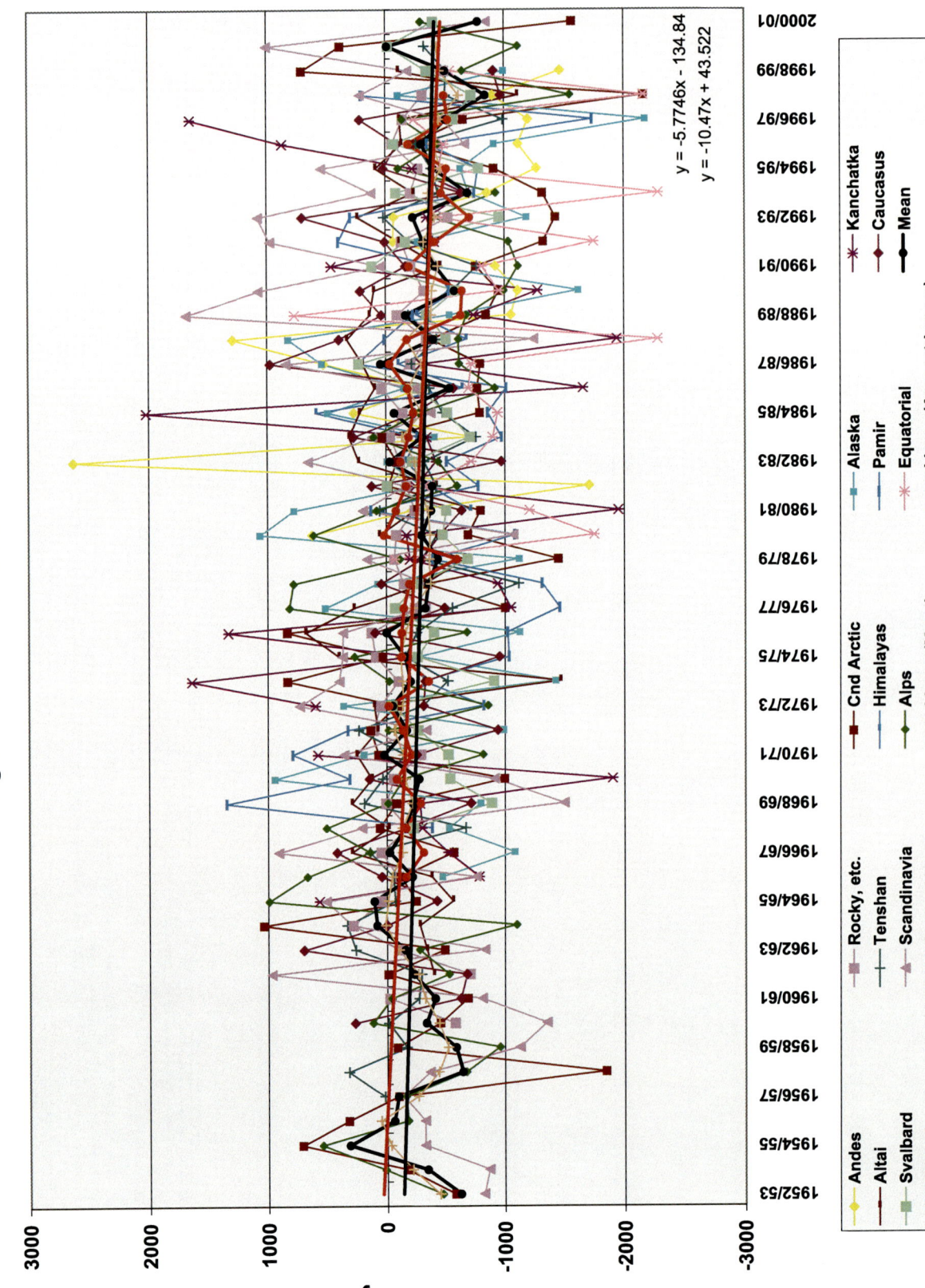

Plate 1: Annual mass balance for 14 regions as defined in Table 3 (While the thick black line indicates the trend of the arithmetic mean of 14 regions, the thick red line shows the trend for the area-weighted means)

Table 4: Regional and Global Mass Balance of Glaciers (for Those Glaciers With Winter, Summer and Annual Balance Measurements)
Bw (Winter Balance)

	Rockies	CND Arctic	Alaska	Kamchatka	Altai	Tenshan	Pamir	Caucasus	Svalbard	Scandinavia	Alps	Mean	Area-Weighted Mean
1956/57						809						809	
1957/58						989						989	
1958/59	3280					749						2014	
1959/60	2210					787						1499	
1960/61	2400					674						1537	
1961/62	2500					716						1608	
1962/63	2230	173				862				1503		1192	
1963/64	3250	252				933				1342		1444	
1964/65	2540	136				840				1572		1272	
1965/66	2048	138	1285			838				1864	1845	1336	829
1966/67	2446	134	1335			897				1423	1370	1267	891
1967/68	2210	151	1685			638	1030	2060	525	2354	865	1280	901
1968/69	2164	148	1205			768	2115	1890	405	2179	1035	1323	812
1969/70	1658	132	2050			825	1516	2410	365	1373	1290	1291	890
1970/71	2300	163	1695			739	1374	2430	675	1364	1220	1329	947
1971/72	2340	116	960			792	1401	2170	965	1996	1010	1306	816
1972/73	1984	106	1510	3690		650	1756	2200	780	1884	950	1551	854
1973/74	2790	180	1035	4130		588	774	2360	725	2391	940	1591	879
1974/75	1886	125	1815	3630		581	1029	2180	805	1885	1285	1522	896
1975/76	2830	136	965	3810		522	571	2220	735	2295	475	1456	837
1976/77	1465	127	2805	3700	1083	584	1942	2140	780	2389	1940	1723	1071
1977/78	2068	124	2175	3310	837	480	1469	2560	780	1311	1555	1515	996
1978/79	1551	101	1390	2970	727	627	1619	2520	785	1675	1693	1423	762
1979/80	1706	133	2870	3620	970	583	1277	2720	790	2200	1665	1685	1105
1980/81	1858	158	3425	2575	753	661	1065	2390	485	1459	1335	1469	1208
1981/82	2501	151	1770	3500	657	447	1046	2530	650	1992	1665	1537	968
1982/83	2150	136	1930	3330	870	514	1305	1650	725	1472	1975	1460	957
1983/84	2038	159	1810	4000	1030	463	1479	2270	665	2233	1585	1612	928
1984/85	1624	152	1630	3610	937	483	1449	1820	955	2073	1720	1496	856
1985/86	2084	181	1610	3000	957	686	1010	2240	1020	1332	1380	1409	930
1986/87	1936	188	2260	3000	973	656	1700	4000	820	1472	1280	1662	1045
1987/88	1828	152	2300	2670	1213	763	1900	2760	585	2046	1520	1613	1011
1988/89	1745	181	1350	3400	1120	642	1190	2630	595	1682	1143	1425	786
1989/90	1970	161	1295	3440	1047	720	1620	2660	810	2972	1180	1625	847
1990/91	3350	155	1725	3250	877	410	1290	2480	900	2924	1355	1701	1119
1991/92	1910	153	1620	3060	810	600	1620	1950	765	1832	1140	1405	877
1992/93	1343	111	1590	2740	1240	547	1640	3180	610	2618	1320	1540	763
1993/94	1644	117	1775	3480	1037	622	1610	2070	825	2436	1610	1566	874
1994/95	1703	206	1775	2740	1263	279	1120	2540	540	1738	1870	1434	856
1995/96	2940	150	1040		890	339	1410	2262	790	2440		1362	865
1996/97	3710	187	1585		990	237	970	2797	530	1310		1368	1079
1997/98	2760	181	2000		635	400	1900		510	2463		1356	1056
1998/99	3590	140	1615		860	47				2021		1379	1034
1999/00	3320					179			500	2035		1509	
Mean	2282	151	1732	3333	947	617	1393	2403	700	1936	1374	1452	917

Rockies, Coastal, Cascase: Helm, Peyto, Place, Ram, Sentinel, South Cascade
Canadian Arctic: Devon, Melville South, Meighen
Alaska: Gulkana, Wolverine
Kamchatka: Kozelskiy
Tenshan/Dzhungariya: Tsentralniy Tuyuksuyskiy, Urumqihe S. No.1, Karabatkak, Golubin, Shumskiy
Altai: Leviy Aktru, Maliy Aktru, No. 125
Pamir: Abramov
Caucasus: Djankuat
Svalbard: Austre Broggerbreen, Midtre Lovenbreen
Scandinavia: Alfotbreen, Engabreen, Grasubreen, Hardangerjokulen, Hellstungubreen, Nigarsbreen, Storbreen, Storglaciären
Alps: Glacier de Sarennes, Vernagtferner

Bs (Summer), Table 4 (cont.)

	Rockies	CND Arctic	Alaska	Kamchatka	Altai	Tenshan	Pamir	Caucasus	Svalbard	Scandinavia	Alps	Mean	Area-Weighted Mean
1956/57						-900						-900	
1957/58						-793						-793	
1958/59	-2580					-974						-1777	
1959/60	-2710					-980						-1845	
1960/61	-3500					-1140						-2320	
1961/62	-2300					-1030						-1665	
1962/63	-3530	-267				-650					-535	-1245	
1963/64	-2050	6				-711					-2168	-1231	
1964/65	-2915	-71				-735					-1272	-1248	
1965/66	-2230	-257	-1755			-994				-1351	-1325	-1319	-991
1966/67	-3234	-116	-2415			-661				-2176	-1553	-1692	-1203
1967/68	-2142	-214	-2210			-1189	-1414	-1960	-590	-1443	-595	-1306	-1064
1968/69	-2510	-180	-1995			-602	-766	-2980	-1290	-1971	-1380	-1519	-1098
1969/70	-3008	-213	-1105			-639	-1204	-2000	-900	-2860	-1650	-1509	-977
1970/71	-2409	-491	-1505			-783	-2167	-2660	-1195	-2279	-2015	-1723	-1155
1971/72	-2144	-14	-1945			-525	-1073	-3310	-1230	-1639	-1125	-1445	-954
1972/73	-2014	-201	-1145	-3090		-1035	-2574	-2480	-830	-2216	-1515	-1710	-876
1973/74	-1997	-302	-2460	-2480		-1071	-1085	-2120	-1630	-1659	-1565	-1637	-1236
1974/75	-2010	-102	-2085	-3850		-838	-2063	-3090	-1065	-1485	-1160	-1775	-1024
1975/76	-1943	-8	-2090	-2480		-816	-1518	-1930	-1135	-1926	-1495	-1534	-968
1976/77	-2571	-297	-2285	-4750	-813	-1065	-1648	-2510	-855	-2021	-1275	-1826	-1221
1977/78	-2525	-155	-2105	-4250	-1160	-1220	-2808	-2120	-1300	-1639	-1130	-1856	-1191
1978/79	-3362	-84	-2510	-3170	-1390	-925	-2006	-2830	-1470	-2016	-1745	-1955	-1358
1979/80	-2359	-164	-1805	-3790	-980	-1033	-2358	-2400	-1265	-2038	-1440	-1785	-1089
1980/81	-2557	-327	-2645	-4700	-967	-715	-936	-3300	-990	-2530	-1345	-1910	-1293
1981/82	-2642	-234	-2065	-3610	-1117	-997	-1820	-2110	-660	-1797	-2150	-1746	-1139
1982/83	-2315	-209	-1940	-3750	-657	-936	-1812	-2620	-945	-1681	-2200	-1733	-1081
1983/84	-2079	-80	-2215	-4340	-723	-1189	-2450	-2060	-1420	-1574	-1605	-1794	-1122
1984/85	-2917	-235	-1135	-1590	-737	-1156	-2304	-2200	-1470	-1958	-2470	-1652	-1089
1985/86	-2924	-21	-1765	-4660	-883	-1164	-2020	-2740	-1285	-1832	-2680	-1998	-1115
1986/87	-2883	-228	-1720	-3300	-790	-982	-1460	-2460	-590	-1718	-1915	-1641	-1080
1987/88	-2584	-530	-1475	-4610	-880	-1186	-1890	-2240	-1090	-1181	-2145	-1801	-1192
1988/89	-2700	-221	-3270	-4140	-1003	-822	-1420	-2590	-940	-2920	-2605	-2057	-1420
1989/90	-2705	-374	-2915	-4720	-940	-1291	-2150	-2320	-1395	-1308	-2565	-2062	-1491
1990/91	-3550	-238	-2115	-2790	-1336	-1280	-1720	-2790	-835	-1866	-2505	-1911	-1320
1991/92	-3920	-224	-1895	-3470	-929	-825	-1220	-2080	-935	-1783	-2235	-1774	-1291
1992/93	-2565	-579	-2775	-3090	-1004	-384	-1340	-2080	-1565	-1571	-2145	-1736	-1472
1993/94	-3085	-231	-2475	-4150	-1262	-1094	-2360	-2910	-975	-1341	-2450	-2030	-1343
1994/95	-2861	-391	-2405	-2970	-1180	-866	-1900	-2500	-1325	-1596	-1695	-1790	-1369
1995/96	-2840	-96	-1955		-1015	-505	-1764	-2434	-865	-1904		-1486	-1061
1996/97	-3080	-224	-3765		-1100	-987	-2710	-2557	-1050	-1954		-1936	-1605
1997/98	-4620	-369	-2110		-1740	-975	-1700		-1235	-2979		-1966	-1552
1998/99	-2570		-2610		-960	-838				-1799		-1755	
1999/00	-2940								-525	-2164		-1876	
Mean	-2723	-221	-2137	-3641	-1025	-919	-1795	-2479	-1089	-1846	-1789	-1688	-1174

presented in Plate 2. On global average the glaciers had annual accumulation and ablation of about 1 m in water equivalent (w.e.), with a slightly negative balance for the second half of 20th century. The trend analysis clearly indicates that the accelerating negative mass balance trend is caused by the increasing ablation in summer, while the winter accumulation has remained almost constant. The rate at which ablation increased during the last 30 years is -13 mm y^{-2} which explains the trend in the annual balance of -12 mm y^{-2}. Since the global mean temperature equivalent of melt is 2.0 x 10^{-3} K mm^{-1} y, the present melt acceleration implies a temperature increase of 0.8 K over the last 30 years. This value compares very well with the global temperature increase of 0.6 K for June, July and August for the period from 1970 to 1997 by Jones et al. (1999).

While the regional mass balance trends fit the globally averaged conditions presented above, the glaciers in Scandinavia often appear to be an exception to this trend. This point deserves a more detailed analysis. First of all, these glaciers are used as an argument for discounting the global warming. Secondly, the quality and the length of the mass balance observations in Norway and Sweden are extremely high, so that more detailed

Bn (Annual Balance), Table 4 (cont.)

	Rockies	CND Arctic	Alaska	Kamchatka	Altai	Tenshan	Pamir	Caucasus	Svalbard	Scandinavia	Alps	Mean	Area-Weighted Mean
1956/57	-100					20						-40	
1957/58	-1847					318						-764	
1958/59	-87					-168						-127	
1959/60	-453					-13						-233	
1960/61	-683					-272						-477	
1961/62	-13				-400	-292						-235	
1962/63	-490	-109			-340	260				970		58	
1963/64	1037	281			-280	333				-827		109	
1964/65	-248	46			-560	-70				289		-109	
1965/66	-117	-95	-470		-380	-56				510	520	-13	-162
1966/67	-566	44	-1080		290	140				-753	-183	-301	-312
1967/68	56	-177	-525		-20	-666	-384	100	-227	911	270	-66	-163
1968/69	-81	3	-790		290	186	1349	-1090	-885	207	-345	-116	-286
1969/70	-999	-62	945		120	31	312	410	-535	-1497	-360	-163	-87
1970/71	66	-291	190		250	-154	793	-230	-520	-911	-795	-160	-208
1971/72	131	109	-985		70	234	328	-1140	-265	358	-115	-128	-138
1972/73	-91	48	365	600	100	-306	-818	-280	-50	-333	-565	-121	-22
1973/74	834	-103	-1425	1640	-1470	-517	-311	240	-905	733	-625	-174	-357
1974/75	28	86	-270	-220	400	-158	-1034	-910	-260	400	125	-165	-128
1975/76	835	123	-1125	1330	680	-254	-1022	290	-400	369	-1020	-18	-131
1976/77	-1002	-243	520	-1050	270	-554	-1459	-370	-75	368	665	-266	-150
1977/78	-333	-149	70	-940	-323	-1116	-1313	440	-268	-328	425	-349	-195
1978/79	-1450	-69	-1120	-200	-663	-297	-387	-310	-685	-341	-53	-507	-596
1979/80	-689	-87	1065	-170	55	-425	-1081	380	-475	163	225	-95	16
1980/81	-795	-241	780	-1950	-243	53	129	-910	-505	-1071	-10	-433	-85
1981/82	-22	-174	-295	-110	-423	-417	-774	420	-10	195	-485	-190	-171
1982/83	-93	-120	-10	-420	230	-317	-507	-970	-220	-165	-225	-256	-124
1983/84	30	-38	-405	-340	313	-757	-971	210	-705	659	-20	-184	-194
1984/85	-790	-144	495	2020	223	-475	-855	-380	-515	115	-750	-96	-233
1985/86	-770	30	-155	-1660	93	-363	-1010	-500	-265	-374	-1300	-570	-185
1986/87	-795	-209	540	-300	195	-219	240	1540	230	-245	-635	31	-35
1987/88	-380	-341	825	-1940	328	-424	10	520	-505	844	-625	-154	-181
1988/89	-844	-99	-535	-740	135	-192	-230	40	-345	-1240	-1463	-501	-634
1989/90	-950	-325	-1620	-1280	98	-556	-530	340	-585	1665	-1375	-465	-644
1990/91	-760	-214	-390	460	-459	-769	-430	-310	115	1058	-1150	-259	-201
1991/92	-1330	-138	-275	-410	-119	-202	400	-130	-170	49	-1095	-311	-414
1992/93	-1433	-525	-1185	-350	236	16	300	1100	-955	1022	-825	-236	-709
1993/94	-1322	-213	-700	-670	-240	-442	-750	-840	-90	1094	-840	-456	-469
1994/95	-912	-278	-630	-230	83	-389	-780	40	-785	141	175	-324	-513
1995/96	-225	-58	-915	600	-140	-262	-354	-172	-75	536	-195	-115	-196
1996/97	-655	-147	-2180	940	-123	-989	-1737	240	-570	-624	-450	-572	-526
1997/98	-2166	-316	-110		-1110	-503	204	-430	-725	-474	-1655	-729	-496
1998/99	705	-494	-995			-113	-511		-1000	-350	222	-663	-355
1999/00	380					-330			-25	-153	-915	-209	
Mean	-441	-127	-365	-216	-78	-269	-409	-114	-382	93	-467	-246	-257

examinations are possible. In Plate 3 the mean annual and seasonal mass balances for 8 Scandinavian glaciers are presented.

The maritime glaciers near the Atlantic coast of Norway receive high winter accumulation. These glaciers therefore tend to be influenced by variations in precipitation. Most glaciers were at equilibrium during the 1960s and moved into a slightly positive mass balance starting in the early 1970s. The mean rate of the annual mass gain, 6.6 mm y^{-2} for the last 40 years was influenced by the strong increase in accumulation (11.6 mm y^{-2}). Although the ablation trend was also increasing at 5.2 mm y^{-2}, the accumulation increase was much greater. Examining individual glaciers it was found that about half of the glaciers showed a decreasing trend of melting over the last 40 years. The linear trend, however, depends on the period chosen. Taking the last 20 year from 1981 to 2000, an increase in ablation was observed on all glaciers except one, the Nigardsbreen which shows a small decreasing rate of melting at 4 mm y^{-2}. The ablation on the Nigardsbreen, however, started to increase rapidly during the last 10 years. The strongest trend in the ablation increase is seen on the Ålfot-

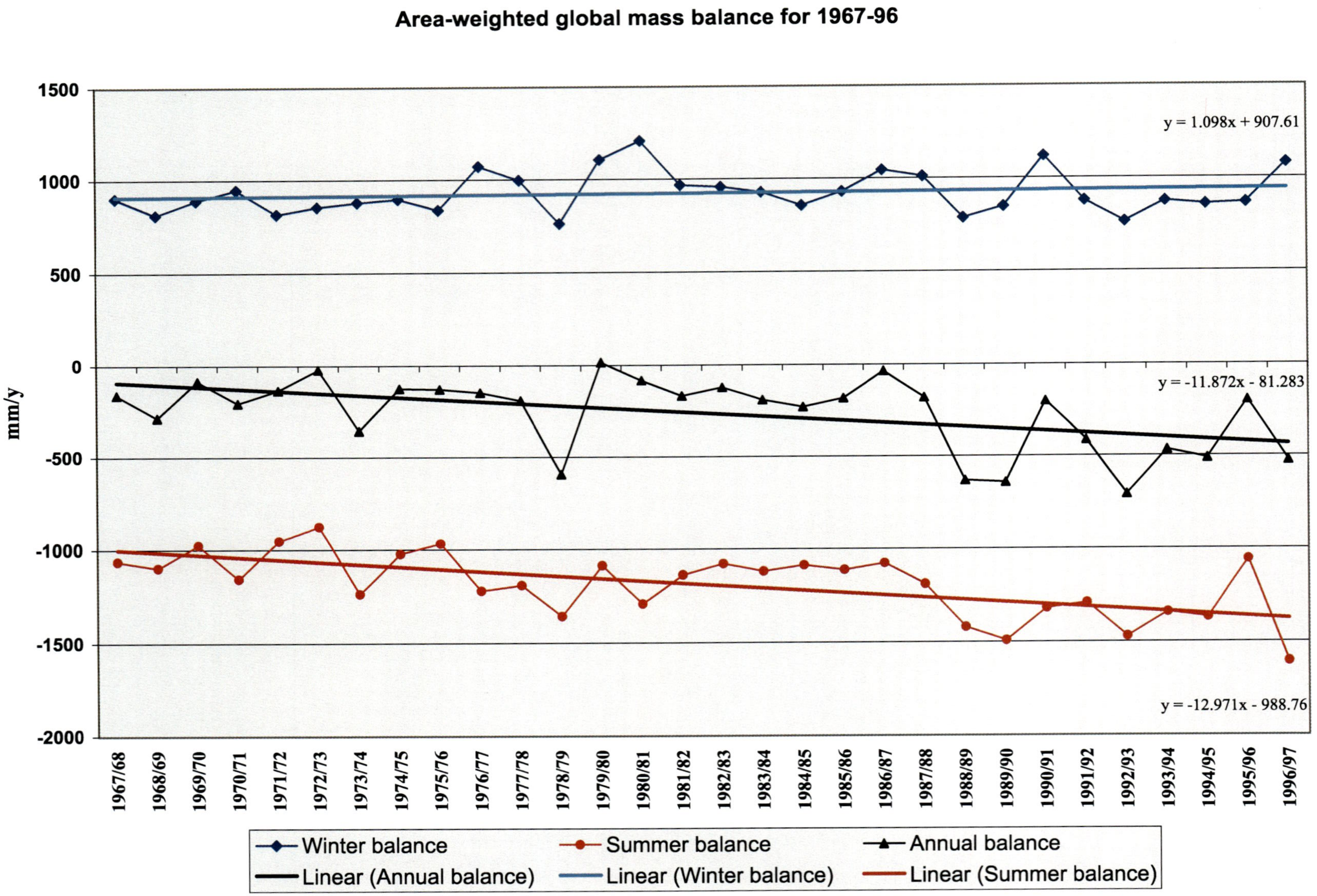

250 CRYOSPHERE DURING THE TWENTIETH CENTURY

Plate 2: Area-weighted global mean winter, summer and annual mass balance for 11 regions as defined in Table 4. (Blue line: winter mass balance (accumulation); Red line: summer mass balance (ablation); Black line: annual mass balance)

breen where the trend for the last 20 years is 48 mm y^{-2}. The average melt acceleration of the eight Scandinavian glaciers for the last 20 years was 14 mm y^{-2}, which is virtually identical to the global trend.

The most significant change in the glacier mass balance is the drastic mass loss observed during the last decade of the 20th century. This feature is clearly visible in Plate 2. Table 5 summarises the area-weighted global mass balance for the last four decades of the 20th century. The earlier quoted annual global mass balance for the last 30 years is -265 mm y^{-1}, which closely resembles the balance of the 1980s. In the last decade, represented by 7 years from 1991/92 to 1997/98, almost twice this amount was lost at a rate of 12 mm y^{-2}. If the present rate continues the majority of glaciers in the Alps will disappear by 2055 A.D. All other glaciers in the world will almost melt away between 2090 and 2140 A.D. This does not mean that all glaciers will completely disappear. Some will survive as small ice patches in favourable depressions at higher altitudes. By this time the total rise of sea-level from the melts of mountain glaciers and small ice caps will be close to 15 cm.

Ice Sheets

With a surface area of 15.85 x 10^6 km^2 and ice volume of 33.09 x 10^6 km^3, 97% of the glacier surface and 99.8% of ice volume are found in Greenland and Antarctica. Strictly speaking 48 x 10^3 km^2 of the ice surface (2.7% of the total ice-covered area) and 20 x 10^3 km^3 of the ice volume (0.8% of the total volume) for Greenland are not part of the Greenland ice sheet but are in the form of valley glaciers, isolated ice caps and glaciers on islands (Weidick, 1995). Similarly, at least 447 x 10^3 km^2 (3% of total ice covered area) of ice surface and 227 x 10^3 km^3 (0.8% of the total volume) of the ice volume are not a part of the Antarctic ice sheet (Drewry, 1983). For the purpose of convenience, in the present section all glaciers on these continents will be included with the ice sheets.

Greenland

The mass balance under the present climate has been a subject of hot debate, because even the plus/minus sign of the mass balance was uncertain. The information in Church and Gregory (2001) is the best summary on this subject up to 2000. This section supplies information that became available after the publication of the last IPCC report and also other useful material that was not included in that volume. Accumulation and melt rates are relatively well known for Greenland. The most serious lack of information is the discharge rate through calving. The author will discuss this subject first for the region of the Greenland ice sheet where the calving rate has been established.

An accurate assessment of the Greenland mass balance is available only for the south-eastern part of the ice sheet. Owing to the detailed calculations of the calving rates of all calving glaciers by Weidick et al. (1992), the direct iceberg discharge on the south-west coast of Greenland west of 44° W and south of 73° N is known with good accuracy. Therefore, we have the best chance to evaluate each term of the mass balance equation to obtain the change in storage. The surface area of the ice drainage basin for this coastal stretch is 510 x 10^3 km^2, which is equivalent to 29% of the total ice sheet. The total discharge presented by Weidick et al. (1992) is 128 ± 20 km^3 in water equivalent (w.e.) y^{-1} or 250 ± 40 mm w.e.y^{-1} in mean specific balance. Recently, the accumulation and melt rates were re-analysed by Ohmura et al. (1999), Ohmura (2001) and Wild et al.(2003) based on the high resolution topography map (Eckholm, 1996), and including more recent core data. For the area of the Weidick et al. (1992) investigation, the author's group obtained an accumulation value of 204 ± 26 km^3 w.e.y^{-1} and melt discharge of 143 ± 26 km^3 w.e. y^{-1}. This yields an annual mass balance of -66 ±42 km^3 y^{-1}. In this part of Greenland, there must therefore be a lowering of the surface by 130 mm w.e. y^{-1} that is geometrically equivalent to 142 mm y^{-1}. This discussion is summarised in Table 6. This area of Greenland contributes 0.18 mm to the sea level rise annually. From this estimate alone it is difficult to extrapolate the mass balance for the entire ice sheet.

There have been attempts to estimate the total calving rate of the Greenland ice sheet. Reeh (1994) and Bigg (1999) presented very different calving rates. Reeh used his extensive experience of ice dynamics and field experiments on the ice sheets and slightly increased his earlier total calving rate from 310 km^3 w.e.y^{-1} (Reeh, 1985) to 316 km^3 y^{-1} (Reeh, 1994). Bigg (1999) used empirical equations for calving and obtained 220 ± 50 km^3 y^{-1}. This is unrealistically low when compared to the discharge for the same region from Weidick et al. (1992). For the same geographical region of the south-west coast Bigg (1999) obtained only 79 km^3 y^{-1} compared with 128 km^3 y^{-1} by Weidick et al. (1992). For estimating the calving rate, the surface velocity is an extremely important variable. Bigg ignored this important variable and depended heavily on the frontal thickness of calving glaciers which are not well known for most glaciers, resulting in this underestimate. Taking Reeh's (1994) estimate of the calving rate and the recently re-evaluated melt rate by the author's team, we propose a new mass balance for the whole of Greenland which is summarised in Table 7. The melt calculation in the table was based on a computational scheme by Ohmura et al. (1996) and a high-resolution digitised topographic map by Eckholm (1996). The melt rate obtained in the present work, 169 mm y^{-1} is considerably smaller than the previous estimation (200 mm y^{-1}: Ohmura et al. 1999). This discrepancy is due to the high-resolution

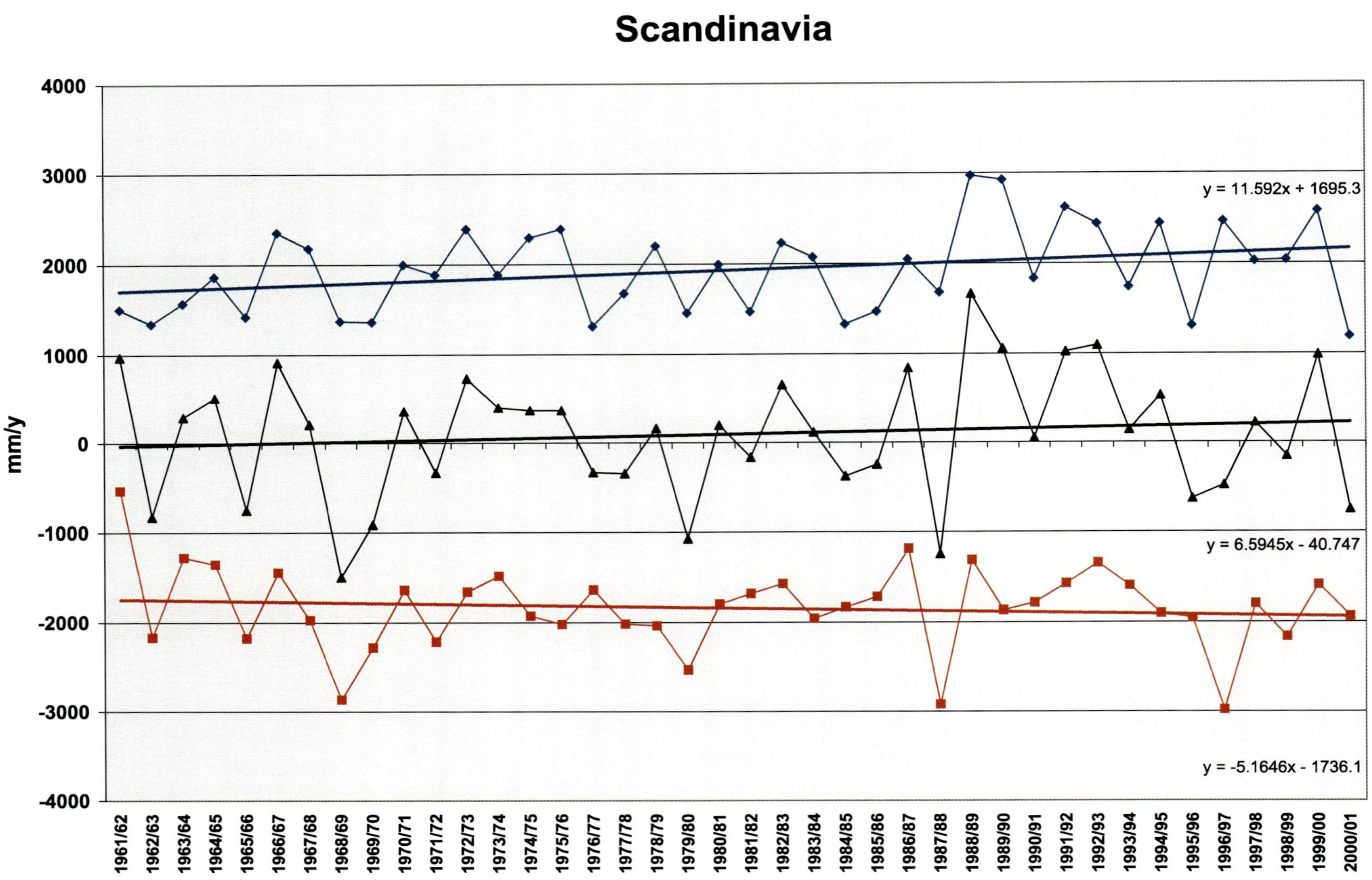

252 CRYOSPHERE DURING THE TWENTIETH CENTURY

Plate 3: Regional mean winter, summer and annual mass balance for 8 Scandinavian glaciers listed in Table 4. (Blue line: winter mass balance (accumulation); Red line: summer mass balance (ablation); Black line: annual mass balance)

Table 5: Global Glacier Mass Balance by Decades (Unit in mm y^{-1})

	Winter Balance	Summer Balance	Annual Balance	Sea Level Contribution
1965/66-1970/71	878	-1081	-203	0.29
1971/72-1980/81	943	-1121	-179	0.25
1981/82-1990/91	945	-1205	-260	0.37
1991/92-1997/98	910	-1385	-475	0.67

cartography which represents the topography of the steep marginal area of the ice sheet more accurately than was previously available. Since most of the discharge takes place within about 50 to 60 km of the ice margin on the west coast, and within 20 km on the east coast, it is absolutely essential to represent the topography of this area accurately for the melt computation. This poses great difficulty for melt estimates based only on satellite remote sensing or GCM simulations. The present calculation gives an annual net loss of 91 km^3 y^{-1} and a mean surface lowering of 57 mm y^{-1} (w.e. 52 mm y^{-1}). This results in an annual sea-level rise of 0.25 mm y^{-1}.

The mass balance of an ice sheet can be calculated by comparing the shape of the surface determined at two stages. Based on terrestrial surveys on the traverse route, EGIG Line (Expédition Glaciologique Internationale au Groenland) in the middle of Greenland, carried out in 1959 and 1968, Seckel (1977) found that in the interior region above 1700 m a.s.l. on the west slope and above 3100 m on the east, the ice sheet was growing at a rate of 90 mm y^{-1}, while it was thinning rapidly at its margins. Based on a third survey of the same traverse line, Kock (1993) obtained a general lowering of the surface for the entire stretch of the EGIG Line at a rate of 150 mm y^{-1} for the period of 1968 to 1992, and also a mean lowering of 90 mm y^{-1} for the period of 1959 to 1992. Seckel (1977) and Kock (1993) therefore came to irreconcilable conclusions. It is unlikely that the surface lowering speed of the ice sheet can fluctuate so much in a short time. The present author believes that something went wrong during the second survey which took place in 1968. Ignoring the results of the second survey, Kock's (1993) mean lowering rate for 1959 to 1992, namely 90 mm y^{-1}, is close to the present result of 57 mm y^{-1}.

The shape of the surface was also surveyed by satellite radar-altimetry. Zwally (1989) and Zwally et al. (1989) reported height increases for the ice sheet of 11 cm y^{-1} for the period 1975 to 1978, 20 cm y^{-1} for 1978 to 1985 and 28 cm y^{-1} for 1985 to 1986 for south Greenland, south of 72° N. This area represents 40% of the ice sheet and contains the EGIG Line used for the terrestrial survey. More recently Krabill et al. (1999; 2000) reported a general balance situation with local thickening and thinning above 2000 m a.s.l. and systematic lowering at the lower altitudes, especially below 1500 m based on an aircraft laser-altimetry. The net loss they reported is 51 km^3 y^{-1}. Rignot and Thomas (2002) compiled satellite- and aircraft-based altimetry and concluded that the ice was more or less in equilibrium above 2000 m. They reported a significant thinning in the coastal regions over the five years from 1993/94 to 1998/99 with an annual net loss of at least 50 km^3 y^{-1}, mostly at the tongue areas of the outlet glaciers. Therefore,

Table 6: Mass balance of South-west Greenland (South of 73° N, West of 44° W)
Surface area: 510 x 10^3 km^2

	Volume in Water Equivalent: km^3 y^{-1}	Mean Specific Mass Balance: mm y^{-1}
Accumulation	204 ± 26	400 ± 50
Melt	143 ± 26	280 ± 50
Calving	128 ± 20	250 ± 40
Balance	-66 ± 42	-130 ± 80

-66 km^3 y^{-1} is equivalent to +0.2 mm y^{-1} sea level rise.

Sources: Ohmura and Reeh (1991), Weidick et al. (1992), Ohmura et al. (1999), Wild et al. (2003)

Table 7: Estimate of the Mass Balance of the Greenland Ice Sheet

	Total Volume km³ w.e. y-1	Mean Specific Balance mm w.e. y^{-1}	Source
Annual precipitation on Greenland	750	346	Ohmura et al. (1999)
Annual precipitation on tundra	156	379	ditto
Annual precipitation on ice-covered area	594	340	ditto
Loss through evaporation/sublimation	61	35	ditto
Loss by run-off as liquid precipitation	14	8	ditto
Annual accumulation	519	297	ditto
Melt	295	169	This work
Calving	315	180	Reeh (1994)
Rate of storage	-91	-52	This work

the annual net loss of 50 to 60 km³ y^{-1} seems to be a likely value for Greenland, based on remote sensing. This rate is not too far removed from the earlier glaciological estimation of -91 km³ w.e. y^{-1}. These estimates and Kock's (1993) terrestrial survey also yield an order of magnitude similar net loss. It is likely that the present mass balance of the Greenland ice sheet is negative and making a sea level contribution at a rate of between 0.15 and 0.25 mm y^{-1}. Since the sign of the mass balance has only recently been clarified, it is still too premature to assess any acceleration of the mass balance. There is, however, evidence for an increasing rate of melting in the upward shift of the dry snow line reported by Abdalati and Steffen (2001).

Antarctica

In terms of surface area, the Antarctic ice sheet is almost eight times larger than the Greenland ice sheet. Additional logistic and technical adverse conditions make assessment of the mass balance of the Antarctic ice sheet much more difficult than for Greenland. Nevertheless, significant progress has been achieved, that has made the range of uncertainty narrower since the publication of the last IPCC report. Rignot and Thomas (2002) assessed the balance for glaciers covering 53% of the entire ice sheet. While the last IPCC consensus gave -376 km³ y^{-1} annual balance (+1.04 mm y^{-1} sea level equivalent), Rignot and Thomas (2002) obtained a substantially smaller value of -26 km³ y^{-1} (+0.07 mm y^{-1} sea level equivalent). Further progress by Rignot and Thomas (2002) is eagerly awaited so that the remaining 47% of the glacier surface can be assessed to complete the total mass balance of the entire Antarctic ice sheet.

Substantial progress has been made in assessing the mass balance of the base of the ice shelves. Melt on the under side of the ice shelves is especially large on their continental sides (Gjessing and Wold, 1986; Grosfeld et al., 1998). The uncertainty of this quantity not only influences the total mass balance of the ice sheet, but impedes the assessment of the stability of ice shelves. Although the stability of ice shelves does not directly influence the global sea level, its influence on the glaciers behind the ice shelves may be significant.

According to Rott et al. (1996) disintegration of ice shelves on Antarctic Peninsula is a continuing process. The Larsen Ice Shelf occupies 1200 km of shoreline from 64° to 74° S on the east coast of the Antarctic Peninsula and it has been slowly retreating since the 1940s. After the disappearance of the ice shelf in the Larsen Inlet (64° S) in 1986 to 1989, the general instability was observed over a much larger area. The ice shelf in Prince Gustav Channel (64° S) disappeared by the beginning of 1995; Larsen A (up to 65° S) lost most of its mass by March 1995; and finally the much larger Larsen B (up to 66° 20′ S) disintegrated from autumn 2002 to 2003. Although Larsen A was known to be in negative mass balance owing to the melt on its underside, the upper surface was an accumulation area as for most ice shelves. However in the early 1990s the surface of Larsen A became an ablation area, producing a large volume of melt water every summer (Rott et al., 1998; Bindschadler et al., 2002). The reason for this change was the steady temperature increase observed in this area since 1960 (Skvarca et al., 1998). A detailed account of the process and dynamics of the Larsen A disintegration was made by Rack (2000). Four years after the collapse of Larsen A, the velocity of major glaciers in the background was found to be accelerating by more than a factor of three (Rott et al., 2002). This observation is crucial in assessing the role of ice shelves in the stabilisation of the ice sheet. The mass balance of one of the major ice shelves, the Ronne Ice Shelf was recently found to be negative (Lambrecht, 1998).

Out of the total coast line of 31,876 km around Antarctica, 44% or 14,110 km is occupied by ice shelves. Many of these ice shelves have ice streams and outlet glaciers on their continental sides. An understanding of the mechanism of ice shelf stability and its influence on the upstream glaciers seems to hold the key to sea level changes in the near future. It is especially important to investigate the West Antarctic ice sheet in this regard. Potential instability of the West Antarctic ice sheet has been argued only on dynamic grounds. The climate of the ice sheet regions in West Antarctica is much milder than East Antarctica. This condition makes the West Antarctic ice sheet respond much faster to global warming than the East Antarctic ice sheet. Earlier considerations on the effect of warming in the Antarctic region have neglected the importance of the mass balance and the melt water production on ice shelves, as the ice shelves are already in hydrostatic balance with the sea. Earlier computations of the sea level contribution from the Antarctic ice sheet rest on the assumption of a gradual increase of the melt volume as the melt line slowly climbs on the ice sheet slopes. It was further projected that warming in the Antarctic would lower the sea level by increasing accumulation, as a result of an increase in water vapour (Ohmura et al., 1996). If the melt water on the ice shelf is the main cause for accelerating the instability of ice shelves and the subsequent increase in the discharge of glaciers, the influence of Antarctica on sea level may become a reality much earlier than previously thought.

CONCLUSION

The cryosphere as a whole preserves various signs of warming during the second half of the 20th century. The snow cover extent showed most rapid recession between the 1970s and 1980s, and especially in spring and early summer. The mean retreat was 1.45×10^6 km^2 in area and 140 km^3 w.e. in volume during the last 30 years. The sea ice receded both in area and thickness in the Arctic. The retreat is most conspicuous in late summer as the sea ice extent reaches its minimum, while the maximum extent in winter showed little variation. This widening gap between the summer and winter ice limits suggests the possibility of increasing annual ice production in the Arctic. If this possibility is real, the present warming trend does not *a priori* create deceleration of the thermohaline circulation in the North Atlantic. Around the Antarctic a slight increase in the sea ice extent was observed after 1978. Although some explanations of this asymmetry in the sea ice trends in the Arctic and Antarctic are offered, this phenomenon must be explained more satisfactorily. Mountain glaciers and small ice caps show a global mass loss at a mean rate of -265 mm y^{-1} for the last 30 years. During this period mean ELA rose at a rate of 4 m y^{-1}. Further, the negative mass balance accelerated at a rate of -12 mm y^{-2}. On glaciers with a positive mass balance there is an increasing trend of melt formation.

Re-evaluation of the mass balance of the Greenland ice sheet shows a negative balance between -50 and -90 km^3 y^{-1}. Most recent work on the Antarctic regional mass balance for the 53% of the total area that is reasonably well studied produced a negative balance of -26 km^3 y^{-1}. The total mass balance of this largest glacier of the world is still unknown.

Based on the above analyses, the total discharge from the terrestrial cryosphere (snow cover and glaciers) is estimated at 260 km^3 y^{-1}, or +0.7 mm y^{-1} in sea level equivalent. The total mass balance of the cryosphere including sea ice is -625 km^3 y^{-1}. The total heat flux consumed to melt the cryosphere, excluding the permafrost during the second half of the 20th century is estimated to be 0.01 to 0.03 W m^{-2}. Compared with the heat stored in ocean reported by Levitus et al. (2001), the storage rate due to the melt is one order of magnitude smaller. The total storage of heat in the ocean and cryosphere during the last decade in the 20th century was much larger than that in the previous decades, and it is estimated to be close to 1 W m^{-2}.

Acknowledgements. In the course of preparing this work, the author received generous support from a number of scientific colleagues. Notably thanks are due to Drs. R. Armstrong (Univ. Colorado), Y. Morinaga-Shinoda (Meiji Univ.) and K. Rikiishi (Hirosaki Univ.) for providing the data on snow distribution and literatures; Drs. H. J. Zwally (NASA), D. A. Rothrock (Univ. Washington) and Dr. H. Huwald (ETH Zurich) for sea ice related data; Dr. A. Roesch (ETH Zurich) for snow cover area in the Southern Hemisphere, Drs. R. Braithwaite (Univ. Manchester), W. Haeberli and M. Hoelzle (Univ. Zurich), B. Kjøllmoen (Norwegian Water Resources and Energy Directorate), M. Dyurgerov (Univ. Colorado), P. Jansson (Stockholm Univ.), C. Vincent (Centre National de la Recherche Scientifique) and L. N. Braun (Bavarian Academy of Sciences) for glacier mass balance data; Drs. H. Rott (Univ. Innsbruck) and K. Grosfeld (Univ. Bremen) for information on ice shelves. Dr. N. Reeh supplied the most recent information on the calving rate of the Greenland ice sheet. Profs. M. Fukuda and A. Blochkov (Hokkaido Univ.) supplied data and literatures for estimating permafrost volume. The author is indebted to three reviewers, Profs. C. Hawkesworth (Univ. Bristol) and P. Jansson (Stockholm Univ.) for suggestions which lead to significant improvement of the text, and M. Kuhn (Univ. Innsbruck) for pointing out crucial errors in Tables 1 and 4 in the original draft.

REFERENCES

Abdalati, W., and K. Steffen, 2001: Greenland ice sheet melt extent: 1979–1999. *J. Geophys. Res.*, **106**, D24, 33,983–33,987.

Bamber, J. L., R. L. Layberry, and S.P. Gogineni, 2001: A new ice thickness and bed data set for the Greenland ice sheet, 1. Measurement, data reduction and errors. *Jour. Geophys. Res.*, **106**, D24, 33,773–33,780.

Bigg, G. R., 1999: An estimate of the flux of iceberg calving from Greenland. *Arctic, Antarctic, and Alpine Research*, **31**, 174–178.

Bindschadler, R., T. A. Scambos, H. Rott, R. Skvarca, and P. Vornberger, 2002: Ice dolines on Larsen Ice Shelf, Antarctica. *Ann Glaciol.*, **34**, 283–290.

Björnsson, H., 1986: Surface and bedrock topography of ice caps in Iceland, mapped by radio-echo-sounding. *Ann. Glaciol.*, **8**, 11–18.

Brown, R. D., 2000: Northern Hemisphere snow cover variability and change. 1915–97. *J. Climate*, **13**, 2339–2355.

Cavalieri, D. J., P. Gloersen, P., C. L. Parkinson, J. C. Comiso, and H. J. Zwally, 1997: Observed hemispheric asymmetry in global sea ice changes. *Science*, **278**, 1104–1106.

Chen, J. and Funk, M., 1990: Mass balance of the Rhonegletscher during 1882/83–1986/87. *Jour. Glaciol.*, **36**, 199–209.

Chen, J. and Ohmura, A., 1990: Estimation of Alpine glacier water resources and their change since 1870s. In Lang, H. and Musy, A.(Eds.): Hydrology in Mountainous Regions I, *IAHS Publ.*, No.**193**, 127–135.

Church, J. A., and J. M. Gregory, 2001: Changes in sea level. Chapt. 11, In Houghton, J. T. et al. (Eds.): *Climate Change 2001: The Scientific Basis, Cambridge.* Univ. Press, Cambridge.

Danilov, I. D., 1996: Basics of Geocryology. Vol. 2, Moscow State University, Moscow, 399pp.

Drewry, D. (Ed.), 1983: *Antarctica: Glaciological and Geophysical Folio.* Cambridge Univ. Press.

Dyurgerov, M., 2002: Glacier Mass Balance and Regime: Data of Measurements and Analysis. *Occasional Paper* No. **55**, Institute of Arctic and Alpine Research, Univ. Colorado, Boulder.

Eckholm, S., 1996: A full coverage, high resolution, topographic model of Greenland computed from a variety of digital elevation data. *J. Geophys. Res.*, **10**(B10), 21,961–21,972

Field, W. O., 1975: *Mountain Glaciers of the Northern Hemisphere,* CRREL, Hanover

Folland, C. K., and T. R. Karl, 2001: Observed climate variability and change. Chapt. 2, In Houghton, J. T. et al. (Eds.): *Climate Change 2001: The Scientific Basis*, Cambridge Univ. Press, Cambridge

Gjessing, Y., and B. Wold, 1986: Absolute movements, mass balance and snow temperature of the Riiser- Larsen Ice Shelf, Antarctica. *Skrifter* (Norsk Polarinstitutt, Oslo), **187**, 23–31.

Grosfeld, K., H. H. Hellmer, M. Jonas, H. Sandhäger, M. Schulte, and D. G. Vaughan, 1998: Marine ice beneath Filchner Ice Shelf: Evidence from a multi-disciplinary approach. *Antarctic Res. Ser.*, **75**, 319–339.

Haeberli, W. 1985: *Fluctuations of Glaciers 1975–1980*, Vol.4, ICSI and UNESCO, Paris.

Haeberli, W.,and P. Müller, 1988: *Fluctuations of Glaciers 1980–1985*, Vol.5, ICSI and UNESCO, Paris.

Haeberli, W., H. Bösch, K. Scherler, G. Østrem, and C. C. Wallén, 1989: *World Glacier Inventory Status 1988.* IAHAS (ICSI)-UNEP_UNESCO, Paris.

Haeberli, W., and M. Hoelzle, 1993: *Fluctuations of Glaciers 1985–1990*, Vol.6, ICSI and UNESCO, Paris.

Haeberli, W., M. Hoelzle, S. Suter, and R. Frauenfelder, 1998: *Fluctuations of Glaciers 1990–1995*, Vol.7, ICSI and UNESCO, Paris.

Hastenrath, S., 1984: *The Glaciers of Equatorial East Africa.* Reidel, Dortrecht.

Hodge, S., M., Trabant, D. C., Krimmel, R. M., Heinrichs, T., A., March, R. S., and Josberger, E. G., 1998: Climate variations and changes in mass of three glaciers in Western North America. *Jour. Climate*, **11**, 2161–2179.

Jones, P. d., M. New, D. E. Parker, S. Martin, and I. G. Rigor, 1999: Surface air temperature and its changes over the past 150 years. *Rev. Geophys.*, **37**, 173–199.

Kasser, P, 1967: *Fluctuations of Glaciers 1959–1965.* ICSI and UNESCO, Paris.

Kasser, P., 1973: *Fluctuations of Glaciers 1965–1970.* ICSI and UNESCO, Paris

Kjøllmoen, B. (Ed.), 1998: *Glaciological Investigations in Norway in 1996 and 1997.* Norwegian Water Resources and Energy Administration. Oslo.

Kjøllmoen, B. (Ed.), 1999: *Glaciological Investigations in Norway in 1998.* Norwegian Water Resources and Energy Administration. Oslo.

Kjøllmoen, B. (Ed.), 2000: *Glaciological Investigations in Norway in 1999.* Norwegian Water Resources and Energy Administration. Oslo.

Kjøllmoen, B. (Ed.), 2001: *Glaciological Investigations in Norway in 2000.* Norwegian Water Resources and Energy Administration. Oslo.

Kjøllmoen, B. (Ed.), 2003: *Glaciological Investigations in Norway in 2001.* Norwegian Water Resources and Energy Administration. Oslo.

Kock, H., 1993: Height determinations along the EGIG line and in the GRIP area. In Reeh, N., and H. oerter (Eds.): Mass Balance and Related Topics of the Greenland Ice sheet. *Open File Ser.* **93**/5, Geol. Surv. Greenland, 68–70.

Krabill, W., E. Frederick, s. Manizade, C. Martin, J. Sonntag, R. swift, R. Thomas, W. Wright, and J. Yungel, 1999: Rapid thinning of parts of the southern Greenland ice sheet. *Science*, **283**, 1,522–1,524.

Krabill, W., W. Abdalati, E. Frederick, S. Manizade, C. Martin, J. Sonntag, R. Swift, R. Thomas, W. Wright, and J. Yungel, 2000: Greenland Ice sheet: high-elevation balance and peripheral thinning. *Science*, **289**, 428–430.

Lambrecht, A., 1998: Untersuchungen zu Massenhaushalt und Dynamik des Ronne Ice shelfs, Antarktis. *Berichte zur Polarforschung*, **265**, Alfred-Wegener-institut für Polarfor- und Meeresforschung, Bremerhaven.

Levitus, S., J. I. Antonov, J. Wang, T. L. Delworth, K.W. Dixon, and A. J. Broccoli, 2001: anthropogenic warming of earth's climate system. *Science*, **292**, 267–270.

Mercer, J. H., 1967: *Southern Hemisphere Glacier Atlas.* Am. Geogr. Soc., US army Natick Lab. Tech. Rep., 67–76-ES.

Meier, M. F.,1984: Contribution of small glaciers to global sea level. *Science*, **226**, 1419–1421.

Meier, M. F., 1993: Ice, climate and sea level: do we know what is happening? In *Ice and Climate System*, Peltier, W. R. (Ed.), NATO ASI Series, Springer Verlag, Heidelberg, 141–160.

Müller, F., 1970: Guide Book for "Perennial Ice and Snow Masses".*UNESCO/IASH Technical Papers in Hydrology*, No.1, 1–23.

Müller, F., 1977: *Fluctuations of Glaciers 1970–1975.* Vol. 3, ICSI and UNESCO, Paris.

Ohmura, A., 1987: Heat budget of the climate system between the Last Glacial Maximum and the present. *Bull. Dept. Geogr., University of Tokyo*, **94**, 109–126.

Ohmura, A., 2001: Physical basis for the temperature-based melt-index method. *J. Appl. Met.*, **40**, 753–761.

Ohmura, A., and N. Reeh, 1991: New precipitation and accumulation maps for Greenland. *J. Glaciol.*, **37**, 140–148.

Ohmura, A., Wild, M, and Bengtsson, L., 1996: A possible change in mass balance of Greenland and Antarctic ice sheets in the coming century. *J.Climate*, **9**, 2124–2135.

Ohmura, A., Calanca, P., Wild, M., and Anklin, M., 1999: Precipitation, accumulation and mass balance of the Greenland ice sheet. *Z. Gletscherkd. Glazialgeol.*, **35**, 1–20.

Ommanney, C. S. L., 1972: Application of the Canadian glacier inventory to studies of the static water balance. *Internt. Geogr.*, **2**, 1266–1268.

Østrem, G., K. D. Selvig, and K. Tandberg, 1988: *Atlas over breer i Sor-Norge*. Norges Vassdrags- og Energiverk, Vassdragsdirektoratet, Medd. 61.

Parkinson, C. L., 2000: Variability of Arctic sea ice. The view from space, an 18-year record. *Arctic*, **53**, 341–358.

Parkinson, C. L.,J. C. Comiso,.H. J. Zwally, D. J. Cavalieri, P. Gloersen, and W. J. Campbell, 1987: *Arctic Sea Ice, 1973–1976: Satellite Passive-Microwave Observations*. NASA, Washington, D.C.

Parkinson, C. L., D. J. Cavalieri, P. Gloersen, H. J. Zwally, and J. C. Comiso, 1999: Arctic sea ice extents, areas, and trends, 1978–1996. *J. Geophys. Res.*, **104** (C9), 20837–20856.

Rack, W., 2000: *Dynamic Behavior and Disintegration of the Northern Larsen Ice shelf, Antarctic Peninsula*.Dissertation, Univ. Innsbruck, Innsbruck.

Reeh, N., 1985: Greenland ice sheet mass balance and sea level change. In *Glaciers, Ice Sheets and Sea Level: Effect of a CO_2-induced Climatic Change*. National Academy Press, Washington, D. C., 155–171.

Reeh, N., 1994: Calving from Greenland glaciers: observations, balance estimates of calving law. In Workshop on the Calving Rate of West Greenland Glaciers in Response to climatic Change, Danish Polar Centre, Copenhagen, 85–102.

Rignot, E., and R. Thomas, 2002: Mass balance of polar ice sheets. *Science*, **297**, 1502–1506.

Rikiishi, K., E. Hashiya, and M. Imai, 2003: Linear trends of the length of snow-cover season in the Northern Hemisphere as observed by the satellites in the period 1972–2000. *Ann. Glaciol.* (in press).

Rothrock, D. A., Y. Yu, and G. A. Maykut, 1999: Thinning of the Arctic sea-ice cover. *Geophys. Res. Lett.*, **26**, 3469–3472.

Rothrock, D. A., J. Zhang and Y. Yu, 2003: The arctic ice thickness anomaly of the 1990s: A consistent view from observations and models. *J. Geophys. Res.*, **108**, C3, 3083, doi: 10.1029/2001JC001208, 2003.

Rott, H., P. Skvarca, and T. Nagler, 1996: Rapid collapse of Northern Larsen Ice shelf, Antarctica. *Science*, **271**, 788–792.

Rott, H., W. Rock, T. Nagler, and P. Skvarca, 1998: Climatically induced retreat and collapse of northern Larsen Ice shelf, Antarctic Peninsula. *Ann. Glaciol.*, **27**, 86–92.

Rott, H., W. Rock, P. Skvarca, H. de Angelis, 2002: Northern Larsen Ice shelf, Antarctica: further retreat after collapse. *Ann. Glaciol.*, **34**, 277–282.

Seckel, H., 1977: Höhenänderungen im grönländischen Inlandeis zwischen 1959 und 1968. EGIG 1967–1968, Vol. 3, No.5, *Medd. om Grønland*, Bd. **187**, Nr. 4, Copenhagen.

Serreze, M. C., J. A. Maslanik, T. A. Scambos, F. Fetterer, J. Stroeve, K. Knowles, C. Fowler, S. Drobot, R. G. Barry, and T. M. Haran, 2003: A record minimum arctic sea ice extent and area in 2002. *Geophys. Res. Lett.*, **30**, 10.1029/2002GL016406.

Skvarca, P., W. Rock, H. Rott, T. I. Donángelo, 1998:Evidence of recent climatic warming on the eastern Antarctic Peninsula: *Ann. Glaciol.*, **27**, 628–632.

Stocker, T. F., and A. Schmittner, 1997: Influence of CO_2 emission rates on the stability of the therohaline circulation. *Nature*, **388**, 862–865.

Vincent, C., 2002: Influence of climate change over the 20th Century on four French glacier mass balances. *Jour. Geophys. Res.*, **107**, D19, 4375, doi:10.1029/2001JD000832.

Vinje, T., N. Nordlund, and Å. Kwambekk, 1998: monitoring ice thickness in Fram Strait. *J. Geophys. Res.*, **103** (C%), 10437–10449.

Vinnikov, K. Y., A. Robock, R. J. Stouffer, J. E. Walsh, C. L. Parkinson, D. J. Cavalieri, J. F. B. Mitchell, D. Garrett, and V. F. Zakharov, 1999: Global warming and Northern Hemisphere sea ice extent. , **286**, 1934–1937.

Wadhams, P., and N. R. Davis, 2000: Further evidence of sea ice thinning in the Arctic Ocean. *Geophys. Res. Lett.*, **27**, 3973–3976.

Walsh, J. E., 1978: Data set on Northern Hemispheresea-ice extent. Glaciology (Snow and Ice), Part 1, 49–51.

Weidick, A., 1995: Greenland. In Williams, R. S., and J. G. Ferrigno (Eds*.)*: *Satellite Image Atlas of Glaciers of the World*, US Geol. Surv. Prof. Paper, 1386-C, US Gov. Ptint. Office, Washington, D. C.

Weidick, A., Egede-Bøggild, C., and Knudsen, N. T., 1992: Glacier Inventory Atlas of West Greenland. *Report* **158**, Geol. Surv. Greenland, Copenhagen.

Wild, M., Calanca, P., Scherrer, S. C., and Ohmura, A., 2003: Effects of polar ice sheets on global sea level in high-resolution greenhouse scenarios. *Jour. Geophys. Res.*, **108** (D5), ACL 5-1–5-10.

Zhang, T., J. A. Heginbottom, R. G. Barry, and J. Brown, 2000: Further statistics on the distribution of permafrost and ground ice in the Northern Hemisphere. *Polar Geogr.* 24, 126–131.

Zwally, H. J., 1989: Growth of Greenland Ice sheet: Interpretation. *Science*, **246**, 1589–1591.

Zwally, H. J., J. C. Comiso, C. L. Parkinson, W. J. Campbell, F. D. Carsey, and P. Gloersen, 1983: *Antarctic Sea Ice, 1973–1976: Satellite Passive-Microwave Observations*. NASA, Washington, D. C.

Zwally, H. J., A. C. Brenner, J. A. Major, R. A. Bindschadler, and J. G. Marsh, 1989: Growth of Greenland Ice sheet: Measurement. *Science*, **246**, 1587–1589.

Zwally, H. J., J. C. Comiso, C. L. Parkinson, D. J. Cavalieri, and P. Gloersen, 2002: Variability of Antarctic sea ice 1979–1998. *J. Geophys. Res.*, **107**, C5, 10.1029/2000JC000733, 2002.

Atsumu Ohmura, Institute for Atmospheric and Climate Science, ETH Zürich, Winterthurerstrasse 190, CH-8057 Zürich, Switzerland

Climate Prediction: The Limits of Ocean Models

Peter H. Stone

Department of Earth, Atmospheric, and Planetary Sciences, Massachusetts Institute of Technology

We identify three major areas of ignorance that limit predictability in current ocean general circulation models (GCMs). One is the very crude representation of subgrid-scale mixing processes parameterized with coefficients whose values and variations in space and time are poorly known. A second problem derives from the fact that ocean models generally contain multiple equilibria and bifurcations, but there is no agreement on how far the current climate is from such a bifurcation. A third problem arises from the fact that ocean circulations are highly nonlinear, but only weakly dissipative and therefore potentially chaotic. The few studies that have dealt with this kind of behavior have not answered fundamental questions, such as: what are the major sources of error growth in model projections, and how large is the chaotic behavior relative to realistic changes in climate forcings. Advances in computers will help alleviate some of these problems; for example, by making it more practical to explore to what extent the evolution of the oceans is chaotic. Be that as it may, models will continue to rely on parameterizations of key small-scale processes such as diapycnal mixing for some time to come. To make more immediate progress here requires the development of physically based prognostic parameterizations and coupling the mixing to its energy sources. Another possibly fruitful area of investigation is the use of paleoclimate data on changes in ocean circulation to constrain more tightly the stability characteristics of ocean circulation.

1. INTRODUCTION

The oceans are of fundamental importance to Earth's climate system. One important role is their transport of heat. Indeed, compared to the total poleward heat transport in the whole climate system, about 5.5 petawatts [*Trenberth and Caron*, 2001], ocean circulations carry about 2 petawatts of heat poleward [*Ganachaud and Wunsch*, 2003]. This heat transport, which profoundly influences latitudinal variations in climate [*Seager et al.*, 2002], also affects the global mean climate by affecting the amount of sea ice in high latitudes. Because of its high reflectivity, sea ice has a substantial effect on the amount of solar energy absorbed by the climate system, and thus changes in the amount of sea ice can cause global warming or cooling. Another important role that the oceans play involves the mixing of heat into deep oceans, which then determines how rapidly surface temperatures change [*Hansen et al.*, 1985]. Within the context of global warming scenerios, strong mixing will retard surface warming rates. Thus any attempt to model or predict climate change requires a good understanding of how the oceans operate.

That our understanding of the climate system as a whole has not yet reached the level where reliable projections can be made is obvious from the lack of robustness of climate change projections made with different state-of-the-art climate models. For example, Cubasch and Meehl [2001] compared pro-

---

[1] This paper was originally presented at the IUGG meeting in Sapporo under the title "The Straits of Ocean Predictability."

jections of changes in the meridional overturning circulation in the North Atlantic from 10 different coupled atmosphere-ocean general circulation models (GCMs) for the same global warming scenario. This circulation is illustrated in Figure 1. The poleward flow near the surface is primarily associated with the Gulf Stream. This circulation is particularly important for climate, because it transports more heat than the circulations in any other ocean basin, and has a substantial warming effect on mid and high latitudes of the Northern Hemisphere [*Seager et al., 2002*]. Estimates of the strength of the overturning circulation range from 16 to 25 Sv [*Macdonald and Wunsch, 1996; Ganachaud, 2003*; Sv = one Sverdrup = 10^6 m^3/s]. However the simulated changes in this circulation by 2100 varied from no change to a decrease of 14 Sv. Since this result comes from coupled models, it is not possible to identify any single component of the climate system, such as the oceans, as being the source of the differences, without further analysis.

An analysis that does implicate the ocean component of the climate models has been carried out by Sokolov et al. [2003]. They found that model differences in projections of changes in global mean surface temperature could be attributed to differences in two model characteristics. One is the model's climate sensitivity, defined as how much the global mean surface temperature would increase if the concentration of CO_2 in the atmosphere were doubled and the climate system were allowed to equilibrate. This sensitivity depends primarily on atmospheric processes such as how clouds change when climate changes. These processes are not well understood and are represented in different ways in different models. The second model characteristic is the rate at which perturbations

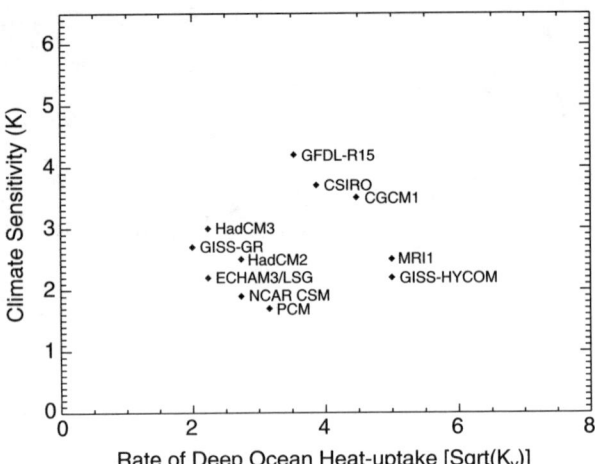

Figure 2. Properties of 11 different coupled GCMs. Vertical axis: climate sensitivity. Horizontal axis: a parameter measuring the depth to which heat has penetrated in the deep ocean (see text). Adapted from Sokolov et al.,[2003].

in the heat flux between the atmosphere and ocean are mixed into the deep oceans.

Figure 2 shows how 11 different coupled atmosphere-ocean GCMs differ with respect to these two characteristics. In the figure, the rate of heat uptake by the deep oceans is measured by the global mean value of a coefficient that describes the effective rate at which heat anomalies are mixed into the deep ocean. In the figure the square root of this coefficient is plotted, since the depth to which heat penetrates at a given time is proportional to the square root of the coefficient. As shown in Figure 2, this depth varies between models by a factor of two and one half. The rate of heat uptake is not well constrained by the available observations [*Forest et al., 2002*], so none of these models can be ruled out by comparing them with the observations. Similarly we cannot be sure that any of them are right.

One likely source of the ocean model differences in the rate of heat uptake is the different representations of small-scale oceanic processes used in different models. The differences reflect our ignorance of these processes, and this is one potential obstacle to our current ability to predict climate change (see section 2 for further discussion here).

Another potential obstacle is the possibility that the circulation in the North Atlantic and its heat transport may be very sensitive to small changes in climate. Since this circulation is coupled to that of the rest of the oceans by the "conveyor belt" circulation, such changes would have global consequences. Uncoupled ocean models that show this possibility include simple box models [*Stommel, 1961; Rooth, 1982; Welander, 1986*], two-dimensional meridional plane models [*Marotzke et al., 1988*], and three-dimensional numerical models [*Bryan, 1986; Marotzke and Willebrand, 1991*]. They all show that

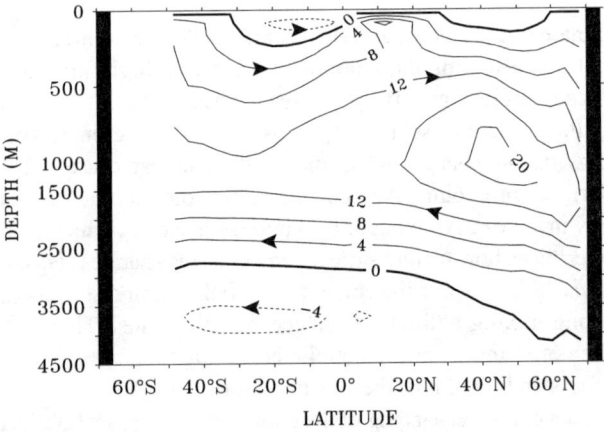

Figure 1. Typical model simulation of the stream function of the zonal mean overturning circulation in the North Atlantic. Depth is given on the vertical axis and latitude on the horizontal axis. Adapted from Huang et al. [2003].

the circulation is very sensitive to salinity perturbations, particularly at high latitudes, and that the circulations can have at least two states. One is like that which currently exists in the North Atlantic Ocean, with a relatively strong poleward heat transport. The other has a much weaker circulation with very little poleward heat transport.

Paleoclimatic evidence also indicates that two states like these with very different climates can exist [*Broecker* et al., 1985; *Boyle and Keigwin*, 1987; *Broecker*, 2003]. Indeed, Broecker et al. [1985] suggest that sudden shifts in climate, such as that associated with the Younger Dryas event some 10,000 years ago, may have been caused by a sudden collapse in the circulation of the North Atlantic. See section 3 for further discussion on how this phenomenon may limit predictions of climate change.

The limits on prediction described above could in principle be overcome if we could acquire data that is sufficiently extensive and accurate, and if our computers were sufficiently fast. However there may be a more fundamental limitation to our ability to predict changes in the oceans. The oceans' circulations are highly nonlinear, but only weakly dissipative. Such systems are potentially chaotic, i.e., unpredictable past a certain time limit, as discussed in section 4. Finally, in section 5, results are summarized and possible paths for improving ocean model predictions and determining the limits of their predictability discussed.

2. SMALL-SCALE OCEANIC PROCESSES

The ocean GCMs used in current climate models have coarse resolution; typical horizontal resolutions are in the range 1° to 3°. Thus there are many subgrid-scale processes that need to be parameterized in these models. In current practice these processes are generally decomposed into four components which are parameterized separately: diapycnal diffusion, isopycnal diffusion, mesoscale eddies, and convection. Diapycnal diffusion refers to diffusion perpendicular to constant density surfaces while isopycnal diffusion refers to diffusion along constant density surfaces. Mesoscale eddies are eddies with typical spatial scales of about 100 km and typical periods of about 100 days. Energy spectra of the oceans show a peak at the frequency of the mesoscale eddies [*Wunsch*, 1981]. The other parameterized processes occur at smaller spatial scales. There are major uncertainties and problems in current parameterizations of all these processes.

Diapycnal diffusion plays a particularly important role in determining the ocean's circulation, since it is the diapycnal mixing of heat and salinity from the ocean's surface into its depths that gives rise to the density gradients that drive the large-scale ocean circulation and its horizontal heat transports [*Munk and Wunsch*, 1998]. In fact, scaling analyses and ocean GCM calculations show that the strength of the ocean circulations and heat transports are sensitive to the value of the diapycnal diffusion coefficient [*Bryan*, 1987; *Martozke*, 1997]. In a basin like the North Atlantic, the strength of the meridional overturning is approximately proportional to the 2/3 power of the coefficient and the poleward heat transport to the 1/2 power [*Marotzke*, 1997]. The strength and heat transport are determined primarily by the values of the diapycnal diffusion at depths of 200 to 500 m in the tropics and subtropics [*Scott and Martozke*, 2002; *Bugnion and Hill*, 2004].

However OGCMs generally treat the diapycnal diffusion coefficients for heat, salinity, and momentum as constants, or as specified functions of depth. These representations are unlikely to be realistic. For example, one would expect the coefficients in general to depend on the shear and/or the stratification. Furthermore the values of the coefficients in the current climate are quite uncertain, with different measurements and estimates giving a range of 10^{-4} to 10^{-5} m^2/s [*Munk and Wunsch*, 1998]. This is at least in part because they have strong spatial variations [e.g., *Polzin* et al., 1997].

OGCM calculations show that vertical mixing by the other three subgrid-scale processes is strongest in high latitudes [*Huang* et al., 2003a and 2003b]. This is because the strong cooling of surface waters in high latitudes favors static instability and a vertical orientation of isopycnals. The former leads to convection; the latter leads both to isopycnal diffusion being predominantly vertical and to large amounts of potential energy being available for mesoscale eddies. The efficiency of all these processes is usually parameterized by specifying a constant diffusion coefficient.

The values of these coefficients are again poorly known. Estimates of the isopycnal diffusivity range from 500 to 2000 m^2/s [*Hirst and Cai*, 1994; *Jenkins*, 1991]. The most popular parameterization of mesoscale eddies is the Gent-McWilliams parameterization, which requires the specification of both an isopycnal diffusion coefficient and a diffusion coefficient parameterizing the effect of the mesoscale eddies on the density field [*Gent and McWilliams*, 1990]. The two diffusivities are commonly (but arbitrarily) taken to be the same. Eddy-resolving simulations show that in fact the mesoscale eddy diffusivity varies over a range of 10 to 10^7 m^2/s [*Nakamura and Chao*, 2000].

There are also theoretical reasons for questioning the adequacy of the parameterizations of high-latitude mixing. A fundamental limitation of the Gent-McWilliams parameterization is its assumption that mesoscale eddies' energy source is potential energy, whereas eddy-resolving simulations show that the kinetic energy of the mean flow is also an important source of eddy energy [*Solovev* et al., 2002]. In the case of parameterizations of convection, current schemes neglect the inhibiting effect of rotation on vertical motions [*Marshall and Schott*, 1999].

Finally we note that the calculation of the large-scale circulations in ocean GCMs is dependent on numerical schemes that are not perfect. Because of their inaccuracies there may be a significant amount of numerical diffusion, i.e., artificial mixing, in a model. Indeed it has been suggested that the unusually rapid mixing of heat into the deep ocean found in a global warming scenario with the GISS-HYCOM model [*Sun and Bleck*, 2001; *Sokolov* et al., 2003; see Figure 2] may be an artifact due to numerical diffusion in the HYCOM model [*R. Bleck*, personal communication].

3. STABILITY OF THE GLOBAL OCEAN CIRCULATION

As noted in the introduction, all ocean models show the possibility that circulation can be very sensitive to salinity perturbations and therefore to changes in surface freshwater fluxes. This sensitivity is closely associated with the fact that ocean models show the existence of more than one equilibrium state under some circumstances. These multiple equilibria arise because of a positive feedback associated with the advection of salinity in a circulation like that illustrated in Figure 1.

In this circulation the sinking is located in high latitudes, because that is where the surface waters are most dense. The density is a maximum there because the surface waters are coldest there. However, the waters in high latitudes are relatively fresh compared to the subtropics because in high latitudes precipitation exceeds evaporation whereas in the subtropics evaporation exceeds precipitation. Thus the poleward flow near the surface in a circulation like that shown in Figure 1 (basically the Gulf Stream) brings saltier water into high latitudes, and this tends to raise the density of the high latitude surface waters. Thus, this advection supplies a positive feedback to perturbations in the strength of the circulation. For example, if the circulation is weakened, the salinity advection weakens, the density of high latitude surface waters is decreased, and this weakens the circulation even more. Given a sufficiently strong initial decrease in the circulation, it will collapse. As noted earlier, paleoclimate evidence does indicate that similar state changes have occurred in the past.

This behavior can be illustrated in a model by tracing out a hysteresis loop [*Stocker and Wright*, 1991; *Rahmstorf*, 1995a]. Two such hysteresis loops, calculated with the Rooth [1982] box model, are shown in Figure 3. The equilibrium strength of the meridional overturning circulation in the Atlantic Ocean is plotted vs. the moisture flux into high latitudes of the North Atlantic, F_1. A positive circulation means that there is a strong poleward heat flux into high latitudes of the North Atlantic, and,

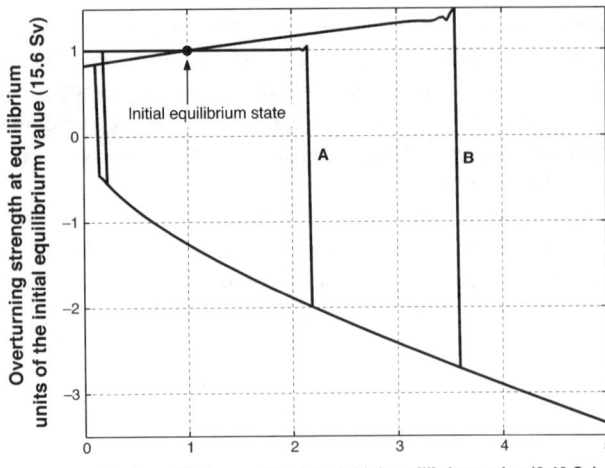

Figure 3. Hysteresis loops calculated from the Rooth [1982] box model with mixed boundary conditions. Vertical axis: strength of the meridional overturning circulation in the Atlantic normalized by its value in the current climate. Horizontal axis: atmospheric moisture flux from low to high latitudes in the Northern Hemisphere, normalized by its value in the current climate. Curve A assumes that the atmospheric moisture flux in the Southern Hemisphere is kept fixed at its value in the current climate. Curve B assumes that Southern Hemisphere flux is increased from its current climate value by 20% of the increase in the Northern Hemisphere.

in this model, a weak poleward heat flux into high latitudes of the South Atlantic. A negative circulation implies the opposite. The former state is the one analogous to that of the Atlantic in the current climate.

As the figure shows, there is a range of values of the moisture flux where two equilibria exist. For smaller values of the moisture flux only the state with strong poleward heat flux in the North Atlantic can exist; for larger values of the moisture flux only the state with strong heat flux in the South Atlantic can exist. If the system is in the former state, a sufficiently large positive perturbation added to the moisture flux will cause this state to collapse to the other equilibrium state, with a consequent large change in the oceanic heat transport and climate. How big a perturbation is required to accomplish this depends on many things. One factor is illustrated by the difference of the two hysteresis loops shown in Figure 2. Curve A is plotted under the assumption that the moisture flux into high latitudes of the South Atlantic does not change when F_1 changes. Curve B shows how the equilibrium state depends on F_1 when there is a simultaneous perturbation of the moisture flux into the high latitudes of the South Atlantic equal to 20% of F_1. As the figure shows, increased moisture flux into southern high latitudes is a stabilizing influence, i.e., it takes larger perturbations in F_1 to shift the system from one equilibrium state to the other.

A question of major importance to our understanding of the sensitivity of climate and its predictability is the question of where on the upper branch of the hysteresis loop the current climate is located. Ideally this question should be addressed with the most sophisticated state-of-the-art coupled GCMs. However to trace out such a curve with one of these models is not computationally feasible. To do so requires either very many integrations with different values of F_1, or a single integration in which F_1 changes very slowly so that the model will evolve through the whole series of possible quasi-equilibrium states. This would require 10,000 or more years of integration, and no coupled GCM has yet been used to calculate such a hysteresis loop.

Recently however hysteresis loops for 11 different models of intermediate complexity have been calculated as part of an intercomparison project for earth models of intermediate complexity (EMICs). EMICs are models which have less detail than state-of-the-art coupled GCMs, but do contain representations of all of the physical processes present in coupled GCMs [*Claussen* et al., 2002]. The results were reported at a workshop at the annual meeting of the European Geophysical Society in April 2003. There was no agreement among the models as to the position of the current climate. All the models did have the position being on the upper branch of the hysteresis loop, as it has to be in order to be consistent with the modern climate, but the locations varied from being far to the left of the hysteresis loop, in the monostable regime, corresponding to a very stable climate, to the position being in the bistable region near the bifurcation at the right side of the loop, corresponding to a state with very weak stability.

Actually the situation appears to be even more complicated than is indicated by the simple hysteresis loops illustrated in Figure 3. EMICs with an ocean GCM and realistic ocean bathymetry indicate the possibility of more than two equilibrium states, with the upper branch of the loop having a more complicated structure than that illustrated. In particular different states with somewhat different strengths for the overturning circulation are possible, depending on the sites of high latitude convection in the North Atlantic [*Rahmstorf*, 1995b].

The diversity of the model results for the state of the ocean circulation ultimately arises from the uncertainties in the input parameters for the climate models. One example is obvious from Figure 3, i.e., one needs to know accurately the values of the freshwater flux into the high latitudes of the Atlantic Ocean. Since these fluxes depend on precipitation and evaporation over the oceans, where measurements are sparse, the errors are large, of order ±30% [*Schmitt* et al., 1989]. In addition we note that the equilibrium states are not steady states, but rather contain fluctuations, presumably about a fixed climate state (see section 4 and Figure 5 below). Also, if the climate forcing is not steady, as for example when greenhouse

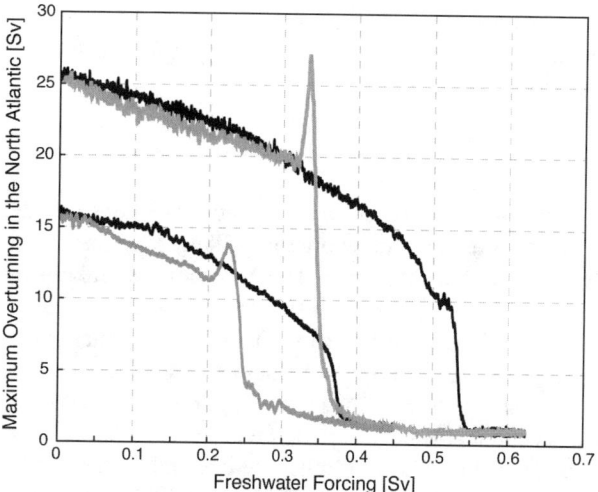

Figure 4. Hysteresis loops calculated with the MIT model of intermediate complexity [*Kamenkovitch* et al., 2002]. Vertical axis: strength of the meridional overturning circulation in the North Atlantic. Horizontal axis: moisture flux into the North Atlantic minus its value in the current climate. The states were traced out by starting with the current climate, then increasing the freshwater flux into the North Atlantic by 0.1 Sv/1000 years, and then after the circulation collapses, reversing the trend and returning to the current climate. The upper curve was calculated with a diapycnal diffusivity of 0.5 cm^2/s, the lower one with 0.2 cm^2/s. Adapted from Dalan, [2003]

gases increase, the equilibrium states and the hysteresis loops will change.

Another major source of uncertainty involves again the presence of uncertainty in small-scale oceanic mixing processes. Figure 4 illustrates two hysteresis loops calculated from an EMIC which includes an ocean GCM [*Kamenkovich* et al., 2002]. In order to complete the calculations in a reasonable amount of time, the moisture flux into the North Atlantic was taken to evolve somewhat more rapidly than required for the plotted states to be precise equilibrium solutions, and thus the forward and return branches of the hysteresis loops do not coincide precisely. Note that in these calculations there was no change in the moisture flux into the South Atlantic, and that in Figure 4 on the horizontal axis is plotted the change in the moisture flux into the North Atlantic from that in the current climate, rather than the actual flux. The two hystersis loops were calculated for different values of the ocean model's diapycnal diffusion coefficient, the upper one being for 0.5 cm^2/s, and the lower one for 0.2 cm^2/s.

As shown in the figure the hysteresis loops are displaced considerably from each other, and correspondingly the stability properties of the system are quite different, with the system being much less stable with the smaller value of the diffusivity. The intersection of the hysteresis curves with the vertical axis gives the strength of the overturning circulation in the North Atlantic in the current climate for the two values of the diapycnal diffusivity. Unfortunately, as we noted earlier, the strength is uncertain.

4. CHAOTIC BEHAVIOR

As noted in the introduction, oceanic circulations are likely to be chaotic, i.e., their evolution is likely to be very sensitive to the initial conditions. This behavior is well known in the atmosphere, and has been studied extensively with atmospheric GCMs. The results show that weather cannot in principle be predicted more than about two weeks in advance because small errors in the initial conditions grow so rapidly. The dynamical time scales in the oceans are much longer than in the atmosphere, of order decades and centuries rather than days, and this makes it much more difficult computationally to assess how chaotic behavior may limit the predictability of ocean circulations. There have only been two studies using ocean GCMs which have attempted to determine if such limits do exist. One by Griffies and Bryan [1997] (hereafter referred to as GB) looked at the predictability of fluctuations in the North Atlantic circulation; the other by Wang et al. [1999] (hereafter referred to as WSM) looked at the predictability of regime changes, i.e., of changes between different branches of the hysteresis loops discussed in the previous section.

GB used a coupled atmosphere-ocean GCM in their study. They carried out a thousand-year integration with fixed forcing corresponding to the current climate. In this integration there were fluctuations in the strength of the meridional overturning circulation of the North Atlantic, as illustrated in the top of Figure 5. They then carried out an ensemble of 12 integrations in which the initial state of the oceans was taken from year 130 of the control run, but the initial state of the atmosphere varied, being picked from 12 different years in the control runs (but all from the same calendar date). Thus only the weather in the initial atmospheric state differed in the 12 runs. The results for the evolution of the strength of the meridional overturning circulation in the North Atlantic are shown in the bottom of Figure 5. We see that the ensemble members diverge, and GB found using a statistical test that there is some reasonable predictability of the circulation strength only for the first 3 years. This result is the oceanic analog (for this model) of the prediction limit for atmospheric weather.

However from the point of view of climate, the GB result is not so relevant. The fluctuations in the circulation strength shown in Figure 5 are analogous to fluctuations in weather, and they all occur within the same climate regime. From the point of view of climate, a more interesting question is, what happens if the forcing changes? Is there a limit on our ability to predict regime changes? WSM examined this question using an ocean GCM with idealized global geometry. The ocean was forced by specified moisture fluxes and wind stresses, and the heat flux was calculated from a relaxation condition for the sea surface temperature. In the control run all these boundary conditions were based on the current climate. In addition a stochastic forcing was added to the wind stress boundary condition in order to mimic atmospheric weather fluctuations.

WSM then carried out an ensemble of runs in which the strength of the hydrological cycle in the Northern Hemisphere increased linearly, at a rate equal to 0.1% of the strength in the control run, per year. Thus the net precipitation in high latitudes of the Northern Hemisphere slowly increases and there is an equivalent increase in the net evaporation in low latitudes of the Northern Hemisphere. Three runs were carried out with three different choices for the initial value of the stochastic component of the wind stress. The results for the evolution of the strength of the meridional overturning circulation in the North Atlantic are shown in Figure 6.

Because of the very slow acceleration of the Northern Hemisphere hydrological cycle, the circulation evolves through a series of quasi-equilibrium states. In these equilibrium states the strength of the circulation does not change because the changes in precipitation and evaporation in the Northern Hemisphere in effect compensate each other. The increased

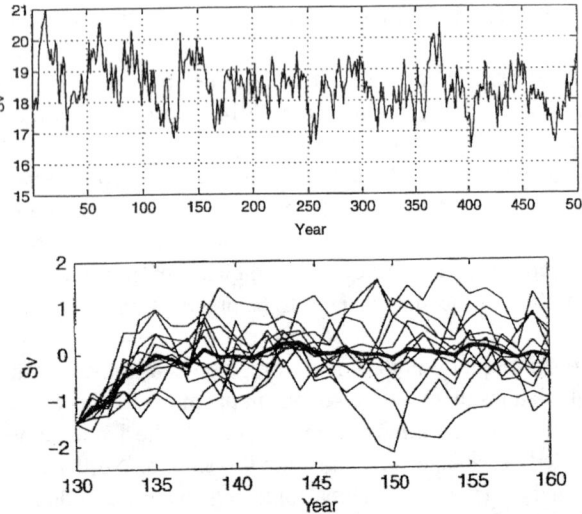

Figure 5. Top: strength of the meridional overturning circulation in the North Atlantic vs. time from a 500-year segment of a control run with the GFDL coupled GCM. Bottom: same as the top figure, except the difference in the strength of the circulation from the mean of the control run is plotted on the vertical axis, and the results are taken from 12 different experiments, all starting from the oceanic state at year 130 in the control run, but with different initial conditions in the atmosphere. The thick line indicates the mean of the 12 experiments. Adapted from GB.

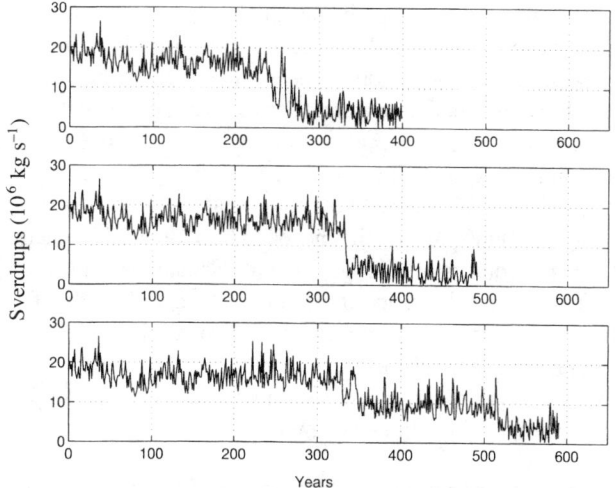

Figure 6. Strength of the meridional overturning circulation in the North Atlantic vs. time from 3 experiments with the WSM model in which the moisture flux into high latitudes of the North Atlantic slowly increased. The only difference between the experiments was the initial value of the atmospheric wind stress. Adapted from WSM.

precipitation in high latitudes reduces the density of the surface water there, but the increased evaporation in the subtropics increases the salinity of the subtropical surface waters, and this increases the advection of salinity into high latitudes. The effect of the latter on the density of the high-latitude surface waters just balances the effect of the former, because there is no net exchange of moisture between the atmosphere and ocean in the Northern Hemisphere as a whole. Thus the system evolves along a hysteresis loop like that shown by curve A in Figure 3.

During the initial phase of the experiments there are interannual fluctuations in the strength of the circulation which are comparable to those in the GB experiments (cf. Figures 5 and 6). However there is a striking difference in the nature of these fluctuations. In the WSM experiments, the fluctuations in all three experiments are identical for about 200 years, i.e., the predictability time is much longer than in the GB experiments. One plausible reason for the difference is that the surface heat flux variations in the GB model were much larger and more realistic. Although GB found that the interannual variations in the ocean circulation were largely controlled by the internal ocean dynamics, surface heat fluxes did play a role, and their variations due to weather could have caused the loss of predictability compared to the WSM experiments. We note however that even the more realistic GB model has significant limitations. For example it has coarse horizontal resolution (~5°), which limits the ability to simulate realistic weather fluctuations, and the model can only reproduce the current climate by introducing large unphysical adjustments to the surface heat fluxes.

The more interesting aspect of the WSM experiments is what happened on the longer time scales. As discussed in the previous section the acceleration of the Northern Hemisphere hydrological cycle must eventually lead to a collapse of the strong North Atlantic circulation (as indicated by curve A in Figure 3). It does in all three experiments but, as shown in Figure 6, the timing of the collapse, and the nature of the transition between the two circulation regimes differ considerably. Evidently the differences in the initial condition do not matter until the system approaches a bifurcation, and then there is a complete loss of predictability.

The two studies just described clearly only touch the surface of the problem of how prediction of changes in the ocean's circulation may be limited by chaotic behavior. For example, it is not clear from these experiments whether fluctuations in the surface heat flux or wind stress are more important in limiting predictability in the ocean circulation on long time scales. In addition neither study looked at how the predictability is affected by perturbations in the initial state of the oceans.

5. POSSIBLE PATHS FORWARD

Forecasts of global warming during the 21st century indicate that the earth is likely to reach global temperatures higher than any it has experienced for at least 100,000 years [IPCC, 2001]. This would take the earth to a situation outside the previous experience of our own species as well as that of many others. Thus one of the most formidable scientific challenges facing society is the need to develop a better understanding of how the climate system operates and to predict, to the extent possible, the changes in climate and the environment that society must cope with in the future. Because of the great complexity of the climate system and the many different disciplines that are required to deal with it, this is arguably the most difficult scientific task that has been undertaken. In addition, because the natural response times of the ocean range from decades to centuries, understanding and predicting ocean behavior is essential for planning over the next few centuries.

In our discussion of the oceans we have focused on three problems that limit our ability to predict the ocean's behavior: (1) our poor understanding of small-scale mixing processes, (2) our inability to characterize the stability characteristics of the ocean circulation, and (3) the presence of chaotic elements in the ocean's behavior. These problems are not independent. For example, the strength and behavior of the mixing properties affect the stability properties, and the stability properties influence the degree of chaotic behavior. In our discussion of the oceans, we also focused on the North Atlantic because that is where ocean heat transports are strongest. However, circulations in the North Atlantic are not exclusive to the North Atlantic, but rather extend throughout the global oceans, as the "conveyor belt" circulation. Because such issues remain unresolved, they have limited our understanding of, and our ability to model, the whole global ocean. Simulations of climate change with current state-of-the-art models are problematic as a result.

With regard to the small-scale mixing processes, advances in computer speeds will considerably alleviate at least the problems associated with parameterizations of mesoscale eddies. Since typical scales of these eddies are of order 100 km, models with horizontal resolutions of order 1/10 degree will have much less need to parameterize their effects. Such resolutions should be achievable for global climate models in the near future. Because oceanic energy spectra peak at the frequency of mesoscale eddies, this should mark a major advance in our models' capabilities.

Unfortunately the other mixing processes occur on a much finer scale and thus ocean models will have to rely on subgrid-scale parameterizations for them for a long time to come. More observational estimates of vertical fluxes of heat and tracers, particularly in high latitudes, would be useful, but obtaining them is difficult and expensive. In this situation theoretical approaches may be the most fruitful. In particular one needs prognostic parameterizations rather than the empirical schemes based solely on the current climate that are commonly used in current ocean GCMs. One promising approach for improving current parameterizations is to use modern turbulence closure models to derive prognostic parameterizations [*Canuto* et al., 2001 and 2002].

However, even these parameterizations still require the specification of the flux of energy into the oceans that drives the mixing. The major sources of this energy are believed to be surface winds and tidal mixing [*Munk and Wunsch, 1998*]. Thus climate changes which lead to changes in the surface winds might change the ocean mixing. Such an interaction has never been included in a climate model. Another potentially valuable step forward would be to couple these processes.

Because the stability characteristics of the ocean circulation depend on the small-scale mixing processes and surface flux climatologies (cf. Figures 3 and 4), improvements in our knowledge of both of these factors would help to determine the stability properties of the current climate. Paleoclimate data could also prove quite useful. There is considerable evidence indicating changes in the ocean's circulation regime in the past [e.g., *Broecker*, 2003, and references therein] and these data could help constrain a fully coupled climate model to have the right stability properties.

The fundamental nature of the ocean's circulations, i.e., their nonlinearity and weak dissipation, make it inevitable that their behavior will contain some chaotic elements. Computers have played a prominent role in advancing our knowledge of chaotic behavior in other systems, and in principle they could also do so for the oceans. The primary obstacle so far has been the inherently long time scales associated with the oceans. Increasing computer speeds, however, are now reaching the point where one can envisage carrying out ensembles of runs over long time scales with EMICS whose ocean component is an ocean GCM. Similar studies using coupled atmosphere-ocean GCMs are likely to be feasible within a decade or so. One key question that needs to be addressed is whether the major sources of error growth are fluctuations in the ocean or in the atmosphere, and if the latter, which surface flux fluctuations lead to the most rapid error growth. From the point of view of climate the key question before us is clear: to what extent does this error growth dominate over changes in forcing in controlling climate change?

Acknowledgments. I am indebted to Chris Forest for Figure 2, to Valerio Lucarini for Figure 3, to Fabio Dalan for Figure 4, and to Anne Slinn for Figures 1 and 6, and for formatting the other figures. Steve Sparks' editorial advice led to significant improvements in the paper.

REFERENCES

Boyle, E. A., and L. Keigwin, North Atlantic thermohaline circulation during the past 20,000 years linked to high-latitude surface temperature, *Nature,* 330, 35–40, 1987.

Broecker, W., D. Peteet, and D. Rind, Does the ocean-atmosphere system have more than one stable mode of operation?, *Nature,* 315, 21–26, 1985.

Broecker, W. S., Does the trigger for abrupt climate change reside in the ocean or in the atmosphere?, *Science,* 300, 1519–1522, 2003.

Bryan, F., High-latitude salinity effects and interhemispheric thermohaline circulation, *Nature,* 323, 301–304, 1986.

Bryan, F., Parameter sensitivity of primitive equation ocean general circulation models, *J. Phys. Oceanogr.,* 17, 970–985, 1987.

Bugnion, V., and C. Hill, Far field regulation of meridional overturning—the role of surface boundary conditions, *J. Climate,* submitted, 2004.

Canuto, V. M., A. Howard, Y. Cheng, and M. S. Dubovikov, Ocean turbulence: Part I: One-point closure model. Momentum and heat vertical diffusivities, *J. Phys. Oceanogr.,* 31, 1313–1426, 2001.

Canuto, V. M., A. Howard, Y. Cheng, and M. S. Dubovikov, Ocean turbulence: Part II: Vertical diffusivities of momentum, heat, salt, mass and passive scalars, *J. Phys. Oceanogr.,* 32, 240–264, 2002.

Claussen, M., et al., Earth system models of intermediate complexity : closing the gap in the spectrum of climate system models, *Clim. Dyn.,* 18, 579–586.

Cubasch, U., and G. A. Meehl, *Projections of future climate change. Climate Change 2002: the Scientific Basis,* J. T. Houghton et al., eds., Cambridge University Press, Cambridge UK.

Dalan, F., Sensitivity of climate change to diapycnal diffusivity in the ocean, Massachusetts Institute of Technology M.S. thesis, Cambridge, MA, 96 pp., 2003.

Forest, C. E., P. H. Stone, A. P. Sokolov, M. R. Allen, and M. D.

Webster, Quantifying uncertainties in climate system properties with the use of recent climate observations, *Science,* 295, 113–117, 2002.

Ganachaud, A., Large-scale mass transports, water mass formation, and diffusivities estimated from World Ocean Circulation Experiment, *J. Geophys. Res.,* 108, No. C7, 3213, doi:10.1029/2002JC001565, 2003.

Ganachaud, A., and C. Wunsch, Large-scale ocean heat and freshwater transports during the World Ocean Circulation Experiment, *J. Climate,* 16, 696–705, 2003.

Gent, P. R., and J. C. McWilliams, Isopycnal mixing in ocean circulation model, *J. Phys. Oceanogr.,* 20, 150–155, 1990.

Griffies, S. M., and K. Bryan, A predictability study of simulated North Atlantic multidecadal variability, *Clim. Dyn.,* 13, 459–487, 1997.

Hansen, J., G. Russell, A. Lacis, I. Fung, D. Rind, and P. Stone, Climate response times: dependence on climate sensitivity and ocean mixing, *Science,* 229, 857–859, 1985.

Hirst, A., and W. Cai, Sensitivity of a world ocean GCM to changes in subsurface mixing parameterization, *J. Phys. Oceanogr.,* 24, 1256–1279, 1994.

Huang, B., P. H. Stone, and C. Hill, Sensitivities of deep-ocean heat uptake and heat content in an OGCM with idealized geometry, *J. Geophys. Res.,* 108(C1), 3015, doi:10.1029/2001JC001218, 2003a.

Huang, B., P. H. Stone, A. P. Sokolov, and I. V. Kamenkovich, The deep-ocean heat uptake in transient climate change, *J. Climate,* 16, 1352–1363, 2003b.

IPCC, 2001: Climate Change 2001: The Scientific Basis (Houghton, J. T., et al., eds.), Cambridge University Press, Cambridge, UK, 881 pp.

Jenkins, W. J., Determination of isopycnal diffusivity in the Sargasso Sea, *J. Phys. Oceanogr.,* 21, 1058–1061, 1991.

Kamenkovich, I. V., A. Sokolov, and P. H. Stone, An efficient climate model with a 3D ocean and statistical-dynamical atmosphere, *Climate Dynamics,* 19, 585–598, 2002.

Ledwell, J. R., E. T. Montgomery, K. L. Polzin, L. C. St. Laurent, R. W. Schmitt, and J. M. Toole, Evidence for enhanced mixing over rough topography in the abyssal ocean, *Nature,* 403, 79–182, 2000.

Levitus, S., J. I. Antonov, J. Wang, T. L. Delworth, K. W. Dixon, and A. J. Broccoli, Anthropogenic warming of earth's climate system, *Science,* 292, 267–274, 2001.

Macdonald, A., and C. Wunsch, An estimate of global ocean circulation and heat fluxes, *Nature,* 382, 436–439, 1996.

Marotzke, J., P. Welander, and J. Willebrand, Instability and multiple steady states in a meridional-plane model of the thermohaline circulation, *Tellus,* 40A, 162–172, 1988.

Marotzke, J., and J. Willebrand, Multiple equilibria of the global thermohaline circulation, *J. Phys. Oceanogr.,* 21, 1372–1385, 1991.

Marotzke, J., Boundary mixing and the dynamics of three-dimensional thermohaline circulations, *J. Phys. Oceanogr.,* 27, 1713–1728, 1997.

Marshall, J., and F. Schott, Open-ocean convection: observations, theory and models, *Revs. Geophys.,* 37, 1–64, 1999.

Munk, W., and C. Wunsch, Abyssal recipes II: energetics of tidal and wind mixing, *Deep-Sea Res.,* 45, 1976–2009, 1998.

Nakamura, M., and Y. Chao, Characteristics of three-dimensional quasi-geostrophic transient eddy propagation in the vicinity of a simulated Gulf Stream, *J. Geophys. Res.,* 105, No. C5, 11,385–11,406, 2000.

Rahmstorf, S., Bifurcations of the Atlantic thermohaline circulation in response to changes in the hydrological cycle, *Nature,* 378, 145–149, 1995a.

Rahmstorf, S., Multiple convection patterns and thermohaline flow in an idealized OGCM, *J. Climate,* 8, 3028–3039, 1995b.

Rooth, C., Hydrology and ocean circulation, *Progress in Oceanography,* 11, Pergamon, 131–149, 1982.

Schmitt, R. W., P. S. Bogden, and C. E. Dorman, Evaporation minus precipitation and density fluxes for the North Atlantic, *J. Phys. Oceanogr.,* 19, 1208–1221, 1987.

Scott, J., and J. Marotzke, The location of diapycnal mixing and the meridional overturning circulation, *J. Phys. Oceanogr.,* 32, 3578–3595, 2002.

Seager, R., D. S. Battisti, J. Yin, N. Gordon, N. Naik, A. C. Clement, and M. A. Cane, Is the Gulf Stream responsible for Europe's mild winters?, *Q. J. Roy. Met. Soc.,* 128 (586), 2563–2586, 2002.

Sokolov, A., C.E. Forest, and P. H. Stone, Comparing oceanic heat uptake in AOGCM transient climate change experiments, *J. Climate,* 16, 1573–1582, 2003.

Solovev, M., P. H. Stone, and P. Malanotte-Rizzoli: Assessment of mesoscale eddy parameterizations for a single basin coarse resolution ocean model, *J. Geophys. Res.,* 107(C9), 9-1–9-19, doi:10.1029/2001JC001032, 2002.

Stocker, T. F., and D. G. Wright, Rapid transitions of the ocean's deep circulation induced by changes in surface water fluxes, *Nature,* 351, 729–732, 1991.

Stommel, H., Thermohaline convection with two stable regimes of flow, *Tellus,* 13, 224–230, 1961.

Sun, S., and R. Bleck, Atlantic thermohaline circulation and its response to increasing CO_2 in a coupled atmosphere-ocean model, *Geophys. Res. Lett.,* 28, 4223–4226, 2001.

Trenberth, K. E., and J. M. Caron, Estimates of meridional atmosphere and ocean heat transports, *J. Climate,* 14, 3433–3443, 2001.

Wang, X., P. H. Stone, and J. Marotzke, Global thermohaline circulation, Part I: Sensitivity to atmospheric moisture transport, *J. Climate,* 12, 71–82, 1999.

Welander, P., Thermohaline effects in the ocean circulation and related simple models, *Large-Scale Transport Processes in Oceans and Atmosphere,* J. Willebrand and D. Anderson, eds., D. Reidel, 163–200, 1986.

Wunsch, C. Low-frequency variability of the sea, *Evolution of Physical Oceanography,* B. A. Warren and C. Wunsch, eds., MIT Press, 342–374, 1981.

Peter H. Stone, Department of Earth, Atmospheric, and Planetary Sciences, Massachusetts Institute of Technology, Cambridge, Massachusetts.

Biosphere Dynamics: Challenges for Earth System Models

I. Colin Prentice[1,2], Corinne Le Quéré[1], Erik T. Buitenhuis[1], Joanna I. House[1], Christine Klaas[1], Wolfgang Knorr[1]

Many lines of evidence have established the role of the biosphere in the Earth's climate. Relevant processes act on time scales from minutes (e.g. stomatal control of evapotranspiration) to hundreds of millions of years (coevolution of the atmosphere and life). Successful attempts have been made to model the dynamics of the terrestrial biosphere on time scales up to that of vegetation dynamics, and comparable efforts in global marine biosphere modelling are under way. A long-term goal of biosphere modelling is the interactive representation of biospheric processes in Earth system models built on coupled models of the ocean and atmosphere. The most advanced climate models already contain sufficient biology to represent in some way the coupling between climate change and the global carbon cycle. This work has shown that the representation of terrestrial carbon cycling is a major source of uncertainty for the future of atmospheric CO_2. However, no climate model as yet includes a representation of land-surface processes and carbon cycling that is consistent with current understanding of plant physiological and ecological processes governing the exchanges of energy, water and carbon between terrestrial ecosystems and the atmosphere. Similarly, ocean biogeochemistry representations in current climate models fall short of representing current understanding of the dynamics of marine ecosystems and their interactions with mixed-layer physics and nutrient cycles. This article illustrates the problem with examples, from both the terrestrial and marine realms, of important processes whose representation in coupled models may be seriously in need of improvement. These examples speak to a need for more effective information exchange and collaboration among the different disciplines involved in Earth system science.

INTRODUCTION

The concept of the Earth system is central to two of the outstanding intellectual challenges in contemporary life. The first is the challenge of predicting the consequences of past, present and possible future fossil-fuel emissions for climate and society. Lacking the ability to make such predictions with sufficient precision for society' needs, the scientific community still cannot satisfactorily quantify either the effectiveness or the economics of strategies to lessen the impacts of human activities on the global environment. The second is the challenge of understanding the ice-core record of changes in the atmosphere and climate over the ice-age cycles, which has emphasized the complexity of "natural" climate dynamics and the close and as yet poorly understood linkages between atmospheric composition (including the production of trace gases of primarily biological origin) and climate. The two challenges are related because confidence in our ability to predict the consequences of actions taken now will be lim-

---

[1] Max Planck Institute for Biogeochemistry, P.O. Box 100164, D-07701 Jena, Germany
[2] QUEST, Dept. Earth Sciences, Univ. Bristol, Wills Memorial Building, Bristol BS8 1RJ, UK

ited, until such time as we can hindcast well-characterized events that have taken place in the past.

Both challenges provide a motivation to develop predictive models that include the set of physical, chemical and biological components that interact to govern Earth system dynamics on the appropriate time scales. The drive to develop interdisciplinary Earth system models has been given further impetus by process analyses, on time scales from "physiological" (minutes) to "geological" (centuries/millennia), which collectively point to a major involvement of life in determining the physical conditions under which life itself occurs. Examples, at opposite ends of the range of time scales, include the partial control of evapotranspiration by stomata (tiny pores in the surface of plant leaves) on a time scale of minutes, and the partial control of the atmospheric content of CO_2 by plant-enhanced chemical weathering and marine photosynthesis and calcite production on a time scale of hundreds of millions of years. In both examples, the word "partial" points to the fact that there are other, non-biological controls too – water supply and solar radiation in the first example, plate tectonics, volcanism and metamorphism in the second. A complete understanding of such phenomena must recognize the close interaction of physical, chemical and biological processes on all time scales.

Our focus is on the time scales up to about a century, which are the most directly relevant to the contemporary political and economic issues surrounding climate change and the global carbon cycle. On this time scale we can safely neglect processes acting on much longer time scales, such as macroevolution and continental drift. On the other hand, we have to consider processes acting on shorter time scales, at least in terms of their aggregated effects at the time scale of immediate interest. Most of the work during the past twenty years concerning the prediction of future climates has been based around general circulation models, which explicitly resolve weather processes. There is scope for more computationally efficient modelling with "intermediate complexity" models [*Schellnhuber*, 1999], but even these must strive to represent adequately the time-averaged behaviour of processes with time constants smaller than the models' smallest time-step. Already, the most advanced models have been developed to include representations of terrestrial and marine biology in order to examine feedbacks to climate from changing land surface conditions and feedbacks to atmospheric CO_2 and climate from changes in the metabolism of terrestrial and marine life. The most advanced models also account for key atmospheric chemistry reactions and the consequences of changing sources and sinks for reactive traces gases, such as CH_4, CO and NO_x, at the Earth's surface. However, these developments pose many challenges due to incomplete information, most importantly about the mechanisms controlling the exchanges of energy, water, CO_2 and reactive gases between ecosystems and the atmosphere.

These challenges are being addressed through the development of separate biosphere models for the land and ocean. Among current biosphere modelling efforts, our major involvements are with the Lund-Potsdam-Jena (LPJ) dynamic global vegetation modelling project [e.g., *Sitch et al.*, 2003; *Gerten et al.*, 2004], and the more recently started Dynamic Green Ocean Modelling (DGOM) project. Both are propelled by international consortia whose aim is to produce models that are state-of-the-art in terms of the way they represent ecosystem processes, and effective in reproducing various kinds of observations at the ecosystem scale from first principles – that is, from observations and experiments on the physiology and biophysics of different types of organisms. Relevant observations for model evaluation and benchmarking include remotely sensed spatial and temporal variation in land and ocean "greenness", and the results of large-scale field experiments, such as CO_2 fertilization and soil warming experiments on land [e.g., *DeLucia et al.*, 1999; *Rustad et al.*, 2001] and Fe fertilization experiments in the ocean [e.g., *Boyd et al.*, 2001].

Based on our experience in developing and evaluating biosphere models, we suggest that the representation of a number of key biogeochemical and biophysical processes in most or all climate models does not match current understanding and may be inadequate for predictive purposes. (This is not intended as a criticism of climate modelling, but rather as call for joint action by the climate modelling and biosphere dynamics communities!) We illustrate this idea with a small selection of examples, two terrestrial and two marine. Each example concerns a process of importance to climate or the carbon cycle or both. In no case can we present a definitive argument to support a particular representation, as these are all areas of active research and even controversy. Nevertheless we believe there is enough information to make it important to re-evaluate conventional wisdom and at least to evaluate the consequences of alternative representations of each process.

RESULTS AND DISCUSSION

Example 1: Controls of Evapotranspiration From the Vegetated Land Surface

Evapotranspiration from the land surface is a process of central importance for the global hydrological cycle. On the one hand, evapotranspiration permits the recycling of moisture, which supports precipitation in inland regions that receive little or no precipitation direct from oceanic sources. On the other hand, evapotranspiration depletes the soil moisture store, which can halt the moisture recycling process by removing the supply of water to be evaporated. Evapotranspiration is also a crucial control point in the physical climate system because of its influence on the partitioning of heat transfer (between sensible and latent heating)—cooling the surface. For example, warming can lead to increased evapotranspiration leading to soil water depletion, causing a positive feedback to further warming and drought.

The greater part of evapotranspiration from the global land surface takes place through transpiration, i.e. water that has been carried from the soil through the xylem (water-conducting tissues) of plants and evaporated from the leaf surfaces at a rate that is tightly controlled by the opening and closing of stomata on fast time scales (seconds to minutes). Early general circulation models used remarkably primitive methods to model evapotranspiration, given its central role in the physical climate system they represent. The situation has been improved over the past decade due to the widespread introduction of so-called "third-generation" land-surface models [*Sellers et al.,* 1997], which explicitly represent the various resistances to water flow in the soil-plant-atmosphere continuum and recognize the key role played by stomata in the regulation of water loss from plants. However, there are several potentially serious problems with current formulations. Most models continue to use empirical parameterizations of stomatal behaviour whose general applicability (for example, under altered ambient CO_2 concentration) is in doubt. Further, the correct representation of evapotranspiration from plant canopies requires not only correct specification of energy inputs and stomatal control, but also – crucially – a sufficient representation of the vertical dynamics of the convective boundary layer (CBL). This requires high vertical resolution in the lower layers of the atmosphere.

Evapotranspiration can also be viewed from a perspective of large-scale emergent properties, which any good model should reproduce. Work by *Monteith* [1995] and *Huntingford and Monteith* [1998] has drawn attention to some general aspects of the control of evapotranspiration. Figure 1 summarizes the adaptive process that Monteith termed "accomodation" between transpiring vegetation and the CBL. Figure 1 consists of three panels representing pairwise functional relationships between evapotranspiration (E), bulk surface conductance (g_c, which in vegetated landscapes is controlled mainly by leaf area and stomatal conductance per unit leaf area), and the vapour pressure deficit ($D = E/g_c$) of the air surrounding the leaves. In each panel, a thin line shows the "biosphere" relationship—that is, the biologically mediated response of surface conductance to changes in evapotranspiration rate. This formulation is a simple and highly general way of accounting for the apparent stomatal response to vapour pressure deficit, and it is consistent with the experimental finding that stomata tend to close in a neg-

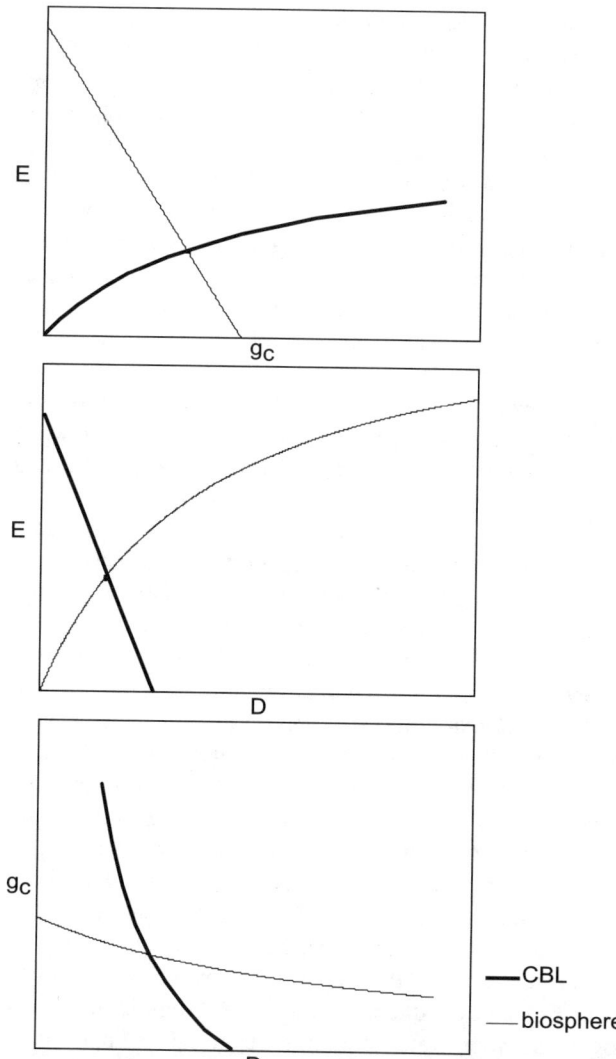

Figure 1. Alternative representations of the *Monteith* [1995] parameterizations of the stomatal response to evapotranspiration rate and the CBL response to the bulk surface conductance. E = evapotranspiration; g_c = bulk stomatal conductance; and D = near-surface vapour pressure deficit (normalized by atmospheric pressure). Panel 1 shows these relations as formulated by Monteith (using the hyperbolic rather than exponential formalism for the CBL response, following *Huntingford and Monteith,* 1998). Here stomatal conductance is shown declining linearly with evapotranspiration, while evapotranspiration increases with stomatal conductance towards an asymptote set primarily by net radiation. Panel 2 shows the same relationships expressed as functions of the near-surface vapour pressure deficit, indicating how this quantity adjusts to be consistent with both the biological and the physical constraints. Panel 3 shows the implied relationships between stomatal conductance and vapour pressure deficit: experimental data on the relationship between these two variables [e.g., *Oren et al.,* 1999] can be difficult to interpret because of potential confusion between the two types of curve.

ative-feedback response to increasing transpiration rate [e.g., *Mott and Parkhurst*, 1991]. A thick line shows the "CBL" relationship—the physically mediated dependence of evapotranspiration on surface conductance. Evapotranspiration is observed to increase linearly at low surface conductance, tending to an asymptotic value (the Priestley-Taylor rate) at high surface conductance. On time scales on the order of a few minutes to hours, both stomatal aperture and atmospheric properties such as the CBL height and water content adjust to one another so that both relationships are approximately satisfied—thus, the variables will tend to values corresponding to the intersections of the curves. The time scale of accomodation is such that disregarding leads and lags in this coupled system is presumably unimportant when the main interest is in the broad features of the diurnal and seasonal cycles.

The key unconventional aspect of this analysis is that it treats vapour pressure deficit as an internal variable of the coupled atmosphere-biosphere system. This approach assumes that over a spatial scale of about 1–100 km, local disequilibria between the atmospheric and biospheric processes are evened out [*Jarvis and MacNaughton*, 1986; *Raupach*, 1998], an assumption that is appropriate for the relatively coarse spatial scales of global atmospheric and biospheric modelling. The approach defines large-scale constraints that effectively put a "cap" on water loss from vegetation to the atmosphere. More precisely there are two caps—one determined by the maximum rate at which stomata will allow transpiration to occur, the other determined by the energy supply for evaporation. Recent experimental evidence points to the former being tightly coupled to the hydraulic properties of plant water conducting tissues. For example, the "biological" cap on water loss from vegetation composed of woody plants is predictable from sapwood conductivity and resistance to embolism (dependent on xylem anatomy, sapwood area and plant height), soil water potential, and leaf characteristics that determine the minimum water potential tolerated by the leaves [e.g., *Ryan and Yoder*, 1997; *Hubbard et al.*, 1999; *Magnani et al.*, 2000; *McDowell et al.*, 2002; *Tyree*, 2003]. These aspects are equally important for modelling vegetation carbon cycling, because the degree of stomatal opening also controls the influx of CO_2.

Modelled evapotranspiration should thus be constrained by well-defined physical and biological limits. These are barely recognizable in most current formulations of land-surface processes. Yet there are large, persistent and largely unexplained differences among models, both in terms of their grid-point simulations of water exchanges between land and atmosphere and in terms of the modelled sensitivity of soil moisture [*Shao and Henderson-Sellers*, 1996] and continental precipitation [*Cubasch et al.*, 2001] to climate change.

We speculate that the differences could be narrowed, and the models made more realistic, by incorporating information from recent advances in the understanding of plant physiology and biophysics. These advances postdate the development of the "third-generation" paradigm in land-surface schemes for climate modelling. Such a development does not necessarily imply further increases in complexity. For example, accounting for the biophysical processes underlying stomatal behaviour should reduce rather than increase the number of parameters that need to be specified, and intelligent simplifications of canopy-CBL interactions may be possible taking into account emergent properties of the coupled system.

Example 2: Temperature-Enhanced Decomposition of Soil Organic Carbon

Soil organic carbon (SOC) has emerged as a major "player" in the global carbon cycle. Not only is SOC the largest C store on land, currently holding at least twice as much C as the entire atmosphere [*Prentice et al.*, 2001]; this store is also potentially susceptible to increased rates of heterotrophic respiration (i.e. remineralization through microbial and fungal decomposer activity, leading to its oxidation to CO_2) in a warmer climate, implying a positive feedback that could enhance the warming due to increased greenhouse gas concentrations in the atmosphere.

Cox et al. [2000] used a coupled atmosphere-ocean general circulation with interactive terrestrial vegetation dynamics and ocean biogeochemistry to simulate the climatic consequences of a continuing exponential increase in anthropogenic CO_2 emission through the 21st century. They found a large feedback, such that the simulated warming climate eventually led to increased emissions of CO_2 from the land. This feedback was in part due to a simulated dieback of forests in some tropical regions, but the largest contribution was due to a general increase in the global rate of decomposition of SOC as indicated above. The effect was to raise atmospheric CO_2 concentration by as much as 200 ppm above what it would be in the absence of this feedback. As in several model studies of terrestrial carbon cycle changes with prescribed CO_2 and climate trends [*Cramer et al.*, 2001], CO_2 fertilization continued to be effective as a cause of terrestrial carbon uptake during the first part of the century, but other, positive-feedback effects—above all the enhanced decomposition rate of SOC—became stronger during the latter part of the century. Further studies by *Dufresne et al.* [2002] and *Friedlingstein et al.* [2003] have shown that another independent coupled atmosphere-ocean-biosphere model produces a similar feedback, although of smaller magnitude. These analyses confirmed that the principal cause of the large feedback found by *Cox et al.* [2000] was in the response of heterotrophic respiration on land. Analyses performed for IPCC [*Prentice et al.,* 2001] with simplified global carbon cycle models also drew attention to this potential feedback as a major source of uncertainty in future projections of atmospheric CO_2.

The picture has been complicated by experimental results [*Giardina and Ryan,* 2000] apparently contradicting the assumption made by *Cox et al.* [2000], in common with other terrestrial biosphere models, that the rate of decomposition of SOC should increase approximately exponentially with temperature. *Giardina and Ryan* [2000] analysed published data from soil samples that had been incubated under constant-temperature conditions for a year, and calculated apparent "turnover times" based on the proportion of initial SOC remaining at the end of the incubation. Contrary to usual expectations, they found no effect of incubation temperature on the apparent turnover time of SOC. This finding was interpreted as evidence against standard model formulations of the response of SOC decomposition to temperature [*Giardina and Ryan,* 2000], and has been cited as evidence against a temperature-SOC feedback in the global carbon cycle [e.g., *Grace and Rayment,* 2000].

Another type of experiment [*Jarvis and Lindner,* 2000; *Oechel et al.,* 2000; *Luo et al.,* 2001] has focused on the consequences of *in situ* warming of terrestrial ecosystems for measured fluxes of CO_2 from the soil. It has been found in such experiments that the total measured respiration flux increases at first, but then declines in one to three years, such that the rate of respiration is by then not significantly different from the pre-warming rate. This process has been referred to as "acclimation" (a poorly defined term that usually refers to a biologically driven process promoting homeostasis in the face of a physical perturbation). Taken at face value, these observations could be thought to contradict the idea of a long-term positive carbon-cycle feedback due to enhanced decomposition of SOC.

Fortunately, there is a straightforward explanation for these observations that does not conflict with other evidence from laboratory incubations with repeated sampling at shorter intervals [e.g., *Holland et al.,* 2000], nor with ^{14}C determinations on SOC from different climates [*Jenkinson et al.,* 1999; *Trumbore,* 1993, 2000; *Bird et al.,* 2002], which provide independent evidence for the dependence of SOC turnover rate on temperature under field conditions. The explanation we propose rests on the extreme chemical heterogeneity of SOC [*Davidson et al.,* 2000]. It has been known for decades that SOC consists of a broad chemical spectrum of materials with differing susceptibilities to microbial or fungal attack. Most models of SOC dynamics [e.g., *Parton et al.,* 1987; *Jenkinson et al.,* 1991] account for this heterogeneity by partitioning SOC into a small number of discrete "pools" with greatly different time constants. In unpublished work, we have shown that a simple model of heterogeneous SOC dynamics can account for both the apparent insensitivity of SOC decay as shown by *Giardina and Ryan* [2000] and the apparent "acclimation" of heterotrophic respiration in warming experiments. This model assumes that the decay rates of two labile SOC pools respond to temperature following the Arrhenius equation [*Lloyd and Taylor,* 1994] while a third pool shows negligible decay over one year (i.e., it has a much longer time constant—in keeping with the measured ^{14}C ages of SOC). The data of *Holland et al.* [2000] yield best-fitting estimates of the turnover time at 25°C of the first (most labile) pool as being only about 5 days, and for the second pool about 200 days. More than 90% of the SOC in these experiments remains at the end of a year, indicating the predominance of the third, non-labile pool.

According to this interpretation, the analysis of *Giardina and Ryan* [2000] tells us nothing about the temperature dependence of SOC decay. By the end of a year, in each experiment, virtually all of the first pool and most of the second pool would have decayed; on the other hand, the third pool would hardly have decayed at all. In effect, *Giardina and Ryan* [2000] were simply measuring the proportion of non-labile SOC in each soil (Plate 1a). The results of soil warming experiments can be explained equally simply. Measured respiratory *fluxes* are dominated by the relatively fast decay of labile SOC, whose abundance would be expected to equilibrate to new environmental conditions within a few years. SOC *stocks*, by contrast, are dominated by non-labile SOC. With a century or more of warming, we would expect to see a reduction in SOC stocks as well, but this cannot be observed in field experiments lasting only a few years (Plate 1b).

In this example, we suggest that the experimental data have been wrongly interpreted as requiring an "acclimation" mechanism because of the adoption of an oversimplified (single-pool) conceptual model to explain the data. The *Cox et al.* [2000] simulation model also treats SOC as a single pool (with temperature sensitivity); this in part accounts for the large magnitude of the simulated terrestrial carbon feedback. Warming in this model promotes release of CO_2 from *all* of the SOC, without regard for the long time constants for decomposition of the more recalcitrant compounds. But the single-pool model also accumulates CO_2 too rapidly in response to CO_2 fertilization. A multi-compartment model thus produces a transition from terrestrial biosphere carbon sink to carbon source at a similar critical CO_2 concentration but the magnitude of the accumulation and subsequent release is less (P.M. Cox, personal communication, 2004). Thus the basic qualitative finding of *Cox et al.* [2000] is robust to different assumptions about the heterogeneity of SOC; whereas the interpretation of the experimental data is not.

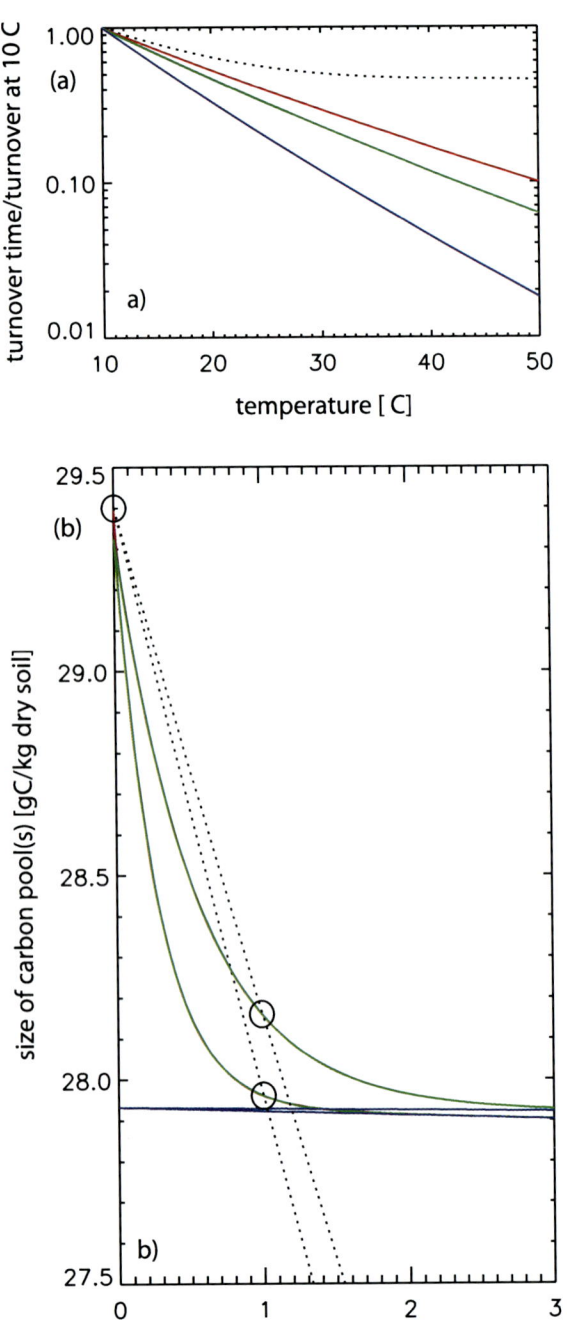

Example 3: Controls on the Sinking of Marine Particulate Organic Carbon

"Export production" by marine ecosystems is the process by which particulate organic carbon (POC) of biological origin sinks below the mixed layer and is thus taken out of contact with the surface waters and the atmosphere. If marine export production was to cease, it has been estimated that atmospheric CO_2 concentration would rise by about 200 ppm in a matter of years or decades [*Sarmiento and Toggweiler*, 1984; *Maier-Reimer et al.*, 1996]. Changes in export production could therefore substantially influence the global carbon cycle. Indeed, increased export production promoted by increased dust and Fe flux to Fe-limited regions of the ocean is one plausible mechanism contributing to the observed lowering of atmospheric CO_2 during glacial maxima, relative to interglacial periods [*Martin*, 1990; *Bopp et al.*, 2003; *Watson et al.*, 2000].

Early compilations of sediment-trap observations indicated that the flux of POC from 100 m to the deep ocean roughly followed an exponentially decreasing profile, suggesting a constant rate of remineralization with depth [*Martin*, 1987]. This simple model of POC sinking is strictly based on the quantity of organic material that sinks out of the mixed layer. It does not take into account the composition of the material. However, *Armstrong* [2002] recently showed that the POC sinking out of the mixed layer can be broadly separated into two components, one of which remineralizes at relatively shallow depths (< 500m), the other which sinks to the deep ocean. Further study by *Klaas and Archer* [2002] has established that the sinking rate of POC depends on the material ("ballast") to which it is attached (Figure 2). Despite the scatter in the observations plotted in Figure 2, it can be generalized that particles attached to $CaCO_3$ and to lithogenic material carry roughly three times more POC than particles attached to SiO_2. As different minerals are produced by different plankton functional types (coccolithophorids and foraminifera produce $CaCO_3$, diatoms produce SiO_2, and lithogenic material comes from atmospheric dust deposition and sediment resuspension), this finding implies a strong control of export production by the functional-type composition of the plankton. It

Plate 1. A re-interpretation of the results of *Giardina and Ryan* [2000]. It is assumed that the SOC consists of three pools, a labile pool with a turnover time of a few days, an intermediate pool with a turnover time of several months, and a slow pool with a turnover time much greater than one year (all at 25°C). (a) Turnover times are strongly temperature dependent (red: fast; green: intermediate; blue: slow pool) while the apparent turnover time as defined by *Giardina and Ryan* [2000] changes only little with temperature (dashed line). (b) Decay of SOC as a function of time for 25°C (lower curves) and 35°C (upper curves), for all three pools (red), the sum of the intermediate and slow pools (green) and the slow pool only (blue). The apparent turnover time of *Giardina and Ryan* [2000] assumes an exponential decay of the entire SOC (dashed lines) and is based on two points, at time=0, and time=1 year (shown as circles). According to the 3-pool model of SOC decay, the apparent turnover time predominently reflects the ratio of the active (fast and intermediate) pools to the slow, quasi-inert pool; it is not related to the turnover time of either of the three SOC pools (see also the discussion by *Davidson* et al., 2000).

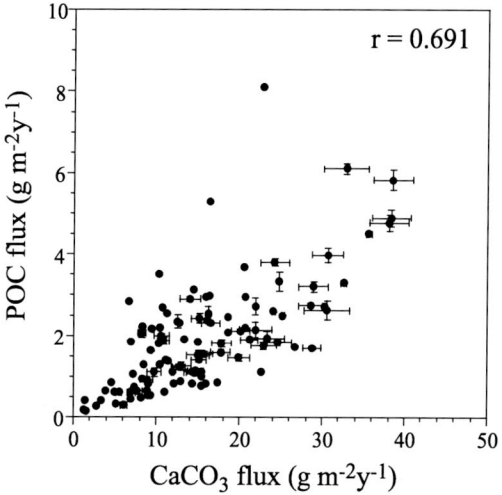 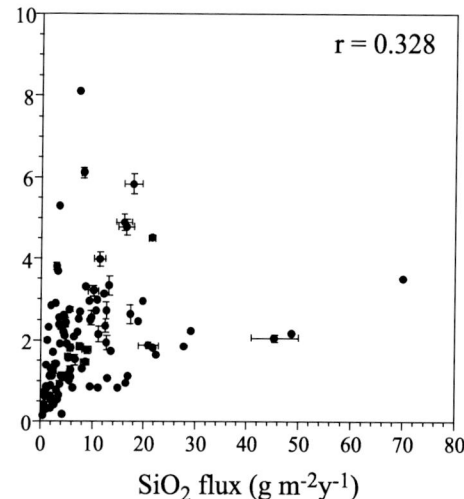

Figure 2. Fluxes of POC relative to $CaCO_3$ (left panel) and SiO_2 (right panel), as observed in sediment traps below 1000 m [from *Klaas and Archer*, 2002]. These data indicate a strong relationship between the composition of the inorganic ballast and the sinking of associated POC.

also casts doubt on the conventional wisdom that diatoms are the "champions" of export production simply because of the large size and density of their shells.

Current global ocean biogeochemical models (OBMs), as included in coupled models, do not pay attention to the production of different types of ballast by different types of plankton. Instead, OBMs use one of two strategies, as follows. (1) Some models instantaneously distribute the POC sinking out of the mixed layer according to the Martin curve [e.g., *Maier-Reimer*, 1993]. (2) Other models use a constant sinking speed of POC and distinguish between export of POC driven by large organisms, and non-sinking small organisms [*Bopp et al.*, 2003]. More detailed 1-D models exist that (3) calculate the sinking speed of POC from the size distribution of the particles [*Kriest and Evans*, 1999], or (4) calculate the sinking speed of POC based on ballasting by $CaCO_3$ [*Buitenhuis et al.*, 2001]. The results of *Klaas and Archer* [2002] show that these advances in the description of sedimentation need to be incorporated in OBMs to explore the potentially large feedback by which changes in marine ecosystem composition (brought about e.g. by changing ocean circulation, climate, external nutrient supply or alkalinity) could influence the global carbon cycle [*Bopp et al.*, 2003].

Example 4: Interannual Variability in Ocean-Surface Chlorophyll Concentration and Export Production

The six years of surface chlorophyll concentration observations recently made available from the SeaWiFS satellite allow us for the first time to make global-scale estimates of the time-variation in the export of POC. Export production can be estimated from satellite-derived chlorophyll *a* (Chl *a*) concentrations using empirical relationships between primary production and Chl *a* [*Behrenfeld and Falkowski*, 1997] that depend on the mixing depth and the temperature, and by estimating the fraction of the primary production that sinks out of the mixing depth [*Laws et al.*, 2000]. Although there are errors associated with both algorithms, the results suggest that the variability in POC export is large, especially in regions where the mean POC export itself is large.

This information on the variability of POC export did not exist at the time when the present generation of OBMs was developed. Thus, these observations provide a particularly revealing test of export production processes as simulated by OBMs. Plate 2 shows that none of the OBMs developed so far even approaches the observed space-time pattern of variability in POC export. The models of lower complexity (HAMOCC3 and NPZD) allocate most variability in the equatorial Pacific, and to other regions where primary productivity is large. However, even the relatively advanced PISCES model of *Aumont* et al. [2003] shows much lower variability than is seen in the data. Of the models examined, only the DGOM shows the right amplitude of variability, while the spatial patterns of variability that it simulates still have considerable discrepancies from the data.

The dramatic differences between the PISCES and DGOM simulations are a consequence of the inclusion of just one additional phytoplankton type (coccolithophorids). This fact draws attention to the complexity of behaviour of ecosystem processes, and the importance of including the main biological "players" in so far as they have different biogeo-

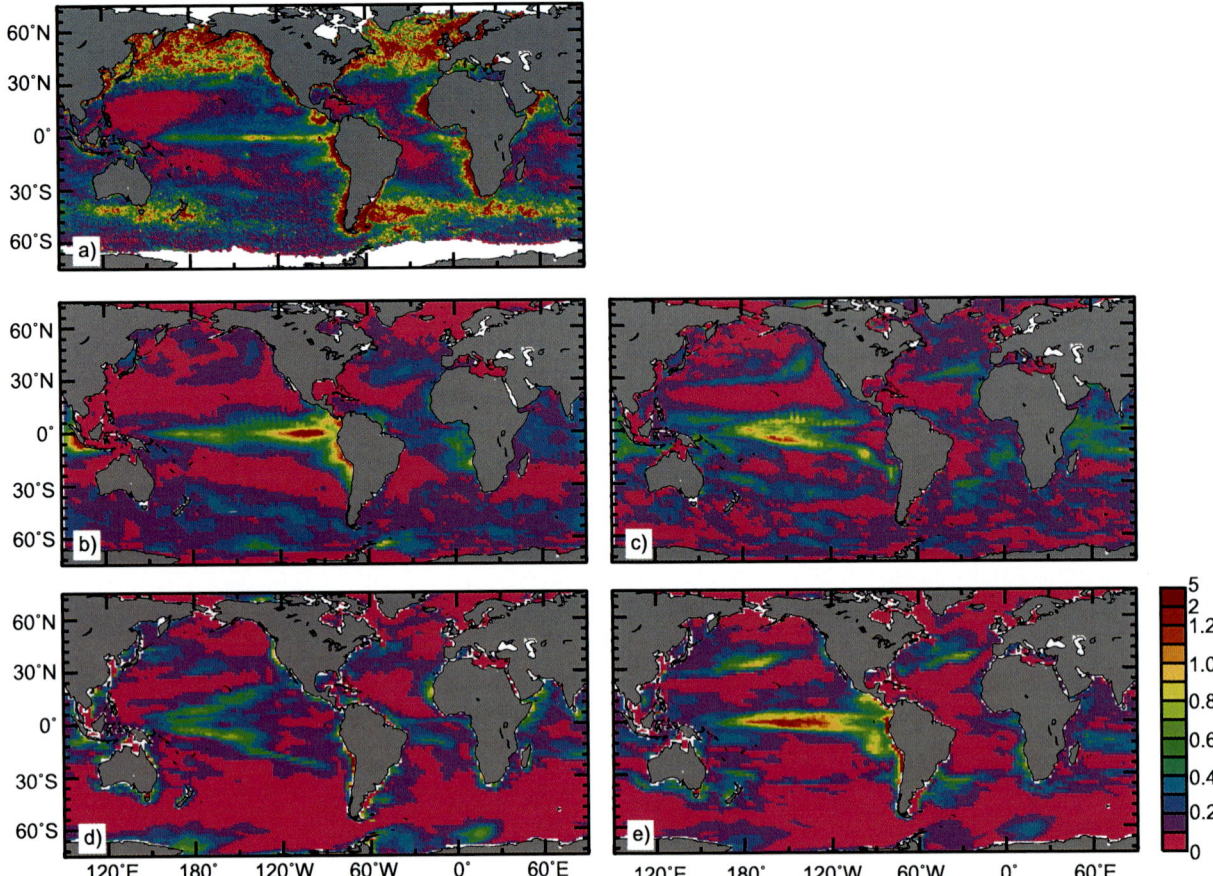

Plate 2. Standard deviation of the interannual variability in export production: (a) estimated from SeaWiFS Chl *a* observations, using *Behrenfeld and Falkowski* [1997] to compute primary production and *Laws* et al. [2000] to compute the fraction of primary production that is exported below 100 m; (b) to (d), simulated using ocean biogeochemistry models of different levels of complexity. Panel (b) shows the HAMOCC3 model, which assumes that export production is proportional to the squared concentration of P [*Maier-Reimer*, 1993]. Panel (c) shows an NPZD model in which just one type of phytoplankton, one type of zooplankton, DOC and detritus are represented [*Aumont* et al., 2002]; as in HAMOCC3 this model considers the cycle of only one nutrient element (P). Panel (d) shows the PISCES model, in which two types of phytoplankton (diatoms and nanophytoplankton) and two types of zooplankton are resolved and the cycles of P, Si and Fe [*Aumont* et al., 2003; *Bopp* et al., 2003] and their interactions are modelled explicitly. Panel (e) shows a current version of the DGOM, which is based on PISCES but models one additional phytoplankton type (coccolithophorids) (E. Buitenhuis and others, unpublished results).

chemical responses and effects. Coccolithophorids are distinguished biogeochemically in part by their exceptionally high affinity for phosphate, such that they can grow rapidly down to very low concentrations of phosphate; they stop growing only when dissolved phosphate concentrations are drawn down below the threshold for measurement [*Riegman et al.*, 2000]. The inclusion of a phytoplankton type with this property in a global ocean model has knock-on effects on the dynamics of all the phytoplankton types and in regions remote from coccolithophorid blooms. The fact that the simulated patterns are still in error indicates the need for further work to improve the DGOM.

CONCLUDING REMARKS

A variety of cultural and communication problems have stood in the way of effective collaboration among the different disciplines involved in Earth system science. We suggest, with support from the examples summarized above and many others, that these problems have hindered both the implementation of current process-level biological understanding in the context of global physical and physical/chemical models and the proper appreciation by biologists working at the process level of the system-level consequences of their findings. In some cases this will necessarily entail an increase in model complexity, but as some of the examples show, this need not always be the case, especially where process-based formulations of biological processes can be substituted for empirical ones. Overall, our recent experience encourages us to believe that such problems can be solved through collaborative work, and indeed that the enterprise of Earth system modelling constitutes an excellent focus for interdisciplinary research.

Acknowledgments. We thank numerous colleagues in the LPJ and DGOM consortia for discussions. ICP thanks the organizers of the IUGG 2003 "State of the Planet" Symposia for the initial impetus to write this article, and the UK National Instiute for Environmental e-Science (NIEeS) for the opportunity to present this work at the UK/Japan collaboration workshop on Earth System Modelling in Cambridge, UK in October 2003. Example 1 also benefited from discussions with Keith MacNaughton; Example 2 from discussions with many scientists including Torben Christensen, Beth Holland, Jari Liski, Jon Lloyd, Dave Schimel, Sue Trumbore and Ricardo Valentini; Example 4 from discussions with Olivier Aumont and Laurent Bopp. The DGOM development has been partially supported by the EU under a Development Host grant to MPI-BGC. We thank the SeaWiFS team for chlorophyll data and Peter Cox for valuable comments on an earlier draft. We thank Gerhard Boenisch and Kerstin Sickel for assistance in designing Figure 1, and Silvana Schott for technical editing of multiple drafts.

REFERENCES

Armstrong, R. A., C. Lee, J. I. Hedges, S. Honjo, and S. G. Wakeham, A new, mechanistic model for organic carbon fluxes in the ocean: based on the quantitative association of POC with ballast minerals, *Deep Sea Res., Part II, 49*, 219–236, 2002.

Aumont, O., S. Belviso, and P. Monfray, Dimethylsulfoniopropionate (DMSP) and dimethylsulfide (DMS) sea surface distributions simulated from a global three-dimensional ocean carbon cycle model, *J. Geophys. Res., 107* (C4), 148–227, 2002.

Aumont, O., E. Maier-Reimer, S. Blain, and P. Pondaven, An ecosystem model of the global ocean including Fe, Si, P co-limitations, *Glob. Biochem. Cycl., 17*, doi:10.1029/2001GB001745, 2003.

Behrenfeld, M. J., and P. G. Falkowski, A consumer's guide to phytoplankton primary productivity models, *Limnol. & Oceanogr., 42*, 1479–1491, 1997.

Bird, M., H. Santruckova, J. Lloyd, and E. Lawson, The isotopic composition of soil organic carbon on a north-south transect in western Canada, *Europ. J. Soil Sci., 53*, 393–403, 2002.

Bopp, L., K. E. Kohfeld, C. Le Quéré, O. Aumont, Dust impact on marine biota and atmospheric CO2 during glacial periods, *Paleoceanogr., 18*, doi: 10.10129/2002/PA000810, 2003.

Boyd, P. W., The role of iron in the biogeochemistry of the Southern Ocean and equatorial Pacific: a comparison of in situ iron enrichments, *Deep Sea Res. Part II, 49* (9–10), 1803–1821, 2002.

Buitenhuis, E. T., P. van der Wal, and H. J. W. de Baar, Blooms of Emiliania huxleyi are sinks of atmospheric carbon dioxide: a field and mesocosm study derived simulation, *Glob. Biogeochem. Cycl., 15*, 577–587, 2001.

Cox, P. M., R. A. Betts, C. D. Jones, S. A. Spall, and I. J. Totterdell, Acceleration of global warming due to carbon-cycle feedbacks in a coupled climate model, *Nature, 408*, 184–187, 2000.

Cramer, W., A. Bondeau, F. I. Woodward, C. Prentice, R. A. Betts, V. Brovkin, P. M. Cox, V. Fisher, J. A. Foley, A. D. Friend, C. Kucharik, M. R. Lomas, N. Ramankutty, S. Sitch, B. Smith, A. White, and C. Young-Molling, Global response of terrestrial ecosystem structure and function to CO_2 and climate change: results from six dynamic global vegetation models., *Glob. Change Biol., 7*, 357–374, 2001.

Cubasch, U., G. A. Meehl, G. J. Boer, R. J. Stouffer, M. Dix, A. Noda, C. A. Senior, C. B. Raper, and K. S. Yap, Projections of future climate change., in *Climate Change 2001: The Scientific Basis. Contribution of Working Group I to the Third Assessment Report of the Intergovernmental Panel on Climate Change*, edited by J. T. Houghton, Y. Ding, D. J. Griggs, M. Noguer, P. J. van der Linden, X. Dai, K. Maskell, and C. I. Johnson, pp. 525–582, Cambridge University Press, Cambridge, 2001.

Davidson, E. A., S. E. Trumbore, and R. Amundson, Biogeochemistry—Soil warming and organic carbon content, *Nature, 408*, 789–790, 2000.

DeLucia, E. H., J. G. Hamilton, S. L. Naidu, R. B. Thomas, J. A. Andrews, A. Finzi, M. Lavine, R. Matamala, J. E. Mohan, G. R. Hendrey, and W. H. Schlesinger, Net primary production of a forest ecosystem with experimental CO_2 enrichment, *Science, 284* (5417), 1177–1179, 1999.

Dufresne, J.-L., Friedlingstein, P., Berthelot, M., Bopp, L., Ciais, P., Fairhead, L., LeTreut, H., and P. Monfray, Effects of climate change due to CO_2 increase on land and ocean carbon uptake, *Geophys. Res. Lett., 29(10)*, doi: 10.1029/2001GL013777, 2002

Friedlingstein, P., J.-L. Dufresne, P. M. Cox, and P. Rayner, How positive is the feedback between climate change and the carbon cycle?, *Tellus (Series B), 55*, 692–700, 2003.

Gerten, D., S. Schaphoff, U. Haberlandt, W. Lucht, and S. Sitch, Terrestrial vegetation and water balance—hydrological evaluation of a dynamic global vegetation model, *J. Hydrol., 286*, 249–270, 2004.

Giardina, C. P., and M. G. Ryan, Evidence that decomposition rates of organic carbon in mineral soil do not vary with temperature, *Nature, 404*, 858–861, 2000.

Grace, J., and M. Rayment, Respiration in the balance, *Nature, 404*, 819–820, 2000.

Holland, E. A., J. C. Neff, A. R. Townsend, and B. McKeown, Uncertainties in the temperature sensitivity of decomposition in tropical and subtropical ecosystems: Implications for models, *Glob. Biogeochem. Cycl., 14*, 1137–1151, 2000.

Hubbard, R. M., J. B. Bond, and M. G. Ryan, Evidence that hydraulic conductance limits photosynthesis in old *Pinus ponderosa* trees, *Tree Physiol., 19*, 165–172, 1999.

Huntingford, C., and J. L. Monteith, The behaviour of a mixed-layer model of the convective boundary layer coupled to a big leaf model of surface energy partitioning, *Boundary-Layer Meteorol., 88*, 87–101, 1998.

Jarvis, P., and S. Linder, Constraints to growth of boreal forests, *Nature, 405*, 904–905, 2000.

Jarvis, P. G., and K. G. McNaughton, Stomatal control of transpiration: scaling up from leaf to region, *Advances in Ecol. Res., 15*, 1–49, 1986.

Jenkinson, D. S., D. E. Adams, and A. Wild, Model Estimates of CO_2 Emissions from Soil in Response to Global Warming, *Nature*, *351* (6324), 304–306, 1991.

Jenkinson, D. S., J. Meredith, J. I. Kinyamario, G. P. Warren, M. T. F. Wong, D. D. Harkness, R. Bol, and K. Coleman, Estimating net primary production from measurements made on soil organic-matter, *Ecology*, *80*, 2762–2773, 1999.

Klaas, C., and D. E. Archer, Association of sinking organic matter with various types of mineral ballast in the deep sea: Implications for the rain ratio, *Glob. Biogeochem. Cycl.*, *16*, #1116, 2002.

Kriest, I., and G. T. Evans, Representing phytoplankton aggregates in biogeochemical models, *Deep-Sea Res. Part I-Oceanogr. Res. Papers*, *46*, 1841–1859, 1999.

Laws, E. A., P. G. Falkowski, W. O. Smith, H. Ducklow, and J. J. McCarthy, Temperature effects on export production in the open ocean, *Glob. Biogeochem. Cycl.*, *14*, 1231–1246, 2000.

Lloyd, J., and J. A. Taylor, On the temperature dependence of soil respiration, *Funct. Ecol.*, *8*, 315–323, 1994.

Luo, Y. Q., S. Q. Wan, D. F. Hui, and L. L. Wallace, Acclimatization of soil respiration to warming in a tall grass prairie, *Nature*, *413*, 622–625, 2001.

Magnani, F., M. Mencuccini, and J. Grace, Age-related decline in stand productivity: the role of structural acclimation under hydraulic constraints, *Plant Cell & Environm.*, *23*, 251–263, 2000.

Maier-Reimer, E., Geochemical cycles in an ocean general circulation model: preindustrial tracer distributions, *Glob. Biogeochem. Cycl.*, *7*, 645–677, 1993.

Maier-Reimer, E., U. Mikolajewicz, and A. Winguth, Future ocean uptake of CO_2—Interaction between ocean circulation and biolgoy, *Clim. Dynam.*, *12*, 711–721, 1996.

Martin, J. H., G. A. Knauer, D. M. Karl, and W. W. Broenkow, VERTEX: Carbon cycling in the northeast Pacific, *Deep Sea Res.*, *34*, 267–285, 1987.

Martin, J. H., R. M. Gordon, and S. E. Fitzwater, Iron in Antarctic Waters, *Nature*, *345* (6271), 156–158, 1990.

McDowell, N. G., N. Phillips, C. Lunch, B. J. Bond, and M. G. Ryan, An investigaiton of hydraulic limitation and compensation in large, old Douglas-fir trees, *Tree Physiol.*, *22*, 763–774, 2002.

Monteith, J. L., Accommodation between transpiring vegetation and the convective boundary layer, *J. Hydrol.*, *166*, 251–263, 1995.

Oechel, W. C., G. L. Vourlitis, S. J. Hastings, R. C. Zulueta, L. Hinzman, and D. Kane, Acclimation of ecosystem CO_2 exchange in the Alaskan Arctic in response to decadal climate warming, *Nature*, *406*, 978–981, 2000.

Oren, R., J. S. Sperry, G. G. Katul, D. E. Pataki, B. E. Ewers, N. Phillips, and K. V. R. Schäfer, Survey and synthesis of intra- and interspecific variation in stomatal sensitivity to vapour pressure deficit, *Plant Cell & Environm.*, *22*, 1515–1526, 1999.

Mott, K. A., and D. F. Parkhurst, Stomatal Responses to Humidity in Air and Helox, *Plant Cell & Environm.*, *14*, 509–515, 1991.

Parton, W. J., D. S. Schimel, C. V. Cole, and D. S. Ojima, Analysis of factors controlling soil organic matter levels in Great Plains Grasslands, *Soil Science Soc. Am. J.*, *51*, 1173–1179, 1987.

Prentice, I. C., G. D. Farquhar, M. J. R. Fasham, M. L. Goulden, M. Heimann, V. J. Jaramillo, H. S. Kheshgi, C. Le Quéré, R. J. Scholes, and D. W. R. Wallace, *The carbon cycle and atmospheric carbon dioxide, in Climate Change 2001: The Scientific Basis. Contribution of Working Group I to the Third Assessment Report of the Intergovernmental Panel on Climate Change*, edited by J. T. Houghton, Y. Ding, D. J. Griggs, M. Noguer, P. J. van der Linden, X. Dai, K. Maskell, and C. A. Johnson, pp. 185–225, Cambridge University Press, Cambridge, 2001.

Raupach, M. R., Influences of local feedbacks on land-air exchanges of energy and carbon, *Glob. Change Biol.*, *4*, 477–494, 1998.

Riegman, R., W. Stolte, A. A. M. Noordeloos, and D. Slezak, Nutrient uptake and alkaline phosphatase (EC 3:1:3:1) activity of Emiliania huxleyi (Prymnesiophyceae) during growth under N and P limitation in continuous cultures, *J. Phycol.*, *36*, 87–96, 2000.

Rustad, L. E., J. L. Campbell, G. M. Marion, R. J. Norby, M. J. Mitchell, A. E. Hartley, J. H. C. Cornelissen, and J. Gurevitch, A meta-analysis of the response of soil respiration, net nitrogen mineralization, and aboveground plant growth to experimental ecosystem warming, *Oecologia*, *126* (4), 543–562, 2001.

Ryan, M. G., and B. J. Yoder, Hydraulic limits to tree height and tree growth, *Biosci.*, *47*, 235–242, 1997.

Sarmiento, J. L., and J. R. Toggweiler, A new model for the roel of the oceans in determining atmospheric pCO_2., *Nature*, *308*, 621–624., 1984.

Schellnhuber, H. J., 'Earth system' analysis and the second Copernican revolution, *Nature*, *402*, C19–C23, 1999.

Sellers, P. J., R. E. Dickinson, D. A. Randall, A. K. Betts, F. G. Hall, J. A. Berry, C. J. Collatz, A. S. Denning, H. A. Mooney, C. A. Nobre, N. Sato, C. B. Field, and A. Henderson-Sellers, Modeling the exchanges of energy, water and carbon between the continents and the atmosphere, *Science*, *275*, 1–35, 1997.

Shao, Y. P., A. Henderson-Sellers, Validation of soil moisture simulation in landsurface parameterisation schemes with HAPEX data, *Glob. & Planet. Change*, *13*, 11–46, 1996.

Sitch, S., B. Smith, I. C. Prentice, A. Arneth, A. Bondeau, W. Cramer, J. O. Kaplan, S. Levis, W. Lucht, M. T. Sykes, K. Thonicke, and S. Venevsky, Evaluation of ecosystem dynamics, plant geography and terrestrial carbon cycling in the LPJ dynamic global vegetation model., *Glob. Change Biol.*, *9*, 161–185, 2003.

Trumbore, S. E., Age of soil organic matter and soil respiration: radiocarbon constraints on belowground C dynamics, *Ecol. Appl.*, *10*, 399–411, 2000.

Trumbore, S. E., Comparison of carbon dynamics in tropical and temperate soil using radiocarbon measuremtns., *Global Biochemical Cycles*, *7*, 275–290, 1993.

Tyree, M. T., Hydraulic limits on tree performance: Transpiration, carbon gain and growth of trees, *Trees*, *17*, 95–100, 2003.

Watson, A. J., D. C. E. Bakker, A. J. Ridgwell, P. W. Boyd, and C. S. Law, Effect of iron supply on Southern Ocean CO_2 uptake and implications for glacial atmospheric CO_2, *Nature*, *407*, 730–733, 2000.

I. Colin Prentice, QUEST, Department of Earth Sciences, University of Bristol, Wills Memorial Building, Bristol BS8 1SS, UK

The UVic Earth System Climate Model and the Thermohaline Circulation in Past, Present and Future Climates

Andrew J. Weaver

School of Earth & Ocean Sciences, University of Victoria, Victoria, B.C., Canada

Over the last few years significant advances have been made towards understanding the mechanisms behind climate variability over the Last Glacial Cycle. This has become possible through the development of a new breed of climate models of intermediate complexity. In this review, the philosophy behind the development of these models is discussed with particular attention given to the UVic Earth System Climate Model. Results are then surveyed from numerous studies using these intermediate complexity, as well as other, climate models aimed at piecing together puzzles buried within the paleo proxy record. Particular attention is given to the climate feedbacks involved in glacial inception 116,000 years ago, as well as modelling efforts aimed at understanding millennial timescale Dansgaard-Oeschger oscillations and their packaging into Bond Cycles in cold climates, their association with Heinrich events, and their dependence on the mean climatic state. In examining the climate over the last 135,000 years, it is apparent that variations in the formation of intermediate waters, both in the Labrador Sea and the Antarctic Circumpolar Current, have important consequences for the stability and variability of the climate system. A discussion of some future challenges for the climate and paleoclimate community is also given.

1. INTRODUCTION

Coupled atmosphere-ocean general circulation models (GCMs) are frequently used to understand both past, present and future climate and climate variability. The computational expense associated with these models, however, precludes their use for undertaking extensive parameter sensitivity studies. While they must ultimately be used as the primary tool for undertaking climate projections on which policy is based, it is important to conduct sensitivity studies, in parallel with the coupled GCM studies, using simpler models. Simple models, or models of intermediate complexity, allow one to explore the climate sensitivity associated with a particular process or component of the climate system over a wide range of parameters. In addition, they allow one to streamline the experiments to be performed by more complicated GCMs. Models of intermediate complexity vary from simple one-dimensional energy balance/upwelling diffusion models [*Wigley and Raper*, 1987], to zonally-averaged ocean/ energy balance atmosphere models [*Stocker et al.* 1992], to models with more sophisticated subcomponents [*Petoukhov et al.* 2000; *Weaver et al.*, 2001].

Simple and intermediate complexity climate models are designed with a particular class of scientific questions in mind [see *Claussen et al.*, 2002 for a review]. In the development of the model, only those processes and parametrisations are included which are deemed important (or are possible to incorporate) in the quest to address the scientific questions of concern. For example *Wigley* [1998] used an upwelling diffusion-energy balance climate model to evaluate Kyoto Protocol implications for increases in global mean temperature and sea level. While such a simple climate model relies on the climate sensitivity and other parameters obtained from coupled GCMs, it nevertheless allows for a first-order analysis. *Stocker and Schmittner* [1997] used a three-basin zonally-averaged ocean model coupled to an energy-balance atmospheric model, in a systematic parameter sensitivity study of the response of the Atlantic meridional overturning (AMO) to both the rate of increase and equilibrium concentration of atmospheric CO_2. Their work clearly illustrated the importance of the rate of CO_2 increase on the AMO, a result difficult to achieve with the computationally expensive present generation GCMs.

The CLIMBER group have taken the approach of building a climate model of intermediate complexity with a sophisticated, albeit highly parametrised, atmospheric component. Their atmospheric model is based on the statistical-dynamical approach without resolving synoptic variability [*Petoukhov et al.* 2000]. Their three-basin, zonally-averaged ocean model is very similar to the one used by *Stocker et al.* (1992) and they also incorporate a simple dynamic/thermodynamic ice model. The CLIMBER-2 model sacrifices resolution and complexity for computational efficiency. This model has been used to investigate the climate of the last glacial maximum [LGM—*Ganopolski et al.* 1998], glacial climate stability and variability [*Ganopolski and Rahmstorf*, 2001] as well as the cause for the collapse of the AMO in climate change experiments [*Rahmstorf and Ganopolski*, 1999]. *Fanning and Weaver* [1996] developed an energy-moisture balance model coupled to an ocean GCM and a thermodynamic sea-ice model. This model was used to undertake a number of sensitivity studies, including the role of sub-grid-scale ocean mixing and flux adjustments in climate change experiments [*Wiebe and Weaver*, 1999; *Fanning and Weaver*, 1997] and steric sea level rise [*Weaver and Wiebe*, 1999], and the climate of the LGM [*Weaver et al.*, 1998].

A quick view of the rest of this paper follows. In the next section I briefly describe the UVic Earth System Climate Model (ESCM): the primary tool I use to illustrate the physical mechanisms behind the key climate events observed during the last glacial cycle (see section 3). In section 4, I discuss the fundamental coupling of water mass formation in the northern and southern hemispheres while, in section 5, I examine the response of the ocean to anthropogenic warming. Finally, I present some future challenges for the paleoclimate community in section 6.

2. THE UVIC EARTH SYSTEM CLIMATE MODEL

The philosophy underlying the development of the UVic ESCM is that on timescales greater than a decade, the ocean, its horizontal gyre structure, and its ability to transport heat and freshwater are key components of the climate system. As such, the model has been built to resolve as many processes and feedbacks as possible that affect climate sensitivity and oceanic heat uptake on long timescales. It is also constructed in a modular fashion so that only a subset of processes or subcomponent models may be included, depending on the particular scientific question of concern.

Since its original conception in *Fanning and Weaver* [1996], the UVic ESCM has undergone significant development. It now consists of an ocean GCM coupled to a thermodynamic/dynamic sea ice model, an ocean carbon cycle model, a dynamic energy-moisture balance atmosphere model, a thermomechanical ice sheet model, a land surface model and a terrestrial dynamic vegetation and carbon cycle model [*Weaver et al.* 2001; *Matthews et al.*, 2003, 2004; *Meissner et al.*, 2003b; *Ewen et al.*, 2004]. A reduced complexity atmosphere model is used for computational efficiency. Atmospheric heat transport is parametrised through diffusion, and moisture transport is accomplished through advection and diffusion, with precipitation occurring when the relative humidity is >85% (see Plate 1). The atmospheric model includes a parametrisation of the water vapour/planetary longwave feedback, while the radiative forcing associated with changes in atmospheric CO_2 is externally imposed as a reduction of the planetary long wave radiative flux. A specified lapse rate is used to reduce the surface temperature over land where there is topography.

The model uses prescribed winds to obtain its present-day climatology, and a dynamical wind feedback is included that exploits a latitudinally-varying empirical relationship between atmospheric surface temperature and density. The ocean component of the coupled model is a 3-D ocean GCM with a global resolution of 3.6° (zonal) by 1.8° (meridional) and 19 vertical levels. It includes a new parameterisation for brine-rejection [*Duffy et al.*, 1999, 2001]. The coupled model incorporates a dynamic/ thermodynamic sea ice model [*Bitz et al.*, 2001, *Holland et al.*, 2001]. Dynamics is represented by an elastic-viscous-plastic rheology and various sea ice thermodynamics and thickness distribution options are included.

The thermomechanical ice sheet model within the UVic ESCM comes from *Marshall and Clarke* [1997]. It employs continuum mixture theory to incorporate ice streams [*Yoshimori et al.*, 2001]. We have also added the Hadley Centre dynamic global vegetation and terrestrial carbon cycle model [TRIFFID—*Cox et al.*, 2000, *Cox*, 2001] in which the relevant land-surface characteristics (vegetation fraction, leaf area index, albedo, etc.) are modelled directly [*Foley et al.*, 1996], and two different land surface models: one being a simple bucket model [*Matthews et al.*, 2003], and the other being a simplified one-layer version of the Hadley Centre MOSES land surface scheme [*Meissner et al.*, 2003b].

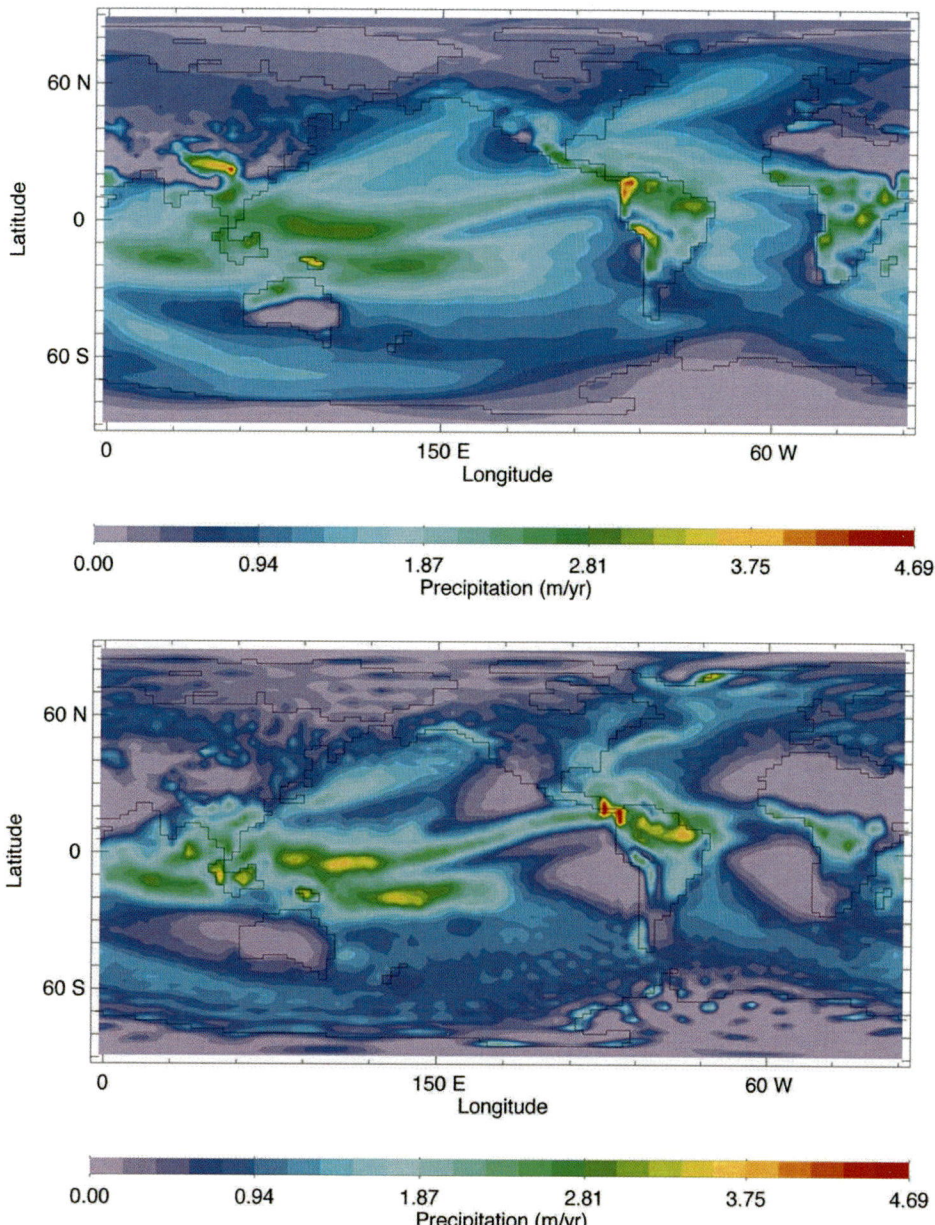

Plate 1. Annual mean precipitation (m/yr) at 1850 as obtained from the UVic ESCM (top) and from the present-day NCEP reanalysis (bottom). The needle leaf tree distribution associated with the model is shown in Plate 2 (top).

The UVic ESCM has not only been used to investigate various scientific questions in contemporary and paleo climate, but also as a laboratory with which to build new subcomponent models, test new parametrisations, and develop intuition and coupling experience/technology. The efficiency of the simple atmospheric component of the ESCM allows for a wide range of parameter space to be explored through long timescale integrations.

One of the virtues of the coupled model is that we do not need to employ explicit flux adjustments to keep the simulation of the present climate stable. It also allows us to conduct many long timescale integrations, in order to investigate climate processes through a wide range of parameter space. Our use of a rotated coordinate system also allows us to examine high latitude processes in detail without having to worry about numerical issues associated with converging meridians. Thus, it is well suited for both climate and paleoclimate modelling, and is especially useful for examining important high latitude processes.

2.1. The Model Evaluation Processs

Of course, before using any model to examine past or future climate variability/change, it is necessary to evaluate it against existing data sets representing both snapshots in time and the transient evolution between these snapshots. In this regard, the UVic model has been extensively and successfully evaluated against both contemporary climate observations [*Weaver et al.*, 2001] as well as paleo proxy records [*Weaver et al.*, 1998; *Schmittner et al.*, 2002b; *Meissner et al.*, 2003a]. For example, Plate 2 shows the preindustrial representation of the boreal forest (obtained from TRIFFID) compared to present-day observations. The model has also been evaluated against the transient 20th century climate (Plate 3) in a manner similar to that of *Stott et al.* [2000]. As in *Stott et al.* [2000], natural forcing agents (solar forcing and volcanic emissions), while necessary to simulate the early century warming, can not account for the warming in recent decades. Similarly, anthropogenic forcing alone (greenhouse gases and sulphate aerosols) is insufficient to explain the

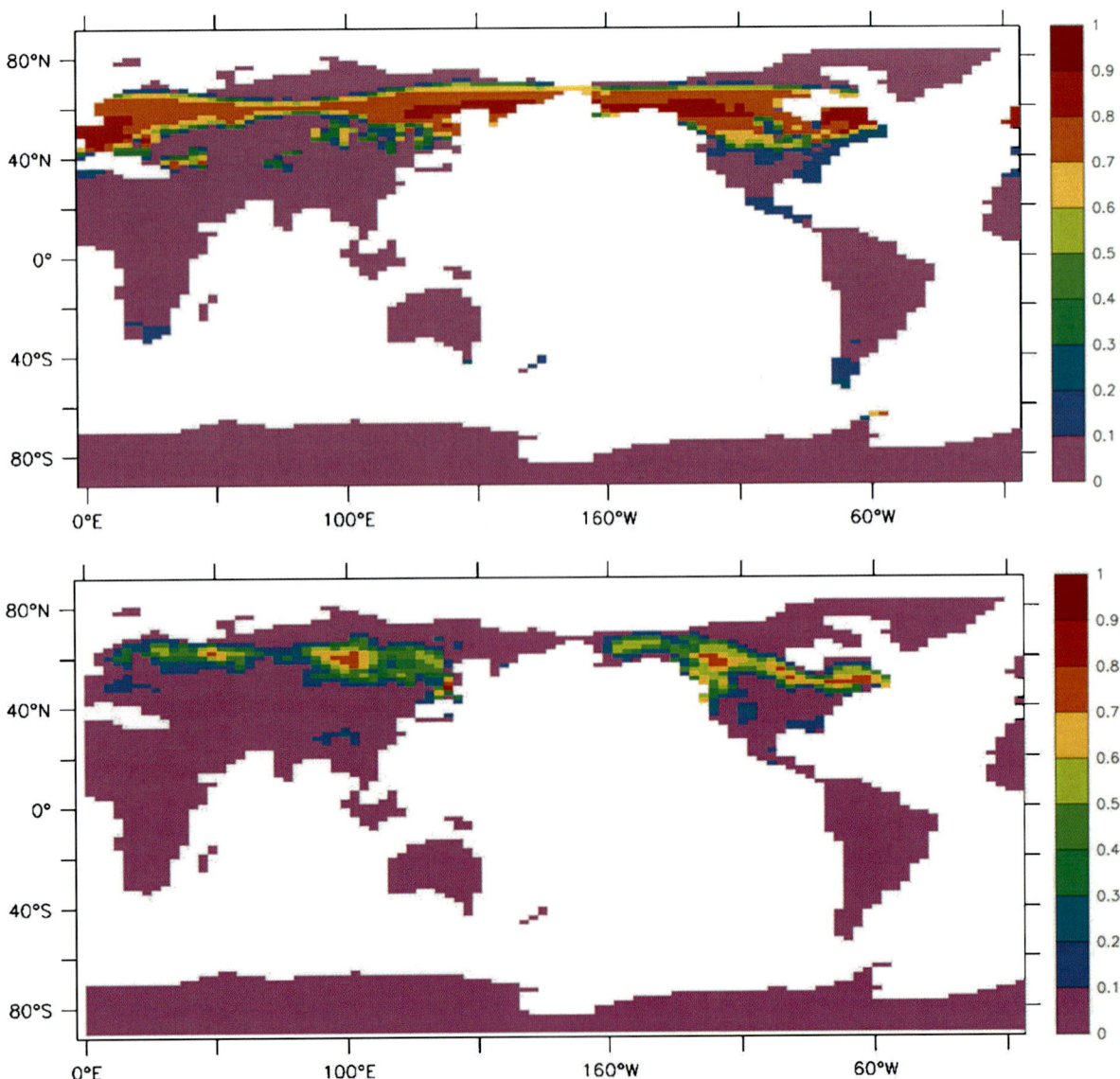

Plate 2. (top) Needleleaf trees at 1850 obtained from the UVic ESCM and (bottom) from present-day IGBP observations. The model has not included the consequences of human activity (deforestation into cropland/pastures etc.) whereas the IGBP data does. The precipitation field associated with the coupled model is shown in the left panel of Plate 1.

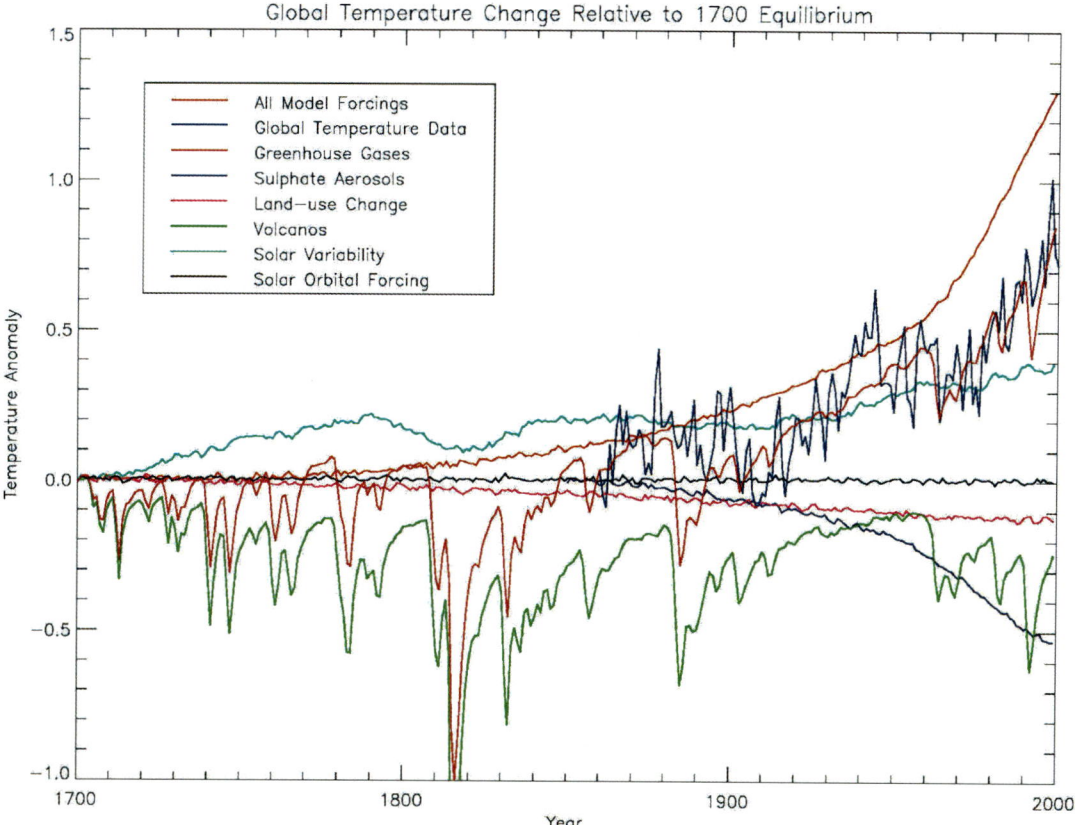

Plate 3. Global mean surface air temperature anomaly (relative to a year 1700 equilibrium) obtained from UVic ESCM simulations. The observed temperature record is given in blue, while the model simulated change including all natural and anthropogenic forcings is given in red. These are plotted in relation to each other according to their 1961–1990 averages. The 20th century surface air temperature response of the UVic model, driven by changes in individual forcings (either natural or anthropogenic), is also shown: greenhouse gases (orange); sulphate aerosols (purple); land-use change (pink); volcanic activity (green); solar intensity (light blue); orbital (Milankovitch) forcing (black). Redrawn from *Matthews et al.* [2003].

1910–1945 warming but is necessary to simulate the warming since 1976.

In *IPCC* [2001] the simulated LGM sea surface and surface air temperatures from the UVic ESCM, the CLIMBER-2 model, and several atmospheric GCM-mixed-layer ocean models were compared with paleo proxy data. The UVic model compares very favourably with the paleo reconstructions and lies in the mid range of estimates (near the model-mean ensemble) obtained from a variety of atmosphere GCM-mixed layer ocean models (Plate 4).

3. CLIMATE OVER THE LAST 135,000 YEARS

North Atlantic Deep Water (NADW) and the intermediate-depth Labrador Sea Water (LSW) are currently formed in the Greenland-Iceland-Norwegian (GIN) and Labrador Seas, respectively. Their rate of formation tightly governs the strength of the AMO and its associated heat transport in the North Atlantic [see, *Weaver et al.*, 1999, for a review]. The Nordic Seas are separated from the North Atlantic by the Greenland-Scotland ridge with a maximum depth of only about 600–800 m, and so the renewal of the deep North Atlantic is fed by an overflow of intermediate-depth water [*Aagaard et al.*, 1985]. As these waters flow southwestward in a deep western boundary undercurrent into the Labrador Sea, they entrain recirculating, relatively cold and fresh Labrador Sea waters [*McCartney*, 1992; *Dickson and Brown*, 1994], that are largely confined to the subpolar gyre of the North Atlantic [*McCartney and Talley*, 1984]. The deep western boundary undercurrent that leaves the Labrador Sea at depths greater than ~2000 m is thought to be about 200–300 km wide and transport about 13–14 Sv of newly formed NADW southward [*Talley and McCartney*, 1982; *Schmitz and McCartney*, 1993] before eventually encountering northward-flowing Antarctic Bottom Water (AABW). Despite the high salinity of the NADW, the colder AABW has higher density and so passes below it.

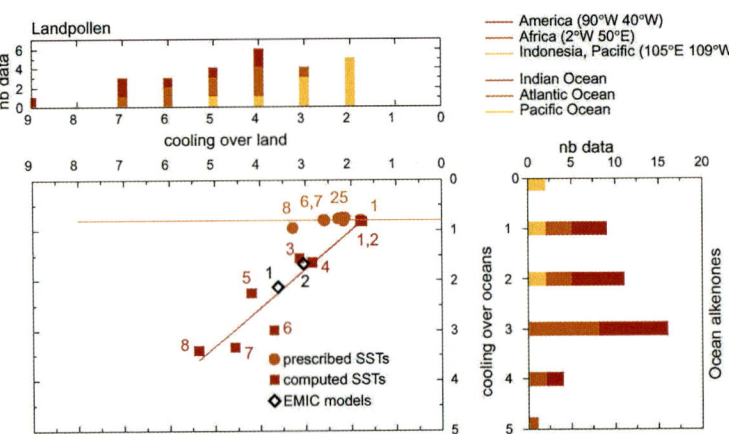

Plate 4. Model simulated (center) versus paleodata (upper and right) comparison of the annual mean land and sea surface tropical cooling (averaged 30°S–30°N) at the LGM. The atmospheric GCMs either used prescribed CLIMAP SSTs (circles) or were coupled to mixed layer ocean models (squares). The numbers refer to models from: [Circles] 1: LMD4, 2-5: MRI2, ECHAM3, UGAMP, LMD5, 6-7: CCSR/NIES1, LMD5, 8: GEN2. [Squares] 1: LMD4, 2: UGAMP, 3: GEN2, 4: GFDL, 5: HADAM2, 6: MRI2, 7: CCM1, 8: CCC2. The results from the UVic (1) and CLIMBER-2 (2) models of intermediate complexity are also displayed (diamonds): The land temperature proxy record (upper) was obtained from various pollen data for altitudes below 1500m ("nb data" refers to the number of data points in three different regions corresponding to the temperature change plotted in the abscissa); The SST proxies (right) were obtained from tropical alkenone reconstructions Model results were averaged over the whole tropical domain and not over proxy-data locations, which may bias the comparison. Taken from *IPCC* [2001].

3.1. Labrador Sea Water Formation

The LSW mass is renewed locally through convective mixing during winter [*Marshall and Schott*, 1999], although the amount of convection varies from year to year and decade to decade [*Lazier*, 1980; *McCartney and Talley*, 1984; *Dickson and Brown*, 1994; *Dickson et al.*, 1996]. *Wood et al.* [1999], using the Hadley Centre coupled atmosphere-ocean GCM with sufficient resolution to resolve Denmark Strait overflow, performed two transient simulations starting with a preindustrial level of atmospheric CO_2 and subsequently increasing it at a rate of 1% or 2% per year. Convection and overturning in the Labrador Sea ceased in both these experiments while deep water formation persisted in the Nordic Seas. As the climate warmed, the Denmark Strait overflow water became warmer and hence lighter so that the density contrast between it and the deep Labrador Sea water was reduced. This made the deep circulation of the Labrador Sea collapse, while DSOW remained unchanged, a behaviour suggested from the paleo-reconstructions of *Hillaire-Marcel et al.* [2001] for the Last Interglacial (Oxygen Isotopic Substage 5e [OIS-5e]; the Eemian—Shackleton, 1987).

The Eemian constitutes an interval that likely experienced warmer conditions than during pre-industrial (PI) times [*White*, 1993]. Lasting from 127 to 118 kyr BP [*Adkins et al.*, 1997], the Eemian was associated with a minimum in ice volume and a sea level, estimated from coral proxies, to be 6 m higher than today [*Ku et al.*, 1974]. Although several earlier studies suggest a high production rate of NADW during OIS-5e [e.g., *Keigwin et al.*, 1994; *Yu et al.*, 1996; *Adkins et al.*, 1997], recent analysis by *Hillaire-Marcel et al.* [2001] has suggested that the modern situation, with active LSW formation, has apparently no analogue throughout the last climatic cycle, and thus appears a feature exclusive to the present interglacial. In particular, the results of *Hillaire-Marcel et al.* [2001] indicate that the LSW/NADW stratification, which characterizes the modern and Holocene Labrador Sea, never developed during the penultimate deglaciation, and that during the Eemian, a single water mass occupied the basin below a low density surface layer. They also indicated the absence of winter convection and hence no LSW formation, but provided evidence for millennial oscillations in surface water conditions, not unlike those of the Holocene [*Bond et al.*, 1997].

In *Cottet-Puinel et al.* [2004] the variation of North Atlantic Deep Water (NADW) formation over the last glacial cycle, from the Eemian through to future climate change projections, was investigated using the UVic model. Equilibrium simulations for the Eemian (125 kyr BP) and the LGM both revealed the absence of Labrador Sea Water (LSW) formation although NADW formation still occured, albeit at a reduced rate relative to the modern times. For the Eemian, the location of convection in the eastern North Atlantic was similar to the present, although it was generally shallower and less extensive. In the case of the LGM, deep convection moved southward to the western coast of Europe and was much more localised. The inferred inception of a modern-like circulation slightly before 7 kyr BP revealed by proxy reconstructions [*Hillaire-Marcel et al.*, 2001] was not captured by the model unless the meltwater forcing from the Laurentide ice sheet was applied in a long 21,000 year transient simulation. This raised questions concerning the value of equilibrium simulations of early Holocene climate. In all global warming projections (see section 4), LSW formation initially ceased as the level of atmospheric CO_2 rose, but recovered once the atmospheric CO_2 level was held fixed. Convection in the north extended further into the Nordic Seas as the sea ice edge retreated. In all simulations convection remained active in the eastern North Atlantic, with its latitude depending on the position of the sea ice edge, suggesting that the formation of lower NADW is a robust feature of Late Quaternary climate. As the Labrador Sea was found to be very sensitive to the freshwater forcing, it suggests that this region represents a good location for the concentration of observational studies to monitor a possible oceanic response to anthropogenic climate change.

3.2. Ice Age Inception

A fundamental and challenging issue in paleoclimate modelling is the failure to capture the last glacial inception [*Rind et al.*, 1989]. Glaciation is indicated by a relatively well dated global ice volume record derived from the oxygen isotope ratio in microfossil shells [*Imbrie et al.*, 1984; *Linsley*, 1996; *Shackleton*, 1987] and glacio-eustatic sea level change derived from uplifted coral reef terraces [*East et al.*, 1999]. While sea level is not strictly proportional to continental ice volume due to concurrent variations of sea surface area [*Marsiat and Berger*, 1990], it gives an approximate idea of globally-integrated glacial evolution. Between 118 and 110 kaBP, the sea level records show a rapid drop of 50–80 m from the last interglacial, which itself had a sea level only 3–5 m higher than today [*East et al.*, 1999; *Cuffey and Marshall*, 2000]. This sea level lowering, as a reference, is about half of the 120 m LGM difference relative to the present [*Fairbanks*, 1989].

As the last glacial inception offers one of few valuable test periods for the validation of climate models, many studies have been conducted. The first attempts to simulate glacial inception using an AGCM were conducted by *Royer and Deque* [1983] and *Royer et al.* [1984]. In their experiments, only orbital parameters were changed with other boundary conditions remaining at modern values, and they were unable to obtain a perennial snow cover. The inability of models to adequately simulate the last glacial inception was also pointed out by *Rind et al.* [1989] wherein several experiments under 116 kaBP and extreme orbital configurations were conducted. In their experiments, glacial inception required the use of full glacial CLIMAP SSTs (reduced by 2°C) and a prescribed initial 10m land ice over the areas of CLIMAP LGM ice sheets [*CLIMAP*, 1981]. The use of full glacial ocean conditions for the simulation of glacial inception, and the resulting inconsistency of the SSTs with applied orbital and CO_2 forcing, restricts the interpretation of these results.

Oglesby [1990] and *Verbitsky and Oglesby* [1992] conducted a series of sensitivity experiments with respect to the level of atmospheric CO_2. They concluded that CO_2 forcing alone could not account for glacial inception, in agreement with the result of *Syktus et al.* [1994]. Similarly, *Phillipps and Held* [1994] and *Gallimore and Kutzbach* [1995], conducted a series of sensitivity experiments with respect to orbital parameters by specifying several extreme orbital configurations. They concluded that although a cool summer orbit brings the most favourable conditions for the development of permanent snow and expansion of glaciers, orbital forcing alone cannot account for glacial inception. This conclusion was also confirmed by *Mitchell* [1993], *Schlesinger and Verbitsky* [1996], and *Vavrus* [1999].

Various other studies have examined combinations of orbital and CO_2 forcing [*Syktus et al.*, 1994; *Schlesinger and Verbitsky*, 1996; *Dong and Valdes*, 1995]. Synthesizing the results from these studies with those listed above leads to the following conclusions: 1) neither orbital nor atmospheric CO_2 forcing alone can account for glacial inception; 2) the results of their combined effects are not consistent; 3) the difficulty in simulating perennial snow cover is not related to the existence of multiple equilibria, but to model resolution or the overlooked role of the biosphere.

Gallimore and Kutzbach [1996], *de Noblet et al.* [1996], and *Pollard and Thompson* [1997] went further by examining biospheric feedbacks in atmospheric GCM studies either by specifying an expansion of tundra at the surface [*Gallimore and Kutzbach*, 1996] or by including interactive vegetation [*de Noblet et al.*, 1996; *Pollard and Thompson*, 1997]. However, none of these studies included an interactive ocean GCM and hence were unable to capture changing SSTs associated with changing ocean dynamics. Nevertheless, these studies pointed to the importance of vegetation feedbacks for capturing glacial inception. In particular, the southward expansion of tundra and its associated vegetation-snow masking effect was found to be an important positive feedback for initiating glaciation. The response of the boreal vegetation during ice age inception is perhaps the most important positive feedback in the climate system able to amplify the small seasonal changes in radiative forcing associated with changing orbital param-

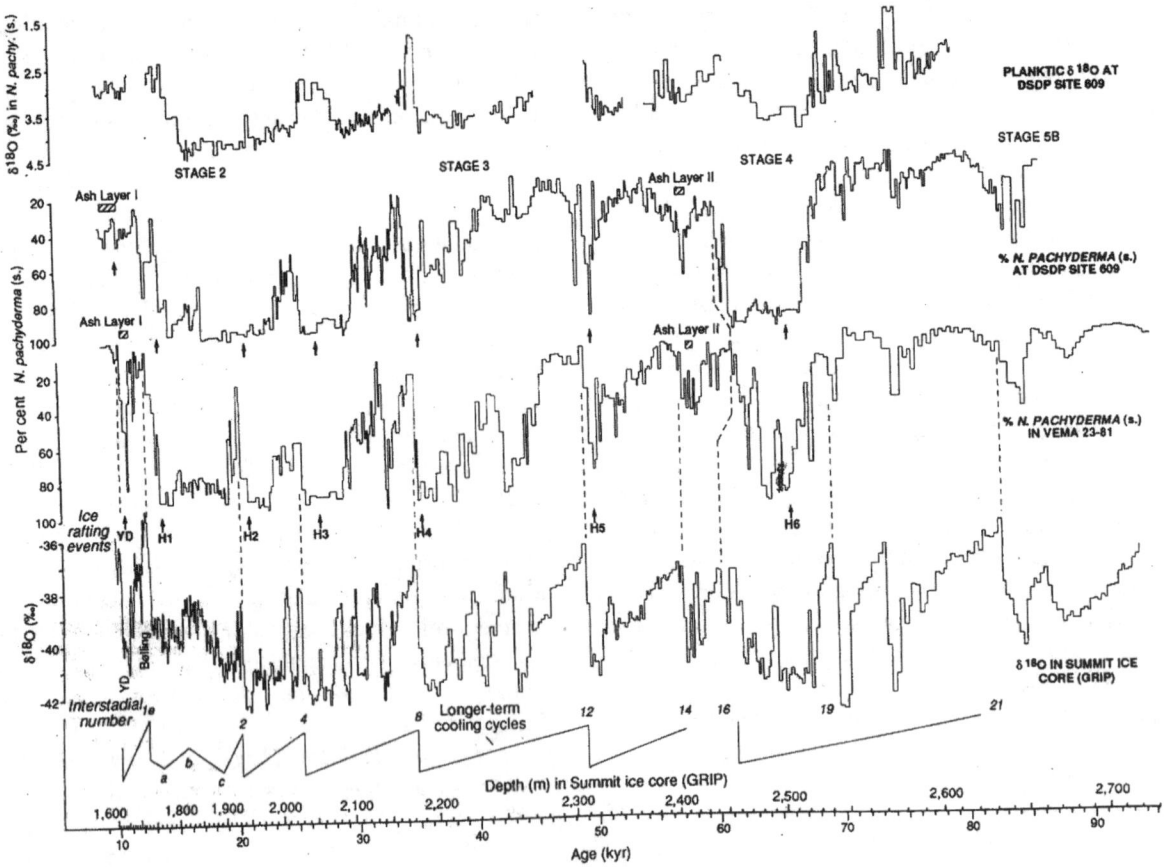

Figure 1. Sediment foraminiferal records from *Bond et al.* [1993] and ice core oxygen isotope records from *Dansgaard et al.* [1993] and *GRIP* [1993] as correlated by *Bond et al.* [1993]. The Younger Dryas (YD) and six Heinrich evens (H1–6) are indicated. Also shown at the bottom is the saw tooth Bond Cycle pattern of successively weaker interstadials associated with a sequence of D-O oscillations following a Heinrich event [see *Bond et al.*, 1993 for a more detailed discussion]. Taken from *Bond et al.* [1993].

eters. As snow covered forests are replaced by snow covered tundra or grass areas the surface albedo is considerably increased from ~0.4 to ~0.8 [*Federer*, 1968].

Recently, *Yoshimori et al.* [2002] examined the issue of glacial inception at 116 kyr BP in both the UVic ESCM as well as the Canadian Centre for Climate Modelling and Analysis (CCCma) atmospheric GCM. Initially, they integrated the UVic ESCM under both present-day and 116 kyr BP orbital forcing and atmospheric levels of CO_2. They then integrated the CCCma AGCM with prescribed SSTs and sea ice mask taken from the UVic ESCM. They examined the sensitivity to specified vegetation changes in the land surface component of the CCCma AGCM, based on climate changes induced at 116kyr BP. In the CCCma model, perennial snow cover occurred over northern Canada under 116 kyr BP orbital and CO_2 forcing with present-day *warm* sea surface conditions, and further expanded when 116 kyr BP *cool* sea surface conditions were applied. Modifying vegetation based on cooling during the summer induced by 116 kyr BP sea surface conditions, led to much larger areas of perennial snow cover. Their results suggested that the capturing of glacial inception at 116 kyr BP required the use of *cooler* sea surface conditions than the present [see also *Khodri et al.*, 2001; *Wang and Mysak*, 2002]. They also showed how feedbacks induced through changes in vegetation type were important in capturing a more realistic representation of glacial inception.

None of the studies mentioned above treated the terrestrial biosphere as a dynamic element of the climate system. Changes in the vegetation cover were either prescribed or obtained from offline integrations of a terrestrial vegetation model. *Crucifix and Loutre* [2002] were pioneers in this regard by including the dynamic vegetation model [VECODE—*Brovkin et al.*, 1997] into the MoBidiC earth system model of intermediate complexity. In MoBidiC, a zonally-averaged atmospheric model is coupled to a sectorially-averaged, multi-level ocean model and a simple thermodynamic-dynamic sea ice model. They showed that during ice age inception, a 14° southward shift of the northern treeline, combined with changes in summer sea ice and snow, lead to surface albedo changes that nearly quadrupled the orbital forcing effect.

The first results of the UVic model coupled to a land surface scheme and the TRIFFID dynamic global vegetation model were presented in *Meissner et al.* [2003b]. A southward shift of the northern treeline and a global decrease in vegetation carbon were observed in their ice age inception run. In tropical regions, up to 88% of broadleaf trees were replaced by shrub and C4 grasses. These changes in vegetation cover had a remarkable effect on the global climate: land related feedbacks doubled the atmospheric cooling during the ice age inception as well as the reduction of the AMO. The introduction of vegetation related feedbacks also increased the surface area with perennial snow significantly.

3.3. Dansgaard-Oeschger Oscillations, Heinrich Events, Bond Cycles

Early ice core records for the last glaciation have revealed large amplitude variability on the millennial timescale characterized by abrupt warming events (interstadials) lasting from several hundred to several thousand years [see *Weaver*, 1999 and *Clark et al.*, 2002b for further reviews]. These oscillations (Plate 5, Figure 1), known as Dansgaard/Oeschger (D-O) oscillations [after the pioneering work of *Oeschger et al.*, 1984 and *Dansgaard et al.*, 1984] are also apparent in North Atlantic sediment records [*Bond et al.*, 1993] suggesting a role or response of the ocean. The last such event, known as the Younger Dryas event, took place between 12,700 and 11,650 years BP and terminated abruptly within a few decades [*Dansgaard et al.*, 1989]. Evidence from the Santa Barbara basin [*Kennett and Ingram*, 1995; *Behl and Kennett*, 1996] and the Northeast Pacific [*Lund and Mix*, 1998] suggests that a signature of these D-O oscillations is also present in the Pacific, while further sediment analyses suggest they may be an inherent part of late [*Oppo et al.*, 1998] and early [*Raymo et al.*, 1998] Pleistocene climate.

In an attempt to provide a mechanism for the observed D-O variability, *Broecker et al.* [1990] proposed that during glacial times, when the northern end of the Atlantic Ocean was surrounded by ice sheets, a stable AMO was not possible. They further suggested that when the AMO was weakened or shut down and there were growing ice sheets, there was little oceanic salt export from the Atlantic to other world basins. Assuming a net evaporation over the North Atlantic, its salinity continued to increase with the moisture being deposited on

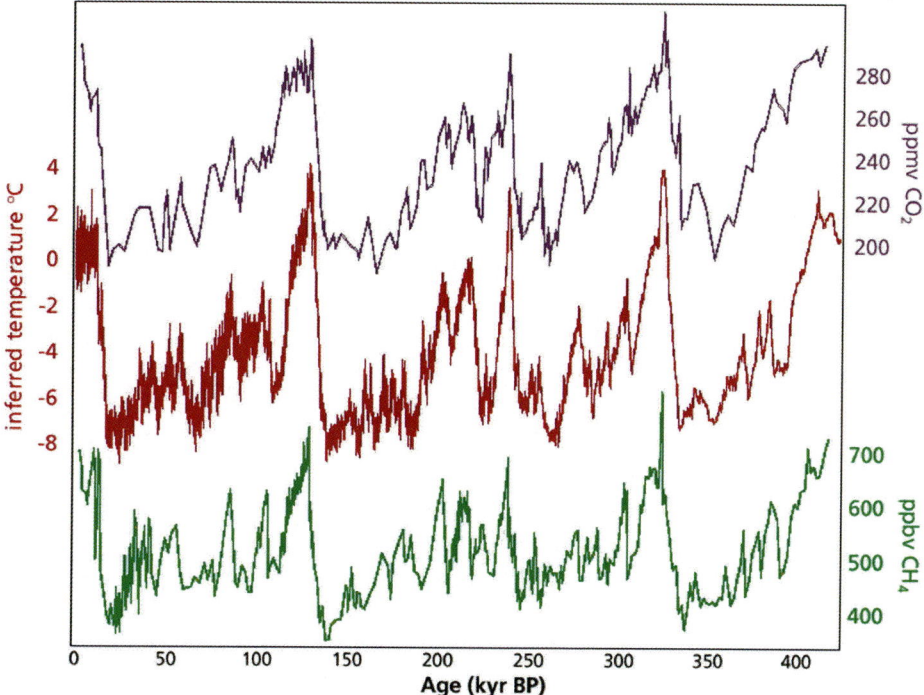

Plate 5. Variations in local Antarctic atmospheric temperature, as derived from oxygen isotope data, as well as concentrations of atmospheric CO_2 and CH_4 from Vostok, Antarctica ice core records. Redrawn from *Petit et al.* [1999].

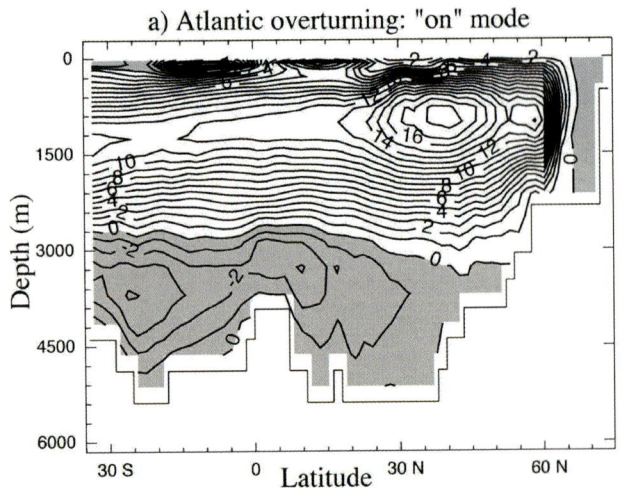

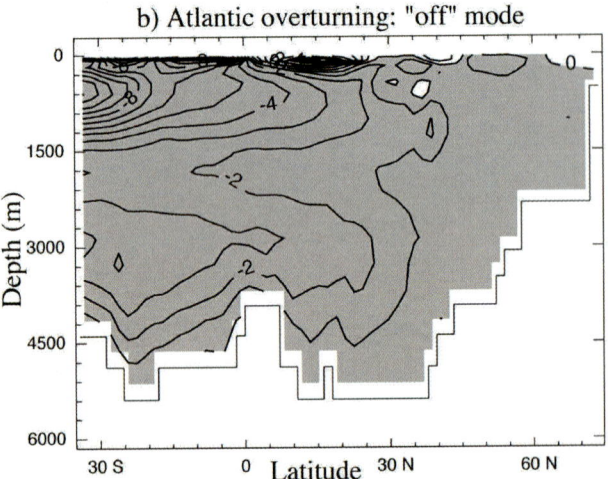

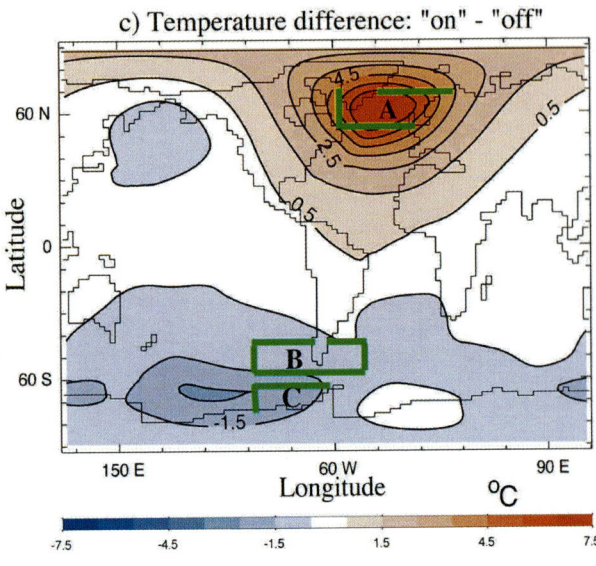

land as snow thereby growing the ice sheets. Upon reaching a critical salinity, deep convection and hence the AMO turned on, transporting and releasing heat to the North Atlantic, thereby melting back the ice sheets. The flux of fresh water into the North Atlantic from the melting ice sheets or enhanced ice berg calving [*Bond and Lotti*, 1995] eventually reduced or shut off the AMO and the process began anew.

Schmittner et al. [2002a], on the other hand, used the UVic model with an interactive continental ice sheet to show that the mechanism proposed by *Broecker et al.* [1990] was in fact opposite to what occurred within the coupled system. During the cold, glacial climate, when the AMO was active, the mass balance over continental ice sheets was positive. That is, rather than melting ice sheets, when the conveyor turned on they grew, since the warmer atmosphere allowed greater precipitation (in the form of snow). They also found a mechanism for D-O oscillations but it involved increased ice berg calving, consistent with *Bond and Lotti* [1995] that provided the freshwater source to the ocean, several hundred years after the conveyor turned on. This mechanism is more intuitively appealing than the stochastic resonance mechanism proposed by *Ganopolski and Rahmstorf* [2002] which relies upon the existence of an unknown 1500 year external periodic forcing.

Heinrich [1988] analyzed marine sediments in three cores from the North Atlantic and noted the presence of six anomalous concentrations of lithic fragments over the last glaciation. Since the source for these fragments was the land [and in particular Canada—*Bond et al.*, 1992], he argued that this provided evidence for six anomalous surges of icebergs into the North Atlantic (Plate 6). *Broecker et al.* [1992] noted that the so-called Heinrich events were even more striking when expressed in terms of the ratio of lithic fragments to the sum of lithic fragments and foraminifera shells, due to the low foraminifera counts in the Heinrich sediment layers. *MacAyeal* [1993] developed a simple model to illustrate the mechanism for Laurentide ice-sheet instability which ultimately gives rise to an ice-sheet surge and a Heinrich event. He argued that the Heinrich cycle consisted of two phases. In the growth phase the Laurentide ice sheet grew through snow accumulation while remaining attached to the bedrock. In the purge phase, he suggested that the high pressures caused by the deep ice sheet caused thawing near the base of the ice sheet thereby

Plate 6. (a) Active and (b) inactive mode of the AMO (Sv). 100 Sv-years of freshwater was added to region A to obtain the state in (b) with enhanced intrusion of AAIW and the absence of NADW. (c) Surface air temperature (°C) difference between the modes shown in (a) and (b). Freshwater was added to region B to get the hysteresis of Plate 7. Similar results are found when freshwater is added to region C. From *Weaver et al.* [2003].

allowing the ice sheet to surge seaward over a lubricated base through Hudson Strait. He pointed out that the resulting freshwater discharge into the North Atlantic would be of the order of 0.16 Sv (1 Sv ≡ $10^6 m^3 s^{-1}$) over a period as short as 250–500 years. Recent evidence [*Hewitt et al.*, 1997] suggests that Heinrich events (and associated ice rafted debris) also have a signature in the northeast Pacific Ocean.

As noted by *Bond et al.* [1993] and *Broecker* [1994] the Heinrich events, appearing about every 10,000 years, occur at the end of a sequence of D-O cycles during a prolonged cold period. *Bond et al.* [1993] further noted that the sequences of D-O oscillations tends to follow a saw-tooth cycle (now termed a Bond Cycle) with successive D-O oscillations involving progressively cooler interstadials (Fig. 1). They argued that this Bond Cycle was terminated by a Heinrich event, after which a rapid warming occurred and the process began anew.

Despite these fundamental advances in paleoclimate modelling and analysis, many challenges remain. In particular, the capturing of millennial timescale (D-O) variability and its packaging into Bond Cycles in cold climates, its association with Heinrich events, and its dependence on the mean climatic state remains one of the greatest challenges for paleoclimate modellers.

3.4. Deglaciation and Melt Water Pulse 1A

The melting of continental ice sheets during the last deglaciation provided a freshwater source to the ocean that also affected global sea level and the strength of the AMO. An exceptionally large melting event, inferred from far-field relative sea level records, occurred ~14,600 yr BP wherein global sea level rose by ~20m in less than 500 years. The ice sheet that served as the source for this event, known as meltwater pulse 1A (mwp-1A) has been the subject of some controversy since it was first identified from Barbados coral records [*Fairbanks*, 1989]. The Laurentide Ice Sheet was commonly cited as the most likely source for mwp-1A, but this raised the apparent conundrum of reconciling a large freshwater forcing (~0.5 Sv over several hundred years) to the North Atlantic Ocean with an active AMO and the associated warm climate of the Bølling-Allerød [*Clark et al.*, 1996]. Until recently, a satisfactory mechanism for the onset of the Bølling-Allerød event, conventionally considered as marking the termination of the last glacial period, had not been identified.

Clark et al. [2002a] provided compelling evidence that the partial collapse of the Antarctic ice sheet was responsible for a substantial component of mwp-1A. They noted that the ice sheet ablation responsible for mwp-1A would lead to a sea level fingerprint, that is a dramatic departure from eustasy, due primarily to a reduction in the gravitational attraction of water towards the location(s) of the source. A comparison of fingerprints predicted for various mwp-1A scenarios with available far-field relative sea level records supported an Antarctic source for mwp-1A and ruled out a sole Laurentide source for the event. Further evidence for an Antarctic source for mwp-1A comes from South Atlantic records of ice-rafted debris (IRD) derived from the Antarctic ice sheet, where *Kanfoush et al.* [2000] document the existence of one such IRD event (SA0) that correlates to mwp-1A and the Antarctic Cold Reversal.

Using the UVic model, *Weaver et al.* [2003] showed that with mwp-1A originating from the Antarctic ice sheet, consistent with the sea-level fingerprinting inferences of *Clark et al.* [2002a], the strength of the AMO increased simultaneously, albeit in a nonlinear fashion (their Figure 4), thereby warming the North Atlantic region and providing an explanation for the onset of the Bølling-Allerød warm interval. The established mode of an active AMO would then be able to respond to subsequent freshwater forcing from the Laurentide and Fennoscandian ice sheets, setting the stage for the Younger Dryas cold period.

4. THE COUPLING BETWEEN AAIW AND NADW

The formation of NADW in the North Atlantic is fed by the less dense Antarctic Intermediate Water (AAIW) and, in part, by the upper thermocline water [e.g., *Rintoul*, 1991]. AABW is formed largely in the Weddell and Ross Seas and spreads below the NADW, whereas the AAIW is thought to form most intensively around the southern tip of South America [*McCartney*, 1977] and spreads into the Atlantic above NADW. Numerous modelling studies have found the existence of another stable mode of the AMO, with no deep water formation in the North Atlantic [e.g., *Manabe and Stouffer*, 1988; *Stocker and Wright*, 1991]. As noted above, the UVic model also supports a stable mode without NADW formation (Plate 6b) and its subsequent evolution is governed by a hysteresis loop [*Schmittner et al.*, 2002a]. In this mode AAIW spreads into the Atlantic below NADW and the absence of active NADW formation leads to the deep North Atlantic being a stagnant, poorly ventilated basin. In the upper 2–3 km of the Atlantic, the modes are characterized by overturning circulations of opposite sign, with either a dominance of the AAIW cell over the NADW cell (inactive or *off* AMO mode; Plate 6b) or vice versa (active or *on* AMO mode; Plate 6a).

Historically, the NADW hysteresis behaviour and the associated transitions between its two stable modes (Plates 6a, b) have normally been found by applying freshwater perturbations within the North Atlantic. *Saenko et al.* [2003b], on the other hand, using the UVic model showed that very similar hysteresis behaviour can be obtained by applying freshwater forcing to the AAIW formation region in the Southern Ocean (Plate 7). This suggests that a tight coupling exists between the formation of NADW and AAIW and hence the existence of (and the transitions between) the two AMO modes. The transition between these modes is controlled by the relationship between the densities in the source regions of the AAIW and NADW water masses (Plate 7a). In particular, the application of a slowly-varying freshwater perturbation to the region of enhanced AAIW formation in the Southern Ocean (Plate 6c) leads to a hysteresis loop of the NADW circulation, with transitions between the active and inactive AMO modes occurring when the surface densities in the source regions of AAIW and NADW become comparable.

5. NADW UNDER A WARMING CLIMATE

Numerous coupled models have examined the transient response of the AMO to increasing greenhouse gases [see *IPCC*, 2001; *Crowley*, 1992; *Broecker*, 1998; *Stocker*, 1998]. Some models find no reduction of the AMO [*Latif et al.*, 2000; *Gent*, 2001] while others find a reduction in the AMO strength over the course of this century [*IPCC*, 2001]. Such a reduction leads to a negative feedback to anthropogenic warming in and around the North Atlantic through reducing the transport of heat from low to high latitudes. That is, SSTs are cooler than they would otherwise be if the AMO remained unchanged, so that warming is reduced over and downstream of the North Atlantic. It is important to note that in all models where the AMO weakens, warming still occurs downstream over Europe (see section 5.1) due to the radiative forcing associated with increasing greenhouse gases. In different models, the AMO-SST feedback is fundamentally determined by the competing effects of differential heat and freshwater flux forcing between low and high latitudes.

Evidence suggests [*Broecker*, 1998; *Stocker*, 1998] that a reduced cross-equatorial heat transport to the North Atlantic, associated with a reduced AMO would, at quasi-equilibrium, lead to enhanced SSTs in the South Atlantic through the seasaw effect. This is turn would suggest a positive feedback to warming in and around the South Atlantic if the AMO were to reduce.

Many future climate change projections show that once the radiative forcing is held fixed, reestablishment of the AMO occurs to a state similar to that for the present day. During this reestablishment phase, the AMO acts as a positive feedback to warming in and around the North Atlantic and, at equilibrium, there is close to zero net feedback. Whether or not reestablishment of the AMO occurs depends on the parameterization of ocean mixing [*Manabe and Stouffer*, 1999], as well as the emission rate and eventual stabilization scenario for atmospheric greenhouse gases [*Stocker and Schmittner*, 1997].

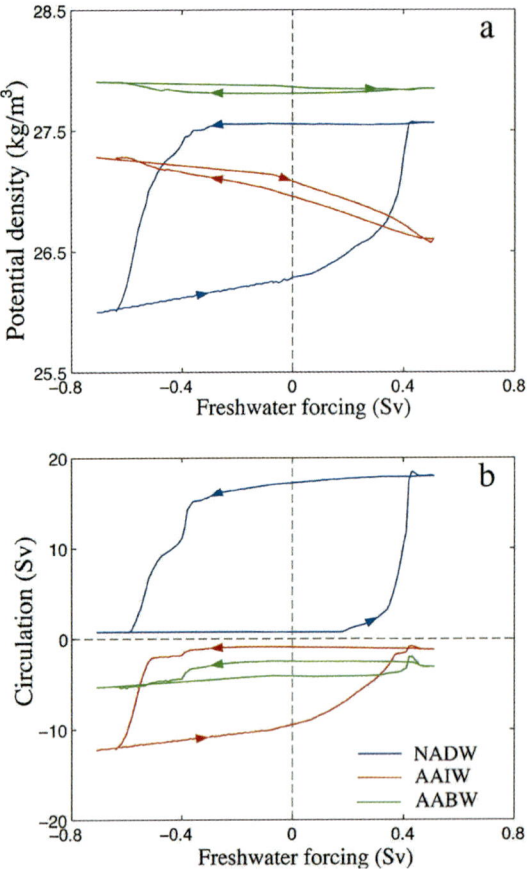

Plate 7. Hysteresis behaviour of AAIW (red), AABW (green) and NADW (blue). (a) surface densities σ_t (kg m^{-3}) in the source region of, and (b), strength of the overturning circulation associated with, these water masses. The hystereses were obtained by starting with an inactive mode of NADW formation (Plate 6b) and then externally-imposing a freshwater flux to region B (Plate 6c) that increased at a rate of 0.2 Sv per 1000 years until NADW formation turned on. The forcing was then reduced at the same rate back to 0 Sv in the 11,000 year total integration. From *Saenko et al.* [2003b].

This fundamental AMO-SST feedback is well understood although different models yield different projections in the strength of the AMO over the 21st century. This is not because the underlying feedback is unknown, but rather because the feedback is ultimately linked to the air-sea exchanges of heat and freshwater. The basic physics of the latter is still a matter of investigation and the present observational network to constrain the physics and its parameterization in models is far from ideal.

Freshwater export from the Arctic to the North Atlantic Ocean, either in the form of freshwater or sea ice, is governed by the net atmospheric moisture transport and runoff into the Arctic. Coupled modelling studies [see *IPCC*, 2001], suggest that a warmer climate is one in which the hydrological cycle, and hence runoff into and precipitation over the Arctic, will be enhanced. All coupled models project an increase in poleward transport of water vapor from low to high latitudes in the atmosphere under enhanced greenhouse conditions. In some, this leads to a freshening of the high latitude North Atlantic which reduces convection and hence the strength of the AMO. In others, no change in the overturning occurs as compensating feedbacks come into play.

Attention has largely been focused on the poleward atmospheric moisture transport in the northern hemisphere as this directly affects the properties of North Atlantic surface waters. Following from the analysis discussed in section 4, *Saenko et al.* [2003a] illustrated the sensitivity of the meridional moisture transport at mid-latitudes of the Southern Hemisphere on the AMO and hence its associated heat transport, using the UVic model. As noted above, an intensified hydrological cycle and subsequent North Atlantic freshening, through enhanced northward moisture transport, has often been evoked as the cause of the AMO weakening in transient climate change experiments. The results of *Saenko et al.* [2003a] point out that an intensified hydrological cycle can also stabilize the AMO by transporting more moisture southward. It is plausible, therefore, that the different AMO response in transient coupled model experiments arises from the relative differences in the poleward freshwater transport in the northern and southern hemisphere. Those models which show an initial weakening of the AMO may have a delayed response to southern hemisphere changes. The subsequent reestablishment would then correspond to the eventual realization of hemispherically symmetric perturbations to the poleward moisture transport.

5.1. Global Warming and the Next Ice Age

Conveying the significance of climate change to the public is a difficult task for scientists. Unfortunately, the media often is of little help here, especially when it sensationalizes the science with eye-catching headlines and confusing reportage. On January 14, 2001, for example, the Victoria Times Colonist published a 'Top Story' on page A3 headlined "*Study deflates global warming.*" Only nine days later, on January 23, 2001, a second front-page lead story offered this headline: "*Global warming severity grows.*" What would the average lay reader make of these apparently contradictory articles save that scientific understanding was swinging like a pendulum from one to another extreme. Their confusion here might even prompt them to dismiss the entire issue, since they would not have the benefit of reading the peer-reviewed literature from which the stories arose.

Another example even more relevant to the topic of this review concerns the erroneous and often cited consequence of anthropogenic fossil fuel emissions: *Global warming will cause the onset of the next ice age.* It is not surprising that such a catchy and sensational hypothesis would spread like a virus through media outlets, and end up as setting the story for a blockbuster movie: "The Day After Tomorrow", which hit movie theatres on May 28, 2004. But where did the virus emerge from?

The first major outbreak of the "Global Warming Induced Ice Age Syndrome" (GWIIAS) virus occurred in 1997 after the publication of two popular survey articles [*Rahmstorf,* 1997; *Calvin,* 1998], which noted the following: that abrupt climate change was prevalent in recent Earth history; in particular, there was some early evidence from the last interglacial, thought to be slightly warmer than preindustrial times [*IPCC,* 2001]; and that abrupt climate change was the norm [e.g., *GRIP,* 1993].

In the media this sequence of events mutated into a startling speculation: If global warming were to increase the hydrological cycle, enhanced freshwater discharge into the North Atlantic could shut down the AMO, such that temperatures would cool downstream over Europe—leading to the slow growth of glaciers and the onset of the next ice age.

Despite the best efforts of several scientists to develop a vaccine to prevent the spread of the GWIIAS virus [*Loutre and Berger,* 2000], a second, particularly virulent strain appeared on April 18th, 2002 following an Opinion Editorial in the New York Times, entitled "*The Heat Before the Cold*". The virus spread to the Discover Magazine in September 2002 [*Lemley,* 2002], and then to national newspapers (*e.g.,* the National Post in Canada published a piece in October, 2002 entitled: *Rumbling of a coming ice age*), and finally local newspapers.

Berger and Loutre [2002] refined their original vaccine although it still didn't eradicate the spread of the virus. They noted the importance of reduced June insolation for ice sheet growth, and suggested that "most CO_2 scenarios led to an exceptionally long interglacial from 5000 years before the present to 50,000 years from now ... with the next glacial maximum in 100,000 years. Only for CO_2 concentrations less than 220 ppmv was an early entrance into glaciation simulated". They further argued that the next glaciation would not likely occur for 50 millennia.

Over the years a relatively solid understanding of glacial inception has emerged (see section 3.2). Periods in the late quaternary that were characterized by reduced summer insolation in the northern hemisphere are those which are conducive to northern hemisphere glacial growth. The changes in seasonal insolation are then amplified through feedbacks (ice/snow albedo, vegetation, carbon cycle) operating within the coupled climate system. As discussed above and demonstrated in *Weaver et al.* [1998], at equilibrium, changes in the strength of the AMO have large local effects on North Atlantic surface air temperature and compensating effects elsewhere (especially the South Atlantic), such that when globally averaged, the surface air temperature is relatively insensitive to the strength of the AMO. Even in transient coupled GCM simulations where the AMO weakens over the 21st century, warming still occurs downstream over Europe.

In summary, despite the IPCC (2001) assessment—that "Most models show weakening of the Northern Hemisphere Thermohaline Circulation (THC), which contributes to a reduction of surface warming in the northern North Atlantic. Even in models where the THC weakens, there is still a warming over Europe due to increased greenhouse gases"—there is still a widespread misunderstanding of the possible consequence of climate change on the AMO. In particular, it is often touted, especially in the media, that global warming will cause the onset of the next ice age. *Weaver and Hillaire-Marcel* [2004] showed how this was impossible. Through analysis of the paleoclimate record as well as a number of climate model simulations, they further argued that it is very unlikely that the AMO will cease to be active in the near future. They also noted that a region where intermediate water formation may cease is in the Labrador Sea, although this has more minor consequences for climate than if deep water formation in the Nordic Seas were to shut down.

6. FUTURE CHALLENGES

Modelling efforts aimed at understanding the ocean's role in paleoclimate variability have evolved substantially over the last decade. Initially, stand-alone component models, such as oceanic or atmospheric GCMs with specified surface boundary conditions, were almost exclusively used. Today, a new breed of intermediate complexity models have evolved that incorporate many interactive subcomponent models, some or all of which are of reduced complexity. Advancements in computer technology over the next decade will likely allow scientists to surpass such intermediate complexity models with Earth System GCMs capable of efficiently undertaking ensembles of millennial timescale integrations.

While fundamental advances in our understanding of the processes involved in climate variability over the last glacial cycle have occurred, many challenges remain—perhaps the greatest of which involves our need to integrate the coupled climate system with fully-interactive ocean, atmosphere, sea ice, land surface, terrestrial vegetation, carbon cycle and continental ice sheet subcomponents over the last glacial cycle. The goal here is a multiple one: to reproduce the last glacial inception, to capture Dansgaard-Oeschger millennial timescale variability, its packaging into Bond Cycles and its association with Heinrich events, and to successfully reproduce the last deglaciation solely as a consequence of changing orbital parameters. While it might never be possible to exactly reproduce the proxy record over the last glacial cycle, reproducing the statistics of millennial timescale variability and Heinrich events, as well as the slow evolution of the glacial-interglacial cycle, is certainly within our grasp.

Throughout this review the importance of intermediate water formation, both in the Labrador Sea and Antarctic Circumpolar Current (and presumably the northwest Pacific), has been emphasized. In the case of the Labrador Sea, proxy records supported by modelling results suggest that LSW formation today is unique in the context of the last glacial cycle. In addition, coupled model studies suggest that changes in, and perhaps the transient cessation of, LSW formation are possible under anthropogenic warming. Combining what we know from the past, and what might possibly occur in the future, suggests that the Labrador Sea represents an ideal location for the concentration of observational efforts to monitor a possible oceanic response to anthropogenic climate change.

In the case of AAIW production, unfortunately, there is a dearth of observations both in the contemporary and proxy record. The model results discussed here suggest a fundamental link between AAIW and NADW formation including their simultaneous hysteresis behaviour. Intuitively this makes sense since AAIW waters affect the near surface water properties of the North Atlantic by becoming thermocline waters as they move northwards in the Atlantic. The historical focus on the North Atlantic and its freshwater budget as a trigger for abrupt climate change, both past and future, needs to be supplemented with more extensive observational and modelling efforts aimed at understanding the stability and variability of AAIW production in the southern hemisphere. This poses a particularly challenging logistical problem, but it is one that will likely reap enormous advances in our further understanding of the ocean's role in climate.

Finally, as Earth System GCMs come on stream over the next decade and focus on the climate of the last glacial cycle, intermediate complexity models will require new niches. In the United Nations Intergovernmental Panel on Climate Change (IPCC) fourth scientific assessment, to be completed in 2007, the leading climate models will include interactive terrestrial and ocean carbon cycles in which anthropogenic greenhouse gas and aerosol emissions, rather than concentration scenarios, will be specified. One might speculate that in the IPCC fifth scientific assessment, probably in ten years time, rather than specifying future emissions of atmospheric aerosols and greenhouse gases, the state of the art climate models will calculate these emissions internally through the interaction of coupled climate/socio-economic models. It is perhaps in this area that the intermediate complexity models will find their new niche.

Acknowledgments. I am grateful to both the Killam Foundation and the Canada Research Chair program for providing both research support and release time. Funding support from the NSERC and CFCAS is also acknowledged.

REFERENCES

Aagaard, K., J. H. Swift, and E C. Carmack (1985), Thermohaline circulation in the Arctic Mediterranean Seas. *J. Geophys. Res., 90*, 4833–4846.

Adkins, J. F., E. A. Boyle, L. Keigwin, and E. Cortijo (1997), Variability of the North Atlantic thermohaline circulation during the last interglacial period. *Nature, 390*, 154–156.

Alley, R., et. al. (1993), Abrupt increase in Greenland snow accumulation at the end of the Younger Dryas event. *Nature, 362*, 527–529.

Behl, R. J., and J. P. Kennett (1996), Brief interstadial events in the Santa Barbara basin, NE Pacific, during the past 60 kyr. *Nature, 379*, 243–246.

Berger, A., and M. F. Loutre (2002), An Exceptionally Long Interglacial Ahead? *Science, 297*, 1287–1288.

Bitz, C. M., M. M. Holland, A. J. Weaver, and M. Eby (2001), Simulating the ice-thickness distribution in a coupled climate model. *J. Geophys. Res., 106*, 2441–2463.

Bond, G. C., and R. Lotti (1995), Iceberg discharges into the North Atlantic on millennial timescales during the last glaciation. *Science, 267*, 1005–1010.

Bond, G., et al. (1992), Evidence for massive discharges of icebergs into the North Atlantic ocean during the last glaciation. *Nature, 360*, 245–249.

Bond, G., et al. (1993), Correlations between climate records from North Atlantic sediments and Greenland ice. *Nature, 365*, 143–147.

Bond, G., et al. (1997), A pervasive millennial-scale cycle in North Atlantic Holocene and glacial climates. *Science, 278*, 1257–1266.

Broecker, W. S. (1994), Massive iceberg discharges as triggers for global climate change. *Nature, 372*, 421–424.

Broecker, W. S. (1998), Paleocean circulation during the last deglaciation: A bipolar seesaw? *Paleoceanogr., 13*, 119–121.

Broecker, W. S., et al. (1990), A salt oscillator in the glacial Atlantic?, 1, The concept. *Paleoceanogr., 5*, 469–477.

Broecker, W. S., et al. (1992), Origin of the northern Atlantic's Heinrich events. *Clim. Dyn., 6*, 265–273.

Brovkin, V., A. Ganopolski, and Y. Svirezhev (1997), A continuous climate-vegetation classification for use in climate-biosphere studies. *Ecol. Modell., 101*, 251–261.

Calvin, W. H. (1998), The great climate flip-flop. *The Atlantic Monthly, 281*(1), 47–64, January.

Chapman, M. R., N. J. Shackleton, M. Zhao, and G. Eglinton (1996), Faunal and alkenone reconstructions of subtropical North Atlantic surface hydrography and paleotemperature over the last 28 kyr. *Paleoceanogr., 11*, 343–357.

Clark, P.U., et al. (1996), Origin of the first global meltwater pulse following the last glacial maximum. *Paleoceanogr., 11*, 563–577.

Clark, P. U., J. X. Mitrovica, G. A. Milne, and M. E. Tamisiea (2002a), Sea-level fingerprinting as a direct test for the source of global meltwater pulse IA. *Science, 295*, 2438–2441.

Clark, P. U., N. G. Pisias, T. F. Stocker, and A. J. Weaver (2002b), The role of the thermohaline circulation in abrupt climate change. *Nature, 415*, 863–869.

Claussen, M., et al. (2002), Earth system models of intermediate complexity: Closing the gap in the spectrum of climate system models. *Clim. Dyn., 18*, 579–586.

CLIMAP (1981), *Seasonal Reconstructions of the Earth's Surface at the Last Glacial Maximum*. Map Chart Series, MC-36, Geological Society of America, Boulder.

Cottet-Puinel, M., et al. (2004), Variation of Labrador Sea water formation over the last glacial cycle in a climate model of intermediate complexity. *Quat. Sci. Rev., 23*, 449–465.

Cox, P. M. (2001), Description of 'TRIFFID' Dynamic Global Vegetation Model. Hadley Centre Tech. Note, HCTN24.

Cox, P. M., et al. (2000), Acceleration of global warming due to carbon-cycle feedbacks in a coupled climate model. *Nature, 408*, 184–187.

Crowley, T. J., (1992), North Atlantic Deep Water Cools the Southern Hemisphere. *Paleoceanogr., 7*, 489–497.

Crucifix, M., and M. F. Loutre (2002), Transient simulations over the last interglacial period (126–115 kyr BP): feedback and forcing analysis. *Clim. Dyn., 19*, 417–433.

Cuffey, K. M., and S. J. Marshall (2000), Substantial contribution to sea-level rise during the last interglacial from the Greenland ice sheet. *Nature, 404*, 591–594.

Dansgaard, W., et al., (1984), North Atlantic climate oscillations revealed by deep Greenland ice cores, in *Climate Processes and Climate Sensitivity, Geophys. Monogr. Ser.*, vol. 29, edited by J. E. Hansen and T. Takahashi, pp. 288–298 AGU, Washington, DC.

Dansgaard, W., J. W. C. White, and S. J. Johnsen (1989), The abrupt termination of the Younger Dryas climate event. *Nature, 339*, 532–534.

Dansgaard, W., et al. (1993) Evidence for general instability of past climate from a 250-kyr ice-core record. *Nature, 364*, 218–220.

de Noblet, N. I., et al. (1996), Possible role of atmosphere-biosphere interactions in triggering the last glaciation. *Geophys. Res. Lett., 23*, 3191–3194.

Dickson, R. R., and J. Brown (1994), The production of North Atlantic deep water: Sources, sinks and pathways. *J. Geophys. Res., 99*, 12319–12341.

Dickson, R. R., et al. (1996), Long-term coordinated changes in the convective activity of the North Atlantic. *Prog. Oceanogr., 38*, 205–239.

Dong, B., and P. J. Valdes (1995), Sensitivity studies of Northern Hemisphere glaciation using an atmospheric general circulation model. *J. Climate, 8*, 2471–2496.

Duffy, P. B., M. Eby, and A. J. Weaver (1999), Effects of sinking of salt rejected during formation of sea ice on results of a global ocean-atmosphere-sea ice climate model. *Geophys. Res. Lett., 26*, 1739–1742.

Duffy, P. B., M. Eby, and A. J. Weaver (2001), Climate model simulations of effects of increased atmospheric CO_2 and loss of sea ice on ocean salinity and tracer uptake. *J. Climate, 14*, 520–532.

East, T. M., et al. (1999), Rapid fluctuations in sea level recorded at

Huon Peninsula during the penultimate deglaciation. *Science, 283,* 197–201.

Ewen, T. L., A. J. Weaver, and M. Eby (2004), Sensitivity of the inorganic carbon cycle to future climate warming in the UVic coupled model. Atmos.-Ocean, in press.

Fairbanks, R. G. (1989), A 17,000-year glacio-eustatic sea level record; influence of glacial melting rates on the Younger Dryas event and deep-ocean circulation. *Nature, 342,* 637–642.

Fanning, A. F., and A. J. Weaver (1996), An atmospheric energy moisture-balance model: climatology, interpentadal climate change and coupling to an OGCM. *J. Geophys. Res., 101,* 15111–15128.

Fanning, A.F., and A.J. Weaver (1997), Temporal-geographical meltwater influences on the North Atlantic Conveyor: Implications for the Younger Dryas. *Paleoceanogr., 12,* 307–320.

Federer, C. A. (1968), Spatial variation of net radiation, albedo and surface temperature of forests. *J. Appl. Meteor., 7,* 789–795.

Foley, J. A., et al. (1996), An integrated biosphere model of land surface processes, terrestrial carbon balance, and vegetation dynamics. *Glob. Biogeochem. Cycl., 10,* 603–628.

Gallimore, R. G., and J. E. Kutzbach (1995), Snow cover and sea ice sensitivity to generic changes in earth orbital parametrs. *J. Geophys. Res., 100,* 1103–1120.

Gallimore, R. G., and J. E. Kutzbach (1996), Role of orbitally induced changes in tundra area in the onset of glaciation. *Nature, 381,* 503–505.

Ganopolski, A., and S. Rahmstorf (2001), Simulation of rapid glacial climate changes in a coupled climate model. *Nature, 409,* 153–158.

Ganopolski, A., and S. Rahmstorf (2002), Abrupt glacial climate changes due to stochastic resonance. *Phys. Rev. Lett., 88(3),* 038501.

Ganopolski, A., S. Rahmstorf, V. Petoukhov, and M. Claussen (1998), Simulation of modern and glacial climates with a coupled global model of intermediate complexity. *Nature, 391,* 351–356.

Gent, P. R. (2001), Will the north Atlantic Ocean thermohaline circulation weaken during the 21st century? *Geophys. Res. Lett., 28,* 1023–1026.

GRIP, (1993), Climate instability during the last interglacial period recorded in the GRIP ice core. *Nature, 364,* 203–207.

Heinrich, H. (1988), Origin and consequences of cyclic ice rafting in the northeast Atlantic Ocean during the past 130,000 years. *Quat. Res., 29,* 143–152.

Hewitt, A. T., D. McDonald and B. D. Bornhold (1997), Ice-rafted debris in the North Pacific and correlation to North Atlantic climatic events. *Geophys. Res. Let., 24,* 3261–3264.

Hillaire-Marcel, C., A. de Vernal, G. Bilodeau, and A. J. Weaver (2001), Absence of deep-water formation in the Labrador Sea during the last interglacial period. *Nature, 410,* 1073–1077.

Holland, M. M., C. M. Bitz, M. Eby, and A. J. Weaver (2001), The role of ice ocean interactions in the variability of the North Atlantic thermohaline circulation. *J. Climate, 14,* 656–675.

Imbrie, J., et al. (1984), The orbital theory of Pleistocene climate: support from a revised chronology of the marine $\partial^{18}O$ record, in M*ilankovitch and climate, Part 1*, edited by A. L. Berger, J. Hays, G. Kukla and B. Salzman, pp 269–305, D. Reidel Dordrecht.

IPCC, (2001), *Climate Change 2001, The Scientific Basis. Contribution of Working Group I to the Third Scientific Assessment Report of the Intergovernmental Panel on Climate Change.* Edited by J. T. Houghton, et al., 881 pp., Cambridge University Press, Cambridge, UK.

Kanfoush, S. L., et al. (2000), Millennial-scale instability of the Antarctic Ice Sheet during the last glaciation. *Science, 288,* 1815–1819.

Keigwin, L., W. B. Curry, S. Lehman, and S. Johnsen (1994), The role of the deep ocean in North Atlantic climate change between 70 and 130 kyr ago. *Nature, 371,* 323–326.

Kennett, J. P., and B. L. Ingram (1995), A 20,000 year record of ocean circulation and climate change from the Santa Barbara basin. *Nature, 377,* 510–514.

Khodri, M., et al. (2001), Simulating the amplification of orbital forcing by ocean feedbacks in the last glaciation. *Nature, 410,* 570–574.

Ku, T.-L., M. A. Kimmel, W. H. Easton, and T. J. O'Neil (1974), Eustatic sea level 120,000 years ago on Oahu, Hawaii. *Science, 183,* 959–962.

Latif, M., E., Roeckner, U. Mikolajewicz, and R. Voss (2000), Tropical stabilization of the thermohaline circulation in a greenhouse warming simulation. *J. Climate, 13,* 1809–1813.

Lazier, J. N. R. (1980), Oceanographic conditions at Ocean Weather Station BRAVO 1964–1974. *Atmos.-Ocean, 18,* 227–238.

Lemley, B. (2002), The new ice age. *Discover, 23(9),* September.

Linsley, B. K. (1996), Oxygen-isotope record of sea level and climate variations in the Sulu Sea over the past 150,000 years. *Nature, 380,* 234–237.

Loutre, M. F., and A. Berger (2000), Future climatic changes: Are we entering an exceptionally long interglacial? *Clim. Change, 46,* 61–90.

Lund, D. C., and A. C. Mix (1998), Millennial-scale deep water oscillations: Reflections of the North Atlantic in the deep Pacific from 10 to 60 ka. *Paleoceanogr., 13,* 10–19.

MacAyeal, D. R. (1993), Binge/purge oscillations of the Laurentide ice sheet as a cause of the North Atlantic's Heinrich events. *Paleoceanogr., 8,* 775–784.

McCartney, M. S. (1977), Subantarctic Mode Water, in *A Voyage of Discovery: George Deacon 70th Anniversary Volume, Deep-Sea Research Supp.*, edited by M. V. Angel, pp. 103–119, Pergamon Press, Oxford.

McCartney, M. S. (1992), Recirculating components to the deep boundary current of the northern North Atlantic. *Prog. Oceanogr., 29,* 283–383.

McCartney, M. S., and L. D. Talley (1984), Warm-to-cold water conversion in the northern North Atlantic Ocean. *J. Phys. Oceanogr., 14,* 922–935.

Manabe, S., and R. J. Stouffer (1998), Two stable equilibria of a coupled ocean-atmosphere model. *J. Climate, 1,* 841–866.

Manabe, S., and R. J. Stouffer (1999), Are two modes of thermohaline circulation stable? *Tellus, 51A,* 400–411.

Marshall, J. and F. Schott (1999), Open-ocean convection; observations, theory, and models *Rev. Geophys., 37,* 1–64.

Marshall, S. J., and G. K. C. Clarke (1997), A continuum mixture model of ice stream thermomechanics in the Laurentide Ice Sheet, I, Theory. *J. Geophys. Res., 102*, 20599–20614.

Marsiat, I. M., A. Berger (1990), On the relationship between ice volume and sea level over the last glacial cycle. *Clim. Dyn., 4*, 81–84.

Matthews, H. D., A. J. Weaver, M. Eby, and K. J. Meissner (2003), Radiative forcing of climate by historical land cover change. *Geophys. Res. Lett., 30*(2), 1055, doi:10.1029/2002GL016098.

Matthews, H. D., et al. (2004), Natural and anthropogenic climate change: Incorporating historical land cover change, vegetation dynamics and the global carbon cycle. *Clim. Dyn., 22*, 461–479.

Meissner, K. J., A. Schmittner, A. J. Weaver, and J. F. Adkins (2003a), The ventilation of the North Atlantic Ocean during the Last Glacial Maximum—A comparison between simulated and observed radiocarbon ages. *Paleoceanogr., 18*(2), 1023, doi:10.1029/2002PA 000762.

Meissner, K. J., A. J. Weaver, H. D. Matthews, and P .M. Cox (2003b), The role of land-surface dynamics in glacial inception: A study with the UVic Earth System Model. *Clim. Dyn., 21*, 515–537.

Mitchell, J. F. B. (1993), Modelling of palaeoclimates: examples from the recent past. *Philos. Trans. Roy. Soc. Lon. B, 341*, 267–275.

Oeschger, H., et al. (1984), Late glacial climate history from ice cores, in *Climate Processes and Climate Sensitivity, Geophys. Monogr. Ser.*, vol. 29, edited by J.E. Hansen and T. Takahashi, pp. 299–306 AGU, Washington, DC.

Oglesby, R. J. (1990), Sensitivity of glaciation to initial snow cover, CO_2, snow albedo, and oceanic roughness in the NCAR CCM. *Clim. Dyn., 4*, 219–235.

Oppo, D. W., J. F. McManus, and J. L. Cullen (1998), Abrupt climate events 500,000 to 340,000 years ago: Evidence from subpolar North Atlantic sediments. *Science, 279*, 1335–1338.

Petit, J. R., et al. (1999), Climate and atmospheric history of the past 420,000 years from the Vostok ice core, Antarctica. *Nature, 399*, 429–436.

Petoukhov, V., et al. (2000), CLIMBER-2: A climate system model of intermediate complexity. Part I: Model description and performance for present climate. *Clim. Dyn., 16*, 1–17.

Phillipps, P. J., and I. M. Held (1994), The response to orbital perturbations in atmospheric model coupled to a slab ocean. *J. Climate, 7*, 767–782.

Pollard, D., and S. L. Thompson (1997), Driving a high-resolution dynamic ice-sheet model with GCM climate: ice-sheet initiation at 116,000 BP. *Ann. Glaciol., 25*, 296–304.

Rahmstorf, S. (1997), Ice-cold in Paris. *New Scientist, 153*, 08 February.

Rahmstorf, S., and A. Ganopolski (1999), Long-term global warming scenarios computeed with an efficient coupled climate model. *Climat. Chang., 43*, 353–367.

Raymo, M. E., et al. (1998), Millennial-scale climate instability during the early Pleistocene epoch. *Nature, **392***, 699–702.

Rind, D., D. Peteet, and G. Kukla (1989), Can Milankovitch orbital variations initiate the growth of ice sheets in a general circulation model? *J. Geophys. Res., 94*, 12851–12871.

Rintoul, S. R. (1991), South Atlantic interbasin exchange. *J. Geophys. Res., 96*, 2675–2692.

Royer, J. F., and M. Deque, 1983: Orbital forcing of the inception of the Laurentide ice sheet? *Nature, 304*, 43–46.

Royer, J. F., M. Deque, and P. Pestiaux (1984), A sensitivity experiment to astronomical forcing with a spectral GCM: simulation of the annual cycle at 125 000 BP and 115 000 BP, in *Milankovitch and climate, Part 2.* edited by A. L. Berger, J. Hays, G. Kukla, and B. Salzman, pp. 733–763, D. Reidel Dordrecht.

Saenko, O. A., A. J. Weaver, and A. Schmittner (2003a), Atlantic deep circulation controlled by freshening in the Southern Ocean. *Geophys. Res. Lett., 30*(14), 1754, doi: 10.1029/2003GL017681.

Saenko, O. A., A. J. Weaver, and J. M. Gregory (2003b), On the link between the two modes of the ocean thermohaline circulation and the formation of global-scale water masses. *J. Climate, 16*, 2797–2801.

Schlesinger, M. E., and M. Verbitsky (1996), Simulation of glacial onset with a coupled atmospheric general circulation/mixed layer ocean - ice-sheet/asthenosphere model. *Paleoclimates, 2*, 179–201.

Schmittner, A., M. Yoshimori, and A. J. Weaver (2002a), Instability of glacial climate in a model of the ocean-atmosphere-cryosphere system. *Science, 295*, 1489–1493.

Schmittner, A., K. J. Meissner, M. Eby, and A. J. Weaver, (2002b), Forcing of the deep ocean circulation in simulations of the Last Glacial Maximum. *Paleoceanogr., 17*(2), 1015, doi:10.1029/ 2001PA000633.

Schmitz, W. J Jr, and M. S. McCartney (1993), On the North Atlantic circulation. *Rev. Geophys.* 31, 29–49.

Shackleton, N. J. (1987), Oxygen isotopes, ice volume and sea level. *Quat. Sci. Rev., 6*, 183–190.

Stocker, T. F. (1998), The seesaw effect. *Science, 282*, 61–62.

Stocker, T. F., and D. G. Wright (1991), Rapid transitions of the ocean's deep circulation induced by changes in the surface water fluxes. *Nature, 351*, 729–732.

Stocker, T. F., and A. Schmittner (1997), Influence of CO_2 emission rates on the stability of the thermohaline circulation. *Nature, 388*, 862–865.

Stocker, T. F., D. G. Wright, and L. A. Mysak (1992), A zonally-averaged, coupled ocean-atmosphere model for paleoclimatic studies. *J. Climate, 5*, 773–797.

Stott, P. A., et al. (2000), External control of 20th century temperature by natural and anthropogenic forcings, *Science, 290*, 2133–2137.

Syktus, J., H. Gordon, and J. Chappell (1994), Sensitivity of a coupled atmosphere-dynamic upper ocean GCM to variations of CO_2, solar constant, and orbital forcing. *Geophys. Res. Lett., 21*, 1599–1602.

Talley, L. D., and M. S. McCartney (1982), Distribution and circulation of Labrador Sea Water. *J. Phys. Oceanogr., 12*, 1189–1205.

Vavrus ,S. J. (1999), The response of the coupled arctic sea ice-atmosphere system to orbital forcing and ice motion at 6 kyr and 115 kry Bp. *J. Clim., 12*, 873–896.

Verbitsky, M. Y., and R. J. Oglesby (1992), The effect of atmospheric carbon dioxide concentration on continental glaciation of the Northern Hemisphere. *J. Geophys. Res., 97*, 5895–5909.

Wang, Z., and L. A. Mysak (2002), Simulation of the last glacial inception and rapid ice sheet growth in the McGill Paleoclimate

Model. *Geophys. Res. Lett.*, *29*(23), 2102, doi:10.10129/2002GL 015120.

Weaver, A. J. (1999), Millennial timescale variability in ocean/climate models, in *Mechanisms of Global Climate Change at Millennial Time Scales. Geophys. Monogr. Ser.*, vol 112, edited by R. S. Webb, P. U. Clark, and L. D. Keigwin, pp. 285–300, AGU, Washington, D.C.

Weaver, A. J., and E. C. Wiebe (1999), On the sensitivity of projected oceanic thermal expansion to the parameterisation of sub-grid scale ocean mixing. *Geophys. Res. Lett.*, *26*, 3461–3464.

Weaver, A. J., M. Eby, A. F. Fanning, and E. C. Wiebe (1998), Simulated influence of carbon dioxide, orbital forcing and ice sheets on the climate of the last glacial maximum. *Nature*, *394*, 847–853.

Weaver, A. J., C. M. Bitz, A. F. Fanning, and M. M. Holland (1999), Thermohaline circulation: high latitude phenomena and the difference between the Pacific and Atlantic. *Ann. Rev. Earth Planet. Sci.*, *27*, 231–285.

Weaver, A. J., et al. (2001), The UVic Earth System Climate Model: Model description, climatology and application to past, present and future climates. *Atmos.-Ocean*, *39*, 361–428.

Weaver, A. J., O. A. Saenko, P. U. Clark, and J. X. Mitrovica (2003), Meltwater pulse 1A from Antarctica as a trigger of the Bølling-Allerød warm interval. *Science*, *299*, 1709–1713.

Weaver, A. J., and C. Hillaire-Marcel (2004), Ice Growth in the greenhouse: A seductive paradox but unrealistic scenario. *Geoscience Canada*, *31*, 77–85.

White, J. W. C. (1993), Don't touch that dial. *Nature*, *364*, 186.

Wiebe, E. C. and A. J. Weaver (1999), On the sensitivity of global warming experiments to the parametrisation of sub-grid scale ocean mixing. *Clim. Dyn.*, *15*, 875–893.

Wigley, T. M. L. (1998), The Kyoto Protocol: CO_2, CH_4 and climate implications, *Geophys. Res. Lett.*, *25*, 2285–2288.

Wigley, T. M. L., and S. C. B. Raper (1987), Thermal expansion of sea water associated with global warming. *Nature*, *330*, 127–131

Wood, R. A., A. B. Keen, J. F. B. Mitchell, J. M. Gregory (1999), Changing spatial structure of the thermohaline circulation in response to atmospheric CO_2 forcing in a climate model. *Nature*, *399*, 572–575.

Yoshimori, M., A. J. Weaver, S. J. Marshall, and G. K. C. Clarke (2001), Glacial termination: Sensitivity to orbital and CO_2 forcing in a coupled climate system model. *Clim. Dyn.*, *17*, 571–588.

Yoshimori, M., M. C. Reader, A. J. Weaver, and N. A. MacFarlane (2002), On the causes of glacial inception at 116KaBP. *Clim. Dyn.*, *18*, 383–402.

Yu, E.-F., R. Francois, and M. P. Bacon (1996), Similar rates of modern and last-glacial ocean thermohaline circulation inferred from radiochemical data. *Nature*, *379*, 689–694.

A. J. Weaver, School of Earth and Ocean Sciences, University of Victoria, PO Box 3055, Victoria, BC, V8W 3P6, Canada. (e-mail: weaver@uvic.ca)

Complexities in the Climate System and Uncertainties in Forecasts

Ronald G. Prinn

Department of Earth Atmospheric and Planetary Science, Massachusetts Institute of Technology, Cambridge, Massachusetts

The global atmosphere is a chemically complex and dynamic system, interacting both internally, mostly within the troposphere and stratosphere, and with the oceans, land, and living organisms. Its composition is changing today, and has also changed markedly over the last 400,000 yr. Current understanding of this complex system resulting from recent observations, theory, and laboratory and modeling studies is reviewed. Also, results are presented from the Integrated Global System Model (IGSM). This is a coupled model of economic development, atmospheric chemistry, climate dynamics and ecosystem processes that explores possible future changes in atmospheric composition and climate. The results of an uncertainty analysis involving hundreds of runs of the IGSM imply that, without specific mitigation policies, the global average surface temperature may rise between 1.0 and 4.9°C from 1990 to 2100 (95% confidence limits). Polar temperatures, absent policy, are projected to rise from about 1 to 12°C (95% limits) with obvious great risks for high latitude ecosystems and ice sheets at the high end of this range. Analysis of the Kyoto Protocol, and a more stringent climate mitigation policy, shows the difficulties in accounting simply for the effects of other greenhouse gases relative to carbon dioxide. Also, the greatest effect of these policies is to lower the probability of extreme changes as opposed to lowering the medians.

1. INTRODUCTION

Global climate change is the subject of policy debate within most nations, and of negotiations within the Conference of Parties (COP) to the Framework Convention on Climate Change (FCCC). Climate is usefully defined as the average of the weather we experience over a ten-or twenty-year time period. Long-term temperature and rainfall changes are typical measures of climate change, and these changes can be expressed at the local, regional, country, or global scale.

Fundamentally, global warming or cooling can be driven by any imbalance between the energy the Earth receives, largely as visible light, from the sun, and the energy it radiates back to space as invisible infrared radiation. The greenhouse effect is a warming influence caused by the presence in the air of gases and clouds, which are very efficient absorbers and radiators of this infrared. The greenhouse effect is opposed by substances at the surface (such as snow and desert sand) and in the atmosphere (such as clouds and white aerosols) that efficiently reflect sunlight back into space, and thus act as a cooling influence.

Easily the most important greenhouse gas is water vapor, which typically remains for a week or so in the atmosphere. Water vapor and clouds are handled internally in climate models, although with considerable uncertainty. Concerns about global warming, however, revolve around somewhat less important but much longer-lived greenhouse gases, especially carbon dioxide. The concentrations of carbon dioxide and some other long-lived gases (methane, nitrous oxide, chlorofluorocarbons, perfluorocarbons, hydrochlorofluorocarbons, sulfur hexafluoride) have increased substantially over the past two centuries due wholly or at least in part to human activity. When the concentration of a greenhouse gas increases (with no other changes occurring), it temporarily lowers the flow of infrared energy to space and increases the flow of infrared energy down toward the surface. The Earth then temporarily receives more energy—for example, 1 percent—than it radiates to space (1 percent is about the radiative energy imbalance caused by the rise of long-lived greenhouse gases from pre-industrial to present times). This small imbalance, which is often called "radiative forcing", tends to raise temperatures at the surface and in the lower atmosphere. The rate of surface temperature rise is slowed significantly by uptake of heat by the world's oceans. The greenhouse effect as quantified by radiative forcing is real and the physics relatively well understood. What is more uncertain, and the cause of much scientific debate, is the magnitude of the warming response of the complex system that determines our climate to radiative forcing. Feedbacks in the system can either amplify or dampen the response in ways that are not fully understood at present (*IPCC*, 2001a, b, c).

There are also complex connections between air pollution, land-use change and climate change that must be taken into consideration [*Prinn*, 1994]. Environmentally-significant chemical processes occurring in the atmosphere include those affecting urban air pollution, the ozone layer and the levels of radiatively active gases and particles. Carbon dioxide levels are governed by a complex set of natural and human activities at the land surface and in the oceans. Methane, another important greenhouse gas, has major natural and anthropogenic sources; it is destroyed largely by reaction with the hydroxyl radical in the troposphere. Both its sources and sinks are strongly influenced by human activity through agriculture and air pollution. Nitrous oxide, with a similar range of sources, and chlorofluorocarbons with purely anthropogenic sources, are also greenhouse gases; however, their potency is offset partially by the ozone they destroy. Ozone, a key chemical and protective ultraviolet shield, has a complex chemistry influenced by many other trace species and is also an important greenhouse gas. Many other trace gases play a key role in air pollution and climate through their influence on the concentrations of ozone, the hydroxyl radical, and methane. Gaseous sulfur compounds, both natural and anthropogenic, are oxidized to particulate sulfates that have an important effect on albedo, counteracting the influence of the greenhouse gases. Naturally and anthropogenically produced organic compounds contribute to the ozone and hydroxyl radical budgets as well as producing carbon-rich aerosols that can absorb and/or reflect sunlight. All these various aerosols can also indirectly change the reflection properties and lifetimes of clouds. Atmospheric chemistry is therefore closely linked to human activity, climate, and land use through numerous interactive environmental processes.

Clearly, integrating and understanding the diverse human and natural components of the problem is a must when informing policy development and implementation. As a result, climate research should focus on predictions of key variables such as rainfall, ecosystem productivity, and sea level that can be linked to estimates of economic, social, and environmental effects of possible climate change. Projections of emissions of greenhouse gases and atmospheric aerosol precursors should be related to the economic, technological, and political forces at play, and to the expected results of international agreements. In addition, such assessments of possible societal and ecosystem impacts, and analyses of mitigation strategies, should be based on realistic representations of the uncertainties of climate science. At MIT, we have developed an Integrated Global System Model (IGSM) to address some of these issues and to help inform the policy process.

2. INTEGRATED GLOBAL SYSTEM MODEL

The IGSM consists of a set of coupled sub-models of economic development and associated emissions, natural biogeochemical cycles, climate, air pollution, and natural ecosystems [*Prinn et al.*, 1999; *MIT*, 2004]. It is specifically designed to address key questions in the natural and social sciences that are amenable to quantitative analysis and are relevant to climate change policy. For example, how do the uncertainties in key component models, like those for ocean circulation and atmospheric convection, affect predictions? Are feedbacks between component models, such as climate-induced changes in oceanic and terrestrial uptake of carbon dioxide, atmospheric chemistry, and terrestrial emissions of methane, important? To answer such questions, and allow examination of a wide variety of proposed policies, the global system model must address not only the major human and natural processes involved in climate change [*Schneider*, 1992; *Prinn*, 1994; *IPCC*, 2001a, b, c], but it must also be computationally feasible for use in multiple 100-year predictions. The IGSM attempts to meet these challenging criteria.

Priorities must be set as to what is and what is not important in the model to ensure computational feasibility. Although we could have simply coupled together the most comprehensive existing versions of component models, the result would have been computationally so demanding that the many runs required to understand inter-model feedbacks, address a wide range of policy measures, and study uncertainty would not have been feasible with current computers. Thus a major challenge in developing our global system model involved determining what processes required inclusion in detail, and what processes could be omitted or simplified. The current structure of the IGSM is shown in Figure 1.

Human activity leads to emissions of chemically and radiatively important trace gases. The Emissions Prediction and Policy Analysis (EPPA) model incorporates the major relevant demographic, economic, world trade, and technical forces involved in this process at the national and international levels. Natural emissions of trace gases must also be predicted, and for this purpose the natural emissions model takes account of changes in both climate and ecosystem states in wetlands and soils around the world.

The coupled atmospheric chemistry and climate model is driven by a combination of anthropogenic and natural emissions. The essential components of this model are chemistry, atmospheric circulation, and ocean circulation, and each of these components by itself can require very large computer resources. The atmospheric chemistry component is modeled in sufficient detail to capture its sensitivity to climate and different mixes of emissions, and to address the effects on climate of policies proposed for control of air pollution, and vice-versa [*Wang et al.*, 1998; *Mayer et al.*, 2000]. For the atmosphere and ocean components, however, a computer has yet to exist that can adequately resolve the important small-scale eddies on a global scale for thousands of century-long integrations. Thus, simplified treatments of these circulations are included in the version of the model used for the Monte Carlo uncertainty and multiple policy applications discussed here [*Sokolov and Stone*, 1998; *Sokolov et al.*, 2003]. Linking these complex models together leads to many challenges, well illustrated by the failure of essentially all existing coupled ocean-atmosphere models (including ours) to simulate current climate over the globe very accurately without arbitrary adjustments to the air-to-sea fluxes of heat, water and (sometimes) momentum. Depending on the model, these non-physical adjustments range from relatively small values to significant fractions of the annual mean fluxes, and indicate deficiencies in the model formulations of air-sea interactions [*IPCC*, 2001a].

Urban air pollution has an impact on global chemistry, and thus on climate. Air pollution is a problem in a steadily growing number of giant cities worldwide. The emissions of chemicals important in air pollution and climate are often highly correlated due to shared generating processes like combustion, which produces CO_2, CO, NO_x, black carbon (BC) aerosols, and SO_x (which forms white sulfate aerosols). Also the atmospheric lifecycles of air pollutants such as CO, NO_x and volatile organic compounds (VOCs), and some climatically important species (e.g. CH_4, sulfate aerosols), both involve the fast photochemistry of the hydroxyl free radical (OH). Indeed, OH removes about 3.7 Pg per year of reactive trace gases from the atmosphere, which is similar to the total mass of carbon removed annually from the atmosphere by the land and ocean [*Ehhalt*, 1999; *Prinn*, 2003]. There is evidence, based on global measurements of the industrial gas methyl chloroform (CH_3CCl_3) which reacts with hydroxyl, that OH levels in the atmosphere are changing, perhaps due to the joint effects of air pollution and climate change [*Prinn et al.*, 2001]. The interconnection of pollution and climate hinders effective policy and decision making. To help unravel the interactions, an urban-scale air chemistry module was added to the IGSM to simulate the chemical reactions occurring in large cities, which are ignored in all other integrated assessment models. In a study of these phenomena [*Mayer et al.*, 2000], the inclusion of urban air chemistry processes led to lower global tropospheric NO_x, ozone, and OH concentrations, and hence to more methane than in forecasts that neglect urban processing. The resulting changes, interestingly, have limited impact on the global mean surface temperature because ozone decreases partially offset the methane increases. However, this advance in the IGSM framework enables the simultaneous consideration of control policies applied to local air pollution and global climate. It also provides the capability to assess the effects of air pollution on ecosystems, and to predict levels of irri-

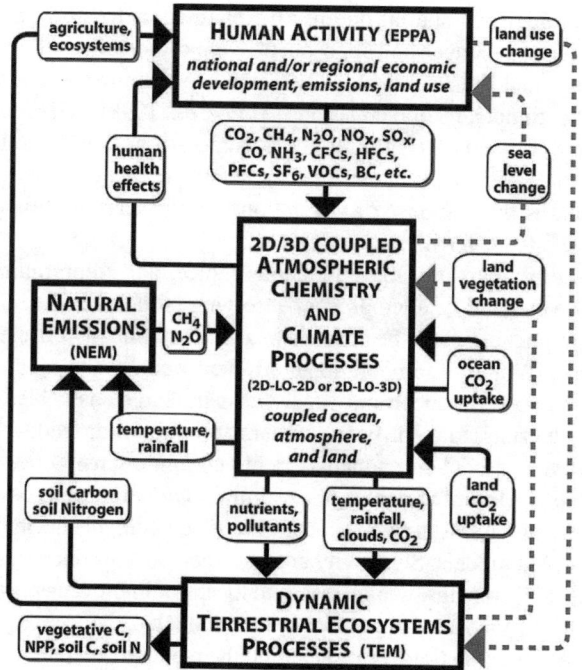

Figure 1. The schematic illustrates the framework and processes of the MIT Integrated Global System Model. Feedbacks between the component models that are currently included, or proposed for inclusion in the next generation, are shown as solid and dashed lines, respectively (adapted from *Prinn et al.*, 1999; MIT, 2004).

tants important to human health, such as ozone, in the growing number of megacities around the world.

The coupled chemistry/climate model outputs then drive a Terrestrial Ecosystems Model [TEM; *Xiao et al.*, 1998], which is capable of predicting vegetation properties including the difference between carbon uptake by plant photosynthesis and carbon loss by plant respiration (net primary production or NPP), land-atmosphere carbon dioxide (CO_2) fluxes, and soil composition. TEM outputs then feed back to the climate model, chemistry model, and Natural Emissions Model (NEM). Finally, NEM, which predicts wetland and soil emissions of methane (CH_4) and nitrous oxide (N_2O), is driven jointly by outputs from the TEM and climate models, and in turn provides inputs to both the atmospheric chemistry and climate models. Not included in the present IGSM, but planned for future versions, are the effects of anthropogenic changes in land cover on ecosystems, hence climate, and the effects of changes in climate and ecosystems on agriculture, hence anthropogenic emissions.

Our IGSM has several capabilities not currently present or simplified in other integrated models of the global climate system. The prediction of global anthropogenic emissions is based on a regionally disaggregated model of global economic growth. This allows for treatment of a shifting geographical distribution of emissions over time and changing mixes of emissions, both of which affect atmospheric chemistry. Also, our model of natural emissions is coupled to climate and land ecosystems models, which provide the needed explicit predictions of temperature, rainfall, and soil organic carbon concentrations.

To attain the necessary computational efficiency, while retaining plausible treatments of key climate processes, we use a longitudinally averaged statistical-dynamical climate model that is two-dimensional (2D) but that also resolves the land and ocean (LO) at each latitude (and so is referred to as the 2D-LO model). It includes a simplified ocean model, which is coupled to the atmosphere with representations of horizontal heat transport in the uppermost ("mixed") layer and heat exchange between the mixed layer and deep ocean [*Sokolov and Stone*, 1998]. It is capable of reproducing many characteristics of the current zonally-averaged climate, and its behavior and predictions are similar to those of coupled atmosphere-ocean three-dimensional general circulation models (GCMs), including the NASA Goddard Institute for Space Studies (GISS) GCM from which it is derived. Specifically, through appropriate choices of the oceanic vertical diffusivity and climate sensitivity in the 2D-LO model, this model can closely mimic the global mean temperature and sea level changes in transient forced runs of 11 GCMs investigated [*Sokolov and Stone*, 1998; *Forest et al.*, 2001; *Sokolov et al.*, 2003]. Climate sensitivity, which is the temperature difference between equilibrium runs of the model with reference and double-reference levels of CO_2, is altered by changing the model cloud parameterization. The 2D-LO model is about 20 times faster than the GISS GCM with similar latitudinal and vertical resolution.

By choosing this climate model, we are able to incorporate detailed atmospheric and oceanic chemistry interactively with climate in sufficient detail to allow study of key scientific and policy issues. However, to better address ocean circulation, the latest version of the IGSM includes a low-resolution three-dimensional (3D) ocean model that can simulate changes in the rate of the deep (thermohaline) oceanic circulation; changes that go untreated in the previous 2D-LO model [*Kamenkovich et al.*, 2002]. In common with several other 3D ocean models (*IPCC*, 2001a), this model shows a slowing down of the thermohaline circulation with rising carbon dioxide levels and is computationally efficient enough to be used in future sensitivity and uncertainty studies with the IGSM [*Kamenkovich et al.*, 2003]. All else being equal, the uptake of heat and carbon by the 3D model should evolve to be less than the 2D model, with higher predicted temperatures, particularly beyond the year 2100.

Fundamental ecosystem biogeochemical processes in 18 globally distributed terrestrial ecosystems are included in the ecosystem model. This model, with its significant biogeochemical and spatial detail, also enables us to study how changes in climate and atmospheric composition affect ecosystems, and the relationships between ecosystems and chemistry, climate, natural emissions [*Xiao et al.*, 1998], and (in the latest version) agriculture. The accuracy of this, and similar models, depends on how adequately they treat key processes, such as the response of vegetation to rising carbon dioxide levels (CO_2 fertilization effect).

In summary, the above judicious choices and compromises have enabled detailed process-resolving models for the relevant phenomena to be coupled in a computationally efficient form. With this computational efficiency comes the capability to perform uncertainty analyses using large ensembles of model runs, to identify and understand important feedbacks between model components, and to compute sensitivities of policy-relevant variables (*e.g.*, rainfall, temperature, ecosystem state) to assumptions in the various sub-components in the coupled models. Sensitivity analysis then enables our assessment of strengths, weaknesses, and means of improvements for future versions of the global system model.

At the same time, I must caution the reader to keep in mind that the various components of the IGSM do contain simplifications when interpreting its climate projections. The climate system contains a number of nonlinearities, feedbacks and critical thresholds that are not present in the IGSM, or most other models [*Rial et al.*, 2004]. In addition to the issues regarding the ocean noted above, the IGSM does not include irreversible conversions of ecosystems and the related releases of greenhouse gases, or the dynamics of the Greenland and Antarctic ice sheets (although it does include the melting of mountain glaciers). These omissions, however, are not expected to be important until after the year 2100.

In the following sections, I will review the application of the IGSM to the science of two policy-relevant issues. Alternative approaches to all these issues exist as reviewed in *IPCC* [2001a,b,c]. These alternative approaches involve the use of scenarios rather than specific economic model outputs to project emissions, and informal expert elicitation rather than large ensemble (Monte Carlo) model runs to estimate warming during the next 100 years. But the IGSM is arguably unique in its combination of scientific and economic detail, climate-atmospheric chemistry-ecosystem feedbacks, and computational efficiency.

3. UNCERTAINTY ANALYSIS

To help decision-makers evaluate how policies might reduce the risk of climate impacts, quantitative assessments of uncertainty in climate projections are very useful. *Webster et al.* [2002, 2003] use several hundreds of runs of the IGSM together with quantitative uncertainty techniques to achieve this assessment. Absent mitigation policies, the median projection in this study shows a global average surface temperature rise from 1990 to 2100 of 2.4°C, with a 95% confidence interval of 1.0°C to 4.9°C (*Webster et al.*, 2003). For comparison, the Third Assessment Report of the Intergovernmental Panel on Climate Change (IPCC) reports a range for the global mean surface temperature rise by 2100 of 1.4 to 5.8°C (1PCC, 2001a). Unfortunately, the IPCC does not provide likelihood estimates for this key finding although it does for others. This omission by the IPCC has been criticized by *Reilly et al.* [2001] and defended by *Allen et al.* [2001].

The IGSM physical climate model is flexible, which enables it to reproduce quite well the global behavior of coupled atmosphere-ocean general circulation models (AOGCMs) [*Sokolov & Stone*, 1998; *Sokolov et al.*, 2003]. This flexibility allows for analysis of the effect of some of the structural uncertainties present in existing AOGCMs (*Forest et al.*, 2001, 2002). The *Webster et al.* [2003] study includes uncertainties in natural and anthropogenic emissions of all climatically important gases and aerosols [*Prinn et al.*, 1999; *Reilly et al.*, 1999, *Webster et al.*, 2002], in critical atmospheric and oceanic interactions, and in carbon-cycle feedbacks in land ecosystems and the ocean. Their estimates of key climate model uncertainties (oceans, clouds or equivalently climate sensitivity, aerosols) are constrained by observations of the climate system for the period 1906-1995 [*Forest et al.*, 2002]. Also, uncertainty in emissions reflect errors in measurement of current emissions, and include expert judgment about variables that influence key economic projections [*Webster et al.*, 2002].

The probability distribution functions (pdfs) for the mean global surface temperature and sea level increases between 1990 and 2100 are shown in Figure 2 for two hypothetical cases: no explicit climate policy, and a stringent policy which is the same as that adopted by *Reilly et al.* [1999]. The stringent policy assumes that the Kyoto Protocol is implemented in 2010 by all the (largely developed) countries that were included in the original 1997 Protocol (the so-called Annex-B countries), including the U.S., which has not subsequently ratified the Protocol. For these Annex B countries, the stringent policy then assumes that they lower their emissions by 5% every 15 years after 2010 so that by 2100 their emissions are 35% below their 1990 levels. For the (largely developing) non-Annex B countries, including China and India, the stringent policy assumes that they lower their emissions in 2025 to 5% below their (unconstrained) 2010 levels. They then continue to lower their emissions by 5% every 15 years so that they are 30% below their 2010 levels in 2100. This stringent policy keeps CO_2 levels in the year 2100 in the median case to be just below 550 ppm (which is about twice the preindustrial CO_2 level).

The peak in the pdf denotes the most probable amount (i.e. mode) of warming or sea level rise. The total area under each pdf is (by definition) unity. The percentage probability of an increase in temperature or sea level being greater or less than

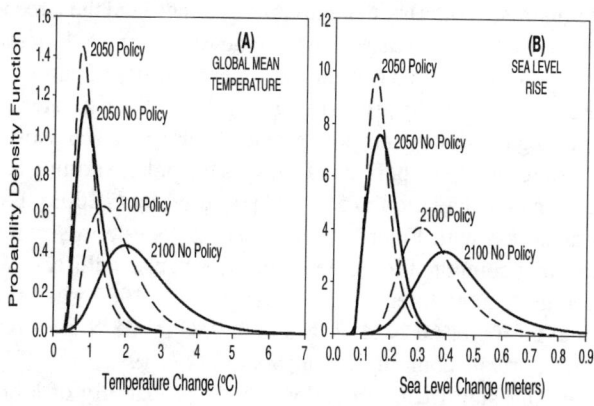

Figure 2. Probability density function (pdf) for the change in global-mean: (A) surface temperature, and (B) sea level rise, from 1990 to 2050 or 2100, estimated as a best-fit to 250 simulations using Latin Hypercube sampling from input pdfs for uncertain variables. The solid lines show the pdfs resulting from no explicit emissions restrictions, and the dashed lines are the pdfs under hypothetical emissions policy leading to steady levels of atmospheric CO_2 of about twice pre-industrial values (adapted from *Webster et al.*, 2003). The *IPCC* (2001, a, b, c) upper estimate of 5.8°C is well beyond the 95% confidence limit. The usefulness of the IPCC estimates is weakened by their omission of confidence limits on their results.

any particular value is given by 100 times the area to the right or left of the value respectively. For example, there is a 2.5% probability (i.e. 1 chance in 40) that the temperature rise from 1990 to 2100 will be greater than 4.9°C or less than 1.0°C (i.e. 95% range) in the no-policy case. When compared to the no-policy case, the stringent policy case has a median of only 1.6°C and 95% range of only 0.8 to 3.2°C. There is a 1 in 2 chance of warming exceeding 2.4°C in the no-policy case and only 1 in 7 chance in the stringent policy case. The policy obviously lowers the probability of large amounts of warming by very significant factors.

Of some interest is the fact that the pdfs for the policy and no-policy cases are not very different in 2050 and only become distinct in 2100. This is caused first by the fact that the world is already committed to significant future warming even at present greenhouse gas levels (due to the delay in warming associated with heat uptake by the ocean), and second because the policy case has its greatest reductions in emissions relative to the no-policy case only after 2050. Because the mitigating effects of the policy only appear very distinctly in the pdf after 50 years, this implies that there is significant risk in waiting for very large warming to occur before taking action.

To better appreciate the risks in the no-policy case, it is also important to examine the latitudinal distribution of the projected warming. In common with other climate models, the computed temperature increases in polar regions are much greater than those in equatorial regions (see Figure 3 for the no-policy case). Polar regions contain vulnerable ecosystems (permafrost, tundra) with large carbon storage, and the Greenland and Antarctic ice sheets with large water storage. Release of some of this stored carbon and water is clearly of concern. The 1 chance in 40 of warming of 8 to 12°C or greater in polar regions for the no-policy case (Figure 3) is worrisome in this respect. The policy case lowers the polar warmings in the 1 in 40 calculation to 5 to 7°C [*Webster et al.*, 2003]. For comparison with Figure 3, the observed long-term surface warming computed from the differences between thirty-year average temperatures centered on 1915 and 1985 was only about 0.6°C for 60°N to 65°N and 0.75°C for 45°S-50°S (C. Forest, private communication; *Jones et al.*, 1999).

Similar significant reductions in the probability of large and risky amounts of sea level rise due to the hypothetical policy are also evident in Figure 2. Emissions reductions will therefore lower the chance of exceeding an extreme climate outcome but not eliminate the risk entirely, and analysis of the reduction in probability is an important policy consideration. We emphasize that due to the simplifications or possible omissions in the various IGSM components, this exercise of determining probabilities has its own implicit uncertainties. Hence it is the qualitative (rather than the exact quantitative) results that should be emphasized. Nevertheless, future cli-

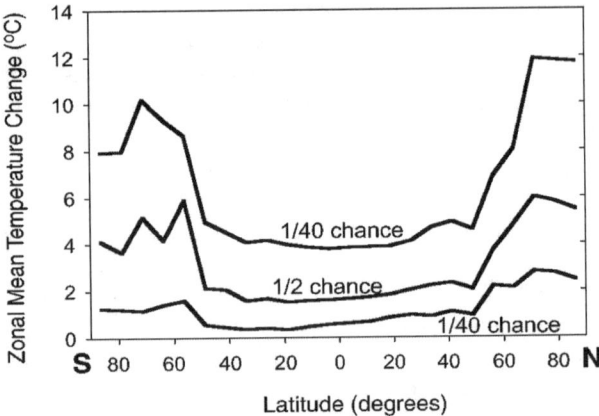

Figure 3. Zonal mean temperature change in surface warming by latitude band between 1990 and 2100 in the case assuming no explicit policy in Figure 2. There is one chance in 40 of being above or below the upper and lower curves and one chance in 2 of being above or below the middle curve respectively (adapted from *Webster et al.*, 2003).

mate assessments would better serve the policy process by including formal analysis of uncertainty for key projections, with an explicit description of the methods used [*Reilly et al.*, 2001; *Allen et al.*, 2001]. The great value of such formal probability analyses for policy decision-making lies in the ability to compare relative risks of various policies, which are less affected by the above climate system model uncertainties. These applications to policy are addressed more explicitly by *Jacoby* [2004].

4. ASSESSMENT OF POLICIES

The IGSM can also be used to assess specific policy proposals in detail. Under the FCCC, the Kyoto Protocol, which addresses the period up to 2010, allows reductions in emissions of several radiative gases to be credited against a CO_2-equivalent emissions cap. For this purpose, the Protocol adopts the concept of Global Warming Potentials (GWPs), which are the masses of CO_2 required to match the radiative forcing of climate changes caused by unit masses of each non-CO_2 gas over a given time horizon (100 years in the Protocol). A study using the IGSM [*Reilly et al.*, 1999] showed that, absent any policy, total CO_2 plus CO emissions in 2100 would be about 19 Pg carbon/year compared to about 8 Pg carbon/year in 1990. They also show that economic analyses that leave out gases other than CO_2 and CO, err in several important ways: reference (no policy) emissions are understated, allowable emissions in the period up to 2010 are too low, and opportunities to reduce emissions of other gases are not considered in abatement options. Although the effects are partially offsetting, the inclusion of other relevant gases (e.g. methane) and carbon sinks (e.g. forests) reduces the costs in 2010 of achieving CO_2 emissions reductions specified by the Kyoto agreement. Specifically, while essentially the same reduction in warming is obtained either by control of fossil CO_2 only, or by control of fossil CO_2 and other gases and sinks, the fossil CO_2-only approach could cost over 60% more in the year 2010. Two extensions of the Kyoto Protocol are considered from 2010 to 2100: a policy where only Kyoto signatories (Annex B countries) keep their emissions constant at 2010 levels resulting in global CO_2 plus CO emissions in 2100 of about 13 Pg carbon/year; and a more stringent policy involving all countries (discussed in the previous section) that results in global CO_2 plus CO emissions in 2100 of only 3 to 7 Pg carbon/year, depending on the amounts of reductions in non-CO_2 and non-CO gases used. Extending the Kyoto Protocol using the more stringent emissions policy, in which reductions in non-CO_2 gases like methane (CH_4) are large, shows that the use of GWPs as defined in the Protocol leads to considerably less mitigation of climate change for CO_2-only control than for the supposedly equivalent multi-gas strategies (i.e. in Figure 4, which shows projected warming from 1990 to 2100, cases 2' and 3' differ significantly). This illustrates the limits of GWPs as a tool for political decision and the need to develop improved methods especially for use in the longer term. The effects of the Kyoto Protocol on temperatures in 2010 are not distinct from the no-policy (reference) case but become very obvious in the extensions after about 2040 (Figure 4). Note once again the significantly greater predicted temperature changes in polar regions compared to the global average in the cases in Figure 4.

5. CONCLUSIONS

Recent research has solidified the need to consider interactions between human activity, atmospheric chemistry and physics, climate dynamics, and ecosystem processes in order to understand how the earth responds to changes in air pollutant and greenhouse gas emissions. With the above two examples, I have attempted to demonstrate that integrated assessments using fully interactive models like the IGSM are invaluable tools for addressing a variety of important issues. These include discovery and elucidation of previously undetected feedbacks between natural and human-related components of the climate system, objective and quantitative assessments of uncertainty in forecasts of key climate and economic variables, and critical analysis of specific policy proposals [see also *Jacoby*, 2004].

However, nonlinear processes are very important in the climate system, and while the IGSM contains some of these (e.g. climatically modulated greenhouse gas fluxes from current ecosystems, sea ice cover, mountain glaciers, and thermohaline circulation [in 3D ocean version]), it does not contain others (e.g. climatically modulated ecosystem transitions, ice sheet ablation, volcanic eruptions, and solar variability). And even if we were to add treatments of all these missing processes, current knowledge of the stability of the great ice sheets, stability of the thermohaline circulation, ecosystem dynamics, connections between climate change and severe storms, future technological innovation, human population dynamics, and political change are all sufficiently inadequate to allow "surprises" not currently evident from our (or indeed all other) model studies to occur. Therefore, as with all investigations and simulations of complex and only partially understood systems, the results presented here must be treated with appropriate caution.

Looking to the future, there are major research challenges to be faced if the uncertainties in climate prediction are to decrease. Recently, a group of climate scientists has compiled a list of recommended research areas, which are useful to paraphrase here [*Rial et al.*, 2004]:

(1) Explore the limits to climate predictability in view of the nonlinearities, feedbacks and critical thresholds in the climate system;

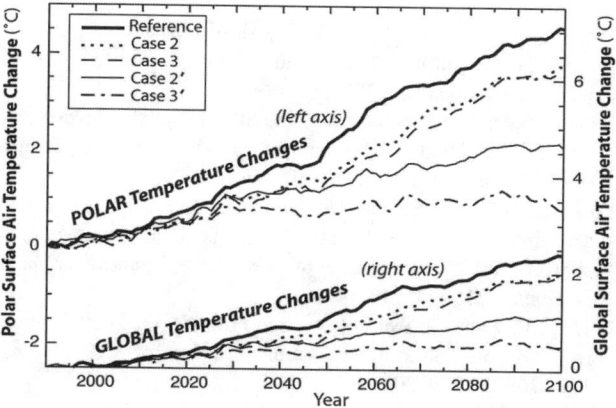

Figure 4. Predicted global and polar temperature changes from 1990 levels for four different policy cases. With the Kyoto Protocol targets simply extended unchanged to 2100 (cases 2 and 3) there is a 17% reduction in warming from the reference. With more stringent policies after 2010 (cases 2' and 3'), which approach a post-2100 steady-state of about twice pre-industrial atmospheric CO_2 concentrations (i.e. 550 ppm), warming is reduced by about 55 and 85% respectively from the reference. Cases 2 and 2' convert non-CO_2 emissions reductions into CO_2 emissions reductions for use in the chemistry and climate model using the approximate approach of Global Warming Potentials (GWPs; *IPCC, 2001a*). Cases 3 and 3' more correctly consider reductions in all gases explicitly in the chemistry and climate model (adapted from *Reilly et al.*, 1999).

(2) Better understand the nonlinear response of climate to changes in solar forcing caused by Earth's orbital variations;

(3) Search for additional measures of past climate changes and use these to develop and test climate models;

(4) Develop three-dimensional climate models which incorporate the dynamics of the cryosphere, hydrosphere, atmosphere and pedosphere, and which are capable of addressing multi-millennial time scales;

(5) Better understand the connections between, and variability of, ocean-atmosphere interactions exemplified by the El Nino-Southern Oscillation and the North Atlantic Oscillation;

(6) Improve techniques for measuring or deducing the spectral variations of the solar output at 10 to 100 year time scales;

(7) Understand better the physics of the deep ocean (thermohaline) circulation and its role in major and rapid climate changes;

(8) Evaluate better the response of climate to biospheric and cloud microphysical variations including those caused by human activity (land-use change and aerosol emissions);

(9) Investigate the benefits and risks of deliberate large-scale interventions in the climate system exemplified by carbon sequestration and water management;

(10) Identify and further investigate those ocean and land areas (e.g. high latitudes) particularly sensitive to climate change;

(11) Investigate nonlinear interactions between atmospheric chemistry, climate and ecosystem fluxes of radiatively and chemically important trace gases.

To these challenges in the natural sciences, we should add the need for substantial research to better understand the economic, technical, and political drivers of human activity affecting, and affected by, climate change.

Acknowledgements. This research was supported by the U.S Department of Energy, U.S. National Science Foundation, and the Industry Sponsors of the MIT Joint Program on the Science and Policy of Global Change (Alstom Power (France), American Electric Power (USA), BP p.l.c. (UK/USA), ChevronTexaco Corporation (USA), DaimlerChrysler AG (Germany), Duke Energy (USA), J-Power (Electric Power Development Co., Ltd.) (Japan), Electric Power Research Institute (USA), Electricité de France, ExxonMobil Corporation (USA), Ford Motor Company (USA), General Motors (USA), Mirant (USA), Murphy Oil Corporation (USA), Oglethorpe Power Corporation (USA), RWE/Rheinbraun (Germany), Shell International Petroleum (Netherlands/UK), Statoil (Norway), Tennessee Valley Authority (USA), Tokyo Electric Power Company (Japan), TotalFinaElf (France), Vetlesen Foundation (USA), We Energies (USA)). I thank Drs. S. Wofsy and S. Sparks and an additional anonymous reviewer for comments which improved the manuscript.

REFERENCES

Allen, M., Raper, S. and Mitchell, J., 2001. Uncertainty in the IPCC's Third Assessment Report. *Science* **293**: 430–433.

Ehhalt, D. H., 1999. Gas phase chemistry of the troposphere. *Topics in Physical Chemistry,* **6**: 21–109.

Forest, C. E., Allen, M.R., Sokolov, A. P. and Stone, P. H., 2001. Constraining climate model properties using optimal fingerprint detection methods. *Climate Dynamics,* **18**: 277–195.

Forest, C. E., Stone, P. H., Sokolov, A. P., Allen, M. R. and Webster, M., 2002. Quantifying uncertainties in climate system properties using recent climate observations. *Science,* **295**: 113–117.

Intergovernmental Panel on Climate Change (IPCC). 2001a. *Climate Change 2001: The Scientific Basis.* Houghton, J. T., Ding, Y., Griggs, D. J., Noguer, M., van der Linden, P. J., and Xiaosu, D. (eds.). Cambridge University Press, U.K. 944 p.

Intergovernmental Panel on Climate Change (IPCC). 2001b. *Climate Change 2001: Impacts, Adaptation and Vulnerability.* McCarthy, J. J., Canzian, O. F., Leary, N. A., Dokken, D. J., and White, K. S. (eds.). Cambridge University Press, U.K. 1000 p.

Intergovernmental Panel on Climate Change (IPCC). 2001c. *Climate Change 2001: Mitigation.* Metz, B., Davidson, O., Swart, R. and Pan, J. (eds.). Cambridge University Press, U.K. 700 p.

Jacoby, H., 2004. Modeling Human-Climate Interaction. In *State of the Planet,* AGU Monograph, Eds. C. Hawksworth and S. Sparks, in press.

Jones, P. D., New, M., Parker, D. E., Martin, S. and Rigor, I. G., 1999. Surface air temperature and its changes over the past 150 years. *Rev. Geophysics,* **37**: 173–199.

Kamenkovich, I. V., Sokolov, A.P. and Stone, P., 2002. An efficient climate model with a 3D ocean and statistical-dynamical atmosphere. *Climate Dynamics,* **19**: 585–598.

Kamenkovich, I. V., Sokolov, A. P. and Stone, P., 2003. Feedbacks affecting the response of the thermohaline circulation to increasing CO_2: A study with a model of intermediate complexity. *Climate Dynamics,* **21**: 119–130.

Mayer, M., Wang, C. Webster, M., and Prinn, R. G., 2000. Linking local air pollution to global chemistry and climate. *J. Geophys. Res.,* **105**: 20,869–20,896.

MIT Joint Program on the Science and Policy of Global Change, 2004. Website: http://web.mit.edu/globalchange/www/.

Prinn, R. G., 2003. The cleansing capacity of the atmosphere. *Ann. Reviews Environ. and Resources,* **28**: 29–57.

Prinn, R. G., 1994. The interactive atmosphere: Global atmospheric-biospheric chemistry, *Ambio,* **23**: 50–61.

Prinn, R. G., Huang, J., Weiss, R., Cunnold, D., Fraser, P., Simmonds, P., McCulloch, A., Harth, C., Salameh, P., O'Doherty, S., Wang, R., Porter, L., and Miller, B., 2001. Evidence for substantial variations of atmospheric hydroxyl radicals in the past two decades. *Science,* **292**: 1882–1888.

Prinn, R. G., Jacoby, H., Sokolov, A., Wang, C., Xiao, X., Yang, Z., Eckaus, R., Stone, P., Ellerman, A. D., Melillo, J., Fitzmaurice, J., Kicklighter, D., Holian, G. and Liu, Y., 1999. Integrated global system model for climate policy assessment: Feedbacks and sensitivity studies. *Climatic Change,* **41**: 469–546.

Reilly, J., Prinn, R., Harnisch, J., Fitzmaurice, J., Jacoby, H., Kicklighter, D., Melillo, J., Stone, P., Sokolov, A. and Wang, C., 1999. Multi-gas Assessment of the Kyoto Protocol. *Nature,* **401**: 549–555.

Reilly, J., Stone, P., Forest, C., Webster, M., Jacoby, H., and Prinn, R., 2001. Uncertainty and climate change assessments. *Science,* **291**: 430–433.

Rial, J. A., Pielke, R. A. Beniston, M., Claussen,M., Canadell, J., Cox, P., Held, H., Noblet-Ducoudre, N. de, Prinn, R. G., Reynolds, J. F. and Salas, J. D., 2004. Nonlinearities, feedbacks and critical thresholds within the Earth's climate system. *Climatic Change,* in press.

Schneider, S. H., 1992. Introduction to climate modeling, in: *Climate System Modeling.* Trenberth, K. (ed.), Cambridge University Press, Cambridge, U.K., pp. 3–26.

Sokolov, A., and Stone, P., 1998. A flexible climate model for use in integrated assessments. *Climate Dynamics,* **14**: 291–303.

Sokolov, A., Forest, C. E. and Stone, P., 2003. Comparing oceanic heat uptake in AOGCM transient climate change experiments. *J. Climate,* **16**: 1573–1582.

Wang, C., Prinn, R. and Sokolov, A., 1998. A global interactive chemistry and climate model: Formulation and testing. *J. Geophysical Research* **103**: 3399–3417.

Webster, M. D., Babiker, M., Mayer, M., Reilly, J. M., Harnisch, J., Sarofim, M. C., and Wang, C., 2002. Uncertainty in emissions projections for climate models. *Atmos. Environ.,* **36**: 3659–3670.

Webster, M. D., C. E. Forest, J. M. Reilly, M. Babiker, D. Kicklighter, M. Mayer, R. G. Prinn, M. Sarofim, A. Sokolov, P. H. Stone and C. Wang, 2003. Uncertainty analysis of climate change and policy response. *Climatic Change,* **61**: 295–320.

Xiao, X., Melillo, J., Kicklighter, D., McGuire, A., Prinn, R., Wang, C., Stone, P. and Sokolov, A., 1998. Transient climate change and net ecosystem production of the terrestrial biosphere. *Global Biogeochemical Cycles,* **12**: 345–360.

Ronald G. Prinn, Department of Earth Atmospheric and Planetary Sciences, Massachusetts Institute of Technology, 77 Massachusetts Avenue, Cambridge, MA 02139-4307, USA.

Modeling Human-Climate Interaction

Henry D. Jacoby

Massachusetts Institute of Technology

If policymakers and the public are to be adequately informed about the climate change threat, climate modeling needs to include components far outside its conventional boundaries. An integration of climate chemistry and meteorology, oceanography, and terrestrial biology has been achieved over the past few decades. More recently the scope of these studies has been expanded to include the human systems that influence the planet, the social and ecological consequences of potential change, and the political processes that lead to attempts at mitigation and adaptation. For example, key issues—like the relative seriousness of climate change risk, the choice of long-term goals for policy, and the analysis of today's decisions when uncertainty may be reduced tomorrow—cannot be correctly understood without joint application of the natural science of the climate system and social and behavioral science aspects of human response. Though integration efforts have made significant contributions to understanding of the climate issue, daunting intellectual and institutional barriers stand in the way of needed progress. Deciding appropriate policies will be a continuing task over the long term, however, so efforts to extend the boundaries of climate modeling and assessment merit long-term attention as well. Components of the effort include development of a variety of approaches to analysis, the maintenance of a clear a division between close-in decision support and science/policy research, and the development of funding institutions that can sustain integrated research over the long haul.

1. ANALYZING HUMAN-CLIMATE INTERACTION

At one level or another almost all of the questions motivating this volume on *The State of the Planet* arise from the interaction between earth's natural systems and the activities of human society. None of these issues, however, presents such a rich combination of complexity and policy immediacy as does anthropogenic climate change. Motivated by this threat, billions of dollars are being spent each year on climate science research. Moreover, the formulation of policy responses occupies political authorities at all levels—from individual cities, to national and regional governments, to the United Nations and summit meetings of the great powers.

Unfortunately, connections remain weak between the efforts of natural scientists to understand potential climate change, and the work of social and behavioral scientists on its human contributors, its economic and environmental consequences, and the formulation of a societal response. As a result, many important efforts on earth observation systems and other data gathering, scientific research, and policy analysis—all needed as guides to action—are diminished in their usefulness for informing policy choice. Here we explore social and behavioral science aspects of global climate change and the ways they are intertwined with the natural science of climate at the frontier of efforts to inform policymakers and the public. As will be argued below, more effective integration of the various components of the issue is both an intellectual and an institutional challenge.

Figure 1 provides a simplified picture of the dynamics of natural/social science and policy aspects of the issue, and can serve as guide to the discussion. Population growth, technology change and economic development are matters of human choice, producing emissions of greenhouse gases (GHGs) that are building-up over time in the atmosphere. In combination with emissions of aerosols and their precursors, and ozone-producing chemical processes in the troposphere—and the resulting effects on land use and land cover—these human activities are changing the radiative balance of the earth. Mediated by complex interactions and feedbacks among the atmosphere, oceans and terrestrial biosphere, the expected result is some level of global climate change over decades to centuries. Because of the slowness of the process and the noisiness of the climate system, this change cannot yet be dependably sensed by casual observation. However, scientific research involving complex and expensive satellite and other observation systems, intensive mining of data from ice cores, corals, tree rings and other records of past change—combined with sophisticated statistical analysis, and complex theoretical and empirical modeling—is producing an ever more convincing picture of a substantial anthropogenic contribution to the warming seen over the past century or so [*Houghton et al.*, 2001].

Through its regional and global manifestations, climate change is expected to have substantial ecosystem effects [e.g., see *Root et al.*, 2003], so another area of complex scientific analysis, shown in Figure 1, concerns the economic and social consequences of such change. It is the public understanding of these, in turn, that motivates action to reduce emissions (or to counter their global effects by geo-engineering) and to devise measures to ease adaptation to levels of climate change that may be unavoidable. In combination with studies of the costs of measures to mitigate human influence, these relations constitute the main inputs to the ongoing debate about the appropriate policy response to the climate change threat.

Other science-based problems of managing the global commons have a similar structure. The buildup of lead and other toxics in the environment and the destruction of stratospheric ozone by chlorofluorocarbons come to mind. However, characteristics peculiar to the climate issue combine to make it the seemingly intractable policy issue it has become. First are the long lags in the system. Any change in global radiative balance results from the buildup of greenhouse gases only over several decades, which means the emissions in the next few years have only a small influence on the long-term risk. Second is our limited understanding of key climate processes—e.g., the behavior of clouds, the response of ocean circulations, and the influence of aerosols—and the fact that the high natural variability of the system greatly complicates efforts to quantify the human influence. Further complicating the decision of what to do now is the fact that some of these uncer-

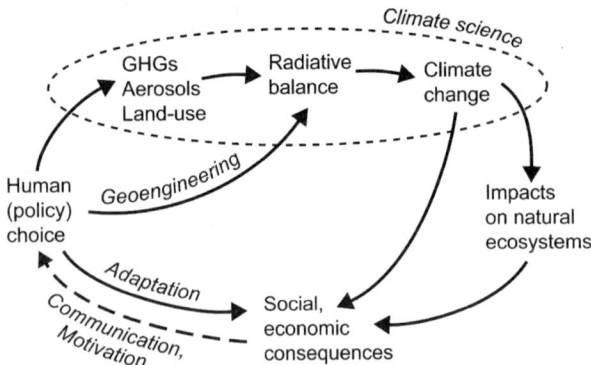

Figure 1. The Expanding Scope of Climate Models. To meet the needs of policymakers and the lay public, conventional boundaries of climate analysis must be expanded to include social and behavioral science aspects.

tainties may be reduced (or perhaps some even increased) over the next decade or two.

These complexities might not be so troublesome if greenhouse gases were a minor byproduct of the modern economy, but they are not. Unlike lead or chlorofluorocarbons, whose control was relatively easy, reducing greenhouse gas emissions will require substantial changes in social organization, impacting all nations and all economic sectors, threatening substantial economic cost, and stirring rancorous controversy over the distribution of the burden.

To gain understanding of the combined human-climate system, with its many and complicated interactions and feedbacks, mathematical models at ever increasing levels of complexity have become essential. Indeed, the history of climate analysis can be charted by the progressive extension of the boundaries of these efforts over the last two or three decades—from atmospheric models drawn from earlier work in meteorology, to a coupling of atmosphere and oceans, to integration with the terrestrial biosphere, hydrosphere and cryosphere. Each increase in model scope has been motivated by the recognition of phenomena that could not be understood through analysis of the disconnected components. It is a process familiar in the natural sciences. Moreover, given the scale and cost of research in these areas, the change in scope naturally is reflected not just in the research and analysis, and model structure, but also in the institutions that fund the work and the disciplinary associations that grow along with them.

Now, in the past dozen years or so, a new challenge has emerged in this process of expanding scope. As nations have begun the difficult task of formulating a response to the climate change threat, a need arises to provide better information about how the climate responds to specific aspects of human activity and the likely effectiveness of proposed control schemes, and about the human and ecological consequences of change that we are unwilling or unable to avoid. These policy choices involve the allocation of human and political effort and economic resources, and the essential question is: what should the nations do *now* given current understanding of the risks and costs? The political task is to find the appropriate mix of effort among three areas of effort: (1) mitigation to be undertaken now and in the next few years, (2) adaptation measures to be taken now in anticipation of possible future change, and (3) research to inform these choices. The research and analysis need, and thus the modeling challenge, is to understand the human-climate system as an integrated whole, with a focus on these choices.

Expansion of the scope of analysis, to include the social and behavioral aspects of these choices as illustrated in Figure 1, involves a reach across disciplinary boundaries that is greater than those needed to achieve collaboration among sub-fields of the earth sciences. Moreover, besides the differences in the nature of the systems studied, and the research methods appropriate to each, the institutional and political barriers are more daunting. Unavoidable tension arises between scientific research for the sake of science, and science constrained to focus on some particular decision problem. Furthermore, because climate change is such a contentious political issue, there is a risk that pressures to meet the objectives of some particular economic or ideological interest may corrupt the scientific enterprise.

From the time the climate issue first gained widespread attention, efforts have been under way to carry out this kind of integrated work, and a substantial capacity and body of literature has been developed. (An impression of these efforts can be gained from the survey by the IPCC [*Bruce et al.*, 1995, Chapter 10]). Nonetheless, the intellectual and institutional challenges to integrated analysis of the human-climate system remain considerable. Some of the more troublesome challenges to conventional policy assessment methods are well summarized by *Morgan et al.* [1999]. If we are to understand this system, and make intelligent decisions about anthropogenic influence over coming decades, an ever-stronger strategic collaboration of the natural, social and behavioral sciences is essential. Otherwise nations will continue to face a pair of dangers: key economic studies and policy analysis could proceed on the basis of a flawed understanding of what is known about the science of climate, and opportunities could be missed to direct scientific efforts to questions of greatest importance to policy analysis and political decision.

2. RESEARCH FRONTIERS: THE INTELLECTUAL CHALLENGE

There are many areas of climate analysis that require close coordination between natural and social/behavioral scientists when pursuing an informed policy decision, but the three that follow will illustrate the need. All are at the forefront of current knowledge, all involve the modeling of human-climate interaction, and all are challenging areas of research. With these examples in hand, a few words can be added on the related task of lay communication.

2.1 Climate Change Projections

Implementing a response to the climate change threat involves potentially costly decisions that must be made with a troubling sense of uncertainty as to their consequences. National economies, and the natural ecosystems on which they depend, will be substantially affected by changes in climate, whether with negative or (for some sectors with a few degrees of warming) positive outcomes. But the main risk is of large negative consequences. Year to year, nations will decide how to manage that risk, taking into account the ability to reconsider decisions in the future, perhaps with better information. Certainly, the foundation of any discussion about emissions control and/or anticipatory adaptation is an analysis of the range of possible climate system outcomes if no action is taken. Development of such projections is a complex task because the analysis must consider uncertainties not only in climate system response, but also in population growth, economic development and technological change, and it must take into account potential feedbacks among these systems over time should climate change occur.

A key difficulty, then, is to combine the natural science and social science analyses of these uncertain systems in order to prepare a useful picture of the nature of the risk. For example, in summaries of the state of the science by the Intergovernmental Panel on Climate Change (IPCC), it has been the practice to present possible outcomes in terms of high and low values of temperature change over the 21st Century—between 1.4°C and 5.8°C as stated in the IPCC Third Assessment Report or TAR [*Houghton et al.*, 2001]. This range is supposed to include uncertainty in both anthropogenic emissions and the response of the climate system to them. Unfortunately, this way of expressing results tends to facilitate the rhetoric of advocates and public misunderstanding. Environmental activists (and much news coverage) emphasize the 5.8°C threat while climate "nay-sayers" argue that the science showing this result is flawed, giving credence to the impression that the lower number is correct. In this debate between polar results, lay observers may misperceive the very nature of the risk, which is not a binary choice (it is a problem or it is not), but some more complex distribution of possible outcomes.

The shortcomings of this way of expressing climate change projections have led researchers to attempt a combined analysis of emissions and climate uncertainty. One example from this family of work is the analysis by *Webster et al.* [2003], which is the basis of Figure 2. The analysis is applied using the MIT Integrated Global System Model [*Prinn et al.*, 1999] and summarized in this volume [*Prinn*, 2003]. An analysis of uncertain greenhouse emissions, as projected by its multi-region, multi-sector model of the world economy [*Webster et al.*, 2002], is combined with an analysis of uncertainty in key parameters of its model of the climate system [*Forest et al.*, 2002]. The result is a representation of the uncertain future behavior of this human-climate system over the 21st Century on the assumption that no greenhouse controls are imposed.

Although a strong effort has been made within the IPCC to incorporate uncertainty analysis in its assessments [*Moss and Schneider*, 2000] controversy surrounds this way of applying social science and natural science models to analysis of uncertainty in the combined human-climate system. Disagreement usually arises in some combination of concerns about conclusions drawn from incomplete science models and/or objection to the methods applied to long-term economic and social processes [*Schneider*, 2001; *Reilly et al.*, 2001; *Allen, Raper and Mitchell*, 2001; *Webster*, 2003]. Also, results must be interpreted with special caution when the coupled systems include models subject to mixes of (perhaps interacting) structural and parameter uncertainties, and/or where the components move out of their ranges of well-understood behavior at different rates with time or scale [*Casman, Morgan and Dowlatabadi*, 1999]. At the very least, therefore, care should be taken to be clear that any such result is conditioned on the model structures applied and the climate processes included and omitted. Further, those assumptions to which results are most sensitive should be transparent. With these qualifications, however, representa-

tion of the threat in the form of Figure 2, or related outputs of formal uncertainty analysis, is an improvement over ranges of outcomes with no confidence bounds. The more complete analysis can serve both as an aid to public understanding of the nature of the threat and as a first step toward analysis of year-to-year decisions about emissions mitigation. Importantly for this discussion, the credibility of the result depends on careful coordination between the natural science and social science contributors to any joint analysis, which in turn requires that each have some minimal understanding of the methods and assumptions applied by the other.

2.2 The "Danger" Level of Atmospheric Concentrations

Given the environmental threat suggested by Figure 2, a natural component of the process of international negotiation and national decision is the formulation of some long-term goal to guide society's response. This goal might be defined at several points in the human-climate system, e.g., at the level of economic and environmental consequences, or some set of climate variables. However, perhaps in an effort to avoid the complicating influence of uncertainties in effects estimates and models of climate system behavior, the drafters of the Framework Convention on Climate Change (FCCC) set the goal of international action on climate as,

> ... stabilization of greenhouse gas concentrations in the atmosphere at a level that would prevent dangerous anthropogenic interference with the climate system ... [and] allow ecosystems to adapt naturally

Since the Convention has been ratified by 188 nations, this notion of stabilized atmospheric concentrations will be an important component of ongoing negotiations.

In gaining a definition with such useful simplicity, however, the diplomats created other puzzles yet to be resolved. When, in the early 1990s, the Convention was negotiated, focus was heavily placed on CO_2 with little consideration given to other substances, and the role they play in the greenhouse effect. Negotiators also seem to have been little concerned about the complicating role of uncertainty in the climate system as it influences the carbon cycle. Thus continuing problems remain in the definition and use of the stabilization goal—problems that require joint analysis of the climate science and the economics of emissions.

For example, Article 2 of the Climate Convention connects to a provision of Article 4 that requires nations to report periodically on the adequacy of efforts, and to continue doing so "until the objective of the Convention is met". Thus the concept of a concentration target provides a basis for debate about whether current and anticipated emissions control efforts are

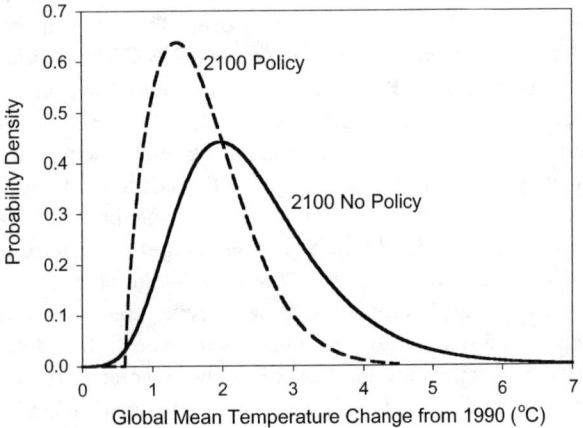

Figure 2. Probability density function (PDF) of global mean temperature change in 2100 under a no-policy case (solid line) and under controls that stabilize atmospheric concentrations at 550 ppmv under a reference projection (dashed line).

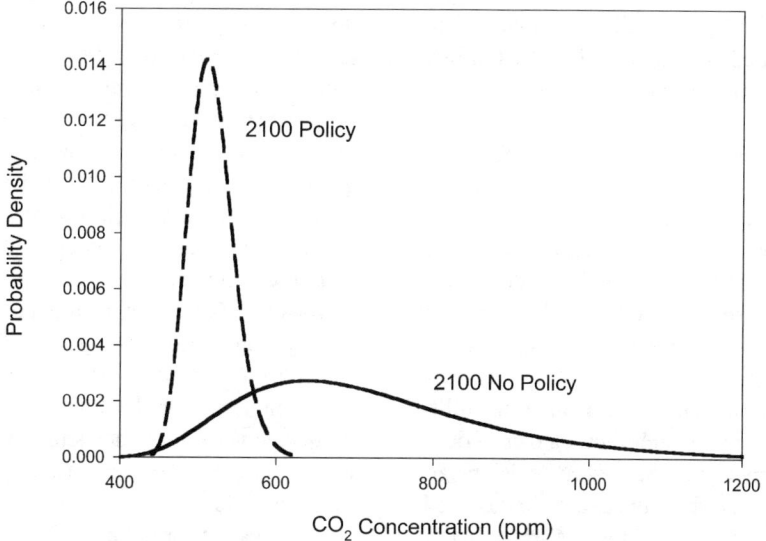

Figure 3. Probability density function (PDF) of atmospheric CO_2 concentration in 2100 under a no-policy case (solid line) and under controls that stabilize atmospheric concentrations at 550 ppmv under a reference projection (dashed line).

consistent with the Convention. Unfortunately, any such judgment must take account of the uncertainties in the carbon cycle, which result from complex processes in the oceans and terrestrial biosphere. There is no deterministic link between emissions paths and atmospheric CO_2 concentrations. At best, any such judgment about the ability of a policy path to attain a particular atmospheric target can only be made in probabilistic terms. This fact is illustrated in Figure 3. It shows PDFs of 2100 CO_2 concentrations for the same two cases that were presented in Figure 2. The solid line shows the no-policy case; the dashed line shows an estimate of the distribution of atmospheric CO_2 concentrations under a profile of global emissions control that would lead to (roughly) a stabilization of greenhouse gases at 550 ppmv under a reference forecast. The insight to be drawn from the figure is that it is not possible to make a one-to-one link between policy action over time and the condition of the atmosphere. Naturally, any projection of the resulting reduction in global temperature change also is uncertain, as shown by the dashed line PDF in Figure 2. At best the policy outcome can be stated in terms of confidence intervals, or the odds that a particular atmospheric concentration or climate result will be achieved.

The non-CO_2 gases introduce other difficulties. The FCCC also covers CH_4, N_2O and a set of industrial gases. Each has a different lifetime and radiative strength, and the lifetime of CH_4 depends on the details of tropospheric chemistry, which changes over time. In this situation even the definition of "stabilization" is not clear. One approach to this puzzle is to seek stabilization in carbon-equivalent terms, with the non-CO_2 gases weighted according to a set of exchange rates (so-called global warming potentials or GWPs) defined by the IPCC, but the result is inconsistent with the intent of the FCCC, which is to stabilize the instantaneous human influence. The target could be defined as stabilization of the concentration of each gas individually, but any strategy to achieve this result would be highly inefficient economically. Finally, some climatically important substances are left out of consideration entirely. Tropospheric ozone is ignored and the influence (both warming and cooling) of aerosols and aerosol precursors remain outside the the FCCC's system of greenhouse accounting [*Reilly, Jacoby and Prinn*, 2003]. Resolution of these issues, to clarify the economic analysis of control schemes and the language of international negotiations, will require joint effort by atmospheric chemists and radiative transfer experts, and those carrying out the economic analysis and policy assessment.

Finally, there is the issue of defining the benefits of restraining atmospheric concentrations, or some other measure of climate change. Such measures are needed for comparison with the expected control costs and to inform discussions of the appropriate long-term atmospheric target. Here again, attaining clarity will require a complex inter-weaving of economic and ecological analysis of the advantages of avoiding change, and the underlying natural science of the change itself. A number of difficult challenges stand in the way of widely accepted estimates of such benefits—including accounting for uncertainty and risk preferences, the difficulties of valuation of non-market impacts, and the puzzles presented by attempts to aggregate effects over very different national circumstances. As a result, it is likely that a portfolio of benefit measures will be used to inform the choice of target, including physical measures, at global and regional scales, computed by global climate models [*Jacoby*, 2003]. Again, if climate model results are to be appropriately applied, natural scientists must inform the related social and political analysis.

2.3. Uncertainty, Learning, and Sequential Decision

The projections of climate change in Figure 2, under a no policy assumption or some imagined path to a stabilization goal, are useful in illustrating the nature of climate change risk. But because of their assumption of a fixed policy path over time they are not realistic representations of the temperature outcomes under possible paths of resolution of climate system uncertainties. The limit to their descriptive capability can be illustrated using Figure 4, which shows a pair of decision trees. The squares represent points where decisions are made; the circles indicate times where uncertainty is resolved. In this example, future climate policy choice is simplified to a set of decisions by a single global authority at a couple of points in time. "No mitigation" or "fixed target" policy scenarios essentially assume that the decision context is the one shown in the upper part of the figure. Choices about a mitigation path are made now, and maintained over time; key uncertainties will not be resolved until the end of the period (for this illustration, the end of the 21st Century). This is the assumption implicit in the results shown in Figures 2 and 3. That is, choices are made today either to take no action over the century, or to follow a trajectory of emissions control that is presumed to lead to 550 ppmv.

Useful as calculations based on this assumption may be for illustrating the nature of the risk, they do not provide an accurate picture of the choice problem. In the jargon of environmental economics, greenhouse gases are a "stock pollutant". As noted earlier, their influence on the environment comes not from emissions today but from their buildup over time—over many decades in the climate case. Furthermore, society does not today decide what response it will take over the long-term future. Even were it wise to commit to such long-term paths, nations are limited in their ability to make far future commitments. Moreover they need not do so: they can decide what to do today knowing that any decision, such as the stringency of mitigation efforts, can be reconsidered in the future (perhaps with some options foreclosed by events along the way). And, since key uncertainties about climate system response to human influence, and the possible economic and ecosystem effects, may be reduced over time by research and observation, there is always a choice between acting now or waiting for more information. Thus the more correct representation appears in the lower part of Figure 4. In a simplified two-stage version, it shows that we will decide and act, then learn over time, then decide again—in a sequence extending far into the future.

Here lies the core of the debate over climate policy and climate research—an issue well illustrated by an early exchange in EOS [*Risbey, Handel and Stone*, 1991a,b; *Schlesinger and Jiang*, 1991a,b]. How much mitigation should be undertaken now given what we may learn over time, what resources should be devoted to ensuring that this learning occurs, and where should resources be directed? The issue of sequential choice under uncertainty, with learning, is an extensive and complex topic in the literature of economics and policy analysis [e.g., *Webster*, 2002], with special complications depending on whether actions are taken by some individual decision maker or as the result of a negotiation involving several parties. In either case, the structure of the human-climate interaction matters, as does the question of which uncertainties are expected to be narrowed, and how soon. Of course, relevance to policy is not the only criterion in deciding climate science research strategy. To the extent it is important, however, those carrying out the social science research and policy analysis need the active involvement of those at the forefront of the natural science work.

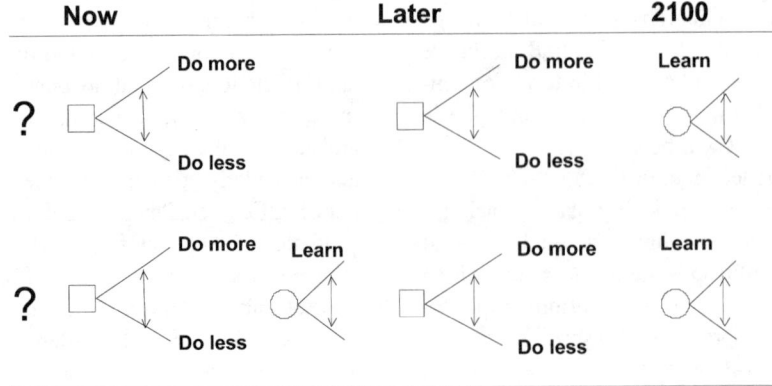

Figure 4. Decision trees for alternative versions of the climate decision problem. The top tree shows the scenario form; the bottom tree represents sequential decision with learning.

2.4 Communicating Results

Up to this point the discussion has proceeded as if the main task was communication and collaboration among natural scientists, social and behavioral scientists, and policy analysts. But there is another key connection that crosses the borders of discipline, and that is communication with policy-makers and the lay public. It is difficult enough for the non-specialist to gain a clear impression of the climate change issue when information is so charged with the agendas of one or another side of disputes over specific policy initiatives. Even more problematic is the fact that natural and social scientists have so little understanding of how their research results—regarding the climate system and its uncertainties, the potential effects of change, and the costs of control efforts—are appropriated and interpreted by non-specialists. Without greater interaction with behavioral scientists, who do try to understand this process, important information may be misperceived or ignored, and policy effort misdirected.

Just two examples will make the point about the need for greater interaction on this disciplinary boundary. Above it is argued that greater care needs to be taken to accurately present uncertainties in projections of emissions and climate outcomes, as if such information (perhaps expressed in PDF's) would be commonly understood, producing a coherent interpretation and reaction. Not so. Studies of attitudes to risk show that they differ among cultures, and across individuals within a culture or nation [*Renn and Rohrmann*, 2000; *Slovic* 2002]. People have different views of the nature of a particular risk, or of what it would be worth to reduce it, even if they agree on the magnitude of the effect under various outcomes. There may be many ways to summarize information about climate change, each with meaning to a particular party, but there may be no way to achieve a uniformly shared impression, and perhaps even a difficulty in achieving a common measuring rod. Moreover, research on cognitive processes shows that people do not absorb new information passively, as if the mind were a blank slate so far as the issue at hand is concerned [e.g., *Kempton, Boster and Hartley*, 1995]. They approach any new phenomenon (like climate change) with a set of existing cultural models and concepts, through which they view the new information. This pattern may help explain why so many lay people think the solution to the climate problem is to control spray cans!

Those of us who work on natural and social science aspects of climate have something important and useful to offer to non-specialists and policymakers, but good work as defined within our own disciplinary standards is not sufficient.

3. INTEGRATED ANALYSIS: THE INSTITUTIONAL CHALLENGE

3.1 Interdisciplinary Modeling and Analysis

For purposes of the work outlined above it is fortunate that strong similarities exist between the analytical methods applied by natural scientists, in models of the climate system and its response to greenhouse gases, and the models used by social scientists (economists mainly) in studies of the origins of these gases and the costs of controlling them. It is thus possible for researchers from these diverse disciplines to come to understand one another's methods, and even to develop mutual respect for the difficulties each faces. Both build models that attempt to capture the structure of a complex system of many components. The natural scientists confront interactions of the atmosphere, oceans and land surface. Economic analysis must integrate multiple nations and economic sectors, each with particular technological characteristics. But both have models that specify these components, their individual behavior, and the interactions among them. Each has some set of natural forcings: existing greenhouse gas concentrations, solar variability, etc. on the climate side, and population change, technological advance, resource depletion and other such phenomena for the economic analysis. Then each discipline forces its model with outside influences: growing greenhouse gas emissions introduced into the climate model, policies such as emissions mitigation imposed on economies.

In both cases the larger models—the atmosphere-ocean general circulation models and multi-region, multi-sector general equilibrium economic models—solve for equilibria period by period, for a range of time steps from minutes to months in climate models and usually a five or ten year period in economic models. Each creates transient projections by introducing vectors of greenhouse forcings or economic processes like technical change. Each is calibrated to match some historical period, and each has parameters—some estimated statistically, some set by expert judgment—that can serve as a basis for exploring within-model uncertainty.

Despite these similarities, of course, substantial differences complicate the integration of social science and natural science modeling efforts. The modeling of human systems must deal with uncertainties that are not faced in the analysis of natural systems. The chemistry and physics of the atmosphere may be uncertain, but modelers need not consider the possibility that future molecules might invent new ways of reacting, or that a forest may anticipate a coming change in soil moisture and start moving a decade ahead of time. In economic models, on the other hand, invention and anticipation are important processes. Furthermore, except in some long-term Darwinian sense processes modeled in the natural sciences don't reflect preferences for ultimate outcomes, or attitudes to risk. These are relevant, however, in modeling the social processes that underlie any response to the climate threat.

These differences in the characteristics of the components of the human-climate system can lead to disagreement over methods, and even to conflict over philosophy of approach to analysis. Thus, the frequentist-subjectivist controversy emerges in the climate debate, and elicitation of expert judgment and application of Bayesian methods are likely to be more familiar and comfortable in some applications than in others [*Webster*, 2003].

The gap is even greater between the natural sciences and social sciences like economics on the one hand, and the behavioral sciences on the other—a phenomenon that can be seen in the paucity of inputs from these fields in existing integrated assessment [*Bruce et al.*, 1995]. The common pool of terminology and techniques, and opportunities for linkage of analytical models, which facilitate collaboration between earth science and economics, are missing in the interaction with behavioral science. The study of human behavior naturally focuses more on survey methods, individual and group experiments and studies of brain function—areas of work unfamiliar to most researchers working on the climate issue. Yet if effective communication is to be achieved, greater understanding is going to be needed about how different groups of people perceive climate information, form judgments, and act.

3.2 Institutional Barriers

Institutional barriers to the strengthening of existing interdisciplinary work are well known. Conventional academic departments, organized along disciplinary lines, are both a blessing and a curse. In their disciplinary orientation they encourage depth of analysis, pushing back the frontiers of the individual field whether it be ocean dynamics, plant biology or economic modeling. But by their laser-like focus on problems in the field they dampen incentives to devote work to topics on the boundaries. The disciplinary journals, at once the arbiters of scientific advance and the key to academic advancement, are major players in this process.

This fact about research, particularly as conducted in academic institutions, leads to other observations about conditions that are helpful if not essential to bringing the best minds to interdisciplinary work. One key requirement is a problem definition that provides a mutual intellectual challenge—one that cannot be successfully attacked by any one discipline alone. Even if such a common challenge is defined, however, top researchers often will resist being diverted completely from their home disciplines. Thus the interdisciplinary work also needs to offer opportunities for contribution to the individual's home field and journals. At least in the academic context, then, the institutional problem of effective interdisciplinary work is to a large degree an intellectual one as well. Problems need to be formulated in ways that attract the needed talent, and standards need to be maintained as to what constitutes quality work in the interdisciplinary domain.

Of course, university departments are not the only sources of work on climate issues. In most countries, government agencies, laboratories and their contractors are the channels for most of the human and financial resources devoted to climate research. Indeed, these agencies often are the funding sources for most of the university-based work. No doubt the disciplinary imperatives apply also here to some degree. In these agencies, moreover, the wide scope of the climate change issue comes again into play as a barrier to integrated work. In normal government organization (using the US as an example) the regulatory and price-incentive policies that might be part of any climate response are spread across one set of agencies (Environmental Protection Agency, Treasury); the major sources of emissions are covered by another set (Agriculture, Transport, Energy, Federal Energy Regulatory Commission); and the major sources of climate research include these agencies, plus several others (National Science Foundation, National Atmospheric and Space Administration, National Oceanographic and Atmospheric Administration). In a recent effort to prepare a strategic plan for US climate science research, over a dozen federal agencies were involved [*US CCSP*, 2003]. Naturally, each has its turf to protect.

As suggested earlier, progress has been made in developing institutions that can integrate the natural science of the atmosphere, oceans and terrestrial biosphere. Unfortunately, despite the efforts cited earlier it has proved more difficult to link this work with analysis critical to public decision about emissions control, adaptation, and research direction. In part the difficulty originates in features already noted: the gulf in terminology and modeling methods is greater than among the natural sciences alone, and the funding agencies are yet more fragmented. But there is yet another source of difficulty. Climate change has become highly charged politically, and existing governments—who are the source of the bulk of the funding—often have policies not only regarding the appropriate policy response but also about the acceptable description of the threat and topics for investigation. The problems that this environment creates for work on human-climate interaction are subtle but important. Some scientists resist bringing their work too close to the human-climate frontier, with its political entanglements, because they fear their efforts will be diverted to what they view as unfruitful areas of work. Even more pernicious is the worry that areas of investigation will be closed off, or at least not encouraged, because they raise issues that have already been decided as a matter of government policy. The concern of individual scientists and research institutions for their reputations, and a long tradition of free inquiry in many countries, are a bulwark against these pressures, but they are nonetheless inevitable in connection with such a contentious social issue.

In an attempt to achieve greater integration, analysis centers have been created in several European countries (with both national and EU funding), in Japan and elsewhere, and they often involve participation by the major climate modeling centers. Although such efforts have a long history in some countries (e.g., the Netherlands), much of this development is relatively new and results are yet to be seen, particularly regarding the integration of policy economics and other social and behavioral science inputs. The same holds for the US where a number of integrated analysis groups are active, some with ties to the federally sponsored climate modeling centers. However, for an issue where the co-evolution of climate learning and policy development is so crucial the level integrated work is inadequate. The Strategic Plan for the US Climate Science Research Program, completed in July 2003, devotes some attention to integration of the type discussed here, under its focus on "decision support" [*US CCSP*, 2003]. Institutional means adequate to achieve this objective remain to be formulated, however. Given the fragmented nature of the disciplines, the large number of agency interests involved, and the high level of political disagreement about the policy response that is warranted at this time, it should be no surprise that the task is so difficult.

4. CLOSING THOUGHTS

As this volume is being prepared the nations are focused on short-term issues including the implementation of Annex B commitments under the Kyoto Protocol, the continuing debate over US policy, and the search for ways to encourage deeper involvement by developing countries. In this discussion, and in the inputs to the process by the IPCC and other assessment efforts, it is discouraging how little effective integration has been attained between the natural sciences and social and behavioral science and policy analysis. However, it should always be kept in mind that climate is a century scale problem, and society is just at the start of an effort to mount a sustained, well-calibrated response. Over and over in the decades to come, nations and sub-national decision makers will revisit their decisions about emissions mitigation (including aid to less developed countries), anticipatory adaptation, and research priorities—at each point seeking guidance from integrated assessments of the then current state of knowledge. An important task today is to create the institutional structure that can facilitate these needed inputs to public and private choice.

Considering the complexity of the human-climate issues illustrated above, and the difficulties in achieving the needed integration of talents and approaches, three suggestions come to mind regarding the organization of this work. First, there is need for a flexible capability to integrate the needed climate science with social and behavioral inputs to studies of different aspects of the climate change issue. Building interdisciplinary teams and carrying out such research and analysis requires years of work, and yet at any one time we do not know what issues will prove most important five or ten years hence. This situation calls for the support of a diverse set of efforts, at least in the research phase, applying a variety of methods. By the same token, it argues against the construction of single, dominant national centers, that attempt to encompass all the needed integrated work in one place or under one organization.

Second, it is important to keep as clear a distinction as possible between the needed research and integrated assessment of policy issues on the one hand, and close-in decision support on the other. All nations use some form of short-term "policy shop" activity to help inform leaders at the point of political decision. It is an essential function, although the institutional form differs substantially from country to country; depending on the circumstance it may be sought from consultants, government laboratories, agency staff, etc. Because of their proximity to political choice, one should always expect that assessments carried out under these conditions will be subject to guidance on the problem definition and the options that can be considered. Moreover, such assessments usually will be on a very short time schedule, as they usually are called into being only when decisions are at hand. From the discussion above of the intellectual and institutional barriers, it can be seen that political direction and a short time scale practically insure that serious integration of natural and social/behavioral science components will not take place.

To achieve the needed joint work, then, research groups need to be created and sustained that are insulated from the short-term political pressures that likely will characterize this issue for the foreseeable future. In some countries it is possible to maintain this type of independence within government agencies and laboratories, and in others not. In keeping with the argument above for diversity, large countries like the US, Japan or major European nations, or the EU itself, can reduce the pressures on any one group by sustaining several. The inputs to short-term decision can then be drawn from them as appropriate by a separate activity for this purpose.

Finally, there is the issue of the funding and organization of the more research-oriented of these two functions. As introduced in the discussion surrounding Figure 1, the extension of modeling and assessment activity outside the conventional boundaries of climate science, to include ecological and societal components, is a much more difficult task than the earlier integration of air chemistry and meteorology, oceanography, and terrestrial biology. Even worse, in some governments (including the US) there is no institutional champion for the integration. To some degree this may be because of a lack of separation between the close-in decision support and policy relevant research; to some degree it may simply reflect the fact that too many governmental agencies have a stake in the problem and in research directed to their particular focus. Solution of these problems thus requires leadership and sustained attention at the highest national level. Given the long-term nature of the issue, it is a goal worth pursuing.

Acknowledgments. The MIT models underlying analysis shown here supported by the US Department of Energy, Office of Biological and Environmental Research [BER] (DE-FG02-94ER61937) the US Environmental Protection Agency (X-827703-01-0), the Electric Power Research Institute, and by a consortium of industry and foundation sponsors.

REFERENCES

Allen, M., S. Raper and J. Mitchell, 2001. Uncertainty in the IPCC's Third Assessment Report, *Science* 293(5529): 430–433.

Bruce, J. et al., 1995. *Climate Change 1995: Economic and Social Dimensions of Climate Change*, University Press, Cambridge, UK.

Casman, E., M. Morgan, H. Dowlatabadi, 1999. Mixed Levels of Unceratinty in Complex Policy Models, *Risk Analysis* 19: 33–42.

Forest, C., P. Stone, A. Sokolov, M. Allen and M. Webster, 2002. Quantifying Uncertainties in Climate System Properties Using Recent Climate Observations, *Science* 295:113–117.

Houghton, J. et al., 2001. *Climate Change 2001: The Scientific Basis*, Cambridge University Press, Cambridge, UK.

Kempton, W., J. Boster and J. Hartley, 1995. *Environmental Values in American Culture*, MIT Press, Cambridge, MA.

Jacoby, H., 2003. Informing Climate Policy Given Incommensurable Benefits Estimates, *Global Environmental Change*, in press.

Morgan, M., M. Kandlikar, J. Risbey and H. Dowlatabadi, 1999. Why Conventional Tools for Policy Analysis Are Often Inadequate for Problems of Global Change, *Climatic Change* 41: 271–281.

Moss, R., and S. Schneider, 2000. Uncertainties in the IPCC TAR: Recommendations to Lead Authors for More Consistent Assessment and Reporting, in Guidance Papers on Cross-Cutting Issues of the Third Assessment Report of the IPCC (eds. R. Pachauri, T. Taniguchi and K. Tanaka, World Meteorological Organization, Geneva.

Prinn, R., 2003. Atmospheric Chemical Change, Air Pollution, and Climate, C. Hawksworth and S. Sparks (eds.) *The State of the Planet* (AGU Monograph), in press.

Prinn, R., H. Jacoby. A. Sokolov, C. Wang, X. Xiao, Z. Yang, R. Eckaus, P. Stone, D. Ellerman, J. Mellillo, J. Fitzmaurice, D. Kicklighter, G. Holian and Y. Liu, 1999. Integrated Global System Model for Climate Policy Assessment: Feedbacks and Sensitivity Studies, *Climatic Change* 41(3/4):469–546.

Reilly, J. P. Stone, C. Forest, M. Webster, H. Jacoby and R. Prinn, 2001. Uncertainty and Climate Change Assessments, *Science* 293(5529): 430–33.

Reilly, J., H. Jacoby and R. Prinn, 2003. Multi-Gas Contributors to Global Climate Change: Climate Impacts and Mitigation Costs of Non-CO_2 Gases, Washington D.C., Pew Center on Global Climate Change.

Renn, O. and B. Rohrmann, 2000. *Cross-Cultural Risk Perception: A Survey of Empirical Studies*, Kluwer, Dordrecht, The Netherlands.

Risbey, J., M. Handel and P. Stone (1991a). Should We Delay Responses to the Greenhouse Issue? *EOS, Transactions* 72(53): 593.

Risbey, J., M. Handel and P. Stone (1991b). Do We Know What Difference a Delay Makes? *EOS, Transactions* 72(53): 596–597.

Root, T., J. Price, K. Hall, S. Schneider, C. Rosenzweig and J. Pounds, 2003. Fingerprints of Global Warming on Animals and Plants, *Nature* **421**(2): 57–60.

Schlesinger, M. and X. Jiang (1991a). Climatic Responses to Increasing Greenhouse Gases, *EOS, Transactions* 72(53): 597.

Schlesinger, M. and X. Jiang (1991b).A Phased-In Approach to Greenhouse-Gas-Induced Climatic Change, *EOS, Transactions* 72(53): 593, 596.

Schneider, S., 2001.What is 'Dangerous' Climate Change?, *Nature* 411: 17–19.

Slovic, P., 2002. "Trust, Emotion, Sex, Politics and Science: Surveying the Risk-assessment Battlefield", P. Slovic (ed) *The Perception of Risk*, Earthscan, London.

[US CCSP] US Climate Change Science Program and the Subcommittee for Global Change Research (2003). Strategic Plan for the US Climate Change Science Program, Washington DC. (http://www.climatescience.gov/.

Webster, M., 2002. The Curious Role of "Learning" in Climate Policy: Should We Wait for More Data?, *The Energy Journal* 23(2): 97–119.

Webster, M., 2003. Communicating Climate Change Uncertainty to Policymakers and the Public: An Editorial Essay, *Climatic Change*, in press.

Webster, M., M. Babiker, M. Mayer, J. Reilly, J. Harnisch, R. Hyman, M. Sarofim and C. Wang, 2002. Uncertainty in Emissions Projections for Climate Models, *Atmospheric Environment* 36(2): 3659–3670.

Webster, M., C. Forest, J. Reilly, D. Kicklighter, M. Mayer, R. Prinn, M. Sarofim, A. Sokolov, P. Stone and C. Wang, 2003. Uncertainty Analysis of Climate Change and Policy Response, *Climatic Change* 61(3): 295–430.

Henry D. Jacoby, Joint Program on the Science and Policy of Global Change, Massachusetts Institute of Technology (E40-439), Cambridge, MA 02139

Uncertainty and Predictability in Geophysics: Chaos and Multifractal Insights

Daniel Schertzer

CEREVE, Ecole Nationale des Ponts et Chaussées and Météo-France, Paris, France

Shaun Lovejoy

Physics Department, McGill University, Montreal, Canada

Uncertainty and error growth are crosscutting geophysical issues. Since the "chaos revolution" the dominant paradigm has been the "butterfly effect": the dependence on initial conditions is so sensitive that errors grow exponentially fast with characteristic times. This was the outcome of studying superficially simple caricatures of more involved systems. We critically analyze the physical relevance of these models and the mathematical generality of this effect. We emphasize that the atmosphere, oceans, rain etc., are spatially extended turbulent systems, with wide ranges of spatial scales. Turbulent phenomenology already shows that errors grow only slowly across these scales; they follow power laws, there are no characteristic times. An important recent realization is that in spite of strong anisotropies the dynamically significant range of scales is much larger than previously thought and that the role of intermittency is drastic and yields much more frequent extremes. The focus is now on time-space geophysical scaling behavior: their multifractality. It is found quite generally - not only for turbulent fields- that an infinite hierarchy of exponents is required to characterize the predictability decay from average to extreme events. Nevertheless, these laws are meaningful over the whole time range from short to long term; we give their explicit expression. This multifractal predictability suggests the advantages of stochastic rather than deterministic sub-grid parametrizations, and makes stochastic forecasting very attractive.

1. INTRODUCTION

Recently, there have been growing societal pressures to provide reliable predictions, in particular of extreme geophysical events (e.g. earthquakes, floods/droughts, cyclones, storms, etc.). Prediction time scales range from nearly zero for "nowcasting" to centuries for global change.

Prediction has several meanings, sometimes called "prediction of the first", "second" and "third kinds": these correspond to a pure initial value problem e.g. a Cauchy problem for differential equations, to a change of boundary value problem e.g. a Diriclet or von Neuman problem for partial differential equations, or to a mixture of both. Practical problems, such as global change, typically correspond to the third type. In spite of advances in prediction techniques, rather little has been done to determine their theoretical limits. The existence of limits are usually accepted at least in principle (e.g. the 'butterfly effect'), but one still hopes to extract some information in the midst of noise. This leads to a fundamental methodological problem, since without an estimate of the limits, it is difficult to evaluate and hence improve prediction skills.

This paper focuses on predictability limits and clarifies the crucial role of spatial scales. Two rather distinct approaches have been followed: one corresponding to the dynamical systems approach ('deterministic chaos') and the other one based on spatial complexity and scaling. Although both approaches share some common features, the types of predictability decay are quite different, corresponding to exponentials and power laws respectively. Although this dichotomy was first investigated by Lorenz in the 1960's, it is still not widely known. Since then, our knowledge of scaling processes has mushroomed and their close connection with intermittency has been clarified. Here we discuss the impact of these ideas showing that the loss of predictability is not smooth in time, but rather occurs by intermittent "puffs".

Although the predictability problem is ubiquitous in geophysics, most early developments and formalizations have occurred in the context of atmospheric or climate dynamics, where they have achieved their highest expression. This paper provides a general review of concepts together with applications to geophysics, e.g. hydrology (Sect. 2.3) atmospheric dynamics (Sect. 3.3). We hope that these examples will stimulate the development of similar applications in other geophysical domains.

2. DYNAMICAL SYSTEM APPROACH

2.1. Chaos Revolution?

The notion of "initial condition sensitivity" became well-known due to the work of [Lorenz 1963a] on his 3-component model (corresponding to the first three Fourier components of convection). By the 1980's such exponential error growth became the hallmark of the "deterministic chaos revolution" and it was widely viewed to be a generic property of nonlinear systems.

2.2. Exponential Error Growth

Exponential error growth emerged from the pioneering work of [Lyapunov, 1907], and was subsequently generalized into the elegant Multiplicative Ergodic Theorem (MET) [Oseledets, 1968], a cornerstone of chaos theory. A key assumption of the theorem is that temporal averages of a single sample of the process are the same as the average at one time over an ensemble of identical processes; i.e. that the process is "ergodic". Let us give some heuristics showing how this theorem follows from this common geophysical ergodicity assumption. Consider the simplest case, of a discrete ($t = 0, 1, 2, 3...$) nonlinear mapping G on real numbers. A well-known example concerns population dynamics [May, 1976] and the approach generally applies to finite difference approximations to scalar differential equations:

$$X(t+1) = G(X(t)) \qquad (1)$$

The amplitude of the infinitesimal separation $\delta X(t)$ of a pair of points $\left(X^1(t), X^2(t) = X^1(t) + \delta X(t)\right)$ is *multiplicatively* modulated by the derivative of the map G at the point $X(t)$, i.e.:

$$\left|\delta X(t+1)\right| \approx \left|D_{X(t)}G\right| \left|\delta X(t)\right| \qquad (2)$$

taking logarithms, one obtains:

$$Log\left[\left|\delta X(t)\right|/\left|\delta X(0)\right|\right] \approx \sum_{t'=0, t-1} Log\left(\left|D_{X(t')}G\right|\right) \qquad (3)$$

If the process defined by Eq. (1) is ergodic, the right hand side of Eq. (3) is determined by replacing the time averages by ensemble averages (square brackets "⟨.⟩"):

$$\sum_{t'=0, t-1} Log\left(\left|D_{X(t')}G\right|\right) \approx t\left\langle Log\left(\left|D_x G\right|\right)\right\rangle \qquad (4)$$

a result which yields an exponential error growth:

$$\left|\delta X(t)\right| \approx e^{\mu t}\left|\delta X(0)\right| \qquad (5)$$

with a Lyapunov exponent:

$$\mu = <Log\left(\left|D_X G\right|\right)> \qquad (6)$$

Equation (5) is valid as long as μ is finite, an assumption usually taken for granted. μ corresponds to the inverse of the characteristic time after which predictions are effectively impossible. Generalizations to finite d-dimensional systems proceed along the same lines. However, if we attempt to generalize this to evolving fields, i.e. to nonlinear partial differential equations (infinite dimensional (functional) spaces), we encounter severe difficulties. In fact, only a few limited extensions have been obtained [e.g. Ruelle, 1982]. Later on (Sect. 3.1), using physical phenomenology, we point out a key difficulty: the likely small scale divergence of μ, which violates the finiteness assumption of the mathematical derivation of the MET.

2.3. Low Dimensional Chaos in Geophysics?

2.3.1. Dimension estimates of geophysical attractors. The exponential growth of error implies that much of the original information is rapidly lost hence that only a few degrees of freedom might be required to specify the state of a system after long times. This raises the possibility that elaborate nonlinear time series analysis techniques might be able to identify the few relevant parameters and greatly simplify the model. This explains the excitement generated, twenty years ago, when the embedding theorem and the correlation dimension algorithm were discovered. Together, they gave some credence to the idea that geophysical systems might not only have finite dimensions, but that the dimensions might be low.

This evolution in thinking started with [*Packard et al.*, 1980] and [*Takens*, 1980] who considered a scalar observable h of a (possibly vector-valued and continuous) time-series $x(t)$ constrained on a (finite dimensional) strange attractor. They showed that in order to obtain a faithful image of this attractor (called a "reconstruction"), it was sufficient to use a (discrete) time series $X_n = h\{x(n\Delta t)\}$. For instance, among the many components of the climate $x(t)$, one may consider the temperature $h\{x(t)\}$ and reconstruct the climate attractor with the help of a discrete time series $h\{x(n\Delta t)\}$. This may be achieved using "delay vectors" $Y_n = (X_n, X_{n-\tau}, X_{n-2\tau}, ..., X_{n-(d-1)\tau})$, for any specified (integer) time delay τ, as soon as the dimension d of the resulting 'embedding space' E_d that they span is large enough. Defining the box dimension D_0 as the the scaling exponent of the number of attractor points N in a box of size ℓ: $N(\ell) \approx \ell^{D_0}$, the precise requirement is $d > 2D_0$. In spite of its simple definition, the box dimension D_0 was numerically difficult to evaluate for large d.

The correlation dimension algorithm [*Grassberger*, 1983] overcame this difficulty [e.g. *Schuster*, 1988; *Tsonis*, 1992]: only the distance between pairs of delay vectors is required, not the embedding space. The correlation dimension D_2 of a set of points is defined as the scaling exponent of the average number of points in a sphere of radius r centered at one of them, i.e.:

$$<N(d,r)> \propto r^{D_2(d)} \qquad (7)$$

The advantage of D_2 is that very large embedding dimensions d can easily be explored numerically. Initially, strange attractors were thought to have a unique (fractal) dimension, therefore the box-dimension and the correlation dimension were considered as equivalent. If this property, as well as the other assumptions of the embedding theorem, is satisfied, then for large d, the estimates $D_2(d)$ should converge towards the theoretical value D_2—at least over a range of r. This is illustrated with the help of Figure 1-a, which displays $Log<N(d,r)>$ vs.

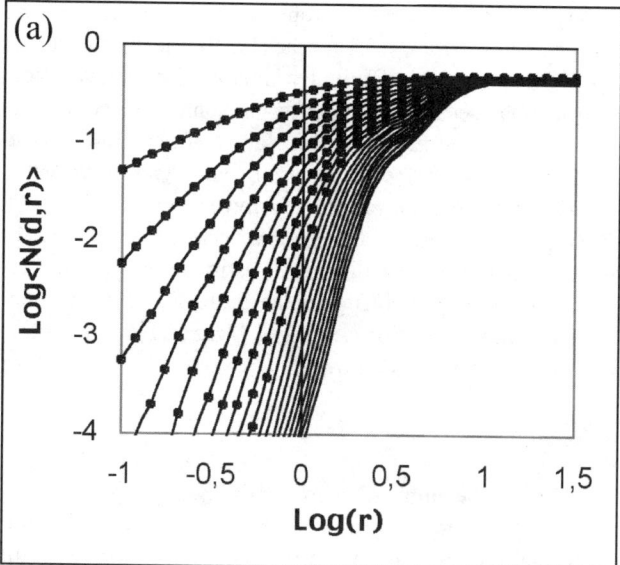

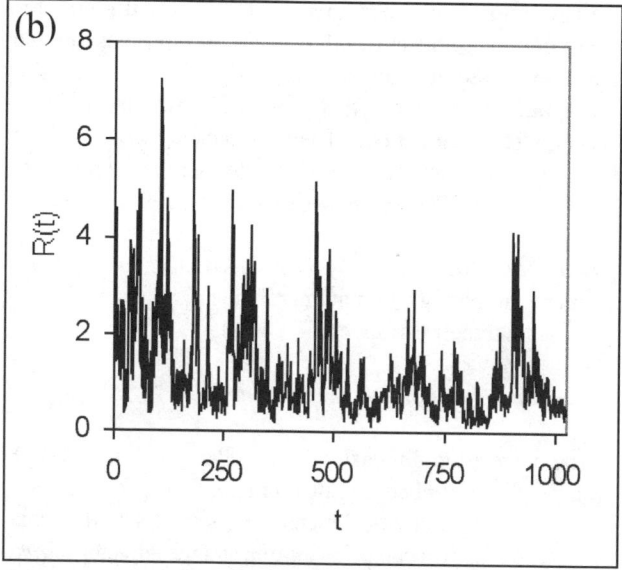

Figure 1: The correlation dimension (D_2) algorithm (a) applied to estimating the dimension D_2 of a synthetic rainfall series (b). (a) shows $Log<N(d, r)>$ vs. $Log(r)$. As d increases, the slopes on the lower left reach an apparently constant value, the correlation dimension $D_2 \approx 2.7$. (b) the time series produced by a lognormal cascade with 12 cascade steps, each of factor 2 in scale, i.e. with $N = 4096$ data points.

$Log(r)$ for a synthetic rainfall time series Figure 1-b. As d increases, the slopes converge to $D_2 \approx 2.7$.

Loosely speaking an attractor dimension measures how its points are concentrated in the phase space. A low dimension corresponds to a very small fraction of the available phase space, implying very constrained dynamics. This explains the excitement following [*Nicolis and Nicolis*,

1984], who analyzed the isotope record of a deep-sea core and estimated that $D_2 \approx 3.1$ for the climate and concluded that 4 ODE's might be sufficient to model climate evolution. Similar analyses have been performed on many geophysical data sets leading to numerous claims of very low dimensionality ($D_2 \approx 5 \div 7$), especially for rain [*Essex et al.*, 1987; *Fraedrich*, 1986; *Hense*, 1986; *Jayawardena and Lai*, 1994; *Rodriguez-Iturbe et al.*, 1989; *Tsonis et al.*, 1993]. Although the proliferation of low dimension estimates gave some credence to the applicability of the butterfly effect to geophysics, many questions emerged [*Marzocchi*, 1997] that we are going to discuss now.

2.3.2. Limitations Due to Sample Size

Although the embedding space is only implicit in the correlation dimension algorithm, an essential problem remains: empirical analyses are performed on finite samples! This leads to an artificial confinement of empirical points to a small fraction of the embedding space and spurious low estimates of correlation dimensions $D_2(d)$, which do not reflect the dynamics. To avoid this problem, a minimal range of scales is required [*Ruelle*, 1990]: a decade yields the celebrated rule of thumb that the minimum number of points for any D is: $N_D \approx 10^D$ [*Nerenberg and Essex*, 1990] and *Essex, 1991*]. [*Grassberger*, 1986] pointed out that the results of [*Nicolis and Nicolis*, 1984] were based on only 184 measurements, implying a maximum reliable dimension $D_2 \approx 2.3$, i.e. less than their estimate of $D_2 \approx 3.1$.

2.3.3. More Questions on the Estimates

The correlation dimension D_2 and the box dimension D_0, required by the embedding theorem are two special cases of the infinite hierarchy of dimensions (Sect. 3.4.1) that characterize the multifractal behavior of a strange attractor [*Grassberger*, 1983; *Hentschel and Procaccia*, 1983]. Physically, this means that systems are not homogeneously constrained. For example, the clustering of point pairs characterized by D_2 is generally less extreme than the clustering of triplets (D_3), quartets (D_4) etc. Perhaps the fundamental point is that the embedding theorem *hypothesizes* that the dynamics are deterministic. One therefore cannot draw any conclusion from a low D_0: this is neither a requisite nor an indication of chaos (see Sect. 2.4). In particular, nonlinear time series analysis techniques are inherently incapable of distinguishing between low-dimensional deterministic systems and high dimensional stochastic systems. The classical example is the reconstruction of the stochastic process known as "Brownian motion" [*Osborne and Provenzale*, 1989; *Theiler*, 1991], which is *linear* and has a box-counting dimension $D_0(d) = 2$ for any $d \geq 2$.

Similar results are obtained for a *nonlinear* stochastic model—a multiplicative cascade—which has been often invoked for rain (Sect. 3.4.2) and which displays an estimate $D_2 \approx 2.7$

Figure 1 a, b). According to the sample size criterion ($N = 4096$ and $\log_{10}(N) \approx 3.6 > 2.7$), this estimate is reliable but nevertheless does not imply a small number of degrees of freedom!

2.4. High Dimensional Chaos?

How to deal with geophysical systems if they do not have low finite dimensions? A radical change that has occurred in weather forecasting during the last decade illustrates the issue. Whereas attention had long been focused on large-scale deterministic modeling, it became clear that small-scale uncertainty must be considered as a first order problem. Deterministic modeling has progressively been replaced by Ensemble Prediction Systems (EPS) [*Molteni et al.*, 1996; *Palmer*, 2000; *Toth and Kalnay*, 1993]. EPS involves a probabilistic study of the trajectories of an ensemble of solutions of a deterministic numerical model started from different initial conditions or from (slightly) different models: e.g. 50 perturbed trajectories are routinely run at the European Center for Medium-Range Weather Forecasts.

Whereas theorists have primarily considered fields as elements of infinite dimensional functional spaces [e.g. *Eckmann and Ruelle*, 1985], meteorologists are interested in the large but finite dimensional projection of meteorological fields onto the phase space (the 'resolved scales') of their numerical models (typically $10^6 – 10^7$ grid-points). There was agreement on the need to dynamically obtain relevant statistics by following the time-evolution of the density of points in phase space. For finite dimensional phase spaces, the equation for this density is the "Liouville equation" [*Liouville*, 1838] and has attracted attention in meteorology [*Ehrendorfer*, 1994; *Epstein*, 1969]. Consider a well-posed finite d-dimensional differential system:

$$\dot{\underline{X}}(t) = \frac{d}{dt}\underline{X} = \underline{F}(\underline{X}, t) \quad (8)$$

The probability density $\rho(\underline{X}, t)$ (with respect to the volume measure $dX_1, dX_2 \ldots dX_d$ of the phase state spanned by $X_1, X_2, \ldots X_d$) satisfies a phase space continuity equation [e.g. *Nicolis*, 1995], the Liouville equation:

$$\frac{\partial}{\partial t}\rho(\underline{X}, t) + \sum_{i=1}^{d} \frac{\partial}{\partial X_i}\left[\dot{X}_i(t)\rho(\underline{X}, t)\right] = 0 \quad (9)$$

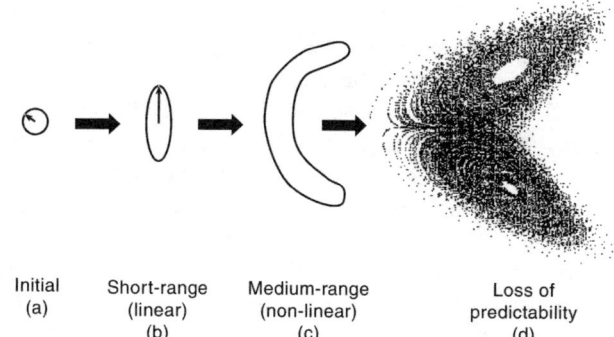

Initial (a) | Short-range (linear) (b) | Medium-range (non-linear) (c) | Loss of predictability (d)

Figure 2: Scheme of the evolution of the empirical pdf in an Ensemble Prediction System according to [*Palmer*, 1999]: from the phase space region occupied by the initial ensemble (a), to (b) linear growth phase, to (c) nonlinear growth phase, to (d) loss of predictability.

Figure 2, reproduced from [*Palmer*, 1999], displays the expected time evolution of the empirical solution of the Liouville equation in an EPS phase space (represented for simplicity as being two dimensional, i.e. spanned by a pair of variables (X_1, X_2)). After an early linear (b) and later nonlinear (c) period of dispersion, the probability converges quickly (exponentially) to an invariant measure (d) of a strange attractor. This is the standard schematic used to illustrate the domain of numerical weather forecasts (b,c) and the loss of predictability (d). Below, we question the physical relevance of this scheme (Sect. 3.4.3).

Due to the scale truncation of the numerical models, the issue of noisy perturbations arising from subgrid processes is fundamental. If these perturbations were gaussian white noises $f(t)$ of intensity ε, then the state of the system would evolve as:

$$\frac{d}{dt}\underline{X} = \underline{F}(\underline{X},t) + \underline{f}(t); \left\langle f_i(t) f_j(t') \right\rangle = \varepsilon \delta_{i,j} \delta(t-t') \quad (10)$$

and the Liouville equation for the probability density would generalize into the Fokker-Planck equation [e.g. *Gardiner*, 1990]:

$$\frac{\partial}{\partial t}\rho(\underline{X},t) + \sum_{i=1}^{d}\frac{\partial}{\partial X_i}\left[\dot{X}_i(t)\rho(\underline{X},t)\right] - \varepsilon \Delta_X \rho(\underline{X},t) = 0 \quad (11)$$

where Δ_x is the Laplacian diffusion operator (in the phase space). However, these perturbations may be strongly non-gaussian and it is better to consider a "fractional" general-

---

[1] Hereafter ρ no longer denotes a probability density, as in a previous section.

ization of the Fokker-Planck equation, which involves fractional derivatives (e.g. [*Schertzer et al.*, 2001] and references therein), in particular, fractional powers of the Laplacian. Furthermore, the noises may be colored rather than white [e.g. *Hasselmann*, 1976]. We will now consider the possibility of taking into account all the scales together.

3. SCALING AND MULTIFRACTAL APPROACHES

3.1. Phenomenology of High Dimensional Systems and Scale Symmetries

3.1.1. Scale symmetry of generating equations. Geophysically relevant equations are nonlinear. They can respect (more or less formal) scale symmetries. This has classically been known as "self-similarity" [e.g. *Sedov*, 1972], but with unnecessary limitations. Consider the Navier-Stokes equations that are used in many geophysical domains:

$$\frac{\partial \underline{u}}{\partial t} + (\underline{u}\cdot\nabla)\underline{u} = -\frac{\nabla p}{\rho} + \nu\nabla^2\underline{u} + \underline{f}; \frac{\partial}{\partial t}\rho + \nabla(u\rho) = 0 \quad (12)$$

with $\underline{u}$ = velocity, t = time, p = pressure, ρ = fluid density[1], ν = viscosity, $\underline{f}$ = forcing density (external forcing, gravity), as well as the associated advection-diffusion equations for a scalar field θ (f_θ = forcing density for the scalar, κ = diffusivity):

$$\frac{\partial \theta}{\partial t} + (\underline{u}\cdot\nabla)\theta = \kappa\nabla^2\theta + f_\theta \quad (13)$$

In geophysics, active scalar fields θ are as important as the velocity field $\underline{u}$, e.g. convection, where ρ sensitively depends on θ, either the temperature (atmosphere or lithosphere) or on the salinity (oceans), contrary to the (academic) passive case (ρ, $\underline{f}$, independent of θ). Although the basic mathematical properties of their solutions (e.g. existence and uniqueness) are unsolved "Hilbert problems" [*Hilbert*, 1902] in both cases, these equations remain formally invariant under any (affine) contraction of the time-space (of scale ratios λ, λ^{1-H}):

$$\underline{x} \to \underline{x}/\lambda \quad t \to t/\lambda^{1-H} \quad (14)$$

as long as the dependent variables are suitably renormalized:

$$\underline{u} \to \underline{u}/\lambda^H; \theta \to \theta/\lambda^{H'}; \rho \to \rho/\lambda^{H''}$$
$$\nu \to \nu/\lambda^{1+H}; \kappa \to \kappa/\lambda^{1+H''}; p \to p/\lambda^{H''-1} \quad (15)$$
$$\underline{f} \to \underline{f}/\lambda^{2H-1}; f_\theta \to f_\theta/\lambda^{H'+H''-1}$$

One may either consider the asymptotic case of fully developed turbulence (with an infinite Reynolds number (Re→∞; equivalently, a vanishing viscosity, ν→0) for the incompressible Navier-Stokes equations [e.g. *Frisch*, 1995]), or consider the case of non-zero eddy (rather than molecular) viscosity and eddy diffusivity, [e.g. *Schertzer and Lovejoy*, 1987]. In numerical models, where scales are split into resolved and (parametrized) sub-grid components this symmetry is unfortunately broken.

3.1.2. Scaling and quantitative laws in turbulence. Scale invariance (or scaling) is also a symmetry directly related to a striking and rather general feature of nonlinear systems: their high variability in space and time. Indeed, this extreme variability can easily be understood as the result of a scale invariant process that is repeated scale by scale, thus multiplicatively amplifying even small variability present at larger scales. This is related to the paradigm of cascades [*Richardson*, 1922] and is exemplified by the basic quantitative turbulence laws: the Richardson law [*Richardson*, 1926] and the Kolmogorov-Obukhov law [*Kolmogorov*, 1941; *Obukhov*, 1941]. The first relates the relative separation $r(t)$ of a pair of particles passively advected by turbulence:

$$\langle r(t)^2 \rangle \propto \overline{\varepsilon}\, t^3 \qquad (16)$$

$\overline{\varepsilon}$ is the spatial average of the energy dissipation rate density, as well of the density of the energy flux to smaller scales. The second relates the shear of the velocity field $\delta u(\ell)$ to the scale ℓ:

$$\langle \delta u(\ell)^2 \rangle \propto \overline{\varepsilon}^{2/3} \ell^{2/3} \qquad E(k) \propto \overline{\varepsilon}^{2/3} k^{-5/3} \qquad (17)$$

where $E(k)$ is the energy spectrum at the wave-number k. Their extensions to a passive scalar field [*Obukhov*, 1949; *Corrsin*, 1951] are:

$$\langle \delta\theta(\ell)^2 \rangle \propto \overline{\chi}\,\overline{\varepsilon}^{-1/3} \ell^{2/3} \qquad E_\theta(k) \propto \overline{\varepsilon}^{2/3} k^{-5/3} \qquad (18)$$

where χ is the flux of scalar variance, $E_\theta(k)$ the spectrum of the scalar field. Surprisingly, these laws are still beyond the reach of analytical techniques, including the Quasi Normal Approximation [*Millionshchikov*, 1941], the Direct Interaction Approximation [*Kraichnan*, 1958; *Kraichnan*, 1959] (for review see [*Leslie*, 1973]), as well from the renormalization group [*Forster et al.*, 1977]. Indeed, rather ad-hoc modifications are required to obtain 'analytical closures' (Eddy-Damped QuasiNormal Model [EDQNM, *Orszag*, 1970] and the Test Field Model [TFM, *Kraichnan*, 1971]), which are compatible with these "mean field" laws. However, closures are unable to account for intermittency [*Frisch et al.*, 1980], see Sect. 3.4.1.

3.1.3. Eddy turnover time and scaling space-time anisotropy. In contrast to the exponential predictability decay law (Eq. (5)) for deterministic chaos (Sect. 2.2), Eqs. (16)–(18) are power laws. This is a consequence of the transformation group, which for any value of the exponents H, H' leaves the generating equations invariant (Eq. (15)). The particular values $H = H' = 1/3$, which correspond to the Kolmogorov-Obukhov (Eq. (17)) and to Corrsin-Obukhov law (Eq. (18)), can be found either by purely dimensional considerations or using the physical notion of "eddy turnover time" $\tau(\ell)$. The latter is the characteristic time for a structure of scale ℓ with a velocity shear across it $\delta u(\ell)$ to "turn over":

$$\tau(\ell) \propto \ell / \delta u(\ell) \qquad (19)$$

Since the characteristic time of destruction of structures of this scale ℓ must be proportional to the eddy turn-over time [e.g. *Robinson*, 1971], one finds that the rate of transfer of energy to smaller scale is:

$$\varepsilon(\ell) \propto \delta u(\ell)^2 / \tau(\ell) \propto \delta u(\ell)^3 / \ell \qquad (20)$$

therefore:

$$\delta u(\ell) \propto \varepsilon(\ell)^{1/3} \ell^{1/3} \qquad \tau(\ell) \propto \varepsilon(\ell)^{-1/3} \ell^{2/3} \qquad (21)$$

By performing a spatial average and considering that $\overline{\varepsilon}(\ell)$ is scale independent and not too fluctuating (i.e. $\overline{\varepsilon^q} \approx \overline{\varepsilon}^q$), Eq. (21) yields the Kolmgorov-Obukhov scaling law Eq. (17) as well as an homogeneous eddy turn-over time $\overline{\tau}(\ell)$:

$$\overline{\tau}(\ell) \propto \overline{\varepsilon}^{-1/3} \ell^{2/3} \qquad (22)$$

This confirms that there is a scaling anisotropy between time and space i.e. the (typical) lifetime of a structure varies as a power of the scale, in agreement with Eq. (14), with the "dynamical exponent": $1 - H_t = 2/3$, to be used in Sects. 3.2, 3.4.2, 3.4.3. It also yields the scaling laws for the passive scalar (Eqs. (18)).

3.2. Predictability in Homogeneous Turbulence

3.2.1. The phenomenology of error growth through scales and the MET. The general phenomenology of error growth through scales is rather straightforward: an error or uncertainty initially confined to small-scales will progressively 'contaminate' large-scale structures through these interactions. This is in sharp contrast with the MET that does not consider the problem of many nonlinearly interacting spatial scales. The problem of the evolution of spatially extended fields was first theoretically investigated by [Thompson,

1957]. Using initial time-derivatives and various meteorological models, Thompson studied the nonlinear uncertainty growth due to errors in the initial conditions resulting from the limited resolutions of the measurement network and of the models. He estimated the root mean square (RMS) doubling time for small errors to be about two days, whereas [Charney and al., 1966], using more elaborate meteorological models, estimated it as five days.

The scale dependency of the predictability times was underlined by [Robinson, 1967; Robinson, 1971]. Indeed, if the notion of characteristic error time τ is still relevant, it should depend on the spatial scale ℓ in a hierarchical manner. For $t > \tau(\ell)$ two fields initially similar at scale ℓ become quite different (e.g. rather decorrelated) at this scale, but may remain similar at larger scales. This is in agreement with the following estimates of the Lyapunov exponent $\mu(\ell) \propto 1/\overline{\tau}(\ell)$ and the characteristic space scale ℓ_e reached by the error at time t (see Eq. (22)):

$$\mu(\ell) \propto 1/\overline{\tau}(\ell) \propto \overline{\varepsilon}^{1/3} \ell^{-2/3} \qquad \ell_e(t) \propto \overline{\varepsilon}^{1/2} t^{3/2} \quad (23)$$

This shows—contrary to the usual assumption (Eq. (6))—that unless a break in the scaling occurs leading to smooth small scale behavior, the Lyapunov exponent μ will diverge at small scales.

3.2.2. Energetics and spectral analysis of the error growth. Let $\underline{u}^1(\underline{x},t)$ and $\underline{u}^2(\underline{x},t)$ be two solutions of a nonlinear system (e.g. velocities for Navier-Stokes equations) initially identical, but with a perturbation (error) $\delta \underline{u}(\underline{x},0) = \underline{u}^2(\underline{x},0) - \underline{u}^1(\underline{x},0)$ at t=0, confined to infinitesimally small spatial scales. The time-evolution of $\delta \underline{u}(\underline{x},t)$ corresponds to the effect of butterflies homogeneously distributed in space, rather than the effect of a single butterfly. When the nonlinear interactions preserve the kinetic energy (e.g. Navier-Stokes equations), it is convenient but not sufficient to consider both the correlated (kinetic) energy (per unit of mass):

$$e^c(\underline{x},t) = \tfrac{1}{2} \underline{u}^1(\underline{x},t) \cdot \underline{u}^2(\underline{x},t) \quad (24)$$

and the decorrelated energy:

$$e^\Delta(\underline{x},t) = \tfrac{1}{2} (\delta \underline{u}(\underline{x},t))^2 = \tfrac{1}{2} (\underline{u}^2(\underline{x},t) - \underline{u}^1(\underline{x},t))^2 \quad (25)$$

as well as the total energy and the energy of each solution:

$$e^T(\underline{x},t) = e^1(\underline{x},t) + e^2(\underline{x},t); \quad e^n(\underline{x},t) = \tfrac{1}{2}(\underline{u}^n(\underline{x},t))^2 \quad (26)$$

This implies the relation:

$$e^T(\underline{x},t) = e^c(\underline{x},t) + e^\Delta(\underline{x},t) \quad (27)$$

hence, if the total energy is statistically stationary (conserved on average), there will be a flux of correlated energy $e^c(\underline{x},t)$ to decorrelated energy $e^\Delta(\underline{x},t)$. This also holds for the corresponding energy spectra $E^T(k,t) = E^c(k,t) + E^\Delta(k,t)$, since the latter corresponds to a linear decomposition of the former with respect to wave number, k. Therefore, the decorrelated energy spectrum $E^\Delta(k,t)$ steadily increases in magnitude from large to small wavenumbers, converging to the total energy spectrum $E^T(k,t) \approx k^{-5/3}$ (Figure 3). The critical wave number $k_e(t)$ of the transition from dominant correlation to dominant decorrelation can be defined by $E^c(k_e(t),t) = E^\Delta(k_e(t),t)$, scales as $1/\ell_e(t)$ (Eq. (23)) and decreases as: $k_e(t) \approx t^{-3/2}$.

3.2.3. Consequences and limitations. If the constant of proportionality in the definition of the eddy-turn over time (Eq. (22)), as well as that relating the latter to the error time, is of order unity, then taking a "typical values" $\overline{\varepsilon} = 10^{-3} \mathrm{m}^2 \mathrm{s}^{-3}$ and $\eta \approx 10^{-3} \mathrm{m}$ (energy flux and the viscous scale), one obtains $\tau_e(\eta) \approx \overline{\varepsilon}^{-1/3} \eta^{2/3} \approx 10^{-1} \mathrm{s}$, as well as $\tau_e(\ell) \approx \tau_e(\eta)(\ell/\eta)^{2/3}$ and therefore $\tau_e(\ell) \approx 10 \mathrm{s}$; 1/2 hr; 28 hr; 5.4 days respectively for $\ell = 1 \mathrm{m}$, 1 km, 10^3 km, 10^4 km. These estimates are close to those obtained by [Lorenz, 1969] (Figure 3), but slightly lower than the numerical (closure, nonintermittent) results obtained by [Leith, 1971; Métais and Lesieur, 1986] and [Kraichnan, 1970; Leith and Kraichnan, 1972].

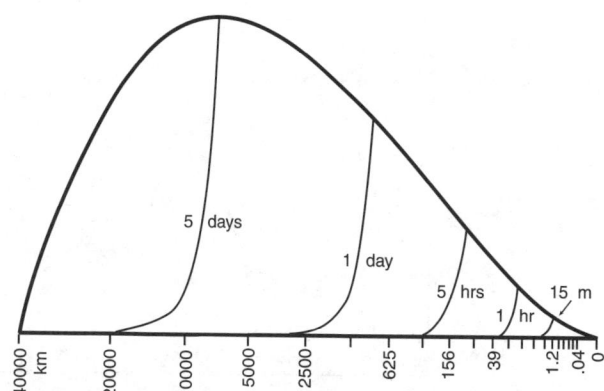

Figure 3: Atmospheric error growth according a quasi-normal closure simulation [Lorenz, 1969]. The decorrelated energy spectrum $E^\Delta(k, t)$ initially confined to a few meters (on the right) "pollute" the larger scales (up to 20,000 km (on the left) after various time intervals (from few minutes to 5 days).

3.3. Questioning the Lorenz-Leith-Kraichnan Approach

3.3.1. Isotropic models of atmospheric motions.

The use by [Lorenz, 1969] of a $k^{-5/3}$ spectrum up to synoptic scales was considered as paradoxical and speculative, e.g. [Robert and Rosier, 2001], in any case as in contradiction with the standard model of atmospheric dynamics that considers large-scale atmospheric motions as quasi-two dimensional (quasi-2D) and small-scale motions as quasi-three dimensional (quasi-3D). This standard model emerges from quasi-linear approximations, in particular the quasi-geostrophic approximation [Charney, 1948] and the related notion of quasi-geostrophic turbulence [Charney, 1971]. Furthermore, it seems eminently sensible since at large scale the atmosphere appears to be a thin film of thickness $h \approx 10$ km with large horizontal scale $L \approx 20{,}000$ km.

However, 2D and 3D turbulence have quite distinct dynamics and transport properties. 2D turbulent flows are fairly smooth, since the main dynamical mechanism of 3D turbulence—vortex stretching—is impossible. Whereas 3D dynamics generate vorticity explosively, 2D dynamics only advect vorticity conservatively. Among various consequences of this vorticity conservation, there is no longer a cascade of energy towards smaller scales, but a cascade of enstrophy (the vorticity squared) with a different spectral slope [Fjortoft, 1953; Kraichnan, 1967] and nearly scale invariant eddy-turn over times.

In order to hold together, the standard model requires a "meso-scale gap", otherwise the 3D turbulence destabilizes the 2D turbulence. The existence of this gap was initially given some empirical support (Figure 4) by estimates of wind spectra [Panofsky and Van der Hoven, 1955; Van der Hoven, 1957]. In spite of strong criticism [Pinus, 1968], it was eventually consecrated by [Monin, 1972; Pedlosky, 1979]. As a consequence, [Leith and Kraichnan, 1972] studied 2D and 3D turbulent flow predictability.

More fundamentally, the atmosphere is buoyancy driven so that we must consider the fact that buoyancy forces generate another conservative flux [Bolgiano, 1959; Obukhov, 1959], related to both the potential energy defined by gravity, and to the large scale stable stratification of the atmosphere. Buoyancy was not considered in the standard model, whereas it is expected to dominate the kinetic energy flux at large scales. Unfortunately, this "buoyancy subrange" was originally hypothesized as an isotropic regime i.e. with the same $k^{-11/5}$ spectrum in both horizontal and vertical directions. However, as it was never observed along the horizontal, the idea languished. As discussed below an anisotropic generalization is required to yield a coherent model of atmosphere.

[Lilly, 1985] argued that empirical findings in the 1980's seriously undermined the standard model. The GASP experiment (Figure 5), the first large-scale campaign to measure the horizontal velocity spectrum [Lilly and Paterson, 1983; Nastrom and Gage, 1983b], found no evidence of a mesoscale spectral gap. This result was confirmed by the more recent MOZAIC experiment [Lindborg, 1999]. Instead, they found a Kolmogorov $k^{-5/3}$ scaling extending to at least hundreds of kilometers. This was confirmed in a variety of climatological and meteorological regimes, including tropical cylonic conditions [Chigirinskaya et al., 1994]. However, the atmospheric dynamics that aircrafts are assumed to measure can induce fractal aircraft trajectories [Lovejoy et al., 2004] and hence possible biases due to mixed (nontrivially correlated) measurements of vertical and horizontal fluctu-

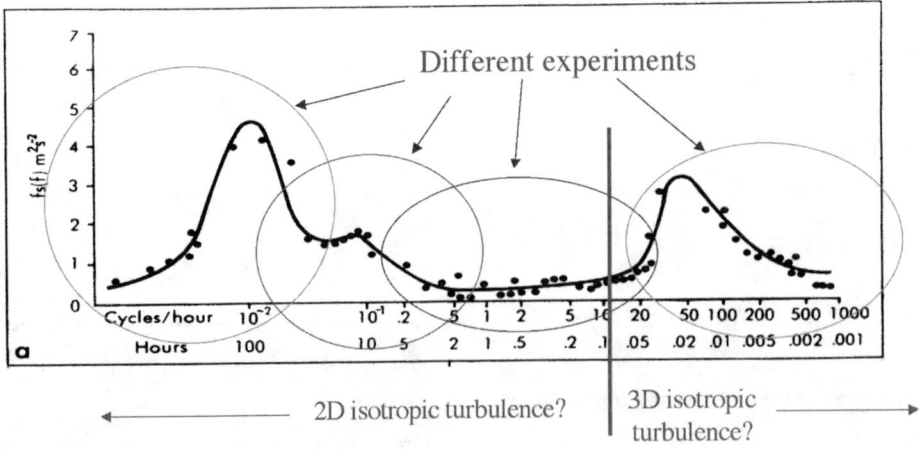

Figure 4: Atmospheric wind spectrum according to [Van der Hoven, 1957]. There are four different experiments at different frequencies. According to [Vinnichenko, 1969], the high frequency data was taken at "hurricane-like conditions", hence the spurious gap.

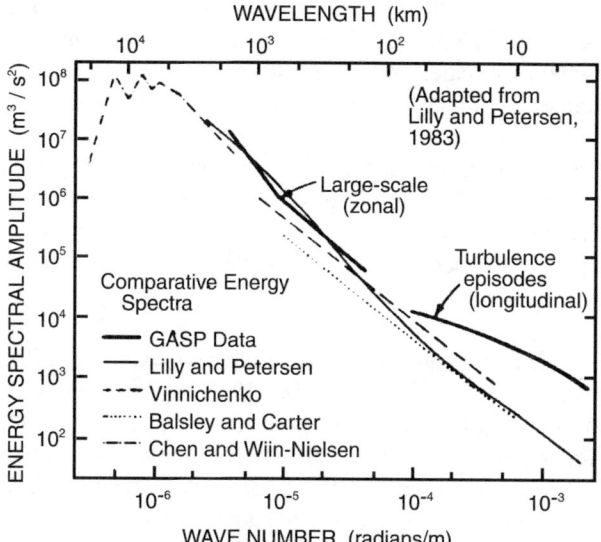

Figure 5: GASP atmospheric wind spectra [*Nastrom and Gage*, 1983a] from instrumented commercial aircraft. The basic spectrum is close to Kolmogorov out to at least several hundred kilometers and there is no evidence for a "meso-scale gap".

ations. As an alternative, [*Lovejoy et al.*, 2001] used the fact that the infra red and visible radiances of cloud fields are strongly coupled to their structures and to the dynamics and therefore should be scaling over the same range: the analysis of nearly 1000 satellite images show that the (multi) scaling of the radiances was respected from planetary scales down to at least kilometer scales.

[*Gage*, 1979] proposed that the meso-scale $k^{-5/3}$ spectrum could be explained by a 2D inverse energy cascade. This modification of the standard model is nevertheless inconsistent with the empirical evidence that atmospheric variability is also scaling along the vertical. Using balloon measurements over heights of 10–20 km the vertical spectrum of the horizontal wind was consistently found (Figure 6) to follow Bolgiano-Obhukov (BO) $k^{-11/5}$ scaling throughout the troposphere [*Adelfang*, 1971; *Endlich et al.*, 1969; *Lazarev et al.*, 1994; *Schertzer and Lovejoy*, 1985].

3.3.2. Anisotropic models of atmospheric motions. [*Schertzer and Lovejoy*, 1985a] proposed that in the horizontal the statistics are dominated by the energy flux (leading to $k^{-\beta_h}, \beta_h \approx 5/3$ in the horizontal), while in the vertical the conservative flux generated by the buoyancy force is dominant leading to $k^{-\beta_v}, \beta_v \approx 11/5$. Contrary to the original BO framework, neither the Boussinesq approximation, nor other stable stratified reference states are required. Fluid particles respond to local gradients, not gradients with respect to theoretical reference values.

In this new model, the atmosphere is anisotropic at all scales and effectively becomes progressively flatter and flatter along the horizontal at larger and larger scales. The differential flattening can be characterized by an intermediate "elliptical dimension" $2 \leq D_{el} \leq 3$ [*Schertzer and Lovejoy*, 1984]: $D_{el} = 2 + (\beta_h - 1)/(\beta_v - 1)$ (where β_h, β_v are the spectral slopes along the horizontal and vertical directions). When the horizontal extent of a structure is increased by λ, its volume increases by $\lambda^{D_{el}}$. The atmosphere is therefore neither 3D isotropic at small scales nor 2D isotropic at large scales ($D_{el} = 3$ and 2 are the 3D and 2D isotropic cases respectively). In terms of D_{el} the current atmospheric debate is between buoyancy driven flows with $D_{el} = 23/9 = 2.555$ and flows resulting from a gravity wave mechanism leading to $D_{el} = 7/3 = 2.333$ ($\beta_v \approx 3$), [*Lumley*, 1963; *Shur*, 1962; *Van Zandt*, 1982; *Weinstock*, 1978]. Recent lidar based aircraft data from pollution (considered as a passive scalar surrogate), using simultaneous vertical and horizontal backscatter measurements, yield $D_{el} = 2.55 \pm 0.02$ over the range 3 m to 120 km, very close to the 23/9 model and incompatible with the 2D, 3D and 7/3D models (Figure 7 and [*Lilley et al.*, 2004]). Note that empirical evidence of anisotropic scaling have been also found for the magnetic susceptibility of the earth crust [*Lovejoy*, 2001] and

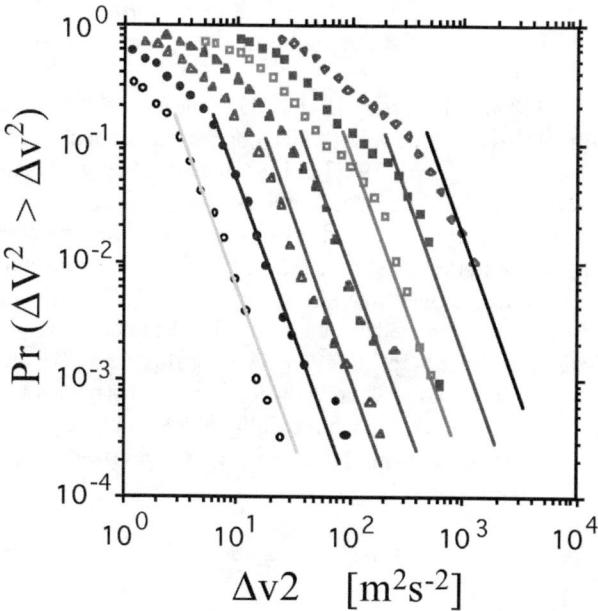

Figure 6: Scaling of the probability distribution of the vertical shear of the horizontal wind [*Schertzer and Lovejoy*, 1985b]. The left curve for 50m thick layers, with thickness increasing by factors of 2 to right (80 radiosonde ascents). The straight lines indicate a reference slope –5 (corresponding to a power-law probability falloff, divergence of moments of order $q>5$), and the line spacing indicates Bolgiano-Obukhov (BO) scaling in the vertical.

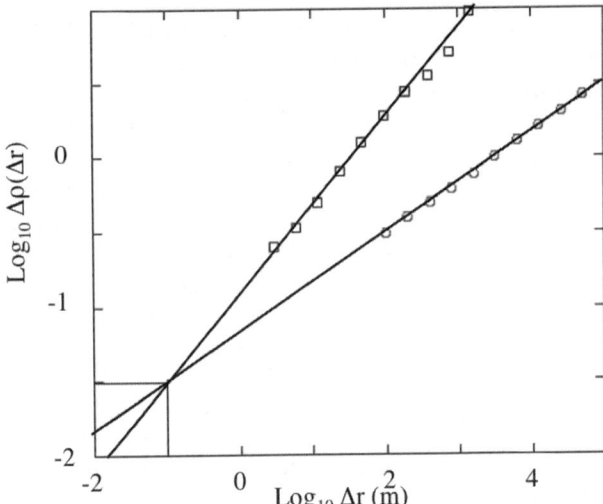

Figure 7. Log-log plot of the first order structure function of the aerosol backscatter ratio [*Lilley et al.*, 2004], a surrogate for the concentration of a passive scalar, obtained from 9 vertical cross-sections (4.5 km thick, 120km long) with vertical resolution 3m, horizontal resolution 100m. The straight lines indicate the theoretical exponents (horizontal), (vertical), their intersection determines point the "sphero-scale" l_s=10cm. Structures larger than l_s are flattened in the horizontal, whereas smaller structures are vertically aligned.

for the soil hydraulic conductivity [*Tchiguirinskaia*, 2002], so that geophysical stratification may be generally scaling.

3.3.3. Which statistics? [*Lilly*, 1985] questioned the quasi-normal closure framework , which implies that the analyses of [*Lorenz*, 1969] [*Leith*, 1971] and [*Kraichnan*, 1970; *Leith and Kraichnan*, 1972] are local as well as global. This is not consistent with the observation that various atmospheric structures (e. g. rotating thunderstorms [*Lilly*, 1983]) maintain a stable identity much longer than their turnover time. [*Schertzer and Lovejoy*, 1984] reported that the probability distributions of vertical wind shear amplitude $|\delta u|$, energy flux density ε and potential temperature θ have power law tails. This means that for example (Figure 6) the probability of the wind shear $|\delta u|$ (as well as it squared δu^2) to exceeding a (large) fixed threshold is a power-law :

$$\Pr(|\delta u| \geq s) \approx s^{-q_D} \quad (28)$$

the power-law exponent q_D is a critical exponent. [*Lilly*, 1985] therefore argued that the error statistics should be similarly divergent and much more variable and extreme than those estimated in the quasi-normal framework of closures. Indeed, power law probability tails, are often considered a hallmark of Self-Organized Criticality [*Bak et al.*, 1987]; the exponent q_D being a critical order of divergence of statistical moments

This means that the (theoretical) statistical moments of order $q \geq q_D$ are infinite. On finite (e.g. empirical) samples the moment estimates are finite but grow (i.e. diverge) as the number of samples increases. This divergence results from the fact that the weights or frequencies of extremes is much higher than usual. This is consistent with the fact that the probability of finding a 10 times larger fluctuation, decreases only by a factor 10^{q_D} (e.g. according to Figure 6 the probability to have a 10 times larger wind shear decreases only by 10^5, not by an exponential factor).

3.3.4. The fundamental role of intermittency. As early as 1942 Landau [*Landau and Lifshitz*, 1987; *Yaglom*, 1994] questioned the assumption of the homogeneity of fluxes used by [*Kolmogorov*, 1941; *Obukhov*, 1941] to derive their scaling law of velocity shears (Eq. (17)). [*Batchelor*, 1953; *Batchelor and Townsend*, 1949] observed that not only does the "activity" of turbulence induce inhomogeneity, but the activity itself is very inhomogeneously distributed: there are "puffs" of active turbulence inside of puffs of (active) turbulence. This inhomogeneity has been termed "intermittency", which may be even more fundamental for active scalars such as rain and cloud fields than for the dynamics. These considerations lead to the general idea [e.g. *Leslie*, 1973] that a turbulent flow is only turbulent in tiny fractions of space and time. A precise meaning to the term "fraction" was achieved with the help of cascade models.

3.4. Strongly Non Gaussian Statistics and Multifractal Modeling

In contrast to the limitations of closure models, multifractal models yield strongly non-gaussian statistics and therefore structures of very different intensities. A key ingredient is a multiplicative property of the models that describe a cascade of instabilities, i.e. incrementally the heterogeneity of fluxes flowing through smaller and smaller structures increases.

3.4.1. Multifractals and the phenomenology of cascades. Stochastic multifractal processes originated from the phenomenological assumption [e.g. *Yaglom*, 1966] that in turbulence successive cascade steps define independent fractions of the flux, F, transmitted to smaller scales and that a cascade is scaling (Figure 8). To be more precise, let $\lambda = L/\ell$ be an intermediate scale ratio ("resolution"), where L is the outer scale and ℓ the scale corresponding to scale ratio λ, and let $\Lambda = L/\ell' = \lambda\lambda'$ be the total scale ratio of the cascade. Scaling means that the cascade from λ to Λ corresponds to a cascade from ratio 1 to λ' contracted by T_λ of scale ratio λ; $T_\lambda(f(\underline{x})) = f(T_\lambda(\underline{x}))$; in the self-similar (isotropic) case $T_\lambda(\underline{x}) =$

$\underline{x}/\lambda$. When combined these two properties imply that the flux is a multiplicative group, with the equality symbol indicating that the random variables on each side have identical probability distributions:

$$F_{\Lambda=\lambda\lambda'} \stackrel{d}{=} F_\lambda \cdot T_\lambda(F_{\lambda'}) \qquad (\lambda, \lambda' \geq 1) \qquad (29)$$

Hence we obtain the following scaling law for the statistical moments:

$$<F_{\lambda\lambda'}^q> = \lambda'^{K(q)} <F_\lambda^q> \qquad (\lambda, \lambda' \geq 1) \qquad (30)$$

where the exponent $K(q)$ is the moment scaling function. The probability distribution for the event $\{F_\lambda \geq \lambda^\gamma\}$ is also scaling (mathematically, this may be obtained using the Mellin transform [*Schertzer and Lovejoy*, 1993; *Schertzer et al.*, 2002a]):

$$\Pr\{F_\lambda \geq \lambda^\gamma\} \propto \lambda^{-c(\gamma)} \qquad (31)$$

the arbitrary exponent γ, which defines a given level of activity or intensity at all resolutions λ, is a "singularity": the larger it is, the faster F_λ grows with resolution/scale λ. The scaling exponent $c(\gamma)$ of the probability is a statistical codimension [*Schertzer and Lovejoy*, 1987] also called the "Cramer" function [*Mandelbrot*, 1991; *Oono*, 1989]. When the embedding dimension $d > c(\gamma)$, it corresponds to a geometric notion of codimension, so that on a given sample of this process, the event $\{F_\lambda \geq \lambda^\gamma\}$ is almost surely a fractal set of dimension: $D(\gamma) = d - c(\gamma)$. In other words a multifractal field can be understood as an infinite hierarchy of embedded fractal sets of dimension $D(\gamma)$ and supporting a given singularity γ, i.e. F_λ grows faster than λ^γ with increasing resolution λ. The highest singularities are the rarest, hence $c(\gamma)$ increases with γ, whereas $D(\gamma)$ decreases. For instance, the schematic Figure 8 displays a unique extreme singularity, three more intermediate ones, and extremely low ones for the rest of the (2D) space. In any case, multifractality cannot be understood as a dimension depending on scale. Whereas scale invariant geometric sets of points are fractals, scale invariant fields (i.e. with a value at each point) are multifractals.

The two scaling functions are related by the Legendre transform [*Parisi and Frisch*, 1985]:

$$K(q) = \max_\gamma \{q\gamma - c(\gamma)\} \quad c(\gamma) = \max_q \{q\gamma - K(q)\} \qquad (32)$$

Thus the main multifractal properties common to all the various formalisms are an infinite hierarchy of statistical exponents [e.g. *Badii and Politi*, 1984; *Grassberger*, 1983; *Grassberger and Procaccia*, 1983; *Hentschel and Procaccia*, 1983; *Schertzer and Lovejoy*, 1984; *Stanley and Meakin*, 1988] and an infinite hierarchy of singularities [e.g. *Benzi et al.*, 1984;

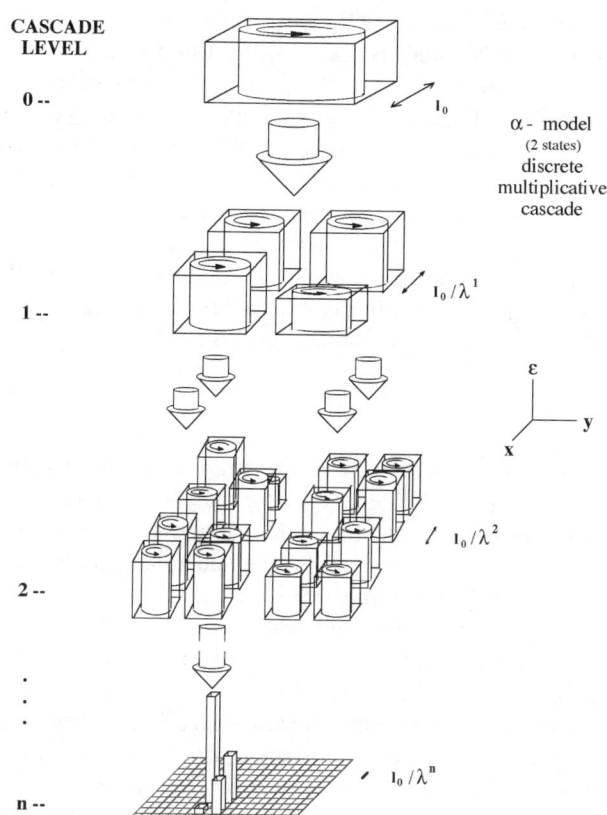

Figure 8. Illustration of a discrete (in scale) cascade process. The displayed first few steps show how the energy flux at large scales multiplicatively modulates the flux at successively smaller scales. Since the energy is conserved on average, the flux is concentrated in a hierarchy of fractal sets whose fractal dimension decreases with intensity threshold.

Halsey et al., 1986; *Parisi and Frisch*, 1985]. However, there are substantial differences. For example, there is an upper bound for singularities of "geometrical" multifractals [*Halsey et al.*, 1986; *Parisi and Frisch*, 1985], where each singularity is assumed to be supported by a well–defined geometrical (fractal) set, and for singularities of "microcanonical" multifractal processes that conserve fluxes on each realization as well as scale by scale [e.g. *Benzi et al.*, 1984; *Meneveau and Sreenivasan*, 1987; *Pietronero and Siebesma*, 1986]. In contrast, more general canonical multifractals, which conserve only ensemble flux averages, do not generally have any upper bound [*Schertzer and Lovejoy*, 1992]. The resulting extreme singularities yield power probability distributions having power-law tails (Eq. (28)). which are discussed in Sect. 3.3.3.

3.4.2. Multifractal modeling. Static multifractal models (pure spatial cuts, without time) have become useful tools for simulations of clouds [e.g. *Arneodo et al.*, 1999; *Naud et al.*,

1996; *Wilson et al., 1991*] and of other geophysical fields [e.g. *Deidda, 2000; Pecknold et al., 1996; Pecknold et al., 1997; Pecknold et al., 1993*]. Their dynamic versions, i.e. space-time processes, have been developed for studying turbulence, rain and the predictability [*Marsan et al., 1996; Over and Gupta, 1996; Schertzer et al., 1997; Marsan, 1998*]. Multiplicative processes, in particular when continuous in scale, can be generated from white noises. Indeed, they can be obtained by the exponential of an additive process $\Gamma_\lambda(\underline{x})$ called the "generator "of the flux ($\underline{x} = (x, y, z)$ for a 3D spatial process, $\underline{x} = (x, y, z, t)$ for a time-space process):

$$F_\lambda = e^{\Gamma_\lambda} \qquad (33)$$

In order to respect the scaling property of the statistical moments (Eq. (30)) the generator must have a logarithmic divergence with resolution λ; $\Gamma_\lambda \propto Log(\lambda); \lambda \to \infty$. This is achieved with an appropriate "fractional integration", i.e. a power-law filtering in the Fourier space [for details e.g. *Schertzer et al., 1997*]) of a white noise, which can be chosen as a Levy stable noise [*Schertzer and Lovejoy, 1987; Schertzer and Lovejoy, 1997*]. One may note that the anisotropy between space and time (Sect. 3.1.3, in particular Eq. (22)), as well as the necessary causality condition (asymmetry between past and future) can easily be taken into account [*Marsan et al., 1996*].

3.4.3. Multifractal predictability limits. In order to generalize the approach followed in the spectral analysis of predictability (Sect. 3.2.2) to multifractals, we consider the time evolution of a pair of fields of common resolution Λ. They are identical up to the time t_0 when one lets the fluxes become independent at small scales [*Schertzer and Lovejoy, 2004*]. For simplicity, consider the scalar rain rate R(x,t) illustrated by Plate 1 (time t along the horizontal, location x along the vertical). Plates 1a–b display a pair of rain rate fields $R^1_\Lambda(x, t)$ and $R^2_\Lambda(x, t)$ and Plate 1d their absolute difference $|\delta R_\Lambda(x,t)|$. One may qualitatively note the role of intermittency: most of the difference $|\delta R_\Lambda(x,t)|$ is due to a small number of extremely large values.

[*Marsan et al., 1996*] checked that the spectral analyses of multifractal simulations of a velocity component are in agreement with homogeneous turbulence results (Sect. 3.2.2, and Figure 3). Bursts of violent fluctuations cannot be accounted for using second order statistical moments, in particular energy spectra; these are evident in Figure 9 which displays an 'elementary' decorrelated/error energy spectrum, i.e. not obtained by ensemble average, but only over a unique sample. It is no longer as smooth as an ensemble averaged decorrelated/error energy spectrum $E^\Delta(k, t)$ (e.g. Figure 3), but rather corresponds to a sequence of decorrelation (more

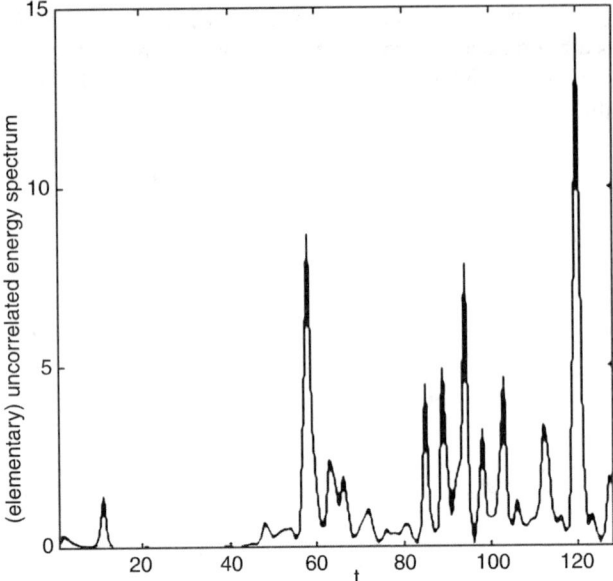

Figure 9. Elementary error energy spectrum displaying decorrelation bursts ([*Schertzer et al., 1997*]). This elementary error energy spectrum is obtained from a unique realization ($\Lambda = 64$ is the resolution of the simulation) rather than from an ensemble average. It is no longer as smooth as Figure 3, but rather corresponds to a sequence of decorrelation bursts at different scales.

generally of independence) bursts at different scales. These bursts result from the fact that although the energetics of the upscale cascade of errors remain basically the same, they do not constrain the largest fluctuations of the errors as much as in the homogeneous turbulence case.

We emphasized that statistics of second order moments, in particular their correlation that corresponds to the correlation energy for a velocity field, are unable to account for the co-evolution of a pair of multifractal fields. Therefore, we need to consider a covariance of order q for different values of q. This is rather simple for fluxes, e.g. the respective energy flux densities ε^i_Λ ($i=1,2$) of a pair of velocities $u^i(\underline{x}, t)$. Up to t_0 the fluxes are identical over the full range of the cascade process (i.e. over the possibly infinite cascade scale ratio Λ). After t_0, they remain rather similar only over a decreasing scale ratio $\lambda(t) \leq \Lambda$, which necessarily follows a power law. More precisely [*Schertzer and Lovejoy, 2004*], the latter is defined by the dynamical exponent $1-H_t$, which defines the scaling space-time-space anisotropy (Sect. 3.1.3, in particular Eq. (22)):

$$t \leq t_0 : \lambda(t) = \Lambda; \ t > t_0 : \lambda(t) \approx Min[\Lambda, (T/(t-t_0))^{\frac{1}{1-H_t}}] \qquad (34)$$

where T is the outer time scale. As a consequence, one obtains for the (normalized) covariance of order q:

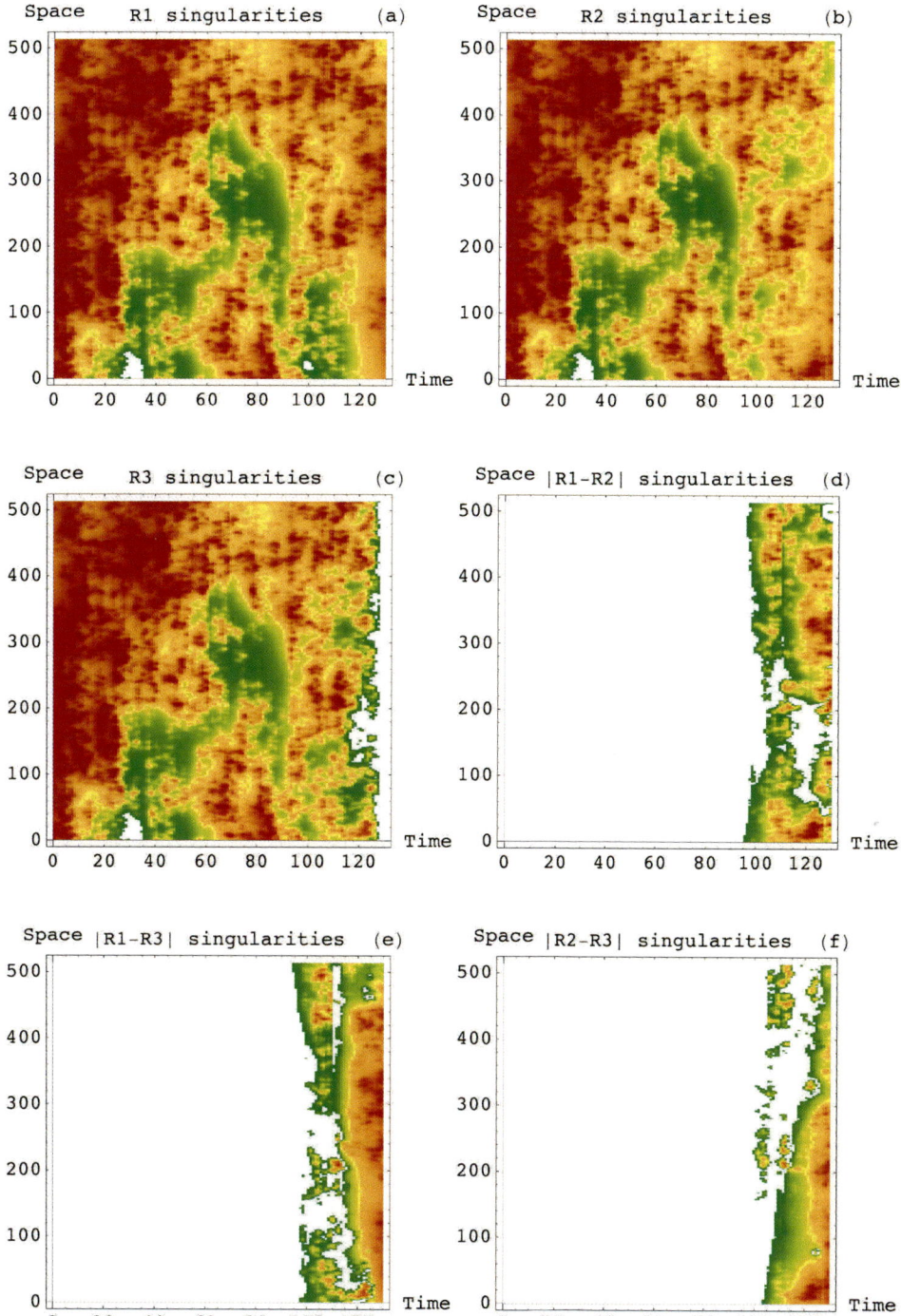

Plate 1.a–f Simulation of multifractal predictability decay for rain field: (a) and (b) are identical up to $t_0=64$, after which their fluxes become independent. (c) displays the forecast based on their common past and the deterministic conservation of the flux afterward. Singularities of the fields (i.e. their log divided by the log resolution), as well as of their absolute differences (d–f), are displayed according to the following palette: white for negative singularities; green to yellow for singularities contributing to statistics up to the mean; red for singularities contributing to second and higher order moments.

$$C^{(q)}(\varepsilon_\Lambda^1,\varepsilon_\Lambda^2)=\left\langle\left(\varepsilon_\Lambda^1\varepsilon_\Lambda^2\right)^q\right\rangle\Big/\left(\left\langle\varepsilon_\Lambda^{1q}\right\rangle\left\langle\varepsilon_\Lambda^{2q}\right\rangle\right)\propto\lambda(t)^{K(q,2)} \quad (35)$$

with $K(q, 2) \equiv K(2q) - 2K(q)$. The multifractality $K(q,2)$ of the joint field $\varepsilon_\Lambda^1\varepsilon_\Lambda^2$ is purely defined by that of ε_Λ^i ($K(q)$). The distinctive feature is that instead of being fixed at Λ (as for $\varepsilon_\Lambda^1=\varepsilon_\Lambda^2$) the range of scale ratios $[1, \lambda(t)]$ also has a power-law decay (Eq. (34)). The same occurs for its probability distribution. It is important to appreciate that these power laws are valid for all time scales, not only the large scales. This is in a sharp contrast with the schematic sequence of predictability behaviors presented by Figure 2 and discussed in Sect. 2.4. The corresponding "Liouville+ MET' scenario is therefore not relevant for multifractal fields.

3.4.4. Forecasts and stochastic parameterizations. We may now explore the question of optimizing forecast procedures so that the decay law of the (normalized) covariance $C^{(q)}(\varepsilon_\Lambda^F,\varepsilon_\Lambda^0)$ of order q of the forecast field ε_Λ^F and of the observed field ε_Λ^0 is as close as possible to the theoretical $C^{(q)}(\varepsilon_\Lambda^I,\varepsilon_\Lambda^2)$ (Eq. (35)). For example, let us point out that the multifractal behavior of meteorological fields theoretically explains and confirms the recent empirical evidence that stochastic parametrizations do better than deterministic ones [*Buizza et al.*, 1999; *Houtekamer et al.*, 1996], in particular in the EPS framework. It suffices to use the fact that a multifractal field may be defined with the help of a white-noise. Indeed, past and future components of a white noise are independent and identically distributed. Therefore, any white-noise identically distributed to the past component is obviously a possible future component. In particular, the resulting process will in the future keep the same statistical properties, as well as the same scale ratio. On the contrary, a future component defined in a deterministic manner cannot have an identical statistical distribution. In particular, its scaling function $K_{det}(q)$ is linear with respect to q, instead of being nonlinear as that of the observations $K(q)$. At best, one can only find a (deterministic) procedure to preserve the statistics of a given order q. This is illustrated by Plate 1 obtained by a numerical simulation, where the (deterministic) future component of the noise of the forecast field $R_\Lambda^3(x, t)$. (Plate 1c) was defined to preserve the mean ($q = 1$) of the flux. Plates 1e–f display the drastic loss of all extreme events ($q \gg 1$) with respect to the samples $R_\Lambda^1(x, t)$ and $R_\Lambda^2(x, t)$. More quantitative statements can be readily obtained with the help of the covariance $C^{(q)}(\varepsilon_\Lambda^F,\varepsilon_\Lambda^0)$ of order q. This should encourage the radical EPS evolution to increasingly account for the randomness of meteorological fields at different scales.

4. CONCLUSION AND PROSPECTS

Prediction in geophysical systems is still in its infancy. Basic questions such as the nature of error growth and limits to predictability must be answered if only to allow the predictions to be seriously evaluated. Geophysics thus faces a situation somewhat analogous to that of celestial mechanics a few centuries ago when the ad-hoc epicycle framework became unmanageable: the definition of a suitable framework for predictability assessment of geophysical fields still requires much theoretical work.

We underlined that geophysical systems are not only complex in time, but also in space and therefore rather complex^{1+D}, where D is the dimension of the spatial extension. This requires a critical assessment of the relevance of concepts that emerged from the study of simple temporally complex systems.

In this direction, we have reviewed, compared and contrasted several frameworks that belong to two broad categories: dynamical systems (or deterministic chaos) and scaling in time and space. We discussed several critical issues related to the transition from (geometrical) finite dimensional phase space to (functional) infinite dimensional ones, in particular the importance of the anisotropic symmetry between time and space, as well as the question of singular behaviors at small scales. We argued that they imply strong limitations on the applicability of the Multiplicative Ergodic Theorem (MET) and of the Liouville equation. We reviewed the evidence brought by homogeneous turbulence phenomenology and statistical closure models that predictability decay laws are algebraic rather than exponential. Unfortunately, the quasi-normal framework of these models prevents them from dealing with intermittency: the strong heterogeneity of the activity of turbulence, i.e. the "bursts" of the energy fluxes through scales. We illustrated the potential of multifractals to quantify the fact that weak and strong events have different predictability limits, in particular with the help of a (normalized) correlation of increasing order. However, the most important single point is that these algebraic laws hold at all times, not only asymptotically. This contrasts to the standard predictability framework that involves a sequence of linear, nonlinear and chaotic regimes. It also shows and explains why stochastic sub-grid modeling in the context of Ensemble Prediction Systems may perform much better than deterministic modeling. In a more general manner, a better understanding of the intrinsic predictability limits of natural phenomena should help us to find alternative modeling strategies approaching the intrinsic predictability limits.

Acknowledgments. We greatly acknowledge Steve Sparks for his careful reading of the manuscript and his many stimulating suggestions.

REFERENCES

Adelfang, S.I., On the relation between wind shears over various intervals, *Journal of Atmospheric Sciences*, *10*, 138, 1971.

Arneodo, A., N. Descoter, and S.G. Roux, Intermittency, Log-Normal Statistics, and Multifractal Cascade Process in High-Resolution Satellite Images of Cloud Structure, *Phys. Rev Letters*, *83* (6), 1255–1258, 1999.

Arnold, L., *Random Dynamical Systems*, 586 pp. pp., Springer, Berlin Heidelberg New York, 1998.

Badii, R., and A. Politi, *Pys. Rev. Lett.*, *52*, 1661, 1984.

Bak, P., C. Tang, and K. Weiessenfeld, Self-Organized Criticality: An explanation of 1/f noise, *Physical Review Letter*, *59*, 381–384, 1987.

Batchelor, G.K., *The theory of homogeneous turbulence*, Cambridge University Press, Cambridge, 1953.

Batchelor, G.K., and A.A. Townsend, The Nature of turbulent motion at large wavenumbers, *Proceedings of the Royal Society of London*, *A 199*, 238, 1949.

Benzi, R., G. Paladin, G. Parisi, and A. Vulpiani, On the multifractal nature of fully developped turbulence, *Journal of Physics A*, *17*, 3521–3531, 1984.

Bolgiano, R., Turbulent spectra in a stably stratified atmosphere, *J. Geophys. Res.*, *64*, 2226, 1959.

Buizza, R., M.J. Miller, and T.N. Palmer, Stochastic Simlulation of Model Uncertainties in the ECWMF Ensemble Prediction System, *Q. J. R. Meteorol. Soc.*, *125*, 28887–2908, 1999.

Charney, J.G., *Geophys. Publ.*, *17*, 1, 1948.

Charney, J.G., *J. Atmos. Sci*, *28*, 1087, 1971.

Charney, J.G., and e. al., The feasibility of a global observation and analysis experiment, *Bull. Amer. Meteor. Soc.*, *47*, 200–220, 1966.

Chigirinskaya, Y., D. Schertzer, S. Lovejoy, A. Lazarev, and A. Ordanovich, Unified multifractal atmospheric dynamics tested in the tropics, part I: horizontal scaling and self organized criticality, *Nonlinear Processes in Geophysics*, *1* (2/3), 105–114, 1994.

Corrsin, S., On the spectrum of Isotropic Temperature Fluctuations in an isotropic Turbulence, *Journal of Applied Physics*, *22*, 469–473, 1951.

Deidda, R., Rainfall downscaling in a space-time multifractal framework, *Water Resour. Res.*, *36*, 1779_1794, 2000.

Eckmann, J.P., and D. Ruelle, Ergodic theory of chaos and strange attractors, *Rev. Mod. Phys.*, *57* (3), 617–656, 1985.

Ehrendorfer, M., The Liouville equation and its potential usefulness for the prediction of forecast skill. Part I: Theory, *Mon. Wea. Rev.*, *122*, 703–713, 1994.

Endlich, R.M., R.C. Singleton, and J.W. Kaufman, Spectral Analyes of detailed vertical wind profiles, *Journal of Atmospheric Sciences*, *26*, 1030–1041, 1969.

Epstein, E.S., Stochastic dynamic prediction, *Tellus*, *21*, 739–759, 1969.

Essex, C., Correlation dimension and data sample size, in *Non-linear variability in geophysics: Scaling and Fractals*, edited by D. Schertzer, and S. Lovejoy, pp. 93–98, Kluwer, Dordrecht, 1991.

Essex, C., T. Lookman, and M.A.H. Nerenberg, The climate attractor over short timescales, *Nature*, *326*, 64, 1987.

Fjortoft, R., On the changes in the spectral distribution of kinetic energy in two dimensional, nondivergent flow, *Tellus*, *7*, 168–176, 1953.

Forster, D., D.R. Nelson, and M.J. Stephen, Large distance and long time properties of a randomly stirred fluid, *Phys. Rev. A*, *16*, 732–749, 1977.

Fraedrich, K., Estimating the dimensions of weather and climate attractors, *J. Atmos. Sci.*, *43*, 419, 1986.

Frisch, U., *Turbulence: The Legacy of A. N. Kolmogorov*, 296 pp., Cambridge University Press, Cambridge, 1995.

Frisch, U., M. Lesieur, and D. Schertzer, Comment on the Quasi-Normal Markovian Approximation for Fully-Developed Turbulence, *J. Fluid Mech.*, *97*, 181–192, 1980.

Gage, K.S., Evidence for $k^{-5/3}$ law inertial range in meso-scale two dimensional turbulence, *Journal of Atmospheric Sciences*, *36*, 1979, 1979.

Gardiner, C.W., *Handbook of Stochastic Methods for Physics, Chemistry and the Natural Sciences.*, 442 pp. pp., Springer, Berlin, 1990.

Grassberger, P., Generalized dimensions of strange attractors, *Physical Review Letter*, *A 97*, 227, 1983.

Grassberger, P., Do climatic attractors exist?, *Nature*, *322*, 609–612, 1986.

Grassberger, P., and I. Procaccia, Measuring the strangeness of Strange atractors, *Physica*, *9D*, 189–208, 1983.

Halsey, T.C., M.H. Jensen, L.P. Kadanoff, I. Procaccia, and B. Shraiman, Fractal measures and their singularities: the characterization of strange sets, *Physical Review A*, *33*, 1141–1151, 1986.

Hasselmann, K., Stochastic climate models. Part I. Theory, *Tellus*, *28*, 473–485, 1976.

Hense, A., On the possible existence of a strange attractor for the southern oscillations, *Contr. to Atmos. Phys.*, *60*, 1987, 1986.

Hentschel, H.G.E., and I. Procaccia, The infinite number of generalized dimensions of fractals and strange attractors, *Physica D*, *8*, 435–444, 1983.

Herring, J.R., D. Schertzer, J.P. Chollet, M. Larchevêque, M. Lesieur, and G.R. Newman, A comparative assesment of spectral closures as applied to passive scalar diffusion., *J. Fluid Mech.*, *124*, 411–437, 1982.

Hilbert, D., Sur les problèmes futurs des mathématiques, in *Comptes Rendus du 2 ème Congrès Internationa des Mathématiciens*, pp. 58–114,, Gauthier-Villars, Paris, 1902.

Houtekamer, P., L. Lefaivre, J. Derome, H. Ritchie, and M. H., A system simulation approach to ensemble prediction, *Monthly Weather Review*, *124*, 1225–1242, 1996.

Jayawardena, A.W., and F. Lai, Analysis and prediction of chaos in rainfall and stream flow time series, *J. Hydrol.*, *153*, 23–52, 1994.

Kolmogorov, A.N., Local structure of turbulence in an incompressible liquid for very large Raynolds numbers, *Proc. Acad. Sci. URSS., Geochem. Sect.*, *30*, 299–303, 1941.

Kraichnan, R.H., A theory of turbulence dynamics, in *Second Symposium on Naval Hydrodynamics*, pp. 29–44, Office of Naval Research, Washington, DC, 1958.

Kraichnan, R.H., The structure of isotropic turbulence at very high Reynolds numbers, *J. Fluid Mech.*, *5*, 497–543, 1959.

Kraichnan, R.H., Inertial ranges in two-dimensional turbulence, *Physics of Fluids*, *10*, 1417–1423, 1967.

Kraichnan, R.H., Instability in fully developped turbulence, *Phys. Fluids*, *13*, 569–575, 1970.

Kraichnan, R.H., An almost-Markovian Galilean-invariant turbulence model, *J. Fluid Mech.*, *47*, 513–524, 1971.

Landau, L.D., and E.M. Lifshitz, *Fluid Mechanics*, Pergamon Press, Oxford, 1987.

Laplace, P.S., *Théorie Analytique des Probabilités*, Gauthier-Villars, Paris, 1886.

Lazarev, A., D. Schertzer, S. Lovejoy, and Y. Chigirinskaya, Unified multifractal atmospheric dynamics tested in the tropics: part II, vertical scaling and Generalized Scale Invariance, *Nonlinear Processes in Geophysics*, *1*, 115–123, 1994.

Leith, C.E., Atmospheric predictability and two-dimensional turbulence, *J. Atmos. Sci.*, *28*, 145–161, 1971.

Leith, C.E., and R.H. Kraichnan, Predictability of turbulent flows, *J. Atmos. Sci*, *29*, 1041–1058, 1972.

Leslie, D.C., *Developments in the theory of turbulence*, 368 pp., Clarendon Press, Oxford, 1973.

Lilley, M., K. Strawbridge, S. Lovejoy, and D. Schertzer, 23/9 dimensional anisotropic scaling of passive admixtures using lidar data of aerosols, *Phys. Rev. E* (in press), 2004.

Lilly, D., and E.L. Paterson, Aircraft measurements of atmospheric kinetic energy spectra, *Tellus*, *35A*, 379–382, 1983.

Lilly, D.K., Dynamics of rotating thunderstorms, in *Meso-scale Meteorology -Theories, Observations and models*, edited by D.K. Lilly, and T. Galchen, D. Reidel, Dordrecht, 1983.

Lilly, D.K., Theoretical predictability of small-scale motions, in *Turbulence and predictability in geophysical fluid dynamics and climate dynamics*, edited by M. Ghil, R. Benzi, and G. Parisi, pp. 281–280, North Holland, Amsterdam, 1985

Liouville, J., Sur la Théorie de la Variation des constantes arbitraires, *Journal des Mathématiques Pures et Appliquées*, *3*, 342–349, 1838.

Lorenz, E.N., Deterministic nonpereodic flow, *J. Atmos. Sci.*, *20*, 130–141, 1963a.

Lorenz, E.N., The predictability of hydrodynamic flow, *Trans. New York Acad. Sc.*, Ser. 2 (25), 409–432, 1963b.

Lorenz, E.N., The predictability of a flow which possesses many scales of motion, *Tellus*, *21*, 289–307, 1969.

Lorenz, E.N., Atmospheric predictability experiments with a large numerical model, *Tellus*, *34* (505–513), 1982.

Lovejoy, S., S. Pecknold, and D. Schertzer, Stratified multifractal magnetization and surface geomagnetic fields – I Spectral analysis and modelling, *Geophys. Inter. J.*, *144*, 1–22, 2001.

Lovejoy, S., D. Schertzer, and J.D. Stanway, Direct Evidence of Multifractal Atmospheric Cascades from Planetary Scales down to 1 km., *Phys. Rev. Letter*, *86* (22), 5200–5203, 2001.

Lovejoy, S., D. Schertzer, and A.F. Tuck, Fractal Aircraft Trajectories and anomalous turbulent statistics, *Phys.Rev. E* (in press), 2004.

Lumley, J., *Journal of the Atmospheric Sciences*, *21*, 99, 1963.

Lyapunov, M.A., Problème général de la stabilité du mouvement, *Annales Fac. Siences Toulouse*, *9*, 1907.

Mandelbrot, B., Random multifractals: negative dimensions and the resulting limitations of the thermodynamic formalism, in *Turbulence and Stochastic Processes*, edited by J.C.R.H.e. al., The Royal Society, London, 1991.

Marsan, D., Multifractals espace-temps, dynamique et prédicibilité; application aux précipitations, Docteur de l'Université thesis, Université Paris 6, Paris, 1998.

Marsan, D., D. Schertzer, and S. Lovejoy, Causal Space-Time Multifractal modelling of rain, *J. Geophy. Res.*, *D 31* (26), 26,333–26346, 1996.

Marzocchi W., F. Mulargia, and G. Gonzato, Detecting low-dimensional chaos in geophysical time series, *J. Geophys. Res.*, *102*, 3195–3209, 1997.

May, R.M., Simple mathematical models with very complicated dynamics, *Nature*, *261*, 459–467, 1976.

Meneveau, C., and K.R. Sreenivasan, Simple multifractal cascade model for fully develop turbulence, *Physical Review Letter*, *59* (13), 1424–1427, 1987.

Métais, O., and M. Lesieur, Statistical predictability of decaying turbulence, *J. Atmos. Sci.*, *43*, 857–870, 1986.

Millionshchikov, M.D., Theory of homogeneous isotropic turbulence, *Izvestiya, Ser. Geogr. Geophys.*, *5* (433–446), 1941.

Molteni, R.R., R. Buizza, and T.N. Palmer, The ECMWF ensemble prediction system: methodology and validation, *Q. J. R. Meteorol. Soc.*, *122*, 73–119, 1996.

Monin, A.S., *Weather forecasting as a problem in physics*, MIT press, Boston Ma, 1972.

Nastrom, G.D., and K.S. Gage, A first look at wave number spectra from GASP data, *Tellus*, *35*, 383, 1983a.

Nastrom, G.D., and K.S. Gage, A first look at wavenumber spectra from GASP data, *Tellus*, 1983b.

Naud, C., Schertzer, D., and S. Lovejoy, Fractional Integration and radiative transfer in multifractal atmospheres, in *Stochastic Models in Geosystems*, edited by W. Woyczynski, and S. Molchansov, pp. 239–267, Springer-Verlag, 1996.

Nerenberg, M.A.H., and E. C, Correlation dimension and systematic geometric effects, *Phys. Rev. A*, *42* (12), 7065–7074, 1990.

Nicolis, C., and G. Nicolis, Is there a climate attractor?, *Nature*, *311*, 529–532, 1984.

Nicolis, G., *Introduction to Nonlinear Science*, Cambridge Univeristy Press, Cambridge, 1995.

Obukhov, A., Structure of the temperature field in a turbulent flow, *Izv. Akad. Nauk. SSSR. Ser. Geogr. I Geofiz*, *13*, 55–69, 1949.

Obukhov, A.M., On the distribution of energy in the spectrum of turbulent flow, *Izvestiya, Geogr. Geophys.*, *5*, 453–466, 1941.

Obukhov, A.N., Effect of Archimedian forces on the structure of the temperature field in a temperature flow, *Sov. Phys. Dokl.*, *125*, 1246, 1959.

Oono, Y., *Progr. theor. phys. Suppl.*, *99*, 165, 1989.

Orszag, S.A., Analytical Theories of Turbulence, *J. Fluid. Mech.*, *41*, 362–386, 1970.

Osborne, A.R., and A. Provenzale, Finite correlation dimension for stochastic systems with power-law spectra, *Physica D*, *35*, 357–381, 1989.

Oseledets, V.I., A multiplicative ergodic theorem.lyapunov characteristic numbers for dynamical systems, *Trans. Mocscow Maths Soc.*, 19, 197–231, 1968.

Over, T.M., and V.J. Gupta, A space time theory of mesoscale rainfall using random cascades, *J. Geophys. Res., D 31*, 26,319–26,331, 1996.

Packard, N.H., J.P. Crutchfield, D. Farmer, and R.S. Shaw, Geometry from a time series, *Phys. Rev. Lett.*, 45 (9), 712–716., 1980.

Palmer, T.M., Predicting uncertainty in forecasts of weather and climate, *Reports on Progress in Physics*, 63, 71–116, 2000.

Palmer, T.N., Predicting uncertainty in forecast of weather and climate, ECWMF, Reading, U.K., 1999.

Panofsky, H.A., and I. Van der Hoven, *Quarterly J. of the Royal Meteorol. Soc.* (**81**), 603, 1955.

Parisi, G., and U. Frisch, On the singularity structure of fully developed turbulence, in *Turbulence and predictability in geophysical fluid dynamics and climate dynamics*, edited by M. Ghil, R. Benzi, and G. Parisi, pp. 84–88, North Holland, Amsterdam, 1985.

Pecknold, S., S. Lovejoy, and D. Schertzer, The morphology and texture of anisotropic multifractals using generalized scale invariance, in *Stochastic Models in Geosystems*, edited by W. Woyinski, pp. 269–311, Inst. for Math. and its Appl., 1996.

Pecknold, S., S. Lovejoy, D. Schertzer, and C. Hooge, Multifractals and the resolution dependence of remotely sensed data: Generalized Scale Invariance and Geographical Information Systems, in *Scaling in Remote Sensing and Geographical Information Systems*, edited by M.G. D. Quattrochi, pp. 361–394, Lewis, Boca Raton, Florida, 1997.

Pecknold, S., S. Lovejoy, D. Schertzer, C. Hooge, and J.F. Malouin, The simulation of universal multifractals, in *Cellular Automata: prospects in astronomy and astrophysics*, edited by J.M. Perdang, and A. Lejeune, pp. 228–267, World Scientific, 1993.

Pedlosky, J., *Geophysical fluid Dynamics*, Springer-Verlag, Berlin, Heidelberg, New York, 1979.

Pietronero, L., and A.P. Siebesma, Self-similarity of fluctuations in random multiplicative processes, *Physical Review Letter*, 57, 1098, 1986.

Pinus, N.Z., The energy of atmospheric macro-turbulence, *Izvestiya, Atmospheric and Oceanic Physics*, 4, 461, 1968.

Richardson, L.F., *Weather prediction by numerical process*, Cambridge University Press republished by Dover, 1965, 1922.

Richardson, L.F., Atmospheric diffusion shown on a distance-neighbour graph, *Proc. Roy. Soc., A110,*, 709–737, 1926.

Robert, R., and C. Rosier, Long range predictability of atmospheric flows, *Nonlinear Processes in Geophysics*, 8, 55–67, 2001.

Robinson, G.D., Some current projects for global meteorological observation and experiment, *Quart. J. Roy. Meteor. Soc.*, 93, 409–418, 1967.

Robinson, G.D., The predictability of a dissipative flow, *Quart. J. Roy. Meteor. Soc.*, 97, 300–312, 1971.

Rodriguez-Iturbe, I., B. Febres de Power, and J.B. Valdés, Chaos in rainfall, *Water Resources Research*, 25, 1667–1675, 1989.

Ruelle, D., Characteristic exponents and invariant manifolds in Hilbert space, *Annals of Mathematics*, 115, 243–290, 1982.

Ruelle, D., *Proc. R. Soc.*, A427, 241–248, 1990.

Sauer, T., J. York, and M. Casdagli, Embedology, *J. Stat. Phys*, 65, 579, 1991.

Schertzer, D., M. Larcheveque, J. Duan, V.V. Yanovsky, and S. Lovejoy, Fractional Fokker–Planck equation for nonlinear stochastic differential equations driven by non-Gaussian Lévy stable noises, *J. Math. Phys*, 41 (1), 200–212, 2001.

Schertzer, D., and S. Lovejoy, On the Dimension of Atmospheric motions, in *Turbulence and Chaotic phenomena in Fluids*, edited by T. Tatsumi, pp. 505–512, Elsevier Science Publishers B. V., Amsterdam, 1984.

Schertzer, D., and S. Lovejoy, The dimension and intermittency of atmospheric dynamics, in *Turbulent Shear Flow 4*, edited by B. Launder, pp. 7–33, Springer-Verlag, 1985a.

Schertzer, D., and S. Lovejoy, Generalised scale invariance in turbulent phenomena, *Physico-Chemical Hydrodynamics Journal*, 6, 623–635, 1985b.

Schertzer, D., and S. Lovejoy, Physical modeling and Analysis of Rain and Clouds by Anisotropic Scaling of Multiplicative Processes, *Journal of Geophysical Research, D 8* (8), 9693–9714, 1987.

Schertzer, D., and S. Lovejoy, Hard and Soft Multifractal processes, *Physica A*, 185, 187–194, 1992.

Schertzer, D., and S. Lovejoy, *Lecture Notes: Nonlinear Variability in Geophysics 3: Scaling and Mulitfractal Processes in Geophysics*, 292 pp., Institut d'Etudes Scientifique de Cargèse, Cargèse, France, 1993.

Schertzer, D., and S. Lovejoy, From scalar to Lie cascades: joint multifractal analysis of rain and clouds processes, in *Space/Time Variability and Interdependance of Hydological Processes*, edited by R.A. Feddes, pp. 153–173, University Press, Cambridge, 1995.

Schertzer, D., and S. Lovejoy, Universal Multifractals do Exist!, *J. Appl. Meteor.*, 36, 1296–1303, 1997

Schertzer, D., and S. Lovejoy, Space-time Complexity and Multifractal Predictability, *Physica A*, 38 (1–2), 173–186, 2004.

Schertzer, D., S. Lovejoy, and P. Hubert, An Introduction to Stochastic Multifractal Fields, in *ISFMA Symposium on Environmental Science and Engineering with related Mathematical Problems*, edited by A. Ern, and W. Liu, pp. 106—179, High Education Press, Beijing, 2002a.

Schertzer, D., S. Lovejoy, F. Schmitt, I. Tchiguirinskaia, and D. Marsan, Multifractal cascade dynamics and turbulent intermittency, *Fractals*, 5 (3), 427–471, 1997.

Schertzer, D., I. Tchiguirinskaia, S. Lovejoy, P. Hubert, and H. Bendjoudi, Which chaos in the rain-runoff process?, *J. Hydrological Sciences*, 47 (1), 139–148, 2002b.

Schuster, H.G., *Deterministic Chaos*, VCH, New York, 1988.

Sedov, L., *Similitudes et Dimensions en Mécanique*, MIR, Moscow, 1972.

Shur, G., *Trudy*, 43 (79), 1962.

Takens, Detecting strange attractors in turbulence, in *Dynamical Systems and Turbulence*, edited by D.A. Rand, and L.S. Young, pp. 366–381, Springer-Verlag, Berlin, 1980.

Tchiguirinskaia, I., Scale invariance and stratification: the unified multifractal model of hydraulic conductivity, *Fractals*, 10 (3), 329–334, 2002.

Theiler, J., Some comments on the correlation dimension of a 1/f-alpha noise, *Phys.Lett., A* (155), 480, 1991.

Thompson, P.D., Uncertainty of intial state as a factor in the predictability of large scale atmospheric flow patterns, *Tellus, 9,* 275–295, 1957.

Toth, Z., and E. Kalnay, Ensemble Forecasting at NMC: the generation of perturbations, *Bull. Amer. Meteor. Soc., 74,* 2317–2330, 1993.

Tsonis, A.A., *Chaos: Frome Theory to Application,* 274 pp., Plenum, New York, 1992.

Tsonis, A.A., J.B. Elsner, and K.P. Georgakakos, Estimating the dimension of weather and climate attractors: important issues about the procedure and interpretation, *J. Atm. Sci., 50,* 2549–255, 1993.

Van der Hoven, I., Power spectrum of horizontal wind speed in the frequency range from .0007 to 900 cycles per hour, *Journal of Meteorology, 14,* 160–164, 1957.

Van Zandt, T.E., A universal spectrum of buoyancy waves in the atmosphere, *Geophysical Research Letter, 9,* 575–578, 1982.

Vinnichenko, N.K., The kinetic energy spectrum in the free atmosphere for 1 second to 5 years, *Tellus, 22,* 158, 1969.

Weinstock, J., *J. Atmos. Sci., 35,* 634, 1978.

Wiin-Nielsen, A., On the annual variation and spectral distribution of atmospheric energy, *Tellus, 19,* 540–559, 1967.

Wilson, J., D. Schertzer, and S. Lovejoy, Physically based modelling by multiplicative cascade processes, in *Non-linear variability in geophysics: Scaling and Fractals,* edited by D. Schertzer, and S. Lovejoy, pp. 185–208, Kluwer, Dordrecht, 1991.

Yaglom, A.M., The influence on the fluctuation in energy dissipation on the shape of turbulent characteristics in the inertial interval, *Sov. Phys. Dokl., 2,* 26–30, 1966.

Yaglom, A.M., A. N. Kolmogorov as a fluid mechanician and founder of a school in turbulence research, *Ann. Rev. Fluid Mech., 26,* 1–22, 1994.

D. Schertzer, CEREVE, Ecole Nationale des Ponts et Chaussées, 6-8 Avenue Blaise Pascal, Cité Descartes, 77455 Marne-la-Vallee Cedex 2, France. Daniel.Schertzer@cereve.enpc.fr

S. Lovejoy, Physics dept., McGill U., 3600 University st., H3A 2T8, Montreal, PQ, Canada. Lovejoy@physics.mcgill.ca

Earthquake Prediction and Forecasting

David D. Jackson

Department of Earth and Space Sciences, UCLA, Los Angeles, California

Prospects for earthquake prediction and forecasting, and even their definitions, are actively debated. Here, "forecasting" means estimating the future earthquake rate as a function of location, time, and magnitude. Forecasting becomes "prediction" when we identify special conditions that make the immediate probability much higher than usual and high enough to justify exceptional action. Proposed precursors run from aeronomy to zoology, but no identified phenomenon consistently precedes earthquakes. The reported prediction of the 1975 Haicheng, China earthquake is often proclaimed as the most successful, but the success is questionable. An earthquake predicted to occur near Parkfield, California in 1988±5 years has not happened. Why is prediction so hard? Earthquakes start in a tiny volume deep within an opaque medium; we do not know their boundary conditions, initial conditions, or material properties well; and earthquake precursors, if any, hide amongst unrelated anomalies. Earthquakes cluster in space and time, and following a quake earthquake probability spikes. Aftershocks illustrate this clustering, and later earthquakes may even surpass earlier ones in size. However, the main shock in a cluster usually comes first and causes the most damage. Specific models help reveal the physics and allow intelligent disaster response. Modeling stresses from past earthquakes may improve forecasts, but this approach has not yet been validated prospectively. Reliable prediction of individual quakes is not realistic in the foreseeable future, but probabilistic forecasting provides valuable information for reducing risk. Recent studies are also leading to exciting discoveries about earthquakes.

INTRODUCTION

Earthquake prediction has long been a central goal of the earth sciences, because it would confirm our understanding of earth processes and help to save life and property. Recently there has been a spirited debate about whether earthquake prediction is even possible, and how wise it is to spend much public money on it. In 1999 the journal *Nature* sponsored an online debate which set out the issues well (see "Is the reliable prediction of individual earthquakes a realistic scientific goal?" at http://www.nature.com/nature/debates/earthquake/equake_frameset.html). Here I take a more general perspective on earthquake prediction and I skip the debate about funding. I define earthquake prediction, review some general properties of earthquakes, describe several approaches to prediction, give a brief history of some high-profile predictions, and assess future prospects.

DEFINITIONS

Much debate about the subject is semantic. The textbook definition of earthquake prediction is specifying in advance the boundaries of location, time, and magnitude of a future quake. Allen [1976] suggested that an earthquake prediction should include a statement of the degree of certainty that the predicted earthquake would occur, the probability that it would occur at random He also stated that earthquake predictions should be stated clearly and archived in such a way that both successes and failures could by identified and analyzed. Unfortunately Allen's definition is general enough to allow some trivial predictions. For example, earthquake prediction could be formally correct but trivial for a region in which earthquakes are already frequent. What the public want, and what we imply when we talk about saving lives, is some timely information beyond what everyone already knows.

In the *Nature* debate, moderator Ian Main defined earthquake prediction as part of a four step progression: (1) long-term or steady-state earthquake probabilities, (2) temporal variations in the long-term rates caused by recent earthquakes, or their absence, (3) probabilities of impending earthquakes, enabling public officials to organize protective action months or weeks before a probable earthquake, and (4) precise location, time, and magnitude with high enough certainty "that a planned evacuation can take place." The fourth step, sometimes called "deterministic prediction," may be closer to what the public has in mind. Main's statement substantially improves the previous ones, but it is a bit complicated. Furthermore, linking prediction to evacuation may be unrealistic, and it fails to recognize many life-saving actions short of evacuation.

My definition uses a two-part progression. Earthquake forecasting means specifying the long-term probability of earthquakes per unit area, magnitude, and time. It may incorporate modest time-dependence. Earthquake prediction, by contrast, means identifying special conditions that make the immediate probability much higher than usual, and high enough to justify unusual action. This definition starts with probability and recognizes the need to describe what is normal. The definition casts forecasting and prediction in a way that both can be formally tested in time. By omitting the comparison to normal conditions, the "textbook" definition implicitly assumed that earthquake prediction would involve such high probabilities that background rates would not matter.

"NORMAL" PROPERTIES OF EARTHQUAKES

A nearly ubiquitous property of earthquakes is their exponential magnitude distribution, frequently written as

$$N(m) = 10^{a-b*m} \quad (1)$$

where N is the number of earthquakes in a closed region of space and time with magnitudes larger than or equal to m, and a and b are constants appropriate to the given region. This relationship is often called the Gutenberg-Richter or simply G-R distribution. If the rate of earthquakes is relatively constant in time (the Poissonian assumption), then equation (1) works as well with dN/dt substituted for N. In general the constant b, often called the "b-value," is near 1, in which case there will be roughly 10 times as many quakes above magnitude 5 as above magnitude 6, etc. Energy considerations require that for very large earthquakes, perhaps above magnitude 8.5, the rate of earthquakes must drop off faster than implied by equation (1), because the energy per earthquake increases more strongly than their frequency decreases with magnitude. Modern versions of (1) have an additional parameter, such as a "corner magnitude," to describe the magnitude range where the extra drop-off begins. Figure 1 shows the magnitude distributions, plotted as log N vs. m, for two separate regions: a 10 degree "square" in latitude and longitude centered about Tokyo, and a similar "square" centered about Los Angeles. These data come from the Harvard Catalog of Central Moment Tensors (http://www.seismology.harvard.edu/CMTsearch.html), and cover the period from Jan 1, 1977 through June 1, 2003. I recommend the Harvard earthquake catalog for anyone who wants to experiment with earthquake data, because it uses a consistent magnitude definition, is globally complete since 1977 for earthquakes above magnitude 5.8, and is easily searchable to produce sub-catalogs such as those in Figure 1. The data in Figure 1 show that California and Japan have slightly different b-values, or slopes, but both are close to 1. Linear fits like those in Figure 1 will be obtained for just about any sample selected for a large enough area. Wesnousky [1994] and many others contend that the G-R distribution holds primarily for large regions with networks of interacting faults, but not for many individual faults. They maintain that the latter have a preferred "characteristic" magnitude at the upper magnitude limit, and that earthquakes of that magnitude are much more frequent than predicted by equation (1). Kagan [1996] questions Wesnousky's conclusions, maintaining that individual faults also obey the G-R rule.

The G-R magnitude distribution of equation (1) has some important consequences for earthquake prediction. First, under certain conditions equation (1) estimates "normal" behavior as required in my definition of earthquake prediction above. The conditions include that the study area be large enough and that the earthquake rate does not vary much with time (the Poissonian assumption). Second, the exponential earthquake rate requires that the magnitude description for any predicted earthquakes must be precise. The background rate of earthquakes approximately doubles for a reduction in magnitude of only 0.2 units, (a typical measurement uncertainty in a well-monitored region). So if one predicts an earthquake of "about" magnitude 6 but would claim success after a 5.8, the odds of random occurrence must be calculated for 5.8 (or whatever is the lowest magnitude for

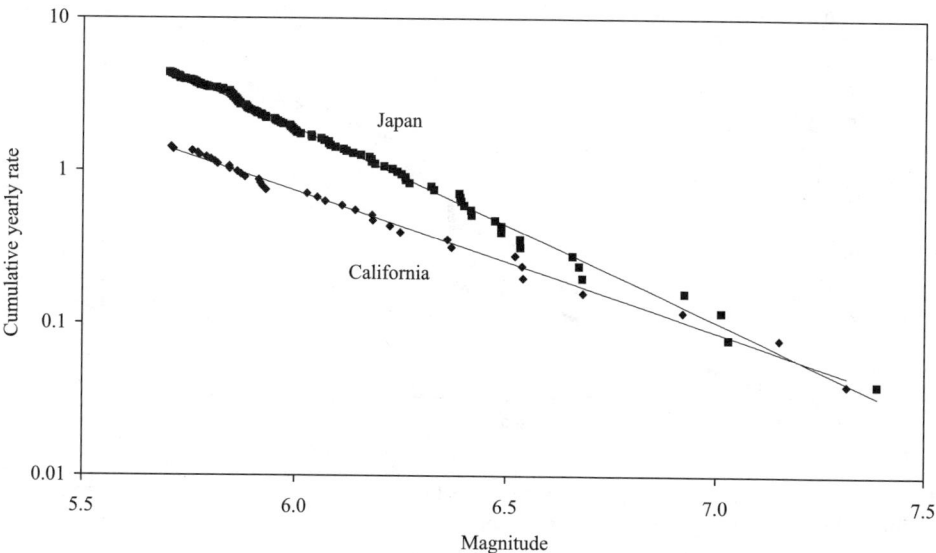

Figure 1. Cumulative numbers of earthquakes over given magnitude for ten degree boxes around Los Angeles and Tokyo.

"success.") In general, be wary of the word "about" in earthquake predictions.

A second general feature of earthquakes is clustering: after an earthquake, many more will follow concentrated in the neighborhood of the recent one. Most of the following earthquakes will have magnitudes smaller than their "parent," in which case the parent will be called the "main shock" and the offspring "aftershocks." However, occasionally the child surpasses the parent in size, in which case the parent is called a foreshock and the large child the main shock. Earthquakes do not have DNA, and labels such as parent, child, foreshock, main shock, and aftershock do not come with the seismic records. Only by waiting for larger quakes can we tell if an event is the main shock. The clustering behavior is fairly regular, when averaged over a large enough region around a previous quake. The total number of offspring N above a given threshold magnitude and within a time t of the previous event is well described by a generalized Omori's law

$$dN/dt = K/(t+c)^p \qquad (2)$$

Here N is the cumulative number of events at time t since a significant earthquake, p is an exponent near 1, K is a constant, and c is a time constant, typically a few minutes to a few hours, depending on the capacity of the seismic network recording the earthquakes. For clustered events an alternate classification is appropriate: the first event in a cluster is "independent," while later events, regardless of size, are "dependent."

An example of clustering is shown in Figure 2 for southern California. The diagram shows the cumulative number of earthquakes greater than magnitude 4 in the southern California catalog (http://www.scecdc.scec.org/data_avail.htm) from 1980 to present. The number of earthquakes tends to increase at a constant (Poissonian) rate until there is a large earthquake, at which time a burst of dependent quakes creates a rounded step in the curve. In most cases the rounded step is well described by (2).

Equation (2) also has some important consequences for earthquake prediction. First, predicting dependent events is easy. Dependent events occur at much higher rates than independent ones, and they are indeed important as they may be even larger than the events that spawned them. However, dependent events do not become predictable until their parent events occur. Many people regard predicting dependent events as not quite earthquake prediction, partly because aftershocks are too easy to predict. However, most damage is done by events that appear independent, and these are properly considered the appropriate target for earthquake prediction. Second, dependent events are to some extent "normal" and they should be included in earthquake forecasts. At present, there is wide agreement that equations like (2) describe the general process well, but there is not good agreement on how to estimate the coefficients, how much they should vary with location and time, etc. In other words the definition of normal behavior is not adequately complete. This fact complicates the definition of prediction, because prediction deals with exceptions to normal behavior.

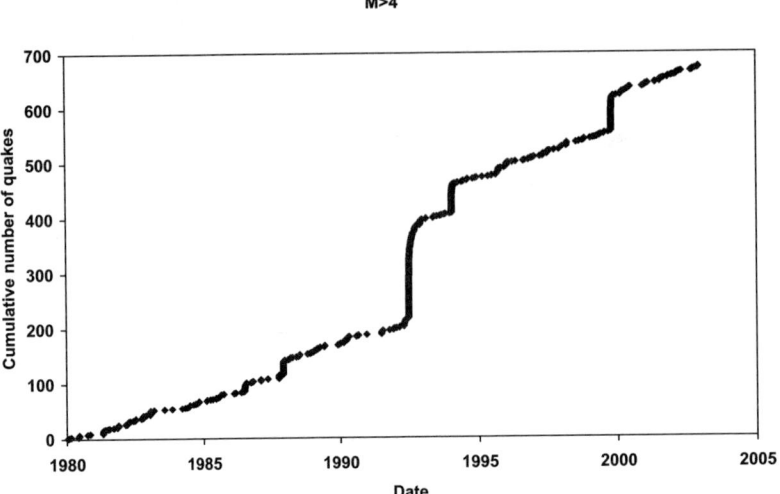

Figure 2. Cumulative numbers of earthquakes over magnitude 4 in southern California since 1980. The rate (the slope of the curve) is higher than normal just after each major event.

EARTHQUAKE PREDICTION AND FORECASTING

Attempts to describe the earthquake future span a wide range, from purely empirical to largely theoretical. Methods fall into four general categories: extrapolating past seismicity, modeling time-dependent variations in past seismicity, searching for precursors, and modeling stress accumulation.

GENERAL COMMENTS ON METHODOLOGY

The scientific method of observing, hypothesizing, testing on independent data, and concluding is a good start, but more is required for earthquake prediction and other endeavors in which public officials might need to act on the results. Rhoades and Evison [1989] introduced a specific scheme appropriate to such sensitive science. An adapted version is shown in Table 1.

The process begins with creative, inductive activity that need not be constrained by too much structure. The final scientific steps are much more objective. When we present a prediction to public officials with the suggestion that they take action in step 8, we must be sure that the methods have proven their validity in similar seismic situations (step 7). The final step, public action, must involve many issues outside the realm of science and is necessarily subjective. One reason for much of the debate in earthquake prediction has to do with where we are in this process. For many scientists, observing strong correlations between earthquakes and other phenomena which show statistical significance in step 3 justifies claiming success, or at least strong optimism. For others, like me, there is no reason to get excited before Step 7. Open thinking in the early steps will generate hypotheses in fine agreement with data that *we* select. However, the correlations may wither in a prospective test in which we cannot select the data in advance.

Figure 3 illustrates why step 7 is so important. It shows 100 randomly selected points on the y-axis plotted against 100 other random points on the x-axis. One could imagine that one axis is time, and the other the location of an earthquake along a fault, or x might be the strength of anomalous electrical field and y the time till a certain earthquake, etc. By construction, there is no systematic relationship between x and y, or between any pair of points. However, inspection shows that local clusters of points, like those in the small boxes, are correlated. It is possible to find groups with either positive or negative correlation. In nature we do not know whether there is a true relationship, or only one produced at random. If we select only the data in a small box for analysis (step 3 in the Rhoades and Evison scheme) we will surely get a strong cor-

Table 1. Scheme for Earthquake Prediction Research

Step	Task	Nature
1	Select data base	Subjective
2	Search for predictive relationships	Subjective
3	Test significance of relationship	Objective
4	Formulate model for hazard estimation	Subjective
5	Estimate parameters of model	Objective
6	Derive hazard estimates	Objective
7	Test model performance on new data	Objective
8	Adopt model for operational use	Subjective

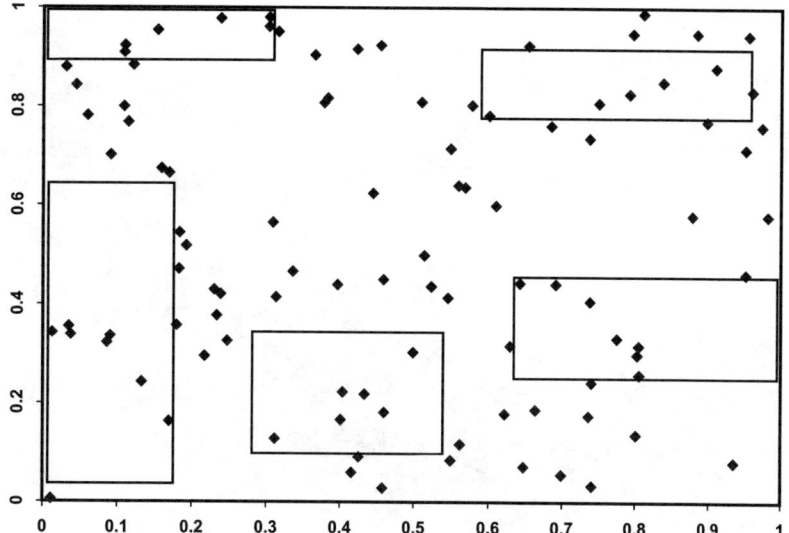

Figure 3. Random values of x (horizontal-) and y (vertical axis). Selected data, shown in boxes, appear correlated even though there is no correlation in the whole data set.

relation. We can generally find a good reason why those data ought to be selected: perhaps they come from a certain location, or a certain type of fault, or faults that have more or less displacement than others, or are straighter or more crooked than others, etc. If we then take these selected data to a statistician, he or she will surely confirm that the correlation is strong and could only be produced at random with small probability. However, the statistician cannot give a meaningful analysis without seeing all the data and the whole process of selection, which is often so complex that we cannot describe it adequately. That is why we need to have fully independent, prospective testing, as described in steps (4) through (7) in Rhoades' and Evison's chart.

EARTHQUAKE FORECASTING FROM PAST SEISMICITY

As in weather prediction, "persistence" is a convenient and practical way to estimate future behavior. In the simplest method, still used extensively, one draws a boundary, compiles a catalog of earthquakes within that boundary, fits the earthquake rate as a function of magnitudes to equation (1), and asserts that earthquakes in the future will have the same rate. Part of the fun is to see how small a region can be used and still have stability. Another argument is whether the relationship obtained for frequent small events can be extrapolated confidently to large earthquakes. A problem of this method, shared with many others in earthquake studies, is the extreme sensitivity caused by fixed boundaries. A large earthquake might be in or out of a region depending on how its location is estimated. That arbitrariness affects both the value of an estimate of future rates and the evaluation of any forecast.

A modification that partly controls the "hard boundary" problem described above is to smooth the seismicity to estimate earthquake rate as continuous functions of position. Jackson and Kagan [1999] have applied this method to forecast earthquakes as a function of magnitude for several seismically active regions of the world. We assumed that the earthquake rate at any point is proportional to a weighted sum of the rate of earthquakes in the region, where the weighting is proportional to magnitude of past events but inversely proportional to the distance away. Several variable parameters, most importantly a distance scaling factor, are adjusted for each region. Jackson and Kagan set out specific probability estimates for the northwest and southwest Pacific regions which contain a major fraction of the world's large earthquakes in any given year. The regions are treated separately because their parameters differ significantly. Figure 4 shows the map for the northwest Pacific, which includes the Japanese Islands. While the map generally indicates that most earthquakes occur along the plate margins (not a great surprise), the expected earthquake rate varies considerably along the boundary. For example, the northeastern tip of Hokkaido is more prone to earthquakes over magnitude 5.8 than is Kyushu. Jackson and Kagan have archived the probability estimates to facilitate testing the model and according to a likelihood ratio test, earthquakes after the 1999 forecast have complied well with the forecast.

TIME-DEPENDENT FORECASTING

The Poisson model discussed above has been widely accepted for simple hazard studies over large areas where it might be

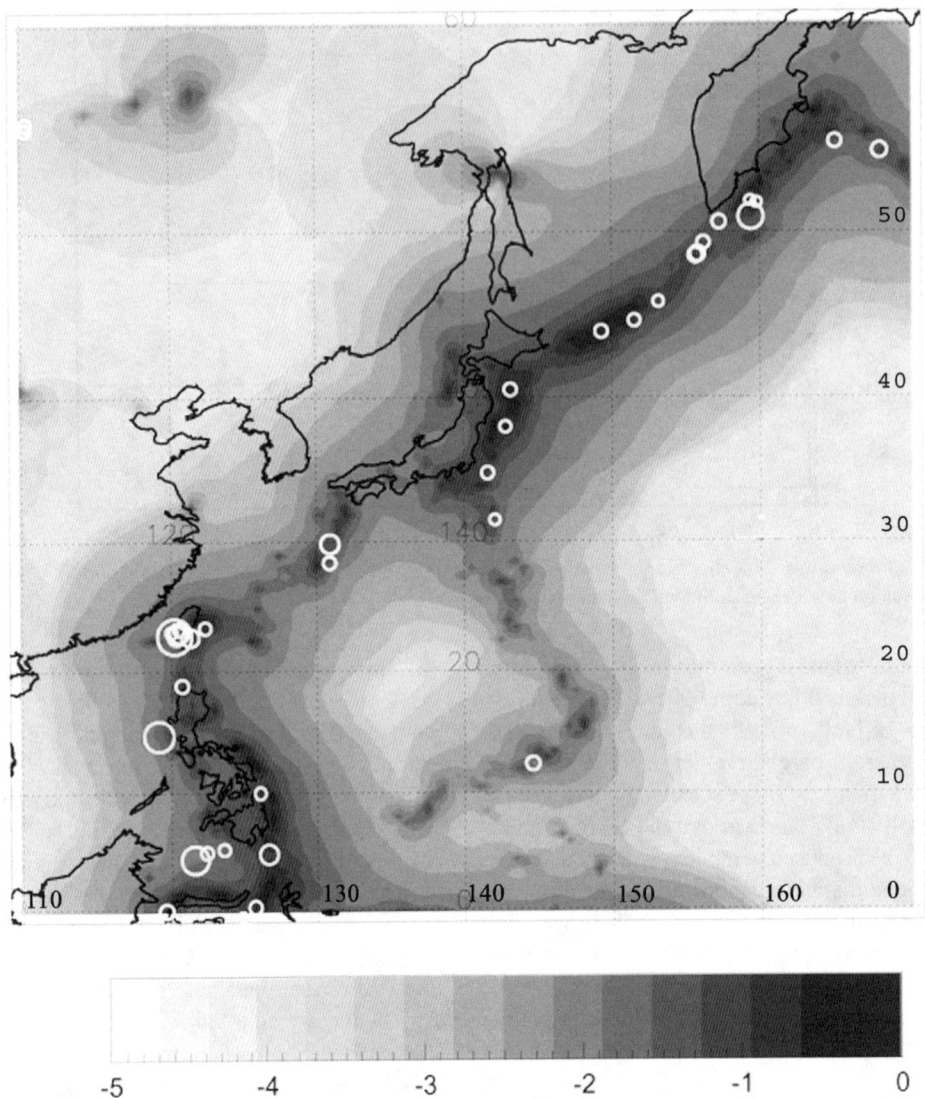

Figure 4. Forecast for the Northwest Pacific region. The grey scale represents the log of the rate density of earthquakes over magnitude 5.8, in units of events per 1000 km^2 per year. The forecast was based on earthquake in the Harvard CMT catalog from 1977 through 1998. Earthquakes in 1999 are shown as white circles with radius proportional to magnitude.

argued that time dependence caused by changes in ambient stress on different faults, for example, may average out. However, stress and other conditions clearly are time dependent, and exploiting this time dependence effectively is at the heart of the quest to transform forecasting into prediction.

The Seismic Gap Model

The seismic gap model quantifies the common wisdom that "the longer it has been since the last earthquake, the sooner the next." We know that earthquakes release stress ultimately caused by plate motions. The motions and the stress buildup are slow, so there must, according to the model, be some time after an earthquake when a similar large earthquake cannot happen. It is an appealing idea, almost self-evident, yet it depends upon a few important assumptions.

First, the seismic gap model assumes that large earthquakes closely resemble previous ones, so that "similar large earthquake" has some meaning. Earthquakes come in all sizes, and it is not necessarily true that any earthquake is essentially a repeat of a previous one. The seismic gap hypothesis depends on the existence of a special or "characteristic earthquake" which is a repeat. It is the largest earthquake that can occur on a particular fault segment or plate boundary segment. It releases the accumulated stress and starts the "clock" ticking till the time of the next characteristic event. A characteristic earthquake may stimulate a cluster of quakes like those shown in Figure 2, but according to the seismic gap model these will be too small to affect the stress environment significantly. Thus, the model posits two classes of earthquakes: the smaller ones that cluster as in Figure 2, and the larger characteristic ones that are instead quasi-periodic. Faults have different sizes, so the characteristic magnitude should differ from place to place. A major problem is how to recognize a characteristic earthquake in a given place.

A second seismic-gap assumption is that a characteristic earthquake releases the stress over a large volume to the extent than it will have to build up slowly by tectonic activity. This is a major assumption, because stress is a tensor and highly variable. At a given place some stress components rise while others fall, and each component varies strongly from place to place. The stress on a fault may have built up over thousands of years, and a single earthquake may not reduce it by a very substantial amount. In fact, small earthquakes may increase some stress components to very high levels in a small region. The seismic gap model requires that these small earthquakes cannot grow to big ones until the regional average stress recovers.

Figure 5 shows a map based on the one used by Nishenko and Sykes [1989] to designate seismic gaps and non-gaps. Nishenko [1991] published essentially the same model in a much more accessible journal. Each polygon on the map shows the projection of a region previously ruptured in a large earthquake assumed to be characteristic. Nishenko forecast that similar sized earthquakes would occur in the future. He estimated "due dates" from the repeat times of historic earthquakes, also assumed characteristic, or by the ratio of the expected displacement to the plate motion rate. With his map Nishenko published a table listing the forecast magnitudes, the expected dates of earthquakes, and the probabilities of occurrence within 5, 10, and 20 years of the 1989 forecast.

Both the five and ten year anniversaries of Nishenko's seismic gap forecast have passed, and the earthquake record has not been kind to the theory. After five years there had only been 2 earthquakes that met the description of a characteristic earthquake in any of the 125 zones, whereas the model implies that there should have been about nine. After ten years, only five had occurred, against an expected total of about 18. If the model were correct, the probability of such a small number of events would be much less than 5% after either the 5 or 10 year anniversary. The results suggest that if there is any such thing as a characteristic earthquake, it must be much larger and much less frequent than Nishenko's estimate. If

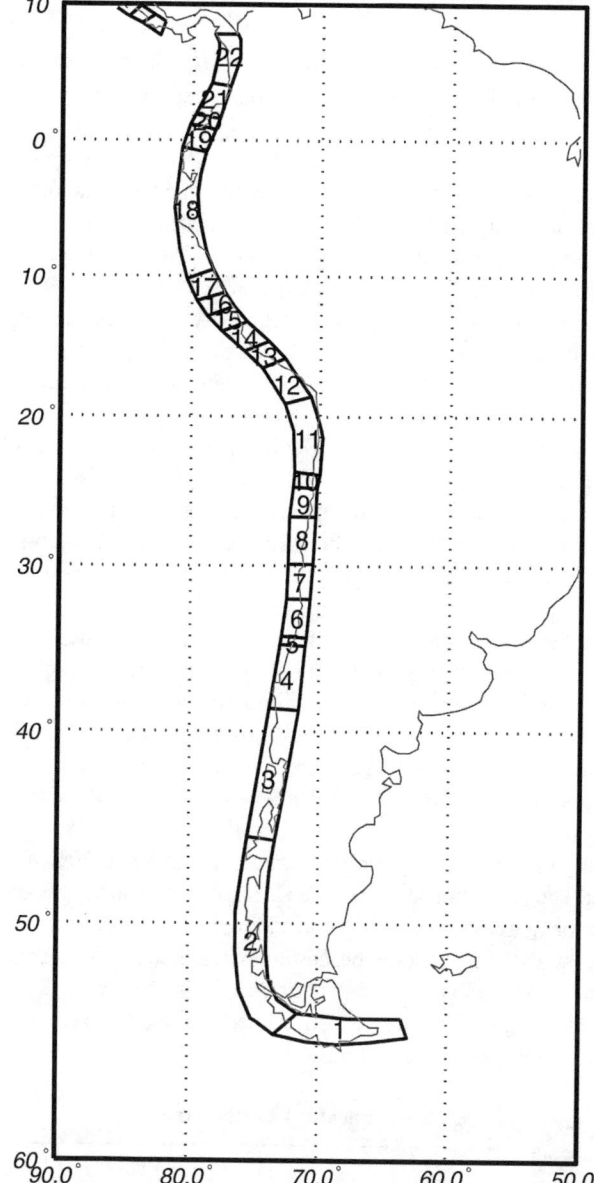

Figure 5. Map showing some of the zones for which Nishenko [1991] estimated characteristic earthquake size and time dependent probabilities in his seismic gap model. These zones are on the west coast of South America, but Nishenko's model covered much of the Pacific Rim. Zone numbers correspond to a table listing expected earthquake sizes and probabilities.

so, we may have no way to identify the characteristic size and test the model for centuries.

While the seismic gap model has not been validated, still the 1989 formulation by Nishenko and Sykes represented a major step in forecast methodology. It was the first earthquake forecast model stated specifically enough for definitive testing.

The Parkfield Prediction.

Another application of the seismic gap model led to the highly publicized prediction of a moderate earthquake on the San Andreas Fault at Parkfield, California. I use the term "prediction" because it is widely called that, although the region and magnitude were not stated specifically, and the announced probability was not high enough to justify emergency response. Historic records indicate that earthquakes of "about magnitude 6" occurred near Parkfield in 1857, 1881, 1901, 1922, 1934, and 1966. Based on the apparent regularity of that sequence, and of course on lots of other data and modeling, Bakun and Lindh [1985] predicted that a similar earthquake should occur before January 1, 1993. Figure 6, adapted from their paper, was so compelling that the U.S. Geological Survey officially adopted the prediction. Bakun and McEvilly [1984] discussed the reasoning behind the prediction. The average time between events, up until 1966, was about 22 plus or minus 2.5 years, suggesting that the next similar event was due in 1998, with a 95% probability of occurring before 1993.

So far no such Parkfield earthquake has occurred: the "prediction" is widely regarded as a failure. What went wrong? A blue-ribbon panel investigated the history of the forecast [*National Earthquake Prediction Evaluation Council Working Group*, 1994], concluding that the calculations were based on an oversimplified model, but that an earthquake at Parkfield was still highly likely. A survey of the literature showed probability estimates ranging from less than 1% to about 20% per year. The low estimate came from Kagan [1997] who suggested that the quakes listed in the Parkfield history were so frequent and the dates so periodic because they had been pre-selected from a much larger universe (an example of the pitfall illustrated in Figure 3, above). Kagan argued that in a large catalog of earthquakes with random times, it is possible that apparently periodic concentrations may occur at random. It is worth noting that the magnitudes and locations of the Parkfield quakes before 1922 are poorly known, some of the earthquakes may not really have been at Parkfield or in the claimed magnitude range, and others may have escaped detection. Also the Parkfield forecast assumes that the past earthquakes were "characteristic," meaning that they avoid the clustering behavior exhibited by other quakes. As in the case of Nishenko's seismic gap forecast, the characteristic earthquake assumption may be invalid.

Clustering

Empirically, earthquakes occur preferentially near each other both in time and space [Jackson and Kagan, 1999]. Specific clustering models can show probability gains of several orders of magnitude for periods of several hours after a major quake (that is, the earthquake rate is thousands of times higher than normal), and the rate can remain modestly above normal for years. However, most of the events occurring in the wake of previous quakes would normally be considered aftershocks. Even though forecasting them can have substantial economic and social value, the term prediction is seldom applied because aftershocks are considered normal. Aftershock models are now being applied with some success to events bigger than their precedents, but again the term prediction is seldom used because the absolute probabilities are very modest (or equivalently, the false-alarm rate is high). Nevertheless, time-dependent forecasts employing clustering models are catching on.

Rhoades and Evison [2003] are pursuing one promising approach which the call "EEPAS," for "every earthquake a precursor according to scale." Their assumption is that each earthquake has a chance of triggering a later one, and that large quakes have more triggering effect than small ones. They have applied this approach retrospectively to earthquakes over magnitude 5.75 in New Zealand for the period 1965–2000. Without adjusting any parameters, they made a pseudo-prospective forecast for California during 1975–2000 based on earthquake data from 1951–1974. Their technique forecasted the probable times and sizes of earthquakes vastly better than a Poisson model with similar spatial clustering. Like the models of Jackson and Kagan [1999] and many other clustering models, EEPAS is primarily empirical and falls somewhere between forecasting and prediction by my definitions. In my view they will set the standard that physics-based models will have to beat to prove their merit.

Chaos

Deterministic chaos is a type of self-organized criticality (SOC) that offers one explanation for the many power law

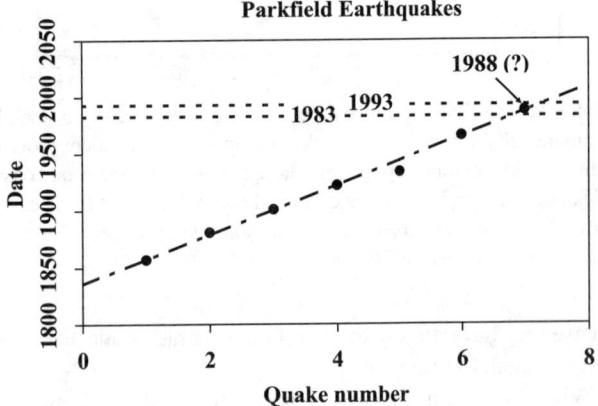

Figure 6. Times of moderate size Parkfield earthquakes as a function of the sequence number. This diagram motivated, in part, the Parkfield earthquake prediction.

distributions of phenomena related to earthquakes. For example, equations (1) and (2) imply that earthquake sizes and interval times have power-law distributions, and independent observations suggest that power laws also describe fault lengths and orientations. The "self-organization" might be accomplished by the stresses propagated from small cracks to slightly larger ones, etc., bringing near-critical stresses to a large part of the crust. SOC might cause profound difficulties for earthquake prediction. SOC implies such extreme sensitivity that any tiny earthquake might, with low probability, cascade into a much larger one [Geller et al., 1997], just as the butterfly might trigger a hurricane in popular conceptions of another complex system. However, mentioning SOC and earthquakes together usually provokes an argument, because there are so many different pictures of SOC and its implications. SOC is clearly defined for solutions to some non-linear differential equations, but in earthquake studies we do not even know the relevant equations. Thus there is no unique mapping from the properties of chaotic systems to earthquakes. Many participants in the Nature debate (http://www.nature.com/nature/debates/earthquake/equake_frameset.html) discussed their views on SOC. Bak (Week 4) commented that the temporal correlation implied by equation (2) and a corresponding power-law spatial correlation between earthquakes provide strong evidence for SOC. The expected behavior conflicts with the independence of fault segments assumed in the characteristic earthquake and seismic gap models. Leon Knopoff (Week 3) reminds us that a few power laws do not prove SOC, and there is evidence for the influence of special and temporal scales not predicted by SOC. Christopher Scholz (Week 2) argues that SOC may describe smaller earthquakes, but that large "system level" earthquakes may behave much differently. Large enough quakes could reduce the stress over a large volume well below the critical state, requiring a period of stress recovery and suggesting quasi-periodic behavior as in the seismic gap model. However, as in the gap model, there is the difficulty of first recognizing the size of the "system level" earthquake, before we could estimate its recurrence time and use it for estimating earthquake potential. Others suggest that a generalized view of SOC could allow, or even require, the existence of seismic or other precursors to earthquakes. Knopoff (Week 3) and Didier Sornette (Week 7) argue that the crust is spatially heterogeneous, and that the communication of stress from one part to another may offer clues to readiness for large events. David Bowman and Charles Sammis (Week 4) suggest "intermittent criticality" in which large volumes of the crust may fall too far below criticality to support large earthquakes for long periods of time. Increasing long-range correlations in seismic behavior might announce the return of SOC and the potential for large earthquakes. Evison [2001] makes a similar argument for the possibility of precursory seismicity increases and swarm activity in the "culmination" of the earthquake generating process.

The arguments above suggest ways that the profound difficulties implied by self-organized criticality need not be fatal for earthquake prediction. However, no one has yet turned these general arguments into a roadmap to success. To get there, we need specific definitions of system level earthquakes, long range correlations, and intermittency; models connecting then with time dependent changes in earthquake probability; and evidence that these models work prospectively for several earthquakes. Given that the inhomogeneity of the crust is an essential part of these models, we need separate specifications and models for each region.

EARTHQUAKE PRECURSORS

What can be done to turn forecasts into predictions? An enduring hope is that measurable phenomena may change noticeably just before an earthquake. Such changes are referred to as "anomalies" when they have no normal explanation. The major reason to expect precursors is that earthquakes are a response to stress, and stress causes cracking and other phenomena in rock that might be identified before the stress reaches the level sufficient to cause a major quake. Recognizing legitimate precursors necessitates associating "anomalies" consistently with earthquakes.

Anomalies that have been observed and published as candidate precursors include seismicity (increased or decreased rate or magnitude distribution), ground deformation (tilt, strain, or apparent sea level changes), water level or pressure, seismic velocity (compressional, shear, or their ratio), gravity, magnetic field, magnetic susceptibility, electromagnetic waves or noise, electrical resistivity, geochemical flux in groundwater and air (e.g. radon, chlorine), heat flux, temperature, infrared radiation, animal behavior, and many more. Each on my list has some possible connection with stress or its effects, and I have listed them in order of how directly they are related to stress. For example, the same stress that leads to big earthquakes may also lead to more or less frequent small ones, depending on the spatial pattern of stress and which stress components are involved. High stress may lead to precursory cracking of the rock just before it breaks. Such cracking might reduce the rock density, leading to uplift or changes in seismic velocity. Similarly, cracks might allow freer flow of fluids, changing the chemistry of ground water or altering the electrical conductivity of the rocks. Animals might be sensitive to high-frequency vibrations from small cracking events or to changes in the electromagnetic environment. Each of the proposed precursors might change in either a positive or negative sense as a result of increasing stress. For example, precursory stresses might cause an increase in the rate or size of earthquakes, but they could also lead to a decrease. The progressive stress accumulation might rupture the easy targets (small cracks) leaving only the hard ones, which will be the source of bigger earthquakes in the future. This seismic lull could then be the quiet before the storm.

Credibility Criteria

If anomalous phenomena relate to the processes leading up to earthquakes, one would expect them to conform to four basic rules: (1) observed anomalies should be strongest at the time of the earthquake itself when stress changes and cracking are far more intense than in the preparation period, (2) anomalies should be strong nearer to the site of the future earthquake and weaker at greater distance, (3) the anomalous changes should be reasonably explained by observed crustal properties and stress dependence, and (4) non-seismic causes of the anomalies should be unlikely.

Over several decades thousands of reports of precursory phenomena have been published and many more reported in scientific meetings. Yet we are still without any consistent precursors. Why? Perhaps it is just the magnitude of the problem: precursory signals might vary from place to place or time to time. The degree to which crack formation and stress variations precede earthquakes, and thus the strength of any precursory anomalies, may depend on local properties of the crust, or even time-dependent properties relating to stresses from past earthquakes, etc. If so, then recognizing the anomalies' true behavior and testing our understanding prospectively will be a huge job. We must observe many strong earthquakes in the same place, or equivalent places, which will require centuries of observation for large earthquakes. Geller [1997] attributes lack of success to another cause: lowered scientific standards allowed due to prediction's perceived social importance. Much research on precursors has lacked rigor, but I believe the problem is more fundamental: our data lack the redundancy needed both to learn how the anomalies are caused and test our understanding. By exhaustive searching we have found many anomalies that may be accidentally correlated with earthquake behavior, an example of the effect illustrated in Figure 3. Many such cases do not satisfy the four credibility rules above, and they should be discarded, but some will satisfy the criteria. However strong the local correlations may be, they are based on selected data, and random variations might be their cause. However, confirming the correlations with independent data has been, and will be, impossible because earthquakes are rare and diverse. We are effectively dealing with irreproducible outcomes.

As stress increases, any response in the region of a future earthquake depends on the preexisting stress pattern, the strength of the rock, the detailed geometry of faults, and the locations and sizes of intermediate and small earthquakes, including those too small to be observed. Whether the ground subsides or uplifts depends on those same quantities, as well as variations in the compressibility and rigidity of the rock, etc. In addition the hypothetical precursors listed above all vary for other reasons besides earthquake stresses. Uplift and subsidence, at the sub-mm level relevant to earthquake prediction, depend on progressive soil motion, expansion of clay minerals in response to rainfall, shrinkage in response to intervening desiccation, earth tides, ocean loading in response to tides and ocean current changes, and other effects. The number and type of observations have been insufficient to resolve the independent variables controlling these phenomena. In principle some of them might be measured; for example, soil expansion can be estimated by measuring the thickness of a soil column in a borehole. The rigor wanting in most prediction studies should be remedied in part by detailed and comprehensive measurements of environmental variables that may affect the "precursory" observations. Thus, the road to earthquake prediction must go through diverse territories where soil properties, water flow, tides, temperature, electrical conductivity, water pressure and chemistry, and many other quantities reign supreme. Furthermore, we cannot guarantee that making all possible environmental measurements will suffice to distinguish tectonic from non-tectonic effects.

Successful Prediction of the Haicheng Earthquake?

Chinese scientists reported that a 1975 earthquake with magnitude 7.3 in Haicheng, China was successfully predicted by analyzing seismic and other geophysical precursors. The observed anomalies included most of those listed in the section above, including strange animal behavior. The earthquake caused extensive damage, but the loss of life was reported to be relatively light because dangerous buildings were evacuated. An international team, including some Chinese-speaking American scientists, visited China to learn the details. Their cautious report [*Haicheng Earthquake Study Delegation*, 1977] generally supported the claims of success and recommended that the US and other countries emulate parts of the Chinese prediction program. Many textbooks [e.g. *Bolt*, 1988; p 170; *Scholz*, 1990, p. 353; *Brumbaugh*, 1999, p.191] now list the Haicheng prediction as an example of success.

According to the Study Delegation and to Ma et al., [1989], the prediction began as a long range forecast that gradually became more focused and more precise. Following a 1969 magnitude 7.4 earthquake in Bohai Bay (about 400 km SW of Haicheng), a National Conference on Seismological Work recommended focused geophysical investigations in northeast China. Some results included mapping of active faults and installation of 17 new seismic stations in Liaoning Province (where Haicheng is located). In 1974 a "Consultative Meeting on the Earthquake Situation" concluded that geophysical monitoring should be intensified in the Beijing–Tianjin–Tangshan–Bohai region, a few hundred km across (and apparently not including Haicheng). After analyzing past earthquakes the group concluded that a quake of magnitude 6 or larger was likely from 1975 through 1977 in the region from 33 to 43 north latitude and 110 to 124 east longitude (which includes Haicheng). In June of 1974 the State Seismological Bureau Conference reviewed data on past earthquakes and variations in geophysical measurements. They suggested that an earthquake with magnitude between 5 and 6 would occur in the northern Bohai region within a year or two. On December 22, 1974, an intense swarm of earthquakes, the largest having a magnitude of 4.8, occurred near Liaoyang, about 70 km northeast of Haicheng. The swarm did not lead to the prediction of larger earthquakes there, but it did sensitize local officials. The Study Delegation reported that "the Liaoning Provincial Revolutionary Committee called an emergency meeting and undertook efforts to have earthquake hazard information reach every family." A State Seismological Bureau conference on January 13, 1975, estimated that a magnitude 5.5 to 6 earthquake would occur in the Yinkou–Dairen–Tantung area in the first 6 months of 1975. The region was not specifically defined in any publications I could find, but the cities mentioned form a triangle outlining the Liaotung Peninsula just south of Haicheng and about 400 km east of Beijing across Bohai Bay. The longest leg of the triangle is about 250 km from Dairen to Tantung. On February 1 an intense swarm of small "foreshocks" began near Haicheng, including 466 events in the 26 hours before the main shock. Most were too small to be felt, but many, including a magnitude 4.8 event 12 hours before the main shock, were easily noticed by the populace. According to the reports, the combination of accelerating earthquake occurrence and the many geophysical anomalies lead to an official prediction.

Given its importance the Haicheng experience merits a second look. The earthquake occurred and the Study Delegation reported during a period in the 1970s and early 1980s when there was great optimism about earthquake prediction but little experience with the practical problems. There are several questions one could ask:

(1) Was there a prediction? According to the study Delegation cited above the Shihpengyu Seismic Observatory reported the small earthquake swarm to the Provincial Earthquake Office at 1830 on February 3, and "the Provincial Revolutionary Committee was apprised of the situation at 0030 on February 4 and given a forecast of a strong earthquake near Haicheng for that day." I have not found evidence that the statement meets either my definition or Allen's [1976] version by stating the limits on location or magnitude, the probability or confidence associated with the prediction, or a complete record of past predictions.

(2) If so, was it fulfilled? Certainly a large earthquak occurred near Haicheng on the specified day. Apparently it violated the State Seismological Bureau's magnitude estimate of 5–6.

(3) What official action was taken in response to the prediction? The Study Delegation reported that emergency responders within 50 km of the epicenter were given special instructions and organized the public for disaster prevention procedures. Emergency shelters and first aid procedures were prepared. "Evacuation was carried out in many areas, and outdoor movies were set up." The extent of evacuations was not mentioned specifically, and the true extent of deaths and injuries from the earthquake are still debated. Geller [1997] discusses this topic in more detail and presents a very skeptical view.

(4) Did the apparent success demonstrate skill? Any prediction might succeed by chance, and of course the more predictions made, the greater chance that one or more will succeed by chance. The team reported that the Chinese had issued many false alarms, but not how many. Allen [1976] suggested that a prediction should be accompanied by complete records of successful and failed attempts. That suggestion was not followed in this case, so we cannot judge now the importance of luck.

Figure 7 shows some of the geophysical data on which the prediction was purportedly based: tilt inferred from leveling data on two short profiles at Jinxiang, about 185 km southwest from the earthquake. The relative elevation changes in late 1974, about 2.5 mm on the WNW profile and 1.5 mm on the NNE profile, were interpreted as earthquake precursors. The WNW and NNE profiles were approximately 560 m and 360 m long, respectively. Inexperienced amateurs, rather than trained professionals, may have collected the leveling data, as that was commonly the case during the Cultural Revolution. While these data were considered fairly convincing at the time, it now seems unlikely that phenomenon related to the earthquake caused the observed changes. First, the anomalies do not meet the credibility conditions listed above: there was no change at the time of the Haicheng earthquake, measurements much closer to Haicheng did not show similar tilts, and there is no reasonable mechanism to explain such deformation at 185 km from the earthquake. Second, the variations before the "anomaly" were too large during the whole period to meet the precision standards of modern leveling. High quality leveling should be accurate to about 0.3 mm for lines of that length [Bomford, 1971, p. 237], and the measurements do not indicate that this standard was ever met. Without demonstrated quality control, any variations must be viewed

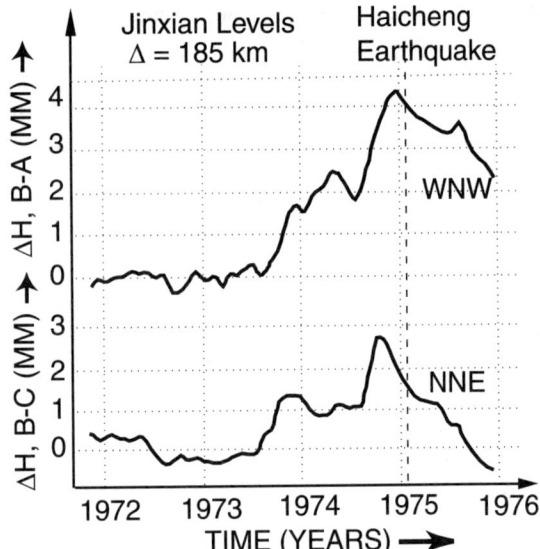

Figure 7: Example of geophysical data reportedly used to justify prediction of 1975 Haicheng earthquake.

as possibly due to experimental error. Gravity, magnetic, and electrical anomalies reported before the Haicheng earthquake have very similar problems.

In summary the claim that scientists successfully predicted the Haicheng has not been documented. The statements made before the earthquake apparently do not qualify as a prediction as commonly defined, the geophysical data on which the reported prediction was based are highly questionable; and an earthquake may well occur within a forecasted region quite by chance, especially if many such forecasts were made.

Shear-Wave Splitting

Several reported precursors are based on observations of apparent temporal variations in seismic wave velocity. Such methods are appealing because growing stress can be expected to change the shape and size of cracks and pores, which in turn affect the velocity. Earthquakes generally nucleate at several km depth, making the relevant velocity measurements challenging. Accurate measurements require that the starting time of seismic waves, and their path through the crust, be known accurately. Controlled artificial sources and seismic recorders are necessarily near the surface, and the times of deeper natural sources (small earthquakes) must be inferred from the same data used to estimate the velocities. For natural sources there is a tradeoff between location and time, seriously degrading the accuracy of velocity estimates. However, relative velocity variations partly circumvent this obstacle. For example, seismic waves include two components of shear waves. In an isotropic medium, these shear waves have particle motion perpendicular to the ray direction and to each other. But cracked rocks are usually anisotropic, with different velocities for the two shear components. Because the two shear waves follow the same path until they encounter a stressed, anisotropic region, natural sources may be used without too much sensitivity to their locations. Crampin et al. [2003] argue that such "shear wave splitting" increases as stress builds before a quake. Crampin et al. [1999] report that they detected enhanced shear wave splitting in Iceland and announced to officials that an earthquake "could occur at any time" within a three month period, sooner for a magnitude 5 and later for a magnitude 6. According to them "specific magnitudes were not suggested" except in a broad range. A magnitude 5 earthquake did occur within their study area three days later. Such observations are intriguing but they have a long way to go before they provide evidence of predictability. The phenomena must be proved to occur systematically, and the earthquake preparation zone must be shown to be the source of the splitting. Then some relationship must be developed to quantify the magnitude, location, and time from the shear-wave observations. Finally the method must be validated with several successful test predictions. Many other phenomena have reached the intriguing level but failed the follow up.

STRESS EVOLUTION

Earthquakes occur when the accumulated stress overcomes the critical strength of rock. Here the word "overcomes" hides some complications, because the strength of rock may be affected by the history of previous stresses and slip on faults. In any case a promising strategy is to calculate the stress accumulation from the long-term slip rate of faults and known earthquakes, and to estimate the strength from the calculated stress just before past earthquakes. Consider two crustal blocks separated by a fault and subject to tectonic forces due to plate motion. Given the displacement rates on the boundary, the location and orientation of the fault, and knowledge of the elastic properties of the two blocks, it is indeed possible to calculate the rate of stress accumulation everywhere in the blocks, and on the fault surface. Should this not be easy? If not, why not? It's not easy because the actual earth is far from the ideal situation described above. First, we do not know the appropriate boundary conditions on the blocks. Geodetic measurements give a good picture of the strain rate on the upper surface, and plate tectonic models give us the relative velocity at a few hundred km from the faults. However, we do not know the forces driving the motion, nor the displacement rates or stresses on the sides or bottoms of the blocks. The stresses differ substantially depending on whether the blocks are riding on a conveyor belt or being pushed over a resistive surface, and we do not know which picture is more accurate. Second, we do not know any appropriate initial conditions. It is tempting to believe that any shear stress on the major fault was erased by the last large earthquake. But there is no consensus and this tempting belief, if true, would still constrain stress only on the fault. Stress may have been extremely heterogeneous since plate tectonics began. Third, we do not know the properties of the medium very well. Measured seismic velocities can tell us average compressibility and rigidity to within a few percent, but un-measurable local variations strongly affect stress. Furthermore, the crust is not perfectly elastic, and we have little information on its viscous and plastic properties. Fourth, we do not know the geometry or even the existence of all the important faults which affect stress profoundly. Even in well studied places like southern California, major earthquakes continue to occur on faults that were not identified in advance. If these faults can slip between major quakes, then stress will concentrate at their boundaries, and we have no way to quantify it. Faults are generally believed to have fractal properties such that their number increases as their size decreases, much like the magnitude frequency relation for earthquakes illustrated in Figure 1. Clearly we can never have a complete fault map, and it is arguable that the many small faults have a stronger effect on stress than the few large ones.

The difficulties cataloged above may possibly be surmounted if three conditions apply. First, large earthquakes must depend predominantly on the average stress over large regions. Local, unknowable stress variations must be more or less irrelevant for the large earthquakes. Small earthquakes might start at local stress concentrations, but they must only grow into large ones when the general level of stress is high over a large region. Thus, stress must be primarily effective at controlling earthquake propagation rather than nucleation. Second, the overall stress level must have significant temporal variations. If it is always at a near critical level, as suggested by some models of self-organized criticality, then the conditions for large earthquakes will be ever-present, and stress calculations based on observable quakes may be misleading. Earthquakes too small to be cataloged might control the regional stress by dint of their large numbers. If so, the second condition might not be satisfied. The third necessary condition is that we must be able to recognize when regional stress is high. We might possibly calculate the stress accumulation from long-term fault slip and known major earthquakes. Of course that would predict the next rupture only if the stress is dominated by large, known earthquakes; if the stress fields from very ancient earthquakes have somehow decayed or been superceded by recent events; and if simple assumptions about boundary conditions, initial conditions, and material properties are valid. Such calculations will not be right in detail, but perhaps the average stress levels would have predictive power. Another hope is that earthquake patterns themselves might reveal regionally high stress. Many have suggested that "long range correlation" of earthquake occurrence could signal that regional stress is high and large earthquakes are more likely. However, reliable measures of long range correlation have not yet been demonstrated and tested for predictive power.

Estimating stress evolution [*Harris*, 1998; *Stein*, 1999] is the most promising avenue for earthquake forecasting and perhaps prediction, but as indicated above it's a difficult road which may not lead to prediction. There are now hundreds of examples of moderate earthquakes that have occurred where stress was increased by previous earthquakes [see Harris, 2000, for a review], but comprehensive studies [*Kagan and Jackson*, 1998; *Parsons*, 2002] show many counterexamples as well. Stress estimates will most likely reveal significant temporal variations in large earthquake probability. However, with all the caveats and uncertainties listed above, the probability gains may not be dramatic.

DISCUSSION

False optimism, unrealized hopes and even outright failures are no reason to surrender. Every science has its failures, and many seemingly impossible challenges have been overcome. Human flight had highly respectable doubters, but now we fly! Space travel and nuclear power were once considered fantasies, yet now they are technically routine. Wyss [2001] makes the point that seismology is young; new technologies like the Global Positioning System (GPS) and interferometric synthetic aperture radar (InSAR) have the potential to measure earth deformation to a precision undreamed of until recently. New data might reveal precursors if we look hard enough, and only the Japanese are exploiting these technologies with the needed vigor. Wyss states that funding for prediction research is "far below what is needed to make significant advances in this difficult problem," and that unrealistic claims by well-meaning but poorly skilled amateurs gives prediction efforts an unfairly bad reputation. However, funding limitations, the youthfulness of our science, and involvement of amateurs are not our only limitations. The facts that earthquakes start in small deep volumes, that weak signals are swamped by strong noise, and that complexity introduces non-uniqueness are not the result of budget limitations. These difficulties may not be impossible to solve, but neither is there any guarantee of success. The fact that people can reach the moon does not mean that they can predict earthquakes reliably.

Earthquake prediction can best be viewed as a highly refined version of forecasting, the specification of long-term probabilities as a function of location, magnitude, and time. So far stochastic approaches based on persistence and earthquake clustering have been most successful in forecasting the rates of earthquakes, but rigorous testing of such techniques is still young. The seismic gap model, which might have led to much sharper temporal forecasting, has not proved effective to date. The sticking point has been whether there are characteristic earthquakes that behave differently than smaller quakes, and if so, how to recognize them. The search for precursors has a checkered history, with no convincing successes. The 1975 Haicheng "prediction" does not withstand scrutiny by modern standards. A general problem with earthquake prediction research is that retrospective correlations are easy to find, but only the most generic models can be tested prospectively. Stress evolution might lead to important refinements in forecasting, but so far the models are not specific enough for prospective testing. Reliable prediction in the popular sense, with high confidence and large probability gain, is not foreseeable at this time. However, long-term forecasting provides reliable information for reducing earthquake risks.

Earthquake research is still a highly valuable endeavor, in spite of the difficulties facing prediction. Scientifically, we can learn a great deal about the causes and effects of earthquakes, the material properties of the medium in which they occur, and their effects on the natural landscape and environment. Modeling the time-dependence of earthquakes more accurately will help us understand better the physics of stress accumulation and release. Practically, we can contribute to public safety and the resiliency of the built environment by exploiting earthquake forecasting more wisely.

Acknowledgments. I thank Harsh K. Gupta and Zhongliang Wu for helpful reviews of an earlier draft, and Zhen Liu for help with illustrations. This research was supported by the Southern California Earthquake Center. SCEC is funded by NSF Cooperative Agreement EAR-0106924 and USGS cooperative Agreement 02HQAG0008. The SCEC contribution number for this paper is 765.

REFERENCES

Allen, C. R. (1976), Responsibilities in earthquake prediction, *Bull. Seis. Soc. Amer.*, 66. 2069–2074.

Bakun, W. H., and A. G. Lindh (1985), The Parkfield, California earthquake prediction experiment, *Science, 229*, 619–624.

Bakun, W. H., and T. H. McEvilly (1984), Recurrence models and Parkfield, California earthquakes, *J. Geophys. Res., 89*, 3051–3058.

Bolt, B. A. (1988), *Earthquakes*, W. H. Freeman, New York, New York.

Bomford, G. (1971), *Geodesy*, 3rd. ed., 731 pp., Oxford U. Press, Oxford, England.

Brumbaugh, D. S. (1999), *Earthquakes: Science and Society*, 252 pp., Prentiss Hall, Upper Saddle River, NJ 07458.

Crampin, S., T. Volti, and R. Stefansson (1999), A successfully stress-forecast earthquake, *Geophys. J. Int., 138*, F1–F5.

Crampin, S., Y. Gao, S. Chastin, S. Peacock, and P. Jackson (2003), Speculations on earthquake forecasting. *Seis. Res. Lett.*, 74, 271–273.

Evison, F. F. (2001), Long-range synoptic earthquake forecasting: an aim for the millennium. *Tectonophysics, 338*, 207–215.

Geller, R. J. (1997), Earthquake prediction, a critical review, *Geophys. J. Int., 131*, 425–450.

Haicheng Earthquake Study Delegation (1977), Prediction of the Haicheng Earthquake, *EOS, Trans. Amer. Geophys. Un., 58*, 236–272.

Harris, R. A. (1998), Introduction to special section: Stress triggers, shadows, and implications for seismic hazard, *J. Geophys. Res., 103*, 24347–24358.

Harris, R. A. (2000), Earthquake stress triggers, stress shadows, and seismic hazard, *Current Science, 79*, 1215–1225.

Jackson, D. D., and Kagan, Y. Y. (1999). Testable earthquake forecasts for 1999, *Seism. Res. Lett.*, 70(4), 393–403.

Kagan, Y. Y. (1996), Comment on "The Gutenberg-Richter or characteristic earthquake distribution, which is it? By S. G. Wesnousky, *Bull Seis. Soc. Amer., 86*, 274–285.

Kagan, Y. Y. (1997), Statistical aspects of Parkfield earthquake sequence and Parkfield prediction experiment, *Tectonophysics, 270*, 207–219.

Kagan, Y. Y., and D. D. Jackson (1998), Spatial aftershock distribution: effect of normal stress, *J. Geophys. Res., 103*, 24453–24467.

Ma, Z., F. Zhengxiang, Z. Yingzhen, W. Chengmin, Z. Guomin, and L. Defu, (1989), *Earthquake prediction: nine major earthquake in China (1966 – 1976)*, 332 pp., Seismological Press, Beijing.

National Earthquake Prediction Evaluation Council Working Group (1994), Earthquake prediction research at Parkfield, 1993 and beyond, *U. S. Geol. Surv. Circular 1116*, 14 pp., U. S. Government Printing Office, Washington, D.C.

Nishenko, S. P. (1989), Circum-Pacific seismic potential 1989–1999, *Open File Rep. 89–85*, 125 pp., U.S. Geol. Surv.

Nishenko, S. P. (1991), Circum-Pacific seismic potential: 1989–1999, *Pure Appl. Geophys. (PAGEOPH) 135*, 169–259.

Parsons, T. (2002), Global Omori law decay of triggered earthquakes: large aftershocks outside the classical aftershock zone, *J. Geophys. Res., 107* (B9) 2199, doi:10.1029/2001JB000646.

Rhoades, D. A., and F. F. Evison (1989), Time variable factors in earthquake hazard, *Tectonophysics, 167*, 201–210.

Rhoades, D. A., and F. F. Evison (2004), Long-range earthquake forecasting with every earthquake a precursor on its own scale, *Pure Appl. Geophys.* 161, 47–72.

Scholz, C. H. (1990), *The mechanics of earthquakes and faulting*, 439 pp., Cambridge University Press, Cambridge, England.

Stein, R. S. (1999), The role of stress transfer in earthquake occurrence, *Nature, 402*, 605–609.

Wesnousky, S. G. (1994), The Gutenberg-Richter of characteristic earthquake distribution, which is it?, *Bull. Seismol. Soc. Amer., 84*, 1940–1959.

Wyss, M. (2001), Why is earthquake prediction research not progressing faster? *Tectonophysics, 338*, 217–223.

Earthquake Prediction, Seismic Hazard And Vulnerability

Seiya Uyeda

Earthquake Prediction Research Center, Tokai University, Shizuoka, Japan

Kimiro Meguro

*International Center for Urban Safety Engineering, Institute of Industrial Science,
The University of Tokyo, Tokyo, Japan*

Mitigation of seismic hazards requires integration of science and human action, namely the science of earthquakes, anti-seismic engineering and socio-political measures. The public, media, policy makers and funding agencies must be constantly reminded that seismic disasters rapidly escalate with civilization's growth and that disasters come when the last tragedy has been "forgotten". Loss of human life is caused overwhelmingly by the collapse of houses and other buildings within less than a few minutes of the main shocks. The most urgent countermeasure is the reinforcement of weak structures. When structural damage is reduced, most other seismic hazards will correspondingly be greatly reduced. If short-term prediction is made, casualties will be further reduced dramatically. Despite general pessimism, short-term prediction research needs to be enhanced because recent research shows real promise. Thus, the reinforcement of existing structures and enhancement of short-term prediction research are the two keys for seismic hazard mitigation.

INTRODUCTION

Earthquakes are caused by sudden fault motion. It has long been known that the global distribution of earthquakes is far from uniform, as shown in Fig. 1. While mid-oceanic ridges are characterized by linear distribution of relatively small earthquakes, large earthquakes occur mainly in the circum-Pacific belt and in the wide zone between Eurasia and the southern continents. The reason why they are distributed in this manner is explained by plate tectonics [e. g., *Stein and Klosko*, 2002]; large earthquakes occur due to plate interactions at convergent and transform plate boundaries. Convergent plate boundaries consist of subduction and collision zones. Statistical properties of earthquake occurrences in these seismic zones are well-known thanks to many decades of seismic observations around the globe, and many seismic hazard maps have been compiled both globally and regionally [e. g., GSHAP program, *Giardini*, 1999].

Plate tectonics has shown with reasonable certainty that plate motions have been essentially steady for the time-scales of a few millions of years, although they have changed over much longer time scales during the earth's history [e.g., *Kumazawa and Maruyama*, 1994; *Uyeda*, 2002]. However, the seismicity of the globe also displays variation on finer time-scales of tens of years, as shown in the upper panel of Fig. 2. The detailed mechanisms for these short-term secular changes are not clear, but it seems plausible that there can be fluctuations even under steady global plate motions. It seems significant that the variations in the loss of life shown in the lower panel of Fig. 2 are very different from variations of seismicity. The reason for this is quite clear. Earthquakes in densely populated regions cause greater disasters [*Utsu*, 2002]. For instance, the two largest giant earthquakes, the 1960 M9.5 Chilean and 1964 M9.2 Alaskan earthquakes, caused much less

350 EARTHQUAKE PREDICTION, SEISMIC HAZARD AND VULNERABILITY

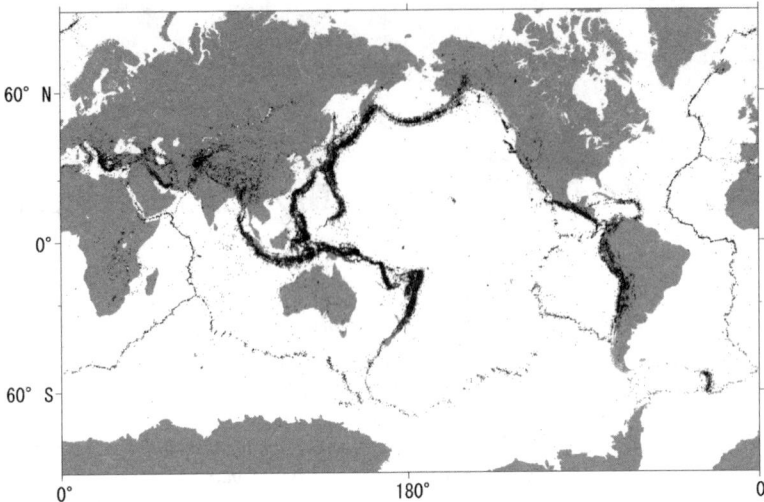

Figure 1. Epicenter distribution of the world for magnitude greater than 4.0 in 1980-2000, after USGS PDE.

loss of life compared to the much smaller 1923 M7.9 Kanto and 1976 M7.8 Tangshan earthquakes. (Hereafter, M stands for the magnitude of an earthquake).

Here we review briefly the state of the art on seismic hazard and present personal views on the priority items related to earthquake hazard mitigation. Specific matters taken up in the text are focused on Japan. However, it is anticipated that what we present will be generally applicable to all earthquake-prone regions of the world.

As well as loss of life, earthquakes cause disasters of all kinds. In terms of monetary loss, this can amount to a substantial fraction of the Gross Domestic Product (GDP) of a nation. For the 1995 Kobe earthquake, monetary loss was estimated at 10 billion US dollars and for the Tonankai-Nankai event, which is a major anticipated future earthquake, it may amount to at least 80 billion US dollars, which would be more than 15% of the GDP according to the Central Disaster Management Council of Japan. Since buildings in Japan are now more seismic resistant than in the past, loss of life in large cities may be less than those in the past. However, huge damage of complicated life-lines and infra-structures can still be expected in modern mega-cities when they are hit by major quakes. Further many mega-cities in the developing world do not have good adherence to building codes and so there can be huge loss of life as illustrated by the 2004 Bam earthquake in Iran. Seismic risk rapidly escalates with population growth and the disasters come when the last experience is "forgotten".

FIRST PRIORITY IS THE REINFORCEMENT OF OUR HOUSES

Many problems arise at and after a disastrous earthquake [*Meguro and Takahashi*, 2001], including:

1) Loss of life at the time of the main shock,
2) Further victims due to fires,
3) Psychological instability of people in the affected areas,
4) Disruption of the community,
5) Building refugee camps for large number of displaced people,
6) Demolition of damaged structures and related environmental effects,
7) Economic, business and societal disruption.

At the time of the main shock, rapid delivery of information on ground motions and damages to the authorities and public are critically important for rescue activity. Thanks to major advances in information technology, there is an emerging new category of hazard mitigation; real-time seismology that has been developed in several areas, including Japan, Mexico, California and Taiwan [e. g., *Nakamura*, 1988; *Epinosa Aranda et al.*, 1995; *Kanamori et al.*, 1997; *Shin et al.*, 2000]. The Seismic Alert System in Mexico made use of the gap of the arrival times of the P- and S-waves. The authorities were able to issue timely alarms so that the subways in the Mexico City were stopped 50 seconds before the arrival of the destructive S-waves from the 1995 M7.3 Guerrero earthquake located at the Pacific coast. However, many of the problems listed above will last for a long time after the event. To cope with the evolution of disaster situations in hours, days, weeks and even months, innovative engineering strategies will be needed. A new kind of strategy can be called the real-time earthquake engineering. After the main shock, continuing data collection of structure damage distribution, rescue activities, lifeline interruptions, debris removal, refugee camps, and power demand fluctuations and other parameters will be critical because of the utmost importance for disaster relief agencies and local and national governments to allocate resources in the

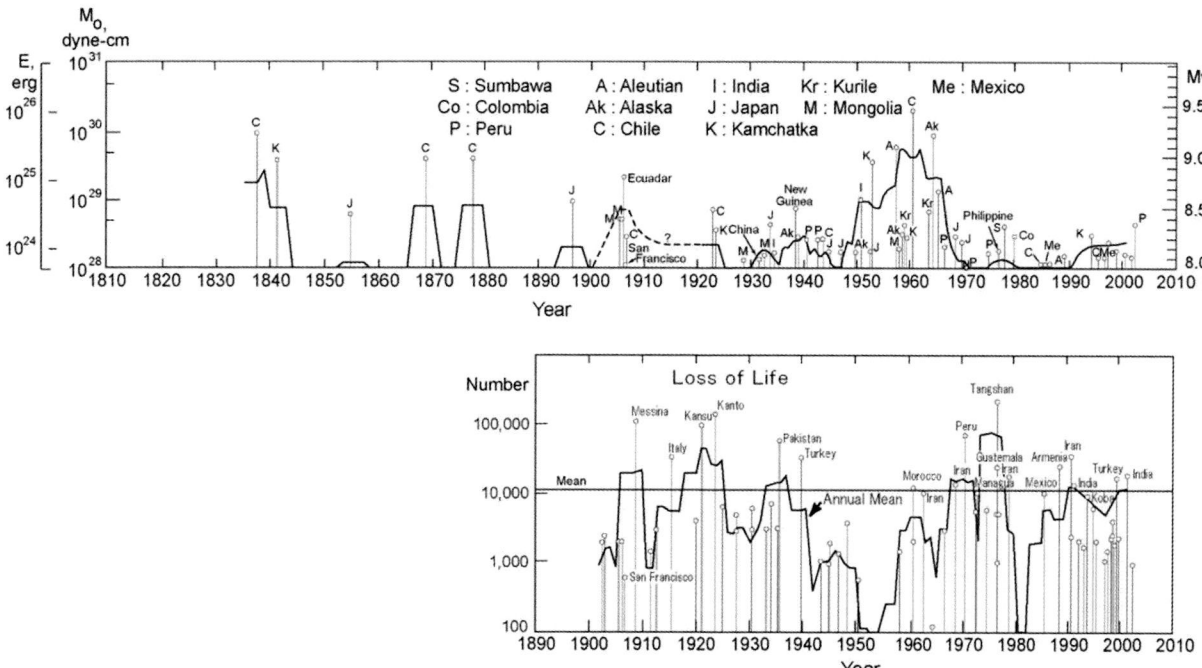

Figure 2. Secular variations of seismic energy release (upper panel) and human loss (lower panel). Ordinate of the upper panel represents the size of earthquakes in seismic moment M0, seismic energy E and the moment magnitude Mw. Ordinate of the lower panel shows the number of victims. In both panels, the vertical bars are for individual event and the solid curve shows the annual average (unlagged 5-year running average) (after H. Kanamori, private communication).

most prompt and efficient manner [*Noda and Meguro*, 1995; *Hada and Meguro*, 2002].

We now focus items 1) and 2) in the above list, namely on how to save human life. Experiences all over the world show that, almost without exception, the majority of victims are killed by the collapse of buildings right at the moment of the main shock (Plate 1, a & b). An example is shown in Fig. 3 for the 1995 M7.3 Hyogoken Nanbu Earthquake, which took over 5,500 human lives. (This earthquake is commonly called the Kobe earthquake after the name of the city, where the damage was most intense.) Almost 90 % of the victims were killed in their own houses. Moreover, medical examinations indicate that 92 % of casualties were killed within less than 15 minutes after the quake before any organized rescue operation had started [*Nishimura et. al.*, 1997a, b].

Many victims of earthquakes can be killed by fires. Although fires can be caused by shaking, most of big fires start from collapsed houses. Fires that started in non-collapsed houses were commonly extinguished by the residents, while those that started in collapsed houses were not. Residents under debris could not fight the fires and the first priority of outside rescuers was to get them out from the fallen houses. Furthermore, roads were often blocked to fire engines by collapsed houses. In Kobe, the Fire Department had a capacity to handle only three to four fires at a time. When many tens of fires started simultaneously, they were completely overwhelmed. To prevent fires from developing to uncontrollable size, the best way is to prevent houses from collapsing.

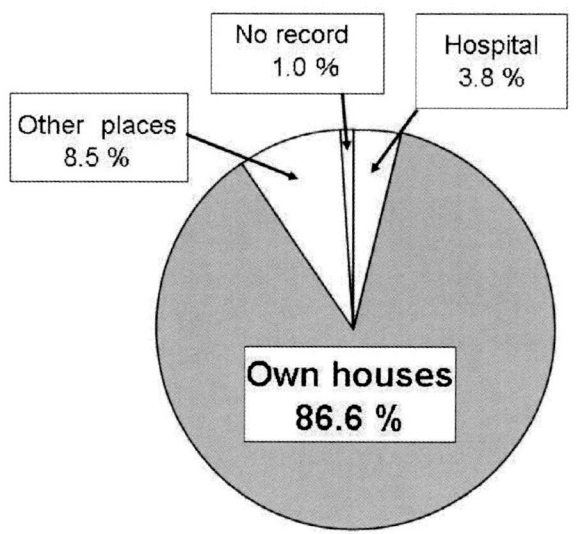

Figure 3. Places where people died (in Kobe City). (after Nishimura et al., 1997b)

Plate 1. Damage to low earthquake resistant structures is major cause of casualties and many problems generated after the earthquake in the world (photo by K. Meguro) (a) Damage to old timber houses due to the Kobe earthquake (M7.3), 1995 in Japan (b)Damage to masonry houses due to the Qayen earthquake (M7.1), 1997 in Northeast Iran.

If the government of Japan were to fund 10 billion US dollars annually, all the necessary reinforcement of houses in Japan would be completed in ten years [*Ohtake*, 2003]. In order to implement the house reinforcement program effectively, the following system, called the Retrofitting Promotion System, has been proposed [*Meguro and Takahashi*, 2001; *Yoshimura and Meguro*, 2003]. The main concept of this system is that the government guarantees to pay a portion of the repair and reconstruction expenses of damaged houses provided that proper retrofitting had been implemented by the owners before the earthquake. With such a system, the overall financial burden of both government and residents can be greatly reduced. The feasibility of this system has been verified for cases in which the retrofitting cost is 10 to 15% of the new construction cost, as is the case in Japan. However, for countries like Turkey, where the retrofitting cost amounts to approximately 75% of the new construction cost, the system is not viable. In such cases, efficient and economic techniques for retrofitting are needed, especially for masonry structures, using locally available and inexpensive materials [*Mayorca and Meguro*, 2003].

If the structural damage is reduced, loss of human life is reduced. Thus, the most effective and highest priority countermeasure against the loss of human life, and all kinds of related seismic hazards and consequences, is the structural issue. In the recent December 26, 2003, M6.7 Bam Earthquake in Iran, there were more than 30,000 casualties due to the collapse of adobe and brick structures which lacked reinforcement. After the disaster, international and local agencies rushed to the affected sites trying to rescue survivors. Although these were valuable efforts, only a few people survived and were rescued, again highlighting the quality of buildings as a key issue.

FIRST PRIORITY IN SCIENCE IS SHORT-TERM PREDICTION

Engineering seismology and earthquake prediction constitute the two key measures that need implementation for major seismic hazard mitigation. It is customary to classify earthquake prediction into three categories: long-term, intermediate-term and short-term predictions. They are different in methodology, accuracy and purposes. For the purpose of saving human life, the short-term (less than a day to months) prediction is most effective. If people had been warned in advance and had escaped adobe buildings in time, casualties in the recent Bam earthquake could have been many times smaller. The impact of short-term prediction on casualties would also be dramatic for cases of large earthquakes that generate tsunamis.

Long-term prediction deals with the probability of earthquake occurrence on time-scales of 10 to 100 years, based mainly on geologic studies of faults and historic records of seismicity, while intermediate-term (1 to 10 years) prediction uses more recent data including seismological observations. One of the most advanced of such efforts is the M8 algorithm which is based on the non-linear dynamics (*Keilis-Borok and Kossobokov*, 1990). It is a pattern recognition approach based on monitoring seismicity and its fluctuations. Long and intermediate-term predictions are typically statistical likelihood estimates, while short-term prediction is based on some definite precursors.

Systematic short-term earthquake prediction research started in the 1960's in several countries including Japan, USA, Soviet Union, and China [*Rikitake*, 1976]. In the 1970's, optimism prevailed due to the encouraging developments, such as dilatancy models [e. g., *Scholz et al.*, 1973] and the successful prediction of the 1975 M7.2 Haisheng earthquake in China [*Chen et al.*, 1990; *Li Hui* 1996]. However, no further generally recognized successes followed, causing the community to become pessimistic. Apparently, the failure of predicting the earthquake in Parkfield, California [*Andrews*, 1992], where the World's best monitoring system was in operation, discouraged American researchers. This is understandable but today's widespread pessimism seems to be unjustified because such a view misses the point that both science and technology are making rapid progress.

In 1978, the Earthquake Prediction Program of Japan designated eight "areas of special observation" and two "areas of intensified observation" (inset of Fig. 4), based on historically disastrous earthquakes, active faults, high seismicity and socio-economic importance [*Hamada*, 1992]. This can be taken as an example of set of nationwide intermediate-term predictions. Comparing the two maps in Fig. 4 indicates that the intermediate-term predictions have been more or less fulfilled, because most of the large earthquakes after 1978 have occurred in or near the "areas of special observation". However, none of these earthquakes, including the Kobe earthquake, was predicted in the short-term. The main reason for this was that the methods for short-term prediction were not applied in any of these "areas of special observation". In fact, the methods, such as densely distributed tilt-meter and strain-meter monitoring, have been applied only in the two "areas of intensified observation" where major earthquakes have not yet occurred. The two "areas of intensified observation" were selected because a great earthquake exceeding magnitude 8 is expected to occur there soon, based on long term predictions. This future earthquake has even been given a name: the Tokai (meaning East Sea in Japanese) earthquake. The postulated epicentral area has lately been extended westward as in the main map of Fig. 4, so that it is now often called the Tonankai/Nankai (East South/South Sea) earthquake (see Fig. 6).

When people assert that the search for earthquake precursors has proven useless, they often overlook the fact that scientific and effective precursor research has seldom been carried out. Certainly, seismic networks have been considerably upgraded, so that even micro-earthquakes can now be detected and located precisely. Precursory changes in local/regional seismicity, such as fore-shock activity and quiescence, may be detected [e. g., *Sobolev et al.*, 2002], but seismic networks are not suited, by definition, for detecting non-seismic precursors.

When an earthquake occurs, it is a mechanical vibration of the ground but since it is caused by breaking of the earth's crust, which, unlike flawless piece of glass, has highly heterogeneous structures, it is reasonable to expect that its preparatory process has various facets which may be observed before the final catastrophe. Therefore, the science of earthquake prediction should be multidisciplinary as depicted in Fig. 5. In seismology, the study of the physical mechanisms of earthquake generation has made significant advances, which include the earthquake dynamics models with pre-seismic slips,

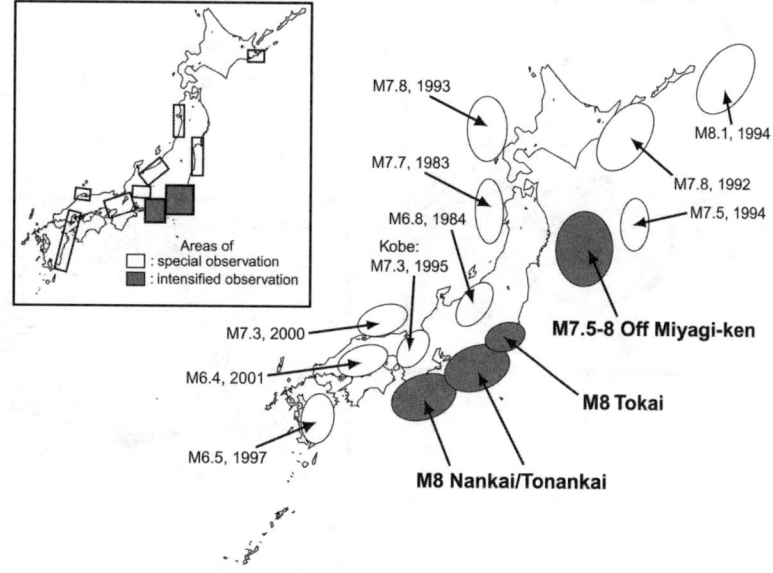

Figure 4. Inset shows the eight "areas for special observation" (empty rectangles) and two "areas for intensified observation" (filled rectangles), selected by the Japanese Earthquake Prediction Program in 1978. Main figure shows the roughly estimated source regions of major earthquakes which occurred afterwards (smaller letters) and expected earthquakes (larger letters).

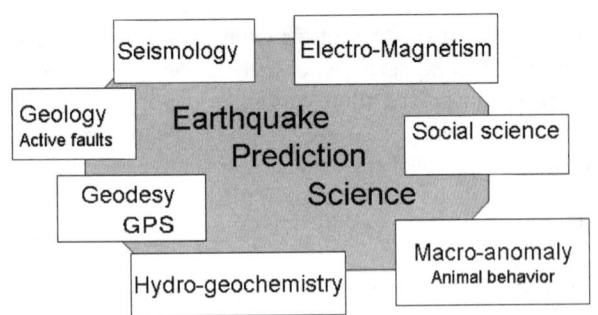

Figure 5. Multidisciplinary nature of earthquake prediction science.

although pre-seismic slip for real earthquakes has yet to be verified. (e. g., *Aochi and Matsu'ura*, 2002).

Pre-seismic changes in levels and chemistry of ground water are providing potentially useful data for short-term prediction [e. g., *Silver and Wakita*, 1996; *Koizumi et al.*, 1999]. Macro-anomalies, such as anomalous animal behaviors may also be useful [e. g., *Ikeya et al.*, 1997], although it may take some time for the scientific community to take them up seriously. Social science should work on establishing better ways for timely dissemination of prediction information to the public without causing undue social unrest. The rapid progress of information technology will play a major role in all these issues.

The introduction of the global positioning system (GPS) has revolutionized geodetic science. Movements of the earth's surface can now be tracked with the precision of 1–0.1 cm. Fig. 6 is an example to illustrate the occurrence of anomalous crustal movements detected by GPS in the Tonankai area, Japan. The displacement after March 2001 shown in (A) has been drastically different from the longer term average shown in (B). The central coastal area, in particular, has moved south-eastward against the north-westward subduction of the Philippine Sea Plate. This might indicate a pre-slip of the feared great Tonankai earthquake cited above or a manifestation of a slow earthquake. Such observations were not possible before GPS.

Another novel direction of research into the mechanics of earthquake generation is focused on the possibility of electromagnetic precursors. Since earthquakes are sudden fault motion that involves fracture of rocks and/or frictional movements of rock masses, it is anticipated that the seismogenic process involves electromagnetic phenomena. The main problem is whether observable electromagnetic pre-seismic manifestations exist or not. There have been a number of positive reports during the last decade. Evidence has been found for electromagnetic phenomena in a wide frequency range from many parts of the world, including Greece, Japan, Russia, China, Taiwan, Armenia and Italy [e. g., *Lighthill* (Ed.) 1996; *Hayakawa and Molchanov* (Eds.), 2002; *Uyeda and Park* (Eds.), 2002, *Balassanian et al.*, 1997 and references therein.].

Pioneering work has been documented in Greece (see for instance, *Varotsos et al.*, 1984: *Lighthill* (Ed.), 1996). A group of Athens University physicists claim that short-term prediction can be made by detecting anomalous pre-seismic transient

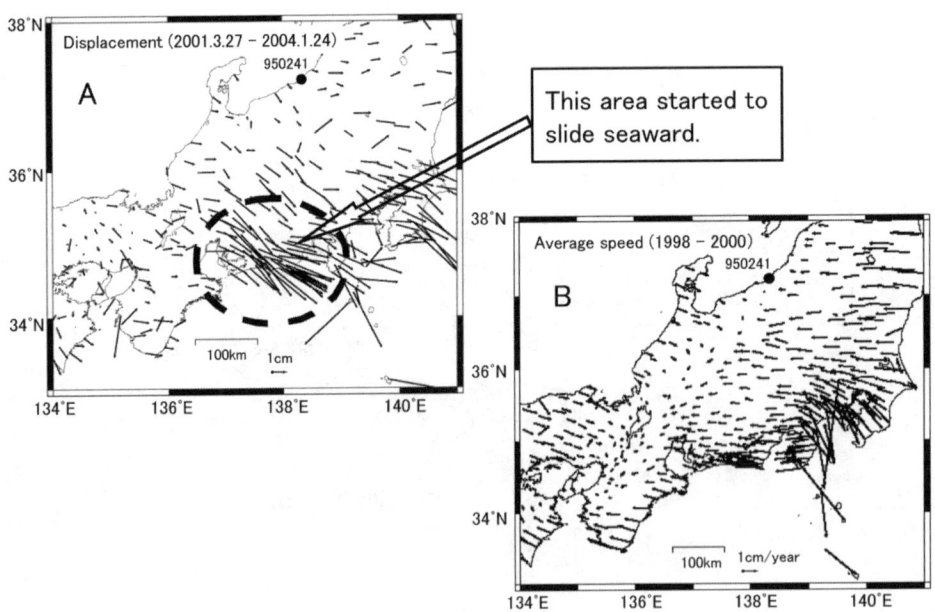

Figure 6. GPS observation of the displacement of central Japan relative to Point 940241 (After website of Geodetic Survey Institute of Japan). A. Displacement in cm from March 27, 2001 to January 24, 2004. B. Average speed in cm/year for 1998–2000.

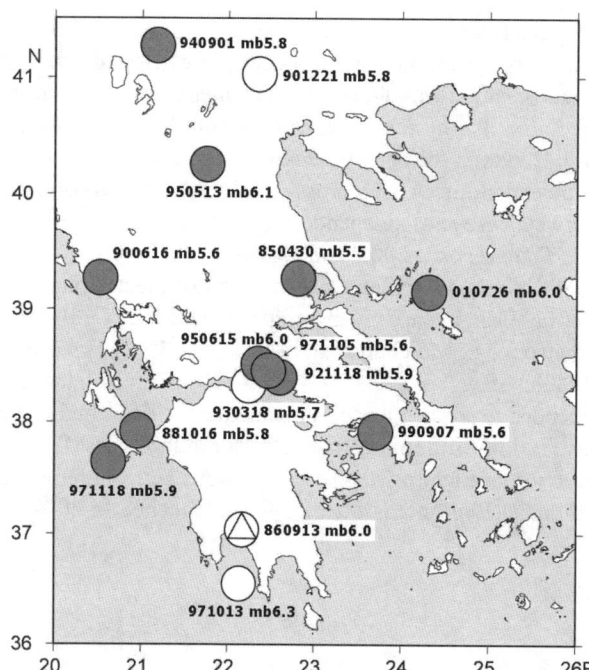

Figure 7. Evaluation of VAN prediction. All the earthquakes with USGS PDE magnitude larger than 5.5 for 1985–2003. Earthquakes are specified next to each circle. For example, 950513 mb 6.1 means year (1995), month (May), day (13th) and magnitude 6.1. The symbol mb is body wave magnitude.
Shaded circles: "successfully" predicted.
White circle with triangle: unsuccessfully predicted.
White circles: missed.

electric currents in the earth. Fig. 7 shows some results of their method, now called the VAN method (after the initials of the members of the founding team) [e. g., *Nagao et al.*, 1996; *Uyeda and Al-Damegh*, 1999, *Uyeda et al.*, 1999; *Kondo et al.*, 2002]. For the period 1985–2003, 12 out of 16 M=5.5 earthquakes in the Greek region were successfully predicted. Successful prediction here means that the errors were less than 0.5 for earthquake magnitude, 100 km in epicentral position and a few weeks in date. The public impact of VAN's predictions has been quite large because citizens' lives have been saved at disastrous earthquakes [*Uyeda*, 2000].

In Japan, several research groups have worked on earthquake related electromagnetic phenomena since the early 1990's. At the time of the January 17, 1995 Kobe earthquake, anomalous electromagnetic phenomena at frequencies ranging from extremely low frequency (ELF; 10^2–10^3 Hz) to very high frequency (VHF; ~10^2 MHz) were observed at many localities [*Nagao et al.*, 2002]. Encouraged by these results, seismo-electromagnetic research in Japan has become very active. Through deploying about 40 monitoring stations for the geoelectric potential changes over the country, it was demonstrated that the VAN-type pre-seismic changes also occur in Japan [*Uyeda et al.*, 2000]. They were observed before M ≥ 5 earthquakes that occurred within 20 km or so of our stations (Fig. 8). Clear co-seismic signals, synchronized with the arrival of seismic waves, have also been observed for nearby earthquakes [*Nagao et al.*, 2000].

Pioneering work has also been made on the precursory phenomena in the ultra low frequency (ULF: 0.01 Hz) geomagnetic field variations. Notable examples are the 1988 M6.9 Spitak [*Kopitenko et al.*, 1993], 1989 M7.1 Loma Prieta [*Fraser-Smith et al.*, 1990], 1993 M8.0 Guam [*Hayakawa et al.*, 1996] and 1997 M6.5 Kagoshima earthquakes [*Hattori et al.*, 2002].

Fig. 9 is the global summary of the ULF investigations, showing the empirical relationship between earthquake magnitude and epicentral distance. White and black marks show the earthquakes with and without observed ULF anomalies, respectively. The slant line indicates the empirical estimate of the threshold for the appearance of ULF signals. For instance, ULF signals of M7 earthquake will be detected within 100 or so km from the epicenter.

In the summer of 2000, there was a large swarm activity in the Izu island region south of Tokyo. In this case, both electric field and magnetic field measured at far apart stations

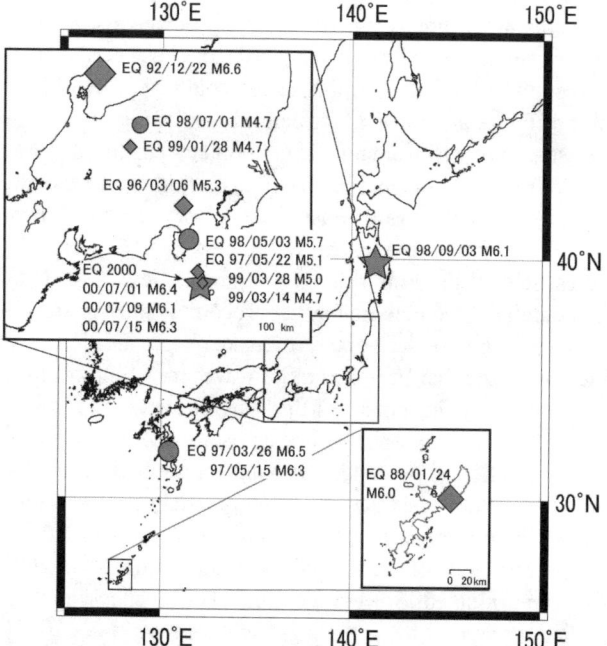

Figure 8. Distribution of M ≥ 5 earthquakes with pre-seismic signals during the period before March 1999. Diamonds and circles showed pre-seismic electric signals and ULF magnetic signals, respectively. Stars showed both electric and magnetic signatures. EQ 98/09/03 M6.1 means M6.1 earthquake which occurred on September 3, 1998.

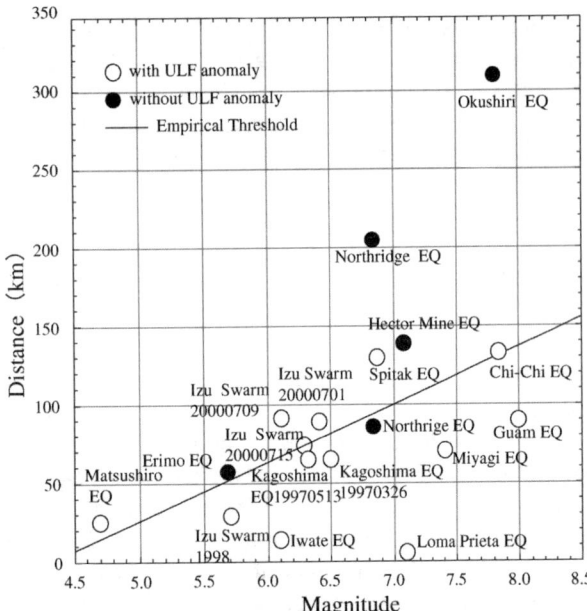

Figure 9. Magnitude—Epicentral distance plot for earthquakes (EQ) with and without ULF signatures. (Revised from Hattori et al., in press).

showed distinct anomalous precursory changes as shown in Fig. 10 [*Uyeda et al.*, 2002]. Remarkably these changes started a few months before the beginning of the swarm activity. Three major earthquakes in this swarm followed distinct precursory signals as indicated in Fig. 10 (b). These earthquakes correspond to the star symbol to the south of Tokyo in the inset of Fig. 8 and those in the central part of Fig. 9.

Research on electromagnetic phenomena has investigated a wide frequency range. For instance, Kushida and Kushida, at the Yatsugatake South Base Observatory, central Japan, were monitoring the transient reflections of VHF FM radio waves beyond the line of sight for meteor detection. They accidentally noted that their system recorded unusual signals on the night before the Kobe earthquake [*Nagao et al.*, 2002]. They proposed that VHF FM radio waves travelling over the focal zone of an imminent earthquake are scattered to reach over-the-horizon distances. They devised a method of short-term prediction, now called the Kushida method. Since the mid-1990's, the Kushidas have been practicing actual short-term prediction experiments [*Kushida and Kushida*, 2002]. The performance of the Kushida method during 2000–2003 has been evaluated for M=5.5 earthquakes by checking their predictions against the actual seismicity [*Uyeda et al.*, 2004]. The criteria for successful prediction are as follows: the error in the date of the earthquake is less than ~10 days, the error in the epicenter is less than ~50 km, and the error in Magnitude is less than 1 Richter unit. About 40% of their predictions were successful and about 30% of M=5.5 earthquakes were

predicted. It was also found that even in unsuccessful predictions, apparently meaningful signals were detected although the interpretation was not correct. The method is still far from perfect and its physical basis is uncertain [*Pilipenko et al.*, 2001]. However, the performance of the method justifies further investigation. Active cross-check experiments are now underway by several independent groups [e. g., *Moriya et al.*, 2003, *Takano et al.*, 2003, *Kamogawa et al.*, 2003].

Pre-seismic anomalous transmission of electromagnetic waves has been intensively investigated in the VLF (~ 10 kHz) band also [e. g., Molchanov and Hayakawa, 1998; Hayakawa and Molchanov (eds.), 2002]. Moreover, the research interest is extended to space science. A micro-satellite dedicated to the investigation of the ionospheric perturbations, DEMETER (Detection of Electro-Magnetic Emissions Transmitted from Earthquake Regions) is planned to be launched in 2004 by

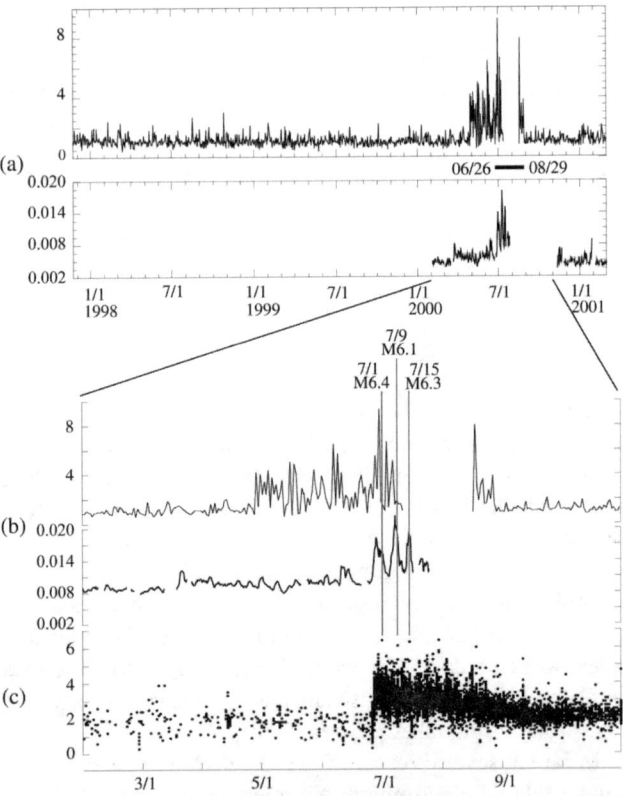

Figure 10. Time change of the 0.01 Hz spectral intensity, in arbitrary unit, of the geoelectric potential at a station in Niijima Island and that of the geomagnetic field at an array station in Izu Peninsula. The data shown are after some data processing process. (a) Three year records of electric (upper panel) and magnetic (lower panel) data. Both plots show the anomalous signals starting a few months before the onset of the swarm activity (June 26 to August 29, 2000). Data gaps are due to system failure by earthquakes and typhoons. (b) Enlarged for January through October, 2000. Three M ≥ 6 earthquakes shown in Figs. 8 and 9 are indicated. (c) Seismicity of the region.

CNES (French Natural Space Agency) [Parrot, 2002]. Thus, the seismo-electromagnetic research involves scientists from diverse branches; not only the conventional earthquake science, such as seismology, geo-electromagnetism, but also the radio, ionosphere and space science. In view of the urgent need for promoting interdisciplinary cooperation and for assisting this new science in the developing world, IUGG established, in 2002, an IAGA/IASPEI/IAVCEI Inter-Association Working Group on Electromagnetic Studies of Earthquakes and Volcanoes (EMSEV: S. Uyeda: Chair).

CONCLUSION

Reinforcement of buildings is the highest priority in the fight against the seismic hazard and to save human life, from the engineering aspect. Earthquake science needs to make utmost efforts in short-term earthquake prediction to further reduce casualties. Disaster prevention measures and prediction research are complementary activities.

Acknowledgments. The authors thank Professors Stephen Sparks, University of Bristol, and Zhongliang Wu, Graduate School of Chinese Academy of Sciences, and Dr. Masashi Kamogawa, Tokyo Gakugei University for their critical and constructive comments. Thanks are also due to Drs. Tom Beer, CSIRO, Australia and G. Heiken, Los Alamos National Laboratory for inviting the authors to Symposium U4 at the 23rd General Assembly of IUGG (2003) in Sapporo. Assistance of Mr. Yoichi Noda, Tokai University in preparing figures is highly acknowledged.

REFERENCES

Andrews, R. A., The Parkfield earthquake prediction of October 1992: the Emergency Services response, *Earthquakes & Volcanoes*, **23**, No. 4, 170–174, 1992.

Aochi, H. and M. Matsu'ura, Slip- and time-dependent fault constitutive law and its significance in earthquake generation cycles, *Pure Appl. Geophys.*, **159**, 2029–2047, 2002.

Aranda, E. et al., Mexico City seismic alert system, *Seimol. Res. Lett.*, **66**, 42–53, 1995.

Balassanian, S., A. Mouradian, A. Sahakian, S. Kalinin, M. Babayan and A. Pogossian, The inveistgation of electromagnetic precursors to earthquakes in Armenia, *Annali di Geofisica*, **XL**, 209–225, 1997.

Chen, Y. T., Z. L. Chen and B. Q. Wang, Seismological studies of earthquake prediction in China: a review, In *"Earthquake Prediction"* (E. Boschi and M. Dragoni, Eds.), 71–109, IL CIGNO GALILEO GALILEI, 1992.

Fraser-Smith, A. C., A. Bernardi, P. R. McGill, M. E. Ladd, R. A. Helliwell, and O. G. Villard, Jr., Low-frequency magnetic field measurements near the epicenter of the Ms7.1 Loma Prieta earthquake, *Geophys. Res. Lett.*, **17**, 1465–1468, 1990.

Giardini, D. (Ed.), The Global Seismic Hazard Assessment Program (SHAP), *Annali di Geofisica*, **42**, No.6, 1999.

Hada, Y. and K. Meguro, Application of Power Demand Changes to Evaluate Building and Dwelling Damages due to Earthquake, *Bulletin of Earthquake Resistant Structure Research Center—Institute of Industrial Science*, **35**, 135–144, 2002.

Hamada, K., Present state of earthquake prediction system in Japan, In *"Earthquake Prediction"* (E. Boschi and M. Dragoni, Eds.), 33–69, IL CIGNO GALILEO GALILEI, 1992.

Hattori, K., Y. Akinaga, M. Hayakawa, K. Yumoto, T. Nagao, and S. Uyeda, ULF magnetic anomaly preceding the 1997 Kagoshima Earthquakes, In *Seismo Electromagnetics*, 19–28, TERRAPUB, Tokyo, 2002.

Hattori, K., I. Takahashi, C. Yoshino, N. Isezaki, H. Iwasaki, M. Harada, K. Kawabata, E. Kopytenko, Y. Kopytenko, P. Maltsev, V. Korepanov, O., Molchanov, M. Hayakawa, Y. Noda, T. Nagao, and S. Uyeda, ULF Geomagnetic Field Measurements in Japan and some recent results associated with Iwateken Nairiku Hokubu earthquake in 1998, *Physics and Chemistry of the Earth*, in press.

Hayakawa, M., R. Kawate and O.A. Molchanov, Ultra-low- frequency signatures of the Guam earthquake on 8 August, 1993 and their implication, *J. Atmos. Electr.*, **16**, 193–198, 1996.

Hayakawa, M. and O. Molchanov (Eds.), *"Seismo Electromagnetics"*, pp. 477, TERRAPUB, Tokyo, 2002.

Ikeya, M., T. Komatsu, Y. Kinoshita, K. Teramoto, K. Inoue, M. Gondou and T. Yamamoto, Pulsed electric field before Kobe and Izu earthquakes from seismically-induced anomalous animal behavior (SAAB), *Episodes*, **20**, 253–260, 1997.

Kamogawa, M., H. Fujiwara, H. Sakata, J. Y. Liu, H. Ofuruton, Y. J. Chuo, Seismo-atmospheric disturbances observed by anomalous transmission of VHF electromagnetic waves, *IUGG 2003 Sapporo Abstracts Week A*, A. 189, JWS01/02P/D-034, 2003.

Kanamori, H., E. Hauksson and T. Heaton, Real-time seismology and earthquake hazard mitigation, *Nature*, **390**, 461–464, 1997.

Keilis-Borok, V. I. and V. G. Kossobokov, Premonitory activation of earthquake flow: algorithm M8, *Phys. Earth and Planet. Inter.*, **61**, 73–83, 1990.

Koizumi, N., E. Tsukuda, O. Kamigaichi, N. Matsumoto, M. Takahashi and T. Sato, Preseismic changes in ground water level and volumetric strain associated with earthquake swarms off the east coast of the Izu Peninsula, Japan, *Geophys. Res. Lett.*, **26**, 3509–3512, 1999.

Kondo, S., S. Uyeda and T. Nagao, The selectivity of the Ioannina VAN station, *J. Geodynamics*, **33**, 433–461, 2002.

Kopytenko, Y.A., T.G. Matishvili, P.M. Voronov, E.A. Kopytenko, and O.A. Molchanov, Detection of ultra-low-frequency emissions connected with the Spitak earthquake and its aftershock activity, based on geomagnetic pulsations data at Dusheti and Vardzia observatories. *Phys. Earth Planet. Inter.*, **77**, 85–95, 1993.

Kumazawa, M. and S. Maruyama, Whole Earth tectonics, *J. Geol. Soc. Japan*, **100**, 81–102, 1994.

Kushida, Y. and R. Kushida, Possibility of earthquake forecast by radio observations in the VHF band, *J. Atmosph. Electricity*, **22**, 239–255, 2002.

Hui, L., China's campaign to predict earthquakes, *Science*, **273**, 1484–1486, 1996.

Lighthill, Sir James (Ed.), *"A Critical Review of VAN"*, pp. 376, World Scientific, 1996.

Mayorca, P., and K., Meguro, Proposal of a new economic retrofitting method for masonry structures, *Proc. 27th Earthquake Engineering Symposium*, CD-ROM, Japan Soc. Civil Engineers, 2003.

Meguro, K., Dynamic evaluation model for large earthquake disaster, *J. Geography*, **110**, 900–914, 2001. (In Japanese with English abstract).

Meguro, K., and T. Takahashi, System for promotion of retrofitting of existing pre code-revision weak structures, *J. Social Safety Science*, 3, 81–86, 2001. (In Japanese with English abstract)

Molchanov, O. A. and M. Hayakawa, Subionospheric VLF signal perturbations possibly related to earthquakes, *J. Geophys. Res.*, **103**, No. A8, 17,489–17,504, 1998.

Moriya, T., T. Mogi, M. Takada and M. Kasahara, Modification of Kushida's method and preliminary result – high possibility method for earthquake forecast based on the exploration for VHF scatterer in the atmosphere, *IUGG 2003 Sapporo Abstracts Week A*, A. 188, JWS01/02P/D-032, 2003.

Nakamura, Y., On the urgent earthquake detection and alarm system (UrEDAS), *Proc. Ninth World Conf. Earthq. Engineering*, **17**, 673–678, 1988.

Nagao, T., Y. Enomoto, Y. Fujinawa, M. Hata, M. Hayakawa, Q. Huang, J. Izutsu, Y. Kushida, K. Maeda, K. Oike, S. Uyeda, T. Yoshino, Electromagnetic anomalies associated with 1995 Kobe earthquake, *J. Geodynamics*, **33**, 401–411, 2002.

Nagao, T., Y. Orihara, T. Yamaguchi, I. Takahashi, K. Hattori, Y. Noda, K. Sayanagi and S. Uyeda, *Geophys. Res. Lett.,* **27**, 1535–1538, 2000.

Nagao, T., M. Uyeshima and S. Uyeda, An independent check of VAN's criteria for signal recognition, *Geophys. Res. Lett.,* **23**, 1441–1444, 1996.

Nishimura, A., Y. Ueno, S. Fujiwara, I. Ijiri, T. Fukunaga, S. Hishida, K. Hatake, A. Tanegashima, H. Kinoshita, Y. Mizoi, and Y. Tatsuno, Medical examination report on the Great Hanshin Earthquake, *J. Advance in Legal Medicine*, **3**, 234–238, 1997a.

Nishimura, A., Y. Ueno, S. Fujiwara, I. Ijiri, T. Fukunaga, S. Hishida, K. Hatake, A. Tanegashima, H. Kinoshita, Y. Mizoi, and Y. Tatsuno, Statistical Report on Casualty of the Great Hanshin Earthquake, *J. Advance in Legal Medicine,* **3**, 346–349, 1997b.

Ohtake, M., Reinforcement is urgently needed for wooden houses, *Japan Society for Disaster Information Studies News Letter*, No. 13, 1, 2003 (in Japanese)

Parrot, M., The miceo-satellite DEMETER, *J. Geodynamics,* **33**, 535–541, 2002.

Rikitake, T., *"Earthquake Prediction"*, pp. 357, Elsevier, 1976.

Scholz, C. H., L. R. Sykes and Y. P. Aggawal, Earthquake Prediction: a physical basis, *Science*, **181**, 803–809, 1973.

Shin, T. C., K. W. Kuo et al., A preliminary report on the 1999 Chi-Chi (Taiwan) earthquake, *Seismol. Res. Lett.*, **71**, 24–30, 2000.

Silver, P. G.. and H. Wakita, A search for earthquake precursors, *Science*, **273**, 77–78, 1996.

Sobolev, G. A., Q. Huang and T. Nagao, Phases of earthquake's preparation and by chance test of seismic quiescence anomaly, *J. Geodynamics*, **33**, 413–424, 2002.

Stein, S. and E. Klosko, Earthquake mechanisms and plate tectonics, in " *International Handbook of Earthquake and Engineering Seismology"* (W. H. K. Lee, H. Kanamori, P. C. Jennings and C. Kisslinger (Eds.), Academic Press, 69–78, 2002.

Takano, T., K. Sakai, A. Yamada, H. Higasa and S. Shimakura, Electromagnetic phenomena possibly associated with earthquakes obtained with broadband observations, *IUGG 2003 Sapporo Abstracts Week A,* A. 188, JWS01/02P/D-030, 2003.

Utsu, T., A list of deadly earthquakes in the world: 1500–2000, in *"International Handbook of Earthquake and Engineering Seismology"* (W. H. K. Lee, H. Kanamori, P. C. Jennings and C. Kisslinger (Eds.), Academic Press, 691–717, 2002.

Uyeda, S., In defense of VAN's earthquake predictions, *EOS, Trans. Amer. Geophys. Un.,* **61**, no.1, Jan. 4, 3 & 6, 2000.

Uyeda, S., Continental drift, sea-floor spreading and plate tectonics/Plume tectonics, in *"International Handbook of Earthquake and Engineering Seismology"* (W. H. K. Lee, H. Kanamori, P. C. Jennings and C. Kisslinger (Eds.), Academic Press, 51–67, 2002.

Uyeda, S. and K. Al-Damegh, Evaluation of VAN method, In: *"Seismo-electromagnetic Phenomena"*, Hayakawa (Ed.), Terra Scientific, Tokyo, 58–69, 1999.

Uyeda, S., K. Al-Damegh, E. Dologlou and T. Nagao, Some relationship between VAN seismic electric signals (SES) and earthquake parameters, *Tectonophys.*, **304**, 41–55, 1999.

Uyeda, S., M. Hayakawa, T. Nagao, O. Molchanov, K. Hattori, Y. Orihara, K. Gotoh, Y. Akinaga and H. Tanaka, Electric and magnetic phenomena observed before the volcano-seismic activity in 2000 in the Izu Island region, Japan, *Proc. Nat. Acad. Sci.*, **99**, 7352–7355, 2002.

Uyeda, S. and A. Kumamoto, Evaluation of the Kushida method of short-term earthquake prediction, *Proc. Japan Acad., Ser. B, Phys. Biolog. Sciences,* **80**, 140–147, 2004.

Uyeda, S., T. Nagao, Y. Orihara, Y. Yamaguchi and T. Takahashi, Geoelectric potential changes: Possible precursors to earthquakes in Japan, *Proc. Nat. Acad. Sci.*, **97**, 4561–4566. Uyeda, S. and S. Park (Eds.), "Recent Investigations of electromagnetic Variations Related to Earthquakes", *J. Geodynamics,* **33**, 377–570, 2002.

Seiya Uyeda, Earthquake Prediction Research Center Tokai University, Shizuoka 424-8610, Japan (suyeda@st.rim.or.jp)

Kimiro Meguro, International Center for Urban Safety Engineering, Institute of Industrial Science, The University of Tokyo, Meguro-ku, Tokyo, 153-8505, Japan

Volcanic Activity: Frontiers and Challenges in Forecasting, Prediction and Risk Assessment

R. S. J. Sparks

Department of Earth Sciences, University of Bristol, Bristol, UK.

W. P. Aspinall

Department of Earth Sciences, University of Bristol, Bristol, UK.
Aspinall & Associates, Beaconsfield, Buckinghamshire, UK.

When a volcano shows signs of unrest scientists are asked to forecast whether an eruption will happen, when it will happen and what kind of eruption it will be. They are also expected to provide information on hazardous volcanic phenomena and their effects, and how long the eruption will last. Eruptions are complex phenomena, however, involving magma ascent to the Earth's surface and interactions with surrounding crust and surface environments during eruption. Magma may change its properties profoundly during ascent and eruption, and many of the governing processes of heat and mass transfer can be highly non-linear. There are both epistemic and aleatory uncertainties involved, which can be large, making precise prediction of a certain event in time and space a formidable or impossible objective; that is, volcanoes can be intrinsically unpredictable. As with other natural phenomena, forecasting is a more achievable goal and needs to be expressed in probabilistic terms that take account of the uncertainties. Ensemble modeling in which uncertainties are sampled with Monte Carlo techniques is likely to become the basis for such forecasting. Despite the limitations, there is significant progress in anticipating volcanic activity and, in favorable circumstances, in making predictions. Data from enhanced monitoring techniques are being combined with advanced numerical models of volcanic flows and their interactions with the environment. Statistical analysis of volcanological data and improvements in methods to treat subjective information are also beginning to provide viable, complementary approaches to basic numerical modeling.

INTRODUCTION

Earth scientists are required to look into the future to advise governments and inform the public about threats from natural hazards. The challenges of prediction become immediate when a volcano is about to erupt. About 500 million people live close enough to volcanoes to be affected by eruptions [*Newhall*, 2000]. Volcanic phenomena can affect climate and are an important contributing factor in forecasting global environmental change. The challenges of forecasting volcanic activity are considerable and come at a time when the demands and expectations of society are increasingly onerous. Some societies are becoming ever more litigious, and prospects of scientists having to justify their judgements in court are emerging.

Volcanic activity has many features in common with other natural hazards, such as extreme weather, earthquakes and landslides. Natural hazards are characteristically complex and involve numerous parameters and processes. Some of the controlling processes are highly non-linear, so that in certain circumstances abrupt changes of behaviour can happen, such as the sudden transition from effusive to explosive eruption. Of particular significance is the need to understand and quantify uncertainty. There is both epistemic and aleatory uncertainty in natural phenomena [*Woo*, 1999], the former being defined as deficiencies in knowledge about natural processes and the latter as natural variability within those processes. Many volcanic processes are stochastic, and need to be characterised by statistical models. Complex nonlinear systems can be very unstable close to critical thresholds between stable states, becoming inherently unpredictable. Such attributes and the importance of quantifying uncertainty are now central in forecasting volcanic phenomena. As a consequence, volcanologists will inevitably shift from deterministic to probabilistic perspectives. This change in focus will require alterations in the conceptual framework within which observations on volcanoes are interpreted. There will be many challenges, because uncertainties in many aspects of volcanic systems are large, and may prove hard to quantify.

Here we consider emerging new approaches to prediction, forecasting, the assessment of volcanic risk, and provision of scientific advice. We also discuss the roles of statistical analysis and computer modelling. This paper complements those of *Sparks* [2003], who reviewed the methods of forecasting volcanic eruptions based around interpretation of monitored data, and of *Newhall and Hoblitt* [2002], who developed parallel themes.

VOLCANIC PROCESSES AND HAZARDS

Volcanism is caused by the buoyant ascent of magmas to the Earth's surface. Magmas change their properties as they ascend, principally due to changes in pressure and temperature. The properties of a magma and its interactions with its surroundings determine whether a given magma erupts or not, and dictate the nature of the activity, if it does erupt. These properties and interactions give rise to a range of physical effects that can be monitored. Important interactions include: fracturing of rocks due to propagation of dykes (*Rubin*, 1995), generating seismicity; escape of pressurised gases and heating of ground waters, leading to 'long-period' volcano-seismic events [e.g. *Chouet*, 1996; *Neuberg*, 2000]; and variations of magma pressure in chambers and conduits leading to ground deformation [e.g. *Voight et al.*, 1999; *Dzurisin*, 2000]. Ascending or erupting magma can cause perturbations of electric, gravity and magnetic fields. Additionally, fluxes of magma and gas, geochemical and petrological variations in erupted products, and observations of surface activity provide important information. Observations from space of heat, strain, topography, gravity, electromagnetic transients and atmospheric emissions are also emerging as a major source of information [*Wadge*, 2003]. Systematic monitoring and observations provide the basis for forecasting, but require understanding of the processes to interpret.

Major types of volcanic hazards, their effects and extents are listed in Table 1. The scale and occurrence of volcanic hazards are inversely related, with small events occurring worldwide at a rate of 10–20 per month, whereas catastrophic eruptions (>10 km^3), that might affect the economy of an entire country, occur every hundred years or so [*Pyle*, 1998]. The very largest volcanic eruptions (>1000 km^3) could threaten civilization [*Rampino*, 2002] and occur about every 50,000 years on average [*Mason et al.*, 2004].

Table 1. Summary of the effects and extents of major volcanic hazards.
L = local; R = regional; N = national; I = international

Hazard	Threat to life	Threat to property	Areas affected
ash and pumice fall	low except near vent high for aviation	depends on thickness: roof collapse, bomb damage, fire	L,R,N,I
pyroclastic flows	very high	very high	L,R
lava flows	low	very high	L
lahars/flooding	high to moderate	high	L,R
gases/dusts/acid rain	low to moderate	moderate	L,R

Volcanic hazards can be entirely caused by the volcanic activity itself, but external factors can be important. Lavas and pyroclastic flows, for example, are strongly influenced by topography. Tephra fall hazards depend on wind strength and direction [e.g. *Bonadonna et al.*, 2002]. Collapse of a lava dome to form pyroclastic flows can be triggered by intense rainfall [*Matthews et al.*, 2002]. Flank collapses can be triggered by earthquakes, as happened on 18th May 1980 at Mount St Helens [*Endo et al.*, 1981]. Separate earthquake–volcano interactions can apparently take place over surprisingly long distances and time-scales [*Linde and Sacks*, 1998; *Hill et al.*, 2002]. Other, subtle complex-systems dynamics [e.g. stochastic resonance; *Weisenfeld and Moss*, 1995] may link volcanic responses to low-level external forcings, such as tides or atmospheric pressure or hydrological cycles [*Jupp et al.*, 2004]. Thus forecasting of hazards in both space and time requires not only understanding of the eruptive processes themselves, but also understanding of the surface and subsurface environment and external processes that interact with volcanic phenomena.

PREDICTION AND FORECASTING

Prediction and forecasting are sometimes used interchangeably, but can be usefully distinguished. A prediction involves a statement about a specific event that is regarded as inevitable within certain defined time limits and, in its most developed form, is found in established laws of physics (e.g. Newton's Laws of motion of bodies). Scientists may judge, for example, that a volcano that is showing unrest will definitely erupt and that lava will reach a village. The prediction will have limited value unless some constraints on time, place and scale are specified, even if these assessments are themselves uncertain to some degree. Forecasting, on the other hand, is a probabilistic statement that a specific event might occur with a certain likelihood in a given time-frame, commonly with associated scales and effects being defined as well.

The requirements for prediction, as defined above, are stringent, because the event must be inevitable. It might be thought that a volcano like Vesuvius, with an historical record of frequent eruptions, would meet the requirements of inevitability of another eruption. However, all volcanoes become extinct and the criteria for recognising that a volcano has had its last ever eruption are very problematic, if not unknowable (and conditional on definitions of 'eruption' and 'extinct'). One cannot be certain that the 1944 eruption of Vesuvius was not its last ever eruption, even though the probability that this is so would be widely judged as diminishingly small. In most eruptions the case for inevitability cannot be easily made.

Many volcanoes erupt in such a way that patterns of behaviour can be established and an eruption can be forecast, and even predicted if these patterns are repeated sufficiently often. In 2000, the eruption of Mount Hekla in Iceland was accurately predicted. Several years earlier borehole strain meters had been deployed in central Iceland and, in the 1991 eruption, large and systematic strain variations (Figure 1), together with seismic data, traced the propagation to the surface of a dyke from a magma chamber about 6 km below Hekla [*Linde et al.*, 1993]. The agreement between the observations and conceptual model gave a high degree of confidence that an eruption was inevitable when the same pattern was repeated in March 2000. Scientists from the Icelandic Meteorological Office informed Icelandic radio that an eruption was imminent in 20 minutes. The prediction was announced at the beginning of a news broadcast and the eruption occurred within 2 minutes of the expected time. In the 1981–1983 activity of Mount St. Helens, patterns of periodic dome growth with precursory ground deformation allowed accurate forecasts to be made of the timing of dome extrusion events [*Swanson et al.*, 1983]. These forecasts became increasingly focussed on narrow time windows to the extent that they, too, might be viewed as predictions.

The above examples show that forecasting and prediction are easier at persistently active volcanoes or in long-lived eruptions with established and repeatable patterns of activity; forecasting and prediction become more difficult for dormant volcanoes with a limited or no historical record. The

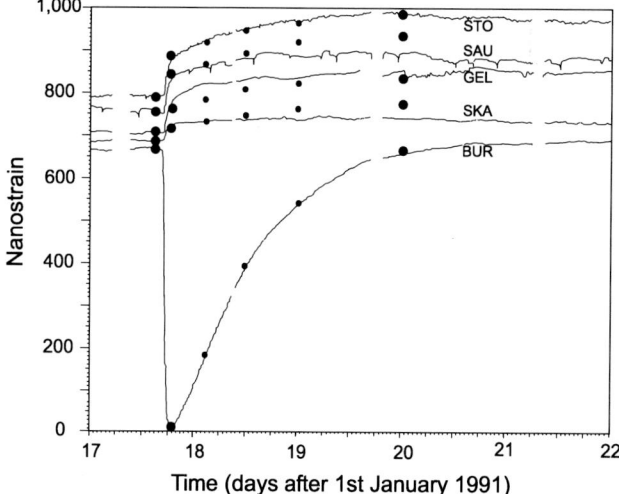

Figure 1. Strain measured in five borehole strainmeters in Iceland in 1991 associated with the eruption of Hekla Volcano [after *Linde et al.*, 1993]. The eruption took place within 20 minutes of the start of the changes in strain at the time when the strain at BUR reached a minimum. Four of the stations show an expansion (positive changes) and the closest station to the volcano (BUR) shows an initial contraction. The data are interpreted as the ascent of a dyke from the magma chamber to the surface.

issue of whether a restless volcano is going to erupt has been a source of angst, controversy and, in some cases, the undermining of scientific credibility. The cause célèbre is the 1976 volcanic crisis at La Soufrière, Guadeloupe, which led to major public disagreements between scientists [*Fiske,* 1984]. One group, led by Haroun Tazieff, judged that the seismic swarm and phreatic activity was not the prelude to a major eruption. Others regarded these disturbances as potential signs of a major eruption, which would endanger over 70,000 people, and so advised an evacuation. The authorities, concerned to incur no public risk in the light of events in Martinique in 1902, accepted the precautionary advice and the consequent three-month evacuation caused hardship and considerable economic cost, running into several hundreds of millions of dollars. A magmatic eruption did not happen and, in some sense, Tazieff's assessment might be construed as "right". A considered appraisal of the Guadeloupe case suggests, however, that the more cautious scientists had better appreciation of scientific uncertainties and the inadequacy of knowledge about how volcanoes work, a view supported by subsequent analysis [*Feuillard et al.,* 1983]. In contrast, the climatic eruption of Mount Pinatubo in June 1991 is a good example of a successful forecast based on a spread of evidence, including precursory seismicity, and variations in SO_2 emissions [*Punongbayan et al.,* 1996]. The prospect of a very substantial explosive eruption was deduced from geological evidence of past eruptions and dating of pyroclastic deposits. The decision to evacuate 250,000 people based on the various strands of evidence was a judgement call and the advice of the Philippine Institute of Volcanology and US Geological Survey to the authorities avoided huge loss of life.

In view of the inherent difficulties in prediction of volcanic phenomena it is often better to address future volcanic activity in terms of probabilistic forecasting. For many eruptions, forecasts are qualitative and expressed in terms such as 'very likely' or more cautious statements that a volcano is showing the signs of unrest and might erupt. Every volcanologist is, however, acutely aware of the problems provoked by calling an evacuation and then nothing happens. The scientists are perceived to have been "wrong", and to have raised a false alarm. For this reason it is better to present forecasts rather than predictions, and to find ways of expressing these in probabilistic terms based on quantitative scientific evidence and analysis.

In the USA public and political decision-makers are used to weather forecasts. In general, weather forecasters have gained public credibility both by a record of improving the accuracy of forecasts and by getting the public accustomed to basic concepts of probability. The statement that there is an 80% chance of rain in a particular town is familiar, and most people will accept that if it doesn't rain then it was not necessarily a wrong forecast. Even in this arena, however, forecasters still have difficulties with extreme conditions and can lose credibility when the weather is much more severe than anticipated. Likewise volcanologists have considerable problems with forecasting the onset and effects of extreme volcanic events.

We now illustrate some of these matters and present practices with the example of the Soufrière Hills volcano, Montserrat, since that ongoing eruption (which started in July 1995) is very familiar to us.

MONTSERRAT: CASE STUDY IN FORECASTING, PREDICTION AND RISK ASSESSMENT

The focus on the Soufrière Hills volcano, Montserrat is selective, but the eruption has displayed a wide range of volcanic phenomena and has proved to be a very good testing ground for development of new approaches to prediction, forecasting and risk assessment. Here we illustrate the principles of probabilistic forecasting with the example of tephra fall hazards. We also consider pyroclastic flow hazards on Montserrat to illustrate methods of risk assessment.

Tephra Fall Hazards

Tephra fall is one of the better understood of volcanic processes (Plate 1). Understanding the dynamics of volcanic plumes is reasonably advanced, as summarised in *Sparks et al.* [1997], with quantitative models for plume ascent, for the interaction of plumes with the wind, and for tephra transport. These models have been calibrated against observed events and consequently there is confidence that the models are simulating natural processes appropriately. There is also information on hazardous effects; for example, mass accu-

Plate 1. An ash-laden volcanic plume at the Soufrière Hills Volcano is blown across Montserrat. Note that the source of the plume is from a pyroclastic flow about 1 km to the east of the crater of the volcano.

mulations of tephra that can lead to roof collapse are well documented [Blong, 1984].

The study of Bonadonna et al. [2002] on assessment of tephra fall hazards on Montserrat illustrates the main features of probabilistic modelling. The model treated the tephra dispersal as an advection–diffusion problem with a source function that reflected the observed origin of the ash plumes from above the pyroclastic flows along the valleys, and the observed approximately exponential decrease of plume height with distance along the valley. The model also considered Vulcanian explosions sourced at the crater. Model simulations were first run for individual volcanic events and for the effects of activity in the sequence from June 1996 to June 1997, under the atmospheric conditions prevailing at the time (wind speeds and directions). Parameters in the model were then adjusted to give best-fits to the observed thickness variations. The volcanic activity of 1996–1998 was then used to build a characteristic scenario for the volcano of combinations of dome collapses, pyroclastic flows and explosions. This scenario was then run several hundred times using re-sampling of the statistical data on daily wind speeds and directions at 1 km intervals up to a height of 20 km for 1992–1996, to generate a statistical distribution of resulting tephra accumulations across the island.

Figure 2 plots the model output in contours of probability (as a percentage) that tephra accumulation will exceed 12 kg/m^2 (Figure 2a) and 120 kg/m^2 (Figure 2b). The first value represents the threshold above which crops commonly fail, and the latter value is the threshold at which roof collapse becomes a problem. This particular model assumes that either erosion or human intervention (e.g. sweeping roofs) removes the ash between each tephra fall event. Such probabilistic models were used on Montserrat to assess vulnerability to roof collapse and to make a risk assessment of the health hazards due to respirable volcanic dust. The forecasts have also been used to guide the UK government in planning the sustainable development of the island.

Tephra fall also poses health hazards. In the case of Montserrat the volcanic dust (<10 μm) contains 10–25 wt% cristobalite, which is a carcinogen and causes silicosis, a chronic lung disease [Baxter et al., 1999]. The model of Bonadonna et al. [2002] was an input into a risk assessment commissioned by the UK Department of Health. In this assessment, simulations of tephra dispersal were combined in a synthesis with other information and processes, including models of ash suspension, epidemiological data and biological data such as in vivo and in vitro experimental results. Multiple model runs were made to assess the exposure of the population to suspended ash. The results of this study have identified outdoor workers (eg gardeners) and children as having relatively high exposure. This study illustrates the

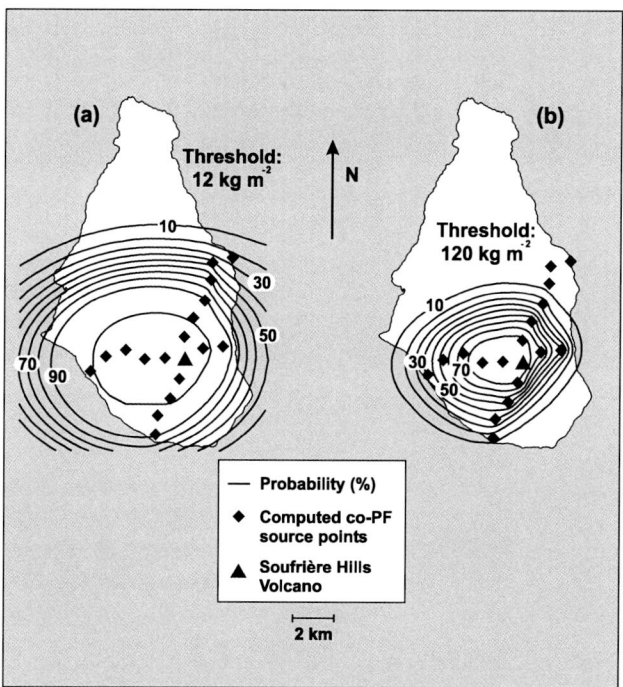

Figure 2. Maps showing contours of probability (as a percentage) across the island of Montserrat of accumulation of 12 and 120 kg/m^2 or more of volcanic ash over a 3-year period [after Bonadonna et al., 2002]. The maps are based on a scenario of activity similar to that experienced on the island. Pyroclastic flows extend in the scenario down the valleys and plumes (co-PF) are source at the points indicated by the solid diamonds. The mass accumulations are equivalent to ash thicknesses of about 1 cm (for 12 kg/m^2) and 10 cm (for 120 kg/m^2).

way in which relatively 'soft' qualitative data can be incorporated into a multi-factor simulation. To produce a probabilistic assessment of human exposure to suspended dusts it is, for example, necessary to parameterise the erosion process, since once ash is eroded it is no longer available for suspension. At the moment, it is beyond present understanding to replicate in a model all the complex processes of tephra erosion and redistribution on a tropical island, but a simple time-dependent removal scheme could be parameterised from empirical information. This was combined with statistical information on rainfall to provide a significant advance on the end member probabilistic maps in Figure 2.

Such modelling can always be improved. An ambitious target might be to produce a probabilistic model of dome growth, collapse and explosion rather than simply adopt a specific, albeit plausible, single scenario as a class example. Monte Carlo sampling of all uncertainties in key parameters that control the underlying volcanic processes could then be incorporated to represent a wide range of potential scenarios. However, the processes of dome growth are not yet well enough understood to do more than develop 'soft' parameterisations, as in the case of erosion. Such models are in the future and will require significant computing power and code development, although adroit use of newly-emerging model inference techniques [*O'Hagan et al.*, 1999] offer the prospect of reduced computing costs. Eventually one can imagine volcano forecasting centres that assist observatory teams with synoptic outputs, just as weather centres have sprung up to underpin local and regional forecasts.

Pyroclastic Flow Hazards

The principal hazard on Montserrat during the eruption is the formation of pyroclastic flows from collapse of the lava dome [*Cole et al.*, 2002]. The assessment of hazards and attendant risks from pyroclastic flows and their accompanying hot turbulent clouds of ash (surges) is described here to illustrate generic issues in relation to prediction and methods of probabilistic forecasting.

The andesite dome has grown in pulses [*Watts et al.*, 2002]. In each pulse a lobe of lava extrudes in a particular direction (Plate 2a). Rockfalls and collapse-induced pyroclastic flows are generated preferentially at the leading edge of a lobe (Plate 2b) and tend to flow away in the same direction as the extrusion direction [*Calder et al.*, 2002]. This behaviour is very useful for forecasting, since the probability of flows going down a particular valley is greatly increased when the dome is extruded in that direction. Pulses in extrusion rate can be marked by the onset of shallow seismicity or by changes in cyclic patterns of ground deformation, as recorded by tiltmeters [*Voight et al.*, 1999], and both symptoms are commonly associated with formation of a new lobe in a new growth direction.

Most major dome collapses on Montserrat (defined here as 3 million cubic metres or greater since only flows of this size or above are large enough to threaten populated areas) occurred within a few hours or days of a pulse in extrusion rate. There have been many such pulses and switches in the extrusion direction, which tend to occur at intervals of a few weeks to a few months. From May to December 1997 the surges and accompanying major dome collapses occurred quite regularly at 6–7 week intervals [*Voight et al.*, 1999; *Sparks and Young*, 2002]. For a while, the regularity of the pattern was sufficiently clear to provide a basis for forecasting. On the other hand, some of the largest dome collapses appear to have been triggered by intense rainfall [*Voight and Elsworth*, 2000; *Matthews et al.*, 2002]. One of these occurred on 3 July 1998 in a period when there was no dome growth. Two large dome collapses occurred on 20 March 2000 and 29 July 2001 (volumes of 20 and 45 million m^3, respectively) and were associated with intense rainfall of over 80 mm/hour.

Plate 2. Activity of the Soufrière Hills volcano, Montserrat illustrating factors determining the directions of pyroclastic flows. In (a) a lobe of lava is extruded towards the north between August 5 and 9, 2002 (view from the east), and in (b) small collapses from the leading edge of a lava lobe generates rock-falls and pyroclastic flows towards the north of the island (view from the north). Images reproduced by permission of the Montserrat Volcano Observatory.

The critical issues in hazard forecasting of pyroclastic flows are the areas that a flow (or series of flows) will inundate and the occurrence of the associated overlying clouds of hot turbulent suspended ash (known as surges) that can spill out of the valleys that confine the main flows. Data on run-out distances and areas affected on Montserrat have led to empirical relationships between flow volume and run-out distance [*Calder et al.*, 1999]. The observed distributions of particular flow deposits can be used to construct a semi-empirical model of run-out [*Wadge et al.*, 1998]. The model incorporates gravitational flow across the observed topography, which is described by a digital elevation model (DEM). The model contains three friction parameters, which can be adjusted by trial and error, or by Monte Carlo techniques, to give a best-fit to the observations. The model can then be run for events that have not yet happened to assess volcanic hazards. Such models have many uncertainties and hidden assumptions. For example, at Soufrière Hills dome collapses typically last for tens of minutes to several hours and may involve numerous individual avalanches. On 25 June 1997, there were three main collapses in 20 minutes [*Loughlin et al.*, 2002]. Using the final total volume of the deposits in a particular collapse episode as a basis for defining mobility by run-out distance, or inundation area, or to estimate friction coefficients might lead to misleading or inappropriate results. There is rarely sufficient information to discriminate the volumes of individual pulses. The fact that dome collapse episodes are almost always pulsed multi-collapse events [*Calder et al.*, 2002] means that there are intrinsic uncertainties and pitfalls in analysing the resulting field data. Another element of semi-empirical models is that the frictional parameters not only reflect the true frictional properties of the flow but implicitly incorporate topographic effects on flow which, as yet, are not explicitly modelled.

The behaviour of the surge clouds is even more problematic because quantitative models of surge generation and dynamics are not yet available. On Montserrat, the development of surge clouds seems to be linked not only with flow volume, but also with the internal pressurisation of the dome, which itself is thought to be related to the extrusion rate [*Cole et al.*, 2002; *Calder et al.*, 2002]. Surge cloud generation and behaviour can also be sensitive to topographic features and can be influenced by wind. Finally, surge clouds can generate dense "secondary" pyroclastic flows, which can move obliquely down valleys away from the main originating flow [*Druitt et al.*, 2002b]. Thus the dispersal of surge clouds is not yet amenable to rigorous modelling and so the assessment of hazards must depend more on qualitative judgements based on observations in circumstances of considerable uncertainty.

Risk Assessment of Dome-Collapse Pyroclastic Flows

We illustrate the development of a quantitative risk assessment of pyroclastic flow hazards, the main hazards issue in the management of the Montserrat crisis, using the recent example of one particular residential area, the Belham Valley. The problem emerged in 2002 when relentless dome growth raised concerns that a large collapse could inundate the lower parts of the Belham Valley, northwest of the volcano (Plate 3). This section describes the modelling procedures that were used to estimate risks and thus inform decisions by the civil authorities. The accompanying map (Figure 3) shows the area evacuated after 8 October 2002 on the basis of scientific advice. This area was re-occupied in August 2003 after a huge dome collapse to the east, on 12

Plate 3. The andesite lava dome of the Soufrière Hills volcano, Montserrat in May 2003, showing the main features pertinent to hazard assessment. If the dome were to collapse down Tyres Ghaut then pyroclastic flows would enter into the lower Belham Valley (see Figure 6), an area that was evacuated after 8th October 2002 because of this threat. (note: 'ghaut' is used locally to denote a valley with a stream or river).

July 2003, removed the threat. For Montserrat, a Risk Assessment Panel has conducted such assessments and involves the scientific staff of MVO and external experts. The authors acted as Chair of the Panel (RSJS) and expert in risk assessment methods (WPA).

The risk to the lower Belham Valley area is a function of the probability that a pyroclastic flow will inundate part or all of the area and the vulnerability of people there. Historical data indicate that 90% of people in areas directly affected by pyroclastic flows and surges are killed [*Baxter*, 1990]. Only those on the fringes of the devastated area might survive and so vulnerability is very high. To estimate risk quantitatively the Montserrat Panel was required to estimate the probability that a flow would happen over a fixed period of time. The time window used was 6 months as this was a useful timescale for decision-makers and was commensurate with the temporal variations in dome-building eruptions.

The procedure for estimating the hazards involved a systematic series of steps in a structured discussion and use of procedures to combine relatively hard and soft information. The risk assessments used, wherever possible, quantitative models of volcanic processes. Thus to assess the run-out distances empirical correlations [*Calder et al.*, 1999] and models (e.g. *Wadge et al.*, 1998) were used. For Belham Valley, the Panel concluded that collapses of 3 million cubic metres or more would have a high likelihood of reaching the lower valley and that collapses of 10 million cubic metres or more would reach the sea and affect most, if not all of the area.

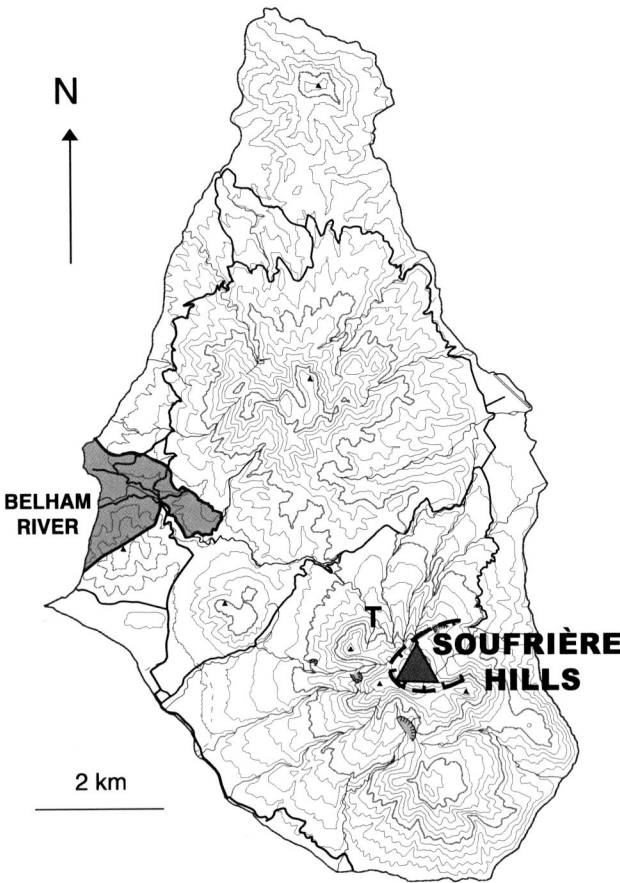

Figure 3. A map showing the location of the Belham Valley, Tyres Ghaut (marked T) and the area (shaded) which was evacuated after 8th October 2002 due to the assessed high level of potential risk from collapse of the lava dome to the northwest. Contour intervals are 200 feet.

The assessment also used a structured method of expert elicitation in which the judgements of the Risk Assessment Panel were pooled. The method, described in *Cooke* [1991], has become a common approach in many scientific and engineering situations that involve significant uncertainty, and it has been applied for the first time in a volcanic crisis on Montserrat. Each expert assesses their judgement of some parameter and his or her confidence limits on that assessment, based on shared, available scientific information. Experts are calibrated by a facilitator, so that the pooled results of the group are weighted according to the individual experts' ability to be informative and knowledgeable. This procedure is designed to give greater weight to those individuals with good judgement in urgent circumstances; that is, a ranking relevant to decision-making capabilities in crisis conditions, and not merely a metric of considered scholarship. The procedure has the advantage of reducing the influence of overconfident, vocal or highly opinionated individuals, while providing a neutral medium for the inclusive incorporation of a spectrum of views, and the outcome can be viewed as a mathematically rational consensus of the opinions of all participants.

A formalised expert elicitation provides a mechanism for structured scientific discussions in an evidence-based approach in which all sources of information (e.g. observations, empirical relationships and theoretical models) are utilised. In the case of the Belham Valley assessment the directionality of dome growth and surge cloud behaviour are examples of components in the estimation of overall probabilities of pyroclastic flow hazards where expert judgement elicitations have proved helpful. For endogenous collapses, directionality of the lava lobe was judged on empirical evidence to be the main determinant of the collapse direction. However, the group discussion concluded that the evidence did not support a random process, since certain directions had not been common during the eruption (possibly related to buttressing effects of older, pre-existing domes) and that switches which might trigger collapses tended to occur away from the direction of previous stagnated lobes, which acted as a barrier [*Watts et al.*, 2002]. The Panel also considered the chances of a collapse triggered by rainfall unrelated to dome growth direction. In this case the previous episodes had all been down the eastern flanks of the volcano. Additionally, the chances of such an event would be greater in the rainy season than in the dry season. An important factor also was the frequency of collapses of 3 million cubic metres or above; not all switches caused a significant collapse. Other factors included the chances of the eruption stopping, and large collapses in directions other than to the northwest over the 6-month period, which would either reduce or entirely remove the threat to the Belham Valley. Thus relatively soft information was integrated into the procedure to produce probabilities of a range of collapse events that might affect the Belham Valley.

The final stage of the risk assessment involved Monte Carlo re-sampling of the probability density functions (pdf) of all controlling factors that contribute to the hazards in different areas, in repeated simulations. The final output evaluates the integrated probability of occurrence of life-threatening events, expressed as the risk of a person being killed in a particular area, or as the associated probability of exceeding a number of casualties in the population at large (Figure 4). The individual risk exposure can be compared with a suitable risk scale: for Montserrat, the comparison was with the UK Chief Medical Officer's Risk Scale. Societal risk diagrams like Figure 4 help illustrate to public officials the consequences of decisions. Volcanic risks can also be compared with other kinds of risk: for instance, casualty exceedance curves in Figure 4 compare the case of the lower Belham Valley being occupied by residents with the case the area is evacuated.

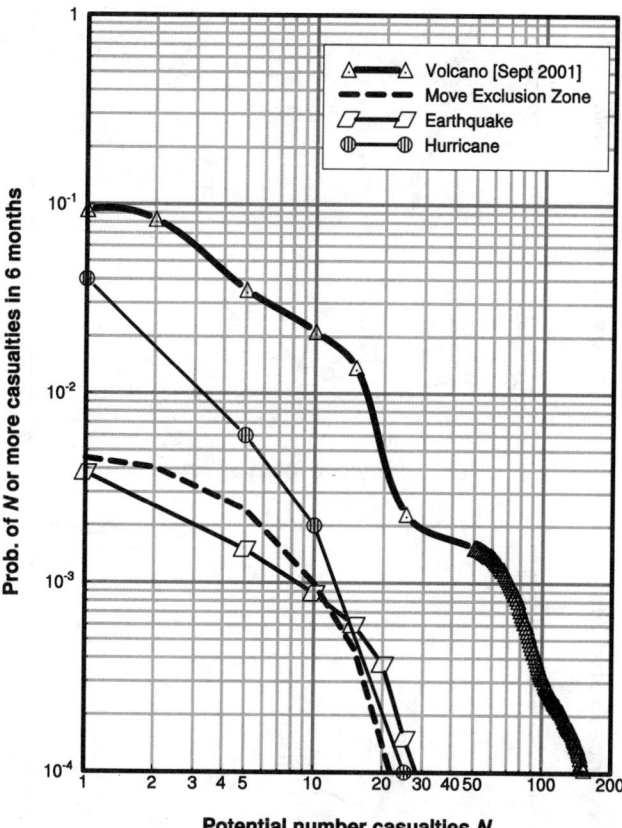

Figure 4. Example of probability curves for societal risk in Montserrat. Each curve shows the probability plotted against number of casualties over a 6-month period, and is the mean of thousands of simulations using Monte Carlo re-sampling from uncertainty distributions on the parameters that influence risk. The upper curve (solid line) is the exposure with the Belham Valley area (Figure 3) populated before the evacuation, and the lower curve (dashed line) shows the reduction in risk with evacuation. Regional risk curves for hurricanes and tectonic earthquakes are shown for comparison. Note that for each curve uncertainties at the 5% and 95% levels were calculated but are not shown for clarity.

Evacuation reduces societal risk by a factor of 10, to levels below those associated with hurricanes in the Caribbean and close to the long-term exposure to earthquakes.

STATISTICS IN VOLCANOLOGY

We now consider the statistical analysis of data, particularly time series data, to extract information on volcanic processes in the context of forecasting and assessment of hazards and risks. With burgeoning amounts of instrumental and other data being acquired statistical analysis is emerging as a major area of research that goes well beyond simply providing measures of uncertainty. Here we give two examples.

Longevity of the Soufrière Hills Volcanic Eruption

The Soufrière Hills eruption has continued for over eight years. The civil authorities have asked the scientists at MVO how long the eruption will last, as an accurate assessment has implications for the sustainable development of the island. To address this issue, duration data on 137 dome-forming eruptions taken from the Smithsonian Institution database [*Simkin and Siebert*, 1994] were gathered, re-interpreted, and fitted to a Generalised Pareto Distribution model (Figure 5). The Generalised Pareto family of distributions have special properties that make them particularly appropriate for analysing extreme-value information in a peaks-over-threshold approach, the mathematical utility of which for 'heavy-tailed' distributions has been recently elucidated (Woo, 1999). An unanticipated outcome was the identification of two distinct groupings in the size distribution of the selected dataset: most dome eruptions (85%) last less than 5 years and fall on a frequency-duration trend (Figure 5) different to those that last more than 5 years. In the latter cases, the eruptions can be very long-lived. Based on this analysis (and no other information) there was, for instance, only a 3% chance of the eruption lasting less than a further 6 months, having already lasted 94 months. There is a 50% chance that the eruption duration will last 20 years or longer, and a 5% chance of lasting more than 180 years.

Such analyses are limited by the quality and nature of the data. There are problems in defining durations because the beginning of an eruption is usually accurately recorded but the ending is often poorly or vaguely recorded. Further, there are

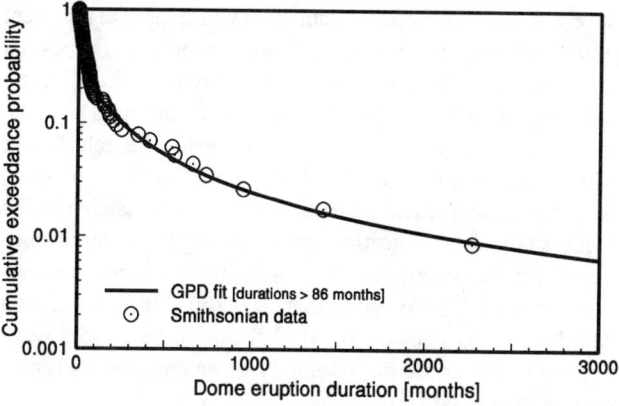

Figure 5. The exceedence probability distribution for the durations of 137 dome-building eruptions, drawn from the Smithsonian database. The data for eruptions lasting longer than 86 months are fitted to a Generalised Pareto Distribution law, which can be used to estimate the likelihood of the duration of an eruption exceeding a given number of months. In the text we have used the distribution for long-lived eruptions (>86 months) to assess the probability of the eruption stopping, given that the eruption had lasted 94 months.

only 15 cases of eruptions lasting more than 5 years so the database is very limited and, in some of these cases, conservative decisions had to be made about what constituted a coherent long-term episode of dome-building when punctuated activity is reported. Nevertheless, the two different distributions are quite distinctive and alternative assumptions on the reliability and uncertainties in the Smithsonian database fail to remove the feature.

This example shows that process information can be extracted by a data-analytic approach, which invites the question of why dome eruptions that last much more than 5 years tend to become very long-lived. A possible answer is that this is sufficiently long for conduits to become mature and stabilise so that heat loss to the walls is balanced or even exceeded by heat advection by magma flow with the longevity of the eruption being controlled by the dynamics of the chamber rather than by gradual freezing of the magma at shallower level. This is certainly not the only possibility; the point is that the data and its analysis provoke scientific enquiry.

Explosion Sequence Time Series

Seventy-five Vulcanian explosions occurred at the Soufrière Hills volcano between 22 September and 22 October 1997. The timings of this sequence were investigated by *Connor et al.* [2003]: on average, there was an explosion every 9.5 hours, but individual intervals varied from as short as 4 hours to as long as 33 hours. *Voight and Cornelius* [1991] had found that, in the run-up to an eruption, time series of volcanic data, such as Real Time Seismic Amplitude Measurement (RSAM) or deformation rate, could fit a Weibull distribution which, in the context of engineering reliability, is widely used to represent times-to-failure in materials. However, the Soufrière Hills explosion data did not fit this form of relationship (Figure 6a), suggesting that the physical controls on intervals between the events were not just an analogue of material mechanics in which strain rate increased with time until failure (explosion) was inevitable. A memory-less (Poisson) process also failed to reflect the data, indicating that the timing to the next explosion had some independence on previous explosions. Instead, *Connor et al.* [2003] found that the explosion interval data fitted a log-logistic statistical distribution extremely well.

They proposed the following dynamic equation as representing the causative processes:

$$t\frac{d\Omega}{dt} = k\Omega - \frac{k}{\Omega_{eq}}\Omega^2 \quad (1)$$

where t is the time since the last explosion, Ω is some state variable, k is a power-law exponent and Ω_{eq} is a characteristic value of Ω when the two right hand terms are equal. A

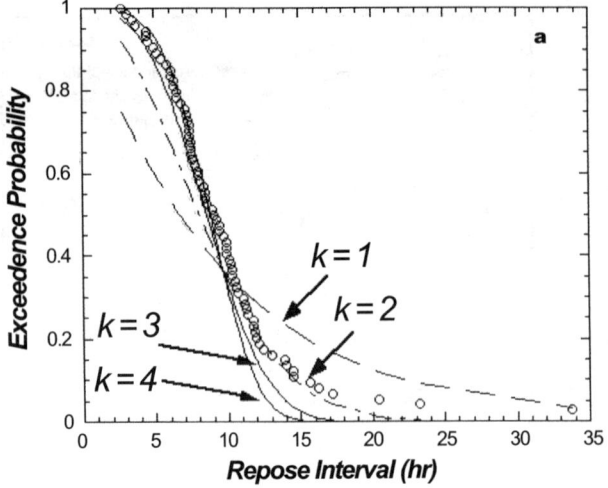

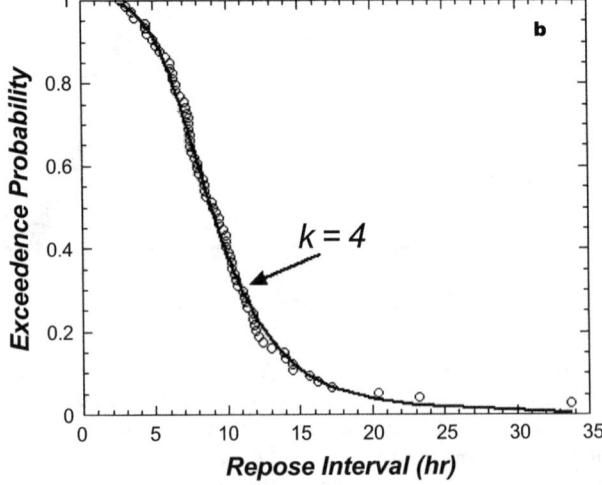

Figure 6. Statistics of intervals between Vulcanian explosions at the Soufrière Hills volcano, Montserrat in September and October 1997. (a) Comparison of the observed interval distribution (open circles) and calculated distributions using the Weibull model with a time constant of $\hat{\mu} = 9.6$ hours and various values of the power exponent k [after *Connor et al.*, 2003]. Using k = 4, estimated from experimental data, gives a good fit to the observed distribution at $t < \hat{\mu}$, but a poor fit at $t > \hat{\mu}$. The exponential (Poisson) model corresponds to k = 1, and does not fit the observed distribution. (b) The observed distribution of repose intervals are fit with > 99% confidence using a log-logistic survivor function with a time constant of $\hat{\mu} = 9.0$ hours (observed distribution median) and an exponent of k = 4 [after *Connor et al.*, 2003].

log logistic survivor function that describes the statistical distribution of repose periods can then be defined [*Connor et al.*, 2003] which fits the observations within 99% confidence limits (Figure 6b). This kind of analysis illustrates that constraints or process information with relevance to hazards can be extracted from such an analysis; in other words, it throws light on the context of what is, otherwise, an abstract statistical model.

For the Montserrat case, the best-fit statistical model suggests that the stochastic dynamics of the system must have certain properties: equation (1) represents two competing processes acting on different time scales which, respectively, increase and decrease internal gas pressure. *Connor et al.* [2003] postulated that after an explosion, gas pressure increases by exsolution from magma, but gas escape by permeable flow through the magma reduces the gas pressure. At early stages, gas exsolution is dominant and gas pressure increases. Later, gas escape plays an increasingly important role and counteracts the exsolution-driven increase in gas pressure. This competing-processes model is consistent with current understanding of the mechanisms of repetitive explosive eruptions at Montserrat [*Voight et al.*, 1999; *Druitt et al.*, 2002a; *Melnik and Sparks*, 2002]. However, a full fluid dynamic model of magma ascent that incorporates all the complex interacting processes involved has not yet been developed. Indeed, a test for the viability of numerical models should be that their outputs mimic closely both the statistical and temporal properties of the natural data.

Such an approach has forecasting relevance, as proposed by *Voight and Cornelius* [1991]. Almost all statistical models can lead to some form of distributional 'hazard function' which, in this case, can be interpreted as the relative probability of an explosion at some definite time after a previous explosion. This probability is constant for a Poisson arrival process, asymptotic after some definite time for a Weibull distribution, but reaches a maximum before declining for a log-logistic model.

This type of analysis of a complex stochastic dynamic system, the analysis of large sample data, and the assessment of uncertainty in wider areas of public concern are somewhat outside the traditional realms of statistical inference [see, e.g., *Chatfield*, 2002], but constitute important propositions for advancing volcanology. As such, they need to be integrated into an overall strategy for modelling volcanic activity, hazards and risks.

MODELLING STRATEGIES IN VOLCANOLOGY

Forecasting of complex volcanic phenomena involves a combination of empiricism, understanding (often at an intuitive level) of the underlying physical processes, and modelling. Monitoring data and observations provide the ingredients for the empirical approach in that patterns may be recognised and interpreted within a framework of physical theory or conceptual models. Monitoring data can also validate quantitative models to increase confidence in their output (although schemes that generate self-fulfilling prophecies must be avoided). However, models should be used with awareness of their limitations as well as their strengths.

The issue of quantifying uncertainty is becoming more prominent. Simplified (scenario-based) deterministic models are exceedingly useful for gaining a good first-order understanding of volcanic processes, but are likely to prove inadequate when it comes to providing models with utility for forecasting and prediction. Given the stochastic and nonlinear nature of many volcanic processes and the uncertainties (which can be large) in the controlling parameters, practical models are likely to follow the approaches now routinely adopted in other natural hazards forecasting, such as floods and extreme weather events. In meteorology, for instance, the integration of results from an array of prediction models, with explicit perturbations to model formulations, initial conditions and parameter probability distributions, generate an ensemble of outcomes that can be treated in a statistical manner and presented as a full probabilistic forecast [see, e.g., *Palmer*, 2000]. This is only just beginning to happen in volcanology. A major challenge is to ensure that such ensemble forecasting encompasses all the key factors involved in the natural processes. In particular, the existence of epistemic uncertainty has to be recognised and assimilated into the procedure, which will entail a suitable suite of alternative models being interrogated jointly. While, in practical terms, this would be a non-trivial undertaking, it would provide a proper rational basis for evaluating the value of such forecasts [*Palmer*, 2000].

In volcanology numerical modelling is becoming an important aspect of forecasting and risk assessment, and evolving approaches have largely focused on numerical simulations. Aided by increasing computer power as well as improving understanding of the physics involved, such models are becoming increasingly sophisticated [e.g. *Neri and Macedonio*, 1996; *Papale*, 1999; *Melnik and Sparks*, 1999]. A presumption is that such models will eventually simulate nature so well that they can be used for forecasting. These expectations may prove to be optimistic, not least because experience has demonstrated with climate modelling that, when uncertainties in such models are disaggregated and appraised individually, the spread of overall uncertainty increases [*Morgan and Keith*, 1995]. Here we discuss the limitations of modelling and consider complementary strategies, while recognizing that numerical models give important insights into volcanic processes.

Most computer modelling of natural phenomena has adopted a strategy of simplification to make matters tractable: those details that are thought to matter most are represented as accurately as possible, and other details, not considered important, are abridged or omitted. Knowing, however, which details matter most can be tricky: models are prone to be incomplete, sometimes leaving out details that could matter under certain conditions. Parametric or even structural uncertainties remain implicit, so that no matter how detailed the model that is created, complete confidence cannot be invested in its predictions about the behaviour of the real system.

Typically, numerical models for volcano dynamics involve several partial differential equations and equations of state that describe the system behaviour. The number of degrees of freedom in the system, and hence number of parameters needed to characterize it adequately, is typically large (commonly a few tens). To make the models computationally manageable and help interpret outputs, simplifications have to be made. For example, published models of pyroclastic flows only involve two particle sizes [*Neri and Macedonio,* 1996]. Certain features of volcanic flow systems are exceedingly computationally demanding: vesiculation processes in explosive eruptions may involve formation of 10^{15} bubbles per cubic metre. Keeping track of what every bubble and melt region does cannot be achieved without making major simplifications that may introduce artificial or unphysical features into the model. As computer codes become more complicated the chances of errors and numerical artifacts increase, and complex codes need substantiating. There may be a limit, however: *Oreskes et al.* [1994] assert that absolute validation is impossible for numerical models in the Earth Sciences.

When numerical models are compared to natural data the issue of uniqueness emerges. It is relatively easy for a skilled modeller to make a model with large numbers of parameters fit the data: the process is one of hindcasting, rather than forecasting. For example, *Barmin et al.* [2002] investigated a model of periodic behaviour of lava dome eruptions and were able to generate simulations similar to the activity of Mount St Helens and Santiaguito lava domes. Such models are very good for gaining insight into how Nature has behaved, but are not unique. There is also inevitably much empiricism in all such models; for example, in high-speed multiphase flows assumptions are made about turbulence based on empirical closure models that have yet to be confirmed in particle-gas mixtures.

It is thus difficult to envisage how deterministic models can be used successfully in forecasting. Interestingly, most of the models that have been developed so far as forecasting tools in volcanology are less physics-based and more overtly empirical, as exemplified by the tephra fall and pyroclastic flow run-out models used on Montserrat. One new direction for numerical models will be to incorporate the quantification of uncertainty. An obvious approach will be to assign uncertainties to every parameter and run ensemble models in which sample the uncertainties using Monte Carlo techniques and repeat calculations a large number of times.

On the topic of uncertainty, an emerging issue for numerical models is the difficult but critical difference between aleatory and epistemic uncertainty, and how to recognize it. Any system has aleatory uncertainty that is real and reflects truly random features like noise and time evolution of magma system properties. Unfortunately model parameters always involve a mixture of both kinds of uncertainty. A good example is conduit dimensions, which are critically important to conduit flow models because of the high powers of flow rate dependence on conduit shape (e.g the fourth power of diameter for a cylindrical conduit). Conduits are non-uniform in Nature and cannot be measured directly, so their shape and dimensions are inevitably poorly constrained with large epistemic uncertainty. Notwithstanding this, conduits are commonly assumed to have a fixed simple geometry (e.g. cylinders of constant diameter). In reality, conduit dimensions might have small aleatory uncertainty (*sensu stricto*), if they could be accessed, but they can evolve with time and be hard to define spatially if there are property gradients between the magma and the wall-rock. So it is unlikely that conduit models can ever be very accurate.

On top of that, many volcanic processes are also highly non-linear. Modelling research in simplified volcanic systems has identified multiple solutions, such that a system can jump suddenly from one state to another [*Jaupart and Allègre,* 1991; *Melnik and Sparks,* 1999; *Slezin,* 2002]. Where systems are close to thresholds for instability, technically known as cusps in catastrophe theory, they can become inherently unpredictable. Worryingly, complex numerical models are unlikely to capture this kind of behaviour easily because of the large epistemic uncertainties in some critical parameters.

Greater use of statistical models and methods should provide an alternative and complementary approach to multivariate, multiparameter forward numerical models. Here we have illustrated how statistical analyses can lead to insights into processes. Statistical models should be given considerable weight in hazards assessments as they are data-based, reflecting how the system actually behaves rather than how it might behave. Such models are effectively free of epistemic uncertainty, except with respect to the measurements themselves. This point has been made strongly by *Young et al.* [2002] in the context of modelling stochastic systems, in particular the climate.

We suggest that the future direction of modelling research for volcanology will involve combining statistical and numerical modeling techniques in a common strategy, with interplay between both. For example, a good test of the validity of a numerical model is that is can reproduce the statistical features of natural data.

THE STATE OF A VOLCANIC PLANET

Over the coming century there are likely to be several major volcanic eruptions (defined as those exceeding 1 km^3), which will affect large numbers of people. With the increasing global population, the dramatic growth of megacities close to active volcanoes, and stresses related to rapid environmental change and globalisation, there is the potential for much larger and more serious volcanic crises and disasters than in the twentieth century. It is certainly plausible that the casualties in a large eruption near an area of dense population could greatly exceed the largest death toll of the last century (30,000 people at Mont Pelée), and might create sufficient destruction to imperil the economies of individual countries and perhaps even continental regions. Human beings are, without doubt, much more vulnerable to volcanic hazards as a consequence of rapid environmental change and globalisation. On the other hand, the advances in understanding of volcanic processes combined with the huge advances in science and technology mean that the scientific community is in a much better position to anticipate volcanic eruptions and, in some circumstances, predict their occurrence, estimate their impact and take steps to protect populations and mitigate the effects.

Many of the scientific tools for dealing with volcanic crises are already available and will continue to improve. The next few decades are bound to produce an increasing number of volcanoes that are adequately monitored. As computer power increases and cheap data storage moves from gigabytes to terabytes there will be huge improvements in the analysis of data from monitoring networks and in the modeling of volcanic processes. There are also likely to be major advances in technology (e.g. satellites, nanotechnology instruments and detectors), which will have significant impact on the ability to analyse signals and monitor volcanoes: remote measurements from space, in particular, offer great promise for enhancing operational forecast models [*Wadge*, 2003].

The approaches to hazards analysis and prediction are likely to move towards probabilistic assessments and, inevitably, to increased use of statistical models. In the latter case, numerical simulations of volcanic processes will be developed in ensemble style, incorporating elements of epistemic and aleatory uncertainty, with comparison of statistical properties of model outputs with real data. A transformation is taking place in the scientific expertise that will be needed to develop these vital quantitative forecasting and prediction tools, and much needs to be done to provide volcanologists with the appropriate knowledge and skills to meet the challenges.

Despite reasons for optimism only a small number of active volcanoes are adequately monitored, and many of these are volcanoes within the Developed World. Most of the world's active volcanoes and a large proportion of the 500 million people living close to volcanic threats are in less developed regions where such volcanoes are either not monitored at all or, at best, have only rudimentary observational or instrumental networks. The challenge is therefore to provide these regions with working access to the technological and scientific advances that can mitigate disaster and assist sustainable development.

Acknowledgements. RSJS acknowledges a Royal Society-Wolfson Award and support from research grants including the EC (MULTIMO EVG1-2000-00574) and NERC. WPA acknowledges a Benjamin Meaker Visiting Fellowship at IAS, Bristol, and support from EC projects MULTIMO EVG1-2000-00574 and EXPLORIS EVR1-CT-2002-40026. The authors acknowledge the tremendous help of staff of the Montserrat Volcano Observatory and many colleagues involved in the hazards and risk assessments on Montserrat. The authors thank Gordon Woo for stimulating discussions, and Colin Wilson, Chris Newhall and Chris Hawkesworth for careful and helpful reviews. Table 1 follows a suggestion from Colin Wilson.

REFERENCES

Barmin, A., Melnik, O and Sparks, R. S. J. Periodic behaviour in lava dome eruptions. *Earth and Planetary Science Letters, 199,* 173–184, 2002.

Baxter, P.J. Medical effects of volcanic eruptions. I. Main causes of death and injury. *Bulletin of Volcanology, 52,* 532–544, 1990.

Baxter, P. J., Bonadonna, C., Dupree, R., Hards, V. L., Kohn, S. C., Murphy, M. D., Nichols, A., Nicholson, R. A., Norton, G., Searl, A., Sparks, R. S. J. and Vickers, B. P. Cristobalite in volcanic ash of the Soufriere Hills Volcano, Montserrat: Hazards implications. *Science, 283,* 1142–1145, 1999.

Blong, R. *Volcanic Hazards: a sourcebook on the effects of eruptions.* Academic Press, Sydney, 424pp, 1984.

Bonadonna, C., Macedonio, G. and Sparks, R. S. J. Numerical modelling of tephra fallout associated with dome collapses and Vulcanian explosions: application to hazard assessment on Montserrat. In: Druitt, T. H. and Kokelaar, B. P. (eds) *The eruption of the Soufrière Hills Volcano, Montserrat 1995 to 1999. Geological Society, London. Memoir 21,* 517–538, 2002.

Calder, E. S., Cole, P. D., Dade, W. B., Druitt, T. H., Hoblitt, R. P., Huppert, H. E., Ritchie, L., Sparks, R. S. J. and Young, S. R. Mobility of pyroclastic flows and surges at the Soufriere Hills Volcano, Montserrat. *Geophysical Research Letters, 26,* 537–540, 1999.

Calder, E. S., Luckett, R., Sparks, R. S. J. and Voight, B. Mechanisms of lava dome instability and generation of rockfalls and pyroclastic flows at Soufrière Hills Volcano, Montserrat. In: Druitt, T. H. & Kokelaar, B. P. (eds) *The eruption of Soufrière Hills Volcano, Montserrat, from 1995 to 1999. Geological Society, London, Memoir 21,* 173–190, 2002.

Chatfield, C. Confessions of a pragmatic statistician. The Statistician, *Journal of the Royal Statistical Society, Series D, 51,* 1–20, 2002.

Chouet, B. A. Long-period volcano seismicity: its source and use in eruption forecasting. *Nature, 380,* 309–316, 1996.

Cole, P. D., Calder, E. S., Sparks, R. S. J., Clarke, A. B., Druitt, T. H., Young, S. R., Herd, R. A., Harford, C. L. and Norton, G. E. Deposits from dome-collapse and fountain-collapse pyroclastic flows at Soufrière Hills Volcano, Montserrat. In: Druitt, T. H. & Kokelaar,

B.P. (eds) *The eruption of Soufrière Hills Volcano, Montserrat, from 1995 to 1999.* Geological Society, London, Memoir 21, 231–262, 2002.

Connor, C. B., Sparks, R. S. J., Mason, R. M. Bonadonna, C., and Young S. R. A "log logistic" volcano: The Soufrière Hills, Montserrat. *Geophysical Research Letters, 30,* 1701 doi:10.1029GL017384, 2003.

Cooke, R. M. *Experts in Uncertainty.* Oxford University Press Oxford, 1991.

Druitt, T. H., Young, S. R., Baptie, B., Bonadonna, C., Calder, E. S., Clarke, A. B., Cole, P. D., Harford, C. L., Herd, R. A., Luckett, R., Ryan, G. and Voight, B. Episodes of repetitive Vulcanian explosions and fountain collapse at Soufrière Hills Volcano, Montserrat. In: Druitt, T. H. & Kokelaar, B. P. (eds) *The eruption of Soufrière Hills Volcano, Montserrat, from 1995 to 1999.* Geological Society, London, Memoir 21, 281–306, (2002a).

Druitt, T. H., Calder. E. S., Cole, P. D., Ritchie, L. J., Sparks, R. S. J., and Voight, B. Small-volume, highly mobile pyroclastic flows formed by rapid sedimentation from pyroclastic surges at Soufrière Hills Volcano, Montserrat: an important volcanic hazard. In: Druitt, T. H. and Kokelaar, B. P. (eds) *The eruption of the Soufrière Hills Volcano, Montserrat 1995 to 1999.* Geological Society, London, Memoir 21, 263–280, 2002b.

Dzurisin, D. Volcano geodesy: challenges and opportunities for the 21st century. *Philosophical Transactions of the Royal Society A, 358,* 1547–1566, 2000.

Endo, E. T., Malone, S. D., Noson, L. L. and Weaver, C. S. Locations, magnitudes and statistics of the March 20–May 18 earthquake sequence. In: Lipman, P. W. and Mullineaux, B. R. (eds) *The 1980 eruptions of Mount St.Helens, Washington. US Geological Survey Professional Paper 1250,* 93–108, 1981.

Feuillard, M., Allegré, C. J., Brandeis, G., Gaulon, R., Le Mouel, J. L., Mercier, J. C., Pozzi, J. P. and Semet, M. P. The 1975–1977 crisis of La Soufrière de Guadeloupe (F.W.I.): a still-born magmatic eruption. *Journal of Volcanology and Geothermal Research, 16,* 317–334, 1983.

Fiske, R. S. Volcanologists, journalists, and the concerned local public: a tale of two crises in the Eastern Caribbean. In: F. R. Boyd, (ed): *Explosive Volcanism: Inception, Evolution and Hazards.* National Academic Press, Washington, DC, 170–176, 1984.

Hill, D. P., Pollitz, F. and Newhall, C. Earthquake–volcano interactions. *Physics Today, 55 (4),* 41–47, 2002.

Jaupart, C. and Allegré, C. Gas content, eruption rate and instabilities of eruption in silicic volcanoes. *Earth and Planetary Science Letters, 102,* 413–429, 1991.

Jupp, T., Pyle, D., Mason, B. and Dade, B. A statistical model for the timing of earthquakes and volcanic eruptions influenced by periodic processes. *Journal of Geophysical Research, 109,* B02206 10.1029/2003 JB002584, 2004.

Linde, A. T. and Sacks, I. S. Triggering of volcanic eruptions. *Nature, 395,* 888–890, 1998.

Linde, A. T., Agustsson, K., Sacks, I. S. and Stefansson, R. Mechanism of the 1991 eruption of Hekla from continuous borehole strain monitoring. *Nature, 365,* 737–740, 1993.

Loughlin, S. C., Calder, E. S., Clarke, A. B., Cole, P. D., Luckett, R., Mangan, M. T., Pyle, D. M., Sparks, R. S. J., Voight, B. and Watts, R. B. Pyroclastic flows generated by the 25 June 1997 dome collapse, Soufrière Hills Volcano, Montserrat. In: Druitt, T. H. & Kokelaar, B. P. (eds) *The eruption of Soufrière Hills Volcano, Montserrat, from 1995 to 1999.* Geological Society, London, Memoir 21, 211–230, 2002.

Mason, B. G., Pyle, D. M. and Oppenheimer, C. The size and frequency of the largest explosive eruptions on Earth. *Bulletin of Volcanology* (in press).

Matthews, A., Barclay, J., Carn, S., Thompson, G., Alexander, J., Herd, R. and Williams, C. Rainfall-induced volcanic activity on Montserrat. *Geophysical Research Letters, 29,* 10.1029/2002GL014863, 2002.

Melnik, O. and Sparks, R. S. J. Nonlinear dynamics of lava extrusion. *Nature, 402,* 37–41, 1999.

Melnik, O. and Sparks, R. S. J. Modelling of conduit flow dynamics during explosive activity at Soufrière Hills Volcano, Montserrat. In: Druitt, T. H. and Kokelaar, B. P. (eds) *The eruption of the Soufrière Hills Volcano, Montserrat 1995 to 1999.* Geological Society, London, Memoir 21, 307–318, 2002.

Montserrat Volcano Observatory. *Dome collapse and explosive activity, 12–15 July 2003.* MVO Open File Report 04/01, 16 pp; 2004.

Morgan, M. G. and Keith, K. W. Subjective judgments by climate experts. *Environmental Science & Technology, 29,* 468–476, 1995.

Neri, A. and Macedonio, G. Numerical simulation of collapsing columns with particles of two sizes. *Journal of Geophysical Research, 101,* 8153–8174, 1996.

Neuberg, J. Characteristics and causes of shallow seismicity in andesite volcanoes, *Philosophical Transactions of the Royal Society Series A, 358,* 1533–1546, 2000.

Newhall, C. G. Volcano Warnings, In: *Encyclopaedia of Volcanoes* (Chief Editor H. Sigurdsson) Academic Press, San Diego, 1185–1197, 2000.

Newhall, C. G. and Hoblitt, R. P. Constructing event trees for volcanic crises. *Bulletin of Volcanology, 64,* 3–20, 2002.

O'Hagan, A., Kennedy, M. and Oakley, J. E. Uncertainty analysis and other inference tools for complex computer codes (with discussion). In: *Bayesian Statistics 6* (eds J. M. Bernardo, J. O. Berger, A. P. Dawid and A. F. M. Smith), pp. 503–524. Oxford: Oxford University Press, 1999.

Oreskes, N., Schrader-Frechette, K. and Belitz, K. Verification, validation, and confirmation of numerical models in the Earth Sciences. *Science, 263,* 641–646, 1994.

Palmer, T. N. Predicting uncertainty in forecasts of weather and climate. *Reports on Progress in Physics, 63,* 71–116, 2000.

Papale, P. Strain-induced magma fragmentation in explosive eruptions. *Nature, 397,* 425–428, 1999.

Pyle, D. M. Forecasting sizes and repose times of future extreme volcanic events. *Geology, 26,* 367–370, 1998.

Punongbayan, R. S., Newhall, C. G., Bautista, M. L. P., Garcia, D., Harlow, D. H., Hoblitt, R. P., Sabit, J. P. and Solidum, R. U. Eruption hazard assessments and warnings, In: *Fire and Mud: eruptions and lahars of Mount Pinatubo, Phillipines,* C. G. Newhall, R. S. Punongbayan, PHILVOLCS, Quezon City and University of Washington Press, Seattle, 67–85, 1996.

Rampino, M. R. Supereruptions as a threat to civilisations on Earth-like planets. *Icarus, 156,* 562–569, 2002.

Rubin, A. Propagation of magma-filled cracks. *Annual Reviews of Earth and Planetary Sciences, 23*, 287–336, 1995.

Simkin, T. and Siebert, L. *Volcanoes of the World: a Regional Directory, Gazetteer, and Chronology of Volcanism During the Last 10,000 Years. (Second edition)*. Geoscience Press, Tucson: 368 pp, 1994.

Slezin, Y. The mechanism of volcanic eruptions (a steady state approach). *Journal of Volcanology and Geothermal Research, 122*, 7–50, 2003.

Sparks, R. S. J. Forecasting volcanic eruptions. *Earth and Planetary Science Letters, 210*, 1–15, 2003.

Sparks, R. S. J. and Young, S. R. The eruption of Soufrière Hills Volcano, Montserrat: overview of scientific results. In: Druitt, T.H. & Kokelaar, B. P. (eds) *The eruption of Soufrière Hills Volcano, Montserrat, from 1995 to 1999. Geological Society London Memoir, 21*, 45–69, 2002.

Sparks, R. S. J., Bursik, M. I., Carey, S. N., Gilbert, J. S., Glaze, L., Sigurdsson, H. and Woods, A. W. *Volcanic Plumes*. Chichester, UK, John Wiley and Sons, 557 pp., 1997.

Swanson, D. A., Casadeall, T. J., Dzurisin, D., Malone, S. D., and Weaver C. S. Predicting eruptions at Mount St. Helens, June 1980 through December 1982. *Science, 221*, 1369–1376, 1983.

Voight, B. and Cornelius, R. R. Prospects for eruption prediction in near-real-time, *Nature, 350*, 695–698, 1991.

Voight, B. and Elsworth D. Instability and collapse of lava domes. *Geophysical Research Letters, 27*, 1–4, 2000.

Voight, B., Sparks, R. S. J., Miller, A. D., Stewart, R. C., Hoblitt, R. P., Clarke, A., Ewart, J., Aspinall, W., Baptie, B., Druitt, T. H., Herd, R., Jackson, P., Lockhart, A. B., Loughlin, S. C., Lynch, L., McMahon, J., Norton, G. E., Robertson, R., Watson, I. M. and Young S. R. Magma flow instability and cyclic activity at Soufriere Hills Volcano, Montserrat, B. W. I. *Science, 283*, 1138–1142, 1999.

Wadge, G. A strategy for the observation of volcanism on Earth from space. *Philosophical Transactions of the Royal Society, London, Series A, 361*, 145–156, 2003.

Wadge, G., Jackson, P., Bower, S. M., Woods, A. W. and Calder, E. S. Computer simulations of pyroclastic flows from dome collapse. *Geophysical Research Letters, 25*, 3677–3680, 1998.

Watts, R. B., Herd, R. A., Sparks, R. S. J. and Young, S. R. Growth patterns and emplacement of the andesite lava dome at the Soufriere Hills Volcano, Montserrat. In: Druitt, T. H. and Kokelaar, B. P. (eds) *The eruption of the Soufrière Hills Volcano, Montserrat 1995 to 1999. Geological Society, London, Memoir 21*, 115–152, 2002.

Weisenfeld, K. and F. Moss. Stochastic resonance and the benefits of noise: from ice ages to crayfish and SQUIDS. *Nature, 373*, 33–36, 1995.

Woo, G. *The Mathematics of Natural Catastrophes*. Imperial College Press, London, 292 pp., 1999.

Young, P. C., Parkinson, S. and M. J. Lees. Simplicity out of complexity: Occam's razor revisited. *Journal of Applied Statistics, 23*, 165–210, 2002.

R. S. J. Sparks, Department of Earth Sciences, University of Bristol, Bristol BS8 1RJ, United Kingdom (Steve.Sparks@bristol.ac.uk)

W.P. Aspinall, Aspinall & Associates, 5, Woodside, Beaconsfield, Buckinghamshire, United Kingdom (willy@aspinall.demon.co.uk)

Geophysical Risk, Vulnerability, and Sustainability

Tom Beer

CSIRO Environmental Risk Network and CSIRO Atmospheric Research, Aspendale, Victoria, Australia

The vulnerability of humans to geophysical hazards (earthquakes, volcanoes, landslides, severe storms, droughts, floods, tsunamis and storm surges, and space weather) affects the sustainability of societies. Risk, over a given time, is the union of a set of likelihoods and a set of consequences of the scenarios under consideration. Consequences depend on the magnitude of the hazard, and the severity of the impacts of the hazard. For humans, vulnerability has both economic and social dimensions, thus a complete study of geophysical risks should incorporate the work of both geophysicists and social scientists. Ecological concepts can categorise vulnerability in terms of the magnitude of the impact, the stability (time to recovery) of the ecosystem, and its resilience. Stability depends on organisation (how well the ecosystem is linked internally), and connectivity—how well the species (humans) is linked to the whole ecosystem. Sustainability is the progression towards improved quality of life, both now and in the future, in a way that maintains the environmental, social and economic processes on which life depends. Because geophysical risk and vulnerability includes all three of these processes, a comprehensive study of geophysical risk and vulnerability inevitably becomes a study of at least one aspect of sustainability. The IUGG Commission on Geophysical Risk and Sustainability issued the Budapest Manifesto to provide a framework for collaboration between scientists and social scientists. The manifesto recommends that scientists examine technical and social issues and contribute to decision-making through a risk management framework.

1. INTRODUCTION

In 1887, an estimated 900,000 people perished in the floods of Henan province in China [*Munich Re*, 2000]. China also suffered 830,000 fatalities in 1556 due to an earthquake in Shaanxi province. Tropical cyclones in Calcutta, India in 1737, in Haiphong, Vietnam in 1881 and in Chittagong, Bangladesh in 1970 each caused 300,000 fatalities. The 1902 eruption of Mont Pelée, Martinique, which destroyed an entire city, caused nearly 30,000 fatalities.

Geophysical hazards, such as floods, earthquakes, volcanoes, and tropical cyclones, all have the potential to kill large numbers of people, destroy property, and disrupt the normal functioning of society. Some societies are at greater risk than others, and some societies are better able to cope than others. A geophysical hazard is a natural phenomenon that has the potential to cause loss of life, or damage to infrastructure with economic loss as a consequence.

To understand the effects of natural hazards on society requires that we examine their intrinsic nature, the risks that they pose, and the character of the society that endures them. One aspect of sustainability is society's vulnerability to natural hazards. A key issue here concerns the dramatic rise in population growth worldwide, and the growing vulnerability of populations to natural hazards. An added complication involves the effects of possible global warming and climate change. Numerical models forecast an increase in extreme events, particularly severe storms, floods, droughts and wildfires, as a consequence of climate change [*IPCC*, 2001]. In recent years, for instance, economic losses from severe storms have risen in North America and in Europe, although there is no consensus on whether this increase is due to population pressure [*Chagnon*, 2003], climate change [*Munich Re*, 2003], or both. We need to find ways to deal with, and resolve, such uncertainties.

2. DEFINITIONS OF RISK AND SUSTAINABILITY

There are many different ways to define risk. The choice of definition is related to the disciplinary background of the individual, and the definition in turn is related to the choice of framework used to assess and manage the risk. At the same time, a consensus has emerged that risk involves not only probability but also a perception of consequences, notwithstanding the US Environmental Protection Agency's definition of risk as "a measure of the probability that damage to life, health, property, and/or the environment will occur as a result of a given hazard" [*USEPA*, 2004]. Indeed, Fournier d'Albe [1979] took risk to be the product of three entities: (i) the probability that a certain area will be affected by an event, (ii) the percentage of lives or goods likely to be lost because of a given event, and (iii) the number of lives or the monetary value of goods potentially at risk.

Beer and Ziolkowski [1995] generalized on the above concepts to define risk as the union of a set of likelihoods and a set of consequences of the scenarios under consideration, over a given time. This definition implies that there are four aspects involved in a consideration of risk: a time scale, the scenarios, the relevant consequences, and the corresponding likelihoods. This paper will discuss each of these four aspects.

There are two principal perspectives on environmental risk: risk to the environment as a result of human activity and risk to people or infrastructure as a consequence of environmental hazards [*Beer*, 2003]. By contrast, geophysical risk is related to natural hazards and refers only to risk as a result of hazardous geophysical events, including: earthquakes, volcanoes, landslides, severe storms, droughts, floods, tsunamis and storm surges, and space weather. In fact, these events cover the disciplinary fields represented by the seven international associations that joined together to form the International Union of Geodesy and Geophysics (IUGG), as shown in Table 1.

Geophysical hazards that pose the most serious problems can commonly be characterised as extreme events. In some cases these extreme events are correlated. Meteorological extreme events, such as severe storms and tropical cyclones, may then

Table 1. Geophysical hazards studied by the seven international associations that comprise the IUGG

Discipline	Name	Acronym	Natural Hazard
Geodesy	International Association of Geodesy	IAG	Landslides
Geomagnetism and Aeronomy	International Association of Geomagnetism and Aeronomy	IAGA	Space Weather
Hydrology	International Association of Hydrological Sciences	IAHS	Floods
Meteorology	International Association of Meteorology and Atmospheric Sciences	IAMAS	Severe Storms Droughts
Oceanography	International Association of the Physical Sciences of the Ocean	IAPSO	Tsunamis and storm surges
Seismology	International Association for Seismology and Physics of the Earth's Interior	IASPEI	Earthquake
Volcanology	International Association of Volcanology and Chemistry of the Earth's Interior	IAVCEI	Volcanoes

produce consequences such as flooding; drought may produce wildfires; extreme rainfall may lead to the collapse of a volcanic dome and resulting pyroclastic flows [*Matthews et al.*, 2002]; and the filling of artificial water reservoirs may trigger earthquakes [*Gupta*, 2002]. There is a well-founded statistical framework, known as the theory of extreme events, explained in Gumbel [1958], that links the concept of an extreme as a tail of a statistical distribution with the concept of an extreme as a rare event in the context of human history or human memory. This theory provides a unifying basis for the mathematical treatment of extreme events. It relies strongly on the "extremal types theorem" [*Leadbetter et al.*, 1983], which shows that, when extreme values are drawn from a probability distribution, the resulting probability distribution of the extreme values is one of only three distributions that are known as the Gumbel, Weibull and Frechet distributions. Kotz and Nadarajah [2000] and Coles [2001] have developed the Generalised Extreme Value (GEV) Distribution, allowing us to combine these three possible distributions into a single family of models.

The theory of extreme events provides the analytical framework for the concept of a return period [or recurrence interval], which Woo [1999: p.176] defined as the inverse of the annual exceedance probability. The theory has found numerous practical applications, especially in civil engineering, where it is used to design the height of dams, spillways, and bridges, the size of culverts, and buildings in areas prone to tropical cyclones [*Beer et al.*, 1993]. In these cases the return periods, for design purposes, are chosen to represent some notion of an acceptable risk and are assigned with some magnitude of hazard; such as flood height, maximum wind speed, and so on. Using this method, McInnes et al. [2003] have calculated sea level heights for various return periods for the Australian city of Cairns. They did so both for the present tropical cyclone climatology, and for the tropical cyclone climatology and the mean sea level (i.e., the storm surge in water level primarily as a result of the decreased pressure) expected to arise from a doubling of carbon dioxide concentrations in the atmosphere. On the basis of their assumptions, they found that the mean sea level heights for a 100-year return period increased from 2.3 m to 2.8 m. The average area inundated by such rises in mean sea level was inferred to more than double under enhanced greenhouse conditions.

There are also risks that do not manifest themselves through extreme events but are associated with continuous or cumulative exposure to a hazard, including certain geophysical risks such as climate change, previously mentioned, or risks inherent in exposure to radon or ultra-violet radiation, long-term exposure to respirable volcanic dust [*Baxter et al.*, 1999], and air pollution [*Beer*, 2001].

The term "sustainability" is now used as shorthand for sustainable development. The concept arose from the findings of the World Commission on Environment and Development [1987], also known as the Brundtland Report, which defined sustainable development as meeting the needs of the present without compromising the needs of future generations. Commission objectives were set by the UN General Assembly, including an attempt to define shared perceptions of long-term environmental issues and efforts appropriate to dealing successfully with the problems of protecting and enhancing the environment. These policy developments were informed and underpinned by parallel scientific activity, which is summarised by Clark and Munn [1986] and Kates et al. [2001].

Sustainability science [*Kates et al.*, 2001], as it has come to be called, is based around the idea of compiling a database or inventory of environmental resources, developing the analytical tools with which to examine the future trajectory of those resources, and then applying the tools. In this paper we explore appropriate means to undertake such science related to geophysical risks.

3. LINKING RISK AND SUSTAINABILITY

Sustainability is a concept that links economics, environment, and society. Our definition of risk brings together likelihoods, consequences, time and scenarios. These entities involve both scientific and societal aspects as depicted schematically in Figure 1. The rectangle represents the domain of science, whereas the area outside the rectangle represents the domain of society. The concepts of time, likelihood and consequences have both scientific and societal aspects, so that they straddle the rectangular boundary. Time, for example, is defined and measured in a scientific way. However, using a particular time-scale in an assessment of risk management (such as the 100 year return period as a design criterion) entails a social choice. The construction of risk scenarios, as explained in the next section, is primarily a social activity, linked to scientific aspects through the time-scales and consequences that are built in to the scenarios.

4. SCENARIO CONSTRUCTION

Scenario construction offers a method with which to examine the complex interactions between hazards, risks and society [*Glenn and Gordon*, 1997, 1998]. Kasperson and Kasperson [2001] note that "imagining sustainable futures is not a matter of wishful thinking but an integral part of risk analysis". They envisage imagining futures, which we shall call scenario construction, as a process that draws on science but primarily involves the search for human values, economic systems, and social structures by which high-risk pathways can switch to lower-risk situations and more sustainable tracks.

The 1997 State of the Future document [*Glenn and Gordon*, 1997] describes two types of scenarios: normative and

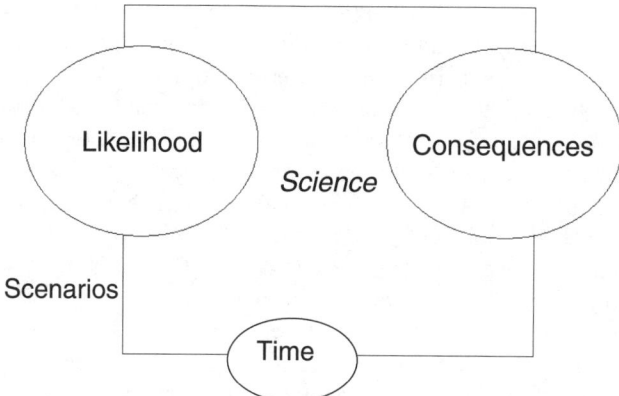

Figure 1. A representation of the definition of risk. The rectangle represents the domain of science, whereas the area outside the rectangle represents the domain of society. The concepts of time, likelihood and consequences have both scientific and societal aspects. Scenario construction is primarily a social activity.

exploratory. Normative scenarios produce images of the hoped-for future. Although this is normally the domain of utopian and science-fiction writers, any group of lay or specialist people can construct a normative scenario. The normative scenario is then the collective response of the group to the question: "What kind of future would you like to see?"

Exploratory scenarios, by contrast, respond to the question: "What do you think the future might be?" They portray images that:

> "seem plausible, given actions or inactions of key players, exogenous developments, chance, and the internal dynamics of the system under study."

The above excerpt comes from the 1997 State of the Future report. Note that this particular ordering of the key factors that influence the future is very much the historical view, "which attaches extreme importance to the exact reconstruction of the actions, words or attitudes of a few personages. . ." [*Bloch*, 1954]. The historian is forced to attach importance to such reconstructions because the internal dynamics of the social system is known, if known at all, only dimly.

By contrast, the empirical approach to exploratory scenario development starts at the other end of the list of actions. A model of the internal dynamics of the system under study is dissected, certain portions enhanced, and then re-assembled. This is most clearly illustrated in weather forecasting and climate modelling where a scenario of the future state of the atmosphere is constructed by setting up a system of partial differential equations and computing their solutions. Lorenz [1963] showed that, under certain conditions, very small changes in the initial conditions used to start the computation can lead to markedly different outcomes. This behaviour has come to be known as chaos [*Lorenz*, 1993], and arises from the non-linear nature of the differential equations. To a true empiricist, chance shows up through the chaotic behaviour inherent in the non-linearity of the model [*Bunde et al.*, 2002] so that the each model run becomes the realisation of an exploratory scenario. Pittock [1993] has used this approach to construct scenarios of climate change and its impacts.

Glenn and Gordon [1997] implement a method of scenario construction that involves, as a first step, determining scenario axes—with the axes referring to important aspects of the scenario. For example, one could develop four scenarios from two issues such as population growth and industrial ecology that, therefore, comprise two scenario axes. Consider the two extremes of population growth (high and low) as points along the first scenario axis, and the two extremes of industrial ecology (clean and dirty) as points along the second axis. The four scenarios then cover the sets (high, clean), (high, dirty), (low, clean) and (low, dirty). Similarly it would be possible to construct scenario axes involving high and low populations around a dangerous volcano and strong versus weak eruptions.

One approach contained within the State of the Future Project [*Glenn and Gordon*, 1997, 1998] was a modified delphi-process of consultation to determine the scenario axes. The delphi-process is a technique of knowledge elicitation that is used when contact between experts is difficult. It works by experts submitting their judgement, with all judgements circulated anonymously to all experts. Each expert then submits a revised judgement. In a full delphi-process the iterations continue until all the judgments converge. There have been methodological criticisms of the Delphi-process that are reviewed in Cooke [1991], and there are more recent developments in methods of expert elicitation [*Morgan and Keith*, 1995; *Cooke*, 1991]. However, the rapid development of internet communication led to the delphi-process being used for the consultation because of its ability to connect experts globally at minimal cost.

Glenn and Gordon [1997] used a two-step Delphi-process. The first step consisted of the identification of key issues (via brainstorming) and their collation. The second step consisted of scoring the issues in relation to the importance and likelihood, determining a single score, and ranking the issues on the basis of the scores. This constitutes a risk-based ranking because the scores are based on consequence (importance) and likelihood.

In studies that examine nuclear facility sites, the scenario axes are determined on the basis of features, events and processes—called FEPs. In one such study [*Brown*, 2000] 46 FEPs were examined and 22 of them were retained as scenario axes (in scenario construction terminology) or as modelled parameters (in modelling terminology).

Once the implicit scenarios have been established, then the consequences that one wishes to examine are determined by adding content to the scenario. In some geophysical applications this is done on the basis of empirical evidence [*Aspinall et al.*, 2002] and in others through computer modelling. Global climate models [*Beer and Foran*, 2000], seismic hazard assessment [*Dunbar et al.*, 2003], coupled ocean-wave circulation model [*Moon et al.*, 2003] and flood routing models [*Chowdhury*, 2000] provide risk-relevant examples of modelling used in the atmospheric, seismological, oceanographic and hydrological sciences respectively. There are important issues related to model construction, parameterization and validation that remain active areas of research. To paraphrase Mahlman [1998], models are imperfect but they are crucial anyway.

5. UNCERTAINTIES IN SCENARIOS

The science involved in determining likelihoods revolves around the concept of uncertainty. Likelihood refers to a verbal or quantitative assignment of certainty or uncertainty.

Woo [1999: p. 74] considers two types of uncertainty: aleatory uncertainty and epistemic uncertainty. Aleatory uncertainty is that associated with natural randomness. Epistemic uncertainty arises from imperfect information and knowledge.

Wynne [1992] considers four different types of uncertainty:
- Wynne's first type of uncertainty is sufficiently tractable that it can be described in terms of statistical probabilities. Wynne [1992] used the term risk to describe this uncertainty, but this mixes usage and meaning of the term risk, so that another term is preferable. Here probabilistic uncertainty is suggested as a suitable term.
- The second type of uncertainty occurs when uncertainty is recognised but it is not possible to quantify it. Here this is termed incertitude, although Wynne [1992] actually called this uncertainty, but this also mixes usage and meaning.
- The third type of uncertainty (called ignorance) occurs when we don't know what we don't know.
- The fourth type of uncertainty (indeterminacy) arises when the social context changes so rapidly that scientific information cannot be sufficiently complete to meet the needs of decision-makers.

Harding [1998] specifically notes that the early history in the investigation of ozone hole was characterised by ignorance. The term "ignorance" is applicable as it was not realised at that time that the chlorofluorocarbons (CFCs) would be chemically active at stratospheric heights and destroy the ozone that protects the biosphere from harmful ultraviolet radiation. The issue was sufficiently complex and large-scale in scope that decisions needed to be made by national governments. Resolving the ambiguity, and making decisions in the face of uncertainty, became the responsibility of government environmental managers. Thus the international process that was set in place, through the Montreal Protocol, was designed to transform epistemic uncertainty (due to ignorance) into aleatory uncertainty.

The whole process of assessing greenhouse gas-induced climate change that led to the development of the Kyoto Protocol can be viewed as a major risk management exercise [Beer, 1997]. Science-based risk assessment is dealt with by the Intergovernmental Panel on Climate Change (IPCC). The IPCC's remit is to assess climate risks in a manner that is useful for policy development without being policy prescriptive. Science-based assessment, in many cases undertaken using Global Climate Models, is concerned with questions such as what are the risks to the environment as a result of human activity? What uncertainties are associated with those risks? What management controls need to be put in place to deal adequately with the risks that have been identified?

The treatment of risk is largely policy-based and is dominated by stakeholders, who include policymakers, planners, industry and the community. The diplomatic processes that led to the Framework Convention on Climate Change and the Kyoto Protocol are aspects of risk treatment. The role of researchers is to work with these stakeholders to communicate the science and its attendant uncertainties and to bring issues relevant to stakeholders into formal assessments such as those for assessing the likelihood of critical thresholds of climate change impact [Jones, 2001, 2003].

Harding and Fisher [1999] make a distinction between prevention and precaution. If uncertainty is of the first type (probabilistic), then a risk management approach is preventive (rather than precautionary), because the problem is fairly clearly bounded and the uncertainties are not likely to involve scientific (or epistemic) uncertainties. The precautionary principle, which is the principle underlying the UN Framework Convention on Climate Change, states that lack of full scientific certainty should not be used as a reason to postpone measures to protect the environment. The principle needs to be invoked when dealing with epistemic types of uncertainty. As Harding and Fisher [1999] say:

> ...not all "risk situations" will require application of the precautionary principle, but all situations warranting application of the precautionary principle can be thought of as imposing *potential* "risks".

Present usage of the precautionary principle envisages "being hazardous" as synonymous with "imposing potential risks".

5.1 Risk and Uncertainty

The first two types of uncertainty mentioned above (probabilistic and incertitude) are amenable to systematic study. In geophysical applications, uncertainty is quantitatively expressed in terms of limits associated with a reading. For example, the solar constant is 1.367 ± 0.003 W m^{-2} (when measured on the scale of weeks). There is a 0.2% uncertainty associated with the average value on a time-scale of the order of weeks, but a 0.4% uncertainty over a 22-year time scale [Beer, 1990: p.21]. This way of expressing uncertainty attempts to represent a probability distribution in a simple form by giving the mean plus or minus some limits. When assessing uncertainty in a more complete manner the nature of the statistical distribution that is appropriate for a given situation needs to be examined. Discussion on this subject is given by Woo [1999] and Vose [2000].

Language has evolved a set of likelihood phrases to allow people to express their certainty and uncertainty (the incertitude). Table 2 reproduces the likelihood statements used by authors of the IPCC Third Assessment Report to represent their judgement in the validity of a conclusion in the report [*IPCC*, 2001: p.44].

There are a number of studies [*Morgan and Henrion*, 1990; *Vose*, 2000] that indicate that scientists (and presumably geophysicists) are poor at estimating the quantitative value of aleatory uncertainties. They tend to be overconfident in their results and ascribe a higher likelihood to them than is warranted. Morgan and Henrion [1990] illustrate this by describing the history of measurement of some of the absolute physical constants, such as the speed of light, for which successive experimental determinations report standard errors that are smaller than they should have been, as revealed by subsequent experiments. Similar observations could be made in the geophysical context in relation to the solar constant [*Beer*, 1990: p.20] and the constant of gravitation [*Jeffreys*, 1948: p. 280].

It is possible to categorise the four types of uncertainty by considering a matrix whose cells are composed of a probability axis, and a consequences axis [*O'Riordan and McMichael*, 2002]. Each axis has two states: well defined, or poorly defined (Table 3).

The addition of ambiguity to Table 3 is based on Anand [2001], who points out that ambiguity refers to a situation, such as bovine spongiform encephalopathy (mad-cow disease), where the consequence is well defined but the associated probabilities are poorly defined.

The precautionary principle has an obvious application to situations that involve ignorance or indeterminancy, or both. The extent to which it applies to situations that involve "incertitude" or "ambiguity" (as defined from Table 3) is less clear according to Ricci et al. [2003]. Nevertheless, if we follow the earlier reasoning, that the precautionary principle relates to epistemic uncertainty, then precaution applies to ambiguity, but prevention applies to incertitude.

Table 2: Verbal and mathematical expression of certainties and probabilities

Verbal expression of certainty [indicating a likelihood measure]	Corresponding mathematical expression of probability
Virtually certain	0.99 to 1.00
Very likely	0.90 to 0.99
Likely	0.66 to 0.90
Medium likelihood	0.33 to 0.66
Unlikely	0.10 to 0.33
Very unlikely	0.01 to 0.10
Exceptionally unlikely	0.00 to 0.01

Table 3: Typology of uncertainty based on whether the uncertainty is associated with likelihoods or consequences

	Well defined consequences	Poorly defined consequences
Well defined probabilities [Aleatory uncertainty]	Probabilistic	Incertitude
Poorly defined probabilities [epistemic uncertainty]	Ambiguity	Ignorance, and Indeterminancy

5.2 Assigning Probabilities to Scenarios

There is disagreement as to whether or not it is possible to assign probabilities to scenarios. The frequentist position, favoured by many of the physical scientists involved in the IPCC process [*Grübler and Nakicenovic*, 2001], claims that this cannot be done. It would be unusual to find a risk analyst who would agree with the frequentist position.

The difficulty is twofold. Firstly, the application of Bayesian statistics may involve estimation or judgement, albeit expert judgement, rather than measurement of the probabilities. Many would deny the validity of such estimates. Although there are difficulties associated with such estimates [*Morgan and Henrion*, 1990], there are techniques that can be used to overcome such difficulties [*Cooke*, 1991; *Morgan and Keith*, 1995; *Vose*, 2000].

Secondly, as pointed out by Pittock et al. [2001], in the earth sciences there will be only one real outcome (or consequence, in the terminology of Figure 1), which cannot be measured beforehand. Probability estimates of future conditions on Earth based on computer modelling appear frequentist, but are essentially Bayesian in that they rely on the same concepts as Bayesian statistics; namely, that they are based on prior knowledge or assumptions embodied in the various models and inputs. The Bayesian position is also the one taken by Morgan and Henrion [1990] and in geophysics extends back in time to at least Jeffreys [1948].

The final aspect of Figure 1 to be discussed relates to the consequences.

6. CONSEQUENCES

Consequences refer to the outcome of an event; for example, loss of life, injury, damage to buildings or other infra-

structure, and traffic congestion. Determining the overall consequences of a geophysical hazard requires consideration of two aspects:
- Magnitude
- Severity

When one considers only a hazard, then it is legitimate to treat the severity of the hazard as being identical to the magnitude of the hazard. There is a rich geophysical literature, exemplified by the journal *Natural Hazards*, which illustrates this.

However, when the focus of interest shifts to geophysical risk, then the consequences can still be considered as a combination of magnitude and severity, but in this case, the severity depends on the vulnerability of the elements that are at risk from the hazard.

Some examples of elements at risk (the receptors) would be agricultural crops, ecosystems, or buildings. The most serious consequences occur when the elements at risk are human beings, and in the subsequent discussion of vulnerability it will be assumed that the situation being considered is that of a population of human beings at risk from one or several geophysical hazards.

7. VULNERABILITY

According to Blaikie et al. [1994], vulnerability is a person's or group's capacity to anticipate, cope with, resist and recover from the impact of a natural hazard. The concept is straightforward, but quantifying it is more difficult [*Brooks and Adger*, in press].

We can consider societal vulnerability in terms of two main classes (Figure 2):
- Societies with an ability to recover from a disaster (Figures 2a, b, c);
- Societies that are unable to recover from a disaster (Figures 2d, e).

Ecologists have long studied ecosystems and the attributes that enable an ecosystem to survive a major disturbance. The two most important attributes, according to Krebs [1978], are stability and resilience.

Resilience is a measure of the ability of a system to persist in the presence of perturbations arising from an impact. In ecology, resilience is measured by the probability of extinction [*Krebs*, 1978: p. 510]. In the context of geophysical risks, resilience is used to refer to the margin of safety between the existing situation, and the threshold at which recovery is no longer possible for a particular impact scenario.

Quantifying such thresholds for recovery is difficult when dealing with human beings and human societies. However, considerable progress is being made in terms of appropriate ecological thresholds related to climate change [*Jones*, 2001, 2003].

7.1 Impact and Resilience

The magnitude of an impact will determine whether or not an individual or a society can recover from the consequences of the impact. However, the magnitude of an impact is not the sole determinant. The key issue is that of the magnitude of the impact in relation to the resources of the individual, or the society. As explained above, we shall use the term "resilience" for this. Rich societies, which have a high per-capita gross domestic product, are more resilient than poor societies. Bangladesh, for example, is less resilient to the consequences of a typhoon than is the United States to the consequences of a hurricane of equal central pressure and wind speed.

Obviously, there will be some impact from which even the wealthiest of societies are unable to recover. In part, the international attention that is being focussed on the problem of climate change is based on a fear that this could be an impact from which wealthy societies may have difficulties in recovering. Similar concerns arise in relation to the possibility of a large asteroid impact that may send excessive quantities of dust into the stratosphere. Here the long term blocking of solar radiation leads to a "nuclear winter" scenario [*Covey et al.*, 1994] and the collapse of agriculture.

7.2 Stability, Connectivity and Organisation

Multiple disasters can push societies below a recovery threshold, so that the return periods and sequences of disasters are also important.

In ecology, stability is a dynamic concept that refers to the ability of a system to recover from a disturbance. By definition, a resilient population is stable. However, the degree of stability is measured by the time that it takes for the population to recover. High stability is characterised by a short recovery time, low stability by a long recovery time.

The ecological literature is replete with studies that illustrate that solitary species and gregarious species have different stabilities [*Krebs*, 1978: p. 506], whereas stability is itself, in ecological terms, a measure of community organisation. These ideas can be directly translated into variables that are of relevance to geophysical vulnerability, noting that stability is a function of both connectivity and organisation.

Organisation, in this instance, refers to the linkages within a society, or to the methodical approach of an individual. Connectivity refers to the linkages that a society makes with other societies, or that an individual makes with other individuals. Connectivity measures the ability of a country, such as Bangladesh, to access aid after a natural disaster. The ability to make optimal use of that aid is determined by the organisation within the society. Large, disparate, diverse societies are

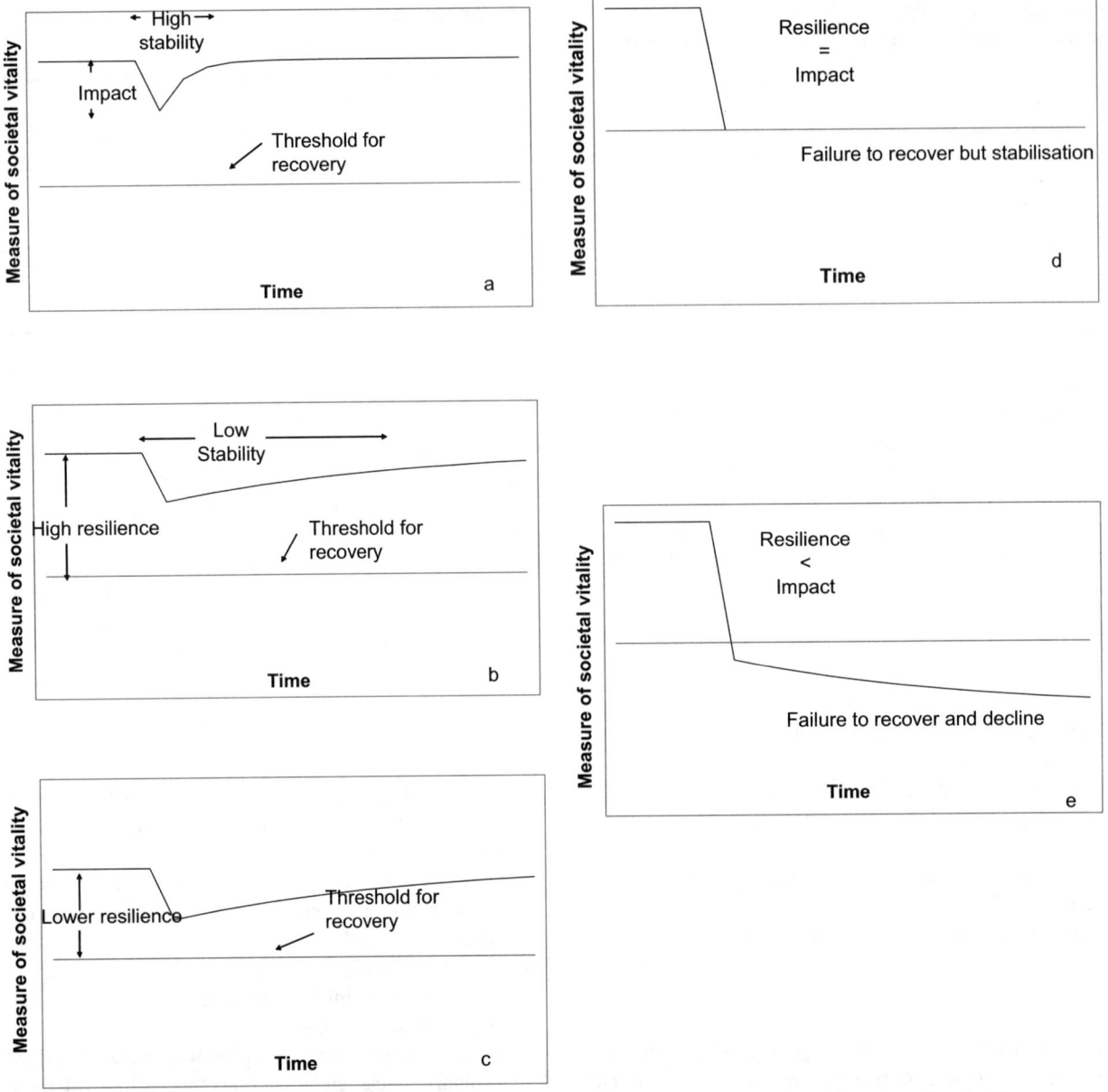

Figure 2. Diagrammatic representation of the variables that determine societal vulnerability. Diagrams a, b and c illustrate the situation in which a society can recover from an extreme event. Diagrams d, e illustrate the situation when the impact is so severe that recovery is no longer possible.

more resilient. Reddy [2000], in a study of long-term recovery following Hurricane Hugo in the United States, found that communities with strong leadership and stakeholder involvement had the highest level of recovery. These are both measures of internal organisation.

The definitions, frameworks and theoretical constructs outlined here are a prelude to the provision of actions designed to minimise the risks associated with geophysical vulnerability. The chance of success improves when scientists and decision-makers can agree on a common vocabulary and a common framework for action.

8. RISK MANAGEMENT AND ASSESSMENT

To undertake appropriate risk management resulting from geophysical risks, both the geophysicist and the decision-maker must be involved. To proceed they both need a framework within which to operate. Beer [2003] examined various environmental risk management frameworks and attempted a typology of frameworks to determine the key attributes that enable a practitioner to decide which framework to use.

There are three attributes. They are:

1. Whether the focus is on episodic (one-off, or oft-repeated) or chronic (quasi-continuous) events. Most natural hazards are examples of episodic events, whereas the chemicals involved in air pollution are examples of chronic events.

2. Whether the focus is on individuals or on populations. Geophysical risk generally focuses on populations.

3. Whether the focus deals with the hazard or with the receptor (e.g. infrastructure or buildings). In this paper, and generally when vulnerability is being examined, interest is in the receptor. This contrasts with geophysical research where the focus is on the hazard. If we ask the question: "what are the myriad risks to which an individual or society is subject?", then we find that the answer to that question may well involve different frameworks yet again.

Castanos and Lomnitz [2003] note that the present paradigm of natural disasters is no longer adequate and they advocate a cross-disciplinary scientific approach to examine such phenomena. The Budapest Manifesto [*Beer and Ismail-Zadeh*, 2003: p. xv], parts of which are reproduced in the next section, was an attempt to find a generic framework that would be suitable for a cross-disciplinary approach in which physical scientists and social scientists have to deal with issues involving risk, vulnerability and sustainability. The Budapest Manifesto was endorsed in late 2002 by the Council of Euroscience and by the Bureau of the International Union of Geodesy and Geophysics [IUGG]. The full text may be found in Beer and Ismail-Zadeh [2003] and on the IUGG web site at www.iugg.org/budapest.pdf . The operationally relevant sections of the manifesto are reproduced below.

9. EXTRACT FROM THE BUDAPEST MANIFESTO ON RISK SCIENCE AND SUSTAINABILITY

An appropriate framework within which to study environmental risk and sustainability needs to be sufficiently flexible to incorporate the diverse aspects included in these terms, yet sufficiently well defined to be able to treat the vulnerabilities to which human and environmental systems are exposed. The methods and tools used to examine natural risk can be applied to the analysis of geo-political risk.

Living in an often turbulent and unpredictable public environment, we scientists can contribute to decision-making through a risk management framework with which to examine technical and social issues related to sustainability that consists of the following:

- Anticipating man-made and natural risks through widespread **consultation**.
- Determining **concerns** by using risk assessment techniques for various scenarios.
- Identifying the **consequences** by systematically cataloguing hazards.
- Undertaking **calculations** with appropriate models.
- Evaluating the **certainties,** uncertainties, and the probabilities involved in the calculations of the vulnerability and of the exposure.
- **Comparing with criteria** to assess the need for further action.
- Determining and acting on options to **control,** mitigate and adapt to the risk.
- **Communicating** the results to those who need to know.
- Promoting and guiding **monitoring** systems to collect, assimilate and archive data relevant to the determination of sustainability and risk, now and in the future.
- Integrating the knowledge and understanding from all relevant disciplines to provide society with the tools to **review** the sustainability and the risks of proposed policies and plans.

Though rational scientific methods hold the promise of an improved science of risk and sustainability, it must be remembered that the priorities for analyses are likely to be heavily influenced by the public and political agenda of the day. This means that implementation of risk management to achieve sustainability can be achieved only through an interaction of theory and praxis.

Acknowledgments. I am grateful to Steve Sparks and his reviewing team for useful and helpful criticism. In particular, an anonymous reviewer pointed out that Wynne's identification of risk with uncertainty was based on a definition and concept of risk that is at variance with the one described in this paper.

REFERENCES

Anand, P., Decision-making when science is ambiguous, *Science*, 295: 1839, 2002.

Aspinall, W. P., Loughlin, S. C., Michael, F. V., Miller, A. D., Norton, G. E., Rowley, K. C., Sparks, R. S. J., and S. R. Young, The Eruption of Soufriere Hills Volcano, Montserrat, from 1995 to 1999, *Mem. Geol. Soc. London*, **21**, 71-91, 2002.

Baxter, P .J., Bonadonna, C., Dupree, R., Hards, V. L., Kohn, S. C., Murphy, M. D., Nichols, A., Nicholson, R. A., Norton, G., Searl, A., Sparks, R. S. J., and B. P. Vickers, Cristobalite in volcanic ash of the Soufriere Hills Volcano, Montserrat, British West Indies, *Science*, **283**, 1142-1145, 1999.

Beer, T., *Applied Environmetrics Meteorological Tables*, Applied Environmetrics, Balwyn, 1990.

Beer, T., Air quality as a meteorological hazard, *Natural Hazards* **23**, 157-169, 2001.

Beer, T., Environmental risk and sustainability, in Beer, T. and A. Ismail-Zadeh, [eds.] *Risk Science and Sustainability*, Kluwer Academic Publishers, Dordrecht, pp. xv–xvi, 2003.

Beer, T., Allan, R., and B. Ryan, Modelling Climatic Hazards, in A. J. Jakeman, M. B. Beck, and M. J. McAleer [eds.], *Modelling Change in Environmental Systems*, John Wiley & Sons, Chichester, pp. 453-477, 1993.

Beer, T. and F. Ziolkowski, Environmental Risk Assessment: an Australian perspective, Report 102, Supervising Scientist, Barton, ACT, 1995.

Beer. T. and B. Foran, Management for the Future: Risk management, future options and scenario analysis, in T. Beer [ed.], *Risk Management and the Future*, Australian Minerals & Energy Environment Foundation, Melbourne, pp. 39-67, 2000.

Beer, T. and A. Ismail-Zadeh, *Risk Science and Sustainability*, Kluwer Academic Publishers, Dordrecht, pp. 39-62, 2003.

Blaikie, P. M., Cannon, T., Davis, I. and B. Wisner, *At Risk: Natural Hazards, People's Vulnerability and Disasters*, Routledge, London, 1994.

Bloch, M., *The Historians Craft*, Manchester University Press, Manchester, 1954.

Brooks, N and W. N. Adger, Country level risk measures of climate-related natural disasters and implications for adaptation to climate change, Tyndall Centre Working Paper 26, University of East Anglia, Norwich, UK. [http://www.cru.uea.ac.uk/%7Ee118/publications/publications-subj.html#climatechange].

Brown, T., *Features, Events and Processes in SZ Flow and Transport*, Report ANL-NBS-MD-000002, Office of Civilian Radioactive Waste Management, US Department of Energy, 2000. [http://www.ymp.gov/documents/amr/22222/22222.pdf]

Bunde, A., Kropp, J. and H. J. Schellnhuber, *The Science of Disasters*, Springer, Berlin, 2002.

Castanos H. and C. Lomnitz, Disasters at the interface of nature and society provoke thought, *EOS Transactions of the American Geophysical Union*, **84**, 521, 2003.

Changnon, S. A., Shifting economic impacts from weather extreme in the United States: a result of societal changes, not global warming, *Natural Hazards*, **29**, 273-290, 2003.

Chowdhury, M. R., An assessment of flood forecasting in Bangladesh: The Experience of the 1998 Flood, *Natural Hazards*, **22**, 139-163, 2000.

Clark, W. C. and R. E. Munn, *Sustainable Development of the Biosphere*, IIASA and Cambridge University Press, Cambridge, 1986.

Coles, S., *An Introduction to Statistical Modeling of Extreme Values*, Springer-Verlag, London, 2001.

Cooke, R. M., *Experts in Uncertainty—opinion and subjective probability in science*, Oxford University Press, New York, 1991.

Covey, C. Thompson, S. L., Weissman, P. R., and M. C. MacCracken, Global climatic effects of atmospheric dust from an asteroid or comet impact on Earth, *Global and Planetary Change* **9**, 263-273, 1994.

Dunbar, P. K., Bilham, R. G. and M. J. Laituri, Earthquake loss estimation for India based on macroeconomic indicators, in T. Beer and A. Ismail-Zadeh [eds.] *Risk Science and Sustainability*, Kluwer Academic Publishers, Dordrecht, pp. 39-62, 2003.

Fournier d'Albe E.M., Objectives of volcanic monitoring and prediction. *Journ.. Geol. Soc. Lond.* **136**, 321-326, 1979.

Glenn, J. C. and Gordon, T. J., *State of the Future*, American Council for the United Nations University, Washington DC, 1997.

Glenn, J. C. and Gordon, T. J., *State of the Future*, American Council for the United Nations University, Washington DC, 1998.

Grübler, A. and Nakicenovic, N., Identifying dangers in an uncertain climate, *Nature* **412**, 15, 2001.

Gumbel, E. J., *Statistics of Extremes*, Columbia University Press, New York, 1958.

Gupta, H. K., A review of recent studies of triggered earthquakes by artificial water reservoirs with special emphasis on earthquakes in Koyna, India, *Earth-Science Reviews,* **58**, 270-310, 2002.

Harding, R., *Environmental decision-making: the roles scientists, engineers and the public*, The Federation Press, Leichardt, NSW, 1998.

Harding, R. & Fisher, E. [eds], *Perspectives on the Precautionary Principle*, The Federation Press, Leichardt, NSW, 1999.

IPCC [Intergovernmental Panel on Climate Change], *Climate Change 2001 Synthesis Report*, Watson, R. T. et al. [eds.] Cambridge University Press, Cambridge, 2001.

Jeffreys, H., *Theory of Probability Second Edition*, Oxford University Press, Oxford, 1948.

Jones, R. N., An environmental risk assessment/management framework for climate change impact assessments, *Natural Hazards*, **23**, 197-230, 2001.

Jones, R. N., *Managing Climate Change Risks*, OECD Report ENV/EPOC/GSP[2003]22/FINAL [http://www.oecd.org/dataoecd/6/12/19519189.pdf], 2003.

Kates, R. W., Clark, W. C., Corell, R., Hall, M. J., Jaeger, C. C., Lowe, I., Schellnhuber, H. J., Bolin, B., Dickson, N. M., Faucheux, S., Gallopin, G. C., Grübler, A., Huntley, B., Jäger, J., Jodha, N. S., Kasperson, R. E., Mabogunje, A., Matson, P., Mooney, H., Moore, B., O'Riordan, T., and U. Svedlin, Sustainability Science, *Science* **292**, 641-642, 2001.

Kotz, S. and S. Nadarajah, *Extreme Value Distributions—Theory and Applications*, Imperial College Press, London, 2000.

Krebs, C. J., *Ecology: The Experimental Analysis of Distribution and Abundance*, 2nd edition, Harper & Row, New York, 1978.

Leadbetter, M. R., Lindgren, G., and Rootzen, H., *Extremes and Related Properties of Random Sequences and Processes*, Springer Verlag, New York, 1983.

Lorenz, E. N., Deterministic nonperiodic flows, *J. Atmos. Sci.*, **20**, 131-141, 1963.

Lorenz, E. N., *The Essence of Chaos*, UCL Press, London, 1993.

Mahlman, J., Science and nonscience concerning human-caused climate warming, *Ann. Rev. Energy Environment*, **23**, 83-105, 1998.

Matthews, A. J., Barclay, J., Carn, S., Thompson, G., Alexander, J., Herd, R., and C. Williams, Rainfall induced volcanic activity on Montserrat, *Geophysical Research Letters*, **29**, [13], 1644, 10.1029/2002GL014863, 2002.

McInnes, K. L., Walsh, K. J. E., Hubbert, G. D. and T. Beer, Impact of sea-level rise and storm surges on a coastal community, *Natural Hazards*, **30**, 187-207, 2003.

Moon , I. J., Oh, I. S., Murthy, T. and Y. H. Youn, Causes of the unusual coastal flooding generated by Typhoon Winnie on the west coast of Korea, *Natural Hazards*, **29**, 485-500, 2003.

Morgan, M. G.., and M. Henrion, *Uncertainty: a Guide to Dealing with Uncertainty in Quantitative Risk and Policy Analysis,* Cambridge University Press, New York, 1990.

Morgan, M. G.., and D. W. Keith., Subjective judgments by climate experts, *Environmental Science and Technolgy,* **29** [10] 468A-476A, 1995.

Munich Re, Topics 2000—Natural catastrophes: the current position, Muenchener RueckversicherungsGesellschaft, Munich, 126p, 2000.

Munich Re, Topics 2002—Annual review: Natural Catastrophes 2002, Muenchener RueckversicherungsGesellschaft, Munich, 126p, 2003.

O'Riordan, T. and A. J. McMichael, Dealing with scientific uncertainties, in Martens, P. and McMichael, A. J. [eds.] *Environmental Change, Climate and Health,* Cambridge University Press, Cambridge, pp. 311-332, 2002.

Pittock, A. B., Climate scenario development, in A. J. Jakeman, M. B. Beck, and M. J. McAleer [eds.], *Modelling Change in Environmental Systems,* John Wiley & Sons, Chichester, pp. 481-503, 1993.

Pittock, A. B., Jones, R. N., and C. D. Mitchell, Probabilities will help us plan for climate change, *Nature* **413**, 249, 2001.

Reddy, S. D., Factors influencing the incorporation of hazard mitigation during recovery from disaster, *Natural Hazards,* **22**, 185-201, 2000.

Ricci, P., Rice, D., Ziagos, J., and L. A. Cox, Precaution, uncertainty and causation in environmental decisions, *Environment International,* **29**, 1-19, 2003.

USEPA, *Terms of Environment* http://www.epa.gov/OCEPAterms/rterms.html, 2004.

Vose, D., *Risk Analysis—a quantitative guide*, John Wiley & Sons Ltd, Chichester, UK, 2000.

Woo, G., The Mathematics of Natural Catastrophes, Imperial College Press, London, 1999.

World Commission on Environment and Development [WCED] and Commission for the Future [1987] *Our Common Future* [Australian edition, 1990][This document is also known as the Brundtland Report], Oxford University Press, Melbourne, 1990.

Wynne, B., Uncertainty and environmental learning: reconceiving science and policy in the preventive paradigm, *Global Environmental Change,* **3**, 111-127, 1992.

Tom Beer, CSIRO Environmental Risk Network, Private Bag 1, Aspendale, Vic. 3195, Australia (tom.beer@csiro.au)

People Induced Geophysical Risks and Urban Sustainability

Ian Douglas

School of Geography, University of Manchester, Manchester, United Kingdom

People-induced geophysical risk is increasing as populations grow, technology advances and urban settlements expand. Some increases in risk, such as those of seismicity induced by reservoirs, underground mining, removal of fluids [oil or water], and deep excavations are well documented. Land reclamation and artificial landform creation lead to new risks. Nonetheless, many aggravated, or apparently new, geophysical risks are more difficult to trace back to specific human activities. Upland land clearance or mining activity is often blamed for flooding in lowland urban areas and deltas. However, while land clearance locally changes river regimes and increases sediment yields, downstream modification of river channels and occupation of floodplains increases human vulnerability. Dams and reservoirs trap sediments and modify flows. Catastrophic floods may involve no more runoff than previous rare floods, but become catastrophic because of the numbers of people at risk and the infrastructure and resources that are now vulnerable when levees are overtopped. Evaluation of geophysical risks must take an integrated approach, to ensure that the resources available for risk reduction and hazard mitigation are spent in the most effective manner.

INTRODUCTION

By 2000 over half the world's people were living in urban areas, most of them in towns and cities of less than one million, but many in huge agglomerations of over 5 million. The world's future environmental and resource security depends on the demands and actions of consumers and industries in those urban areas. If the well-being of future generations is to be assured, the environment has to be managed more sustainably. Such management requires cities to reduce their environmental impacts, both within the city and on the rest of the world with regard not only to the present, but also for the future. Urban sustainability, a community's ability to develop and/or maintain a high quality of life in the present in a way that provides for the same in the future [Shafer et al., 2000], is thus a process of recognising likely consequences of human action and introducing alternative strategies for using land, and managing soils, water and biota. It needs careful use of earth resources, sound hillslope engineering, and integrated management of hydrologic systems and soils. Avoiding or minimising geophysical hazards is a key part of this urban sustainability process.

Geophysical hazards broadly include all forms of earthquakes, landslides, floods and associated erosion and sedimentation, land subsidence, volcanic activity, extreme weather events and tsunami. In and around cities, some of these hazards occur as direct consequences of human action or as a result of major modifications of natural features by people and machines. All activities that change the stresses within, and stability of, the rocks, weathered regolith and soils that support urban areas have the potential to induce geophysical risks. The processes linked to urban development, and to the mining and quarrying that supports urban life, now involve a deliberate movement of earth surface materials that may exceed the natural transport of material to oceans by rivers by 3 to 4 times [Douglas and Lawson, 2002]. This enormous shift of materials involves excavations and void creation in one set of places and accumulation and stocking of materials in others. Local changes of pressure on the earth's surface and substrates result from this activity. The stocked materials include water in reservoirs and waste in dumps, landfill and at landraise sites. Each stock creates potential risks. Most are minor. Some are life threatening.

Urban activities intensify or re-activate many natural geophysical risks, for example when hillsides are terraced and water flows altered to allow for housing development. Paving and roofing the land surface, building drains, modifying stream channels and straightening river channels all alter flood risks. The impacts of these changes vary with environmental, geologic, social and economic conditions. Seismic activity related to surface loading or subsurface void creation is more likely to occur in fractured rocks than in little disturbed strata. Poor people in ill-constructed dwellings suffer more than wealthy people in well-built houses. The ability to apply and assure appropriate earth science and engineering solutions to these risk reduction issues varies greatly from community to community and country to country. Although the bulk of the world's urban people live in Asia, Latin America and Africa, the majority of the published, accessible case-studies of these urban geophysical risks come from North America, Europe and a few Asian countries. Good documentation on places like Mexico City, San Paulo, Rio de Janeiro, Beijing, Shanghai, Bangkok and Jakarta hides our lack of knowledge about the issues in the smaller, but equally rapidly growing urban centres, where most of the urban people live.

Improved methods of assessing these risks distinguish assessment of individual risks at specific locations from multi-criteria assessments of societal and individual risks [Luria and Aspinall, 2003]. Assessment of individual risks, such as landslide hazard, usually involves ranking likely frequency and magnitude of occurrence and mapping the distribution of risk classes. Geologic hazard models combine several types of risk into integrated assessments, often using Geographic Information Systems (GIS) [Mejía-Navarro et al., 1994, Zerger, 2002].

Risk management plans take account of design, security (protection of the public, operators and facilities), monitoring and inspection systems, maintenance programs, and staff training and supervision [ICOLD Committee on Tailings Dams and Waste Lagoons, 2001]. Effective risk assessment and management is a key component of making urban areas more sustainable. Good environmental and earth resource management is a major contributor to improving the quality of lives of all urban people. Greater sustainability requires reduction of the losses and social disruption from hazards, especially those that are induced by human activities that good applications of earth science would help to avoid.

SEISMICITY INDUCED BY HUMAN ACTIVITIES

Several types of engineering and extractive industry activities can trigger limited localized earthquakes known as induced-seismicity. Examples include increased or accelerated activity in seismically active areas or the onset of seismicity in aseismic areas. Earthquakes occur whenever the stresses acting on a point in the lithosphere exceed the strength of rock at that point. The scientific study of induced-seismicity began when earthquakes started to be felt in Johannesburg in 1894. By 1908 it was recognized that the gold mining initiated in 1886 was causing these events. The earth movements in this naturally highly stable area increased as the mines became deeper and disturbed the stress field in the Archean rocks [Guha, 2000].

The main causes of induced seismicity are:

1. Reservoir storage: the impounding of large amounts of water behind dams

2. Extraction of minerals: quarrying at the surface, shallow mine workings, deep underground workings

3. Detonation of underground explosions

4. Fluid removal: the extraction of oil or the pumping of water from an aquifer

5. Fluid injection: the injection of high-pressure fluids into the ground, as in solution mining, waste disposal, geothermal power exploitation, and secondary oil recovery.

Most reservoir seismicity is small and as yet there have been only 4 cases which have exceeded Moment Magnitude scale level [M]=6 [Table 1]. The 1967 Koyna earthquake in Maharashtra state, India, claimed over 200 lives, injured over 1,500, rendered thousands homeless, and destroyed much of the infrastructure in the Koyna Nagar town. The region south of the Shivaji Sagar reservoir has the highest reservoir induced seismicity (RIS) in the world, with more than 150 events of magnitudes M_L 4.0 occurring from 1962 to 1998 [Gupta et al., 2000, 2002]. The activity is still continuing. Remarkably, most of the earthquakes are concentrated in a 25 km ? 40 km area [Gupta, 1992; Gupta et al., 1997]. In the 1990s, a second reservoir, the Warna, one-third the size of the Shivaji Sagar reservoir, was completed some 30 km south of the Koyna Dam. It also appears to have created considerable RIS [Gupta

Table 1. The ten largest reservoir-induced earthquakes

Date	Origin Time UTC	Latitude	Longitude	Magnitude M	Location
11 December 1967	22:51:19	17.400 N	73.740 E	6.3	Koyna, India
05 February 1966	02:01:45	39.050 N	21.750 E	6.2	Lake Kremasta, Greece
19 March 1962		23.780 N	114.580 E	6.2	Xinfengjiang, China
20 July 1938	00:23:35	38.290 N	23.790 E	6.0	Oropos, Greece
22 April 1983		14.400 N	99.130 E	5.8	Srinagarind, Thailand
01 August 1975	20:20:12	39.440 N	121.530 W	5.8	Oroville, U.S.A.
23 September 1963		16.930 S	27.930 E	5.8	Kariba, Zimbabwe
13 April 1969	17:58:39	18.100 N	80.500 E	5.7	Kinnersani, India
14 September 1941	18:39:12	37.570 N	118.730 W	5.4	Lake Crowley, U.S.A.
04 October 1978	16:42:48	37.520 N	118.660 W	5.4	Lake Crowley, U.S.A.

et al., 1997]. Although underlain by approximately 1.5 km of basalt, this 400 km-long area between Mumbai and Goa, has a long record of natural seismicity with at least 15 earthquakes being felt between 1594 and 1832 [Gupta et al., 2000]. The reservoirs appear to have intensified the frequency of seismicity in the area.

The start of water storage behind the 105m high Hsinfengkiang dam in China in 1959 led to new seismicity a month later. During the first few months, seismicity increased as the water level rose. The main shock of 1962 occurred in an area where NNW faults intersected an ENE fault system creating cracks in the dam itself, the riverbank and local buildings. The local rocks have well developed joints. All events occurred within 5 km of the reservoir.

In 1958 Lake Kariba became the largest sustained artificial increase in loading on the earth's crust as it filled with water slowly from 1958 to 1963. In 1963 several large earthquakes occurred near the deepest portion of the reservoir [Meade, 1991]. The earthquake mechanism was dip-slip on normal faults. Later earthquakes in the area, in 1966–68, appear to have been unconnected with the reservoir.

The level of the Kremasta reservoir in Greece repeatedly went up and down between 1967 and 1972. Careful observations by the project engineer showed that these fluctuations corresponded with levels of increased seismic activity [Therianos, 1974]

EXTRACTION OF MINERALS

The removal of large masses of rock in mining and surface quarrying induces subsidence and seismicity primarily by changing the elastic stress in the surrounding rocks. The total stress state around a mine excavation is the sum of the ambient stress state in the rock mass and the stresses induced by mining. The ambient stress is lithostatic, corresponding to the weight of the overburden. Hence there is an increase in stress with depth. Mining induced earthquakes of $M = 5.7$ have been associated with potash workings, while subsidence occurs in areas of underground coal and salt extraction. Mining induced earthquakes occur fairly often, the US Geological Survey reporting 6 of magnitude $M = 4$ or greater between 1973 and 2000 [Table 2].

Mining Induced Earthquakes

Crustal unloading due to quarrying has triggered minor earthquakes in New York State along with horizontal compressive stress [Guha, 2000]. At a quarry in Wappingers Falls, New York,

Table 2. Mining related earthquakes in the USA January 1973-March 2000 that exceeded $M = 4.0$ [after: http://neic.usgs.gov/neis/mineblast/induced_pde.html]

Date	Time	Latitude	Longitude	Magnitude
7 May 1986	0227	33.233	-87.361	4.2
14 Apr 1988	2337	37.238	-81.987	4.1
03 Feb 1995	1526	41.259	-109.640	5.3
29 Jun 1996	1930	37.187	-81.950	4.1
18 Jan 1999	0700	33.405	-87.255	4.8
30 Jan 2000	1446	41.464	-109.679	4.4

a shallow, 0.5 to 1.5 km deep earthquake sequence occurred in 1974 with a main shock of magnitude M = 3.3. These thrust earthquakes were located directly below the floor of the quarry which had been worked for 75 years down to a depth of 50 m, providing a stress change, albeit over many years, of 1.5 MPa [Megapascals]. As this region is an area of high horizontal compression, removal of surface material equated to reducing the normal stress, so leading to the appearance of pop-up structures in the quarry floor [Pomeroy et al. 1977].

Rockbursts

Rockbursts in deep mines in the ancient shield areas of South Africa, India and Canada have sometimes caused widespread damage to mining structures and surface residential buildings [Fairhurst, 1990; Young, 1993]. For example, in the Kolar goldfield, near Bangalore, India major rockbursts occurred in 1952, 1956, 1960, 1962, 1963, 1966, 1971, 1972, 1983 and 1985 [Guha, 2000]. Mine rockbursts may be divided into two types, those influenced by mining operations, extraction rate and blasting schedule [Type 1 Table 2], and those depending on the interaction of a mine-induced stress field with a tectonic stress field and pre-existing faults [Type II Table 3].

Mining Induced Subsidence

Most mining areas have suffered some degree of surface subsidence. Mine subsidence also can cause foundation damage to buildings, disrupt underground utilities, and be a potential risk to human life (Figure 1). Coal mining subsidence has affected large urban areas in much of northern Europe and North America [Bishop et al., 1993; Siriwardane, 1989], while that due to gold mining in South Africa remains a considerable problem. In Pennsylvania USA alone, there are over 800 incidents of mine subsidence each year [Voros et al., 2001]. This has led to much monitoring, analysis and mitigation of the associated risks. In Ohio, for example, the history of coalmine subsidence problems dates back to at least 1923 [Crane, 1931; Crowell, 1995; DeLong, 1988]. Several Ohio communities, such as Wellston (Jackson County) and North Canton (Stark County), have been plagued with numerous mine-subsidence problems. Generally, mine subsidence affects very few people. But, if a mine collapses under an interstate highway, many lives and industries are affected. Mine subsidence in March 1995 caused a portion of the eastbound lane of Interstate Route 70 in Guernsey County to collapse. This subsidence event and the ensuing repair work closed the eastbound and westbound lanes of I-70 for several months, and the cost of the repair work was estimated at $3.8 million [Crowell, D. L., 1995].

SUBSIDENCE DUE TO FLUID REMOVAL

Probably the most pervasive and widespread induced urban geophysical risks are those associated with the extraction of fluids. Pumping of groundwater beneath urban areas can induce land subsidence (as in Bangkok [Rau and Nutalaya, 1982], Mexico City [Durazo and Farvolden, 1989], Shanghai, and Venice) and cause flooding, especially of coastal communities near sea level. The classic USA examples of subsidence include the Houston–Galveston area of Texas where falling land levels have created severe and costly coastal-flooding hazards and affected a critical environmental resource, the Galveston Bay estuary [Holzer, 1984], and the Las Vegas Valley, Nevada, where ground-water depletion and associated subsidence have accompanied the conversion of a desert oasis into a thirsty and fast-growing metropolis [Péwé, 1990].

The European classic example of Venice is part of the wider central–eastern Po Plain system in northern Italy, a rapidly subsiding sedimentary basin that hosts about 30% of the Italian population and 40% of Italy's total productive activities. Subsidence rates range from 0 to -70 mm/year, the maximum

Table 3. Types of mine rockburst [after Guha, 2000]

Type I	Type II
Rate of occurrence generally is a function of mining activity	Insufficient data available to determine relationship to mining rate
Location is generally within 100 m of mining face or of some pre-existing zone of weakness or geological discontinuity near the mine	Location is on a pre-existing fault surface that may be as much as 3 km from the mine
Intact rock may be broken in the rupture if mining stresses exceed the shear strength of the original material. Orientation of the rupture planes may vary	All occur in pre-existing, possible pre-stressed, tectonic faults. Mining may simply trigger these events on faults of preferred orientations
Often high stress drops are recorded	Stress drops are similar to those in natural earthquakes
Low to medium magnitudes	Potential for high magnitude events

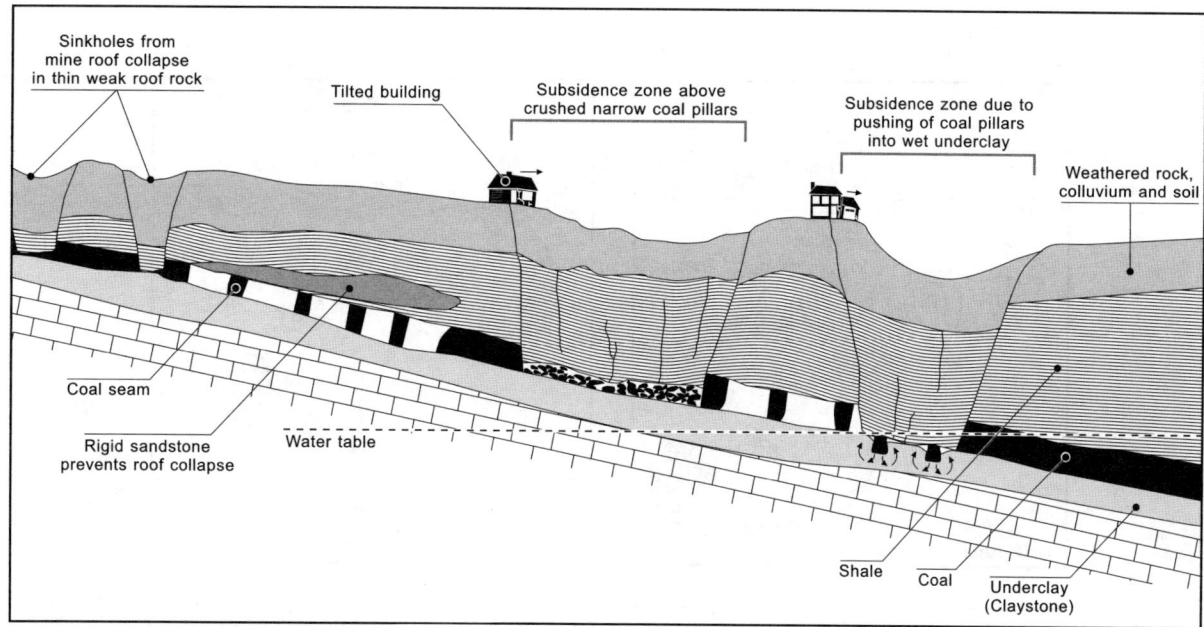

Figure 1. Patterns of sinkhole development over coal mineworkings, showing variations with mining method and the local stratigraphy (after DeLong, 1988).

occurring in synclinal areas in the Po Delta and near Bologna, and the minimum above buried, probable tectonically active, anticlines. Present-day subsidence is at least an order of magnitude higher than that due solely to long-term natural processes, suggesting that most subsidence in the Po Plain has been induced by human activities (Figure 2). The main factor leading to accelerated subsidence is water withdrawal, which was particularly intense during rapid economic development in the second half of the 20th century. Comparison of subsidence rate maps with flood maps reveals a clear correlation between flood frequency and rapid subsidence [Carminati and Martinelli, 2002].

Greater Tianjin, China, a conurbation of 9 million people occupies part of a flat coastal plain overlying a thick alluvial sequence. Groundwater extraction has lead to some 7000 km^2 being affected by land subsidence, with rates of over 100 mm y^{-1} in places, and a maximum total land surface decline of 2.6 m. Since 1982 groundwater pumping has been reduced, cutting the subsidence rates to 10 to 20 mm y^{-1}. Shanghai subsided as much as 2.8 m in the 20th century due to groundwater withdrawal.

Groundwater tables in the Greater Jakarta metropolitan area of West Java have been sinking rapidly for two decades, with a host of related problems, including subsidence and saltwater penetrations affecting the sustainability of urban water supplies. One of the problems in many of these Asian cities is that even if the municipal supply authorities stop pumping from the affected aquifers, illegal private pumping may continue. Where existing public supplies are inadequate and unreliable, it is often difficult to close down private sources. It is equally difficult for such inadequately resourced authorities to regulate groundwater use. Strengthening good governance and urban management in local authorities thus becomes a key part of sustainable environmental policy implementation and geophysical risk avoidance.

In extreme cases, subsurface changes following the withdrawal of fluids have caused earthquakes. Since 1970 several small earthquakes have occurred in the Western Canada Sedimentary Basin. Many of the epicenters are shallow and are situated near areas of oil and gas production in the western side of the basin. The earthquake of October 19th, 1996, in the Strachan Field, in the Alberta foothills southwest of Rocky Mountain House was most probably triggered by gas extraction.

Although subsidence seldom causes loss of life, the cumulative costs of all forms of subsidence are high, of the order of $ 125 million per year for the U.S.A. alone. Damage in U.S. urban areas from mining-related subsidence is around $25 million per year, while oil extraction under Long Beach California caused £5 million worth of damage per year from 1937 to 1966. Shanghai suffers a loss of more than 10 million US$1.2 million for every millimeter it sinks. China's total land subsidence losses are about exceed US$12.1 million annually.

Subsidence is monitored by a combination of microseismic surveys and detailed ground level survey, commonly

392 PEOPLE INDUCED GEOPHYSICAL RISKS AND URBAN SUSTAINABILITY

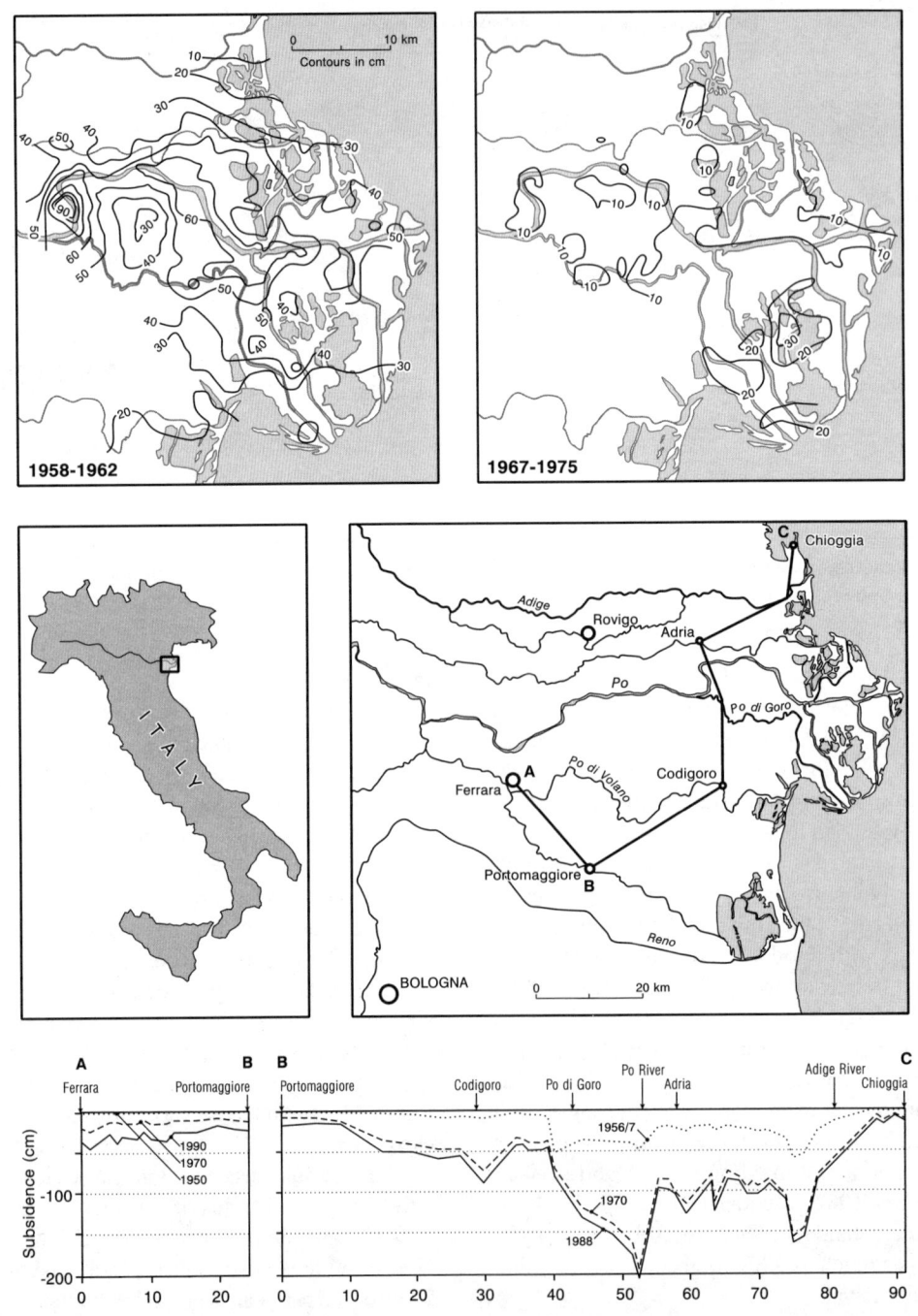

Figure 2. Subsidence in the Po Valley of Italy. The two uppermost diagrams show isopleths of surface lowering during (1958–1962) and after (1967–75) methaniferous-water exploitation (after Caputo, 1970 and Bondesan and Simeoni, 1983). The two lowest diagrams show the amount of subsidence in the second half of the twentieth century (after Bondesan et al., 2000) along the transect A-B-C shown on the centre right map of the Po delta.

Figure 3. Potentially unstable fill on a new housing development site in Kuala Lumpur in 1987. Note the land clearance and disturbance on the hillsides in the background.

Table 4. Loss of life from landslides on people made slopes (compiled from varied sources)

Date	Location	Deaths
	MODIFIED HILLSLOPES	
1966	Rio de Janeiro, Brazil (heavy rains on steep slope occupied by shack dwellers)	c. 1000
18 Jun 1972	Sau Mau Ping, Hong Kong (failure of cut and fill slope)	71
18 Jun 1972	Po Shan Road, Kong Kong (failure of built slope above new excavation)	67
1976	Sau Mau Ping, Hong Kong (failure of ill-compacted fill slope)	18
Oct 1985	The Mamayes, Puerto Rico (catastrophic block slide; drains channeling water into the ground, densely populated slope))	129
Sep 1987	Medellín, Colombia (failure of occupied slopes after heavy rains)	207
1988	Rio de Janeiro, Brazil (heavy rains on steep slope occupied by shack dwellers)	289
1993	Highland Towers, Ampang, Kuala Lumpur, Malaysia	48
20 Nov 2002	Taman Hillview, Ampang, Kuala Lumpur, Malaysia	8
	MINE WASTE TIPS AND TAILINGS DAMS	
21 Oct 1966	Aberfan, Wales, UK (failure of mine waste dump above town)	144
Sep 1970	Mufilira, Zambia (68,000 m^3 of waste entered mine workings)	89
Feb 1972	Buffalo Creek, U.S.A.	125
Nov 1974	Bafokeng, South Africa (3 million m^3 of slurry flowed 45 km)	12
July 1985	Stava, Italy (tailings flowed up to 8 km)	269
16 Oct 2003	Lake Wanagon, Grassberg Mine, West Papua, Indonesia	8
	MUNICIPAL WASTE DUMPS	
9 Jul 2000	"Promised Land" Payatas, Manila, Philippines	At least 205

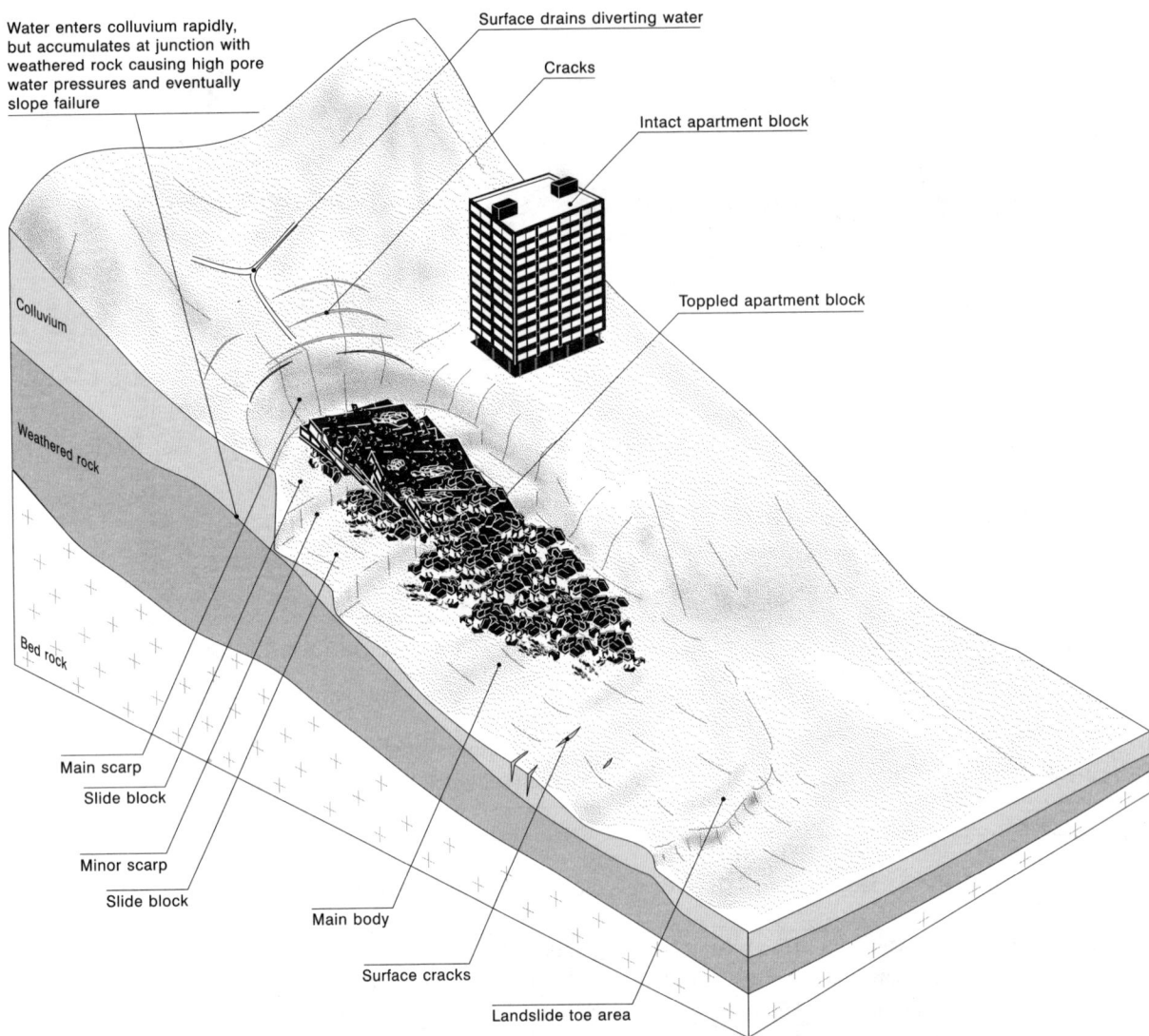

Figure 4. Diagram of an urban landslide typical of deeply weathered rocks in the humid tropics. The excavations for the tower block of apartments and the drainage diversion upslope of the building have concentrated water inflows to the surface soil and underlying colluvium. On reaching the contact with the in-situ weathered rock which is less permeable than the colluvium, water builds up and lubricates the junction, with a landslide occurring after an exceptionally heavy rainstorm. The tower block toppled as its foundations slipped downhill.

employing interferometric synthetic aperture radar (InSAR), which uses repeat-pass radar images from Earth-orbiting satellites to measure subsidence and uplift at high spatial detail (80 m x 80 m) and measurement resolution (sub-centimeter) [Prince and Galloway, 2000].

URBAN LANDSLIDES

Modification of hillslopes for urban development changes their stability, as in cut and fill slopes, often increasing the risk of landsliding (Figure 3). Retaining walls may divert water flows. Drains may lead to water entering the soil at defined points, rather than in a diffuse, dispersed manner. Disruption of the natural flow of water alters pore water pressures in the slope and leads to failure, particularly at the interfaces between made ground and colluvium, and between colluvium and in-situ weathered rock (Figure 4). Poorly consolidated fill is often liquefied following heavy rain, making poorly designed slopes likely to fail during downpours associated with hurricanes (Figure 5).

Losses of life from individual urban landslides are small (Table 4), but annual damage to property and social disruption

Figure 5. Remedial work on a failed slope in Damansara, Kuala Lumpur in 1986.

are cumulatively high. Urban landslide mitigation techniques are highly developed. The Hong Kong Geotechnical Control Office, for example, assesses the risk of landsliding through detailed mapping of slope, soils and weathering mantles and a careful appraisal of the engineering techniques applied to slopes. The Office's long-term "Slope Safety" programme aims to achieve the quickest possible reduction in landslide risk to the community in Hong Kong. Geotechnical practitioners in Hong Kong increasingly use Quantitative Risk Assessment (QRA) to evaluate and managing landslide risk. Successful application of these techniques in Hong Kong has halved the risks and losses from landslides since 1979.

CONCLUSIONS

Any form of urban land development alters the form of the slopes and the passage of water over the ground and into the weathering profile. Exposure and compaction of the ground surface increases the amount of surface runoff. Paving of roads and construction of buildings and parking areas renders large surfaces impermeable and redirects water to other places. The weight of artificial structures changes the stresses within the soil, altering slope stability. Earthmoving results in cut slopes whose internal pore water pressures may change and fill areas with less compact material than natural weathered slopes and a higher degree of erodibility. The extraction of minerals and the dumping of wastes creates new landforms that are often less stable than their natural equivalents (Figure 6). Removal of subsurface liquids and materials leads to subsidence, earth surface fissures, sinkhole formation through the collapse of the roofs of voids below ground and, sometimes, earthquakes.

As urban areas grow, particularly in Asia, Latin America and Africa, these problems are likely to occur with greater frequency and magnitudes. Much is known about the causes, consequences and mitigation of the people-induced geophysical risks, but the ability to apply that knowledge varies greatly from one city to another. Major challenges for the earth sciences as a whole are:

1. to increase the importance of urban earth science in the training of planners, architects, civil engineers and urban administrators;

2. to increase public awareness of the importance of understanding the dynamics of the environment, especially knowledge of the ground on which they build;

3. to assist and strengthen the application of urban earth science in towns and cities where municipal human and financial resources are small and many people have no option but illegally to build homes on unsuitable land;

4. to encourage lateral thinking that recognizes that the consequences of modifications of the earth surface may not be on site and immediate, but may occur offsite, downstream, or some distance away, years or decades in the future,

Figure 6. Potentially unstable outer slope of a large tailings dam near Ranau, Sabah, Malaysia. An exceptional rain event could disrupt such a slope and release a mudflow of fine-grained mine waste on to the surrounding area.

5. to set earth science applications in the context of increasing sustainability, especially through the recycling of materials, the better use of land, the design of ecological flood mitigation and slope stabilization schemes, and the integration of future sustainable land use alternatives into all mineral resource development and rehabilitation schemes.

REFERENCES

Bishop, I. Styles, P. and Allen, M. 1993 Mining-induced Seismicity in the Nottinghamshire Coalfield, *Quarterly Journal of Engineering Geology* 26 253–279.

Bondesan, M.. Gatti, M., and Russo, P. 2000 Subsidence in the Eastern Po Plain *Proceedings of the Sixth International Symposium and Land Subsidence (SISOLS). Ravenna (Italy) 24–29 September 2000*, Vol. II, 193–204

Bondesan, M. and Simeoni, U, 1983 Dinamica e analisi morfologica statistica dei litorali del delta del Po e alle foci dell'Adige e del Brenta. *Memoria società geologica d'Italia.*, 36, 1–48.

Caputo M., Pieri L. and Unghendoli, M. 1970- Geometric investigation of the subsidence in the Po Delta. *Bollettino di Geofisica Teorica ed Applicata*, 14 (47), 187–207

Carminati, E. and Martinelli, G. 2002 Subsidence rates in the Po Plain, northern Italy: the relative impact of natural and anthropogenic causation. Engineering Geology, 66, 241–255

Crane, W. R., 1931, Essential factors influencing subsidence and ground movement: *U.S. Bureau of Mines Information Circular 6501*, 14 p.

Crowell, D. L., 1995, The hazards of mine subsidence: *Ohio Division of Geological Survey, Ohio Geology*, Fall, p. 1–5.

DeLong, R. M., 1988, Coalmine subsidence in Ohio: *Ohio Division of Geological Survey, Ohio Geology*, Fall, p. 1–4.

Douglas, I. and Lawson, N. 2001 The human dimensions of geomorphological work in Britain. *Journal of Industrial Ecology,*4, 9–33.

Durazo, J. and Farvolden, R.N. 1989 The groundwater regime of the Valley of Mexico from historic evidence and field observations. *Journal of Hydrology*, 112, 171–190.

Fairhurst, C. (ed) 1990 *Proceedings 2nd International Symposium on Rockbursts and Seismicity in Mines, Minneapolis, Minnesota June 8–10 1988.* Balkema, Rotterdam.

Galloway, D.I., Jones, D.R., Ingebritsen, S.E. 2000, Land Subsidence in the United States. USGS Fact Sheet-165-00.

Guha, S.K. 2000 *Induced Earthquakes*. Kluwer, Dordrecht

Gupta, H.K. 1992 *Reservoir-induced earthquakes*. (Developments in Geotechnical Engineering Vol. 64) Elsevier, New York.

Gupta, H.K. 2002 A review of recent studies of triggered earthquakes by artificial water reservoirs with special emphasis on earthquakes in Koyna, India. Earth-Science Reviews, 58, 279–310

Gupta, H.K., Rastogi, B.K., Chadha, R.K., Mandal, P., and Sarma, C.S.P., 1997: Enhanced reservoir-induced earthquakes in Koyna region, India, during 1993–95. *Journal of. Seismology*, 1, 47– 53.

Gupta, H.K., Radhakrishna, I., Chadha, R.K., Kümpel, H.J., and Grecksch, G. 2000 Pore pressure studies initiated in area of reservoir induced earthquakes in India. *Eos, Transactions, American Geophysical Union*, 81 [14], 145,151.

Holzer, T.L. 1984 Ground failure induced by ground-water withdrawal from unconsolidated sediment. *Geological Society of America, Reviews in Engineering Geology*, VI, 67–105.

ICOLD Committee on Tailings Dams and Waste Lagoons. 2001 *Tailings Dams: Risk of Dangerous Occurrences: Lessons Learnt from Practical Experiences*. Commission Internationale des Grands Barrages, Paris.

Luria, P. and Aspinall, P.A. 2003 Evaluating a multi-criteria model for hazard assessment in urban design. The Porto Marghera case study. *Environmental Impact Assessment Review.* 23 625–653.

Meade, R.H. 1991 Reservoirs and earthquakes. *Engineering Geology,* 30, 245–262 (Reprinted in Goudie, A. 1997 *The Human Impact Reader: Readings and Case Studies*, Blackwell, Oxford, 33–46).

Mejía-Navarro, M., Wohl, E.E. and Oaks, S.D. 1994 Geological hazards, vulnerability, and risk assessment using GIS: model for Glenwood Springs, Colorado. *Geomorphology.* 10 331–354.

Péwé, T.L. 1990 Land subsidence and earth-fissure formation caused by groundwater withdrawal in Arizona; A review. In Higgins, C.G. and Coates, D.R. (eds) *Groundwater geomorphology; The role of subsurface water in earth-surface processes and landforms*. Boulder, Colorado, Geological Society of America Special Paper 252, 219–233.

Pomeroy, P.W., Nowak, T.A., and Fakundiny, R.H., 1977, Claredon-Linden Fault System Of Western New York: A Vibroseis Seismic Study: *New York State Geological Survey Open File No.* 4iC008, 36p

Prince, K.R. and Galloway, D.L. 2003 *U.S. Geological Survey Subsidence Interest Group Conference, Proceedings of the Technical Meeting, Galveston, Texas, November 27–29, 2001*. U.S. Geological Survey Open File Report 03–308

Rau, J. L. and Nutalaya, P. 1982 Geomorphology and land subsidence in Bangkok, Thailand. In Craig, R. G. & Craft, R. L. (eds.) *Applied Geomorphology*, Allen & Unwin, London 181–201.

Shafer, C.S., Koo Lee, B., and Turner, S. 2000 A tale of three greenway trails: user perceptions related to quality of life. *Landscape and Urban Planning.* 49, 163–178.

Siriwardane H.J. 1989 *Mine Induced Subsidence: Effects on Engineered Structures: Proceedings* (Geotechnical Special Publication, No 19) American Society of Civil Engineers

Therianos, A.D. 1974 The Seismic Activity of the Kremasta Area, Greece, between 1967 and 1972. *Engineering Geology,* 8 49–56.

Voros, A.S., Sands, S.C., and Linnan, J.P. 2001 The Use of Dredged Materials in Abandoned Mine Reclamation. *Paper presented at the 2001 National Association of Abandoned Mine lands National Conference, August 19–22, 2001,* Athens, Ohio,

Young, P.R. 1993 *Proceedings 3rd International Symposium on Rockbursts and Seismicity in Mines, Kingston Ontario, August 16–18, 1993*. Balkema, Rotterdam.

Zerger, A. 2002 Examining GIS decision utility for natural hazard risk modeling. *Environmental Modelliing & Software.* 17 287–294.

Ian Douglas, School of Geography, University of Manchester, Manchester, M13 9PL, United Kingdom.

Urban Climate, Weather and Sustainability

Gerald Mills

Department of Geography, University College Dublin, Belfield, Dublin 4, Ireland.

As concentrated areas of human activities, urban areas and urbanization are key drivers of global environmental change and pose a challenge to the achievement of sustainability. One of the key goals of sustainable development is to separate increases in non-renewable resource use (particularly fossil fuels) from economic growth. This is to be accomplished by modifying individual practices, encouraging technological innovation and redesigning systems of production and consumption. Settlements represent a scale at which significant advances on each of these can be made and where there is an existing management structure. However, urban areas currently consume a disproportionate share of the Earth's resources and urbanization has modified local climate and weather significantly, usually to the detriment of urban dwellers. There is now a lengthy history of urban climate study that links existing settlement form to climatic consequences yet, there is little evidence that climate information is incorporated into urban designs or that the climatic impact of different plans is considered. Consequently, opportunities for planning sustainable urban forms that are suitable to local climates and promote energy conservation and healthy atmospheres are not taken and much effort is later expended in 'fixing' problems that emerge. This paper will outline the links between urban climate and sustainability, identify gaps in our urban climate knowledge and discuss the opportunities and barriers to the application of this knowledge to urban design and planning.

1. INTRODUCTION

Sustainable development has been defined as development that meets the needs of the current generation while leaving sufficient resources for future generations to meet their needs. It is distinguished from conventional development by its recognition of the fundamental link between the economy and the environmental base that underpins it. A central objective of sustainability is to 'decouple' the relationship between economic growth and increases in the consumption of resources and the generation of wastes. In particular, prosperity is strongly associated with access to, and use of, fossil fuels, which have a limited lifetime and produce pollutants. Decoupling will require technological innovation, redesigning systems of production and consumption and modifying individual behaviour. The rhetoric of sustainability has been incorporated into planning at all scales, which now considers environmental goals alongside traditional economic and social goals. Adoption of the tenets of sustainability has occurred for a number of reasons, including a concern about global climate change related to anthropogenic emissions of greenhouse gases.

Sustainability has particular relevance to urban settlements. The majority of the world's population is now classified as 'urban': 50% currently occupies less than 3% of the Earth's inhabitable land at densities exceeding 300 people per km^2 [*Small*, 2001]. Urban areas and urbanisation are key drivers of economic development and consume a disproportionate share of the Earth's resources. They are a 'dominant factor in the world's social, economic, cultural and political matrix' [*U.N.*, 2001, p10]. Rates of urbanization are greatest in the developing world [Table 1] where urban growth is occurring

Table 1. Global statistics on urbanisation, energy and resource use for the Developed and Developing countries and for the world. Compiled from tables published by the *World Resource Institute* [www.wri.org].

	Developed	Developing	World
Urban population, 2000 (millions)	903	1986	2890
Percent urban, 2000	76	41	47
Growth rate 2000-2005	0.5	2.9	2.2
Passenger Cars, 1996 (per 1000 persons)	326	15	84
Gasoline consumption, 1997 (litres per person)	626	55	182
CO_2 Emission, 1999 (metric tons per capita)	10.8	1.8	3.9
Energy consumption, 1997 (kg of oil equivalent per capita)	4505	803	1635
Industry, 1997 (percent of energy consumption)	30.7	34.9	32.2
Transport, 1997 (percent of energy consumption)	29.8	17.8	24.8
Residential, 1997 (percent of energy consumption)	21.4	34.4	27.1
Fossil fuels, 1997 (percent of energy consumption)	84	72	79

in megacities that have immense, and basic, management problems [*Hardoy et al*, 2001]. If these settlements follow the established route to economic prosperity, any improvements in the patterns of consumption in the developed world are likely to be overwhelmed. On the other hand, sustainable practices may be best implemented at the urban scale where economic activity is concentrated and effective planning systems already exist.

There is no agreement on what the sustainable settlement should be. Currently, most settlements are considered inefficient as they utilize large amounts of energy and materials, much of which is wasted. A 'strong' view of sustainability holds that resources should only be consumed at a rate consistent with the ability of the Earth to renew these resources. The conventional urban area that has little or no ecologically productive land cannot be sustainable, as it must be supported by a larger area that produces for example, its food and building materials. A model for the sustainable settlement is one that is linked to an appropriate, local, ecological unit [such as a river basin] from which it would draw its resources. The extent to which an urban area is sustainable in this view is evaluated by assessing its metabolism [e.g. *Warren-Rhodes and Koenig*, 2001], that is, its food, energy, material and waste assimilation requirements. An effective means of integrating this information is to convert all resource use into energy units and compare this with the ability of the Earth to supply this energy via the conversion of available solar radiation to biomass via photosynthesis. These analyses allow an assessment of the extent to which an urban area is reliant on non-renewable resources. Thus, *Girardet* [1999] estimates that London requires the equivalent of 90% of the agriculturally useful land of Britain. Applying this 'strong' model would require radical changes to existing cities [and their inhabitants], which are increasingly linked by a global flow of resources. A 'weak' view of sustainability focuses on strategies that make settlements more efficient while maintaining (or improving) living standards. This approach concentrates on the conservation of existing resources and expects that technological innovations will create new resources. Thus, while it may not be possible to make settlements self-sustaining, making them more efficient is a key part in achieving global sustainability [*Rees and Wackernagle*, 1996]. In this view, monitoring the rates of consumption and waste is central to urban environmental management.

In this paper, I review work on urban sustainability from the perspective of climate and consider the potential for urban design to achieve environmental objectives. The review is organised into three main sections. The first considers the relationship between settlements, climate and sustainability generally. The second examines research and knowledge on the urban climate and its bioclimatic effect. The final section links urban design approaches to climate management. The paper concludes by identifying gaps in our existing knowledge base. The review concentrates primarily on energy issues, which links urban climates most closely with global sustainability and anthropogenic climate change.

2. URBAN SUSTAINABILITY AND CLIMATE

Urban climate, urban design and sustainability intersect in three related areas:

1. The consumption of resources,
2. the generation of wastes and
3. the alteration of natural cover.

The design of the settlement at all scales influences the rate at which it consumes resources. Energy use is divided between industry, commercial-residential and transport [Table 1]. While energy spent on residential heating/cooling is closely related to climate, it is also dependent on building design (its materials, dimensions and orientation] and the expectations of dwellers. The energy consumed by transportation depends on a number of factors including, the distance between home and work and availability of private/public transport systems. These parameters are governed by policies on, for instance, land-use and dwelling density that affect the spread of the urban area, population density and the viability of mass transport systems. An efficient city should be organised to minimise energy use and maximise opportunities for gathering energy. As an example, at the scale of a neighbourhood, buildings can be arranged to ensure equitable access to solar energy for passive heating, day-lighting or active energy generation via photovoltaic cells. Planning guidelines can apply these principles by creating building plots that vary in size according to their solar gain [*Matus*, 1988]. Alternatively, where plots already exist, existing boundaries can be used to generate a solar 'envelope' within which a building must be contained [*Knowles*, 2002]. The precise form of the envelope would be based on the period of 'useful' solar access [Figure 1]. *Steemers* [2003] suggests that, in Britain, an obstruction of 30° to the south-facing side of a building can allow sufficient access to daylight. Employing this criterion, he estimates that a density of 200 dwellings per hectare is possible with good design; this is eight times Britain's current building density.

A by-product of resource consumption is the generation of waste, much of which degrades air quality. At the global scale, fossil fuels contribute to global carbon dioxide concentrations. However, at a local scale, pollutants are an immediate health risk, the severity of which depends on both the type of pollutant and the level and duration of exposure [*Brunkreef and Holgate*, 2002]. While the former depends on the nature of activity, the latter is governed by pollutant concentration at the level of breathing, which is controlled by the emission rates, the manner of release and the mixing in the atmosphere.

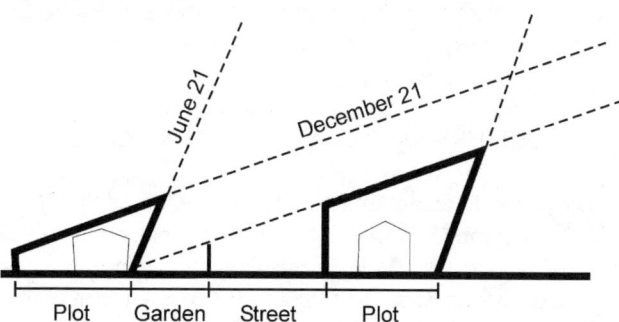

Figure 1. Solar volumes (or envelopes) created for building plots on an east/west street. The dashed lines represent the solar beam at 9am (and 3pm) for the winter solstice (December 21) and for the summer solstice (June 21). Envelopes are constructed by ensuring that the solar needs of individual plots of land are each met. Building within these envelopes ensures access to 'useful' solar energy throughout the year for all plots within the group. Redrawn from *Matus* [1988].

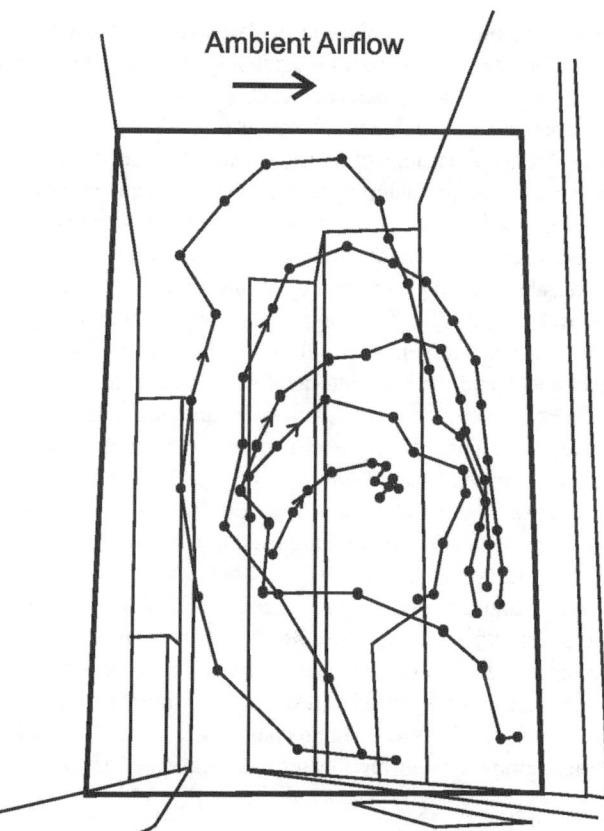

Figure 2. A recirculating vortex in a city street limits pollution dispersal. This diagram shows the paths of balloons released into a narrow city street from the perspective of an individual looking along the street axis. The ambient airflow above the building is perpendicular to the street axis and drives a vortex. The position of the balloons was recorded at regular time intervals (indicated by •) and the lines trace the paths (shown by arrows] taken by individual baloons. Redrawn from *de Paul and Sheih* [1986].

The local climate influences concentrations (by regulating the depth of the mixing volume and its ventilation) and creates conditions for chemical reactions [*Oke*, 1987]. However, the design of settlements governs the geography of emissions and the near ground concentrations [Figure 2]. Air quality management strategies rely on health guidelines, an inventory of emissions (and an understanding of their transformations] and an observational network [*Joumard et al*, 1996]. For cities in the developed world, there has been considerable improvement in air quality, despite their increased resource use [e.g. *Lents and Kelly*, 1993]. However, for many poorer cities, emissions are uncontrolled and, while there is little doubt that air quality is very poor, there is little information on either emissions or air quality [*Mage et al*, 1996]. Their experience is repeating the pollution history of the developed world cities – for example, recent observations show that the concentrations of suspended particulate matter in Karachi and Islamabad can be an order of magnitude greater than the WHO health guidelines [*Parekh et al*, 2001].

Urban areas have dramatic impacts on the local energy and water budgets because of the replacement of natural surface cover. Most urban materials have a low albedo and are impermeable (so that little water is available for evaporation). As a result, much of the energy available at the surface is transferred as sensible heat to the underlying ground and overlying atmosphere [Figure 3]. In addition, the three-dimensional geometry of the settlement has a dramatic impact on radiation exchanges and airflow patterns. Consequently, a distinctive urban climate is both easily observed and is bioclimatically significant. This climate is characterised by higher surface and air temperatures [Figure 4], lower average wind speed but greater variability, and decreased direct solar radiation when compared to surrounding rural areas [*Oke*, 1987]. Climatic extremes can be amplified by these urban 'effects' and result in unhealthy conditions; *Matzarakis and Mayer* [1991] show that the bioclimatic stresses that characterise heatwaves are enhanced by Athens' urban climate. The local water balance is also transformed by these surface changes as much of the natural cover is sealed and channels are provided for the rapid removal of surface water. Consequently, urban river

Figure 4. One of the most recognizable features of the urban climate is its heat island (UHI), which is strongest at night under clear and calm skies. This figure shows the UHI of Mexico City (shown in grey) as revealed by the mean minimum temperature (in °C) for November 1981. Contours are in meters. Redrawn from *Jauregui* [1986].

systems become very responsive to rainfall events [*Hough*, 1995]. An important theme in sustainability is a 'design-with-nature' philosophy that emphasises the potential of natural management systems; floodplains, for example, should be incorporated into the design of urban areas and designated for robust, inexpensive land uses that can respond to the vagaries of weather events.

3. THE URBAN CLIMATE AND BIOCLIMATE

The uniquely urban 'effect' is a modification of the 'natural' or background climate (a product of global, regional and local effects) and is readily detectable across a range of meteorological parameters. Over the last 100 years, urban climatology has been concerned with describing and understanding this effect, with an emphasis on particular phenomena, such as the urban heat island [Figure 4]. Over the last 40 years, studies have adopted a process-based approach to understanding urban climates by concentrating on the exchanges of mass, momentum and energy. To cope with the complex urban environment, this research has selected common urban elements for study, often represented in simple geometrical

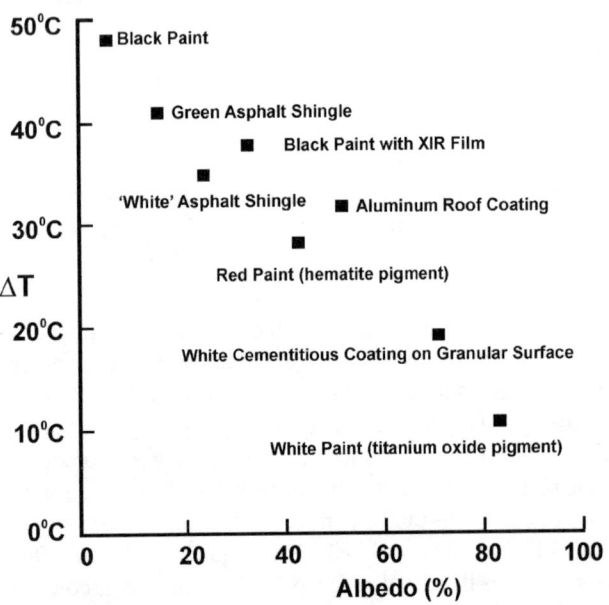

Figure 3. Sensible heat exchange between the urban surface and the overlying atmosphere can be managed through selection of materials with appropriate reflective properties. This diagram illustrates the relationship between the surface–air temperature difference (?T), which is a measure of the potential for sensible heat exchange, and the albedo of selected paints and roofing materials facing the sun. Redrawn from *Rosenfeld et al* [1995].

terms. Thus, streets are represented as very long, symmetrical 'canyons'. Nevertheless, many fundamental aspects of the urban effect are not fully understood and our existing knowledge base is largely confined to middle-latitude settlements of the developed world. Rather than describe the range of urban effects, I will concentrate on concepts and findings that have wide applicability and relevance for urban sustainability.

An important distinction can be made between urban scales that has relevance for observing and understanding urban climates [Figure 5]. Air traversing the city interacts with a generalized 'surface' at roof-level and the depth of this modified air grows from the upwind urban edge. Within this urban boundary layer (UBL), the distinctive contributions of the underlying land-uses are rapidly blended together. Below roof-level is the urban canopy layer (UCL) where human activities are based. This is composed of the enclosed volumes (buildings), and the open volumes between buildings. The climate of buildings is moderated by controlled exchanges of mass and energy across the building surfaces and energy added within the structure. The climates of the open spaces are extremely varied and are a product of exchanges at the walls of buildings, at street level, with the overlying UBL and of advection within the open urban canopy. Energy, materials and momentum are also added by vehicles. Design decisions on building materials, dimensions, orientation, etc. have implications on the outdoor climate and on the climates experienced by other buildings. Similarly, the interior building climate is partly governed by outdoor air quality, landscaping and the relative disposition of neighbouring buildings. Sustainable urban design focuses on managing the canopy layer climates that are, to varying degrees, the outcome of interactions between elements within the UCL and with the overlying UBL.

Two generalisations that establish a relationship between UCL geometry and climate effects have widespread applicability [Figure 6]. The urban heat island (UHI) is commonly identified as a difference between the near-surface air temperatures recorded in urban and surrounding rural areas (ΔT_{u-r}). The magnitude of ΔT_{u-r} is greatest at night, under clear, calm skies when radiation exchange with the sky is the dominant means of surface heat loss. Compared to rural areas, the night-time cooling rate within the UCL is slowed as its 'view' of the sky is obscured by buildings. Figure 6a shows the relationship between the maximum ΔT_{u-r} observed in several cities and a measure of sky access (the ratio of building height to

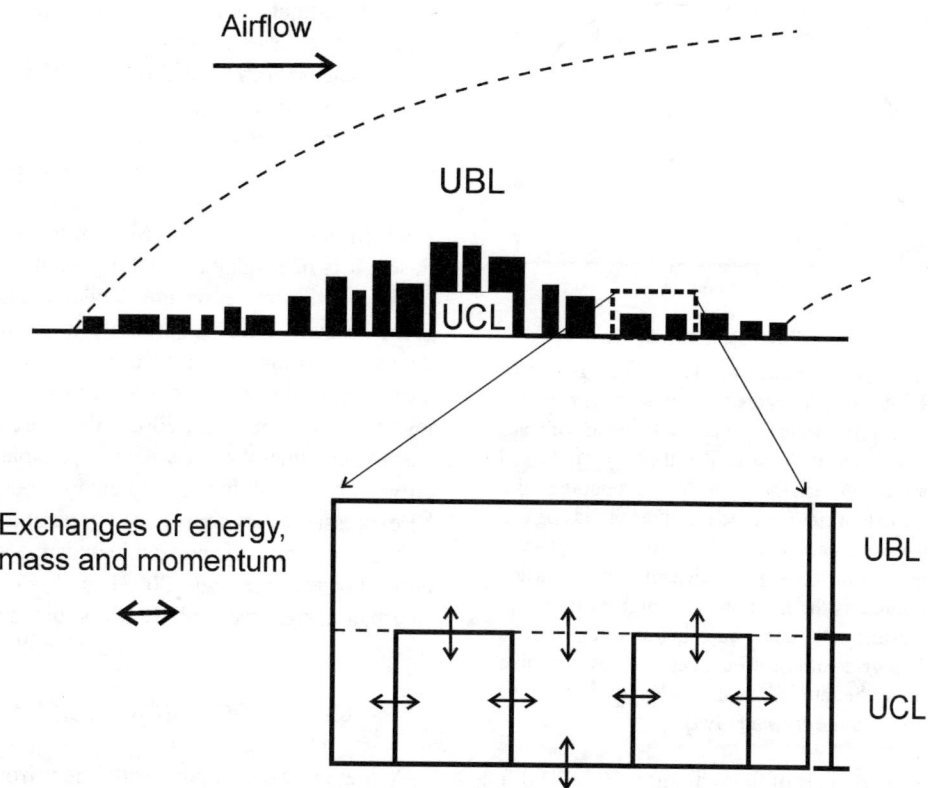

Figure 5. The vertical organisation of the city atmosphere. The urban boundary layer (UBL) forms above the buildings and grows in depth (at a slope of less than 1:100) from the upwind urban edge. Below roof-level lies the urban canopy layer (UCL) with closed (buildings) and open volumes.

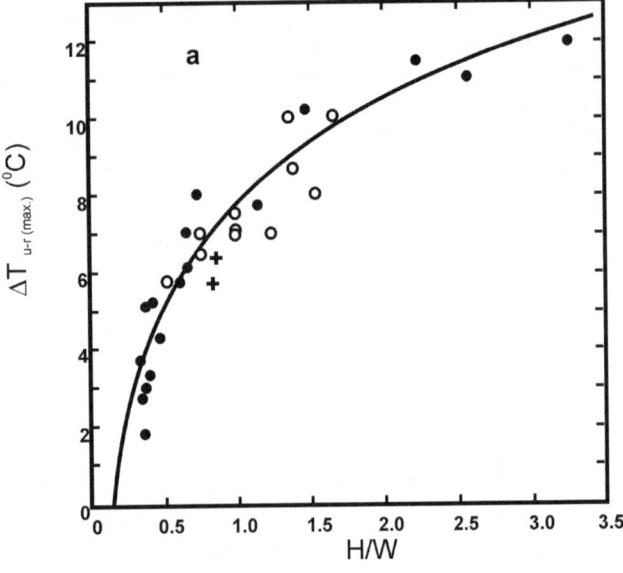

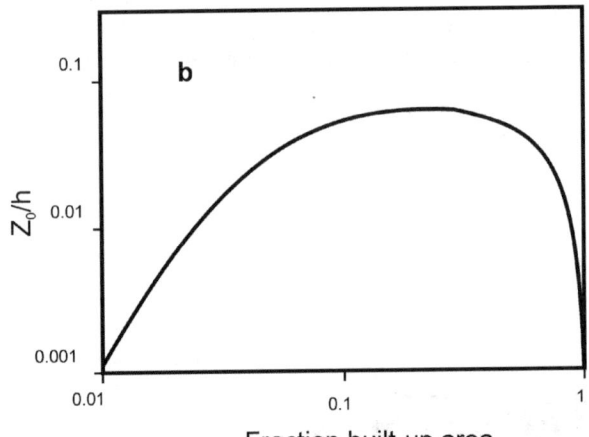

Figure 6. The relationship between elements of urban geometry and climate effect. Figure 6a illustrates the relationship between street dimensions (height (H) to width (W) ratio) in the centre of cities and maximum heat island intensity (measured as the largest observed air temperature difference between the urban and surrounding rural area, $\Delta T_{u-r(max)}$). These data were observed for cities in Europe (•), North America (○) and Australasia (+). Redrawn from *Oke* [1987]. Figure 6b shows the relationship between aerodynamic roughness and building density measured as the fraction of urban area occupied by buildings. These results were obtained using an array of cube buildings arranged in a grid pattern. Roughness is presented in a non-dimensional form as the ratio of roughness length (z_0) to cube dimension (h). Redrawn from *Bottema* [1999].

street width) in the densest part of the settlement [*Oke*, 1987]. The geometry of the UCL will also affect the exchange of momentum with the overlying atmosphere with consequences for canopy level wind speed and pollution dispersion. In a general sense, the effect of decreasing building separation on surface roughness is non-linear [Figure 6b]. When the gaps between buildings are large compared to the building dimensions, the urban area offers little aerodynamic resistance (other than spatially isolated circulations) to the ambient flow. At closer distances, the flow patterns around individual obstacles begin to interact and the near-surface flow becomes highly turbulent. As the separating distance decreases further, UBL airflow flow effectively 'skips' over the gaps between buildings and the roof-level surface becomes progressively smoother. In these cases, the UCL becomes aerodynamically separated from the overlying flow and develops highly complex flows driven by momentum from the overlying flow yet confined and modified by solid (buildings) and porous (trees) obstacles. Thus, design decisions on the geometry of the UCL produce a strong climatic response.

The basis for climate-conscious design is a concern for the total atmospheric environment experienced by the individual; sustainability places this concern within the contexts of resource use and urban planning. Evaluating levels of human (dis)comfort requires meteorological data (e.g. wind speed) and information on individuals (e.g. clothing). For design use, it is necessary to link the bioclimate response of individuals to the properties of the environment experienced. At the building scale, the climate goal is clearest: to create an indoor climate appropriate to its use. For many buildings, including dwellings, this climate objective is described as 'comfortable' and is to be achieved by establishing near steady-state conditions that minimise environmental stress. In outdoor, variable, environments few assumptions can be made and assessing levels of stress, except in instances of weather extremes, has proved very difficult [*Soligo et al*, 1998]. Moreover, for urban design purposes, it is necessary to establish relationships between the background climate, design decisions and bioclimatic outcome. However, this requires very specific meteorological data and information on the urban physical environment that are rarely available. While there have been outdoor case-studies [e.g. *Capeluto et al*, 2003], there are few general urban design guidelines that are widely applicable. Ideally, the cumulative effect of all the relevant environmental variables could be expressed as an equivalent temperature value that would be easily determined and readily understood. This is an area of current research [*Hoppe*, 2002] and has obvious applications to urban design, particularly if combined with an equivalent measure to assess the net effects of air pollution [*Mayer*, 1999].

CLIMATE-CONSCIOUS URBAN DESIGN

Achieving a sustainable settlement from a climatic viewpoint will require a coherent strategy for all scales of urban form and activity. Table 2 outlines three levels at which planning and design decisions are implemented. Each level can be associated with distinctive climate objectives that are linked directly and indirectly to resource use. Moreover, each has a distinctive perspective that is related to its needs, activities and decision-makers.

At the settlement level, decisions create the gross characteristics of the urban area. These are made primarily to allow its efficient operation and establish the overall form, land-use geography, and transportation network. Climate management at this level is based on measures of resource use and air quality. Strategies for minimizing energy use (and preserving air quality) focus on reducing transportation costs and building heating and cooling needs. In the former case, energy use is linked to population density [Figure 7], the placement of land-uses relative to each other and the provision of mass transit systems. Only at this level can comprehensive measures be taken to manage the urban air resource by observing its quality and controlling pollutant sources. At the scale of the individual building, energy use is related to its function and its design. A desirable indoor climate can be achieved by isolating the indoor from the outdoor and ensuring a near constant, monitored, interior climate. An alternative, less energy-intensive, approach is to rely on a design that maximizes passive energy gain and minimizes heat loss by controlling exchanges between the indoor and outdoor. At the scale of the building group, the climatic objectives are less clear. Ideally, design should moderate outdoor climate stresses while ensuring that individual buildings can satisfy their needs. The design tools available at this scale include the organization of building elements, the layout of streets, selection of materials, etc.

A successful urban design will consider the climate impact of decisions across these scales. At each higher level, decisions limit the parameters for design at lower levels. For example, the possible design freedom at the building scale is controlled by plot size and street orientation. Similarly, guidelines on residential densities at the settlement scale set parameters on the design of building groups. Crucially, one of the biggest obstacles to making informed design decisions is the general absence of relevant weather/climate data that would allow one to identify periods of climate stress and to select the best overall design to counteract these.

4.1 Urban Design Tools

Climatic design tools may be categorised into those that manage the physical character of the urban area and those that control its functioning. These aspects of settlement design are not separate, for example, high-density mixed-use developments are advocated as a means of limiting the need for private vehicles. The physical design tools can be classified according to their ability for controlling radiation, wind, evaporation or pollution emission and can be regarded as fixed (e.g. street geometry) or responsive (e.g. vegetation) elements of the urban landscape. The potential roles for some of these physical tools in climate management are discussed below.

The selection of construction material and surface coatings will affect the urban thermal environment and building energy needs. In hot climates, the selection of low albedo materials (e.g. asphalt) for roads and parking lots will raise outdoor

Table 2. A summary of the tools (gray diagonal) employed at the building, building group and settlement scales to achieve climatic objectives at those scales. The application of tools at each scale has a climate impact at (below the diagonal), and places limits on decisions made at (above the diagonal), the other scales.

Objective	Impacts	Limits		
		Buildings	Building Groups	Settlement
Indoor comfort. Shelter	Buildings	Location. Materials. Design (e.g. shape, orientation, etc.)	Access to light, solar energy, wind. Air quality	Building codes
Outdoor comfort. Outdoor health.	Building groups	Local climate change: Emissions, Materials/surfaces, Building dimensions – flow interference, & shadow areas	Building placement. Outdoor landscaping, materials and surfaces. Street dimensions & orientation	Guidelines on densities, heights, uses and green-spaces
Energy use. Air quality. Protection from extremes.	Settlement	Energy efficiency. Air quality. Urban climate effect	Mode and intensity of traffic flows. Energy efficiency. Air quality. Urban climate effect.	Zoning. Overall extent and shape. Transport policy

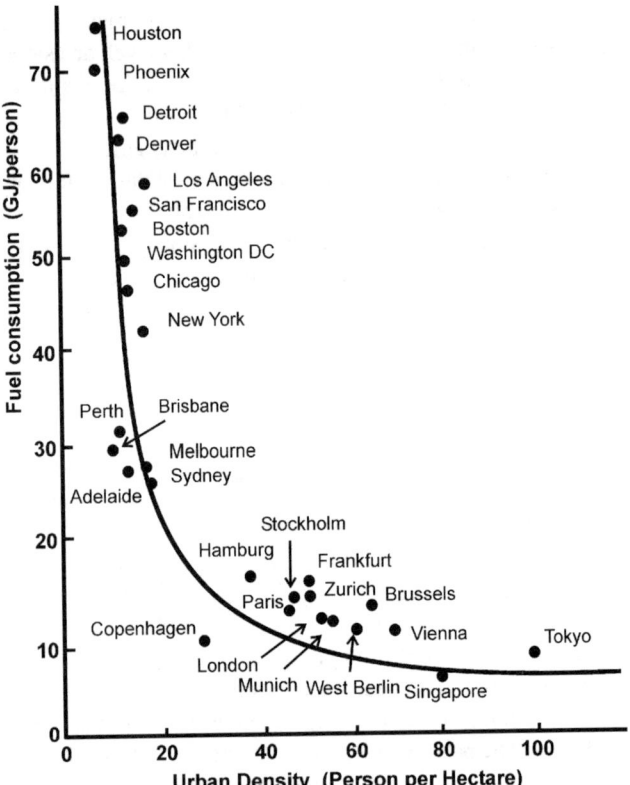

Figure 7. The design of cities has an impact on its energy consumption [*Kenworthy and Laube*, 1996]. This diagram illustrates a relationship between urban population density and per capita gasoline consumption (in giga-Joules] based on data compiled for 30 cities in 1980. Redrawn from *Newman* [1999].

surface and air temperatures [Figure 3] and increase building cooling needs. *Rosenfeld et al* [1995] estimate that increasing the albedo of a single air-conditioned dwelling in such circumstances could yield direct energy savings of 20–40%. Moreover, modelling experiments carried out for the Los Angeles basin suggest that raising the albedo of the urban area (from 0.13 to 0.26) can double these savings by reducing outdoor air temperature. On the other hand, in cold climates, a low albedo surface that increases energy absorption is preferable. The physical design of streets will affect radiative and wind patterns most directly. In cold climates, heat loss can be minimised by ensuring access to direct sunshine and shelter from the wind; both can be managed through street alignment and geometry. Generally, narrow streets that are aligned perpendicular to the path of airflow will provide shelter, while those that are parallel allow ventilation [*Bottema*, 1999]. The design of a community can support these decisions by placing taller and wider buildings on the pole-ward side of the development. These will provide shelter and not overshadow lower buildings.

Air quality management focuses on pollutant sources and dispersal. While technological means of controlling emissions has proven an effective means of limiting large-scale sources, it has had limited success with traffic sources, where technological improvements are often overwhelmed by increases in the number of vehicles [*Lents and Kelly*, 1993]. An alternative approach is to manage emissions by modifying individual behaviour through inducements (e.g. multiple occupancy lanes) or penalties (e.g. road pricing). However, there is ample evidence that the physical design of the urban area itself can affect behaviour [Figure 7]. *Lyons et al* [2003] examined the average daily vehicle kilometres travelled for 84 international cities and found that 88% of the variation could be statistically explained by urban area alone. Thus, the form of the city and the placement of activities (work, home and services) relative to each other can affect modes of transport usage. Two urban layout strategies have been proposed to reduce reliance on private vehicles [*Rogers*, 1997]:

1. The promotion of mixed land-uses within urban areas where appropriate commercial/industry and residential development co-exist.
2. The promotion of fixed-line mass transit systems with high-density developments at designated access points.

Design will also affect pollution concentration and personal exposure within the UCL. Research has shown that in narrow, long streets, ambient airflow that is perpendicular to the street axis will drive a vortex within the street that circulates air pollution with little dilution [Figure 2]. Even moderate traffic flows, in such circumstances, may generate an unhealthy atmosphere.

Vegetation is the most versatile 'tool' for climate modification. It can simultaneously affect change in radiation exchange, airflow and ventilation, evaporation and temperature, erosion and runoff, pollution concentration and noise levels. The magnitude of these effects varies with its canopy form, foliage density, branch and root systems and physiologic properties. These attributes will themselves vary with species, age and health [*Bernatzky*, 1978]. Their effectiveness as a design tool depends on the selection of appropriate species and their placement. However, there has been little research on the total impact of vegetation—rather, there are general guidelines [*McPherson*, 1992] and some specific information of the effect of a vegetation type on a particular property. For example, rows of dense vegetation with low canopies are a suggested solution for a dwelling that needs solar access and to regulate near-surface airflow. By comparison, for hot and humid environments where shade and ventilation are required, tall vegetation with an elevated canopy is ideal. Unlike other design tools, vegetation can be responsive; deciduous trees shed their leaves during winter, thus providing both summer shade and winter solar access. At the scale of the settlement, urban parks can be an integral component of climate management. Their presence is readily detectable in measurements of urban surface and air temperature where they appear as cool (and humid) islands within the warmer urban setting. Their impact on the surrounding urban area is limited however, and is related to the width of the park and the length of the park in the along-wind direction [*Barradas*, 1991]. This suggests that spatially distributed parks of a limited size are preferable to large isolated parks as a means of moderating urban climates.

Best design practice ensures that guidelines for one climate property are compatible with those for other properties and those at different scales. Inevitably, this will have to resolve the, often contradictory, effects of implementing a set of best practices to achieve a given outcome. As an example, *Zrudlo* [1988] presents a design strategy for planning a new settlement in a cold, life-threatening environment. Three designs were developed, each representing an optimal solution to ensure solar access, control wind speed and prevent snow accumulation. Streets should be oriented east to west and of sufficient width to guarantee solar access. To minimise heat loss due to low temperatures and wind, buildings should be arranged to provide mutual shelter and minimize time spent outdoors—a long building located on the north side of the settlement would act as a windbreak. Finally, to ensure adequate ventilation to prevent snow accumulation, streets within the village should be aligned along the path of airflow. The composite design satisfies as many of the requirements as possible [Figure 8]. However, this approach is difficult to repeat. Most settlements are more complex, are already in place and have inherited properties. Decisions are now designed to ameliorate or modify the existing urban climate, which includes the effect of the urban area itself. Moreover, in less extreme environments, the nature of, and role for, climate knowledge in urban design will be different. For example, the issue of the local contribution to global scale climate change may take precedence over aspects of local climate management.

4.2 The sustainable city and climate

There is no real world model for a sustainable urban area that implements climate guidelines across all scales. Designs for climatically sustainable urban settlements have been presented by a number of researchers [e.g. *Bitan*, 1992; *deSchiller and Evans*, 1998; *Emmanuel*, 1995; *Givoni*, 1998; *Yannas*, 2001]. These provide a valuable pedagogic service by drawing upon current best practice for a number of urban elements such as individual buildings, material selection and street characteristics and providing design solutions for current urban problems. Moreover, although there is a limited scope for the unfettered design of entirely new settlements, the 'ideal city'

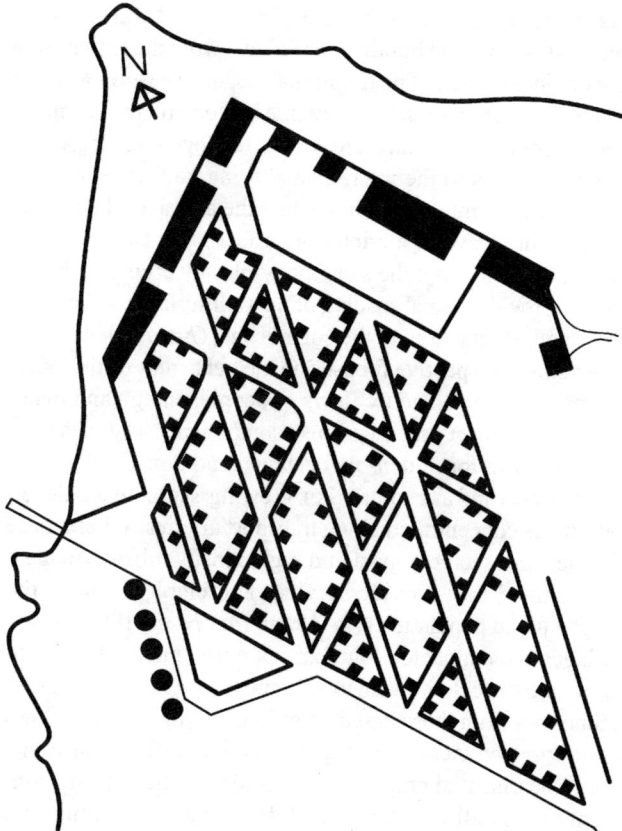

Figure 8. The 'best overall' urban design achieves a compromise between a set of solutions to individual issues. In this diagram, a new settlement design for a harsh climatic environment (the Arctic) is proposed that simultaneously considers access to sunshine, protection from the wind and snow accumulation. Redrawn from *Zrudlo* [1988].

provides a template for re-design of settlements.

Generally, one can distinguish between two types of idealised, sustainable settlements that represent distinct perspectives. The first sees the settlement as a place of habitation and focuses attention on meeting building needs. The strong sustainability model limits energy consumption by selecting local, non-manufactured materials and relying on design to manage passively the interior climate. Another approach sees the advantages of modern materials (e.g. light-sensitive glass) and technologies (e.g. solar heaters) and integrates these into building design. Both recognise the value of traditional building forms that are adjusted to local climates and are distinguished by their analysis of energy cost. The strong sustainability model accounts for both the energy required for building operation and that 'embodied' in its physical structure (that is, the energy required to extract materials, manufacture and transport elements and assemble the building). Building group design should manage relationships between buildings to ensure for example, solar access or mutual shading [e.g. *Littlefair*, 1998].

As an example, the ideal building for a hot and arid climate is one that isolates the building air volume from the extremes of the outdoor climate. The traditional response ensures a cooler indoor environment during daytime by constructing compact forms (minimise surface area) with few openings (blocking sunlight access to the interior) and using a massive, earthen envelope (that moderates and delays the arrival of the diurnal temperature wave to the interior surface). A sustainable settlement would adopt the same principles by arranging buildings to provide mutual shading of the wall surfaces and present the smallest total area to the solar beam [*Gupta*, 1984].

Another perspective focuses on the behaviour of the urban system as represented by flows of people, energy and materials that connect urban land-use components. The ideal city is compact in form, densely occupied and organized around its transport infrastructure. For growing settlements, development is concentrated along transport arteries. Where these arteries intersect, compact and high-density urban 'villages' are planned that provide services and employment to the nearby urban population [*The Urban Task Force*, 1999]. These villages are connected together via mass transit systems to allow urban scale movements. Some envisage a shift in city economies from those based on movements of people to flows of information thus, obviating the need for much private transport. The essential properties of this ideal settlement are constant across climate types and decisions at building and building group scales must both fit the grand urban design and be adapted to the local climate. An exemplar for sustainable settlement practices is Curitiba, Brazil, which preserves floodplain functions and organizes urban development around public transport axes [*Rabinovitch and Leitman*, 1996].

Good design at the scale of building groups is critical if these perspectives are to be compatible. Figures 6b and 7 caution against simple prescriptions for achieving sustainability. Figure 7 supports the argument that high-density, compact cities are more efficient. However, Figure 6b indicates that, if higher population densities are achieved by increasing building density, the exchange of air between the UBL and UCL will decrease unless 'roughness' is consciously designed into the UCL structure. Otherwise, although energy use and waste emissions decrease, pollutant concentrations within the UCL (the layer of human exposure) increase. Unfortunately, there have been few guidelines developed that link general aspects of design at this scale with climatic outcomes. An exception is *Oke's* [1988] work on street design in middle-latitude cities for which there is a considerable body of data. The simultaneous need for winter shelter, solar and daylight access, urban warmth and pollutant dispersal was assessed and an east–west oriented street with a height-to-width ratio of about 0.5 was considered best overall. However, this solution is unlikely to be appropriate in hot and humid climates where protection from the sun and ventilation are key issues. More generally, outdoor climate control at the building group scale has received little attention until recently [e.g. *Mertens*, 1999]. This may be attributed to the absence of clear objectives [e.g. *Ahmed*, 2003] and the weakness of design control at this scale.

Existing settlements are the artefacts of decisions accumulated over a lengthy period. Making them 'more' sustainable will take considerable time, particularly where liberal urban planning, plentiful space and the availability of private cars have produced very low-density cities. A quicker solution for some urban problems may be to alter urban surface cover. *Rosenfeld et al* [1995] suggest that the daytime urban heat island, chemical smog production and residential cooling demand in hot, sunny climates can be offset by settlement-wide planting of shade trees and introduction of higher albedo surfaces. For air pollution, there are few 'easy' design solutions. In those settlements that occupy sites with limited natural ventilation, there is little choice but to limit emissions. In some circumstances, this strategy can be aided by protecting existing resources. For example, the city of Graz, Austria, relies on cold and clean katabatic flows that drain out of nearby valleys and ventilate the urban area. To conserve this air 'reservoir', *Lazar and Podesser* [1999] propose controlling development in these valleys based on their potential for producing and transporting clean air. At a building group scale, urban spaces may be similarly protected by establishing guidelines for future development that ensure access to sunshine and protection from winds [*Arens et al*, 1989]. Similarly, urban spaces can be preserved or 'rehabilitated' for outdoor use if solar guidelines are employed to ensure access to sunshine for selected sites such as parks [e.g. *Bosselmann et al*, 1995].

5. DISCUSSION AND CONCLUSIONS

There are two chief obstacles to achieving climatically sustainable cities. Firstly, climate knowledge is not incorporated into routine planning practice. While there is a global impetus for changing current settlement practices, implementing these is primarily a political decision. It is not obvious that the inherent contradictions in sustainable development, particularly the simultaneous desire for economic growth and environmental protection, are resolvable. Secondly, much of the knowledge for making informed decisions that are consistent and coherent for all urban scales is either inadequate or absent.

The lack of knowledge on design decisions and climate outcomes can be attributed to several causes, including a traditional division between the fields of scientific discovery and applications. Most current urban climate research is carried out with the purpose of discovering the processes responsible for localised climate change and is carried out by a relatively small number of scientists, often working on isolated projects. While there is general agreement on the need for research in certain areas (such as cities in the developing world], there is rarely the critical numbers of scientists and equipment available to carry out such research in a comprehensive manner. As a result, our information on urban climates is largely confined to those that follow European and North American city traditions.

With few exceptions [e.g. *Bottema*, 1999], little climate research is carried out to assess different urban designs. Most research tries to understand processes or observe climate impacts. Consequently, only a small part of the available urban climate knowledge is presented in a manner readily employable by designers [e.g. *Buckland and Middleton*, 1999]. For example, most outdoor design research on building groups is concerned with altering urban components, such as building placement, to achieve comfortable and/or safe outdoor, daytime environments [e.g. *Capeluto et al*, 2003]. On the other hand, much climate research is undertaken when the urban effect is maximized, often at night when outdoor spaces are generally unoccupied. In addition, researchers have adopted an experimental approach where the role of specific urban features on climate modification is isolated and studied. The advantage of this approach is that, where sufficient study has been done (as is the case for streets) general guidelines can be offered. By comparison, design studies are characterised by a case-study approach that give practical expression (and illustration) to climate-based design. There remains a need for a comprehensive manual of applied urban climatology that includes information on meteorological data and the methods for assessing design options. This manual would be supported by case-studies that illustrate best practice in a range of climates. In contrast to the field of architecture, an equivalent bank of knowledge on links between urban design and the bioclimate effect does not yet exist. There is no accepted methodology for assessing the climates experienced by individuals in outdoor situations and much work needs to be done on common outdoor spaces like courtyards and parks.

At all scales, the lack of appropriate information to base decisions is pervasive. Standard meteorological data are valuable but are often inappropriate for the design task. The process of developing guidelines for assessing and adjusting information collected at a standard site to an urban site has just begun. On the settlement side, there is a dearth of information on the geography and structure (building heights, street orientation, etc.) of individual settlements. *Steemers et al* [1997] have demonstrated that this basic structural data can be incorporated into widely available software and usefully employed to assess microclimatological outcomes. This offers the opportunity for assessing the climates of existing settlements and the effect of modifying elements without extensive measurement programmes. At the same time, there remain conceptual problems associated with implementing best-practice at different scales. Planning at the urban scale tends to be two-dimensional, focussed on urban activities and gathers information accordingly. However, the climate experienced by dwellers is also a product of its three-dimensional structure. Good design at the scale of the building group is required if the broad objectives at the settlement scale are to be compatible with the specific needs of indoor spaces and the wider needs of outdoor spaces.

The widespread incorporation of the tenets of sustainability into planning offers an opportunity for including climate/weather issues into urban design on a routine basis. The global concern for climate change and resource use provides a mandate for the development of a coherent and broad-based applied urban climatology, which has not existed previously. In particular, it encourages research that is guided by the needs of planners/designers. Thus far, studies on urban design-climate links have been limited in scope and much of the available research does not translate into clear guidelines that are supported by scientific knowledge and illustrative case-studies. While more studies that explicitly compare the climatic performance of different design solutions is needed, research in this area has increased as the interest in planning sustainable settlements has grown. As the global population urbanizes, managing settlements and their climates will be critical to achieving sustainability.

Acknowledgements. The author is grateful for the comments of anonymous reviewers and those of the editor that have significantly improved the manuscript. Thanks are also due to Stephen Hannon who helped in drawing the figures.

REFERENCES

Ahmed K.S. 2003. Comfort in urban spaces: defining the boundaries of outdoor thermal comfort for the tropical urban environments. Energy and Buildings 35, 103–110.

Arens E., Ballanti D., Bennett C., Guldman S. and White B. 1989. Developing the San Francisco wind ordinance and its guidelines for compliance. Building and Environment 24, 297–303.

Barradas V.L. 1991. Air temperature and humidity and human comfort index of some parks of Mexico City. International Journal of Biometeorology 35, 24–28.

Bernatzky A. 1978. Tree Ecology and Preservation. Elsevier, New York.

Bitan A. 1992. The high climatic city of the future. Atmospheric Environment 26B, 313–329.

Bosselmann P., Arens E., Dunker K. and Wright R. 1995. Urban form and climate: Case study, Toronto. American Planning Association Journal, Spring 1995, 226–239.

Bottema M. 1999. Towards rules of thumb for wind comfort and air quality. Atmospheric Environment 33, 4009–4017.

Brunekreef B. and Holgate S.T. 2002. Air pollution and health. The Lancet 360, 1233–1242.

Buckland A.T. and Middleton D.R. 1999. Nomograms for calculating pollution within street canyons. Atmospheric Environment 33, 1017–1036.

Capeluto I.G., Yezioro A. and Shaviv E. 2003. Climatic aspects in urban design—a case study. Building and Environment 38, 827–835.

dePaul F.T. and Sheih C.M. 1986. Measurements of wind velocities in a street canyon. Atmospheric Environment 20, 455–459.

deSchiller S. and Evans J.M. 1998. Sustainable urban development: design guidelines for warm, humid cities. Urban Design International 4, 165–184.

Emmanuel R. 1995. Energy-efficient urban design guidelines for warm–humid cities: Strategies for Colombo, Sri Lanka. Journal of Architectural and Planning Research 12, 58–79.

Girardet H. 1999. Sustainable cities: A contradiction in terms?, 413–425 in Satterthwaite D. (Ed.) 1999. Sustainable Cities. Earthscan, London.

Givoini B. 1998. Climate Considerations in Building and Urban Design. John Wiley & Sons, New York.

Gupta V.K. 1984. Solar radiation and urban design for hot climates. Environment and Planning B 11, 435–454.

Hardoy J.E., Mitlin D. and Satterthwaite D. 2001. Environmental problems in an urbanizing world. Earthscan.

Hoppe P. 2002. Different aspects of assessing indoor and outdoor thermal comfort. Energy and Buildings 34, 661–665.

Hough M. 1995. Cities and Natural Process. Routledge, London.

Jauregui E. 1986. The urban climate of Mexico city, p63–86 in Oke T.R. (Ed.) Urban Climatology and its Applications with special regard to Tropical Areas. World Meteorological Organization, Switzerland. WMO– No. 652.

Joumard R., Lamure C., Lambert J., and Tripiana F. 1996. Air quality and urban space management. The Science of the Total Environment 189/190, 57–67.

Kenworthy J.R. and Laube F.B. 1996. Automobile dependence in cities: An international comparison of urban transport and land use patterns with implications for sustainability. Environmental Impact Assessment Review 16, 279–308.

Knowles R.L. 2002. The solar envelope: its meaning for energy and buildings. Energy and Buildings 35, 15–25.

Lazar R. and Podesser A. 1999. An urban climate analysis of Graz and its significance for urban planning in the tributary valleys east of Graz (Austria). Atmospheric Environment 33, 4195–4209.

Lents J.M. and Kelly W.J. 1993. Clearing the air in Los Angeles. Scientific American 269(4), 32–39.

Littlefair P. 1998. Passive solar urban design: ensuring the penetration of solar energy into the city. Renewable and Sustainable Energy Reviews 2, 303–326.

Lyons T.J., Kenworthy J.R., Moy C. and dos Santos F. 2003. An international model for the transportation sector. Transportation Research D 8, 159–167.

Mage D., Ozolins G., Peterson P., Webster A., Orthofer R., Vandeweerd V. and Gwynne M. 1996. Urban air pollution in megacities of the world. Atmospheric Environment 30, 681–686.

Matus V. 1988. Design for Northern Climates. Von Nostrand Reinhold, New York.

Matzarakis A. and Mayer H. 1991. The extreme heat wave in Athens in July 1987 from the point of view of human biometeorology. Atmospheric Environment 25B, 203–211.

Mayer H. 1999. Air pollution in cities. Atmospheric Environment 33, 4029–4037.

McPherson E.G. 1992. Shading urban heat islands in U.S. desert cities. Wetter und Leben 44, 107–123.

Mertens E. 1999. Bioclimate and city planning – open space planning. Atmospheric Environment 33, 4115–4123.

Newman P. 1999. Transport: Reducing automobile dependence, p173–198 in Satterthwaite D. (ed.) Sustainable Cities. Earthscan, London.

Oke T.R. 1987. Boundary Layer Climates. 2nd edition Routledge, London.

Oke T.R. 1988. Street design and the urban canopy layer climate. Energy and Buildings 11, 103–113.

Parekh P.P., Khwaja H.A., Khan A.R., Naqvi R.R., Malik A., Shah S.A., Khan K. and Hussain G. 2001. Ambient air quality of two metropolitan cities of Pakistan and its health implications. Atmospheric Environment 35, 5971–5978.

Rabinovitch J. and Leitman J. 1996. Urban planning in Curitiba. Scientific American 274(3), 46–53.

Rees W. and Wackernagel M. 1996. Urban ecological footprints: Why cities cannot be sustainable–and why they are a key to sustainability. Environmental Impact Assessment Review 16, 223–248.

Rogers R. 1997. Cities for a Small Planet. Faber & Faber, London.

Rosenfeld A.H., Akbari H., Bretz S., Fishman B.L., Kurn D.M., Sailor D. and Taha H. 1995. Mitigation of urban heat islands: materials, utility programs, updates. Energy and Buildings 22, 255–265.

Small C. 2001. Global analysis of urban population distributions and the physical environment. Proceedings of the Open Meeting of the Human Dimensions of Global Environmental Change Research Community. Rio de Janeiro.

Soligo M.J., Irwin P.A., Williams C.J. and Schuyler G.D. 1998. A comprehensive assessment of pedestrian comfort including thermal effects. Journal of Wind Engineering and Industrial Aerodynamics 77&78, 753–766.

Steemers K., Baker N., Crowther D., Dubiel J., Nikolopoulou M-H., and Ratti C. 1997. City texture and microclimate. Urban Design Studies 3, 25–50.

Steemers K. 2003. Energy and the city: density, buildings and transport. Energy and Buildings 35, 3–14.

Stoll M.J. and Brazel A.J. 1992. Surface–air temperature relationships in the urban environment of Phoenix, Arizona. Physical Geography 13, 160–179.

The Urban Task Force 1999. Towards and Urban Renaissance. E & F N Spon, London.

U.N. 2001. The State of the World's Cities, 2001. United Nations Centre for Urban Settlements, Nairobi, Kenya.

Warren-Rhodes K. and Koenig A. 2001. Escalating trends in the urban metabolism of Hong King: 1971–1997. Ambio 30, 429–438.

Yannas S. 2001. Toward more sustainable cities. Solar Energy 70, 281–294.

Zrudlo L.R. 1988. A climatic approach to town planning in the Arctic. Energy and Buildings 11, 41–63.

Gerald Mills, Department of Geography, University College Dublin, Belfield, Dublin 4, Ireland.